Principles of Biotechnology

Principles of Biotechnology

Editor
Christina A. Crawford, MS Ed

SALEM PRESS

A Division of EBSCO Information Services
Ipswich, Massachusetts

GREY HOUSE PUBLISHING

Cover image: DNA sequence by Gio_tto (iStock)

Copyright ©2018, by Salem Press, A Division of EBSCO Information Services, Inc., and Grey House Publishing, Inc.

All rights reserved. No part of this work may be used or reproduced in any manner whatsoever or transmitted in any form or by any means, electronic or mechanical, including photocopy, recording, or any information storage and retrieval system, without written permission from the copyright owner. For permissions requests, contact proprietarypublishing@ebsco.com.

∞ The paper used in these volumes conforms to the American National Standard for Permanence of Paper for Printed Library Materials, Z39.48 1992 (R2009).

Publisher's Cataloging-In-Publication Data
(Prepared by The Donohue Group, Inc.)

Names: Crawford, Christina A., 1983- editor.
Title: Principles of biotechnology / editor, Christina A. Crawford, MS Ed.
Description: [First edition]. | Ipswich, Massachusetts : Salem Press, a division of EBSCO Information Services, Inc. ; [Amenia, New York] : Grey House Publishing, [2018] | Series: Principles of | Includes bibliographical references and index.
Identifiers: ISBN 9781682176788 (hardcover)
Subjects: LCSH: Biotechnology.
Classification: LCC TP248.2 .P75 2018 | DDC 660.6--dc23

FIRST PRINTING
PRINTED IN THE UNITED STATES OF AMERICA

TABLE OF CONTENTS

Publisher's Note vii
Editor's Introduction ix
Contributors xiii

~A~
Alternative energy sources 1
Animal breeding 4
Animal testing 6
Animals as a medical resource 9
Anthrax and biological warfare 11
Anthropogeomorphology 14
Antibiotic-resistant bacteria 17
Antibiotics as defense against biological warfare 19
Archaebacteria 21
Artificial intelligence 25
Artificial organs 30
Audio engineering 34

~B~
Bioassays 39
Biochemical engineering 40
Biodetectors 46
Bioenergy technologies 47
Bioengineering 51
Biofertilizers 57
Biofuels 58
Biofuels and synthetic fuels 62
Bioinformatics 66
Biological terrorism 71
Biological weapon identification 75
Biomathematics 77
Biomechanical engineering 80
Biomechanics 84
Biometric eye scanners 87
Biometric identification systems 88
Bionics and biomedical engineering 91
Biopesticides and the environment 95
Bioprocess engineering 97
Bioremediation 100
Biosensors 102
Biostratigraphy 105
Biosynthetics 108
Biotechnology and genetic engineering 112
Biotoxins 116
Botany and genetic engineering 118
Botulinum toxin as a biological weapon 120
Bubonic plague as a biological weapon 121

~C~
Cell and tissue engineering 123
Cloning 127
Cloning of plants 133
CRISPR-Cas9 135
Cryogenics 136

~D~
Desalination plants and technology 140
Detection and prevention of food poisoning . 142
Diamond v. Chakrabarty 144
DNA analysis 145
DNA banks for endangered animals 149
DNA database controversies 150
DNA extraction from hair, bodily
    fluids, and tissues 152
DNA fingerprinting as evidence 154
DNA isolation methods 156
DNA profiling 158
DNA recognition instruments 160
DNA sequencing and crime scenes 161
DNA typing 163
DNA: recombinant technology 164
Dolly the sheep 166
Drug testing 169

~E~
Engineering 174
Environmental biotechnology 179
Enzyme engineering 182
Estrogens from plants 186

~F~
Fiber technologies 189

~G~
Genetic engineering 193
Genetic resources 199
Genetically engineered pharmaceuticals 203
Genetically modified food production 205
Genetically modified organisms 209
Genetically altered bacteria 212
Genomics 214

~H~
Human genetic engineering 220
Human-computer interaction 226
Hybridization (botany) 230

~I~
Industrial fermentation 233
Intelligence 237
Intensive farming 241

~M~
Medicinal plants 243
Metabolic engineering 245
Microscopy 250
Mitochondrial DNA analysis and typing 257
Model organisms 258
Molecular systematics 261

~N~
Nanotechnology 264
Nanotechnology and the environment 268
Neural engineering 270
Night vision technology 275

~P~
Pasteurization and irradiation 279
Pathogen genomic sequencing 283
Performance-enhancing drugs 284
Plant biotechnology 287
Plant breeding and propagation 294
Plant cells: molecular level 297
Plants as a medical resource 301
Polymerase chain reaction 303
Prokaryotes 304
Proteomics and protein engineering 308

~R~
Radiocarbon dating 313
Refuse-derived fuel 319
Renewable and nonrenewable resources 320
Reproductive science and engineering 322

~S~
Scanning probe microscopy 327
Science of cloning 330

Seed banks 336
Stem cell research and technology 337
Synthetic fuels 340

~T~
Tularemia as a bioweapon 343

~Z~
Zygomycetes 345

~Important Figures in Biotechnology~
David Baltimore 348
Françoise Barré-Sinoussi 351
Paul Berg 354
J. Michael Bishop 355
Herbert Wayne Boyer 358
Erwin Chargaff 361
Stanley Cohen 364
Francis S. Collins 368
Carl F. Cori 371
Erasistratus 374
Alexander Fleming 377
Camillo Golgi 379
Carol W. Greider 383
Alfred D. Hershey 385
Edward B. Lewis 387
Konrad Lorenz 390
Barbara McClintock 394
Christiane Nüsslein-Volhard 396
Frederick Sanger 400
Hans Spemann 403
Jack W. Szostak 405
J. Craig Venter 407
James D. Watson 410
Edmund Beecher Wilson 413
Norton David Zinder 417

~Appendixes~
Time Line of Inventions and Scientific
Advancements in Biotechnology 421
Glossary 425
Bibliography 427
Subject Index 459

PUBLISHER'S NOTE

Salem Press is pleased to add *Principles of Biotechnology* as the ninth title in the *Principles of* series that includes *Chemistry, Physics, Astronomy, Computer Science, Physical Science, Biology,* and *Scientific Research*. This new resource introduces students and researchers to the fundamentals of biotechnology using easy-to-understand language that gives readers a solid start and deeper understanding and appreciation of this complex subject.

- The 134 articles include 109 entries that explain basic principles of biotechnology, ranging from Alternative energy sources to Zygomycetes, with attention paid to Cloning; Synthetic fuels; Medicinal plants; Stem cell research and technology; Genetically modified organisms; and more. All of the entries are arranged in A to Z order, making it easy to find the topic of interest.
- The volume also features 23 biographies of key figures in biotechnology that include a description of each individual's significant contributions to the field, ranging from David Baltimore to David Norton Zinder.

Entries related to basic principles and concepts include the following:
- Fields of study to illustrate the connections between the topic and the various branches of science related to biotechnology;
- An Abstract that provides brief, concrete summary of the topic and how the entry is organized;
- Text that gives an explanation of the background and significance of the topic to biotechnology as well as describing the way a process works, how a procedure or technique is applied to achieve important goals related to the environment, health, nutrition, industry, and agriculture.
- Illustrations that clarify difficult concepts via models, diagrams, and charts of such key topics as biomechanical engineering and DNA fingerprinting;
- Further reading lists that relate to the entry.

Entries related to important figures in biotechnology include the following:
- A brief overview of the individual and his or her contributions;
- Key dates and biographical data;
- Primary field(s) and specialties;
- Sidebars explaining the individual's significant advances, inventions, or discoveries;
- Text that provides information about the scientist's Early Life, Life's Work, and Impact;
- Further reading lists that relate to the entry.

This reference work begins with a comprehensive introduction to the field, written by volume editor Christina A. Crawford, Assistant Director for Science and Engineering at the Rice Office of STEM Engagement (R-STEM) at Rice University in Houston, Texas.

The book includes helpful appendixes as another valuable resource, including the following:

- Time Line of Inventions and Scientific Advancements in Biotechnology
- Glossary;
- General Bibliography; and
- Subject Index.

Salem Press and Grey House Publishing extend their appreciation to all involved in the development and production of this work. The entries have been written by experts in the field. Their names and affiliations follow the Editor's Introduction.

Principles of Biotechnology, as well as all Salem Press reference books, is available in print and as an e-book. Please visit www.salempress.com for more information.

Editor's Introduction

It's Monday, around 6 p.m., and like any other night, it is time to sit down and have dinner. This simple event happens every night for millions of people around the world with such consistency and regularity that it is often taken for granted. One out of every eight people, about 795 million worldwide, however, is malnourished, meaning that he or she does not have enough food to maintain his or her health and be active. As of 2016, that number represented about 12.9 percent of the world's population. Making food more accessible to everyone in the world is a serious concern as populations continue to grow while the land available to produce food to feed those populations stays the same, or even decreases. Many people believe that this issue is one of the many problems facing our globe that can be addressed by means of biotechnology, along with other vital concerns including access to routine medications such as vaccinations or vitamin supplements. For many biotechnologists, the goal is to save the world, one invention at a time.

What is Biotechnology?

Biotechnology is defined as the use of living organisms, or substances obtained from living organisms, to produce products or processes of value to humankind.

While it is true that biotechnology has made tremendous advances in the areas of human and veterinary medicine, agriculture, food production, and other fields, these advances have also raised some genuine concerns about how biotechnology should be monitored. Ongoing debates continue the effort to balance the potential of biotechnology, in particular genetic engineering, and its ability to produce organisms that may benefit humans, against growing concerns about the potential for these advances and inventions to disrupt ecosystems, negatively affect human health, or be used in ethically inappropriate ways.

Many countries have different interpretations of the field. Australia, for example, defines biotechnology as any technological application that uses biological systems, including living organisms or derivatives thereof, to make or modify products or processes for specific use. Canada defines biotechnology as the application of science and engineering to the direct or indirect use of living organisms, or parts or products taken from living organisms, in their natural forms.

In the United States, an entity, whether a research facility or a business, is considered as working in the field of biotechnology if any of the following types of research take place:

- **Researching the coding or sequencing of DNA**: Genomics, pharmacogenetics, gene probes, DNA sequencing/synthesis/amplification, genetic engineering.
- **Researching proteins and other biomolecules**: Protein/peptide, sequencing/synthesis, lipid/protein glycoengineering, proteomics, hormones and growth factors cell receptors/signaling/pheromones.
- **Researching cell and tissue culture and engineering**: Cell/tissue culture, tissue engineering, hybridization, cellular fusion, vaccine/immune stimulants, embryo manipulation.
- **Researching process biotechnologies**: Bioreactors, fermentation, bioprocessing, bioleaching, bio-pulping, bio-bleaching, biodesulphurization, bioremediation, and biofiltration.
- **Researching subcellular organisms**: Gene therapy, viral vectors.

Each of the research areas above can be studied in any of the various fields of biotechnology.

Branches of Biotechnology

Researchers who study biotechnology generally organize themselves among five different branches of biotechnology: animal, bioinformatics, environmental, medical, and industrial.

- **Animal biotechnology**: This branch of biotechnology deals with the development of transgenic animals. The research is focused in areas such as the increase of milk or meat production with high resistance to various diseases. It also deals with in vitro fertilization of an egg and the transfer of an embryo to the womb of female animal for further development.
- **Bioinformatics**: Bioinformatics is a branch of biotechnology that is a combination of both computer science and biotechnology.

Bioinformatics helps us to develop tools to analyze data related to biotechnology. Bioinformatics is used in various research studies aimed at goals such as the development of new and more effective medicines. It is also used to increase the fertility of plants and to provide them with a defense against pest, drought and diseases. Bioinformatics is vital in a number of areas that are key contributors to biotechnology and the pharmaceutical sector.
- **Environmental biotechnology**: This branch of biotechnology is dedicated to promoting a healthy environment. With the help of the process called micropropagation (a practice of producing larger number of plants through the existing stock of plants), scientists are better able to select the right quality of plants and crops for specific growing conditions and environments. Transgenic plants (plants whose DNA is modified) can also be designed to grow in a specified environment with the help of certain chemicals.
- **Medical biotechnology**: This branch of biotechnology focuses on improving the health of humans and animals by developing the technology necessary to produce medicines. It also helps to create or design organisms meant to more effectively treat diseases and conditions. Through the process of genetic manipulation, it can be used to cure genetic issues in organisms. Medical biotechnologists research diseases in organisms in order to develop new ways to achieve an accurate diagnosis through the use of more accurate testing protocols. Stem cell therapy is one area that continues to show great potential to either generate new organs or repair the damaged tissues in organisms.
- **Industrial biotechnology**: This kind of biotechnology is used and applied in various industries and their processes. The development of biopolymer (plastics) substitutes has allowed the automobile industry to invent improved parts and fuels for the vehicles of the future. The creation of "high-tech" fibers has had an impact on the clothing industry as well as the sports and fitness sectors. There may be other broad impacts across all industries, from energy to entertainment, as industrial biotechnology develops new chemicals and production processes.

STEPPING STONES IN BIOTECHNOLOGY

The term biotechnology itself is relatively new, but the practice of biotechnology is as old as civilization. In fact, it played a key role in the initial development of societies and civilizations, since they could only start to evolve once humans learned to apply selective breeding techniques–a basic biotechnological skill–to improve and increase the food crops they grew and livestock they raised.

Artificial insemination, the process in which semen is collected from the male animal and deposited into the female reproductive tract through artificial techniques rather than natural mating, emerged as a practical procedure roughly a century ago, although as early as 1784, Italian biologist Lazzaro Spallanzani successfully inseminated a dog.

While genetic engineering is often considered a phenomenon of the late twentieth century, the building blocks for such technology began with the first isolation of DNA in 1869 and the subsequent awareness of its relevance to heredity starting in 1928. The first accurate double-helix model of DNA was developed in 1953 by James D. Watson and Francis Crick, and the first gene sequence and recombinant DNA was created in 1972 by researchers from Stanford University. The latter discovery truly heralded the beginning of the biotechnological industry and the development of genetically modified organisms (GMOs).

In 1963, Margaret Oakley Dayhoff, an American physical chemist and a major figure in the evolving science of bioinformatics, began compiling protein sequences into a series of books titled *Atlas of Protein Structure and Function.* By 1978, however, the *Atlas of Protein Structure and Function* had grown too large and cumbersome to permit comparisons and analyses to be easily performed. Her second major contribution was to create a database infrastructure to convert the atlas to the first online biological database, making it accessible to researchers who used it to sort, manipulate, and align multiple protein sequences. The database developed by Dr. Dayhoff and her successors, Dr. Winona Barker and Dr. Robert Ledley, became know as the Protein Information Resource (PIR). In 2002 PIR, along with its international partners, EBI (European Bioinformatics Institute) and SIB (Swiss Institute of Bioinformatics), were awarded a grant from NIH to create UniProt, a single worldwide database of protein sequence and function.

Biotechnology has played a significant role in treating serious medical conditions, including diabetes. The first effective treatment called for the extraction of insulin from cattle and pigs and while it saved millions of lives, it wasn't perfect, as it caused allergic reactions in many patients. The first recombinant DNA-produced synthetic "human" insulin was produced in 1978 using *E. coli* bacteria to produce the insulin. Eli Lilly went on in 1982 to sell the first commercially available biosynthetic human insulin under the brand name Humulin. Genetic engineering has also been used to synthesize protropin, a human growth hormone (HGH) employed in the treatment of growth failure conditions such as hyposomatotropism.

Biotechnology has played an important role in helping to effectively remediate environmental disasters. In March, 1989, the Exxon Valdez oil tanker spilled millions of gallons of crude oil in Prince William Sound, Alaska. On many beaches, the Environmental Protection Agency authorized the use of simple bioremediation techniques, such as stimulating the growth of indigenous oil-degrading bacteria by adding common inorganic fertilizers. Beaches cleaned by this method did as well as beaches cleaned by mechanical methods. In another instance of successful bioremediation, selenium-contaminated soil in the Kesterson National Wildlife Refuge in California was partially decontaminated in the 1980s through the method of supplying indigenous fungi with organic substrates such as casein and waste orange peels. This promoted as much as 60 percent selenium volatilization in less than two months.

The cloning of Dolly the sheep in Scotland in 1996 opened a whole new avenue in the use of biotechnology for livestock production. The use of cloning technology in conjunction with surrogate mothers provides the means to produce a whole herd of genetically superior animals in a short period of time. As well as offering a new way to raise livestock, it also brought serious questions about the ethics of cloning and genetic engineering to the forefront of a public debate that continues to this day.

CURRENT ADVANCES IN BIOTECHNOLOGY

Over the past few years there have been many advances in the field of biotechnology that border on the realm of science fiction. From cloned animals to nerve regeneration, this field appears to be limited only by the imagination of the scientist working within it. Cloning, for example, was a subject only talked about in science fictions books and movies. Now, as cloning technologies continue to advance, scientists in South Korea have managed to not only clone a dog, but also change its genetic make-up such that the dog actually glows in the dark.

If you have ever played or been a fan of football, biotechnology is helping there as well. Most football fans are aware of the horrific injuries that happen on the football field. One wrong tackle could cause nerve damage that might impair a player for the rest of his or her life. Now a company called Regenxx is attempting to regenerate damaged nerves by using stem cells in hopes that they can give injured athletes the chance to regain mobility.

Biotechnology is also changing the way we use prescription drugs in America. In 2017, Abilify Mycite became the first smart pill that that has been approved by the Federal Drug Administration (FDA). This smart pill is able to transmit information back to doctors once it has been ingested, giving caretakers vital data on medication regimens of the patient.

Biotechnology breakthroughs are driving advances in far more areas than health care alone; they are improving environmental quality all over the world. Global warming is widely perceived as one of the biggest problems of the twenty-first century, playing a role in widespread climate change that has changed weather patterns, growing conditions, and sea levels. To help combat the problem, many countries are researching ways that biotechnology might make the switch to renewable sources of energy such as biofuels possible The potential of biofuels made from ethanol and algae as fuels for vehicles may help reduce levels of carbon emissions produced by fossil fuels, thereby reducing or eliminating one of the root causes of global warming. Biofuels are derived primarily from plants, a renewable resource, making them sustainable rather than nonrenewable. As the demand for biofuel has grown, biotechnology has been applied to improve the methods of producing fuel. Biotechnologists have created microorganisms that can ferment biomass materials derived from sugars (sugar cane, sugar beet and molasses), starch (corn, wheat, grains) or cellulose (forest products). They have also learned to modify strains of algae in order to dramatically increase the amount of oil that the algae produce without significantly inhibiting growth.

The fundamental aim of environmental biotechnology is to use organisms to control contamination and treat waste by means of the following four concepts: bioremediation; prevention; detection and monitoring; and genetic engineering. The use of biotechnology in the treatment of waste and pollution control is not a new idea. For more than a century, many communities have relied on natural processes and microbes to break down and treat sewage. In the process called bioremediation, microorganisms including fungi and bacteria along with their enzymes are used to return a contaminated environment to its original condition. Naturally occurring biological degradation processes are purposely employed to remove contaminants from areas where they have been released. The use of such processes requires a solid scientific understanding of the contaminant, its impact, and the affected ecosystem. The aim of environmental biotechnology is to provide a natural approach to tackling environmental issues that begins with the identification of biohazards—determining which contaminants are present, for how long, and in what quantity— and concludes with the successful restoration of industrial, agricultural, and natural areas that have been affected by contamination.

With the help of two promising gene editing tools—Cluster Regularly Interspaced Short Palindromic Repeats (CRISPR) and the RNA-guided DNA endonuclease enzyme Cas9—scientists are hoping to change the world, especially for people born with a genetic disorder. CRISPR-Cas9 is fast and inexpensive, as well as being the most accurate technique for editing DNA thus far. This unique technology allows scientists to edit a person's DNA sequence by removing, adding, or altering sections of the sequence. One promising aspect is the potential to cure genetic diseases like sickle cell anemia or more complex diseases including cancer and HIV. In 2016, a Chinese group lead by oncologist Lu You at Sichuan University in Chengdu was the first research team to conduct human trials using CRISPR-Cas9 on immune cells to improve the body's attack against cancerous tissue. In 2017, China announced that there will be over 20 human trials underway by 2018 that will utilize CRISPR-Cas9 as a treatment for conditions ranging from HIV infection, human papillomavirus, and even high blood cholesterol. This is an exciting and thrilling time to be involved with medical biotechnology. It is also one that has started to sound some alarms. Scientists, politicians, and religious leaders are voicing worries about the ethical issues of using a genetic tool of this nature, especially if it is used to edit reproductive cells, which could affect generation after generation of humans.

In this Book

In this book, you will find a collection of biotechnology articles that explain the basics, history, and applications of biotechnology. This volume provides readers with the important information they need to understand the basic concepts, ethical arguments, possibilities, and consequences of biotechnology. The text provides students and researchers with an easy-to-understand introduction to the fundamentals of biotechnology.

——*Christina A. Crawford, MS Ed*

Work Cited

Cyranoski, David. "CRISPR Gene-Editing Tested in a Person for the First Time." *Nature News*, Nature Publishing Group, 15 Nov. 2016, www.nature.com/news/crispr-gene-editing-tested-in-a-person-for-the-first-time-1.20988.

Dahms, A. Stephen. "Biotechnology: What It Is, What It Is Not, and the Challenges in Reaching a National or Global Consensus." *Biochemistry and Molecular Biology Education*, vol. 32, no. 4, 2004, pp. 271–278., doi:10.1002/bmb.2004.494032040375.

Kumar, Srinibas. "Biotechnology: Scope and Branches of Biotechnology." Biology Discussion, 26 Oct. 2015, www.biologydiscussion.com/biotechnology/branches-biotechnology/biotechnology-scope-and-branches-of-biotechnology/15653.

CONTRIBUTORS LIST

Kenneth H. Brown

Michael A. Buratovich, PhD

Christina Capriccioso

Richard P. Capriccioso, MD

Michael W. Cheek

Dennis W. Cheek

Mark Coyne

D. R. Gossett

Patrick Norman Hunt

Carly L. Huth, JD

April D. Ingram

Micah L. Issitt

Karen N. Kähler

Narayanan M. Komerath

Padma P. Komerath

Jeanne L. Kuhler, PhD

M. Lee, MA

Spencer G. Lucas

Joel P. MacClellan

Sergei A. Markov, PhD

Eric Metchik

Ralph R. Meyer

Edward C. Nwanegbo

Henry R. Owen

Cynthia Racer

Diane C. Rein, PhD, MLS

James L. Robinson

John Richard Schrock

Dwight G. Smith

Bruce L. Stinchcomb

Bethany Thivierge, MPH

Ruth N. Udey

Oluseyi A. Vanderpuye

Christine Watts, PhD

George M. Whitson III, PhD

Ming Y. Zheng

A

ALTERNATIVE ENERGY SOURCES

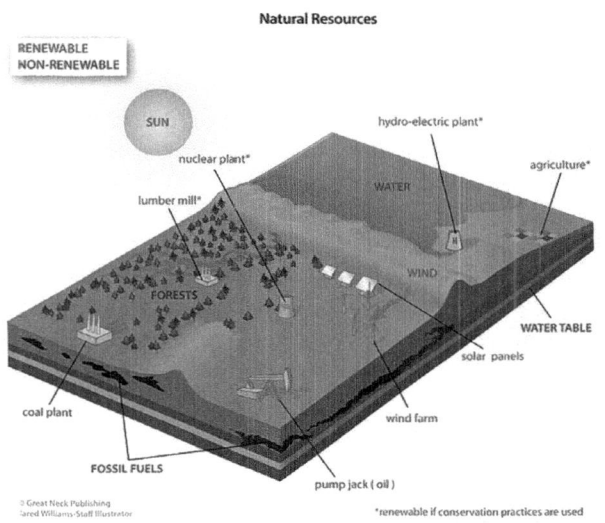

Natural Resources: The earth has many natural resources, some are renewable and others are not. Conservation practices will ensure that resources do not become depleted.
© EBSCO.

FIELDS OF STUDY

Biochemistry; Biotechnology; Molecular biology; Chemical engineering; Industrial biotechnology (or White biotechnology)

ABSTRACT

Energy sources that offer alternatives to the burning of fossil fuels such as coal and petroleum are urgently needed to address rising demand for energy in ways that will not contribute to air pollution and climate change. The ideal alternative energy source is renewable or inexhaustible and causes no lasting environmental damage.

BACKGROUND

Both the extraction and the burning of fossil fuels have caused severe and growing damage to the environment, contributing to such problems as air pollution, the release of greenhouse gases (which retain heat and contribute to climate change), and sulfuric acid in rainfall. Nuclear energy sources are very limited in supply and expensive, require extreme amounts of processing, and produce long-lasting radioactive waste. In the long term, energy release from nuclear fusion has been proposed as a limitless supply of power, but industrial-scale production of fusion power continues to pose large and uncertain obstacles and hazards.

ALTERNATIVE SOURCES

Solar Power. The sun powers winds, ocean currents, rain, and all biomass growth on the earth's surface. Because the availability and extraction means for each of these secondary sources of solar power are diverse, each forms a different field of alternative energy technology. Where solar power is extracted and converted to energy directly, the capture can be by means of flat-plate receivers that collect at the incident intensity but can operate in diffuse light, or by means of concentrators that can achieve intensities of several hundred suns but work poorly in diffuse light.

In solar photovoltaic power (PV) technology, solar radiation is directly converted to useful power through PV cell arrays, which require semiconductor mass-production plants. PV cell technologies have evolved from using single-crystal silicon to using thinner polycrystalline silicon, gallium arsenide, thin-film amorphous silicon, cadmium telluride, and copper indium selenide. The needed materials are believed to be abundant enough to meet projected global growth. The process of purifying silicon requires large inputs of energy, however, and it generates toxic chemical waste. Regeneration of the energy required to manufacture a solar cell requires about three years of productive cell operation.

Solar cell technology continues to evolve. Broadband solar cell technologies have the potential to make cells sensitive to as much as 80 percent of the energy in the solar spectrum, up from about 60 percent. High-intensity solar cells could enable operation at several hundred times the intensity of sunlight, reducing the cell area required when used with concentrator mirrors and enabling high thermal efficiency.

Direct solar conversion is another option. Laboratory tests have shown 39 percent conversion from broadband sunlight to infrared laser beams using neodymium-chromium fiber lasers. Direct conversion of broadband sunlight to alternating-current electricity or beamed power through the use of optical antennae is projected to achieve 80 to 90 percent conversion. Such technologies offer hope for broadband solar power to be converted to narrowband power in space and then beamed to the earth by satellites.

Another way of harnessing solar power is through solar thermal technology. Solar concentrators are used with focal-point towers to achieve temperatures of thousands of kelvins and high thermal efficiency, limited by containment materials. The resulting high-temperature electrolysis of water vapor generates hydrogen and oxygen in an efficient manner, and this technology has demonstrated direct solar decomposition of carbon dioxide (CO_2) to carbon monoxide (CO) and oxygen.

Wind Power. Winds are driven by temperature and pressure gradients, ultimately caused by solar heating. Wind energy is typically extracted through the operation of turbines. Power extraction is proportional to the cube of wind speed, but wind-generated forces are proportional to the square of wind speed. Wind turbines thus can operate safely only within a limited range of wind speed, and most of the power generation occurs during periods of moderately strong winds. Turbine efficiency is strongly dependent on turbine size and is limited by material strength. The largest wind turbines have reached 8 megawatts in capacity. Denmark, the Netherlands, and India have established large wind turbine farms on flat coastal land, and Germany and the United Kingdom have opted for large offshore wind farms. In the United States, wind farms are found in the Dakotas, Minnesota, and California, as well as on Colorado and New Mexico Mountain slopes and off the coasts of Texas and Massachusetts.

Because of wind fluctuations and the cubic power relation, wind power is highly unsteady, and means must be established for storing and diverting the power generated before it is connected to a power grid. In addition, offshore and coastal wind farms must plan for severe storms. Smaller wind turbines are sometimes used for power generation on farms and even for some private homes in open areas, but these tend to be inefficient and have high installation costs per unit power. They are mainly useful for pumping irrigation water or for charging small electrical devices.

Environmentalists have raised some concerns about large wind turbines. The machinery on wind farms causes objectionable noise levels, and many assert that the wind turbine towers themselves constitute a form of visual pollution. Disturbances to wildlife, particularly deaths and injuries among bird populations, are another area of concern. In addition, the construction of wind farms often requires the building of roads through previously pristine areas to enable transportation of the turbines' large components.

Hydroelectric Power and Tidal Power. Large dams provide height differences that enable the extraction of power from flowing water using turbines. Hydroelectric power generated by dams forms a substantial percentage of the power resources in several nations with rivers and mountains. However, the building of large dams raises numerous technical, social, and public policy issues, as damming rivers may displace human inhabitants from fertile lands and may result in the flooding of pristine ecosystems, sometimes the habitats of endangered species. Increased incidence of earthquakes has also been associated with the existence of very large dams.

In some of the world's remote communities, micro hydroelectric (or micro hydel) plants provide power, generating electricity in the 1–30 megawatt range. Very small-scale systems, known as pico hydel, extract a few kilowatts from small streams; these can provide viable energy sources for individual homes and small villages, but the extraction technology has to be refined to bring down the cost per unit of power.

Although tidal power is abundant along coastlines, the harnessing of that power has been slow to gain acceptance, in part because of the difficulties of building plants that can survive ocean storms. Tidal power is extracted in two principal ways. In one method, semipermeable barrages are built across estuaries with high tidal ranges, and the water collected in the barrages is emptied through turbines to generate power. In the second, offshore tidal streams and currents are harnessed through the use of underwater equivalents of wind turbines.

Tidal power plants typically use pistons that are driven up and down by alternating water levels or the action of waves on turbines. A rule of thumb is that a tidal range of 7 meters (23 feet) is required to produce enough hydraulic head for economical operation. One drawback is that the 12.5-hour cycle of tidal operation is out of synchronization with daily peak electricity demand times, and hence some local means of storing the power generated is desirable. In many cases, impellers or pistons are used to pump water to high levels for use when power demand is higher.

China, Russia, France, South Korea, Canada, and Northern Ireland all have operational tidal power stations. The first U.S. tidal power station became operational in 2012 in Cobscook Bay, near Eastport, Maine.

Biomass Power. Biomass, which consists of any material that is derived from plant life, is composed primarily of hydrocarbons and water, so it offers several ways of usage in power generation. Combustion of biomass is considered to be carbon-neutral in regard to greenhouse gas emissions, but it may generate smoke particles and other pollution.

One large use of biomass is in the conversion of corn, sugarcane, and other grasses to ethyl alcohol (ethanol) to supplement fossil petroleum fuels. This use is controversial because the energy costs associated with producing and refining ethanol are said to be greater than the savings gained by using such fuel. It is argued that subsidies and other public policies and rising energy prices entice farmers to devote land to the production of ethanol crops, thus triggering shortages and increases in food prices, which hurt the poorest people the most. Brazil has advanced profitable and sustainable use of ethanol extracted from sugarcane to replace a substantial portion of the nation's transportation fossil-fuel use.

Jatropha plants, as well as certain algae that grow on water surfaces, offer sources of biodiesel fuel. Biodiesel from *Jatropha* is used to power operations on several segments of India's railways, and vegetable oil from peanuts and groundnuts, and even from coconuts, has been used in test flights of aircraft ranging from strategic bombers to jetliners.

Biogas and Geothermal Energy. Hydrocarbon gases from decaying vegetation form large underground deposits that have been exploited as sources of energy for many years. Technology similar to that used in extracting energy from these natural deposits, which are not considered a renewable energy source, can be used to tap the smaller but widely distributed emissions of methane-rich waste gases from compost pits and landfills. Creating the necessary infrastructure to capture these gases over large areas poses a difficult engineering challenge, however. In addition, care must be taken to avoid the release of methane from these deposits into the atmosphere, as methane is considered to be twenty times as harmful as carbon dioxide as a greenhouse gas.

Geothermal energy comes from heat released by radioactive decay inside the earth's core, perhaps augmented by gravitational pressure. Where such heat is released gradually through vents in the earth's surface, rather than in volcanic eruptions, it forms an abundant and steady, reliable, long-term source of thermal power. Hot springs and geothermal steam generation are used on a large scale in Iceland, and geothermal power is used in some American communities and military bases.

———*Narayanan M. Komerath and Padma P. Komerath*

FURTHER READING

Charlier, R. H., and C. W. Finkl. *Ocean Energy: Tide and Tidal Power*. London: Springer, 2009.

Edwards, Brian K. *The Economics of Hydroelectric Power*. Northampton, Mass.: Edward Elgar, 2003.

Klass, Donald L. *Biomass for Renewable Energy, Fuels, and Chemicals*. San Diego, Calif.: Academic Press, 1998.

"Largest WTG Yet Makes Its Debut." *Modern Power Systems* 34.3 (2014): 39–40. *Energy & Power Source*. Web. 15 Jan. 2015.

Pollan, Michael. *The Omnivore's Dilemma: A Natural History of Four Meals*. New York: Penguin Press, 2007.

"Tidal Power." *Chemical Business* 28.10 (2014): 53. *MasterFILE Premier*. Web. 15 Jan. 2015.

Traynor, Ann J., and Reed J. Jensen. "Direct Solar Reduction of CO_2 to Fuel: First Prototype Results." *Industrial and Engineering Chemistry Research* 41, no. 8 (2002): 1935-1939.

Vaitheeswaran, Vijay. *Power to the People: How the Coming Energy Revolution Will Transform an Industry, Change Our Lives, and Maybe Even Save the Planet.* New York: Farrar, Straus and Giroux, 2003.

Walker, John F., and Nicholas Jenkins. *Wind Energy Technology.* New York: John Wiley & Sons, 1997.

Wenisch, A., R. Kromp, and D. Reinberger. *Science or Fiction: Is There a Future for Nuclear?* Vienna: Austrian Ecology Institute, 2007.

ANIMAL BREEDING

Since the 1940's, dairy cows in the United States, such as these in Sacramento, California, have been intensively bred for milk production, which has tripled since that time. Photo courtesy of USDA NRCS. [Public domain], via Wikimedia Commons

FIELDS OF STUDY

Animal Science; Biology; Genetics; Reproductive technology

ABSTRACT

Animal breeding has been used to produce animals more useful to humankind since animals were first domesticated. Traditionally it involved selecting individual animals for desired traits and mating them, with the intent of producing improved offspring. In the second half of the twentieth century, extensive performance records and computer-aided analysis permitted superior animals to be identified more accurately and, via reproductive technologies, to be utilized more rapidly for improving the major livestock species. In the future, molecular biology and biotechnology promise to expedite this process by identifying desirable genes from the same or different species and incorporating them into domesticated animals. Animal breeding will continue to augment the value of domesticated animals as a renewable resource.

GENETIC INHERITANCE

Animal breeding is predicated on two principles: that the genes of an animal are inherited from its parents and that its genes are an important determinant of its appearance, structure, behavior, and productivity. In animal species, almost all genes are located in the nucleus of an organism's cells; these nuclear genes are inherited from both parents. A few genes, located outside the nucleus in subcellular structures called mitochondria, are derived only from the mother. The full complement of genes, known as the genome, directs the development of an individual animal, the synthesis of all body tissues, including metabolic machinery, and to a large extent the characteristics or traits exhibited. Different forms of genes, referred to as alleles, are responsible for the individuality of living things.

Some characteristics are determined by alleles of one gene—for instance, the absence of horns or the occurrence of a metabolic disease. In such cases a single mutation can lead to a deleterious condition. However, most traits of significance involve alleles of more than one gene. Superior characteristics for growth rate or milk yield, so-called polygenic traits, result from the combination of alleles of many genes. Animal breeding seeks to improve genetically the future population of a particular species by increasing the proportion of desirable alleles or the appropriate combination of such alleles. Genetic improvement requires selection of appropriate breeding animals and a mating plan for such animals.

SELECTION AND MATING SYSTEMS

Selection is the process of determining which animals are to be used as breeding stock. The simplest form of selection considers only traits of the individual, whereas more complex selection takes account of additional information on relatives, such as siblings, parents, and offspring. The accuracy of predicting genetic progress is improved by considering relatives. This process requires reliable measures for desired traits, acquisition of records from numerous animals, and analysis of the records, which has been aided by advances in statistical theory and computational power. The result is a ranking of animals based on their genetic merit for single or multiple traits.

Several systems have been used for mating selected animals. One involves complementarity, whereby individuals with high genetic merit for different traits are mated. It has been used to improve livestock in developing countries by mating animals adapted to local conditions with highly productive ones from developed countries. The beef cattle, swine, and poultry industries make heavy use of crossbreeding, in which animals from different breeds are mated. One of its advantages is the "hybrid vigor" that results. Another system is mating the best to the best. One of its hazards is inbreeding, or the mating of relatives, which often results in decreased fertility and viability.

Environmental Factors

Performance or productivity is determined not only by genetics but also by environmental factors. Climate, nutrition, and management can affect the extent to which the genetic potential of an animal is realized. Because the productivity of an animal can be affected deleteriously by heat and disease, climate and other environmental factors can influence animals' performance. Similarly, the management system used, whether intensive or extensive, can also affect productivity. Accordingly, the most productive animal under one set of conditions may not necessarily be the most productive under another. Interactions between genetics and the environment must be considered in animal breeding.

Post-1940's Developments

Beginning in the mid-twentieth century, reproductive technologies, most notably artificial insemination, contributed to rapid improvement in animal performance. These technologies permit animals with the best genetics to be used widely, resulting in numerous offspring from which to select the best breeding stock for the next generation. As a result of intensive selection and management in the United States beginning in the 1940's, milk production per cow has more than tripled. The growth rate of chickens has more than doubled, as has egg production. Such increases have occurred concurrently with a higher efficiency in raising animals for human food.

Molecular biology and biotechnology hold the potential to alter animal breeding processes significantly in the early twenty-first century. Further understanding of the genomes of livestock species should permit identification of specific genes that will increase the productivity of these animals. One approach, known as marker-assisted selection, would use genetic markers associated with desirable production characteristics to enhance genetic improvement. If such markers prove to be accurate predictors, they will allow selection of desirable animals long before performance records are available. Transfer of desirable genes, within or between species, may also expedite the generation of superior animals. The goal of animal breeding can be expected to remain similar to that of the past—namely the improvement of animal species to better meet human needs—but the precise nature of the improvements desired and the methodologies used to achieve them could be vastly different.

—*James L. Robinson*

Further Reading

Bourdon, Richard M. *Understanding Animal Breeding.* 2d ed. Upper Saddle River, N.J.: Prentice Hall, 2000.

Falconer, D. S., and Trudy F. C. Mackay. *Introduction to Quantitative Genetics.* 4th ed. New York: Longman, 1996.

Field, Thomas G., and Robert E. Taylor. *Scientific Farm Animal Production: An Introduction to Animal Science.* 10th ed. Upper Saddle River, N.J.: Prentice Hall, 2008.

Sandøe, Peter, and Stine B. Christiansen. *Ethics of Animal Use.* Oxford, England: Blackwell, 2008.

Schatten, Heide, and Gheorghe M. Constantinescu, eds. *Comparative Reproductive Biology.* Ames, Iowa: Blackwell, 2007.

Van der Werf, Julius, Hans-Ulrich Graser, Richard Frankham, and Cedric Gondro, eds. *Adaptation and Fitness in Animal Populations: Evolutionary and*

Breeding Perspectives on Genetic Resource Management. London: Springer, 2009.

Weaver, Robert F., and Philip W. Hedrick. *Genetics.* 3d ed. Dubuque, Iowa: W. C. Brown, 1997.

U.S. Department of Agriculture. Animal Breeding, Genetics, and Genomics. www.csrees.usda.gov/animalbreedinggeneticsgenomics.cfm

ANIMAL TESTING

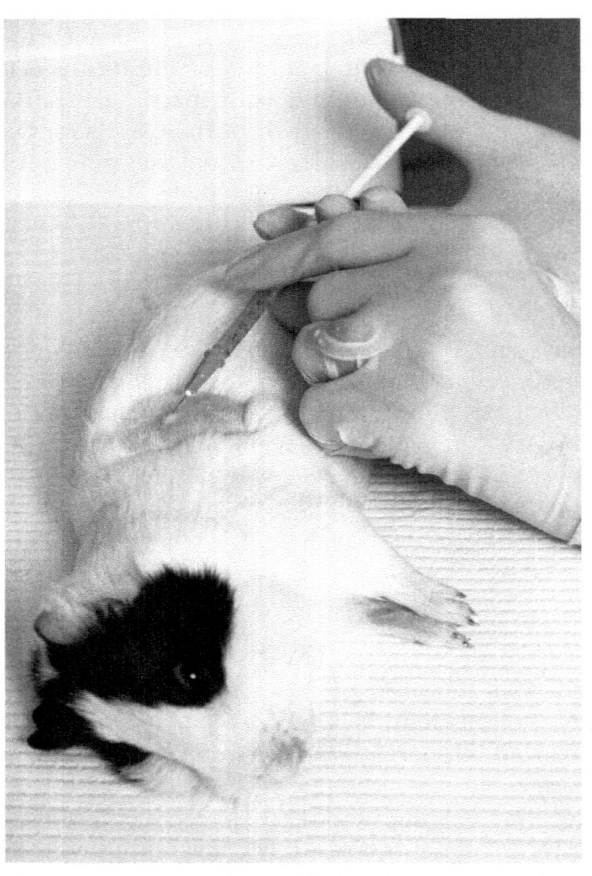

Guinea pig getting an experimental injection. Photo courtesy of National Institutes. Taken by Linda Bartlett. [Public domain], via Wikimedia Commons wikicommons

FIELDS OF STUDY

Animal Science; Biochemistry; Biology; Bio pharmaceutics; Philosophy and Religious Studies

ABSTRACT

Animal testing is an integral component of modern science, product testing, and education. Most significant developments in medicine directly or indirectly rely on animal testing. Public debate about the moral and legal status of animals in society has resulted in numerous regulations on animal testing, yet it continues to be a controversial subject.

BACKGROUND

Animal testing, also known as animal experimentation, animal research, in vivo testing, and vivisection, is used to advance pure and applied research. Behavior, development, evolution, reproductive cloning, and genetic engineering research are all forms of pure research involving animals. Applied research includes medical research, defense research, and toxicology studies for drugs, food additives, pesticides, and cosmetics. The use of animals in medical education and training is typically considered to be a form of animal testing as well.

RATIONALE, SCOPE, AND REGULATIONS

The underlying rationale for the use of animal testing is that living organisms provide interactive, dynamic systems that scientists can observe and manipulate in order to understand normal and pathological functioning as well as the effectiveness of medical interventions. The vast majority of animal testing is for human benefit and relies on the physiological and anatomical similarities between humans and other animals. The term "animal model" refers to the use of live animals to study particular biological processes with the end of extrapolating that information to other animals, particularly humans.

Many species are used in animal testing. Nematode worms and fruit flies are commonly used invertebrates. Zebra fish and mice are commonly used vertebrates. While it is difficult to ascertain the exact number of animals used in research worldwide, the U.S. Department of Agriculture (USDA) reported that in 2015 over 767,000 laboratory animals were used for research, testing, and educational purposes.

This was approximately an 8 percent drop from the USDA's 2014 figures, which stated that over 834,000 animals were used. (In 2013, the USDA noted that over 891,000 animals were used in testing, which was approximately 6 percent greater than was reported in 2014.) These statistics, however, do not include animals not protected by the 1966 Animal Welfare Act, a federal law that has been amended numerous times and regulates the treatment of animals in exhibition, research, and transport. Many rats, mice, and fish are not protected by the act, and it is impossible to know exactly how many are used in the United States for testing. In 2013, the European Union (EU) provided 2011 statistics in its seventh report on the number of animals used for scientific purposes. The report stated that over 85 percent of research animals in the EU were rats, mice, fish, and cold-blooded species such as reptiles. In 2015, the European Commission voted to strike down a proposed ban to stop laboratory experiments on animals, although animal testing of cosmetics was banned in the EU in 2004.

A chief moral concern raised in regard to animal testing is the pain and suffering it involves. Sentience, or subjective awareness, particularly of pain and pleasure, is common to all vertebrates. Evidence for sentience in invertebrates is generally absent, however, and for this reason research on invertebrates is largely unregulated. Cephalopods (the class of animals that includes octopi and squid) are notable exceptions and are covered by regulations in several countries owing to evidence of their sentience. Animal welfare regulations require those who use animals for research or educational purposes to report whether the animal experiences pain and whether pain-relieving medication was provided during procedures. Reporting on mice and rats is not required.

A societal consensus exists that animal testing for the advancement of science and medicine is justified, provided that there are no alternatives the use of animals is kept to a minimum, and that animal pain and distress is minimized. Supporters of animal testing commonly cite the number of major medical advances that have resulted from the practice.

Animal testing is heavily regulated in many countries. Regulations have changed significantly since the mid-twentieth century, and they differ in the numbers of species covered, the kinds of animal welfare protections offered, and the regulatory approaches taken. In the United States, animal testing is governed by two federal statutes: the Animal Welfare Act of 1966 (AWA) and the Health Research Extension Act of 1985, the provisions of which are carried out in the Public Health Service Policy on Humane Care and Use of Laboratory Animals (PHS Policy). AWA establishes the minimum acceptable standards of care and treatment for certain animals in research, testing, experimentation, exhibition purposes, and use as pets. AWA covers all warm-blooded animals yet specifically excludes birds, mice, and rats bred for research purposes, as well as animals used for food, fiber, or many forms of agricultural research. PHS Policy, which applies to all research funded by the National Institutes of Health, applies to all live vertebrates used for research purposes.

Opposition and Alternatives

Opposition to animal testing is diverse and the pros and cons of the use of animals for scientific testing are heavily debated. Disagreement with the practice is based on both scientific and ethical grounds, and it varies according to species and to purpose. Research using primates, monkeys, cats, and dogs is particularly controversial.

Cosmetics testing on animals is especially controversial because many consider the benefit of yet another cosmetic product to be of dubious value when weighed against animals' interests. The practice remains legal in the United States, but in June of 2014, the California Senate passed Joint Resolution 22, which urged the U.S. Congress to pass federal legislation banning the use of all animals in cosmetics testing. Several countries, including Israel, the United Kingdom, and the European Union Cosmetics have outlawed testing on animals. In July of 2014, China removed a requirement for "ordinary cosmetics," such as hair products and fragrances, to be tested on animals. The country still requires, however, that imported and special-use products such as sunblock be animal tested. The next year, South Korea followed in the European Union's footsteps by proposing a five-year plan to phase out the use of cosmetics testing on animals, starting with the prohibition of the use of animals to test finished cosmetic products. New Zealand banned animal testing in cosmetics in 2015. The U.S. Humane Cosmetics Act was introduced to Congress by Rep. Don Beyer and Martha McSally in the summer of 2017, but skeptics believed the bill would not pass due to limited alternatives.

Since federal laws, for the most part, do not protect farm animals from mistreatment in agricultural research experiments, the U.S. Meat Animal Research Center came under fire in early 2015 as a *New York Times* article exposed cruel studies being conducted on animals such as cows and lambs in an effort to increase the meat industry's bottom line. This investigation led to a review by the USDA, which funds the program, and the subsequent appointment of the USDA's Agricultural Research Service's first animal welfare ombudsman.

Those opposed to animal testing on scientific grounds cite the unreliability of predicting effects in humans based on animal models. Some argue that animal testing is not cost-effective; they assert that, given the substantial costs of conducting animal tests, which often last years and cost millions of dollars, the goal of improving human health would be more fully and efficiently realized through a reallocation of funding to implement existing medical technologies more widely. Some argue that much animal testing is immoral because the animal suffering caused is greater than the expected benefits to humans. The stronger animal rights view is that each animal has inherent moral worth, which prohibits humans from using them as experimental subjects for any reason.

First articulated by scientists William M. Russell and Rex L. Burch, the three *R*s—replacement, reduction, and refinement—are influential guiding principles for the humane use of animals in research. Replacement involves seeking to increase alternatives to animal testing that generate the desired research data without the use of sentient animals. Examples of replacement include the use of computer models, epidemiological data, tissue cultures, isolated organs, and nonsentient animals. Reduction is the effort to obtain comparable data using fewer animals or to obtain more data using the same number of animals. Refinement involves favoring research protocols that alleviate or minimize animal pain and distress through the use of analgesics, veterinary care, improved living quarters, and enrichment. Further development and increased implementation of alternatives to and refinement of animal testing is an area of common ground between animal advocates and animal researchers.

———*Joel P. MacClellan*

FURTHER READING

"Animal Research Numbers Continue Downward Trend According to Newly-Released Report." *National Anti-Vivisection Society.* Natl. Anti-Vivisection Soc., 6 July 2015. Web. 9 July 2015.

Aubrey, Allison. "Outrage Over Government's Animal Experiments Leads to USDA Review." *NPR.* NPR, 6 Feb. 2015. Web. 9 July 2015.

Bayne, Kathryn A. L., et al. *Laboratory Animal Welfare.* London: Academic Press, 2014. Print.

Carbone, Larry. *What Animals Want: Expertise and Advocacy in Laboratory Animal Welfare Policy.* New York: Oxford UP, 2004. Print.

"Ending Cosmetics Testing on Animals." *Cruelty Free International.* Cruelty Free International, n.d. Web. 14 June 2016.

Engebretson, Monica. "Will South Korea Be the Next Country to Beat the United States in Ending Cosmetics Testing on Animals?" *Huffington Post.* HuffingtonPost.com, 6 Jan. 2015. Web. 9 July 2015.

Fisher, Elizabeth. "Why We Should Accept Animal Testing." *Huffington Post.* AOL (UK), 17 July 2013. Web. 14 Aug. 2014.

Goodman, Justin R., et al. "Mounting Opposition to Vivisection." *Contexts* 11.2 (2012): 68–69. Print.

Guerrini, Anita. *Experimenting with Humans and Animals: From Galen to Animal Rights.* Baltimore: Johns Hopkins UP, 2003. Print.

Liebsch, Manfred, et al. "Alternatives to Animal Testing: Current Status and Future Perspectives." *Archives of Toxicology* 85 (2011): 841–58. Print.

Monamy, Vaughan. *Animal Experimentation: A Guide to the Issues.* 2nd ed. New York: Cambridge University Press, 2009. Print.

Newton, David E. *The Animal Experimentation Debate.* Santa Barbara: ABC-CLIO, 2013. Print.

"Opening the Doors on Animal Testing." *BBD.* BBC, 15 May 2014. Web. 15 June 2016.

Paul, Ellen Frankel, and Jeffrey Paul, eds. *Why Animal Experimentation Matters: The Use of Animals in Medical Research.* New Brunswick: Transaction, 2001. Print.

Wolfensohn, Sarah, and Maggie Lloyd. *Handbook of Laboratory Animal Management and Welfare.* 4th ed. Oxford: Wiley, 2013. Print.

Animals as a Medical Resource

This image shows a 0.1 x 0.03 inch (2.5 x 0.8 mm) small *Drosophila melanogaster* fly. By André Karwath. (Own work.) [CC BY-SA 2.5 (https://creativecommons.org/licenses/by-sa/2.5)], via Wikimedia Commons

FIELDS OF STUDY

Animal Science; Biochemistry; Biology; Bio pharmaceutics; Medical biotechnology (or Red biotechnology); Philosophy and Religious Studies

ABSTRACT

The use of animals has been a critical component of both human medical research and veterinary research. Although animal research has become a source of controversy among the public, nearly all modern medical advances have been based on some form of animal research.

BACKGROUND

Animals have served purposes related to medicine for centuries. They provided medical products for apothecaries in medieval Europe and for traditional Chinese medicine. Most applications, such as the use of ground rhinoceros horn as an aphrodisiac, were based on nonscientific concepts that have been discarded by modern medicine. However, some techniques, such as the use of spiderweb to stop bleeding, functioned until more effective products became available.

More recently, insulin used to treat diabetes was first harvested from the pancreatic glands of cattle or pigs used in the meat industry; however, the foreign animal proteins sometimes caused allergic reactions. Today, genetically engineered bacteria can provide many hormone products that previously were extracted from animals. Animals have also provided transplant organs. Tissue rejection has been a major problem, but medications can reduce rejection dramatically.

Through genetic engineering, genes that code for pharmaceutical proteins can be incorporated into animals, and medicinal drugs can then be produced from the animal's milk. In 2009, the U.S. Food and Drug Administration (FDA) approved an anticoagulant drug harvested from the milk of genetically engineered goats for the treatment of hereditary antithrombin deficiency; the drug, sold under the brand name ATryn, was the first ever medication produced from genetically engineered animals, and the first FDA-approved biological product derived from such animals. Transgenic animals have also been proposed for the production of drugs to treat cystic fibrosis, cancer, and other disorders. The uses of animals for products and tissues, however, have been minor compared with the use of animals as test subjects in medical research.

ANIMAL MEDICAL RESEARCH

Phenomenal advances in the treatment of human diseases occurred in the century following the U.S. Civil War. The development of the germ theory of disease by Louis Pasteur and Robert Koch, as well as the conquest of most major infectious diseases, was based on extensive animal research. Pasteur's studies of chicken cholera formed a basis for his work, and Koch's breakthrough work with anthrax involved studies with sheep. Most Nobel Prizes in Physiology or Medicine (first awarded in 1901) have involved some form of animal research. The first half of the twentieth century was an era of widespread public support for medical and scientific research. Animal research underlay basic studies in the development of penicillin and other major antibiotics, as well as insulin, surgical techniques, and vaccinations. Partly because people had recent memories of the severity of such major diseases as smallpox and polio, animal research engendered little protest or controversy.

Not all animals are equally useful or appropriate in medical research, because some have systems that differ significantly from human physiology. The closer an animal is to humans evolutionarily, the more likely it is that it will respond to drugs and medical interventions in the same manner that humans will. Most new drugs are first screened on laboratory rats or mice; those drugs that show promise and have no toxic effects may then be tested on primates. Approximately 95 percent of medical research uses mice and other rodents, and nearly all the mice and rats used are "purpose bred" for research. (Cats, dogs, and nonhuman primates make up less than 1 percent of animals used in research.) The protocols for FDA approval of new drugs, as well as agricultural and environmental standards, are based on substantial animal testing to ensure the safety and effectiveness of new medications.

OPPOSITION AND CONTROVERSY
Beginning in the late 1970s, opposition to animal research began to gain national attention. The books *Animal Liberation* (1975), by Peter Singer, and *The Case for Animal Rights* (1983), by Tom Regan, provided a rationale to activists who questioned humans' use of other animals for medical research as well as for food, fur, and other uses. Organizations opposed to some or all animal use in research range from radical groups allegedly responsible for vandalism of research laboratories (the Animal Liberation Front is among the most radical groups) to milder animal protectionist organizations. Probably the best-known animal-rights group is People for the Ethical Treatment of Animals (PETA). Well-known defenders of animal research for medical science include the National Association for Biomedical Research (NABR), the Incurably Ill for Animal Research, and Putting People First.

Some anti-animal-research activists have contended that all animal research can be replaced with alternatives such as research involving tissue culture and computer simulation. Activists also object to the use of animals taken from animal shelters and complain that current care regulations for research animals, particularly under the Animal Welfare Act, are not rigorously enforced by the U.S. Department of Agriculture (USDA).

The scientific community has defended animal research for a number of reasons. Biological systems are much more complex than any computer model devised, so at present computer simulation has severe limitations. New drugs rarely respond in tissue culture exactly as they do in a whole living organism. Researchers point out that although animals are taken from shelters for research use, they constitute a minuscule amount of the dogs and cats that are euthanized annually. By far, most of the animals used in research and teaching are mice and rats. Research facilities are inspected by agencies such as the USDA's Animal and Plant Health Inspection Service, which enforces Animal Welfare Act criteria. The FDA and the Environmental Protection Agency (EPA) also have laboratory practice regulations. The research community also states that approximately 95 percent of laboratory animals are never subjected to pain and that the remaining animals are provided pain-relieving drugs or anesthetics as soon as the study permits.

——*John Richard Schrock*

FURTHER READING
"Animals in Laboratories." *The Humane Society of the United States,* www.humanesociety.org/about/departments/animals_research.html. Accessed 3 Nov. 2016.

"Animal Testing & Cosmetics." *US Food & Drug Administration,* 29 July 2014, www.fda.gov/Cosmetics/ScienceResearch/ProductTesting/ucm072268.htm. Accessed 3 Nov. 2016.

Birke, Lynda, et al. *The Sacrifice: How Scientific Experiments Transform Animals and People.* Purdue UP, 2007.

Carbone, Larry. *What Animals Want: Expertise and Advocacy in Laboratory Animal Welfare Policy.* Oxford UP, 2004.

Haugen, David M., editor. *Animal Experimentation.* Greenhaven Press, 2007. Opposing Viewpoints.

Monamy, Vaughan. *Animal Experimentation: A Guide to the Issues.* 2nd ed., Cambridge University Press, 2009.

Morrison, Adrian R. *An Odyssey with Animals: A Veterinarian's Reflections on the Animal Rights and Welfare Debate.* Oxford UP, 2009.

National Research Council of the National Academies. *Science, Medicine, and Animals: A Circle of Discovery*. National Academies Press, 2004. *The National Academies Press*, www.nap.edu/catalog/10733/science-medicine-and-animals. Accessed 3 Nov. 2016.

Paul, Ellen Frankel, and Jeffrey Paul, editors. *Why Animal Experimentation Matters: The Use of Animals in Medical Research*. Transaction Publishers, 2001.

Pollack, Andrew. "FDA Approves Drug from Gene-Altered Goats." *The New York Times*, 6 Feb. 2009, www.nytimes.com/2009/02/07/business/07goatdrug.html. Accessed 7 Nov. 2016.

Regan, Tom. *The Case for Animal Rights*. Updated ed., U of California P, 2004.

Rudacille, Deborah. *The Scalpel and the Butterfly: The War between Animal Research and Animal Protection*. Farrar, Straus and Giroux, 2000.

Singer, Peter. *Animal Liberation: The Definitive Classic of the Animal Movement*. Updated ed., Harper Perennial, 2009.

Anthrax and biological warfare

Pasteur inoculating a sheep against anthrax. [CC BY 4.0 (http://creativecommons.org/licenses/by/4.0)], via Wikimedia Commons

FIELDS OF STUDY

Biosystems engineering; Biophysics; Chemical engineering; Industrial biotechnology (or White biotechnology); Philosophy and Religious Studies; Political Science

ABSTRACT

Because anthrax is capable of debilitating and killing people and animals quickly, it is an attractive agent for use in biological warfare. The abilities to detect, treat, and neutralize anthrax efficiently are thus necessary to ensure public safety.

BACKGROUND

The bacterium *Bacillus anthracis* resides in soil, and, like other members of the bacterial genus *Bacillus*, can make a highly resistant resting cell known as an endospore. Endospores can withstand heat, desiccation, harsh chemicals, and ultraviolet radiation and can last in soils for centuries. Anthrax, the disease caused by *B. anthracis*, afflicts herbivorous animals, but human anthrax infections result from contact with infected animals or animal products.

Types of Anthrax Infections

Anthrax is caused by the inhalation or ingestion of *B. anthracis* endospores or, in the case of cutaneous anthrax, by contact between damaged skin and *B. anthracis*. Inhalation of endospores causes inhalation anthrax, which typically occurs among workers in textile or tanning industries who handle contaminated animal products such as wool, hair, and hides. The incubation period of inhalation anthrax ranges from one to six days, and the disease follows a two-stage progression. After infection, the patient develops a dry cough, muscle weakness, tiredness, fever, and pressure in the middle of the chest. The second stage begins with the onset of respiratory distress and typically culminates in death within twenty-four hours. Inhalation anthrax has a mortality rate of 95 percent if untreated.

Gastrointestinal anthrax results from the ingestion of undercooked, contaminated meat. Two to

Powell holding a model vial of anthrax while giving a presentation to the United Nations Security Council in February 2003. By United States Government. [Public domain], via Wikimedia Commons

seven days after ingestion, abdominal pain and fever occur, followed by vomiting, nausea, and diarrhea. Gastrointestinal bleeding is observed in some severe cases, and dissemination of the disease throughout the body also results. Fluid loss can result in shock and kidney failure. Approximately 50 percent of cases of gastrointestinal anthrax are lethal.

Cutaneous anthrax results from invasion of the skin by *B. anthracis*. If the skin is damaged by scrapes, cuts, or insect bites, endospores can breach the outer layers of the skin and infect it. After an incubation period of two to five days, small solid and conical elevations of the skin devoid of pus called papules form; these papules then swell, rupture, and blacken. Without treatment, anthrax skin infections can disseminate to other systems, and death occurs about 20 percent of the time.

DETECTION OF ANTHRAX

Growing *B. anthracis* from a blood sample is the best way to demonstrate an anthrax infection in patients who have not yet been given antibiotics. In patients who have begun antibiotic therapy, serological methods that detect antibodies made by the immune system against the bacterium are efficacious. Blood samples from a person who has died from anthrax should yield copious quantities of relatively large, rod-shaped bacteria that are encapsulated and easily visualized with polychrome methylene blue stains.

Automated detection systems (ADSs) can determine whether *B. anthracis* endospores have been released into a setting. The BSM-2000 (Universal Detection Technology), for example, continuously samples the air and heats it. Captured, heated spores release dipicolinic acid (DPA), a compound unique to bacterial endospores. DPA binds to terbium ions (Tb^{3+}), which, together, fluoresce green under ultraviolet light. Other ADSs use Polymerase chain reaction (PCR) to test the air for DNA (deoxyribonucleic acid) sequences specific to *B. anthracis*.

TREATMENT AND PREVENTION OF ANTHRAX

Several antibiotics are effective in the treatment of anthrax infections. High-dose intravenous penicillin G, ciprofloxacin, and doxycycline are typically quite effective. Preventive treatments with oral ciprofloxacin or doxycycline for six weeks are also effective. Anyone exposed to anthrax should begin treatment immediately because the disease can become untreatable with the passage of time.

BioThrax (made by Bioport Corporation) is a vaccine against anthrax. It consists of an extract prepared from a non-disease-causing strain of *B. anthracis*. It is administered as three inoculations given under the skin at two-week intervals, followed by booster injections at six, twelve, and eighteen months, after which yearly boosters are necessary to maintain immunity. BioThrax vaccinations are 93 percent effective in preventing anthrax infections.

When bodies or clothes are contaminated with *B. anthracis* endospores, personal contact can spread the disease. Washing with antibacterial soap and water and treating the wastewater with bleach can rid contaminated bodies of all endospores. Burning contaminated clothing and the corpses of those who have died from anthrax is an effective means of liquidating anthrax from the environment. Burial does not kill endospores. Endospores of *B. anthracis* released into the air are easily removed by means of high-efficiency particulate air (HEPA) or P100 filters.

Decontamination of areas that have been exposed to *B. anthracis* presents several challenges because the bacterial endospores are rather difficult to destroy. Ethylene oxide, chlorine dioxide, liquid bleach, and a decontamination foam created by Sandia National Laboratories kill *B. anthracis* endospores slowly. A cleanup method approved by the Environmental Protection Agency (EPA) that utilizes liquid bleach, water, and vinegar requires contact with a surface for at least sixty minutes. If chlorine dioxide is used in combination with an iron-based catalyst, sodium carbonate, and bicarbonate, disinfection requires only thirty minutes.

ANTHRAX AS A BIOLOGICAL WEAPON
Many nations have examined the potential of *B. anthracis* as a biological weapon. Growing *B. anthracis* is extremely easy, but processing the endospores into a form that is easily disseminated is extremely difficult. The first attempts to use anthrax as a biological weapon utilized rather crude methods. During World War II, the British military experimented with anthrax on Gruinard Island. This experiment so thoroughly contaminated the site that it was quarantined for the next fifty years. Britain then manufactured some five million "N-bombs," which were anthrax-laced explosive devices, to attack German livestock, but the bombs were never used. In 1986, the British government hired a private company to disinfect the soil of Gruinard Island. The company first carted away the island's topsoil in sealed containers and then used 280 tons of formaldehyde mixed with 2,000 tons of seawater to disinfect the soil that remained. In 1990, the British defense minister declared the island safe.

At Fort Detrick in Frederick, Maryland, the U.S. Army developed a special form of anthrax endospores for use as a biological weapon. Such weaponized endospores lack the ionic charges that ordinarily cause them to stick together. Consequently, the spores are easily dispersed as a fine powder that can float for miles on the wind. On November 25, 1969, an executive order from President Richard M. Nixon outlawed offensive biological weapons research in the United States. All existing U.S. stockpiles of biological weapons were subsequently destroyed.

Despite the fact that it was a signatory to the international Biological Weapons Convention of 1972, which was intended to end the production of biological weapons, the Soviet Union produced extensive quantities of weapons-grade anthrax endospores. On April 2, 1979, more than one million people in Sverdlovsk (now Yekaterinburg), Russia, were exposed to an accidental release of anthrax organisms from the local biological weapons plant. More than sixty people died from inhalation anthrax. An extensive KGB-sponsored cover-up from 1979 to 1992 prevented the international community from learning the truth of what happened until Russian president Boris Yeltsin admitted Soviet involvement in this incident. In Africa, South African intelligence services helped the Rhodesian government of Ian Smith use anthrax against humans and the cattle of the black nationalists who were fighting against his government during the late 1970s.

Weaponized endospores were used in the United States during the final four months of 2001, when spores of *B. anthracis* were mailed within the continental United States. Eleven cases of inhalation anthrax and eleven cases of cutaneous anthrax resulted from these attacks, and five people died.

ANTHRAX AND MICROBIAL FORENSICS
Microbial forensics is concerned with the isolation and identification of any microbes used during bioterrorist attacks. Upon arrival at the site of an attack, the microbial forensics team must remove all persons from the site and decontaminate them. Sample collections taken from the air, vents, countertops, sinks, floors, and other surfaces can help the scientists to determine the source of the infection. All samples collected must be properly identified and stored in tamper-proof containers to preserve the chain of custody.

By identifying the exact strain of *B. anthracis* involved in an anthrax outbreak, experts can determine whether the disease has occurred as the result of a bioterrorism attack or as a naturally acquired infection. Various strains of *B. anthracis* show very little DNA sequence variation, but because the entire genome of this organism has been completely sequenced, scientists are able to use PCR to detect single base differences between strains, called single nucleotide polymorphisms (SNPs), and thus provide a fingerprint for each *B. anthracis* strain. If the strains found at the scene of an attack and in the infected individuals are the same, then the agent used in the bioterrorism attack is confirmed. This information can be used in determining both the source of the biological weapon employed and the best treatment options. Molecular forensics identified the strain used in the 2001 postal attacks on American soil as the Ames strain of *B. anthracis*, which was, ironically, developed at Fort Detrick.

——*Michael A. Buratovich, PhD*

FURTHER READING

Alibek, Ken, with Stephen Handelman. *Biohazard: The Chilling True Story of the Largest Covert Biological Weapons Program in the World—Told from Inside by the Man Who Ran It.* London: Hutchinson, 1999. Print.

Decker, Janet. *Anthrax.* New York: Chelsea, 2003. Print.

Guillemin, Jeanne. *Anthrax: The Investigation of a Deadly Outbreak.* Berkeley: U of California P, 2001. Print.

Holmes, Chris. *Spores, Plague, and History: The Story of Anthrax.* Dallas: Durban, 2003. Print.

Miller, Judith, Stephen Engelberg, and William Broad. *Germs: Biological Weapons and America's Secret War.* New York: Simon, 2001. Print.

Wheelis, Mark, Lajos Rózsa, and Malcolm Dando, eds. *Deadly Cultures: Biological Weapons Since 1945.* Cambridge: Harvard UP, 2006. Print

ANTHROPOGEOMORPHOLOGY

The municipality of Dubai has built a global reputation for large-scale developments and architectural works. Among the most visible of these developments are three human-made archipelagos. The two Palm Islands—Palm Jumeirah (image lower left) and Palm Jebel Ali) appear as stylized palm trees when viewed from above. The World Islands evoke a rough map of the world from an air- or space-borne perspective. Photo courtesy of NASA, taken by a member of the Expedition 22 crew. [Public domain], via Wikimedia Commons

FIELDS OF STUDY

Civil engineering; Economics; Environmental biotechnology (or Green biotechnology); Marine biology (or Blue biotechnology); Political Science; Sociology; Urban planning

ABSTRACT

Anthropogeomorphology as a by-product of human activity is a growing concern, as evidenced by a burgeoning global environmental response. Quantitative analyses of climatic changes resulting from anthropogeomorphology have yet to be produced, but growing attention to the subject will render such analyses of great scientific and political interest. Once the phenomenon is better understood, it may be possible not only to prevent further projects from creating negative climatic effects but also to launch projects to mitigate local and global climate trends.

BACKGROUND

Human civilizations alter natural landscapes, both land and water. They partially remove or scrape away natural landforms (subtractive processes), and they build up new topographic features (additive processes). Humans have long reshaped existing bodies of water, created new bodies of water, redirected water in canals or aqueducts, changed the paths of existing rivers, scraped away surfaces, and reclaimed land from seas or marshes behind dikes and seawalls. There is some evidence that these modifications of the landscape affect local climates.

Anthropogenic changes in the landscape easily surpass the scope of natural erosional processes globally. Overconsumption of water resources in such places as the Jordan Valley in the Levant causes bodies of water such as the Sea of Galilee and the Dead Sea to shrink. It also increases desertification by changing the watershed and reducing evaporation with increased land temperature. Replanting trees in arid zones can mitigate surface temperatures and increase orographic precipitation at lower elevations, where evaporation is reduced with dew point and condensation is reached more easily. This has occurred, for example, in Israel's Judaean Hills, where rainfall increased dramatically over thirty documented years after reforestation.

Anthropogeomorphology and Climate Change

One anthropogenic subtractive process has been documented for millennia. Plato noted that severe soil erosion in fourth century BCE Greece was causing much lost topsoil. Ancient observers were sometimes able to determine causes of topsoil loss, such as aggressive deforestation to clear farmland or destruction of native plant roots by animal grazing. Sheep and goat herds were particularly destructive, because they consumed root systems as well as surface plants, effectively killing the plant cover.

Once there were no longer roots to hold the soil, erosion often removed devastating volumes of surface soil. As soil cover moved downward, the resulting alluviation filled river bottoms, lakes, and harbors, which were eventually silted up, changing surface landscapes. Coastlines also changed when alluviated river deltas encroached into bodies of water, as, for example, the Rhone River encroached into the Mediterranean Sea. Erosion is normally a subtle, gradual process, but it can be accelerated by extreme deforestation through human agency. Erosion was exacerbated in the late Roman Empire and afterward in North Africa, where Atlas Mountain deforestation by humans resulted in alluviated coastal watersheds and silted up harbors. This erosion combined with rising land temperatures when rainfall ceased, partly as a result of deforestation, to render the climate much less hospitable. Ultimately, once great North African cities such as Sabratha and Leptis Magna were abandoned as a result of these climate changes.

Alteration of Seacoasts

Another example of ancient anthropogenic land change along seacoasts occurred at Tyre (now in Lebanon), beginning in 332 BCE, when Alexander the Great built a stone causeway out to the then-island city in a siege. Over millennia, the causeway trapped enough marine-transported sandy alluvium that the land bridge—originally around 7 meters across—widened into today's peninsula, which is so much broader that a casual observer would not guess that there was once an open water channel of almost 400 meters between the mainland and island. Prevailing currents from the north built up far more seaborne sandy alluvium on the curved northern side of the peninsula, whereas the southern side of the artificial peninsula remains more contiguous to the original causeway. The volume of sand and eventual structures added over time now approximate about 200 hectares and millions of metric tons of alluvium. This land extension has changed local water and air circulation patterns along and over the coast of Lebanon.

Dams and Reservoirs

Humans have also created many artificial, interior bodies of water, such as reservoirs and artificial lakes. Water storage in artificial lakes fills millions of hectares of land surface on every continent, with concomitant climatic impacts ranging from temperature changes to increased evaporation. Additional anthropogenic aquatic change includes the construction of major canals, such as the Suez Canal linking the Mediterranean and Red Seas and the Panama Canal linking the Pacific Ocean and the Caribbean Sea. Construction of such canals includes the creation or exploitation of connecting lakes and locks to accommodate sea-level differences.

One dramatic human engineering project, China's Three Gorges Dam, is already threatening to offset potential hydroelectric economic gains. Water has seeped into steep lands along the dam's perimeter, causing more than 35 kilometers of banks to cave in, resulting in more than 20 million cubic meters of rockslide since 2003. In addition, there is mounting evidence than regional rainfall has been decreased, leading to drought and loss of biodiversity, while fault activity has increased where the dammed lake sits across two major, active fault lines.

Land Reclamation

Conversely, the Zuider Zee's extensive dikes in the Netherlands reclaimed from the North Sea millions of hectares of land slightly below or just at sea level. The reclaimed land was used extensively for farming. The major dike (Afsluitdijk) created new land polders that gradually reduced the former early twentieth century Zuider Zee by about 38 percent.

A similar phenomenon exists around urban New Orleans, where former swamps and Mississippi River delta wetlands were drained, and dense human settlements were generally protected by extensive levees. Hurricane Katrina's surge in summer, 2005, however, emphasized the fragility of such land reclamation and urbanization. Levee failure caused devastating flooding of 80 percent of New Orleans at great local, regional, and national cost.

Channeling former coastal rivers into stone or concrete storm drains is a major surface change, mostly aimed at reducing flooding. Precedents for such projects can be found in pre-Columbian Inca Peru, along the Urubamba River in the Yucay Valley. Around 1400, Inca engineers not only took out natural oxbows and straightened the river but also created farmland from the natural floodplain, where there had been no prior extensive agriculture, thus humidifying the air in the Yucay Valley.

Anthropogeomorphology and the Mining Industry

Perhaps the most extensive modern anthropogeomorphologic surface change in North America has been produced by the Canadian oil sands industry, which engages in open pit mining of the Athabaska River Valley in Alberta. Boreal forest and earth are scraped away to depths of about 30 meters in many places over a 390-square-kilometer area. This mining is creating vast, toxic, mine-tailing sludge lakes and growing pollutant containment problems. Naphthenic acid and polycyclic aromatic hydrocarbons, which are not easily degradable for centuries, leak from the mines into water tables, as well as the Athabaska River.

The oil sands industry extracts enormous quantities of bitumen-laced sands, heats them, and cleans them using hot water and other agents. This process has resulted in a huge spike in aerosol CO_2 emissions over northern Canada, at a much higher rate than the emission rate of conventional oil production. Oil produced in Alberta's oil sands is mostly consumed by the United States; this single source supplies 10 percent of total U.S. foreign oil.

It takes 3.6 metric tons of earth to produce one barrel of oil, and 750,000 barrels of synthetic crude oil are produced per diem in Alberta. This non-stop daily process mixes 907,000 metric tons of crushed oil sands with 181,000 metric tons of water to be heated to boiling temperatures for steam. It requires considerable energy, generated from natural gas, and yields a vast landscape of stored toxic waste. Environmental scientists are beginning to link accelerated Arctic ice-cover loss with the northern Canadian oil sand industry. The industry may raise local temperatures through deforestation, as well as adding steam to the air.

—*Patrick Norman Hunt*

Further Reading

Atlas of Israel: Cartography, Physical and Human Geography. New York: Macmillan, 1985. Documents and maps all geophysical data available in Israel for over forty years of quantitative measurements from meteorology, oceanography, demography, and other disciplines.

Hvistendahl, Mara. "China's Three Gorges Dam: An Environmental Catastrophe?" *Scientific American*, March 25, 2008. Assessment of China's initial failure to consider the dam's ecological impacts and recent acknowledgment of its possible consequences for the environment.

Kunzig, Robert. "Scraping Bottom: The Canadian Oil Boom." *National Geographic*, March, 2009. Fair but controversial report on the Canadian oil industry's bringing about both an economic boom and environmental degradation.

LaFreniere, Gilbert. *The Decline of Nature: Environmental History and the Western Worldview.* Bethesda, Md.: Academica Press, 2008. Examines human policies that have affected nature through history, detailing both philosophical attitudes underlying those policies and their possible catastrophic results.

Nikiforuk, Andrew. *Tar Sands: Dirty Oil and the Future of a Continent.* Berkeley, Calif.: Greystone Books, 2008. Examines potential long-term environmental effects of Canada's oil sands industry in light of possible climatic changes and the ethics of fossil fuel consumption.

Antibiotic-resistant bacteria

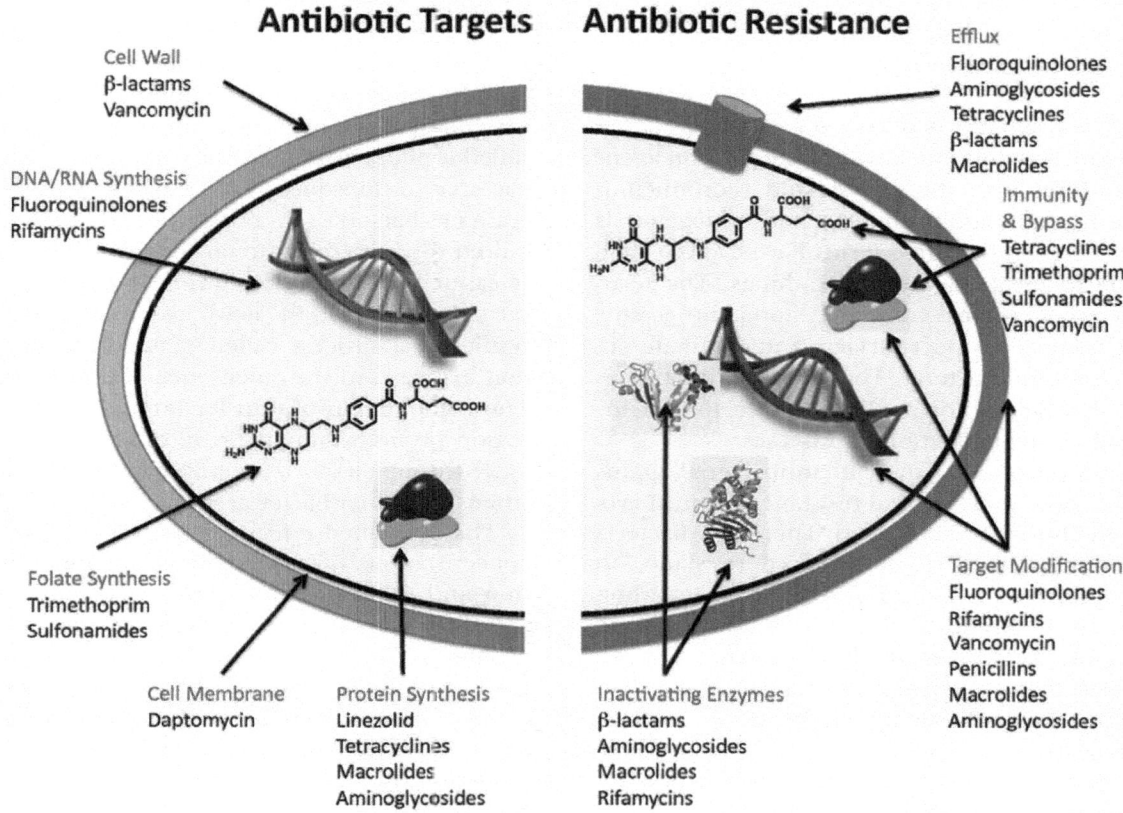

A number of mechanisms used by common antibiotics to deal with bacteria and ways by which bacteria become resistant to them. By Gerard D Wright. [CC BY 2.0 (http://creativecommons.org/licenses/by/2.0)], via Wikimedia Commons

FIELDS OF STUDY

Biochemistry; Biology; Biopharmaceutics; Biophysics; Biotechnology; Medical biotechnology (or Red biotechnology); Molecular biology

ABSTRACT

Although most bacteria are benign, a small percentage are pathogenic, or disease-causing. Bacteria rank among the most important of all disease-causing organisms in humans, and bacterial infections are countered by a wide variety of antibiotic and antibacterial agents. Repeated use of such agents results in bacterial resistance, necessitating the development of stronger antibacterial agents. Increasing fears that antibiotic-resistant strains of bacteria may be used as bioweapons add urgency to efforts to develop new antibacterial agents.

BACKGROUND

Less than 10 percent of all bacteria threaten human health. These disease-causing species are notorious for such diseases as cholera, typhus, and syphilis. The most common and some of the most deadly forms of bacterial diseases are respiratory infections, such as tuberculosis, which kill millions of people every year. Countries around the world have used antibiotic drugs to treat bacterial infections for more than fifty years. The initial introduction of antibiotics was markedly successful, but continued and widespread use has resulted in a phenomenon in which microbial adaptation is making targeted bacteria increasingly difficult to control. This bacterial resistance to antibiotics is of special concern, as ever more powerful antibiotics must be developed.

ANTIBIOTICS AND ANTIBACTERIALS
In its broadest definition, an antibacterial is an agent that interferes with the growth and reproduction of bacteria. Although antibiotics and antibacterials both attack bacteria, these terms have evolved over the years to mean two different things. The term "antibacterials" is most commonly applied to agents that are used to disinfect surfaces and eliminate potentially harmful bacteria. The term "antibiotics" is commonly reserved for medicines given to humans or animals to treat infections or diseases.

Bacteria become resistant to antibacterial agents in one of three ways: natural resistance, vertical evolution, and horizontal evolution. Therefore, bacteria exhibit either inherited or acquired resistance to antibacterial agents. Natural resistance occurs when bacteria are inherently resistant to an antibacterial. For example, a gram-negative bacterium has an outer membrane that establishes an impermeability barrier against the antibiotic it manufactures, so it does not self-destruct.

Acquired resistance occurs when bacteria develop resistance to an antibacterial agent to which the population has been exposed. This may occur through mutation and selection (vertical evolution) or exchange of genes between strains and species (horizontal evolution) of the bacteria exposed to the antibacterial agent.

Vertical evolution represents an example of Darwinian evolution driven by principles of natural selection. Genetic mutations in the bacteria population create new genes or combinations of genes that are resistant to the antibacterial agent. While the nonmutant, sensitive bacteria are killed, bacteria containing the mutated genes survive, and their progeny populate the increasingly resistant colony.

Another form of acquired resistance, horizontal evolution, is the transfer of resistant genes from one bacterium to another in the population. For example, *Escherichia coli* or *Shigella* may acquire a gene from a streptomycete that is resistant to the antibiotic streptomycin. Following this transfer, the population contains a mutant *E. coli* bacterium now resistant to streptomycin. Then, through the process of selection, it donates these genes to further generations, creating a resistant strain.

Transfer of genes in bacteria occurs in one of three ways: conjugation, transduction, or transformation. In conjugation, the gene-containing DNA (deoxyribonucleic acid) crosses a connecting structure, called a pilus, from a donor bacterium to recipient bacteria. In transduction, a virus may transfer genes between bacteria. In transformation, DNA is acquired directly from the environment, having been released from another bacterium. Following transfer, the combination of the newly acquired gene or genes results in a process called genetic recombination that may lead to the emergence of a new genotype. The combination of transfers and genetic recombination promotes rapid spread of antibacterial resistance through a species population and also between strains and other bacterial species.

The combined effects of fast growth rates, high concentrations of cells, genetic processes of mutation and selection, and genetic recombination account for the extraordinary rates of adaptation and evolution observed in bacteria populations. For these reasons, bacterial resistance to antibacterials is a common occurrence and one that promises to be of increasing concern in the future. In fact, high levels of antiobiotic-resistent bacteria are found in numerous common infections, including pneumonia and urinary tract infections, and there are an increasing number of cases of antibiotic resistant tuberculosis.

BACTERIAL RESISTANCE AND FORENSIC SCIENCE
The importance of bacteriology in forensic science is recognized in diverse areas, including DNA profiling, toxicology studies, fingerprinting, and the tracing of violence stemming from or potentially relating to murders. Bacteria have been used as weapons and can be the causes of violence, but they may also serve as tools in the investigation of crimes.

The most serious threat posed by bacteria is their possible use in biological warfare, especially in acts of bioterrorism. For example, *Bacillus anthracis*, which causes anthrax, has become a preferred bacterial strain used by terrorists. Strains of deadly bacteria selected especially for their antibody resistance can pose health threats of enormous proportions at both local and global levels.

Some research has suggested that bacterial infections can lead to criminal behavior. For example, *Streptococcus* infections have been linked to hyperactivity, and hyperactivity has been linked to criminal behavior. Some defense lawyers have used such research findings in attempts to explain their clients' actions, connecting criminal behavior with infection-caused states of delirium.

In some cases, the bacteria present at the site of a crime can give important clues about the crime itself. For instance, bacteria can reveal how long a person has been dead or the temperature the body was subjected to after death. Heart and spleen blood cultures may be taken at autopsy to identify any possible infections or diseases the deceased may have had.

—*Dwight G. Smith*

FURTHER READING

"Antimicrobial Resistance." *WHO*. World Health Organization, Apr. 2014. Web. 10 Mar. 2013.

Bartelt, Margaret A. *Diagnostic Bacteriology. A Study Guide*. Philadelphia: Davis, 2000. Print.

Breeze, Roger G., Bruce Budowle, and Steven E. Schutzer, eds. *Microbial Forensics*. Burlington: Elsevier, 2005. Print.

Cummings, Craig A., and David A. Relman. "Microbial Forensics: When Pathogens Are 'Cross-Examined.'" *Science* 296 (2002): 1976–79. Print.

Larkin, Marilynn. "Microbial Forensics Aims to Link Pathogen, Crime, and Perpetrator." *The Lancet Infectious Diseases* 3.4 (2003): 180–81. Print.

Tsokos, Michael, ed. *Forensic Pathology Reviews*. Vol. 4. Totowa: Humana, 2006. Print.

Antibiotics as defense against biological warfare

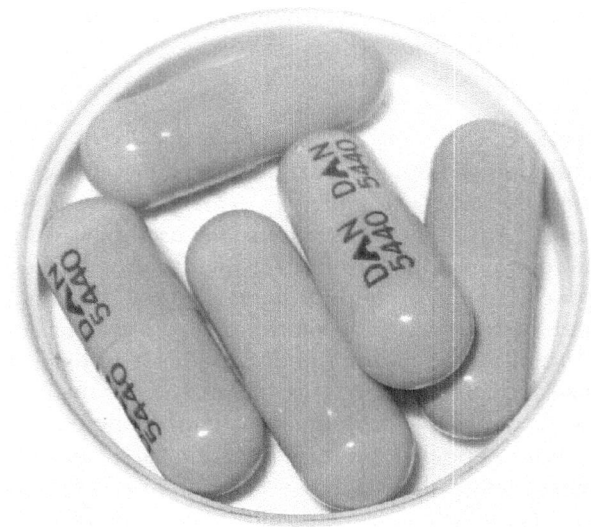

Generic 100 mg doxycycline capsules. By Shorelande. (Own work.) [GFDL (http://www.gnu.org/copyleft/fdl.html), CC-BY-SA-3.0 (http://creativecommons.org/licenses/by-sa/3.0/) or CC BY 2.5 (http://creativecommons.org/licenses/by/2.5)], via Wikimedia Commons

FIELDS OF STUDY

Biochemistry; Biology; Biopharmaceutics; Biophysics; Biotechnology; Medical biotechnology (or Red biotechnology); Molecular biology

ABSTRACT

Antibiotics kill certain types of bacteria that cause diseases without severely hurting the patients; they can thus abate the progression of some diseases and extensively reduce the effects of those diseases on human populations. Because of increasing threats of terrorism in the modern world, law-enforcement agencies are interested in the effective use of antibiotics for blunting the potential threat of microorganisms as biological weapons.

BACKGROUND

Microbial infections cause illnesses that diminish the quality of life and productivity and can eventually cause death. Effective early treatment can reverse the progression of some diseases, decrease the convalescence time, and potentially prevent the spread of infection from one person to another. Treatment can also check the onset of particular undesirable aftereffects caused by some diseases. Antibiotics are the first-line treatments against infectious diseases.

CLASSIFICATION

Most antibiotics are derived from compounds made by various microorganisms to kill competing bacteria. Many antibiotics, however, are completely synthetic in their composition, even though their chemical structures are variations of naturally produced antibiotics.

Antibiotics are classified according to their chemical structures, and drugs with similar chemical structures are classified in a common group. The largest group of antibiotics, the beta-lactams, consists of the penicillins (such as ampicillin, amoxicillin, and carbenicillin), cephalosporins (such as cephalexin, ceflaclor, and ceftizoxime), monobactams (aztreonam), and carbapenems (such as imipenem and meropenem). Other antibiotic groups include the macrolide antibiotics (such as erythromycin, clarithromycin, and azithromycin), tetracyclines (such as doxycycline and minocycline), aminoglycosides (such as streptomycin, kanamycin, tobramycin, and neomycin), sulfanilamides (such as sulfadiazine, sulfamethoxazole, and sulfamethizole), trimethoprim (similar to sulfanilamides but does not contain sulfur atoms), fluoroquinolones (such as ciprofloxacin, levofloxacin, moxifloxacin, and norfloxacin), and glycopeptide antibiotics (vancomycin).

Several antibiotic groups consist of only one drug; these include bacitracin, clindamycin, chloramphenicol, cycloserine, and fosfomycin. Streptogramin A and dalfopristin are given as a combination, and these drugs are the only members of the streptogramin group. The oxazolidinones group contains only one member, linezolid. A handful of antibiotics called antimycobacterials are specifically used to treat tuberculosis: isoniazid, ethambutol, pyrazinamide, and rifampin.

MODE OF ACTION
Several chemically unrelated groups of antibiotics can target similar biochemical processes in bacterial cells. The abilities of distinct antibiotics to kill particular bacterial species vary extensively. Some antibiotics can kill only a few bacterial species (narrow-range antibiotics), whereas others can eradicate many different types of bacteria (broad-spectrum antibiotics).

Several antibiotics inhibit bacterial protein synthesis, which quickly kills bacterial cells. The protein synthesis-inhibiting antibiotics include the macrolides, tetracyclines, aminoglycosides, clindamycin, streptogramins, oxazolidinones, and chloramphenicol. Clindamycin is used to treat infections with anaerobic bacteria, and streptogramins and oxazolidinones are used for infections that resist other antibiotic treatments.

Many antibiotics inhibit the synthesis of the bacterial cell wall, which surrounds the bacterium and protects it. Without their cell wall, bacterial cells succumb to the host's immune system. Antibiotics that inhibit bacterial cell wall synthesis include the beta-lactams, glycopeptides, bacitracin, cycloserine, and fosfomycin.

Some antibiotics interfere with the synthesis of essential molecules. Folic acid is an exceedingly vital cofactor for bacterial metabolism, and without it, bacteria die. The sulfanilamides, trimethoprim, and the drug dapsone (used to fight Hansen's disease, or leprosy) obstruct the synthesis of folic acid. The fluoroquinolones inhibit bacterial DNA (deoxyribonucleic acid) replication.

Of the antituberculosis drugs, rifampin inhibits gene expression, and isoniazid and ethanbutol hamper the synthesis of the waxy cell wall of *Mycobacterium tuberculosis*, the bacterial agent that causes tuberculosis. Pyrazinamide inhibits the synthesis of fatty acids, which are used for the construction of biological membranes.

CLINICAL USE
Antibiotic treatment can have great benefits even when the exact causative agent of a disease is unknown. Antibiotics thus are usually used before the microorganism responsible for the illness is defined. However, because continuous exposure of bacteria to antibiotics allows the evolution of bacteria that are resistant to antibiotics, the overuse of these drugs is ill-advised, and judicious use of antibiotics is the rule. For example, given that more than 90 percent of sinus and upper-respiratory infections are caused by viruses rather than by bacteria, immediate prescription of antibiotics for such conditions is unwarranted.

Prescribing heath care professionals use a protocol known as empirical antimicrobial therapy (EAT) to guide their choices of antibiotics. Using EAT, the prescribing professional attempts to identify the bacterium most likely responsible for the illness through collection of a medical history, physical examination, and laboratory analyses of infected tissues. Because certain bacterial species have a tendency to infect certain organs, specific antibiotics are typically recommended for particular infections. Typically, certain drugs are considered to be the first choice for particular infections, and alternative drugs are used if the first-choice drugs fail to achieve the desired results. For example, the bacterial organisms *Streptococcus pneumoniae*, *Moraxella catarrhalis*, and *Haemophilus influenzae* cause the vast majority of middle-ear infections (otitis media), so the first-choice treatment for these infections is amoxicillin or a combination of trimethoprin and sulfamethoxazole (in a one-to-five ratio); the second-choice treatment is amoxicillin in combination with clavulanate or cefurxime axetil.

ANTIBIOTICS AND FORENSICS

The presence of antibiotics in bodily fluids or tissue samples obtained after death usually indicates the presence of an infection in the deceased. The techniques for detecting antibiotics or their breakdown products in postmortem tissues exploit the unique chemical structure of each antibiotic. Cephalosporins, for example, are detected in postmortem tissues by means of high-performance liquid chromatography (HPLC), which separates compounds according to their differing rates of movement through a porous support material.

The prescription of antibiotics to prevent an impending infection is called antibiotic prophylaxis. In one example of the use of antibiotic prophylaxis, ciprofloxacin was given to approximately ten thousand people who had potentially been exposed to *Bacillus anthracis*, the causative agent of anthrax, as the result of bioterrorism attacks in New York City, Washington, D.C. and Boca Raton, Florida, in the fall of 2001. This step probably saved many lives. The aggressive prophylactic use of antibiotics has the potential to thwart a bioterrorism attack.

——*Michael A. Buratovich*

FURTHER READING

Gilbert, David N., et al. *Sanford Guide to Antimicrobial Therapy 2015*. 45th ed. Sperryville: Antimicrobial Therapy, 2015. Print.
Sachs, Jessica Snyder. *Good Germs, Bad Germs: Health and Survival in a Bacterial World*. New York: Hill, 2007. Print.
Scholar, Eric M., and William B. Pratt, eds. *The Antimicrobial Drugs*. New York: Oxford UP, 2000. Print.
Smith, Frederick P., ed. *Handbook of Forensic Drug Analysis*. Burlington: Elsevier, 2005. Print.
Walsh, Christopher. *Antibiotics: Actions, Origins, Resistance*. Washington: ASM, 2003. Print.

ARCHAEBACTERIA

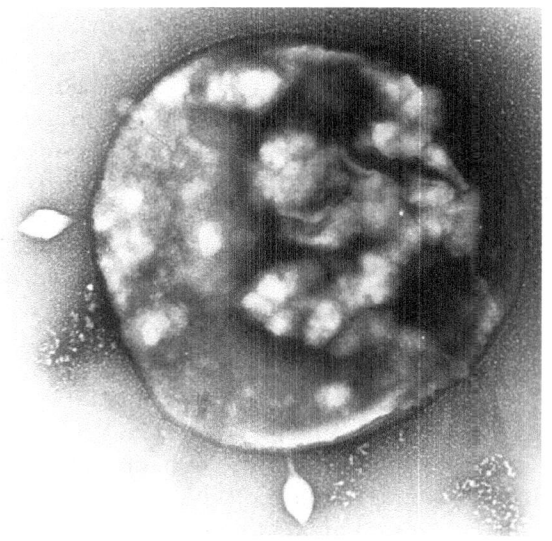

Sulfolobus infected with the DNA virus STSV1.[133] Bar is 1 micrometer. By Xiangyux. (English Wikipedia.) [Public domain], via Wikimedia Commons

FIELDS OF STUDY

Biology; Botany; Chemistry

ABSTRACT

Archaebacteria are primitive, one-celled life-forms without a distinct nucleus, different from bacteria in their genetic components. They have been found to be genetically unique and are probably one of the earth's earliest life-forms.

EARLY LIFE-FORMS

The nature of the earth's earliest life-forms has always been an intriguing question for both the earth sciences (paleontology) and biology. The fossil record shows that one-celled organisms are very ancient, their oldest-known fossils being almost 3.5 billion years old. The long fossil record of prokaryotes consists of both preserved fossil cells and distinctive layered mineral structures called stromatolites and microbialites; these structures were produced from the cell metabolisms of colonies of prokaryotes. From at least 3.5 billion years ago to around 1 billion years ago, microscopic prokaryotes were the earth's only organisms. They included, as they do today, a diversity of forms commensurate with their long evolutionary history. The appearance of the eukaryotic cell (more than 1 billion years ago) ushered in the age of multicellular organisms (metazoans and metaphytes) some 700 million years ago, and these have become the dominant lifeforms on the earth. What the much earlier prokaryotic organisms were like and which type of one-celled organisms produced the microbialites is unclear. Usually these oldest of fossils are attributed to the life activities of photosynthetic organisms, particularly the cyanobacteria, referred to in many works as the blue-green algae. A number of other types of one-celled life-forms could have been responsible for some of them, particularly the photosynthetic bacteria and possibly the archaebacteria.

Molecular biologists, utilizing ribonucleic acid (RNA) nucleotide sequencing and other biochemical methods, believe that the nature of early life-forms can be discovered. RNA nucleotide sequences of amino acids can be regarded as a sort of chemical "historical document" that is capable of being "read." The closer two RNA nucleotide sequences are to each other, the smaller is the evolutionary distance between them and the more recently in geologic time they separated from each other. The further one nucleotide RNA sequence is from another, the greater is the evolutionary distance that separates the two organisms. Utilization of nucleotide sequencing can thus produce an evolutionary tree, or phylogeny, for even the most primitive of organisms, and therefore it becomes possible to determine which organisms out of the great variety of primitive life-forms currently living were some of the first to appear. Through information obtained by such sequencing, archaebacteria have been recognized by molecular biologists as some of the most primitive and biochemically unique of organisms. Archaebacteria RNA sequences turn out to be distinctly different from those of other bacteria, even though the various organisms that comprise archaebacteria look like and were previously placed with the bacteria.

Archaebacteria is a polyphyletic group that once included species from two domains, Archaea and Bacteria. There are now considered to be three domains: Archaea, Bacteria, and Eucaryota. Although there are similarities in cell structure and function among Bacteria and Archaea, and some trees group the Archaea and Bacteria, the relationship is currently unclear. In both their tolerance of extreme ecological conditions and their metabolism, the archaebacteria differ from all other prokaryotes, a condition that has led biologists to consider these organisms as particularly well suited to the adverse conditions of the early earth. The very earliest eras of geologic time may well have been the age of archaebacteria.

MOLECULAR CHARACTERISTICS

On a fundamental molecular level, archaebacteria are different in their biochemistry from the other prokaryotes. Nucleotide RNA sequences and other biochemical differences that exist between the various types of eubacteria (bacteria exclusive of the archaebacteria) are minor when compared with the differences between eubacteria and the archaebacteria. Nongenetic differences include such features as cell walls, those of all eubacteria being composed of a complex polymer called peptidoglycan, which is a sugar derivative. In contrast, cell walls of the various types of archaebacteria are composed of a variety of other materials, none of which is peptidoglycan. The lipids (fats) in archaebacteria cells are also fundamentally distinct from the lipids in the cells of both eubacteria and eukaryotes. Ribosomal RNA is what ultimately distinguishes the archaebacteria, for it is markedly different in its sequences of bases from any eubacteria. In higher eukaryotic organisms, where the fossil record is good, greater RNA ribosomal differences exist between those organisms that are separated by long periods of geologic time than between those that are separated by shorter periods of time. Ribosomal RNA differences and the biochemical differences that exist between the eubacteria and the archaebacteria suggest that an evolutionary distance of great magnitude separates them.

The eukaryotic cell has long been observed to be a sort of combination between prokaryotic-type cells, functioning within the cell as chloroplasts and mitochondria, and another cell type that "ingested" the prokaryotes and incorporated them to become a more complex entity in symbiotic collaboration.

Archaebacteria have some genetic characteristics that suggest a link with the eukaryotes, which has led some scientists to propose a predecessor to them both: The "other" cell type that linked up with prokaryotes underwent substantial further evolution and eventually became the eukaryotic cell. The pre-eukaryotic other cell type is known as urkaryote. All three of these cell types—the prokaryote, the urkaryote, and the eukaryote—are hypothesized to have arisen from a common cell ancestor, the progenote. The progenote may have been biochemically simpler than any of the three fundamental life-forms that arose from it, an event that might have taken place during the first 1 billion years of earth history.

THERMOACIDOPHILES

Archaebacteria are represented by three classes: the thermoacidophiles, the extreme halophiles, and the methanogens. The thermoacidophiles occupy hot, acid environments, often rich both in metallic ions and in sulfur compounds, such as hot springs and fumaroles. These organisms are viable under the acidic hot conditions intolerable to other life-forms, with temperatures as high as 75 degrees Celsius. The extreme halophiles, or halobacteria, live only in extremely salty environments. The methanogens are anaerobes that metabolize organic material to form methane; they were the first of the archaebacteria to be discovered.

Igneous activity of various sorts appears from the geologic record to have been much more intensive and widespread on the early earth (between 2 and 4 billion years ago) than during more recent geologic times or at present. A terrestrial geologic record for the first 1 billion years of the earth's history is unknown, as the widespread igneous and tectonic activity during this time seems to have destroyed the evidence; an actual record begins at nearly 4 billion years ago. During the next 1.5 billion years (the Archean eon), igneous phenomena and massive tectonism were still dominant. Hot-spring and fumarolic activity would have been more commonplace during these early times than during later geologic time, and these environments favor the thermoacidophiles. Although the fossil and sedimentational record of the Archean eon does not negate the possibility that archaebacteria were some of the most widespread and dominant life-forms of that time, determination that a particular organic— or presumed organic—structure of the early earth was produced by archaebacteria is quite difficult and may well be impossible. A number of puzzling structures, seemingly of biogenic origin, have been reported from Archean strata. Some of these stromatolite-like or microbialite-like structures are associated with what appear to be hot-spring deposits. These structures may well represent minerals deposited as a consequence of life activities of thermoacidophiles associated with geothermal activity. Like so many stromatolite-like and microbialite-like structures, unequivocal proof as to their biogenic origin is difficult to obtain. Structures similar to them, however, are produced today in hot springs, in the hottest waters of which live communities of thermoacidophile archaebacteria.

Sometimes hot-spring deposits and structures contain carbon-rich sediments or graphite. In addition, some hydrothermal veins of various ages have associated carbonaceous or graphitic material. It is possible that such material might have originated from thermophyllic archaebacteria. Stratified metallic element deposits are known from Archean strata, some of which have been thought to be of biogenic (possibly archaebacterial) origin. Some of these deposits yield microbialites exhibiting distinctive dome, finger, or layered structures containing metallic oxides or carbonates. Today, such structures are produced by the cyanobacteria and by the photosynthetic bacteria, but these younger stromatolites lack components like the oxidized metals. Other Archean stromatolites or microbialites and stromatolite-like structures associated with geothermally active environments have a distinctive "signature" different from later forms, and their origin by thermoacidophile archaebacteria cannot be ruled out.

EXTREME HALOPHILES

The second group of archaebacteria, the extreme halophiles, requires an intensely saline environment. Shallow, marginal marine areas, evaporite basins, and salt flats are the niches in which these organisms generally flourish. Physiologically, the extreme halophiles are photosynthetic; however, the photosynthetic pigment is not chlorophyll, but rather a light-sensitive red pigment, bacterial rhodopsin. The cell walls of the extreme halophiles differ from those of other bacteria in the presence of compounds that prevent destruction of the walls in the high salt concentration conditions under which they live. The chemical similarities of ribosomes and lipids of both the extreme halophiles and the methanogens suggest a common origin.

Again, the fossil record of these organisms is difficult to interpret; some biologists suggest that the halophiles were more prevalent early in the earth's history than they are today. Fossil rod-shaped bacterial cells have been found as far back as the mid-Archean (3.2 billion years ago); however, as the gross morphology of archaebacteria differs little from that of eubacteria, the evidence remains inconclusive.

Peculiar and distinct microbialites of Archean age that are associated with radial sprays of gypsum crystals were described from western Ontario in the 1910's by Charles Doolittle Walcott. Walcott, a pioneer North American paleontologist who concentrated on the early (Precambrian and Cambrian) fossil record, made many finds of peculiar structures resembling fossils in Precambrian strata, many of which remain a mystery. Walcott thought that the radiating gypsum crystals were the rays, or spicules, of a type of spongelike organism he called atikokania. Associated with Walcott's atikokania are distinctive microbialites that contain "lenses" of gypsum that almost certainly originated in a very saline environment. These microbialites could possibly represent the product of physiological activity of the extreme halophiles when, during the process of photosynthesis, they locally removed carbonic acid from the saline water. The white lenses that characterize these distinctive microbialites are gypsum fillings between the black calcium carbonate bands, possibly precipitated by photosynthesis of halophilic archaebacteria.

METHANOGENS

Methanogens produce their metabolic energy either from the breaking down of organic compounds incorporated into sediments or from the reduction of carbon dioxide in the presence of elemental hydrogen, with the consequent release of methane. Were it not for the methanogens, organic carbon would eventually become incorporated into the sediments of the earth's crust, where it would accumulate and could not be recycled back into the biosphere; the methanogens facilitate this recycling of carbon. Methanogens, like the other archaebacteria, have biochemical features distinct from all other bacteria, suggesting that they evolved separately from them.

Like the other archaebacteria, methanogens differ from other prokaryotes in the sequences of nucleotides that make up the RNA in their ribosomes and protein. Fossil methanogens are more difficult to distinguish within the geologic record than other archaebacteria, however, as they leave no distinctive chemical "footprint," as the others can. The abundance of black, carbon-rich sediments in strata of the Archean eon suggests that the oxygen-free, anaerobic environment in which the methanogens flourish was commonplace during that time. The methanogens' biochemical uniqueness, and thus presumed great geologic age, along with the anaerobic Archean Earth environment, suggests that they may have been a dominant part of the Archean biosphere and not restricted as they are today.

———*Bruce L. Stinchcomb*

FURTHER READING

Garrett, Roger A., and Hans-Peter Klenk, eds. *Archaea: Evolution, Physiology, and Molecular Biology.* Malden, Mass. Wiley-Blackwell, 2007. A compilation of scientific reviews and specialist articles discussing the Archaea domain. Discusses the phylogenetic research conducted on these organisms as well as the physiology and specifically metabolic pathways of Archaea. Written for a scientific audience with a biology background.

Gunde-Cimerman, Nina, Aharon Oren, and Ana Plemenita. *Adaptation to Life at High Salt Concentrations in Archaea, Bacteria, and Eukarya.* New York: Springer, 2011. This text provides information on high-salinity habitats and the halophilic organisms found there. The environment overview is followed by sections devoted specifically to Archaea, bacteria, fungi, protozoa, and viruses.

Howland, John L. *The Surprising Archaea: Discovering Another Domain of Life.* New York: Oxford University Press, 2000. Provides an all-encompassing knowledge of Archaea, from their evolution to their ecology. The author captures the excitement of their discovery within the scientific community.

Kandler, Otto, and Wolfram Zillig, eds. *Archaebacteria Eighty-five: Proceedings of the EMBO Workshop on Molecular Genetics of Archaebacteria.* Forestburgh, N.Y.: Lubrecht and Cramer, 1987. A proceedings volume on molecular genetics, biology, and biochemistry of archaebacteria. Two papers are concerned with the geologic and paleontologic record of archaebacteria: "Traces of Archaebacteria in Ancient Sediments," by J. Hahn and Pat Haug, and "Morphological and Chemical Record of the Organic Particles in Precambrian Sediments," by H. D. Pflug. The latter paper illustrates a wide variety of microstructures from Precambrian sediments and discusses possible pathways by which archaebacteria and other prokaryotes could have been responsible for the concentration of many metallic ore deposits throughout various parts of the Precambrian. Other papers such as "Archaebacterial Phylogeny: Perspectives on the Ur Kingdoms," by C. R. Woese and G. J. Olsen, present biochemical reasons substantiating the uniqueness of archaebacteria from both the eubacteria and the eukaryotes. Other papers probe the biochemical similarities with eukaryote cytoplasm and peculiar substrate requirements.

McMenamin, Mark. *Discovering the First Complex Life: The Garden of the Ediacara.* New York: Columbia University Press, 1998. This entertaining study of the earliest complex life-forms on the planet details the author's work on these organisms. Written for the interested student but understandable by the general reader.

Schopf, J. William, ed. *Major Events in the History of Life.* Boston: Jones and Bartlett, 1992. An excellent overview of the origin of life, the oldest fossils, and the early development of plants and animals. Written by specialists in each field but at a level that is suitable for high school students and undergraduates. Although technical language is used, most of the terms are defined in the glossary.

Woese, Carl R. "Archaebacteria." *Scientific American* 244 (June 1981): 98-122. One of the most comprehensive articles available on the archaebacteria. Distinctive attributes characteristic of the archaebacteria as determined through molecular biology are enumerated. The author was one of the workers originally involved in the discovery of the biochemical uniqueness of archaebacteria.

ARTIFICIAL INTELLIGENCE

FIELDS OF STUDY

Bioinformatics; Biorobotics; Computer engineering; Electrical engineering

ABSTRACT

Artificial intelligence is the design, implementation, and use of programs, machines, and systems that exhibit human intelligence, with its most important activities being knowledge representation, reasoning, and learning. Artificial intelligence encompasses a number of important subareas, including voice recognition, image identification, natural language processing, expert systems, neural networks, planning, robotics, and intelligent agents. Several important programming techniques have been enhanced by artificial intelligence researchers, including classical search, probabilistic search, and logic programming.

BACKGROUND

Artificial intelligence is a broad field of study, and definitions of the field vary by discipline. For computer scientists, artificial intelligence refers to the development of programs that exhibit intelligent behavior. The programs can engage in intelligent planning (timing traffic lights), translate natural languages (converting a Chinese website into English), act like an expert (selecting the best wine for dinner), or perform many other tasks. For engineers, artificial intelligence refers to building machines that perform actions often done by humans. The machines can be simple, like a computer vision system embedded in an ATM (automated teller machine); more complex, like a robotic rover sent to Mars; or very complex, like an automated factory that builds an exercise machine with little human intervention. For cognitive scientists, artificial intelligence refers to building models of human intelligence to better understand human behavior. In the early days of artificial intelligence, most models of human intelligence were symbolic and closely related to cognitive psychology and philosophy, the basic idea being that regions of the brain perform complex reasoning by processing symbols. Later, many models of human cognition were developed to mirror the operation of the brain as an electrochemical computer, starting with the simple

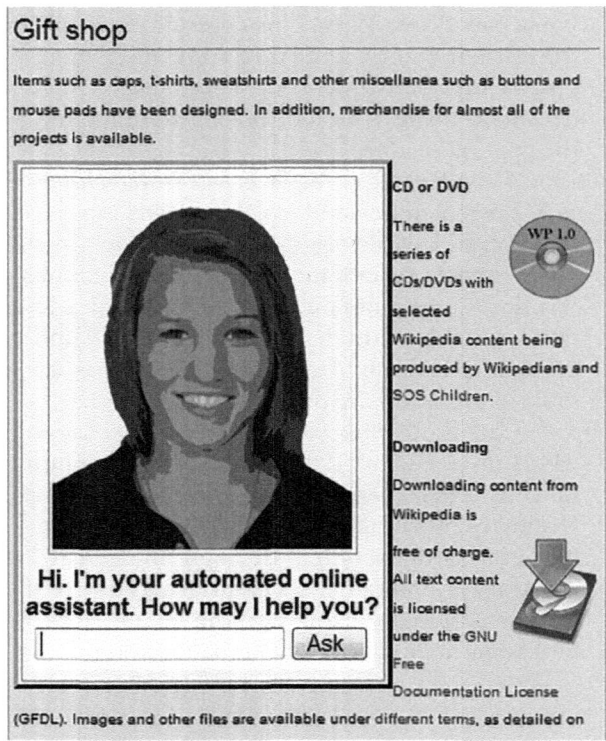

An example of an automated online assistant. Attributed to Bemidji State University. [Public domain], via Wikimedia Commons

Perceptron, an artificial neural network described by Marvin Minsky in 1969, graduating to the backpropagation algorithm described by David E. Rumelhart and James L. McClelland in 1986, and culminating in a large number of supervised and nonsupervised learning algorithms.

When defining artificial intelligence, it is important to remember that the programs, machines, and models developed by computer scientists, engineers, and cognitive scientists do not actually have human intelligence; they only exhibit intelligent behavior. This can be difficult to remember because artificially intelligent systems often contain large numbers of facts, such as weather information for New York City; complex reasoning patterns, such as the reasoning needed to prove a geometric theorem from axioms; complex knowledge, such as an understanding of all the rules required to build an automobile; and the ability to learn, such as a neural network learning to recognize cancer cells. Scientists continue to look for better models of the brain and human intelligence.

HISTORY

Although the concept of artificial intelligence probably has existed since antiquity, the term was first used by American scientist John McCarthy at a conference held at Dartmouth College in 1956. In 1955–56, the first artificial intelligence program, Logic Theorist, had been written in IPL, a programming language, and in 1958, McCarthy invented Lisp, a programming language that improved on IPL. *Syntactic Structures* (1957), a book about the structure of natural language by American linguist Noam Chomsky, made natural language processing into an area of study within artificial intelligence. In the next few years, numerous researchers began to study artificial intelligence, laying the foundation for many later applications, such as general problem solvers, intelligent machines, and expert systems.

In the 1960s, Edward Feigenbaum and other scientists at Stanford University built two early expert systems: DENDRAL, which classified chemicals, and MYCIN, which identified diseases. These early expert systems were cumbersome to modify because they had hard-coded rules. By 1970, the OPS expert system shell, with variable rule sets, had been released by Digital Equipment Corporation as the first commercial expert system shell. In addition to expert systems, neural networks became an important area of artificial intelligence in the 1970s and 1980s. Frank Rosenblatt introduced the Perceptron in 1957, but it was *Perceptrons: An Introduction to Computational Geometry* (1969), by Minsky and Seymour Papert, and the two-volume *Parallel Distributed Processing: Explorations in the Microstructure of Cognition* (1986), by Rumelhart, McClelland, and the PDP Research Group, that really defined the field of neural networks. Development of artificial intelligence has continued, with game theory, speech recognition, robotics, and autonomous agents being some of the best-known examples.

HOW IT WORKS

The first activity of artificial intelligence is to understand how multiple facts interconnect to form knowledge and to represent that knowledge in a machine-understandable form. The next task is to understand and document a reasoning process for arriving at a conclusion. The final component of artificial intelligence is to add, whenever possible, a learning process that enhances the knowledge of a system.

Knowledge Representation. Facts are simple pieces of information that can be seen as either true or false, although in fuzzy logic, there are levels of truth. When facts are organized, they become information, and when information is well understood, over time, it becomes knowledge. To use knowledge in artificial intelligence, especially when writing programs, it has to be represented in some concrete fashion. Initially, most of those developing artificial intelligence programs saw knowledge as represented symbolically, and their early knowledge representations were symbolic. Semantic nets, directed graphs of facts with added semantic content, were highly successful representations used in many of the early artificial intelligence programs. Later, the nodes of the semantic nets were expanded to contain more information, and the resulting knowledge representation was referred to as frames. Frame representation of knowledge was very similar to object-oriented data representation, including a theory of inheritance.

Another popular way to represent knowledge in artificial intelligence is as logical expressions. English mathematician George Boole represented knowledge as a Boolean expression in the 1800s. English mathematicians Bertrand Russell and Alfred Whitehead expanded this to quantified expressions in 1910, and French computer scientist Alain Colmerauer incorporated it into logic programming, with the programming language Prolog, in the 1970s. The knowledge of a rule-based expert system is embedded in the if-then rules of the system, and because each if-then rule has a Boolean representation, it can be seen as a form of relational knowledge representation.

Neural networks model the human neural system and use this model to represent knowledge. The brain is an electrochemical system that stores its knowledge in synapses. As electrochemical signals pass through a synapse, they modify it, resulting in the acquisition of knowledge. In the neural network model, synapses are represented by the weights of a weight matrix, and knowledge is added to the system by modifying the weights.

Reasoning. Reasoning is the process of determining new information from known information. Artificial intelligence systems add reasoning soon after they have developed a method of knowledge representation. If knowledge is represented in semantic nets, then most reasoning involves some type of tree search. One popular reasoning technique is to traverse a decision tree, in which the reasoning is represented by a path taken through the tree. Tree searches of general semantic nets can be very time-consuming and have led to many advancements in tree-search algorithms, such as placing bounds on the depth of search and backtracking.

Reasoning in logic programming usually follows an inference technique embodied in first-order predicate calculus. Some inference engines, such as that of Prolog, use a back-chaining technique to reason from a result, such as a geometry theorem, to its antecedents, the axioms, and also show how the reasoning process led to the conclusion. Other inference engines, such as that of the expert system shell CLIPS, use a forward-chaining inference engine to see what facts can be derived from a set of known facts.

Neural networks, such as backpropagation, have an especially simple reasoning algorithm. The knowledge of the neural network is represented as a matrix of synaptic connections, possibly quite sparse. The information to be evaluated by the neural network is represented as an input vector of the appropriate size, and the reasoning process is to multiply the connection matrix by the input vector to obtain the conclusion as an output vector.

Learning. Learning in an artificial intelligence system involves modifying or adding to its knowledge. For both semantic net and logic programming systems, learning is accomplished by adding or modifying the semantic nets or logic rules, respectively. Although much effort has gone into developing learning algorithms for these systems, all of them, to date, have used ad hoc methods and experienced limited success. Neural networks, on the other hand, have been very successful at developing learning algorithms. Backpropagation has a robust supervised learning algorithm in which the system learns from a set of training pairs, using gradient-descent optimization, and numerous unsupervised learning algorithms learn by studying the clustering of the input vectors.

APPLICATIONS AND PRODUCTS

There are many important applications of artificial intelligence, ranging from computer games to programs designed to prove theorems in mathematics. This section contains a sample of both theoretical and practical applications.

Expert Systems. One of the most successful areas of artificial intelligence is expert systems. Literally thousands of expert systems are being used to help both experts and novices make decisions. For example, in the 1990s, Dell developed a simple expert system that allowed shoppers to configure a computer as they wished. In the 2010s, a visit to the Dell website offers a customer much more than a simple configuration program. Based on the customer's answers to some rather general questions, dozens of small expert systems suggest what computer to buy. The Dell site is not unique in its use of expert systems to guide customer's choices. Insurance companies, automobile companies, and many others use expert systems to assist customers in making decisions.

There are several categories of expert systems, but by far the most popular are the rule-based expert systems. Most rule-based expert systems are created with an expert system shell. The first successful rule-based expert system shell was the OPS 5 of Digital Equipment Corporation (DEC), and the most popular modern systems are CLIPS, developed by the National Aeronautics and Space Administration (NASA) in 1985, and its Java clone, Jess, developed at Sandia National Laboratories in 1995. All rule-based expert systems have a similar architecture, and the shells make it fairly easy to create an expert system as soon as a knowledge engineer gathers the knowledge from a domain expert. The most important component of a rule-based expert system is its knowledge base of rules. Each rule consists of an if-then statement with multiple antecedents, multiple consequences, and possibly a rule certainty factor. The antecedents of a rule are statements that can be true or false and that depend on facts that are either introduced into the system by a user or derived as the result of a rule being fired. For example, a fact could be red-wine and a simple rule could be if (red-wine) then (it-tastes-good). The expert system also has an inference engine that can apply multiple rules in an orderly fashion so that the expert system can draw conclusions by applying its rules to a set of facts introduced by a user. Although it is not absolutely required, most rule-based expert systems have a user-friendly interface and an explanation facility to justify its reasoning.

Theorem Provers. Most theorems in mathematics can be expressed in first-order predicate calculus. For any particular area, such as synthetic geometry or group theory, all provable theorems can be derived from a set of axioms. Mathematicians have written programs to automatically prove theorems since the 1950s. These theorem provers either start with the axioms and apply an inference technique, or start with the theorem and work backward to see how it can be derived from axioms. Resolution, developed in Prolog, is a well-known automated technique that can be used to prove theorems, but there are many others. For Resolution, the user starts with the theorem, converts it to a normal form, and then mechanically builds reverse decision trees to prove the theorem. If a reverse decision tree whose leaf nodes are all axioms is found, then a proof of the theorem has been discovered.

Gödel's incompleteness theorem (proved by Austrian-born American mathematician Kurt Gödel) shows that it may not be possible to automatically prove an arbitrary theorem in systems as complex as the natural numbers. For simpler systems, such as group theory, automated theorem proving works if the user's computer can generate all reverse trees or a suitable subset of trees that can yield a proof in a reasonable amount of time. Efforts have been made to develop theorem provers for higher order logics than first-order predicate calculus, but these have not been very successful.

Computer scientists have spent considerable time trying to develop an automated technique for proving the correctness of programs, that is showing that any valid input to a program produces a valid output. This is generally done by producing a consistent model and mapping the program to the model. The first example of this was given by English mathematician Alan Turing in 1931, by using a simple model now called a Turing machine. A formal system that is rich enough to serve as a model for a typical programming language, such as C++, must support higher order logic to capture the arguments and parameters of subprograms. Lambda calculus, denotational semantics, von Neuman geometries, finite state machines, and other systems have been proposed to provide a model onto which all programs of a language can be mapped. Some of these do capture many programs, but devising a practical automated method of verifying the correctness of programs has proven difficult.

Intelligent Tutor Systems. Almost every field of study has many intelligent tutor systems available to assist students in learning. Sometimes the tutor system is integrated into a package. For example, in Microsoft Office, an embedded intelligent helper provides popup help boxes to a user when it detects the need for assistance and full-length tutorials if it detects more help is needed. In addition to the intelligent tutors embedded in programs as part of a context-sensitive help system, there are a vast number of stand-alone tutoring systems in use.

The first stand-alone intelligent tutor was SCHOLAR, developed by J. R. Carbonell in 1970. It used semantic nets to represent knowledge about South American geography, provided a user interface to support asking questions, and was successful enough to demonstrate that it was possible for a computer program to tutor students. At about the same time, the University of Illinois developed its PLATO computer-aided instruction system, which provided a general language for developing intelligent tutors with touch-sensitive screens, one of the most famous of which was a biology tutorial on evolution. Of the thousands of modern intelligent tutors, SHERLOCK, a training environment for electronic troubleshooting, and PUMP, a system designed to help learn algebra, are typical.

Electronic Games. Electronic games have been played since the invention of the cathode-ray tube for television. In the 1980s, games such as Solitaire, Pac-Man, and Pong for personal computers became almost as popular as the stand-alone game platforms. In the 2010s, multiuser Internet games are enjoyed by young and old alike, and game playing on mobile devices has become an important application. In all of these electronic games, the user competes with one or more intelligent agents embedded in the game, and the creation of these intelligent agents uses considerable artificial intelligence. When creating an intelligent agent that will compete with a user or, as in Solitaire, just react to the user, a programmer has to embed the game knowledge into the program. For example, in chess, the programmer would need to capture all possible configurations of a chess board. The programmer also would need to add reasoning procedures to the game; for example, there would have to be procedures to move each individual chess piece on the board. Finally, and most important for game programming, the programmer would need to add one or more strategic decision modules to the program to provide the intelligent agent with a strategy for winning. In many cases, the strategy for winning a game would be driven by probability; for example, the next move might be a pawn, one space forward, because that yields the best probability of winning, but a heuristic strategy is also possible; for example, the next move is a rook because it may trick the opponent into a bad series of moves.

SOCIAL CONTEXT, ETHICS, AND FUTURE PROSPECTS

After artificial intelligence was defined by McCarthy in 1956, it has had a number of ups and downs as a discipline, but the future of artificial intelligence looks good. Almost every commercial program has a help system, and increasingly these help systems have a major artificial intelligence component. Health care is another area that is poised to make major use of artificial intelligence to improve the quality and reliability of the care provided as well as to reduce its cost by providing expert advice on best practices in health care. Smartphones and other digital devices employ artificial intelligence for an array of applications, syncing the activities and requirements of their users.

Ethical questions have been raised about trying to build a machine that exhibits human intelligence. Many of the early researchers in artificial intelligence were interested in cognitive psychology and built symbolic models of intelligence that were considered unethical by some. Later, many artificial intelligence researchers developed neural models of intelligence that were not always deemed ethical. The social and ethical issues of artificial intelligence are nicely represented by HAL, the heuristically programmed algorithmic computer, in Stanley Kubrick's 1968 film *2001: A Space Odyssey*, which first works well with humans, then acts violently toward them, and is in the end deactivated.

Another important ethical question posed by artificial intelligence is the appropriateness of developing programs to collect information about users of a program. Intelligent agents are often embedded in websites to collect information about those using the site, generally without the permission of those using the website, and many question whether this should be done.

As more complex AI is created and imbued with general, humanlike intelligence (instead of concentrated intelligence in a single area, such as Deep Blue and chess), it will run into moral requirements as humans do. According to researchers Nick Bostrom and Eliezer Yudkowsky, if an AI is given "cognitive work" to do that has a social aspect, the AI inherits the social requirements of these interactions. The AI then needs to be imbued with a sense of morality to interact in these situations. If an AI has humanlike intelligence and agency, Bostrom has also theorized that AI will need to also be considered both persons and moral entities. There is also the potential for the development of superhuman intelligence in AI, which would breed superhuman morality. The questions of intelligence and morality and who is given personhood are some of the most significant issues to be considered contextually as AI advance.

—George M. Whitson III, PhD

FURTHER READING

Basl, John. "The Ethics of Creating Artificial Consciousness." *American Philosophical Association Newsletters: Philosophy and Computers* 13.1 (2013): 25–30. *Philosophers Index with Full Text*. Web. 25 Feb. 2015.

Berlatsky, Noah. *Artificial Intelligence*. Detroit: Greenhaven, 2011. Print.

Bostrom, Nick. "Ethical Issues in Advanced Artificial Intelligence." *NickBostrom.com*. Nick Bostrom, 2003. Web. 23 Sept. 2016.

Bostom, Nick, and Eliezer Yudkowsky. "The Ethics of Artificial Intelligence." *Machine Intelligence Research Institute*. MIRI, n.d. Web. 23 Sept. 2016.

Giarratano, Joseph, and Peter Riley. *Expert Systems: Principles and Programming*. 4th ed. Boston: Thomson, 2005. Print.

Minsky, Marvin, and Seymour Papert. *Perceptrons: An Introduction to Computational Geometry*. Rev. ed. Boston: MIT P, 1990. Print.

Rumelhart, David E., James L. McClelland, and the PDP Research Group. *Parallel Distributed Processing: Explorations in the Microstructure of Cognition*. 1986. Rpt. 2 vols. Boston: MIT P, 1989. Print.

Russell, Stuart, and Peter Norvig. *Artificial Intelligence: A Modern Approach*. 3rd ed. Upper Saddle River: Prentice, 2010. Print.

Shapiro, Stewart, ed. *Encyclopedia of Artificial Intelligence*. 2nd ed. New York: Wiley, 1992. Print.

ARTIFICIAL ORGANS

In an artificial heart, the lower two chambers of the heart are replaced. © EBSCO

FIELDS OF STUDY

Bioengineering; Biosystems engineering; Biotechnology; Biophysics; Medical biotechnology (or Red biotechnology)

ABSTRACT

Artificial organs are complex systems of natural or manufactured materials used to supplement failing organs while they recover, sustain failing organs until transplantation, or replace failing organs that cannot recover. Some whole organs have artificial counterparts: heart, kidneys, liver, lungs, and pancreas. Smaller body parts also have artificial counterparts:

blood, bones, heart valves, joints, skin, and teeth. In addition, there are mechanical support systems for circulation, hearing, and breathing. Artificial organs are composed of biomaterials, biological or synthetic materials that are adapted for use in medical applications.

BACKGROUND

Artificial organs are complex systems that assist or replace failing organs. The human body is composed of ten major organ systems: nervous, circulatory, respiratory, digestive, excretory, reproductive, endocrine, integumentary (skin), muscular, and skeletal. The nervous system transmits signals between the brain and the body via the spinal cord and nerves. The circulatory system transports blood to deliver oxygen and nutrients to the body and to remove waste products. Its organs are the heart, blood, and blood vessels. It works closely with the respiratory system, in which the lungs and trachea perform oxygen exchange between the body and the environment. The digestive system breaks down food and absorbs its nutrients. Its organs include the esophagus, stomach, intestinal tract, and liver. The excretory system rids the body of metabolic waste in the forms of urine and feces. The reproductive system provides sex cells and in females the organs to develop and carry an embryo to term. The endocrine system consists of the pituitary, parathyroid, and thyroid glands, which secrete regulatory hormones. The integumentary system is the body's external protection system. Its organs include skin, hair, and nails. The muscular system recruits muscles, ligaments, and tendons to move the parts of the skeletal system, which consists of bones and cartilage.

HISTORY

While he was still a medical student in 1932, renowned cardiac surgeon Michael DeBakey introduced a dual-roller pump for blood transfusion. It has since become the most widely used type of clinical pump for cardiopulmonary bypass and hemodialysis. Physician John H. Gibbon, Jr., of Philadelphia, developed the first clinically successful heart-lung pump. He initially demonstrated it in 1953, when he closed a hole between the atria of an eighteen-year-old girl.

In 1954, American physician Joseph Murray performed the first successful human kidney transplant from one identical twin to the other in Boston. In 1962, he performed the first kidney transplant in unrelated persons. In 1967, surgeon Christiaan Barnard performed the first successful human heart transplant in Cape Town, South Africa. The patient, a fifty-four-year-old man, lived another eighteen days.

Physician Willem J. Kolff is considered to be "the father of the artificial organ." In 1967, he emigrated from the Netherlands and spent a good deal of his career at the University of Utah, where he became a distinguished professor emeritus of internal medicine, surgery, and bioengineering. He led the designing of numerous inventions, including the modern kidney dialysis machine, the intra-aortic balloon pump, an artificial eye, an artificial ear, and an implantable mechanical heart.

American physician Robert K. Jarvik refined Kolff's design into the Jarvik-7 artificial heart, intended for permanent use. In 1982, at the University of Utah, American surgeon William C. DeVries implanted it into retired dentist Barney Clark, who survived 112 days.

HOW IT WORKS

The existence and performance of artificial organs depend on the collaboration of scientists, engineers, physicians, manufacturers, and regulatory agencies. Each of these groups provides a different perspective of pumps, filters, size, packaging, and regulation.

Hemodynamics. The human heart acts as a muscular pump that beats an average of 72 times a minute. Each of the two ventricles pumps 70 milliliters of blood per beat or 5 liters per minute. Blood pressure is measured and reported as two numbers: the systolic pressure exerted by the heart during contraction and the diastolic pressure, when the heart is between contractions. Hemodynamics is the study of forces related to the circulation of the blood. The hemodynamic performance of artificial organs must match that of the natural body to operate efficiently without resulting in damage. Calculations may be made using computational fluid dynamics (CFD); relevant parameters include solute concentration, density, temperature, and water concentration. In addition to artificial hearts, which are intended to perform all cardiac functions, there is a mechanical circulatory implement called a ventricular assist device (VAD) that supports the function of the natural heart while it is recovering from a heart attack or surgery. Its pumping action may be pulsatile, in rhythmic waves matching those of the beating heart, or continuous.

Mass Transfer Efficiency. The human kidney acts as a filter to remove metabolic waste products from the blood. A person's kidneys process about 200 quarts of blood daily to remove two quarts of waste and extra water, which are converted into urine and excreted. Without filtration, the waste would build to a toxic level and cause death. Patients with kidney failure may undergo dialysis, in which blood is withdrawn, cleaned, and returned to the body in a periodic, continuous, and time-consuming process that requires the patient to remain relatively stationary. Portable artificial kidneys, which the patient wears, filter the blood while the patient enjoys the freedom of mobility. Filtration systems may involve membranes with a strict pore size to separate molecules based on size or columns of particle-based adsorbents to separate molecules by chemical characteristics. Mass transfer efficiency refers to the quality and quantity of molecular transport.

Scale. The development of artificial organs requires that biological processes that can be duplicated in the laboratory be scaled up to work within the human body without also magnifying the weaknesses. Biological functions occur at the organ, tissue, cellular, and molecular levels, which are on micro- and nanoscales. In addition, machines that work in the engineering laboratory must be scaled down to work within the human body without crowding the other organs. Novel power sources and electronic components have facilitated miniaturization. Size must also be balanced with efficiency and cost. Computer-aided design software is being used to create virtual three-dimensional models before fabrication.

Biomaterials. Artificial organs are made of natural and/or manufactured materials that have been adapted for medical use. The properties of these materials must be controlled down to the nanometer scale. The biological components may serve in gene therapy, tissue engineering, and the modification of physiological responses. The synthetic materials must be biocompatible, which means that they do not trigger an adverse physiological reaction such as blood clotting, inflammatory response, scar-tissue formation, or antibody production. The biomechanics of the artificial organ, such as friction and wear, must be known and parts must be sterile before use. Biomaterials have been developed for subspecialties such as orthopedics and ophthalmics.

Regulation. The body has natural feedback systems that allow the exchange of information with the brain for optimal regulation. Artificial organs that communicate directly with the brain are still in development. The present models require sensors and data systems that may be monitored by physicians. Implanted devices must be able to be inspected without direct observation. Another aspect of regulation is the uniform manufacturing of artificial organs in compliance with performance and patient safety specifications.

APPLICATIONS AND PRODUCTS

The collective knowledge of scientists, engineers, physicians, manufacturers, and regulatory agencies has produced the applications and products in the interdisciplinary realm of artificial organs.

Hemodynamics. Knowledge of hemodynamics, the study of blood-flow physics, has led to the development of artificial circulatory assistance. The ventricular assist device (VAD) supplements the contraction of the two lower chambers of the heart so the heart muscle does not have to work as hard while it is healing. The cardiopulmonary bypass pump, also known as a heart-lung machine, provides blood oxygenation and circulating pressure during open-heart surgery when the heart is stopped. A similar application called extracorporeal membrane oxygenation (ECMO) is used to assist neonates and infants in the intensive care unit and to maintain the viability of organs pending transplantation. The natural pressure generated by a healthy heart is used to send blood through versions of artificial lungs and kidneys without batteries.

Mass Transfer Efficiency. Information about molecular transport and delivery, known as mass transfer efficiency, has been applied to separation and secretion functions of artificial organs. In hemodialysis, toxins are removed from circulating blood that passes through a filter called a dialyzer. This process also removes excess salts and water to maintain a healthy blood pressure. The dialyzer is composed of a semipermeable membrane or cylinder of hollow synthetic fibers that separates out the metabolic-waste solutes in the incoming blood by diffusion into dialysate solution, leaving cleaner outgoing blood. Hemofiltration is a similar process; however, the filtration occurs without dialysate solution because instead of diffusion, the solutes are removed more quickly by hydrostatic pressure. Another separation technique in medical applications is apheresis, in which the constituents of blood are isolated. This may be achieved by gradient density centrifugation or absorption onto specifically coated beads. The therapeutic application is the absorptive removal of a specific blood component that is causing an adverse reaction in a patient, with the remaining components returned to the patient's circulatory system. The pathogenic blood component might be malignant white blood cells, excess platelets, low-density lipoprotein, autoantibodies, or plasma. The second application of apheresis is the separation of components following blood donation. Concentrated red blood cells are administered in the treatment of sickle-cell crisis or malaria. Plasmapheresis is used to collect fresh frozen plasma as well as rare antibodies and immunoglobulins.

Scale. Miniaturization of artificial organs has been facilitated by the application of smaller, more efficient batteries, transistors, and computer chips. For example, hearing aids once had to be worn with cumbersome amplifiers and batteries disguised in a purse or camera case with a carrying strap. Existing models fit completely in the ear canal and a computer chip facilitates digital rather than analogue processing for crisper sound. The artificial kidney has evolved into a wearable model that weighs 10 pounds and is seventeen times smaller than a conventional dialysis machine. Its hollow-fiber filter must be replaced once a week and its dialysate solution must be replenished daily. However, this maintenance is a trade-off that many patients are willing to make for freedom of movement. On the horizon is an artificial retina that depends on a miniature camera to transmit images. Conversely, research is under way to produce large-scale cultures of tissues on biohybrid matrices and scaffolding for transplantation.

Biomaterials. Synthetic materials are used in artificial organs. Dacron (polyethylene terephthalate) is a polyester fiber with high tensile strength and resistance to stretching whether wet or dry, chemical degradation, and abrasion. Patches of it are sewn to arteries to repair aneurysms. When tubing of it is used as an aortic valve bypass, the patient will not require subsequent blood-thinning medications. Gore-Tex (expanded polytetrafluoroethylene) is an especially strong microporous material that is waterproof. Vascular grafts made from it are supple and resist kinks and compression. It is also used for replacing torn anterior and posterior cruciate ligaments in the knee.

Perfluorocarbon fluids are synthetic liquids that carry dissolved oxygen and carbon dioxide with negligible toxicity, no biological activity, and a short retention time in the body. These features make them ideal for medical applications. One of these fluids, perfluorodecalin, is typically used as a blood substitute (also called a blood extender) because it mixes easily with blood without changing the hemodynamics. It increases the oxygen-carrying capacity of the blood and penetrates ischemic (oxygen-deprived) tissues especially easily because of its small particle size. This makes it particularly useful in the healing of ulcers and burns. It is also used in conjunction with ECMO in the life support of preterm infants to increase oxygenation and to keep the lungs inflated, reducing exertion. Furthermore, it is used in the preservation of harvested organs and cultured tissue for transplantation, extending their viable storage time.

Regulation. The application of regulatory systems has allowed artificial organs to be adjusted while they are in use. Artificial cardiac pacemakers, which supplement the natural electrical pace-making capabilities of the heart to normalize a slow or irregular heartbeat, are externally programmable so that cardiologists are able to establish the optimal pacing parameters for each patient. Adjustments are made with radio frequency programming, so no further surgery is required. Contemporary hearing aids have volume controls that the wearer can adjust to suit changing surroundings. The inability to detect high- or low-pitch sounds is not a function of volume, yet pitch range can be adjusted in a hearing aid by an audiologist. Other parameters are also adjustable and the audiologist can reprogram the hearing aid as a person's hearing loss changes.

SOCIAL CONTEXT AND FUTURE PROSPECTS

The number of Americans older than sixty-five years of age is expected to double within the next twenty-five years. The fastest growing age group is people older than eighty-five years of age. Increasing life span of the general population is a direct result of improved health care. The shortage of donor organs is also increasing. As of 2010, more than 16,000 people were waiting for liver transplants, but each year, only about 6,500 kidneys become available. For the 93,000 people waiting for kidney transplants in 2010, only 17,000 donated kidneys became available for transplant. In 2012 there were 117,040 people waiting for transplants overall, but only 28,053 transplants were made. At the end of 2017, according to the United Network for Organ Sharing (UNOS), there were 115,873 people on the transplant waiting list with only around fifteen thousand donors registered. The need for artificial organs as a bridge to transplantation or even as a permanent substitute for failed organs remains urgent.

Once only made of synthetic components, artificial organs are becoming biohybrid organs: a combination of biological and synthetic components. Examples include functionally competent cells enveloped within immuno-protective artificial membranes and tissues cultured on chemically constructed matrices. Experiments are underway to develop an antibacterial agent that can be incorporated into biomaterials to reduce the risk of infection from these organ surfaces. Emerging technologies also involve sensors and intelligent control systems, biological batteries and alternate power sources, and innovative delivery systems.

Other areas of research include the miniaturization of artificial organs for pediatric use and the development of smaller and more efficient batteries and sensors that will be capable of more accurate communication between the artificial organ and the brain. Another goal is to incorporate wireless capabilities so the artificial organ may be programmed, monitored, and recharged remotely so the patient has increased freedom of mobility.

—*Bethany Thivierge, MPH*

FURTHER READING

"Data." *UNOS*, 2016, www.unos.org/data/. Accessed 26 Oct. 2016.

Fox, Renée C., and Judith P. Swazey. *Spare Parts: Organ Replacement in American Society*. Oxford UP, 1992.

Hench, Larry L., and Julian R. Jones, eds. *Biomaterials, Artificial Organs, and Tissue Engineering*. CRC Press, 2005.

McClellan, Marilyn. *Organ and Tissue Transplants: Medical Miracles and Challenges*. Enslow, 2003.

Sharp, Lesley A. *Bodies, Commodities, and Biotechnologies: Death, Mourning, and Scientific Desire in the Realm of Human Organ Transfer*. Columbia UP, 2008.

U.S. Dept. of Health and Human Services. "Need Continues to Grow." *Organ Procurement and Transplantation Network*. U.S. Dept. of Health and Human Services, optn.transplant.hrsa.gov/need-continues-to-grow/. Accessed 27 Feb. 2015.

AUDIO ENGINEERING

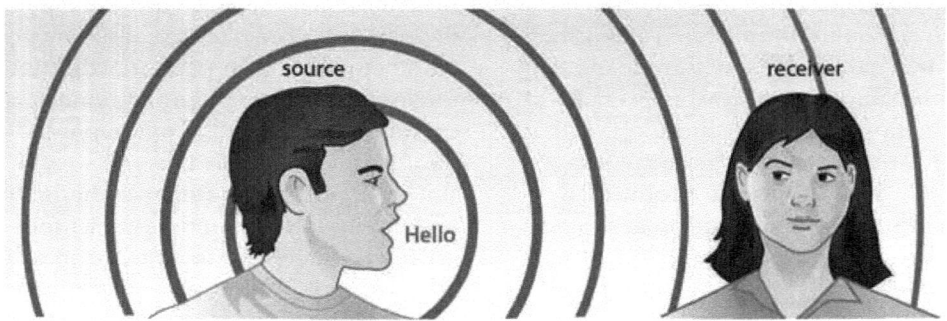

Sound is transmitted through the air as a sound wave. © EBSCO

FIELDS OF STUDY

Bioengineering; Biosystems engineering; Biotechnology; Biophysics; Medical biotechnology (or Red biotechnology)

ABSTRACT

Audio engineering is the capture, enhancement, and reproduction of sounds. It requires an aesthetic appreciation of music and sound quality, a scientific understanding of sound physics, and a technical familiarity with recording equipment and computer software. This applied science is essential to the music industry, film, television, and video game production, live television and radio broadcasting, and advertising. In addition, it contributes to educational services for the visually impaired and to forensic evidence analysis.

BACKGROUND

A recording studio is a specialized environment designed to capture sounds accurately for enhancement and reproduction. The acoustic sounds produced by instruments and vocalists are picked up by strategically placed microphones and transmitted as analog electrical signals to recording equipment, where they may be converted into digital data. Signals may be modified by the use of a mixing console, also called a mixing board or sound board, which changes the characteristics and balance of the input, which may be coming from multiple microphones or signals recorded in different sessions. The final product then undergoes mastering for commercial reproduction and compression for distribution in a digital format.

Audio engineers run recording sessions, work the equipment, and collaborate on the finished product. They are simultaneously technicians, scientists, and creative advisers. They work in recording studios, producing the following: instrumental and vocal music recordings; film and television soundtracks, syncing, and sound effects; music and voice-overs for radio and television commercials; and music and sound effects for video games.

Audio engineers are not acoustic engineers: graduates of a formal university program in engineering who work with architects and interior designers to plan and install audio systems for large venues such as churches, school auditoriums, and concert halls. Audio engineers are engineers in the sense that they are needed to devise a creative solution to a complex sound challenge and oversee its implementation.

HISTORY

Sound capture and reproduction, and thus audio engineering, began with the invention of the phonograph by Thomas Alva Edison in 1877. Sound was recorded on cylinders; the first were wrapped in tin foil and later ones in wax. By 1910, cylinders were replaced with disks, which held longer recordings, were somewhat louder, and could be more economically mass-produced. The disks were spun on a turntable at standard speeds—initially 78 revolutions per minute (rpm). Larger disks were played at 33 rpm and smaller disks were played at 45 rpm. Discs were originally made of shellac and later made of vinyl. They were played with needles (styli) made of industrial diamond, which held a point.

Concurrently, RCA was creating microphones that improved recorded sound quality. In the 1940's, sound began being recorded on magnetic tape and could be reproduced in stereo and as mixed multiple tracks. Digital technology appeared in the 1980's and by the turn of the century, digital recordings were produced with computer technology. Using data compression, digital audio recordings can produce quality replication of the original music in a format that requires less data storage (computer memory); MP3 is one such format and is popular for portable consumer music systems.

HOW IT WORKS

Sound. Sound is the waves of pressure a vibrating object emits through air or water. The three most meaningful characteristics of sound waves are wavelength, amplitude, and frequency. The wavelength is the distance between equivalent points on consecutive waves, such as peak to peak. A short wavelength means that more waves are produced per second, resulting in a higher sound. The amplitude is the strength of the wave; the greater the amplitude, the greater the volume (loudness). The frequency is the number of wavelengths that occur in one second; the greater the frequency, the higher the pitch because the sound source is vibrating quickly.

Hearing. Hearing is the ability to receive, sense, and decipher sounds. To hear, the ear must direct the sound waves inside, sense the sound vibrations, and translate the sensations into neurological impulses that the brain can recognize. The outer ear funnels sound into the ear canal. It also helps the brain determine the direction from which the sound is coming.

When the sound waves reach the ear canal, they vibrate against the eardrum. These vibrations are amplified by the eardrum's movement against three tiny bones (the malleus, incus, and stapes) located behind it. The stapes rests against the cochlea, and when it transmits the sound, it creates waves in the fluid of the cochlea.

The cochlea is a coiled, fluid-filled organ that contains 30,000 hairs of different lengths that resonate at different frequencies. Vibrations of these hairs trigger complex electrical patterns that are transmitted along the auditory nerve to the brain, where they are interpreted.

The frequency of sound waves is measured in hertz (Hz). Humans have a hearing range from 20 to 20,000 Hz. Another name for the frequency of a sound wave is the musical pitch. Pitches are often referred to as musical notes, such as middle C.

The relative loudness of a sound compared with the threshold of human hearing is measured in decibels (dB). Conversation is usually conducted at 40 to 60 dB, while a car passing at 10 meters may be 80 to 90 dB, a jet engine 100 meters away may be 110 to 140 dB, and a rifle fired 1 meter away is 150 dB. Long-term (not necessarily continuous) exposure to sounds greater than 85 dB may cause hearing loss.

The human ear can discern between two musical instruments playing the same note at the same volume by the recognition of a sound characteristic called timbre. Often described by adjectives such as bright versus dark, smooth versus harsh, and regular versus random or erratic, timbre is often what distinguishes music from noise.

Sound Capture. Transducers are devices that change energy from one form into another. A microphone changes acoustical signals into electrical signals, while a speaker changes electrical signals into acoustical signals. The source of the incoming electrical signals may be immediate, such as a microphone or electrical musical instrument, or it may be a recording, such as a compact disc or an MP3 file.

Microphones come in many varieties, such as dynamic, ribbon, condenser, parabolic, and lavaliere. They also vary by their polar patterns, that is, their area of sensitivity to sounds coming in from different directions relative to the receiving membrane. They may be omnidirectional (sensitive to sounds coming from all directions), unidirectional (intended for directed sound reception), or cardioid (having a heart-shaped area of sensitivity). The choices of variety, polar pattern, and placement affect the quality and quantity of sound capture.

Signal Processing. The auditory electrical signal from a microphone is relatively weak, so it must be amplified before the sound can be deliberately modified through signal processing. Incoming sound may be modified in its analogue form or converted to digital data before alteration. Sound mixing is the process of blending sounds from multiple sources into a desired end product. It often starts with finding a balance between vocal and instrumental music or dissimilar instruments so that one does not overshadow the other. It involves the creation of stereo or surround sound from the placement of sound in the sound field to simulate directionality (left, center, or right). Equalizing adjusts the bass and treble frequency ranges. Effects such as reverberation may be added to create dimension. Signals may undergo gating and compression to remove unwanted noise and extraneous data selectively.

Sound Output. Auditory electrical signals may then be sent to speakers, where they are converted into acoustical signals to be heard by a live audience. Digital signals may be broadcast in real time over the Internet as streaming audio. Otherwise, the processed signals may be stored for future reproduction and distribution. Analogue signals may be stored on magnetic tape. Digital signals may be stored on a compact disc or subjected to MP3 encoding for storage on a computer or personal music player.

APPLICATIONS AND PRODUCTS

Instrumental and Vocal Music. As specialists in the capture, enhancement, and reproduction of sound, audio engineers are crucial to successful recording sessions. They collaborate with producers and performers technically to generate the shared artistic vision. They determine the choice and placement of microphones and closely scrutinize the parameters of the incoming signals to collect sufficient data with which to work. They manage the scheduling of studio sessions to keep all participants working efficiently, especially when multiple tracks are being recorded and mixed at different times. They act professionally and deliver the finished product with the highest quality possible.

Audio engineers are also responsible for the restoration of classic recordings that would otherwise be lost. They rescue the raw data that was captured in the first recording, strengthen the sound while preserving the style of the original period, and return it to audiences in a contemporary format.

Because musicians go on concert tours, audio engineers accompany them to provide optimum live sound quality in each different venue. They conduct sound checks before performances and make adjustments for conditions such as wind on outdoor stages.

Film and Television. In the recording studio, audio engineers oversee the production of music soundtracks for films and television shows. Unlike songs that stand alone, the music must be carefully synchronized to the action of the film. It must also swell and ebb with precision to arouse audience emotion.

Foley recording is the production of sound effects that are inserted into videos after they are filmed to add realism and dramatic tension. Foley recording can be synchronized efficiently to video footage because the sound effects are produced in real time, not modified from stock recordings. In addition, sounds that do not exist in reality and so would not be catalogued in a prerecorded audio library must be created.

Live Broadcasting. Audio engineers may be seen sitting at mixing consoles or computers monitoring and adjusting the audio input and output quality at church services, lectures, theatrical performances, and events held in large auditoriums. They may similarly be found as part of a broadcasting team at live sporting events held outdoors, such as football or baseball games, golf tournaments, and the Olympic Games.

Radio and Television Commercials. Audio engineers are instrumental in the production of radio and television commercials, not only for their recording and sound-processing skills but also for their production skills. Because advertising time is sold in specific brief allotments, engineers must encourage the actors to perform at an accelerated pace and later edit the audio to fit within the time allowed. They may also be asked to recruit or audition competent musicians and voice actors to meet the client's needs.

Video Games. The skills of audio engineers enhance the production of popular video games. In addition to providing sound effects such as explosions and gunfire, engineers must create appropriate imaginary sounds such as spaceships landing, ambient sound effects such as slot machine bells and crowd murmurs, and realistic situational sounds, such as footsteps going from grass to gravel. In some cases, they may be called on to provide minor character voices or record spoken instructions.

Forensic Evidence Analysis. Police may seek the assistance of an experienced audio engineer to remove unnecessary background noise from covert recordings of suspected criminals and to make voiceprint comparisons with known exemplars. Voiceprints, vocal qualities that can be demonstrated on a sound spectrograph, are personal because each person's oral and pharyngeal anatomy is distinctive; however, they are not unique like fingerprints because children often sound like their parents and share similar voiceprints. Research has shown that the error rates of misidentifying suspects (false positives) and improperly eliminating suspects (false negatives) are respectably low.

Audio Books. Recordings of books originated in 1932 under the auspices of the American Foundation for the Blind as educational tools for the visually impaired. Books were recorded on shellac discs and played on a turntable. Books on audio cassettes came along twenty years later, and later audio books could be listened to on CDs or portable digital music devices. Audio engineers are responsible for processing the audio signal to optimize the clarity of human speech and editing numerous recitations into one continuous, flawless performance.

SOCIAL CONTEXT AND FUTURE PROSPECTS

Audio engineering is a combination of technology, science, and art. Advancements in audio engineering will come in all three areas. On the technical front, classic (especially pre-1920) recordings will continue to be found, researched, and digitally restored. Improved transducer materials are being sought and new computer software applications for signal processing are being developed. Surround sound is being refined to accompany three-dimensional and high-definition television programs and films as well as video games.

Scientific research into psychoacoustics, the study of sound perception, is expanding. The eventual understanding of how music affects a person's brain will advance the field of music therapy, which seems to touch every facet of a person's being to restore and maintain health. Researchers are also exploring the connections between sound characteristics and the perceptions of timbre and spatial placement and between these perceived attributes and listening preference.

The artistic manipulation of sound is broadening the definition of music and musical instruments. Computer-mediated music has inspired the creation of mobile phone and laptop orchestras. Music enhancement by selectively masking undesired frequencies of instruments and highlighting others is introducing new sound combinations previously not experienced.

——*Bethany Thivierge, MPH*

FURTHER READING

Dittmar, Tim. *Audio Engineering 101: A Beginner's Guide to Music Production*. Waltham: Focal, 2012. Print.

Friedman, Dan. *Sound Advice: Voiceover from an Audio Engineer's Perspective*. Bloomington: AuthorHouse, 2010. Print.

Hampton, Dave. *The Business of Audio Engineering*. 2nd ed. New York: Hal Leonard, 2013. Print.

Hampton, Dave. *So, You're an Audio Engineer: Well, Here's the Other Stuff You Need to Know*. Parker: Outskirts, 2005. Print.

Powell, John. *How Music Works: The Science and Psychology of Beautiful Sounds, from Beethoven to the Beatles and Beyond*. New York: Little, 2010. Print.

Talbot-Smith, Michael. *Sound Engineering Explained*. 2nd ed. Woburn: Focal, 2001. Print.

Talbot-Smith, Michael, ed. *Audio Engineer's Reference Book*. 2nd ed. Woburn: Focal, 1999. Print.

B

Bioassays

EPA Gulf Breeze laboratory: Biologists are seining for fish to be used in bioassays. Photo courtesy of the National Archives and Records Administration. [Public domain], via Wikimedia Commons

FIELDS OF STUDY

Bioengineering; Biosystems engineering; Biotechnology; Biophysics; Environmental biotechnology (or Green biotechnology)

ABSTRACT

Bioassays enable environmental scientists to evaluate the effects of the chemicals used in pesticides as well as the resistance of plants to particular pests.

BACKGROUND

In many instances, a scientist may suspect that a certain chemical is present in a given environment but may not have access to a specific piece of equipment designed to measure the presence of the chemical. In some cases, an experimental protocol for the detection of the chemical may not exist. In either of these cases, the scientist may be able to detect the presence of the chemical by using a biological organism that responds in a specific manner when exposed to that particular chemical agent. At other times, a scientist may know that a certain chemical is present but not know how a particular organism will respond when exposed to the agent. In this case, the scientist will expose the test organism to the chemical and measure a particular physiological response.

HOW IT WORKS

Bioassays are utilized in many different areas of the biological sciences, including environmental studies. Some bioassay methods work better than others. A good bioassay meets two basic criteria. First, it is specific for a given physiological response. For example, if a given chemical is responsible for inhibiting the feeding response of a particular insect, then the bioassay for that chemical should measure only the inhibition of feeding of that insect and not some other physiological response to the chemical. Second, a good bioassay measures the same response in the laboratory that is observed in the field. Again, if a particular chemical inhibits the feeding response in the field, then the laboratory bioassay for that chemical should also inhibit feeding. An ongoing need exists for the development of accurate bioassay methods as well as the improvement of existing techniques.

ENVIRONMENTAL USES

Many different bioassays are used in environmental studies. One of the most common is the measure of the median lethal dose (LD_{50})—the concentration or dose of a chemical that will result in the deaths of one-half of a population of organisms—of a new pesticide on species of pest and nonpest organisms. To conduct this bioassay, the test species is exposed to a wide range of different concentrations of the chemical. The concentration of the pesticide that kills one-half of the test organisms represents the LD_{50}.

Another common environmental bioassay is the measure of resistance of plants to a particular insect pest. In order to reduce the dependence on chemical insecticides, plant breeders are continually trying to develop insect-resistant plants, either through traditional breeding programs or by using biotechnology to transfer resistance genes to susceptible crop strains. Bioassays are used to measure the degree of success of these attempts. In these bioassays, the same numbers of susceptible and resistant plants are subjected to infestation by equal numbers of the insect pest for which the breeder is trying to develop resistance. The two groups of plants are observed, and the degree of resistance, if any, is recorded.

—D. R. Gossett

FURTHER READING

Ohkawa, H., H. Miyagawa, and P. W. Lee, eds. *Pesticide Chemistry: Crop Protection, Public Health, Environmental Safety.* New York: Wiley-VCH, 2007.

Rand, Gary M., ed. *Fundamentals of Aquatic Toxicology: Effects, Environmental Fate, and Risk Assessment.* 3d ed. Boca Raton, Fla.: CRC Press, 2008.

BIOCHEMICAL ENGINEERING

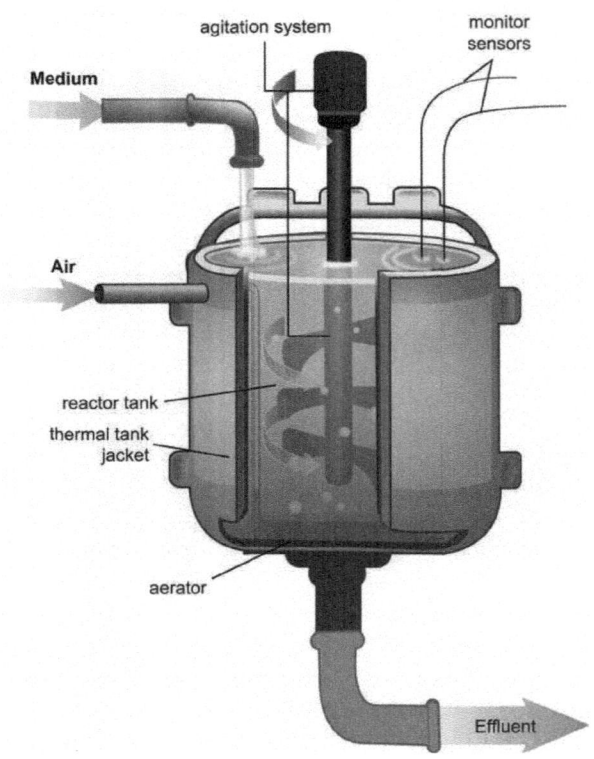

Chemical reactors are designed to contain chemical reactions such as emulsifying, solid suspension, and gas dispersion. © EBSCO

FIELDS OF STUDY

Bioengineering; Biochemistry; Biotechnology; Biophysics; Chemical engineering; Chemistry

ABSTRACT

Biochemical engineers are responsible for designing and constructing those manufacturing processes that involve biological organisms or products made by them. Biochemical engineers take commercially valuable biological or biochemical commodities and design the means to produce those commodities effectively, cheaply, safely, and in mass quantities. They do this by optimizing the growth of organisms that produce valuable molecules or perform useful biochemical processes, establishing the most effective way to purify the desired molecules, and designing the operation systems that execute these processes, while adhering to a high standard of quality, purity, worker safety, and environmental cleanliness.

BACKGROUND

One of the main tasks of bioengineers is to optimize the production of commercially valuable molecules by genetically engineered microorganisms. Biochemical engineers design culture containers known as bioreactors that accommodate growing cultures and maintain an environment that keeps growth at optimal levels. They also create the protocols that separate the cultured cells and their growth medium from the molecule of interest and purify this molecule from all contaminating components. Biochemical engineers do not make the genetically engineered organisms that produce or do valuable things, but instead they maximize the capacities of such organisms in the safest and most cost-effective ways.

Biochemical engineers also design systems that degrade organic or industrial waste. In these cases, bioreactors house biological organisms that receive and decompose waste. They select the right organism or mix of organisms for the job at hand, establish environments that allow these organisms to thrive, and design systems that feed waste to the organisms and remove the degradation products.

A branch of biochemical engineering called tissue engineering combines cultured cells with synthetic materials and external forces to mold those cells into organs that can serve as a replacement for diseased or damaged organs. Biochemical engineers determine the forces, materials, or biochemical cues that drive cells to form fully functional organs and then design the bioreactor and associated instrumentation to provide the proper environment and cues.

HISTORY

Biochemical engineering is a subspecialty of chemical engineering. Chemical engineering began in 1901 when George E. Davis, its British pioneer, mathematically described all the physical operations commonly used in chemical plants (distillation, evaporation, filtration, gas absorption, and heat transfer) in his landmark book, *A Handbook of Chemical Engineering*.

Biochemical engineering emerged in the 1940s as advancements in biochemistry, the genetics of microorganisms, and engineering shepherded in the era of antibiotics. World War II created shortages in commonly used industrial agents; therefore, manufacturers turned to microorganisms or enzymes to synthesize many of the chemicals needed for the war effort. Growing large batches of microorganisms presented scaling, mixing, and oxygenation problems that had never been encountered before, and biochemical engineers solved these problems.

During the 1960s, advances in biochemistry, genetics, and engineering drove the creation of biomedical engineering, which is the application of all engineering disciplines to medicine, and separated it from biochemical engineering. During this decade, biochemical engineers developed new types of bioreactors and new instrumentation and control circuits for them. They also made breakthroughs in kinetics (the science that mathematically describes the rates of reactions) within bioreactors and whole-cell biotransformations.

The 1970s saw the development of enzyme technologies, biomass engineering, single-cell protein production, and advances in bioreactor design and operation. From 1930 to 2000 there was a virtual explosion in biochemical-engineering advances that had never been seen before. The advent of recombinant DNA and hybridoma technologies, cell culture, molecular models, large-scale protein chromatography, protein and DNA sequencing, metabolic engineering, and bioremediation technologies changed biochemical engineering in a drastic and profound way. These technologies also presented new challenges and problems, many of which are still the subject of intense research and development.

HOW IT WORKS

Bioreactors. Bioreactors that utilize living cells are typically called fermenters. There are several different types of bioreactors: mechanically stirred or agitated tanks; bubble columns (cylindrical tanks that are not stirred but through which gas is bubbled); loop reactors, which have forced circulation; packed-bed reactors; membrane reactors; microreactors; and a variety of different types of reactors that are not easily classified (such as gas-liquid reactors and rotating-disk reactors). Biochemical engineers must choose the best bioreactor type for the desired purpose and outfit it with the right instrumentation and other features.

Bioreactor operation is either batch-wise or continuous. Batch-wise operation or batch cultures include all the nutrients required for the growth of cells prior to cultivation of the organisms. After inoculation, cell growth commences and ceases once the organisms have exhausted all the available nutrients in the culture medium. A modification of this type of operation is a fed-batch or semi-batch operation in which the reactants are continuously fed into the bioreactor, and the reaction is allowed to go to completion, after which the products are recovered. Continuously operated bioreactors, use "continuous culture systems" that continuously feed culture medium into the bioreactor and simultaneously remove excess medium at the same rate. Batch-culture bioreactors work best for fast-growing biological organisms. Slow-growing organisms usually require continuous-culture bioreactors.

Several factors influence the success of bioreactor-based operations. First, choosing the right strain to make the desired product is essential. Second, the culture medium and growth conditions must optimize the growth of the chosen organism. Third, supplying the culture with adequate oxygen requires the use of agitators or stirring equipment that must operate at high enough levels to aerate the culture without severely damaging the growing cells. Fourth, the bioreactor must have sensors to measure accurately the physical properties of the culture system, such as temperature, acidity (pH), and ionic strength. Fifth, the bioreactor should also be equipped with the means to adjust these physical properties as needed. Finally, the bioreactor must be integrated into a network of peripheral equipment that allows automated monitoring and adjustment of the culture's physical factors.

Separation. Once a bioreactor makes a product, separating this molecule or group of molecules from the remaining contaminants, byproducts, and other components is an integral part of preparing that molecule for market.

There are several different separation techniques. Filtration separates undissolved solids from liquids by passing the solid-liquid mixture through solids perforated by pores of a particular size (like a membrane). If the liquid is viscous or the particle size of the solid is too small for filtration, centrifugation can separate such solids from liquids. The liquid samples are loaded into centrifuges, which spin rotors at very high speeds. This process creates pellets from the solids and separates them from liquids. Neither filtration nor centrifugation can separate dissolved components from liquids.

Adsorption and chromatography can effectively separate dissolved molecules. Adsorption involves the accumulation of dissolved molecules on the surface of a solid in contact with the liquid. The solid in most cases consists of a resin made of porous charcoal, silica, polysaccharides (complex chains of sugars), or other molecules. Chromatography runs the liquid through a stationary medium packed into a cylindrical column that has particular chemical properties. The interaction between the desired molecules and the stationary medium facilitates their isolation. Other types of separation techniques include crystallization, in which the molecule of interest is driven to form crystals. This effectively removes it from solution and facilitates "salting out," in which gradually increased salt concentrations precipitate the molecules of interest, or contaminating molecules, from a liquid solution.

Sterilization. If a culture of genetically engineered organisms is used to produce a commercially useful product, contamination of that culture can decrease the amount of product or cause the production of harmful byproducts. Therefore, all tubes, valves, the bioreactor container, and the air supplied to it during operation must be effectively sterilized before the start of any production run.

Heat, radiation, chemicals, or filtration can sterilize equipment and liquids. One of the most economical means of sterilization is moist steam. Calculating the time it takes to sterilize something depends on the initial number of organisms present, the resilience of those organisms to killing with the chosen agent, the ability of the air or liquid to conduct the sterilizing agent, and length of time the organisms are exposed to the sterilizing agent.

APPLICATIONS AND PRODUCTS

Pharmaceuticals. Hundreds of pharmaceuticals are proteins made by genetically engineered organisms. Because these reagents are intended for clinical use, they must be produced under completely sterile conditions and are usually grown in disposable (plastic), prepackaged, sterile bioreactor systems. A variety of wave bioreactors, hollow-fiber membrane bioreactors, and variations on these devices help grow the cells that make these products.

Some of the proteins made by genetically engineered cells are enzymes. Genentech, for example, makes dornase alfa, an enzyme that degrades DNA. This enzyme is made by genetically engineered Chinese hamster ovary (CHO) cells and is purified by filtration and column chromatography. Dornase alfa is administered as an inhalable aerosol to allay the symptoms of cystic fibrosis. Other therapeutic enzymes include clotting factors such as Helixate FS (native clotting factor VIII made by CSL Behring), NovoSeven (clotting factor VII made by Novo Nordisk) to treat hemophilia, and Fabrazyme or Replagal (agalsidase alfa) to treat Anderson-Fabry disease.

Other pharmaceuticals are peptide hormones. Serostim and Saizen are commercially available versions of recombinant human growth hormone. Both products are made with cultured mouse C127 cells in bioreactors. Human growth hormone is used to treat children with hypopituitary dwarfism or those who experience the chronic wasting associated with AIDS.

Therapeutic proteins are normally made in the human body under certain conditions, and synthetic versions of these proteins that are made in labs can be used as medicine. For example, human cells make a protein called interferon in response to viral infections, but synthetic interferon can also be used to treat multiple sclerosis. Two synthetic forms of interferon-1, Rebif, which is made in CHO cells by EMD Serono, and Avonex, also made in CHO cells by Biogen Idec, serve as treatments for multiple sclerosis. Alefacept (brand name Amevive), which is made by Astellas Pharma, is a fusion protein that blocks the growth of specific T cells (immune cells). No such protein exists in the human body, but alefacept is used to treat psoriasis and various cancers.

These are only a few examples of the hundreds of pharmaceutical compounds made by genetically engineered organisms in bioreactors designed by biochemical engineers.

Monoclonal Antibodies. Monoclonal antibodies are Y-shaped proteins secreted by specific cells of the immune system that precisely bind to specific sites (epitopes) on the surface of foreign invaders, and act as guided missiles that facilitate the destruction or neutralization of the foreign invaders.

Immune cells called B lymphocytes secrete antibodies, and the fusion of these antibody-producing cells with myelomas (B-cell tumor cells) produces a hybridoma, an immortal cell that grows indefinitely in culture and secretes large quantities of a particular antibody. Antibodies made by hybridoma cells can bind to one and only one site on a specific target and are known as monoclonal antibodies.

Monoclonal antibodies are powerful clinical and industrial tools, and by growing hybridoma cell lines in bioreactors, biotechnology companies can produce large quantities of them for a variety of applications.

Mouse monoclonal antibodies end with the suffix "-omab." Tositumomab (brand name Bexxar) was approved by the Food and Drug Administration (FDA) for treatment of non-Hodgkin's lymphoma in 2003.

Chimeric are humanized monoclonal antibodies, and have the suffixes "-ximab" (chimeric antibodies that are about 65 percent human) or "-zumab" (humanized antibodies that are about 95 percent human). Cetuximab (Erbitux) is a chimeric antibody that was approved by the FDA in 2004 for the treatment of colorectal, head, and neck cancers. Bevacizumab (Avastin) is a humanized antibody approved by the FDA in 2004 that shrinks tumors by preventing the growth of new blood vessels into them.

Human monoclonal antibodies are made either by hybridomas from transgenic mice that have had their mouse antibody genes replaced with human antibody genes, or by a process called phage display. Human monoclonal antibodies end with the suffix "-mumab." The first human monoclonal antibody developed through phage display technologies was adalimumab (Humira), which was approved by the FDA to treat several immune system diseases.

Tissue Engineering. Making artificial organs for transplantation represents a unique challenge. Bioreactors tend to grow cells in two-dimensional cultures, but organs are three-dimensional structures. Thus, biochemical engineers have designed synthetic scaffolds that support the growth of cultured cells and mold them into structures that bear the shape and properties of organs. They have also designed special bioreactors that subject cells to the physical conditions that induce the cells to form the tissues that compose particular organs.

People often need cartilage repair or replacement, but bone and cartilage form only when their progenitor cells are subjected to mechanical stresses and shear forces. Biochemical engineers have grown bone by seeding bone marrow stem cells on a ceramic disc imbued with zirconium oxide and loading these discs into bioreactors with a rotating bed. Cartilage biopsies are taken from the nose or knee and grown in a bioreactor in which the cells are perfused into a complex sugar called glycosaminoglycan (GAG). This engineered cartilage is then used for transplantations. Such experiments have established that nasal cartilage responds to physical forces similarly to knee cartilage and might substitute for knee cartilage.

Heart muscle is grown in bioreactors that pulse the liquid growth medium through the chamber under high-oxygen tension. Blood vessels are grown in two-chambered bioreactors and contain a reservoir of smooth muscle cells and a chamber through which culture medium is repeatedly pulsed.

Food Engineering. Companies making foods that require fermentation by microorganisms or digestion of complex molecules by enzymes use bioreactors to optimize the conditions under which these reactions occur. Biochemical engineers design the industrial processes that manufacture, package, and sterilize foods in the most cost-effective manner.

Starch is a polymer of sugar made by plants and is a very cheap source of sugar. To convert starch into glucose, enzymes called amylases are employed. These enzymes are often isolated from bacteria or fungi, and some are even stable at high temperatures. Degrading starch at high temperatures often clarifies it and rids it of contaminating proteins.

Lactic acid fermentation metabolizes simple sugars to lactic acid and is commonly used in the production of yogurt, cheeses, breads, and some soy products. Cheese production begins with curdling milk by adding acids such as vinegar that separate solid curds from liquid whey and an enzyme mixture called rennet that comes from mammalian stomachs and coagulates the milk. Starter bacterial cultures then ferment the milk sugars into lactic acid. Yogurt is made from heat-treated milk to which starter cultures are added. The acidity of the culture is monitored, and when it reaches a particular point, the yogurt is heated to sterilize the culture for packaging.

Ethanol fermentation converts simple sugars to ethyl alcohol and is used in the production of alcoholic beverages. The most common organism utilized for ethanol fermentation is the baker's yeast, *Saccharomyces cerevisiae*. Malted barley is the sugar source in beer production, and grapes are used to make wine. Beer production involves the extraction of wort, a sugar-rich liquid from barley, which is treated with hops to add aroma and flavor and is then fermented by yeast to form beer. For wine production, the juice from crushed grapes is fermented by yeast for from five to twelve days to generate ethanol. For most red wines and some white wines, the mixture is fermented a second time by malolactic bacteria that degrade the malic acid in the wine, which has a rather harsh, bitter taste, to lactic acid. This lowers the acidity of the wine.

Biofuel Production. Burning of fossil fuels as an energy source is not sustainable, since the supply of these fuels is finite and their combustion generates greenhouse gases such as carbon dioxide (CO_2), sulfur dioxide (SO_2), and nitrogen oxides. First-generation biofuels (biodiesel and bioethanol) utilize biomass from cultivated crops such as corn, sugar beets, and sugar cane. This results in the unfortunate consequences of tying up large swathes of farmland for fuel production and raising food prices. Second-generation biofuels come from grasses, rice straw, and bio-ethers, which are economically superior to first-generation biofuels. Third-generation biofuels show the most ecological and economic promise and come from microalgae. The oil content of some microalgae can exceed 80 percent of their dry weight, and since they use sunlight as their energy source and atmospheric CO_2 as their carbon source, microalgae can produce substantial amounts of oil with little material investment.

Microalgae can be grown in open ponds, which ties up land, or special bioreactors called photobioreactors. The fast-growing microalgae are harvested and then liquefied by microwave high-pressure reactors. Oils extracted from the algal species *Dunaliella tertiolecta* at 340 degrees Celsius for sixty minutes had physical properties comparable to fossil fuel oil.

Waste Management. The removal of pollutants from air and water provides a large global challenge to environmental engineers. While there are nonbiological ways to degrade pollution, biological strategies represent some of the most innovative and potentially effective ways to remediate pollution.

To treat polluted air, it is piped through a biofilter, which consists of an inert substance called a carrier. Nutrients are trickled over the carrier, and consequently the carrier is colonized by biological organisms that can degrade the pollutant. Devices called bioscrubbers eliminate pollutants such as hydrogen sulfide (H_2S), which smells like rotten eggs, or SO_2, by dissolving the air pollutants in water and running the water into a bioreactor where the pollutants are degraded. For air pollutants that are poorly soluble in water, such as methane (CH_4) or nitric oxide (NO), hollow-fiber membrane bioreactor (HFMB) systems that house a robust population of biological organisms that can degrade gas-phase pollutants effectively treat air polluted with such molecules. Many of these same strategies can also treat polluted water.

Bioreactor landfills were designed to accelerate the degradation of municipal solid waste (MSW) in landfills. Bioreactor landfills use microorganisms to degrade solid wastes, but they also drain the water (leachate) that moves through the landfill, clean it, and recycle it back through the landfill in a process called leachate recirculation. The design of a bioreactor landfill requires extensive knowledge of the surroundings, the nature of the MSWs to be treated, and the quality of the water that becomes the leachate.

SOCIAL CONTEXT AND FUTURE PROSPECTS

Two aspects of biochemical engineering can be cause for concern to the general public. First, biochemical engineers work with genetically engineered organisms. Many people have never completely made peace with the use of such organisms, despite the fact that many of the items people consume on a regular basis, from seasonal flu vaccines and other medicines to the foods they eat, are made by genetically engineered organisms. Nevertheless, fear of genetically engineered organisms remains. For example, despite repeated tests establishing that genetically engineered foods are as safe as food from nongenetically modified crops, some people still feel the need to label genetically engineered food as Frankenfood. As long as this fear persists, the work of biochemical engineers will make some people uncomfortable. Second, biochemical engineers tend to work for large industries that are sometimes painted as inveterate polluters by environmental groups or as greedy, unconcerned capitalists by consumer-advocate groups. Since many companies abide by strict environmental standards and engage in humanitarian work, these accusations are somewhat unfair.

The development of new technologies in fields like genetic engineering, biomedicine, bioinstrumentation, biomechanics, waste management, and alternative energy development are driving new employment opportunities for biochemical engineers. According to the U.S. Department of Labor, biochemical engineers are expected to have 2 to 4 percent employment growth during the period of 2014 through 2024, which is slower than the average for all occupations. Greater demands for more sophisticated medical equipment, procedures, and medicines will increase the need for greater cost-effectiveness.

——*Michael A. Buratovich, PhD*

FURTHER READING

Katoh, Shigeo, and Fumitake Yoshida. *Biochemical Engineering: A Textbook for Engineers, Chemists, and Biologists.* John Wiley & Sons, 2009.

McNamee, Gregory. *Careers in Renewable Energy: Get a Green Energy Job.* PixyJack, 2008.

Mosier, Nathan S., and Michael R. Ladisch. *Modern Biotechnology: Connecting Innovations in Microbiology and Biochemistry to Engineering Fundamentals.* John Wiley & Sons, 2009.

Murphy, Kenneth M., Paul Travers, and Mark Walport. *Janeway's Immunobiology.* 7th ed., Taylor & Francis, 2007.

Pahl, Greg. *Biodiesel: Growing a New Energy Economy.* 2nd ed., Green, 2008.

"Summary Report for: Biochemical Engineers." *O*Net OnLine*, www.onetonline.org/link/summary/17-2199.01. Accessed 27 Oct. 2016.

Vasic-Racki, Durda. "History of Biotransformations: Dreams and Realities." *Industrial Biotransformations.* Edited by Andreas Liese, Karsten Seelbach, and Christian Wandrey, Wiley, 2000.

Walker, Sharon. *Biotechnology Demystified.* McGraw-Hill, 2006.

BIODETECTORS

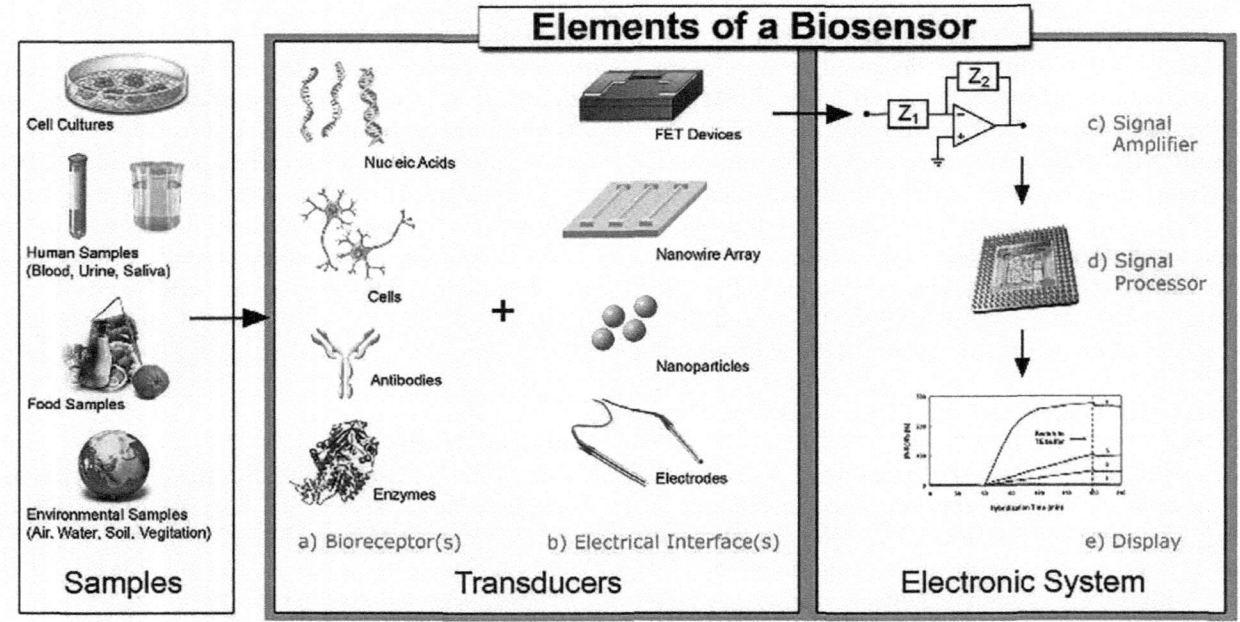

Biosensor system and components. By Dorothee Grieshaber. [CC-BY-SA-3.0 (http://creativecommons.org/licenses/by-sa/3.0)], via Wikimedia Commons

FIELDS OF STUDY

Bioengineering; Biochemistry; Biotechnology; Biophysics; Chemical engineering; Chemistry; Environmental biotechnology (or Green biotechnology)

ABSTRACT

Combining the ability to process data with the selectivity of biological systems, biodetectors are powerful analytical tools employed in forensic science. They can be used to counter the growing threat of biocrimes or acts of bioterrorism because of their ability to detect even minute levels of colorless and odorless harmful agents (such as pathogenic viruses, fungi, bacteria, and other noxious substances) days before concentrations of the agents are high enough to cause medical symptoms.

SIGNIFICANCE

Following a biocrime, responses based on data obtained from biodetection may include forensic investigation, medical diagnoses, and crisis management. In 2001, the importance of timely forensic investigation of surface contamination was demonstrated following identification of the anthrax bacterium found in letters sent to the Hart Senate Office Building in Washington, D.C.; early detection allowed for prophylactic treatment with antibiotics, thus saving the lives of those exposed to the pathogen. For highly contagious diseases such as smallpox, it may be crucial to institute immediate measures such as vaccination or quarantine to halt the spread of the disease.

The significance of early detection of harmful biological agents cannot be overemphasized. At first, medical symptoms may seem mild, and outbreaks may be mistaken for ordinary influenza; this can delay necessary remedial actions that could lessen, or even prevent, morbidity and mortality. The greatest benefit of biodetectors may be to protect against highly lethal pathogens such as Ebola and Marburg viruses, for which no vaccines, treatments, or cures have been developed.

In the mid-1960's, Leland C. Clark, considered the "father of biosensors," developed the first enzyme electrodes, which eventually led to creation of more advanced versions for applications in biotechnology and forensic science, especially as the latter pertains to countering acts of bioterrorism. Biosensors of this type, employed to detect DNA and related biomolecules, are also known as biodetectors; they are key players in the investigation of events leading up to and following exposure to such pathogenic agents as ricin (a highly toxic protein derived from the castor bean) and *Bacillus anthracis*, the bacterium that causes anthrax. Biodetectors may also be employed for continuous monitoring of the environment, surveillance of medical symptoms, and ancillary intelligence activities that may be put in place to mitigate or prevent the aftereffects associated with biocrimes and acts of bioterrorism.

Ideally, biodetectors should be networked—that is, decentralized—during an attack involving biological weapons so that they can be used to define the perimeter of the assault. Portability is another desirable characteristic for biodetectors; such devices could be moved quickly to the locations of biocrimes to perform evaluation and monitoring. Although the task of building a system of networked biodetectors is fraught with complexity, the future of emerging biosensor technology lies in scientists' ability to develop networks of sophisticated alarm-bearing biodetectors that can differentiate between harmful and benign entities and can be used anywhere, with wireless and remote capabilities.

—*Cynthia Racer*

FURTHER READING

Behnisch, Peter A. "Biodetectors in Environmental Chemistry. Are We at a Turning Point?" *Environment International* 27 (December, 2001): 441-442.

Cooper, Jon, and Tony Cass, eds. *Biosensors: A Practical Approach.* 2d ed. New York: Oxford University Press, 2004.

Malhotra, Bansi D., et al. "Recent Trends in Biosensors." *Current Applied Physics* 5 (February, 2005): 92-97.

Bioenergy technologies

Stirling engine capable of producing electricity from biomass combustion heat. By User:Wtshymanski. (Own work.) [Public domain], via Wikimedia Commons

FIELDS OF STUDY

Bioengineering; Biochemistry; Biotechnology; Biophysics; Botany; Forestry; Horticulture

ABSTRACT

The introduction of green bioenergy technologies may reduce pollution and dependence on finite fossil fuels. Bioenergy derived from biomass has the potential to provide renewable and sustained energy on both a local and a global scale. According to the International Energy Agency Bioenergy, the fundamental objective of bioenergy technology is to increase the use and implementation of ecologically sound, economically viable, and sustainable bioenergy that will help meet the world's increasing energy demands.

BASIC PRINCIPLES

Although they are closely related terms, bioenergy should not be confused with biomass. Fundamentally, bioenergy is energy derived from biomass, that is, energy derived from living (or recently living) biological organisms. Although fossil fuels are naturally occurring substances formed through the decomposition of biological organisms, the creation of these types of fuels takes millions of years, which means they are, on a human history scale, nonrenewable and unsustainable. The global demand for fossil fuels such as oil continues to increase. Oil has allowed human society to thrive because it has supplied seemingly endless cheap energy, creating diverse industries and employment. Scientific evidence continues to mount, however, in regard to its environmental impact. Oil is also a finite resource, and research has indicated that peak oil has either already occurred or will occur by the mid-twenty-first century. A growing number of environmental and scientific organizations state that once the point of peak oil is reached, demand will far outstrip supply, and therefore, it is imperative that an alternative and sustainable source of fuel is found and implemented.

BACKGROUND

The history of bioenergy is as long as the history of human civilization itself. The most basic form of bioenergy from biomass—the burning of wood for heat and light—has been used for thousands of years. Human society might rely heavily on fossil fuels for its energy needs, but bioenergy has been the world's primary energy source for most of human history and is still in use.

Wood and other combustibles such as corn husks are not the only sources of bioenergy with a long history of human use. One of the most popular biofuels is ethanol. Ethanol was first developed in the early to mid-nineteenth century, before Edwin Drake's 1859 discovery of petroleum. The push for alternative fuels during that time was driven by the need to replace the whale oil used in lamps, as supplies dwindled and prices increased. By the late 1830's, ethanol mixed with naturally derived turpentine was the preferred and cheaper alternative.

In 1826, Samuel Morey invented the internal combustion engine. His engine was powered by a composite fuel of ethanol and turpentine, and although he was unable to find a suitable investor, his engine is considered to be his greatest and most progressive invention. It was not until 1860, however, that the German inventor Nicholas Otto independently invented the internal combustion engine (again fueled by ethanol) and achieved financial backing to develop the engine.

The next considerable leap in interest in bioenergy and in technology development occurred with the invention of the automobile in the early 1900's and Henry Ford's vision of an ethanol-fueled vehicle. As early as 1917, scientists knew that ethanol and other alcohols could be used as fuels. Because ethanol and other alcohols could be derived easily from any vegetable matter that undergoes fermentation, they could be cheaply and easily produced. Despite this, however, ethanol fuels were not widely embraced as the fuel of choice for automobiles, particularly in counties such as the United States, where an ethanol tax made the alternative fuel more expensive than petrol. By 1906, when the tax was removed, gasoline fuel had developed an extensive infrastructure, and ethanol could not compete.

Because of rationing and shortages of petrol during World War II, ethanol and other vegetable-based fuels were used extensively. In particular, the use of vegetable-based fuels such as palm oil became common in European colonies in Africa so as to increase fuel self-sufficiency. During the oil crises of the 1970's, oil prices skyrocketed and oil-dependent countries, such as the United States, became desperate to find a replacement for fossil fuels and to reduce their dependence on oil-producing countries. The 1974 oil embargo was influential in renewing interest in alternative fuels, such as ethanol, particularly in the United States.

Despite this, however, fossil fuels have continued to be used to a much greater degree than ethanol fuels. Although many automotive fuels are mixed with ethanol, there is still significant room for growth and technological and infrastructure development of bioenergy and biofuels. The potential of biofuels as a sustainable alternative to fossil fuels has once again increased interest in bioenergy. However, there are arguments against the use of bioenergy, particularly the fuel-versus-food debate. Some people believe that as long as large numbers of people worldwide do not get enough to eat, using crops for fuel instead of food is at best misguided and at worst highly unethical, and it contributes to the rates of starvation and malnutrition seen in many developing countries.

HOW IT WORKS

Although scientific debate still surrounds human-influenced climate change and the timing of peak oil, many experts believe that developing sustainable fuels from renewable sources and implementing technology that helps reduce pollution is important and necessary. Bioenergy technologies play an important role in accelerating the adoption of environmentally sound bioenergy at a reasonable cost and in a sustainable manner, thereby helping meet future energy demands.

Bioenergy can be produced from many different biological materials, including wood and various crops, as well as human and animal waste. All these materials can be used to produce electricity and heat, and after coal, oil, and natural gas, biomass is a major energy resource in the world. The International Energy agency reports that about 1.5 percent of the world's electricity and 10 percent of the world's energy was produced from biomass sources in 2012. This percentage differs from country to country and is greater in developing countries, than in developed countries such as the United States. The United States is, however, one of the world's largest biopower generators and possesses much of the world's installed bioenergy capacity.

Since the 1990's, bioenergy technologies have experienced continuous development. Generally, these technologies can be divided into two main groups in relation to bioenergy production: energy crops and waste energy. Energy crops, which include trees, sugarcane, and rapeseed, are either combusted or fermented to produce high-energy alcohols such as ethanol and biodiesel that can be used as a replacement for petrol and liquid fuels. Certain waste, including organic waste from agriculture, human and animal effluent, food and plant waste, and industrial residue, can be used to produce methane gas. This methane gas can be combusted to produce steam, which can then be used to turn turbine generators and produce heat and electricity.

Most forms of bioenergy require combustion and thus the release of carbon dioxide into the atmosphere at some stage during their production. This release, however, is offset by the initial absorption of carbon dioxide by the fuel crops during the growing process. According to research, even accounting for all carbon dioxide released as a result of the planting, harvesting, producing, and transporting of bioenergy, net carbon emissions are significantly reduced.

APPLICATIONS AND PRODUCTS

A number of different types of domestic biomass resources (also referred to as feedstocks) are used to produce bioenergy. These include biomass processing residues such as paper and pulp, agricultural and forestry wastes, urban landfill waste and gas, animal (including human) sewage and manure waste, and land and aquatic crops. There are basically two very important and useful applications of bioenergy: the production of electricity and the replacement of liquid fuels such as petrol.

Electricity Production. Biomass is capable of producing electricity in many different ways. The most commonly used methods include pyrolysis, cofiring, direct-fired/conventional stream method, gasification, anaerobic digestion, and landfill gas collection.

The term "pyrolysis" is derived from the Greek words *pyro* (meaning fire) and *lysys* (meaning decomposition). Pyrolysis is a thermochemical conversion technology that involves the combustion of biomass at very high temperatures and its decomposition without oxygen. Although this process is energy consumptive and expensive, it can be used to produce electricity through the creation of pyrolysis oil, biochar, and syngas (oil, coke, and gas). These three products can be used for electricity production, as soil fertilizer, and for carbon storage. There are two types of pyrolysis—fast and slow. Fast, or flash, pyrolysis, which uses any organic material as a biomass source, takes place within seconds at temperatures of 300 to 550 degrees Celsius with rapid accumulation of biochar. In slow, or vacuum, pyrolysis, which uses any organic material as a biomass source, the combustion of the biomass occurs within a vacuum to reduce the boiling point and adverse chemical reactions.

Cofiring basically involves the combustion of solid biomass such as wood or agricultural waste with a traditional fossil fuel such as coal to produce energy. Many consider this form of bioenergy to be the most efficient in terms of the existing fossil fuel infrastructure, with its high dependence on coal. In addition, because the combustion of biomass is carbon neutral (the carbon absorbed during growth is equal to the amount released during combustion), mixing biomass with coal can assist in reducing net carbon emissions and other pollutants such as sulfur. The production of electricity by this method is considered advantageous because it is inexpensive and makes use of already existing power plants.

The direct-fired/conventional stream method is the most commonly used method of producing bioenergy. This process involves the direct combustion of biomass to produce steam, which then turns turbines that drive generators to produce electricity. Gasification is another type of thermochemical conversion technology, in which any organic biomass is converted into its gaseous form (known as syngas) and used to produce energy. The process relies on biomass gasifiers that heat solid biomass until it forms a combustible gas, which is then used in power production systems that merge gas and steam turbines to generate electricity. Although this technology is still evolving, it is hoped that gasification of biomass may lead to more efficient bioenergy production. Biomass gasification technologies basically can be categorized as fixed-bed gasification, fluidized-bed gasification, and novel-design gasification.

Anaerobic digestion, a natural biological process, has a long history and involves the decomposition of organic biomass material such as manure and urban solid wastes in an air-deficient environment. Fundamentally, anaerobic digestion involves the production of methane gas through bacteria and archaea activity. This methane gas, also known as a type of biogas, is captured and then used to power turbines and produce electric and heat energy, as well as a soil-enhancement material called digestate. The advantage of this method is that it uses waste, such as wastewater sludge, to produce renewable energy and reduces the amount of greenhouse gas being released into the atmosphere.

The process of landfill gas collection is closely related to anaerobic digestion and involves the capture of gas from the decomposition of landfill urban wastes. Landfill gas, which is about 50 percent methane, 45 percent carbon dioxide, 4 percent nitrogen, and 1 percent other gases, is then used to produce energy.

Fuel Production. Liquid biofuels are a significant alternative to petroleum-based vehicle and transportation fuels. Biofuels are attractive because they can be used in already existing vehicles with little modification required and also in the production of electricity. As of 2014 the International Energy Agency estimated that biofuels accounted for about 4 percent of the transportation fuels used in the world's vehicles.

Bioenergy production is generally divided into first-generation and second-generation fuels. The main distinction between these two types of fuels relates to the feedstock used. First-generation fuels are already in commercial production in many countries and are primarily made from edible grains, sugars, or seeds. The two most common biofuels are ethanol and biodiesel, both of which are already used in large quantities in many countries. Second-generation fuels are considered superior to first-generation fuels because they are primarily made from nonedible whole plants or waste from food crops such as husks and stalks.

Ethanol, also known as ethyl alcohol, is an alcohol that can be used as a vehicle or transportation fuel. It is a renewable energy produced from sustainable agricultural feedstocks, particularly sugar and starch crops such as sugarcane, potatoes, and maize. Although it can be used as a direct replacement fuel, it is more commonly used as a fuel additive. Many vehicles use ethanol-petrol blends of 10 percent ethanol, which improve octane and decrease emissions. Ethanol is particularly popular in Brazil and the United States, which together produce most of the world's ethanol. Brazil has been at the forefront of ethanol use and as early as 1976 mandated the use of ethanol with fuels, eventually requiring a blend containing 25 percent ethanol (the most of any country). Although using arable land to grow crops to produce ethanol is somewhat controversial, the development of ethanol from cellulosic biomass (obtained from the cellulose of trees and grasses) is considered promising and may play a large role in the future of ethanol as a biofuel.

Biodiesel is a renewable energy produced from sustainable animal fat and vegetable oil feedstocks, such as soy, rapeseed, sunflowers, palm oil, hemp, and algae, and can be used as a vehicle or transportation fuel. As with ethanol, however, biodiesel is more often used as a diesel additive to reduce the levels of pollution emitted by traditional diesel engines. It is primarily produced through a process known as transesterification, which is the exchange or conversion of an organic acid ester into another ester.

Biobutanol can be used as fuel in internal combustion engines. It is usually produced from the fermentation of biomass, and because of its chemical properties, it is actually more similar to petrol than ethanol is. It can be produced from the same feedstocks used for ethanol production, such as corn, sugarcane, potatoes, and wheat. Despite its possible applications, however, the U.S. Department of Energy reported in

April 2016, biobutanol had not yet been sold commercially as a transportation fuel in the United States because of regulatory hurdles.

SOCIAL CONTEXT AND FUTURE PROSPECTS

The concept of bioenergy relies on the fact that such energy is sustainable. Although many believe bioenergy is synonymous with green energy, this is not always the case. Individual types of bioenergy produced in different ways and from various biomasses can have very diverse environmental impacts. Although the goal of bioenergy is to reduce the world's dependence on nonrenewable (to all intents and purposes) fossil fuels and thereby reduce greenhouse gas emissions and pollution, some forms of bioenergy can be equally harmful in terms of pollution or the energy expended to produce the bioenergy. As such, there have been significant movements since the 1980's to develop cleaner, greener, and move advanced biofuels and technologies. Many countries are investigating the potential of bioenergy, and many researchers believe that bioenergy will become a key contributor to sustainable global energy use.

Biofuels are, however, controversial, and some are calling for a moratorium on their use and advancement because of environmental and social concerns, particularly the loss of biodiversity and habitat destruction and the use of crops for fuel rather than food. First-generation fuels, already in commercial production in many countries, have been criticized because they are obtained from edible seeds and plants. However, because of the increasing problems of fossil fuel dependence, many of the world's governments and international organizations are stepping up research into bioenergy as a viable and sustainable alternative fuel. Many researchers believe that investigation and implementation of second-generation fuels, which are made from nonedible whole plants or waste from food crops, is the way of the future in terms of both efficiency and social responsibility. Biomass does offer many countries the opportunity to use fuels that are both sustainable and domestically sourced.

———*Christine Watts, PhD*

FURTHER READING

Geller, Howard. *Energy Revolution: Policies for a Sustainable Future.* Washington, D.C.: Island Press, 2003.

Rosillo-Calle, Frank, et al., eds. *The Biomass Assessment Handbook: Bioenergy for a Sustainable Environment.* Sterling, Va.: Earthscan, 2008.

Scragg, Alan. *Biofuels: Production, Application, and Development.* Cambridge, Mass.: CAB International, 2009.

Silveira, Semida, ed. *Bioenergy: Realizing the Potential.* San Diego, Calif.: Elsevier, 2005.

Sims, Ralph. *The Brilliance of Bioenergy: In Business and Practice.* London: James and James, 2002.

Singh, Om V., and Steven P. Harvey, eds. *Sustainable Biotechnology: Sources of Renewable Energy.* London: Springer, 2009.

BIOENGINEERING

FIELDS OF STUDY

Bioengineering; Biotechnology

ABSTRACT

Bioengineering is the field in which techniques drawn from engineering are used to tackle biological problems. For example, bioengineers may use mechanics principles—knowledge about how to design and construct mechanical objects using the most ideal materials—to create drug delivery systems. They may work on developing efficient ways to irrigate and drain land for growing crops, or they may be involved in building artificial environments that can support life even in the harsh climate of outer space. A highly interdisciplinary, collaborative field that synthesizes expertise from multiple research areas, bioengineering has had a significant impact on many fields of study, including the health sciences, technology, and agriculture.

BACKGROUND

In many contexts, the term bioengineering is used to refer solely to biomedical engineering. This is the application of engineering principles to medicine, such as in the development of artificial organs or limbs.

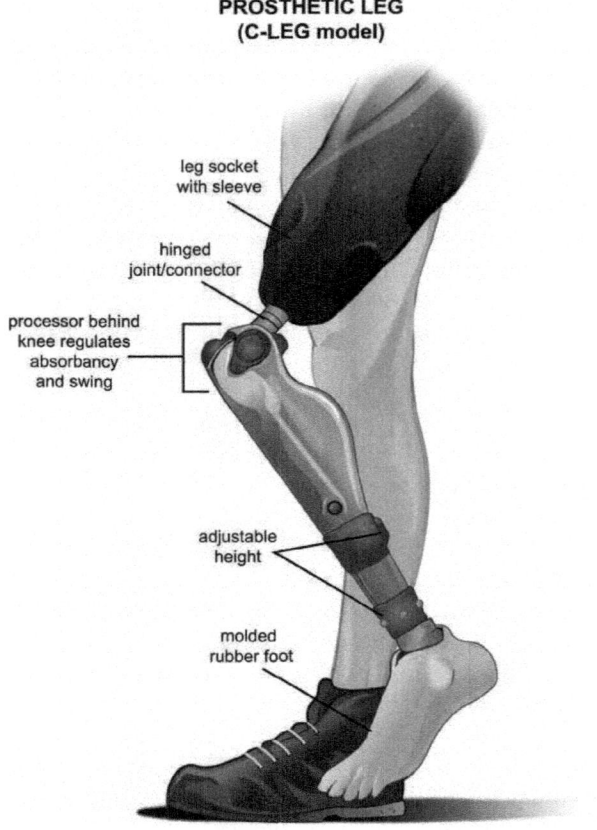

Bioengineers have made significant advances in the field of prosthetics; the C-Leg prosthesis was introduced in 1997. © EBSCO

However, the field of bioengineering has many applications beyond the field of health care. For example, genetically modified crops that are resistant to pests, suits that protect astronauts from the ultra-low pressures in space, and brain-computer interfaces that may allow soldiers to exercise remote control over military vehicles all fall under the wide umbrella of bioengineering.

Each of the subdisciplines within bioengineering relies on different sets of basic engineering principles, but a few fundamental approaches can be said to apply broadly across the entire field. From an engineering perspective, three basic steps are involved in solving any problem: an analysis of how the system in question works, an attempt to synthesize the information gathered from this analysis and generate potential solutions, and finally an attempt to design and test a useful product. Bioengineers apply this three-stage problem-solving process to problems in the life sciences. What is somewhat novel about this approach is that it is a holistic one. In other words, it treats biological entities as systems—sets of parts that work together and form an integrated whole—rather than looking at individual parts in isolation. For example, to develop an artificial heart, bioengineers need to consider not just the structure of the heart itself on a cellular or tissue level but also the complex dynamics of the organ's interactions with the rest of the body through the circulatory system and the immune system. They must build a device whose parts can mimic the functionality of a healthy heart and whose materials can be easily integrated into the body without triggering a harmful immune response.

HISTORY

Principles of chemical and mechanical engineering have been applied to problems in specific biological systems for centuries. For example, bioengineering applications include the fermentation of alcoholic beverages, the use of artificial limbs (which are documented as far back as 500 BCE), and the building of heating and cooling systems that regulate human environments.

Bioengineering did not emerge as a formal scientific discipline, however, until the middle of the twentieth century. During this period, more and more scientists began to be interested in applying new technologies from electronic and mechanical engineering to the life sciences. As the United States, Japan, and Europe began to enter a period of economic recovery and growth following World War II, governments increased funding for bioengineering efforts. The cardiac pacemaker and the defibrillator, both developed during this postwar period, were two of the earliest and most significant inventions to come out of the quickly developing field. In 1966, the Engineers Joint Council Committee on Engineering Interaction with Biology and Medicine first used the term "bioengineering." At about the same time, academic institutions began to form specialized departments and programs of study to train professionals in the application of engineering principles to biological problems. In the twenty-first century, rapid technological advances continue to produce growth in the field of bioengineering.

HOW IT WORKS

Because bioengineering is such a large and diverse field, it would be impossible to enumerate all the processes involved in creating the totality of its applications. The following are a few of the most significant examples of the types of technological tools used in bioengineering.

Materials Science. One of the most important areas of bioengineering is the intersection of materials science and biology. Scientists working in this field are charged with developing materials that, although synthetic, are able to successfully interact with living tissues or other natural biological systems without impeding them. (For example, it is vital that biocompatible materials not allow blood platelets to adhere to them and form clots, which can be fatal.) Depending on the specific application in question, other properties, such as tensile strength, resistance to wear, and permeability to water, gases, and small biological molecules, are also important. To manipulate these properties to achieve a desired end, engineers must carefully control both the chemical structure and the molecular organization of the materials. For this reason, biocompatible materials are generally made out of some kind of synthetic polymer—substances with simple and extremely regular molecular structures that repeat again and again. In addition, additives may be incorporated into the materials, such as inorganic fillers that allow for greater mechanical flexibility or stabilizers and antioxidants that keep the material from becoming degraded over time.

Biochemical Engineering. Since living cells are essentially chemical systems, the tools of chemical engineering are especially applicable to biology. Biochemical engineers study and manipulate the behavior of living cells. Their basic tool for doing this is a fermenter, a large reactor within which chemical processes can be carried out under carefully controlled conditions. For example, the modern production of virtually all antibiotics, such as penicillin and tetracycline, takes place inside a fermenter. A central vessel, sealed tight to prevent contamination and surrounded by jackets filled with coolants to control its temperature, contains propellers that stir around the nutrients, culture ingredients, and catalysts that are associated with the reaction at hand.

Genetic engineering is a subfield of biochemical engineering that is growing increasingly significant. Scientists alter the genetic information in one cell by inserting into it a gene from another organism. To do this, a vector such as a virus or a plasmid (a small strand of DNA) is placed into the cell nucleus and combines with the existing genes to form a new genetic code. The technology that enables scientists to alter the genetic information of an organism is called gene splicing. The new genetic information created by this process is known as recombinant DNA. Genetic engineering can be divided into two types: somatic and germ line. Somatic genetic engineering is a process by which gene splicing is carried out within specific organs or tissues of a fully formed organism; germ-line genetic engineering is a process by which gene splicing is carried out within sex cells or embryos, causing the recombinant DNA to exist in every cell of the organism as it grows.

Electrical Engineering. Electrical engineering technologies are an essential part of the bioengineering tool kit. In many cases, what is required is for the bioengineer to find some way to convert sensory data into electric signals and then to produce these electric signals in such a way as to enable them to have a physiological effect on a living organism.

The cochlear implant is an example of one such development. The cochlea is the part of the brain that interprets sounds, and a cochlear implant is designed for people who are profoundly deaf. A cochlear implant uses electronic devices that capture sounds and relay them to the cochlea. The implant has four parts: a microphone, a tiny computer processor, a radio transmitter, and a receiver, which surgeons implant in the user's skull. The microphone picks up nearby sounds, such as human speech or music emerging from a pair of stereo speakers. Then the processor converts the sounds into digital information that can be sent through a wire to the radio transmitter. The software used by the processor separates sounds into different channels, each representing a range of frequencies. In turn, the radio transmitter translates the digital information into radio signals, which it relays through the skull to the receiver. The receiver then turns the radio signals into electric impulses, which directly stimulate the nerve endings in the cochlea. It is these electric signals that the brain is able to interpret as sounds, allowing even profoundly deaf people to hear.

Another example of how electric signals can be used to direct biological systems can be found in brain-computer interfaces (BCIs). BCIs are direct channels of communication between a computer and the neurons in the human brain. They work because activity in the brain, such as that produced by thoughts or sensory processing, can be detected by bioinstruments designed to record electrophysiological signals. These signals can then be transmitted to a computer and used to generate commands. For example, BCIs allow stroke victims who have lost the use of a limb to regain mobility; a patient's thoughts about movement are transmitted to an external machine, which in turn transmits electric signals that precisely control the movements of a cradle holding his or her paralyzed arm.

APPLICATIONS AND PRODUCTS

Biomedical Applications. Biomedical engineering is a vast subdiscipline of bioengineering, which itself encompasses multiple fields of interest. The many clinical areas in which applications are being developed by biomedical engineers include medical imaging, cell and tissue engineering, bioinstrumentation, the development of biocompatible materials and devices, biomechanics, and the emerging field of bionanotechnology.

Medical imaging applications collect data about patients' bodies and turn that data into useful images that physicians can interpret for diagnostic purposes. For example, ultrasound scans, which map the reflection and reduction in force of sounds as they bounce off an object, are used to monitor the development of fetuses in the wombs of pregnant women. Magnetic resonance imaging (MRI), which measures the response of body tissues to high-frequency radio waves, is often used to detect structural abnormalities in the brain or other body parts.

Cell and tissue engineering is the attempt to exploit the natural characteristics of living cells to regenerate lost or damaged tissue. For example, bioengineers are working on creating viable replacement heart cells for people who have suffered cardiac arrests, as well as trying to discover ways to regenerate brain cells lost by patients with neurodegenerative disorders such as Alzheimers disease. Genetic engineering is a closely related area of biomedicine in which DNA from a foreign organism is introduced into a cell so as to create a new genetic code with desired characteristics.

Bioinstrumentation is the application of electrical engineering principles to develop machines that can sense and respond to biological or physiological signals, such as portable devices for diabetics that measure and report the level of glucose in their blood. Other common examples of bioinstrumentation include electroencephalogram (EEG) machines that continuously monitor brain waves in real time, and electrocardiograph (ECG) machines that perform the same task with heartbeats.

Many biomedical engineers work on developing materials and devices that are biocompatible, meaning that they can replace or come into direct contact with living tissues, perform a biological function, and refrain from triggering an immune system response. Pacemakers, small artificial devices that are implanted within the body and used to stimulate heart muscles to produce steady, reliable contractions, are a good example of a biocompatible device that has emerged from the collaboration of engineers and clinicians.

Biomechanics is the study of how the muscles and skeletal structure of living organisms are affected by and exert mechanical forces. Biomechanics applications include the development of orthotics (braces or supports), such as spinal, leg, and foot braces for patients with disabling disorders such as cerebral palsy, multiple sclerosis, or stroke. Prostheses (artificial limbs) also fall under the field of biomechanics; the sockets, joints, brakes, and pneumatic or hydraulic controls of an artificial leg, for example, are manufactured and then combined in a modular fashion, in much the same way as are the parts of an automobile in a factory.

Bionanotechnology. Nanotechnology is a fairly young field of applied science concerned with the manipulation of objects at the nanoscale (about 1^{-100} nanometers, or about one-thousandth the width of a strand of human hair) to produce machinery. Bionanotechnological applications within medicine include microscopic biosensors installed on small chips; these can be specialized to recognize and flag specific proteins or antibodies, helping physicians conduct extremely fast and inexpensive diagnostic tests. Bioengineers are also developing microelectrodes on a nanoscale; these arrays of tiny electrodes can be implanted into the brain and used to stimulate specific nerve cells to treat movement disorders and other diseases.

Military Applications. Bioengineering applications are making themselves felt as a powerful presence on the front lines of the military. For example, bioengineering students at the University of Virginia designed lighter, more flexible, and stronger bulletproof body armor using specially created ceramic tiles that are inserted into protective vests. The armor is able to withstand multiple impacts and distributes shock more evenly across the wearer's body, preventing damaging compression to the chest. Others working in the field are creating sophisticated biosensors that soldiers can use to detect the presence of potential pathogens or biological weapons that have been released into the air.

One of the most significant contributions of bioengineering to the military is in the development of treatments for severe traumas sustained during warfare. For example, stem cell research may one day enable military physicians to regenerate functional tissues such as nerves, bone, cartilage, skin, and muscle—an invaluable tool for helping those who have lost limbs or other body parts as a result of explosives. The United States military was responsible for much of the early research done in creating safe, effective artificial blood substitutes that could be easily stored and relied on to be free of contamination on the battlefield.

Agriculture. Agricultural engineering involves the application of both engineering technologies and knowledge from animal and plant biology to problems in agriculture, such as soil and water conservation, food processing, and animal husbandry. For example, agricultural engineers can help farmers maximize crop yields from a defined area of land. This technique, known as precision farming, involves analyzing the properties of the soil (factors such as drainage, electrical conductivity, pH [acidity] level, and levels of chemicals such as nitrogen) and carefully calibrating the type and amount of seeds, insecticides, and fertilizers to be used.

Farm machinery and implements represent another area of agriculture in which engineering principles have made a big impact. Tractors, harvesters, combines, and grain-processing equipment, for example, have to be designed with mechanical and electrical principles in mind and also must take into account the characteristics of the land, the needs of the human operators, and the demands of working with particular agricultural products. For example, many crops require specialized equipment to be successfully mechanically harvested. Thus a pea harvester may have several components—one that lifts the vines and cuts them from the plant, one that strips pea pods from the stalk, and one that threshes the pods, causing them to open and release the peas inside them. Another example of an agricultural engineering application is the development of automatic milking machines that attach to the udders of a cow and enable dairy farmers to dispense with the arduous task of milking each animal by hand.

The management of soil and water is also an important priority for bioengineers working in agricultural settings. They may design structures to control the flow of water, such as dams or reservoirs. They may develop water-treatment systems to purify wastewater coming out of industrial agricultural production centers. Alternatively, they may use soil walls or cover crops to reduce the amount of pesticides and nutrients that run off from the soil, as well as the amount of erosion that takes place as a result of watering or rainfall.

Environmental and Ecological Applications. Environmental and ecological engineers study the impact of human activity on the environment, as well as the ways in which humans respond to different features of their environments. They use engineering principles to clean, control, and improve the quality of natural spaces, and find ways to make human interactions with environmental resources more sustainable. For example, the reduction and remediation of pollution is an important area of concern. Therefore, an environmental engineer may study the pathways and rates at which volatile organic compounds (such as those found in many paints, adhesives, tiles, wall coverings, and furniture) react with other gases in the air, causing smog and other forms of air pollution. They may design and build sound walls in residential areas to cut down on the amount of noise pollution caused by airplanes taking off and landing or cars racing up and down highways.

The life-support systems designed by bioengineers to enable astronauts to survive in the harsh conditions of outer space are also a form of environmental engineering. For example, temperatures around a space shuttle can vary wildly, depending on which side of the vehicle is facing the Sun at any given moment. A complex system of heating, insulation, and ventilation helps regulate the temperature inside the cabin. Because space is a vacuum, the shuttle itself must be filled with pressurized gas. In addition, levels of oxygen, carbon dioxide, and nitrogen within the cabin must be controlled so that they resemble the atmosphere on Earth. Oxygen is stored on board in tanks, and additional supplies of the essential gas are produced from electrolyzed water; in turn, carbon dioxide is channeled out of the shuttle through vents.

Geoengineering. Geoengineering is an emerging subfield of bioengineering that is still largely theoretical. It would involve the large-scale modification of environmental processes in an attempt to counteract the effects of human activity leading to climate change. One proposed geoengineering project involves depositing a fine dust of iron particles into the ocean in an attempt to increase the rate at which algae grows in the water. Since algae absorbs carbon dioxide as it photosynthesizes, essentially trapping and containing it, this would be a means of reducing the amount of this greenhouse gas in the atmosphere. Other geoengineering proposals include the suggestion that it might be possible to spray sulfur dust into the high atmosphere to reflect some of the Sun's light and heat back into space, or to spray drops of seawater high up into the air so that the salt particles they contain would be absorbed into the clouds, making them thicker and more able to reflect sunlight.

SOCIAL CONTEXT AND FUTURE PROSPECTS

Bioengineering is a field with the capacity to exert a powerful impact on many aspects of social life. Perhaps most profound are the transformations it has made in health care and medicine. By treating the body as a complex system—looking at it almost as if it were a machine—bioengineers and physicians working together have enabled countless patients to overcome what once might have seemed to be insurmountable damage. After all, if the body is a machine, its parts might be reengineered or replaced entirely with new ones—as when the damaged cilia of individuals with hearing impairments are replaced with electro-mechanical devices. Some aspects of bioengineering, however, have drawn concern from observers who worry that there may be no limit to the scientific ability to interfere with biological processes. Transgenic foods are one area in which a contentious debate has sprung up. Some are convinced that the ecological and health ramifications of growing and ingesting crops that contain genetic information from more than one species have not yet been fully explored. Stem cell research is another area of controversy; some critics are uncomfortable with the fact that human embryonic stem cells are being obtained from aborted fetuses or fertilized eggs that are left over from assisted reproductive technology procedures.

One aspect of bioengineering that has been the subject of both fear and hope in the twenty-first century is the question of whether it might be possible to stop or even reverse the harmful effects of climate change by carefully and deliberately interfering with certain geological processes. Some believe that geoengineering could help the international community avoid the devastating effects of global warming predicted by scientists, such as widespread flooding, droughts, and crop failure. Others, however, warn that any attempt to interfere with complex environmental systems on a global scale could have wildly unpredictable results. Geoengineering is especially controversial because such projects could potentially be carried out unilaterally by countries acting without international agreement and yet have repercussions that could be felt all across the world.

—M. Lee, MA

FURTHER READING

Artmann, Gerhard M., and Shu Chien, editors. *Bioengineering in Cell and Tissue Research.* Springer, 2008.

Enderle, John D., et al., editors. *Introduction to Biomedical Engineering.* 3rd ed., Academic, 2012.

Huffman, Wallace E., and Robert E. Evenson. *Science for Agriculture: A Long-Term Perspective.* 2nd ed., Blackwell, 2006.

Madhavan, Guruprasad, et al., editors. *Career Development in Bioengineering and Biotechnology.* Springer, 2008.

Nemerow, Nelson Leonard, et al., eds. *Environmental Engineering: Environmental Health and Safety for Municipal Infrastructure, Land Use and Planning, and Industry.* John Wiley & Sons, 2009. 3 vols.

Biofertilizers

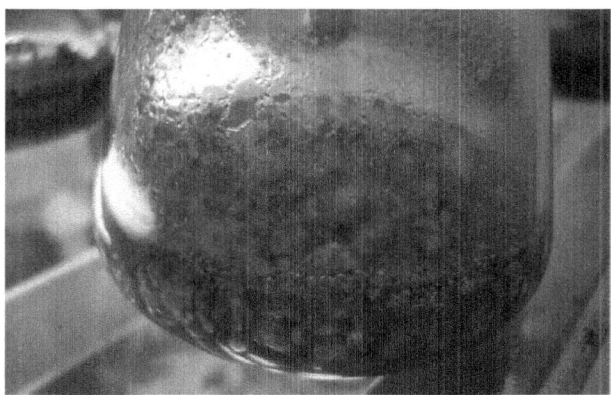

Cyanobacteria cultured in specific media. Cyanobacteria can be helpful in agriculture as they have the capability to fix atmospheric nitrogen to soil. This nitrogen is helpful to the crops. Cyanobacteria is used as a bio-fertilizer. By Joydeep. (Own work.) [CC-BY-SA-3.0 (http://creativecommons.org/licenses/by-sa/3.0)], via Wikimedia Commons

FIELDS OF STUDY

Agricultural engineering; Bioengineering; Biochemistry; Biotechnology; Biophysics; Botany; Forestry; Horticulture

ABSTRACT

Biofertilizers provide a means by which biological systems can be utilized to supply plant nutrients such as nitrogen to agricultural crops. The use of biofertilizers could reduce the dependence on chemical fertilizers, which are often detrimental to the environment.

BACKGROUND

Plants require adequate supplies of the thirteen mineral nutrients necessary for normal growth and reproduction. These nutrients, which must be supplied by the soil, include both macronutrients (those nutrients required in large quantities) and micronutrients (those nutrients required in smaller quantities). As plants grow and develop, they remove these essential mineral nutrients from the soil. Since normal crop production usually requires the removal of plants or plant parts, the nutrients are continuously removed from the soil. Therefore, the long-term agricultural utilization of any soil requires periodic fertilization to replace lost nutrients.

Nitrogen is the plant nutrient that is most often depleted in agricultural soils, and most crops respond to the addition of nitrogen fertilizer by increasing their growth and yield; therefore, more nitrogen fertilizer is applied to cropland than any other fertilizer. In the past, nitrogen fertilizers have been limited to either manures, which have low levels of nitrogen, or chemical fertilizers, which usually have high levels of nitrogen. The excess nitrogen in chemical fertilizers often runs off into nearby waterways, causing a variety of environmental problems.

HOW IT WORKS

Biofertilizers offer a potential alternative: They supply sufficient amounts of nitrogen for maximum yields yet have a positive impact on the environment. Biofertilizers generally consist of either naturally occurring or genetically modified microorganisms that improve the physical condition of soil, aid plant growth, or increase crop yield. Such fertilizers provide an environmentally friendly way to increase plant health and yields with reduced input costs, new products and additional revenues for the agricultural biotechnology industry, and cheaper products for consumers.

While biofertilizers could potentially be used to supply a number of different nutrients, most of the interest thus far has focused on enhancing nitrogen fertilization. The relatively small amounts of nitrogen found in soil come from a variety of sources. Some nitrogen is present in all organic matter in soil; as this organic matter is degraded by microorganisms, it can be used by plants. A second source of nitrogen is nitrogen fixation, the chemical or biological process of taking nitrogen from the atmosphere and converting it to a form that can be utilized by plants. Bacteria such as *Rhizobia* can live symbiotically with certain plants, such as legumes, which house nitrogen-fixing bacteria in their roots. The *Rhizobia* and plant root tissue form root nodules that house the nitrogen-fixing bacteria; once inside the nodules, the bacteria use energy supplied by the plant to convert atmospheric nitrogen to ammonia, which nourishes the plant. Natural nitrogen can also be supplied by free-living microorganisms, which can fix nitrogen without forming a symbiotic relationship with plants. The primary objective of biofertilizers is to enhance any one or all of these processes.

One of the major goals in the genetic engineering of biofertilizers is to transfer the ability to form nodules and establish effective symbiosis to nonlegume plants. The formation of nodules in which the *Rhizobia* live requires plant cells to synthesize many new proteins, and many of the genes required for the expression of these proteins are not found in the root cells of plants outside the legume family. Many research programs have been devoted to efforts to transfer such genes to nonlegume plants so that they can interact symbiotically with nitrogen-fixing bacteria. If this is accomplished, *Rhizobia* could be used as a biofertilizer for a variety of plants.

There is also much interest in using the free-living, soil-borne organisms that fix atmospheric nitrogen as biofertilizers. These organisms live in the rhizosphere (the region of soil in immediate contact with plant roots) or thrive on the surface of the soil. Since the exudates from these microorganisms contain nitrogen that can be utilized by plants, increasing their abundance in the soil could reduce dependence on chemical fertilizers. Numerous research efforts have been designed to identify and enhance the abundance of nitrogen-fixing bacteria in the rhizosphere. Soil microorganisms primarily depend on soluble root exudates and decomposed organic matter to supply the energy necessary for fixing nitrogen; hence there is also an interest in enhancing the biodegradation of organic matter in the soil. This research has largely centered on inoculating the soil with cellulose-degrading fungi and nitrogen-fixing bacteria or applying organic matter to the soil, such as straw that has been treated with a combination of the fungi and bacteria.

—D. R. Gossett

FURTHER READING
Akinyemi, Okoro M. *Agricultural Production: Organic and Conventional Systems.* Enfield, N.H.: Science Publishers, 2007.
Black, C. A. *Soil-Plant Relationships.* New York: John Wiley & Sons, 1988.
Chrispeels, M. J., and D. E. Sadava. *Plants, Genes, and Agriculture.* Sudbury, Mass.: Jones and Bartlett, 1994.
Lynch, J. M. *Soil Biotechnology: Microbiological Factors in Crop Production.* Malden, Mass.: Blackwell, 1983.
Salisbury, F. B., and C. W. Ross. *Plant Physiology.* Pacific Grove, Calif.: Brooks/Cole, 1985.
Yadav, A. K., S. Ray Chaudhuri, and M. R. Motsara, eds. *Recent Advances in Biofertilizer Technology.* New Delhi: Society for Promotion and Utilisation of Resources and Technology, 2001.

BIOFUELS

FIELDS OF STUDY

Agricultural engineering; Bioengineering; Biochemistry; Biotechnology; Biophysics; Industrial biotechnology (or White biotechnology); Industrial fermentation

ABSTRACT

Biofuels are made mainly from plant material such as corn, sugarcane, or rapeseed. Theoretically, biofuels can be generated anywhere on Earth where living organisms can grow. Biofuels such as ethanol and biodiesel are excellent transportation fuels that are used as substitutes or supplements for gasoline and diesel fuels. Biofuels can also be burned in electrical generators to produce electricity. Two biofuels are used in vehicles: ethanol and biodiesel. Biogas and methane are used mainly to generate electricity. Biomass was used traditionally to heat houses.

DESCRIPTION, DISTRIBUTION, AND FORMS
Over millions of years, dead organic matter—both plant and animal organisms—played a crucial role in the formation of fossil fuels such as oil, natural gas, and coal. Since the nineteenth century, humans have increasingly depended on fossil fuels to meet energy needs. As the supply of fossil fuels has diminished, humankind has begun looking for alternative energy sources. Thus, the use of biofuels—including

> **Biofuel Energy Balances**
>
> *The following table lists several crops that have been considered as viable biofuel sources and several types of ethanol, as well as each substance's energy input/output ratio (that is, the amount of energy released by burning biomass or ethanol, for each equivalent unit of energy expended to create the substance).*
>
> | Biomass/Biofuel | Energy Output per Unit Input |
> |---|---|
> | Switchgrass | 14.52 |
> | Wheat | 12.88 |
> | Oilseed rape (with straw) | 9.21 |
> | Cellulosic ethanol | 1.98 |
> | Corn ethanol | ~1.13–1.34 |
>
> *Source:* Data from the British Institute of Science in Society.

© EBSCO

ethanol, biodiesel, methane, biogas, biomass, biohydrogen, and butanol—is increasing.

Ethanol is a colorless liquid with the chemical formula C_2H_5OH. Another name for ethanol is ethyl alcohol, grain alcohol, or simply alcohol.

Biodiesel is a diesel substitute obtained mainly from vegetable oils, such as soybean oil or restaurant greases. It is produced by the transesterification of oils, a simple chemical reaction with alcohol (ethanol or methanol), catalyzed by acids or bases (such as sodium hydroxide). Transesterification produces alkyl esters of fatty acids that are biodiesel and glycerol (also known as glycerin).

Methane is a colorless, odorless, nontoxic gas with the molecular formula CH_4. It is the main chemical component (70 to 90 percent) of natural gas, which accounts for about 20 percent of the U.S. energy supply. Methane was discovered by the Italian scientist Alessandro Volta, who collected it from marsh sediments and showed that it was flammable. He called it "combustible air."

Biogas is a gas produced by the metabolism of microorganisms. There are different types of biogas. One type contains a mixture of methane (50 to 75 percent) and carbon dioxide. Another type comprises primarily nitrogen, hydrogen, and carbon monoxide (CO) with trace amounts of methane.

Biomass is a mass of organisms, mainly plants that can be used as an energy source. Plants and algae convert the energy of the Sun and carbon dioxide into energy that is stored in their biomass. Biomass, burning in the form of wood, is the oldest form of energy used by humans. Using biomass as a fuel source does not result in net CO_2 emissions, because biomass burning will release only the amount of CO_2 it has absorbed during plant growth (provided its production and harvesting are sustainable).

Molecular hydrogen (H_2) is a colorless, odorless, and tasteless gas. It is an ideal alternative fuel to be used for transportation because the energy content of hydrogen is three times greater than in gasoline. Also, it is virtually nonpolluting and a renewable fuel. Using H_2 as an energy source produces only water; H_2 can be made from water again. A great number of microorganisms produce H_2 from inorganic materials, such as water, or from organic materials, such as sugar, in reactions catalyzed by enzymes. Hydrogen produced by microorganisms is called biohydrogen.

Butanol (butyl alcohol) is a four-carbon alcohol with the molecular formula C_4H_9OH. Among other types of biofuels, butanol has been the most promising in terms of commercialization. It is another alcohol fuel but has higher energy content than ethanol. It does not pick up water as ethanol does and is not as corrosive as ethanol but is more suitable for distribution through existing pipelines for gasoline. However, compared to ethanol, butanol is considered toxic. It can cause severe eye and skin irritation and suppression of the nervous system.

HISTORY

The concept of biofuels is not new. People have been using biomass such as plant material to heat their houses for thousands of years. The idea of using hydrogen as fuel was expressed by Jules Verne in his novel *L'Île mystérieuse* (1874-1875; *The Mysterious Island*, 1875). In 1900, Rudolf Diesel, the inventor of the diesel engine, used peanut oil for his engine during the World Exhibition in Paris, France. Henry Ford's first (1908) car, the Model T, was made to run on pure ethanol. Later, the popularity of biofuels as a fuel source followed the "oil trouble times." For example, biofuels were considered during the 1970's oil embargo. Early in the twenty-first century, concerns about global warming and oil-price increases reignited interest in biofuels. In 2005, the U.S. Congress passed the Energy Policy Act, which included several sections related to biofuels. In particular, this energy bill required more research on biofuels, mixing ethanol with gasoline, and an increase in the production of cellulosic biofuels.

Since they are renewable, biofuels are considered by many as potential future substitutes for fossil fuels, which are nonrenewable and dwindling. Moreover, pollution from fossil fuels affects public health and has been associated with global climate change, because burning them in engines releases carbon dioxide (CO_2) into the atmosphere. Using biofuels as an energy source generates fewer pollutants and little or no carbon dioxide. In addition, the utilization of biofuels reduces U.S. dependence on foreign oil.

Obtaining Biofuels

Ethanol is produced mainly by the microbial fermentation of starch crops (such as corn, wheat, and barley) or sugarcane. In the United States, most of the ethanol is produced by the yeast (fungal) fermentation of sugar from cornstarch. Ethanol can be produced from cellulose, the most plentiful biological material on Earth; however, current methods of converting cellulosic material into ethanol are inefficient and require intensive research and development efforts. Ethanol can also be produced by chemical means from petroleum. Therefore, ethanol that is produced by microbial fermentation is commonly referred to as "bioethanol."

In the United States, biodiesel comes mainly from soybean plants; in Europe, the world's top producer of biodiesel, it comes from canola oil. Other vegetative oils that have been used in biodiesel production are corn, sunflower, cottonseed, jatropha, palm oil, and rapeseed. Another possible source for biodiesel production is microscopic algae (microalgae), the microorganisms similar to plants.

Methane is produced by microorganisms and is an integral part of their metabolism. Biogas is produced during the anaerobic fermentation of organic matter by a community of microorganisms (bacteria and archaea). For practical use, methane and biogas are generated from wastewater, animal waste, and "gas wells" in landfills. Biomass is produced naturally, in the forest, and agriculturally, from agricultural residues and dung.

No commercial biohydrogen production process exists. The most attractive for industrial applications is H_2 production by photosynthetic microbes. These microorganisms, such as microscopic algae, cyanobacteria, and photosynthetic bacteria, use sunlight as an energy source and water to generate hydrogen.

Butanol can be produced by the fermentation of sugars similar to the ethanol production. The most well-known pathway of butanol generation is fermentation by bacterium *Clostridium acetobutylicum*. Substrates utilized for butanol production—starch, molasses, cheese whey, and lignocellulosic materials—are exactly the same as for ethanol fermentation. The biological production by fermentation is not economically attractive because of low levels of product concentrations and high cost of product recovery compared to the chemical process.

Applications

With increasing energy demands and oil prices, ethanol has become a valuable option as an alternative transportation fuel. The Energy Policy Act of 2005 included a requirement to increase the production of ethanol from 15 to 28 billion liters by 2012. Beginning in 2008, a majority of fuel stations in the United States were selling gasoline with 10 percent ethanol in it. Nearly all cars can use E10, fuel that is 10 percent ethanol. Blending ethanol with gasoline oxygenates the fuel mixture, which burns more completely and produces fewer harmful CO emissions. Another environmental benefit of ethanol is that it degrades in the soil, whereas petroleum-based fuels are more resistant to degradation and have many damaging effects when accidentally discharged into the environment. However, a liter of ethanol has significantly less energy content than a liter of gasoline, so vehicles must be refueled more often. Ethanol is also more expensive than gasoline, although rising prices of gasoline could cancel that disadvantage. In addition, carcinogenic aldehydes, such as formaldehyde, are produced when ethanol is burned in internal combustion engines. Carbon dioxide, a major greenhouse gas, forms as well. Moreover, the widely used fuel mix that is 85 percent ethanol and 15 percent gasoline (the E85 blend) requires specially equipped "flexible fuel" engines. In the United States, only a fraction of all cars are considered "flex fuel" vehicles. By comparison, however, most cars in Brazil have flex engines. Beginning in 1977, the Brazilian government made using ethanol as a fuel for cars mandatory. Brazil has the largest and most successful "ethanol for fuel" program in the world. As a result of this successful program, the country reached complete self-sufficiency in energy supply in 2006.

Biodiesel performs similarly to diesel and can be used in unmodified diesel engines of trucks, tractors, and other vehicles, and it is better for the environment. Burning biodiesel produces fewer emissions than petroleum-based diesel; it is essentially free of sulfur and aromatics and emits less CO. Additionally, biodiesel is less toxic to the soil. Biodiesel is often blended with petroleum diesel in different ratios of 2, 5, or 20 percent. The most common blend is B20, or 20 percent biodiesel to 80 percent diesel fuel. Biodiesel can be used as a pure fuel (100 percent or B100), but pure fuel is not suitable for winter because it thickens in cold temperatures. In addition, B100 is a solvent that degrades engines' rubber hoses and gaskets. Moreover, biodiesel energy content is less than in diesel. In general, biodiesel is not used as widely as ethanol. However, biodiesel users include the United States Postal Service; the U.S. Departments of Defense, Energy, and Agriculture; national parks; school districts; transit authorities; and public-utilities, waste-management, and recycling companies across the United States. In January, 2009, Continental Airlines successfully demonstrated the use of a biodiesel mixture from plants and algae (50 percent to 50 percent) to fly its Boeing 737-800.

In the 1985 Mel Gibson movie *Mad Max Beyond Thunderdome*, a futuristic city was run on methane that was generated by pig manure. In reality, methane can be a very good alternative fuel. It has a number of advantages over other fuels produced by microorganisms. First, it is easy to make and can be generated locally, which does not require distribution. Extensive natural gas infrastructure is already in place to be utilized. Second, the utilization of methane as a fuel is an attractive way to reduce wastes such as manure, wastewater, or municipal and industrial wastes. In local farms, manure is fed into digesters (bioreactors) where microorganisms metabolize it into methane. Methane can be used to fuel electrical generators to produce electricity. In China, millions of small farms have simple small underground digesters near the farm houses. There are several landfill gas facilities in the United States that generate electricity using methane. San Francisco has extended its recycling program to include conversion of dog waste into methane to produce electricity and to heat homes. With a dog population of 120,000 this initiative promises to generate a significant amount of fuel with a huge reduction of waste at the same time. Methane was used as a fuel for vehicles for a number of years. Several Volvo car models with bi-fuel engines were made to run on compressed methane with gasoline as a backup. Biogas can also be compressed, like methane, and used to power motor vehicles.

In many countries, millions of small farms maintain a simple digester for biogas production to generate energy. Currently, there are more than five million household digesters in China, used by people mainly for cooking and lighting, and there are more than one million biogas plants of various capacities in India.

Utilization of methane and biogas as an energy source in place of fossil fuels is providing significant environmental and economic benefits. Biofuels are essentially nonpolluting, although their utilization results in production of CO_2 and contributes to global warming, though with less impact on Earth's climate than methane itself as a greenhouse gas. Even though the use of methane and biogas as energy sources releases CO_2, the process as a whole can be considered "CO_2 neutral" in that the released CO_2 can be assimilated by their producers, archaea and bacteria.

Some examples of biomass use as an alternative energy source include burning wood or agricultural residue to heat homes. This is an inefficient use of energy—typically only 5-15 percent of the biomass energy is actually utilized. Using biomass that way produces harmful indoor air pollutants such as carbon monoxide. Yet biomass is an almost "free" resource costing only labor to collect. Biomass supplies more than 15 percent of the world's energy consumption. Biomass is the top source of energy in developing countries; in some countries it provides more than 90 percent of the energy used.

Hydrogen powered U.S. rockets for many years. Today, a growing number of automobile manufacturers around the world are making prototype hydrogen-powered vehicles. Only water is emitted from the tailpipe—no greenhouse gases. The car is moved by a motor that runs on electricity generated in the fuel cell via a chemical reaction between H_2 and O_2. Hydrogen vehicles offer quiet operation, rapid acceleration, and low maintenance costs. During peak time, when electricity is expensive, fuel-cell hydrogen cars could provide power for homes and offices. Hydrogen for these applications is obtained mainly from natural gas (methane and propane) via steam reforming. Biohydrogen is used in experimental applications only. Many problems need to be overcome before biohydrogen can be easily available. One of the reasons for the delayed acceptance of biohydrogen is the difficulty of its production on a cost-effective basis. For biohydrogen power to become a reality, tremendous research and investment efforts are necessary.

Butanol can be used as transportation fuel. It contains almost as much energy as gasoline and more energy than ethanol for a particular volume. Unlike 85 percent ethanol, a butanol/gasoline mix (E85 blend) can be used in cars designed for gasoline without making any changes to the engine.

—*Sergei A. Markov*

Further Reading

Chisti, Yusuf. "Biodiesel from Microalgae." *Biotechnology Advances* 25, no. 3 (2007): 294-306.

Glazer, Alexander N., and Hiroshi Nikaido. *Microbial Biotechnology: Fundamentals of Applied Microbiology*. New York: W. H. Freeman, 2007.

Service, Robert F. "The Hydrogen Backlash." *Science* 305, no. 5686 (August 13, 2004): 958-961.

Wald, Matthew L. "Is Ethanol for the Long Haul?" *Scientific American* 296, no. 1 (January, 2007): 42-49.

Wright, Richard T., and Dorothy F. Boorse. *Environmental Science: Toward a Sustainable Future*. 11th ed. Boston: Benjamin/Cummings, 2011. This textbook describes several bioprocesses used in waste treatment and pollution control.

AE Biofuels. http://www.alternative-energy-news.info/technology/biofuels/

Biofuels and synthetic fuels

A bus fueled by biodiesel. United States Department of Energy. By Vincecate at en.wikipedia. [Public domain], from Wikimedia Commons

Fields of Study

Agricultural engineering; Bioengineering; Biochemistry; Biotechnology; Biophysics; Industrial biotechnology (or White biotechnology); Industrial fermentation

Abstract

The study of biofuels and synthetic fuels is an interdisciplinary science that focuses on development of clean, renewable fuels that can be used as alternatives to fossil fuels. Biofuels include ethanol, biodiesel, methane, biogas, and hydrogen; synthetic fuels include syngas and synfuel. These fuels can be used as gasoline and diesel substitutes for transportation, as fuels for electric generators to produce electricity, and as fuels to heat houses (their traditional use). Both governmental agencies and private companies have invested heavily in research in this area of applied science.

Background

Around the world, concerns about global climate change due to the emission of greenhouse gases from human use of fossil fuels, as well as concerns over energy security, have ignited interest in biofuels and synthetic fuels. A large-scale biofuel and synthetic fuel industry has developed in many countries, including the United States. A number of companies in the United States have conducted research and development projects on synthetic fuels with the intent to begin commercial production of synthetic fuels. Although biofuels and synthetic fuels still require long-term scientific, economic, and political investments, investment in these alternatives to fossil fuels is expected to mitigate global warming, to help protect the global climate, and to reduce U.S. reliance on foreign oil.

HISTORY

People have been using biofuels such as wood or dried manure to heat their houses for thousands of years. The use of biogas was mentioned in Chinese literature more than 2,000 years ago. The first biogas plant was built in a leper colony in Bombay, India, in the middle of the nineteenth century. In Europe, the first apparatus for biogas production was built in Exeter, England, in 1895. Biogas from this digester was used to fuel street lamps. Rudolf Diesel, the inventor of the diesel engine, used biofuel (peanut oil) for his engine during the World Exhibition in Paris in 1900. The version of the Model T Ford built by Henry Ford in 1908 ran on pure ethanol. In the 1920's, 25 percent of the fuels used for automobiles in the United States were biofuels rather than petroleum-based fuels. In the 1940's, biofuels were replaced by inexpensive petroleum-based fuels.

Gasification of wood and coal for production of syngas has been done since the nineteenth century. Syngas was used mainly for lighting purposes. During World War II, because of shortages of petroleum, internal combustion engines were modified to run on syngas and automobiles in the United States and the United Kingdom were powered by syngas. The United Kingdom continued to use syngas until the discovery in the 1960's of oil and natural gas in the North Sea.

The process of converting coal into synthetic liquid fuel, known as the Fischer-Tropsch process, was developed in Germany at the Kaiser Wilhelm Institute by Franz Fischer and Hans Tropsch in 1923. This process was used by Nazi Germany during World War II to produce synthetic fuels for aviation.

During the 1970's oil embargo, research on biofuels and synthetic fuels resumed in the United States and Europe. However, as petroleum prices fell in the 1980's, interest in alternative fuels diminished. In the twenty-first century, concerns about global warming and increasing oil prices reignited interest in biofuels and synthetic fuels.

HOW IT WORKS

Biofuels and synthetic fuels are energy sources. People have been using firewood to heat houses since prehistorical time. During the Industrial Revolution, firewood was used in steam engines. In a steam engine, heat from burning wood is used to boil water; the steam produced pushes pistons, which turn the wheels of the machinery.

Biofuels and synthetic fuels such as ethanol, biodiesel, butanol, biohydrogen, and synthetic oil can be used in internal combustion engines, in which the combustion of fuel expands gases that move pistons or turbine blades. Other biofuels such as methane, biogas, or syngas are used in electric generators. Burning of these fuels in electric generators rotates a coil of wire in a magnetic field, which induces electric current (electricity) in the wire.

Hydrogen is used in fuel cells. Fuel cells generate electricity through a chemical reaction between molecular hydrogen (H_2) and oxygen (O_2). Ethanol, the most common biofuel, is produced by yeast fermentation of sugars derived from sugarcane, corn starch, or grain. Ethanol is separated from its fermentation broth by distillation. In the United States, most ethanol is produced from corn starch. Ethanol can also be produced from cellulose, found in the inedible parts of corn plants and other crops. Because using the waste from the production of food crops is better for the environment than growing more crops so that some can be used for fuel, the U.S. government and various private companies have invested significant money into improving the methods of producing cellulosic ethanol, but as of 2016, distilling ethanol from cellulose remained more difficult and costly than distilling it from corn starch or sugarcane. Biodiesel, another commonly used biofuel, is made mainly by transesterification of plant vegetative oils such as soybean, canola, or rapeseed oil. Biodiesel may also be produced from waste cooking oils, restaurant grease, soap stocks, animal fats, and even from algae. Methane and biogas are produced by metabolism of microorganisms. Methane is produced by microorganisms called *Archaea* and is an integral part of their metabolism. Biogas produces by a mixture of bacteria and archaea.

Industrial production of biofuels is achieved mainly in bioreactors or fermenters of some hundreds gallons in volume. Bioreactors or fermenters are closed systems that are made of an array of tanks or tubes in which biofuel-producing microorganisms are cultivated and monitored under controlled conditions.

Syngas is produced by the process of gasification in gasifiers, which burn wood, coal, or charcoal. Syngas can be used in modified internal combustion engines. Synfuel can be generated from syngas through Fischer-Tropsch conversion or through methanol to gasoline conversion process.

APPLICATIONS AND PRODUCTS

Transportation. Biofuels are mainly used in transportation as gasoline and diesel substitutes. As of the early twenty-first century, two biofuels—ethanol and biodiesel—were being used in vehicles. In 2005, the U.S. Congress passed an energy bill that required that ethanol sold in the United States for transportation be mixed with gasoline. The U.S. Department of Energy (DOE) reports that by 2014, 95 percent of gasoline sold in the United States had up to a 10 percent ethanol content. The federal government has allowed the ethanol content to be raised from 10 to 15 percent in some gasoline, which can be used in cars produced after the year 2000. Most cars in Brazil can use an 85 percent/15 percent ethanol-gasoline mix (E85 blend). These cars must have a modified engine known as a flex engine. In the United States, only a small fraction of all cars have a flex engine.

Biodiesel performs similarly to diesel and is used in unmodified diesel engines of trucks, tractors, and other vehicles and is better for the environment. Biodiesel is often blended with petroleum diesel in ratios of 2, 5, or 20 percent. According to the DOE, the most common blend is B20, or 20 percent biodiesel to 80 percent diesel fuel in the United States. Biodiesel can be used as a pure fuel (100 percent or B100), but pure fuel is a solvent that degrades the rubber hoses and gaskets of engines and cannot be used in winter because it thickens in cold temperatures. The energy content of biodiesel is less than that of diesel. In general, biodiesel is not used as widely as ethanol, and its users are mainly governmental and state bodies such as the U.S. Postal Service; the U.S. Departments of Defense, Energy, and Agriculture; national parks; school districts; transit authorities; public utilities; and waste-management facilities. Several companies across the United States (such as recycling companies) use biodiesel because of tax incentives.

Hydrogen power ran the rockets of the National Aeronautics and Space Administration for many years. A growing number of automobile manufactures around the world are making prototype hydrogen-powered vehicles. These vehicles emit only water, no greenhouse gases, from their tailpipes. These automobiles are powered by electricity generated in the fuel cell through a chemical reaction between H_2 and O_2. Hydrogen vehicles offer quiet operation, rapid acceleration, and low maintenance costs because of fewer moving parts. During peak time, when electricity is expensive, fuel-cell hydrogen automobiles could provide power for homes and offices. Hydrogen for these applications is obtained mainly from natural gas (methane and propane), through steam reforming, or by water electrolysis. Many problems need to be overcome before hydrogen-powered vehicles become widely used and readily available. Nonetheless, auto manufacturers began introducing fuel-cell production vehicles in 2014 and 2015.

Methane was used as a fuel for vehicles for a number of years. Several Volvo automobile models with Bi-Fuel engines were made to run on compressed methane with gasoline as a backup. Biogas can also be used, like methane, to power motor vehicles.

Airlines have also begun to use biofuels to power their planes. In 2015, United began using a biofuel blend for its flights from Los Angeles, and in 2016 Lufthansa and JetBlue made agreements with biofuel companies. JetBlue, in particular, vowed to use biofuel for about 20 percent of its fuel needs, which was the largest commitment any airline had made at that time.

Electricity Generation. Biogas and methane are mainly used to generate electricity in electric generators. In the 1985 film *Mad Max Beyond Thunderdome*, starring Mel Gibson, a futuristic city ran on methane generated by pig manure. While the use of methane has not reached this stage, methane is a very good alternative fuel that has a number of advantages over biofuels produced by microorganisms. First, it is easy to make and can be generated locally, eliminating the need for an extensive distribution channel. Second, the use of methane as a fuel is a very attractive way to reduce wastes such as manure, wastewater, or municipal and industrial wastes. In farms, manure is fed into digesters (bioreactors), where microorganisms metabolize it into methane. There are several landfill gas facilities in the United States that generate electricity using methane. San Francisco has extended its recycling program to include conversion of dog waste into methane to produce electricity and to heat homes. With a dog population of more than 100,000, this initiative promises to generate a significant amount of fuel and reduce waste at the same time.

Heat Generation. Some examples of biomass being used as an alternative energy source include the burning of wood or agricultural residues to heat homes. This is a very inefficient use of energy, because typically only 5 to 15 percent of the biomass energy is actually used. Burning biomass also produces harmful indoor air pollutants such as carbon monoxide. On the positive side, biomass is an inexpensive resource whose costs are only the labor to collect it. Biomass supplies about 8 percent of the energy consumed worldwide, according to a 2014 report by research firm Worldwatch Institute. Biomass is the number-one source of energy in developing countries; in some countries, it provides more than 90 percent of the energy used.

In many countries, millions of small farmers maintain a simple digester for biogas production to generate heat energy. For instance, according to the Global Methane Initiative, by 2010, nearly 40 million household digesters were being used in China, mainly for cooking and lighting.

SOCIAL CONTEXT AND FUTURE PROSPECTS

The field of biofuels and synthetic fuels expanded greatly between 2000 and 2012. Demands for biofuels and synthetic fuels were driven by environmental, social, and economic factors and governmental support for alternative fuels.

The use of biofuels and synthetic fuels reduces the U.S. dependence on foreign oil and helps mitigate the impact of increases in the price of oil, which reached a record $140 per barrel in 2008. The production and use of biofuels and synthetic fuels reduces the need for oil and thus helps keep the price of oil lower. Many experts believe that biofuels and synthetic fuels may one day replace oil.

Pollution from oil use affects public health and contributes to global climate change because of the release of carbon dioxide. Using biofuels and synthetic fuels as an energy source generates fewer pollutants and little or no carbon dioxide.

The biofuel and synthetic fuel industry in the United States was affected by the economic crisis in 2008 and 2009. Several ethanol plants were closed, some plants were forced to work below capacity, and other companies filed for Chapter 11 bankruptcy protection. Such events led to layoffs and hiring freezes. Nevertheless, overall, the industry was growing and saw a return to profitability in the second half of 2009. One segment of the biofuel and synthetic fuel industry, the biogas industry, was not affected by recession at all. More than 8,900 new biogas plants were built worldwide in 2009. Research and development efforts in biofuels and synthetic fuels actually increased during the economic crisis. However, the falling price of petroleum and the boom in shale gas in North America in the mid-2010s posed challenges to the future of biofuels and synthetic fuels, at least in the short term.

—*Sergei A. Markov, PhD*

FURTHER READING

Bart, Jan C. J., and Natale Palmeri. *Biodiesel Science and Technology: From Soil to Oil.* Cambridge, England: Woodhead, 2010.

Biello, David. "Whatever Happened to Advanced Biofuels?" *Scientific American,* 26 May 2016, www.scientificamerican.com/article/whatever-happened-to-advanced-biofuels/. Accessed 25 Oct. 2016.

Bourne, Joel K. "Green Dreams." *National Geographic* 212, no. 4 (October, 2007): 38–59.

Cardwell, Diane. "JetBlue Makes Biofuels Deal to Curtail Greenhouse Gases." *New York Times,* 19 Sept. 2016, www.nytimes.com/2016/09/20/business/energy-environment/jetblue-makes-biofuels-deal-to-curtail-greenhouse-gases.html. Accessed 25 Oct. 2016.

Glazer, Alexander N., and Hiroshi Nikaido. *Microbial Biotechnology: Fundamentals of Applied Microbiology.* New York: Cambridge University Press, 2007.

Mikityuk, Andrey. "Mr. Ethanol Fights Back." *Forbes,* November 24, 2008, 52–57.

Probstein, Ronald F., and Edwin R. Hicks. *Synthetic Fuels.* Mineola, N.Y.: Dover Publications, 2006.

Service, Robert F. "The Hydrogen Backlash." *Science* 305, no. 5686 (August 13, 2004): 958–961.

Syngellakis, S., ed. *Biomass to Biofuels.* WIT Press, 2015.

Wall, Judy, ed. *Bioenergy.* Washington, D.C.: ASM Press, 2008.

BIOINFORMATICS

FIELDS OF STUDY

Bioinformatics; Biology; Biorobotics; Computer engineering; Medical biotechnology (or Red biotechnology)

ABSTRACT

Bioinformatics is simultaneously a relatively new type of research practice and a rapidly emerging discipline. As research, bioinformatics is defined as the manipulation and the varied analyses performed by laboratory-based researchers on massive biological datasets residing in thousands of Internet-based databases, each with a distinct set of data and a specific purpose. Originating from molecular biology, bioinformatics has rapidly spread to cell biology, chemistry, statistics, computer sciences, physics, biomedical engineering, psychology, and even anthropology. As a discipline, bioinformatics draws on those professionals with advanced skills from the computer sciences, information sciences, and mathematics disciplines to bear on biological problems posed by laboratory-based bioresearchers. Collectively, these specialists are referred to as "bioinformaticians" to distinguish them from the laboratory-based science researchers carrying out experiments with the products bioinformaticians have created.

BASIC PRINCIPLES

Researchers often define bioinformatics from within the perspective of their specific discipline or individual research efforts. Some biologists view bioinformatics as only involving DNA or protein sequencing. Chemists and physicists tend to view bioinformatics as involving protein molecular structures. Computer scientists describe bioinformatics from a programming or information infrastructure perspective. Pharmacologists often define bioinformatics from the viewpoint of drug-protein interactions. All of these variations share the concept of applying computational analyses to biological processes.

A definition that encompasses bioinformatics both as a profession and a research practice and also takes into account the multitude of disciplines involved is still very much a work in progress as bioinformatics continues to evolve. A unified definition views bioinformatics as the convergence of the biological sciences and computer technologies and the integration of statistics and probability mathematics to understand biological processes of molecules on a very large scale. In turn, collecting, cataloging, classification, storage, organization, management, and retrieval of these massive biodatasets requires information theory and practice (informatics) from the information sciences disciplines to make them available for problem solving.

BACKGROUND

Bioinformatics originates from within the fields of genetics and molecular biology. The computational, mathematical, and biodatabase origins of bioinformatics arose not from within the biological, computer, or mathematical sciences but rather from two individuals who had a fascination with the computer technologies being introduced in the 1960's: Robert Ledley, a dentist turned theoretical physicist, and Margaret Dayhoff, a quantum chemist. Ledley, the inventor of the whole-body computerized tomography machine, founded the National Biomedical Research Foundation (NBRF) in 1960 to research and discover possible uses of computers in biomedical research. He recruited Dayhoff to apply her knowledge and skills at data entry and processing toward protein sequencing, which, at that time was taking more than a year to sequence a single protein by traditional laboratory methods.

Using computational analyses, Dayhoff discovered sequence patterns that identified similar proteins and predicted possible functions. She created a series of mathematical scoring matrices and defined a set of mathematical expressions that accurately reflected these similarities across evolutionary distances. In so doing, she created the first bioinformatics algorithms. Her sequence similarity matrices and rules still provide the basis for contemporary sequence similarity searching algorithms, most notably the suite of Basic Local Alignment Sequence Tools (BLAST) created by the National Center for Biotechnology Information (NCBI) in 1997.

In 1963, Dayhoff began compiling protein sequences into a series of books titled *Atlas of Protein Structure and Function*. By 1978, the *Atlas of Protein Structure and Function* had grown too large to make comparisons and perform analyses. Her second major contribution was to create a database infrastructure to convert the atlas to the first online biological database accessible to researchers who could use it to sort, manipulate, and align multiple protein sequences. The database created, the Protein Information Resource (PIR), has become the major Internet UniProt protein bioinformatics resource at the European Bioinformatics Institute (EBI).

Although the National Institutes of Health (NIH) founded the DNA bioinformatics database, GenBank, in 1982 to specifically accelerate nucleic acid sequence experimentation, progress in DNA sequencing and gene cloning technologies lagged behind protein sequencing. In 1985, then Chancellor of the University of California, Santa Cruz and molecular biologist Robert Sinsheimer convened a workshop of prominent scientists and made what was considered a radical and controversial proposal. He proposed to sequence the entire human genome and then use computational analyses to discover unknown genes and their functions and interrelationships. Thus, the Human Genome Project was initiated in 1990 by the National Institutes of Health. It soon became obvious that existing computational power and hardware were insufficient to process or hold the data being generated. Major engineering innovations were needed to process larger sample numbers, faster. The computer sciences and engineering disciplines responded. Within a few years, specialized robotics, miniaturization of samples, faster computers and processors, larger data storage capacity, and new kinds of software engineering tools were in use, greatly accelerating DNA sequencing.

HOW IT WORKS

How bioinformatics is practiced depends on whether the research is conducted on small data as typified by individual research laboratories or on a mega-scale. Small-scale data handling is often called low throughput, while large-scale is always called high throughput.

Low-Throughput Bioinformatics. In a simple sequencing scenario, researchers working to identify a protein or a gene perform "wet research" experiments ultimately yielding DNA or protein candidates. The candidates are sequenced. The researcher accesses the appropriate bioinformatics databases over the Internet and searches for similar sequences using sequence similarity algorithms, analyzing the results to provide clues to the function and identity of the candidate sequences. Once the researcher has clues to possible functions, additional bioinformatics databases are searched to aid in the development of the next experiment to be performed. In the process, many different bioinformatics databases and tools are used. Learning what databases and tools exist and are best is part of the process of learning bioinformatics research. There are times when the tool may not exist or existing databases are not sufficient. The bench researcher may ask the local bioinformatician to help design a more specific programming tool. If this becomes a critical problem for this area of research in general, bioinformaticians develop new tools and/or databases. These are published in the peer-reviewed literature and tried out by the scientific community. Those that work eventually become established as bioinformatics resources.

Sequences recovered by laboratory researchers with federally funded grants must be uploaded to a sequence repository, along with any information discovered. In the United States, this is NCBI GenBank. Data uploaded in the United States, Europe, and Asia are shared among the countries daily, permitting rapid access to the biodata generated worldwide. NCBI curators then work on the uploaded sequences to integrate and incorporate them into any of the thirty-plus databases at the National Center for Biotechnology Information. When new types of data are being uploaded as research progresses into new areas, the center or its European and Asian counterparts design new kinds of databases and algorithms or fund others to do so.

High-Throughput Bioinformatics. This research typically involves massive generation of data, such as large-scale genome sequencing efforts, or the simultaneous analyses of very large datasets. An example of the latter would be clinical data arising from the identification of proteins unique to a specific cancer isolated from many patients. In these scenarios, millions of sequences need to be processed daily. This kind of bioinformatics requires robotic bioinstrumentation and different algorithms to process. It typically is carried out by supercomputing facilities supported by bioinformaticians with experience in parallel computing, networking, grid computing, advanced algorithms, statistical programming skills, and advanced database modeling and design. Any sequence data recovered from research supported from federal funds must be uploaded to the National Center for Biotechnology Information. In this case, since the functions of the sequences are unknown, the center computationally processes these data to different databases than GenBank, making them available for others to search and identify the function of the sequences.

APPLICATIONS AND PRODUCTS

Biological Databases. Biosequence databases are at the very foundation of bioinformatics research and discovery. The National Center for Biotechnology Information, the European Bioinformatics Institute, and the DNA Database of Japan are the major biosequence spaces; each has an extensive suite of hyperlinked protein, genomes, nucleotide, genes, gene expressions, disease, and chromosome databases. Scientific organizations, government agencies, and research institutes have collaborated to create other databases.

The major protein databases are UniProt of EBI and the three-dimensional structural protein resource at Protein DataBank. Online Mendelian Inheritance of Man (OMIM) and Animal (OMIA) correlate mutations and their inheritance patterns with disease phenotypes. PharmGKB is a major pharmacogenetics database that monitors human genetic variations to specific patient drug reactions and their symptoms. Biological pathway databases, including BRENDA, Reactome, and KEGG, enable researchers to locate proteins that interact with each other and determine how protein sequence alterations could give rise to abnormal biological processes.

Genomes of many different organisms have been sequenced, each representing biodata that detail a model biological system or disease process. Finally, there are thousands of smaller "boutique" biodatabases for specific diseases, the different functional or structural components of genes or proteins, and similar topics.

Algorithms. Although there are many mechanisms to search biodatabases, the most critical and extensively used is sequence similarity searching. Needleman-Wunsch, Smith-Waterman, FASTA, and BLAST represent the major similarity algorithms. They differ in algorithmic mechanism and computational speed, with Needleman-Wunsch being the most accurate but also the most computationally intense. At the time of its publication in 1970, it took days to return results. BLAST is the least accurate of the set but computationally the fastest, taking only minutes to return results. BLAST supported laboratory bench research in real time and is the major sequence similarity algorithm in use. However, as personal computers have advanced to faster processors, the Needleman-Wunsch and Smith-Waterman algorithms have been reengineered and made available at the National Center for Biotechnology Information and the European Bioinformatics Institute.

Molecular Visualization and Modeling. A combination of software engineering and sequence algorithm, three-dimensional molecular viewers enable researchers to manipulate and computationally model proteins. They are particularly important in drug design and analyses of mutant proteins involved in disease, as researchers can introduce changes *in silico* and view how they alter drug interaction or structure or compare a mutant protein directly with a normal protein superimposed in three dimensions. The two most important molecular viewers are Cn3D at NCBI and RasMol for the other protein and nucleic acid databases.

Biodiversity. Microbes (bacteria, fungi, protozoans, and viruses) represent half of the Earth's biomass. It is estimated that there are at least 10 million bacterial species alone, with only a few thousand described. Since the 1990's, it has become clear that most microbes live in mixed communities with other microbes, with any given species present in a small number, none of which can be cultured in a laboratory.

Metagenomics is that part of bioinformatics that determines and then studies what organismal communities are present in various environmental samples such as soil or oceans. It can also detect the organisms present in animal organs or tissues such as the digestive tract or skin. The present-day state of bioinformatics technology and data acquisition and storage permits the identification of microorganism communities only. This includes mixed communities containing bacteria, fungi, and viruses. The samples are collected and all the organisms present in the sample are recovered. Without an attempt to isolate, culture, or identify any of the organisms present, all the DNA from all the organisms is extracted in mass, sequenced, and reassembled into the original genomes, thereby identifying what organisms are present.

Initiated in 2008, the Human Microbiome Project aims to identify all the microorganisms present in five areas of the human body: the digestive tract, the mouth, the skin, the nose, and the vagina from samples taken from healthy human volunteers. Once the healthy human microbiome has been characterized, the human microbiome will be studied in different disease, nutritional, or treatment states. The aim is to use the human microbiome to identify particular diseases and to study the effectiveness of probiotics, pharmaceutical drug treatments, and other therapies. In 2010, 900 microbial genomes had been identified as components of the human microbiome, of which 178 have been fully sequenced. The data indicate that the human microbiome is massive and at least one hundred times larger than the human genome itself. It contains nearly twice the microbial diversity already identified in public domain databases

Understanding the oceanic microbiome and how it responds to climate and human impact is an important step to oceanic conservation. In addition, adding to the catalog of known proteins enhances the ability to discover new proteins that could be reengineered or repurposed for medicinal or bioremediation uses. In a metagenomics approach similar to that taken by the Human Microbiome Project, oceanic samples have been collected, all the microorganisms recovered, DNA extracted, sequenced, and genomically reassembled. This Global Ocean Sampling expedition has identified at least 400 new microbial species and 6 million predicted proteins, doubling the total number of proteins previously identified.

Bioinformatics' contribution to biodiversity is not limited to the present day. Museums worldwide contain unique specimens (both plant and animal) that can be sequenced, genomically cataloged, and characterized. Ancient DNA, the DNA recovered from fossil organisms trapped in underground ancient lake beds or water droplets trapped in various geological samples, is also available for genomic analyses, adding to the publicly available Neanderthal and *Mastodon* genomes.

Personal Genomes. In less than seven years, the cost of sequencing a human genome dropped from almost $3 billion for the Human Genome Project to less than $30,000 in 2010 because of rapid advances in computational and engineering technologies in bioinformatics. By the end of 2011, it is estimated that more 30,000 different human genomes will have been sequenced by various genomic centers and institutes worldwide. As costs continue to drop, sequencing a human genome will be within the reach of individual research laboratories in several years and affordable by many private citizens in possibly five more years. Several private companies are already advertising (at a cost ranging from $400 to $1,500) to scan people's genome for common DNA sequence variations that are associated with specific diseases or conditions such as diabetes or high cholesterol or are known to reduce or enhance the metabolism of pharmaceutical drugs. Some of these companies offer services that trace an individual's ancestry through his or her inheritance of specific DNA patterns now known to be specific to particular ethnicities or to have originated in distinct geographical areas around the world. This new branch of genomics, in which individual human genomes are sequenced and analyzed, is called personal genomics and carries with it evolving ethical issues that are themselves undergoing rapid debate and analyses.

At the academic research level, large consortiums are being formed to analyze vast numbers of individual human genomes to first catalog and then study all the known genetic differences among both individuals and different kinds of populations. The 1000 Genomes project is a consortium of more than seventy-five universities, institutes, and companies worldwide. Regardless of its name, it aims to sequence the genomes of 2,300 individuals with ancestry from Europe, East Asia, South Asia, West Africa, and the Americas. Each genome will be independently analyzed as well as compared to genomes within the same populations. Early studies indicate that each person may carry 250 to 300 mutations in genes known to cause disease, as well as 50 to 100 sequence variations known to be implicated in inherited disorders. Not all genetic variations give rise to disease. The 1000 Genomes catalog has already identified several candidate genetic differences in two genes inherited within one family group that may be responsible for this family having very low cholesterol levels. The hope is that the study of these genes can lead to new cholesterol-lowering strategies.

IMPACT ON INDUSTRY

Bioinformatics is creating new industries and services. Industrial applications are very much in their infancy and in the research and development phase for the most part.

Government and University Research. The National Institutes of Health is a major source of research funding for basic science, biomedical, and clinical research. The National Science Foundation funds more in the environmental and educational sectors. Both are active policy setters and enforcers of data sharing and integration, cyberinfrastructure, supercomputing, and bioethical issues. Because of a federal mandate, the National Institutes of Health focuses on issues related to health and disease, which is the reason that the majority of bioinformatics research in the United States is related to medical and clinical research. The Department of Energy, through the Office of Biological and Environmental Research (BER), funds research on selected organisms, as well as on environmental genomics and proteomics projects related to bioremediation. BER also studies the ethical, legal, and social ramifications of genome projects through the Ethical, Legal, and Social Issues Program. The Department of Energy's Advanced Scientific Computing Research office funds computer science, networking, and mathematics research.

Industry and Business Sectors. The Howard Hughes Medical Institute (HHMI) is a nonprofit independent research institution that both funds and performs bioinformatics research. Its funding program entails appointing scientists as Hughes Investigators, providing them with long-term funding at their home institutions with the freedom to explore research projects as they choose. It is influential in recommending policies and standards for undergraduate science and medical education, including incorporating bioinformatics into the curriculum.

Many independent research institutions (including the Broad Institute, the J. Craig Venter Institute, and the Sanger Institute) carry out bioinformatics research. Several are large biodata producers, performing only high-throughput genomic and metagenomic computationally intense bioinformatics research.

In the for-profit sector, the biotechnology and pharmaceutical companies are significantly invested in protein engineering and modeling for pharmaceutical drug and diagnostic kit development. Bioproduct companies produce the enzymes, reagents, and kits needed to support the laboratory-based molecular biology research related to bioinformatics. Bioinstrumentation companies research, develop, and provide the highly specialized automated genomics and proteomics sequence analyzers and processors needed by both for-profit and nonprofit research efforts. They are an important source of innovation in sequencing methodologies that is largely responsible for continuing to advance bioinformatics into practical applications including metagenomics and personalized medicine initiatives.

SOCIAL CONTEXT AND FUTURE PROSPECTS

Metagenomics, personalized medicine, and future bioinformatics initiatives yet to be discovered will rapidly affect nearly all individuals. It is not surprising that with these major advances in bioinformatics technologies comes a caution by many for careful introspection and debate of their implications. Interest in bioethics is on the rise and has been added to the curricula of not only bioinformatics educational programs but those of many other disciplines as well.

——*Diane C. Rein, PhD, MLS*

FURTHER READING

Baxevanis, Andreas D., and B. F. Francis Ouellette, eds. *Bioinformatics: A Practical Guide to the Analysis of Genes and Proteins.* 3d ed. Hoboken, N.J.: John

Wiley & Sons, 2005. Covers bioinformatics from the database and searching perspective. Contains chapters on various biological databases and their search interfaces.

Gu, Jenny, and Phillip E. Bourne, eds. *Structural Bioinformatics*. 2d ed. Hoboken, N.J.: John Wiley & Sons, 2009. Combination textbook and manual covering all aspects of protein bioinformatics, including the major protein databases, visualization, mass spectrometry, and protein modeling.

Lesk, Arthur M. *Introduction to Bioinformatics*. 3d ed. New York: Oxford University Press, 2008. Comprehensive overview of genomes, proteomics, protein structure, databases, phylogenetics, programming languages, and more.

Zvelebil, Marketa, and Jeremy Baum. *Understanding Bioinformatics*. New York: Garland Science, 2008. Intermediate text with detailed descriptions on sequence alignments, phylogenetics, genomics, proteomics, and protein structure and modeling.

Fascinating Facts about Bioinformatics
- During the 2009 H1N1 swine flu epidemics, more than 24,000 individual virus genomes were sequenced immediately from infected patients worldwide. These data played a major role in the autumn 2009 vaccine development.
- GenBank has become a major sequence resource containing 150 million sequence records in 2009 and giving rise to hundreds of secondary, specialized databases worldwide. More than 500 million records in thirty-plus secondary databases exist at the National Center for Biotechnology Information alone. Worldwide, the number of bioinformatics records based on GenBank is in the billions.
- The genomes between different humans are 99.9 percent similar. The 0.1 percent difference is due to single DNA nucleotide variations at very specific points within the human genome, making each person different from all others physically, behaviorally, and physiologically.
- Before the Human Genome Project, scientists thought that the human genome contained up to 100,000 genes because of the large number of proteins that are known to exist in humans. Scientists now know that there are fewer than 30,000 genes in humans, with each gene estimated to give rise to 3 to 8 different proteins. There are exceptions. The *dscam* gene is involved in the development of neural circuits. In humans, this gene codes for more than 16,000 variations of the dscam protein, and in the fruit fly, 38,016 isoforms of the dscam protein have been proven to exist.
- Human DNA is 98 percent similar in sequence to chimpanzees. However, the genetic difference between any two chimpanzees is four to five times greater than the difference between any two humans.

Of the greater than 3 billion nucleotides in the human genome, less than 3 percent actually codes for protein molecules. The functions of the vast majority of the human genome and what it does or does not do remain unknown.

BIOLOGICAL TERRORISM

FIELDS OF STUDY

Biology; Medical biotechnology (or Red biotechnology); Philosophy and Religious Studies; Political Science; Sociology

ABSTRACT

A bioterrorist attack is perhaps one of the events most feared by emergency responders and government officials in the field of counterterrorism, in large part because, although the probability of a wide-scale attack is rather low, in the event of such an attack, the potential for catastrophic results is high.

BACKGROUND

Ever since the influenza pandemic of 1918–19 (a natural event), which killed some forty million people around the world, a heightened awareness has existed of the potential for the spread of harmful, even lethal, biological cultures among human populations. Among the purposeful biological attacks that have been perpetrated, perhaps the one with which the

M-17 nuclear, biological and chemical warfare mask and hood. By Senior Airman Walker, Kadena Air Force Base. (U.S. Department of Defense (DOD). [Public domain], via Wikimedia Commons

most Americans are familiar is the case in which letters containing the bacterium that causes anthrax were sent to addresses in New York City, Washington, D.C., and Boca Raton, Florida, in October and November of 2001, shortly following the September 11 terrorist attacks on the World Trade Center and the Pentagon. This case greatly increased awareness of the need for government agencies (including the U.S. Postal Service) to learn how to identify and respond effectively to any biological crisis. The outbreak of severe acute respiratory syndrome (SARS) in Canada in 2002–3, which quickly spread from one to more than two hundred persons in Toronto-area hospitals and resulted in thirty-three deaths among patients and health workers, also demonstrated the need for improvements in government and health care responses to epidemic and pandemic disease outbreaks. The investigation and prevention of biological terrorism have become foremost components of nations' efforts to improve their homeland security.

Bioterrorist attacks can target human populations directly or indirectly, through food and water supplies. Agroterrorism—biological terrorism that targets agricultural food sources—is a very real threat to national security in some countries because modern agricultural systems are tightly integrated, and many points in the harvesting, processing, and distribution systems represent potentially "soft" targets for terrorists and difficult targets to defend from terrorist acts. The routine transport and commingling of production and processing systems greatly aid the dissemination of any biological pathogens. It is estimated that about 70 percent of the value production in U.S. agriculture occurs on just 5 percent of U.S. farms, so a successful attack on any of these locations would be catastrophic.

HISTORY

The use of biological weapons can be easily traced back to ancient times. Soldiers used to dip their weapons in animal excrement or known plant toxins before battle so as to cause infection in whomever they stabbed or shot with arrows. In both ancient and medieval times, poisoning water supplies with dead animals was a favorite tactic, as was slinging or firing dead animal or human carcasses over defender walls in the hopes of spreading disease. Although few records exist to prove that European settlers in the New World purposely spread disease among Native Americans, sufficient evidence is found in the form of a letter from Colonel Henry Bouquet to Lord Jeffrey Amherst in 1763 to suggest that the British attempted to spread smallpox to their Native American opponents during the French and Indian War. Emperor Napoleon I drew on the expertise of French scientists to visit swamp fever on his opponents in the eighteenth century, and Confederate soldiers were known to poison ponds as they retreated from the advancing Union Army during the American Civil War.

By the time World War I began in 1914, science was sufficiently advanced that the mechanisms of the spread of disease were understood, and serious consideration was given to making use of biological agents during this global conflict. The German government formally and repeatedly refused to deploy biological agents against humans during the war, however, and the Allied Powers followed Germany's lead in this regard. Nevertheless, German saboteurs deployed anthrax against horses and mules that were to be sent to Allied soldiers on the front lines. During World War II, as ample surviving film footage and written evidence shows, the Japanese tested biological agents extensively on Chinese prisoners and Chinese civilians. Whether the Japanese employed these agents as weapons of war, as some scholars allege, has not been proved. The Geneva Protocol, signed by various nations in 1925, outlaws the use of biological weapons, but such prohibitions are only as good as the resolve of nations to follow the protocol.

A number of terrorist organizations have at least discussed the use of biological weapons, including the Italian Brigate Rosse (Red Brigades) and the German Rote Armee Fraktion (Red Army Faction), earlier known as the Baader-Meinhof Gang. Members of cults in the United States have poisoned restaurants with agents such as salmonella to cause sickness. The Japanese group Aum Shinrikyo (known as Aleph since 2000) actively acquired and cultured *Bacillus anthracis* (the bacterium that causes anthrax) and Ebola virus, both of which were found in significant quantities when police raided the group's headquarters in 1995 following its sarin gas attack on the Tokyo subway. The group purportedly released botulinum toxin as well as anthrax in the same period, but these attempts were not successful. Experts are not sure why these attacks failed; possible reasons include the method of delivery, manufacturing problems, and that the group may have released an anthrax vaccine and a slowly reproducing botulinum toxin rather than more potent varieties of these pathogens.

Since 1996, the Federal Bureau of Investigation (FBI) has opened numerous cases involving the potential use of biological agents. Many have amounted to mere threats, but some have included attempts to produce such pathogens as botulinum toxin, anthrax, and ricin.

TYPES OF AGENTS

Because the variety of biological agents available for use in terrorist acts is quite extensive, stockpiling vaccines that may be needed in the event of biological attacks is extremely difficult; it is virtually impossible to have safeguards in place against every potential type of biological agent. Some of the most dangerous pathogens that may potentially be used by bioterrorists, as categorized by the Centers for Disease Control and Prevention, are anthrax, pneumonic plague, botulinum toxin, smallpox, and ricin.

Anthrax is perhaps the biological pathogen most likely to be used in a bioterrorist attack. It is relatively easy to cultivate the spores of *B. anthracis*, and the spores are fairly stable under a variety of conditions, so dissemination of the pathogen is not particularly difficult. When inhaled, the agent works into the lungs and causes fever, shock, and, ultimately, death. Anthrax can also cause sores on the skin of people working with infected livestock, which can result in other bodily infections. Approximately ten thousand spores of *B. anthracis* must be inhaled to prove deadly, but a mere gram of the bacterium contains millions of lethal doses.

The possibility of the use of pneumonic plague in a biological attack is high on the list of such threats maintained by first responders because this disease is incredibly virulent. Its killing potential in an uninoculated population is extremely high, close to 90 percent, and lethal exposure requires far fewer spores (around three thousand) than does anthrax. Pneumonic plague first appears as a fever accompanied by coughing, which progresses into hemorrhaging in the lungs. If left untreated for a relatively short period, the disease is almost always fatal.

Botulinum toxin is also fairly easy to cultivate. The potential of this toxin for use in aerosol form makes it very attractive as a biological weapon because the pathogen can be spread rapidly over a wide area. Botulinum toxin attacks the muscle nerves, paralyzing the nerve endings and preventing the muscles from responding to the brain. The paralysis begins near the head and works its way down through the body.

Smallpox is considered to be high on the list of potential bioterrorism pathogens because many people in the United States and around the world are no longer immunized against the disease, ever since aggressive vaccination programs let to its global eradication, which was verified and announced in December, 1979. The *Variola major* virus, which causes the most deadly form of smallpox, is relatively easy to cultivate and is easily spread using aerosols. Smallpox is contracted through inhalation, and after it incubates, the infected person normally experiences headache, fever, and other common signs of the flu. Next a rash develops, followed by pus-filled bumps on the skin.

The mortality rate is approximately 30 percent for victims who have not been inoculated.

Ricin is a toxic protein found in castor beans; it is extracted from the waste produced in the manufacture of castor oil. Ricin is relatively easy to acquire and also much easier to stockpile than most other biological pathogens. A large dose is required to kill, but the toxin can be either ingested or inhaled. When employed in conjunction with other pathogens, ricin can enable other pathogens to attack an already afflicted body. Ricin can cause respiratory problems, fever, cough, abdominal pain, and, when ingested, damage to organs such as the liver and kidneys. Ricin prevents cells in the body from making protein, which causes the cells to die off.

METHODS OF INVESTIGATION

Perhaps the greatest difficulty in the investigation of biological attacks is the fact that many of the initial symptoms caused by intentionally introduced agents are very similar to the symptoms of common diseases, such as influenza. Most often, the only way first responders are even aware that a biological attack has potentially taken place is the presence of a massive influx of people with the same symptoms. Such attacks are not usually discovered until after the pathogens have been widely disseminated and have infected large numbers of people.

The teams that investigate biological attacks need to include persons with knowledge of both biology and chemistry, who can understand the interplay between the body and the pathogen. Other areas of knowledge that are extremely important in the investigation of such attacks include the disciplines of anthropology and geography. An understanding of human living, interaction, and moving patterns, combined with meteorological data, can help investigators to track a disease back to where it may have originated, particularly in the case of aerosol dissemination.

Much of the investigative strategy used in determining whether biological agents have been intentionally spread involves the review of medical diagnoses and the employment of effective vaccines against the various agents. Investigators usually trace such agents back to their sources by comparing strains of genetic material with a database that catalogs various strains and the laboratories or environments in which the strains originated. Many materials used in the manufacture of biological agents are sold commercially, and investigators try to track where such materials may have been purchased and by whom. Scientists have been working on developing a system of biological agent detection that will be able to identify pathogens through size, nucleic acid sequence, and antigen recognition.

It is clear that the modern world has seen neither the end of bioterrorist activities nor the full range of bioterrorism possibilities yet displayed. It is equally certain that just as formal counterterrorism measures evolve and successfully propagate, so will the methods, means, and modes of bioterrorism.

—*Michael W. Cheek and Dennis W. Cheek*

FURTHER READING

Anderson, Burt, Herman Freedman, and Mauro Bendinelli, eds. *Microorganisms and Bioterrorism.* New York: Springer, 2006. Print.

Cordesman, Anthony H. *Terrorism, Asymmetric Warfare, and Weapons of Mass Destruction: Defending the US Homeland.* Westport: Praeger, 2002. Print.

Foster, George T., ed. *Focus on Bioterrorism.* New York: Nova Science, 2006. Print.

Katz, Linda B., ed. *Agroterrorism: Another Domino?* New York: Novinka, 2005. Print.

Pilch, Richard F., and Raymond A. Zilinskas, eds. *Encyclopedia of Bioterrorism Defense.* Hoboken: Wiley, 2005. Print.

Ursano, Robert J., Anne E. Norwood, and Carol S. Fullerton, eds. *Bioterrorism: Psychological and Public Health Intervention.* New York: Cambridge University Press, 2004. Print.

Wagner, Viqi, ed. *Do Infectious Diseases Pose a Serious Threat?* New Haven: Greenhaven, 2005. Print.

Wheelis, Mark, Lajos Rózsa, and Malcolm Dando, eds. *Deadly Cultures: Biological Weapons since 1945.* Cambridge: Harvard UP, 2006. Print.

Biological weapon identification

Bacillus anthracis Bioterrorism Incident, Kameido, Tokyo, 1993, Fluid collected from the Kameido site cultured on Petri dishes to identify potential *Bacillus anthracis* isolates. By Centers for Disease Control, United States. [Public domain], via Wikimedia Commons

FIELDS OF STUDY

Biology; Forensics; Medical biotechnology (or Red biotechnology); Microbiology; Molecular biology; Philosophy and Religious Studies; Political Science

ABSTRACT

Heightened concerns regarding the possibility of bioterrorist attacks have led to increased emphasis on microbial forensic science. Microbial forensic data may be presented in court as evidence in cases of terrorist attacks.

BACKGROUND

Virtually all disease-causing microorganisms are potentially useful as biological weapons. The most important candidates for biological weapons are microorganisms that cause diseases with high human mortality rates, such as anthrax, smallpox, plague, encephalitis, and hemorrhagic fever. In addition, biological weapons that are designed to wipe out crops or kill livestock could cause mass starvation and devastating economic losses.

In 2001, the general public in the United States became aware of biological weapons as a result of a series of attacks involving mail containing *Bacillus anthracis* (the bacterium that causes anthrax). Since that time, the possibility that terrorists might employ biological weapons, many of which are easily produced and spread, has been a growing concern. In response to the threat of terrorist attacks, the U.S. government has led efforts to develop quick and efficient methods of biological weapon identification, ultimately leading to the establishment of the new scientific discipline of microbial forensics. In general, the identification of microorganisms is based on techniques that rely on microscopic examination, analysis of the growth and metabolic functions of the microbes (growth-dependent and biochemical tests), and immunological and genetic tests.

GROWTH-DEPENDENT AND BIOCHEMICAL TESTS

Classic methods of microbial identification involve preliminary examination of stained specimens under a microscope, followed by growth-dependent tests. Growth-dependent tests are based on the growth patterns of microorganisms on artificial food sources (media). Particular media can be selected that will produce microbial populations—known as colonies—that have distinctive appearance and color. By comparing the reactions on these media with the known characteristics of different species of microorganisms, scientists can usually identify which microbe is present. However, most growth-dependent tests do not provide results that are extremely specific; that is, they may not distinguish among closely related microorganisms.

To aid in definitive microorganism identification, scientists have developed a series of biochemical tests that can be used to differentiate even the most closely related microbes. These tests are based on the identification of various metabolic reactions and products of different microbes. Microbial species can often be identified on the basis of fermentation patterns and the production of different chemical compounds, such as indole or hydrogen sulfide. Microorganisms are not easily identified by a single biochemical test, so it is usually necessary to perform several tests. A number of rapid identification systems are available that allow several (approximately twenty) biochemical tests to be performed quickly on a particular microorganism.

IMMUNOLOGICAL TESTS

Immunological tests utilize antibodies that are produced in response to the presence of a specific microorganism; actually, they respond to the presence of specific molecules, called antigens, on the microorganism cell surfaces. Antibodies are proteins produced by the body that recognize and bind to those antigens. Specific antibodies for many known disease-causing microorganisms are commercially available. Immunological tests vary in the ways they make the antigen-antibody reaction visible; some show obvious clumps and precipitates, whereas others show color changes or the release of fluorescence.

An example of an immunological test is the agglutination test, which is performed routinely in hospitals to determine blood types. In an agglutination test, antibody-antigen complexes form visible clumps on a test glass slide. Extremely sensitive immunological tests called immunoassays permit rapid and accurate measurement of trace bioweapon agents. These methods are being used increasingly in criminology. A good example of such an immunoassay is the enzyme-linked immunosorbent assay (ELISA). A positive result in this immunoassay is the appearance of a colored product. ELISA is a common screening test for the antibodies to toxins and bacteria that may be used as bioweapons. In radioimmunoassays, antibodies are labeled with radioactive isotopes and traced. Immunological methods are especially important for the identification of viruses, as other identification methods are not suitable to them, and growth times are long.

GENETIC TESTS

Genetic tests of microorganisms are based on the detection of the unique DNA (deoxyribonucleic acid) sequences of potential weapon microorganisms. Certain viruses maintain their genetic material in the form of RNA (ribonucleic acid), which can be converted into corresponding DNA for detection purposes. One particular technique has been widely used for identifying microorganisms based on their DNA sequences: polymerase chain reaction (PCR).

Two variations of the PCR technique have been adopted for identification: PCR and real-time PCR. Both utilize specific sets of primers (short DNA sequences) to amplify and detect DNA sequences unique to a particular microorganism. In PCR, amplified DNA sequences are subjected to separation by electrophoresis, where negatively charged DNA fragments move toward the positive pole. Separated DNA fragments can be classified by the distance they traveled depending on their molecular size. Each microorganism exhibits a characteristic DNA moving pattern by which it can be identified. In real-time PCR, detection of a microorganism's amplified DNA and confirmation of that microorganism's presence are sensed by activation of a fluorescent dye. Officials of the United Nations used portable PCR detectors when they conducted their 2002-2003 inspections of Iraqi facilities for weapons of mass destruction. These detectors can identify a single *B. anthracis* bacterium in an average kitchen-sized room.

ONGOING CHALLENGES

Although, in most cases, agents used as biological weapons could be identified easily within twenty-four hours, prosecutors may have difficulty proving that microorganisms identified in the homes or laboratories of suspects are in fact the same microorganisms used as weapons or intended for such use. One problem with making legal arguments based on weapon microbe identification is that some potentially dangerous microorganisms, such as *B. anthracis*, are found widely in soil. A prosecutor thus must prove that the microbes submitted as evidence in a given case are the same microbes used in the attack in question, and not simply microorganisms that have been transported into the suspect's home or lab accidentally.

—*Sergei A. Markov, PhD*

FURTHER READING

Cowan, Marjorie Kelly, and Kathleen Park Talaro. *Microbiology: A Systems Approach.* 2d ed. Boston: McGraw-Hill, 2008. General microbiology text focuses on the health sciences. Includes a chapter devoted to description of microbial identification techniques.

Fritz, Sandy, comp. *Understanding Germ Warfare.* New York: Warner Books, 2002. Collection of materials describes twenty-first century bioterrorism and germ weapons, including anthrax, smallpox, plague, viral fevers, and toxins. Also discusses methods of delivery of biological agents and their identification, symptoms, and treatment.

Lindler, Luther E., Frank J. Lebeda, and George W. Korch, eds. *Biological Weapons Defense: Infectious Diseases and Counterbioterrorism.* Totowa, N.J.: Humana Press, 2005. Prominent experts in biodefense research—many from the U.S. Army Medical Research Institute of Infectious Diseases—describe how to identify the presence of biological weapons through proteomic and genomic analysis.

Madigan, Michael T., John M. Martinko, Paul V. Dunlap, and David P. Clark. *Brock Biology of Microorganisms.* 12th ed. Upper Saddle River, N.J.: Pearson Prentice Hall, 2008. Widely respected basic microbiology textbook includes information about biological weapons and methods of microbial identification.

Peruski, Anne Harwood, and Leonard F. Peruski, Jr. "Immunological Methods for Detection and Identification of Infectious Disease and Biological Warfare Agents." *Clinical and Diagnostic Laboratory Immunology* 10 (July 2003): 506-513. Technical article describes immunological methods of biological weapon identification.

BIOMATHEMATICS

FIELDS OF STUDY

Biology; Biomathematics; Biopharmaceutics; Medical biotechnology (or Red biotechnology); Microbiology

ABSTRACT

Biomathematics is a field that applies mathematical techniques to analyze and model biological phenomena. Often a collaborative effort, mathematicians and biologists work together using mathematical tools such as algorithms and differential equations in order to understand and illustrate a specific biological function. Biomathematics is used in a wide variety of applications from medicine to agriculture. As new technologies lead to a rise in the amount of biological data available, biomathematics will become a discipline that is increasingly in demand to help analyze and effectively utilize the data.

Biologists have used different ways to explain biological functions, often employing words or pictures. Biomathematics allows biologists to illustrate these functions using techniques such as algorithms and differential equations. Biological phenomena vary in both scale and complexity, encompassing everything from molecules to ecosystems. Therefore, the creation of a model requires the scientist to make some assumptions in order to simplify the process. Biomathematical models vary in length and complexity and several different models may be tested.

The use of biomathematics is not limited to modeling a biological function and includes other techniques, such as structuring and analyzing data. Scientists may use biomathematics to organize data or analyze data sets, and statistics are often considered an integral tool.

DEFINITION AND BASIC PRINCIPLES

biomathematics: A discipline that quantifies biological occurrences using mathematical tools. Biomathematics is related to and may be a part of other disciplines including bioinformatics, biophysics, bioengineering, and computational biology, as these disciplines include the use of mathematical tools in the study of biology.

HISTORY

As early as the 1600's, mathematics was used to explain biological phenomena, although the mathematical tools used date back even farther. In 1628, British physician William Harvey used mathematics to prove that blood circulates in the body. His model changed the belief at that time that there were two kinds of blood. In the mid-1800's, Gregor Mendel, an Augustinian monk, used mathematics to analyze the data he obtained from his experiments with pea plants. His experiments would become the basis for genetics. In the early 1900's, British mathematician R. A. Fisher applied statistical methods to population biology, providing a better framework for studying the field. In 1947, theoretical physicist Nicolas Rashevsky argued that mathematical tools should be applied to biological processes and created a group dedicated to mathematical biology. Despite the fact that some dismiss Rashevsky's work as being too theoretical, many view him as one of the founders of mathematical biology. In the 1950's, the Hodgkin-Huxley equations were developed to describe a cellular function known as ion channels. These equations are still used. In the 1980's, the Smith-Waterman algorithm was created to aid scientists in comparing DNA sequences. While the algorithm was not particularly efficient, it paved the way for the BLAST (Basic Local Alignment Search Tool) software, a program that has allowed scientists to compare DNA sequences since 1990. Despite the fact that mathematical tools have been applied to some biological problems during the second half of the twentieth century, the practice has not been all-inclusive. In the twenty-first century, there has been a renewed interest in biology becoming more quantitative, due in part to an increase in new data.

HOW IT WORKS

Basic Mathematical Tools. Biomathematicians may use mathematical tools at different points during the investigation of a biological function. Mathematical tools may be used to organize data, analyze data, or even to generate data. Algorithms, which use symbols and procedures for solving problems, are employed in biomathematics in several ways. They may be used to analyze data, as in sequence analysis. Sequence analysis uses specifically developed algorithms to detect similarity in pieces of DNA. Specifically developed algorithms are also used to predict the structure of different biological molecules, such as proteins. Algorithms have led to the development of more useful biological instruments such as specific types of microscopy. Statistics are another common way of analyzing biological data. Statistics may be used to analyze data, and this data may help create an equation to describe a theory: Statistics was used to analyze the movement of single cells. The data taken from the analysis was then used to create partial equations describing cell movement.

Differential equations, which use variables to express changes over time, are another common technique in biomathematics. There are two kinds of differential equations: linear and nonlinear. Nonlinear equations are commonly used in biomathematics. Differential equations, along with other tools, have been used to model the functions of intercellular processes. Differential equations are utilized in several of the important systems used in biomathematics for modeling, including mean field approaches. Other modeling systems include: patch models, reaction-diffusion equations, stochastic models, and interacting particle systems. Each modeling system provides a different approach based on different assumptions. Computers have helped in this area by providing an easier way to apply and solve complex equations. Computer modeling of dynamic systems, such as the motion of proteins, is also a work in progress.

New methods and technology have increased the amount of data being obtained from biological experimentation. The data gained through experimentation and analysis may be structured in different ways. Mathematics may be used to determine the structure. For example, phylogenetic trees (treelike graphs that illustrate how pieces of data relate to one another) use different mathematical tools, including matrices, to determine their structure. Phlyogenic trees also provide a model for how a particular piece of data evolved. Another way to organize data is a site graph, or hidden Markov model, which uses probability to illustrate relationships between the data.

Modeling a Biological Function. The scientist may be at different starting points when considering a mathematical model. He or she may be starting with data already analyzed or organized by a mathematical technique or already described by a visual depiction or written theory. However, there are several considerations that scientists must take into account when creating a mathematical model. As biology covers a wide range of matter, from molecules to ecosystems, when creating a model the scale of phenomena must be considered. The time scale and complexity must also be considered, as many biological systems are dynamic or interact with their environment. The scientist must make assumptions about the biological phenomena in order to reduce the parameters used in the model. The scientist may then define important variables and the relationships between them. Often, more than one model may be created and tested.

Applications and Products

The field of biomathematics is applicable to every area of biology. For example, biomathematics has been used to study population growth, evolution, and genetic variation and inheritance. Mathematical models have also been created for communities, modeling competition or predators, often using differential equations. Whether the scale is large or small, biomathematics allows scientists a greater understanding of biological phenomena.

Molecules and Cells. Biomathematics has been applied to various biological molecules, including DNA, ribonucleic acid (RNA), and proteins. Biomathematics may be used to help predict the structure of these molecules or help determine how certain molecules are related to one another. Scientists have used biomathematics to model how bacteria can obtain new, important traits by transferring genetic material between different strains. This information is important because bacteria may, through sharing genetic material, acquire a trait such as a resistance to an antibiotic. To model the sharing of a trait, scientists have combined two of the ways to structure data: the phylogenetic tree and the site graph. The phylogenetic tree illustrates how the types of bacteria are related to one another. The site graph illustrates how pieces of genetic material interact. Then, scientists use a particular algorithm to determine the parameters of the model. By using such tools, scientists can predict which areas of genetic material are most likely to transfer between the bacterium.

Biomathematics has been used in cellular biology to model various cellular functions, including cellular division. The models can then be used to help scientists organize information and gain a deeper understanding about cellular functions. Cellular movement is one example of an application of biomathematics to cellular biology. Cellular movements can be seen as a set of steps. The scientists first considered certain cellular steps or functions, including how a cell senses a signal and how this signal is used within the cell to start movement. Scientists also considered the environment surrounding the cell, how the signal was provided, and the processes that occurred within the cell to read the signal and start movement. The scientists were then able to build a mathematical function that takes these steps into account. Depending on the particular question, the scientists may chose to focus on any of these steps. Therefore, more than one model may be used.

Organisms and Agriculture. Biomathematics has been used to create mathematical models for different functions of organisms. One popular area has been organism movement, where models have been created for bacteria movement and insect flight. A more complete understanding of organisms through mathematical models supports new technologies in agriculture. Biomathematics may also be used to help protect harvests. For example, biomathematics has been used to model a type of algae bloom known as brown tide. In the late 1980's, brown tide appeared in the waters near Long Island, New York, badly affecting the shellfish population by blocking sunlight and depleting oxygen. Four years later, the algae blooms receded. Both mathematicians and scientists collaborated in order to create a model of the brown tide in order to understand why it bloomed and whether it will bloom in the future. To create a model, the collaborators used differential equations. They focused on the population density, which included factors such as temperature and nutrients. The collaborators had to consider many variables and remove the ones they did not consider important. For instance, they hypothesized that a period of drought followed by rain may have affected growth. They also considered fertilizers and pesticides that were used in the area. A better understanding of the brown tide may help protect the shellfish harvest in future years.

Medical Uses. Biomathematical models have been developed to illustrate various functions within the human body, including the heart, kidneys, and cardiac and neural tissue. Biomathematics is useful in modeling cancer, enabling scientists to learn more about the type of cancer, thereby allowing them to study the efficacy of different types of treatment. One project has focused on modeling colon cancer on a genetic and molecular level. Not only did scientists gain information about the genetic mutations that are present during colon cancer, but they also developed a model that predicted when tumor cells would be sensitive to radiation, which is the most common way to treat colon cancer. Studies such as this can be built on in future experimentations, the results of which may someday be used by doctors to create more effective cancer treatments.

Biomathematics has also been used to organize and analyze data from experiments dealing with drug efficacy and gene expression in cancer cells. Using matrices, statistics, and algorithms, scientists have been able to understand if a particular drug is more likely to work based on the patient's cancer cell's gene expression. Biomathematics has also been integral in epidemiology, the field that studies diseases within a population. Biomathematics may be used to model various aspects of a disease such as human immunodeficiency virus (HIV), allowing for more comprehensive planning and treatment.

SOCIAL CONTEXT AND FUTURE PROSPECTS

While mathematical tools have been applied to biology for some time, many scientists believe there is still a need for increased quantitative analysis of biology. Some call for more emphasis on mathematics in high school and undergraduate biology classes. They believe that this will advance biomathematics. As more universities develop biomathematics departments and degrees, more mathematics classes will be added to the curriculum. A concern has been raised in the biomathematics field about the assumptions used to create simplified mathematical models. More complex and accurate models will likely be developed.

Important future applications for biomathematics will be in the bioengineering and medical industries. The development of mathematical models for complex biological phenomena will aid scientists in a deeper understanding that can lead to more effective treatments in such areas such as tumor therapy. As new tools and methods continue to develop, biomathematics will be a field that expands to sort and analyze the large influx of data.

—*Carly L. Huth, JD*

FURTHER READING

Hochberg, Robert, and Kathleen Gabric. "A Provably Necessary Symbiosis." *The American Biology Teacher* 72, No. 5 (2010): 296-300. This article describes some mathematics that can be taught in biology classrooms.

Misra, J. C., ed. *Biomathematics: Modelling and Simulation.* Hackensack, N.J.: World Scientific, 2006. This book provides an in-depth guide to several modern applications of biomathematics and includes many helpful illustrations.

Schnell, Santiago, Ramon Grima, and Philip Maini. "Multiscale Modeling in Biology." *American Scientist* 95 (March-April, 2007): 134-142. This article gives an overview of how biological models are created and provides several modern examples of biomathematicalapplications.

BIOMECHANICAL ENGINEERING

FIELDS OF STUDY

Biology; Biomechanical engineering; Bioengineering; Civil engineering; Computer engineering

ABSTRACT

Biomechanical engineering is a branch of science that applies mechanical engineering principles such as physics and mathematics to biology and medicine. It can be described as the connection between structure and function in living things. Researchers in this field investigate the mechanics and mechanobiology of cells and tissues, tissue engineering, and the physiological systems they comprise. The work also examines the pathogenesis and treatment of diseases using cells and cultures, tissue mechanics, imaging, microscale biosensor fabrication, biofluidics, human motion capture, and computational methods. Real-world applications include the design and evaluation of medical implants, instrumentation, devices, products, and procedures. Biomechanical engineering is a multidisciplinary science, often fostering collaborations and interactions with medical research, surgery, radiology, physics, computer modeling, and other areas of engineering.

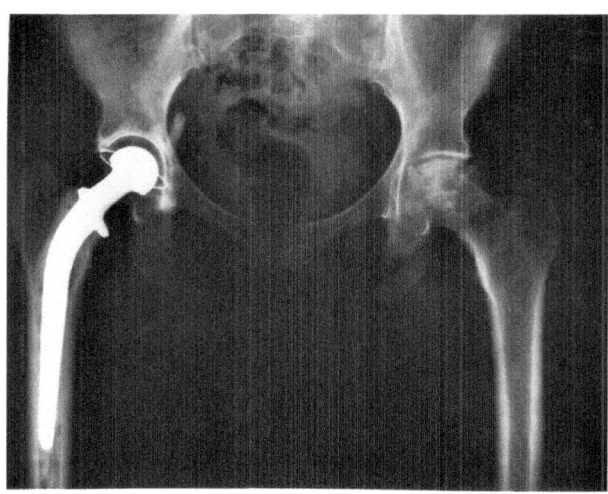

X-ray Image ID: 3684. Photographer: Unknown. Photo courtesy of the National Institutes of Health. By [Public domain], via Wikimedia Commons

HISTORY

The history of biomechanical engineering, as a distinct and defined field of study, is relatively short. However, applying the principles of physics and engineering to biological systems has been developed over centuries. Many overlaps and parallels to complementary areas of biomedical engineering and biomechanics exist, and the terms are often used interchangeably with biomechanical engineering. The mechanical analysis of living organisms was not internationally accepted and recognized until the definition provided by Austrian mathematician Herbert Hatze in 1974: "Biomechanics is the study of the structure and function of biological systems by means of the methods of mechanics." Aristotle introduced the term "mechanics" and discussed the movement of living beings around 322 BCE in the first book about biomechanics, *On the Motion of Animals*. Leonardo da Vinci proposed that the human body is subject to the law of mechanics in the 1500's. Italian physicist and mathematician Giovanni Alfonso Borelli, a student of Galileo's, is considered the "father of biomechanics" and developed mathematical models to describe anatomy and human movement mechanically. In the 1890's German zoologist Wilhelm Roux and German surgeon Julius Wolff determined the effects of loading and stress on stem cells in the development of bone architecture and healing. British physiologist Archibald V. Hill and German physiologist Otto Fritz Meyerhof shared the 1922 Nobel Prize for Physiology or Medicine. The prize was divided between them: Hill won "for his discovery relating to the production of heat in the muscle"; Meyerhof won "for his discovery of the fixed relationship between the consumption of oxygen and the metabolism of lactic acid in the muscle."

The first joint replacement was performed on a hip in 1960 and a knee in 1968. The development of imaging, modeling, and computer simulation in the latter half of the 1900's provided insight into the smallest structures of the body. The relationships between these structures, functions, and the impact of internal and external forces accelerated new research opportunities into diagnostic procedures and effective solutions to disease. In the 1990's, biomechanical engineering programs began to emerge in academic and research institutions around the world.

HOW IT WORKS

Biomechanical engineering science is extremely diverse. However, the basic principle of studying the relationship between biological structures and forces, as well as the important associated reactions of biological structures to technological and environmental materials, exists throughout all disciplines. The biological structures described include all life forms and may include an entire body or organism or even the microstructures of specific tissues or systems. Characterization and quantification of the response of these structures to forces can provide insight into disease process, resulting in better treatments and diagnoses. Research in this field extends beyond the laboratory and can involve observations of mechanics in nature, such as the aerodynamics of bird flight, hydrodynamics of fish, or strength of plant root systems, and how these findings can be modified and applied to human performance and interaction with external forces.

As in biomechanics, biomechanical engineering has basic principles. Equilibrium, as defined by British physicist Sir Isaac Newton, results when the sum of all forces is zero and no change occurs and energy cannot be created or destroyed, only converted from one form to another.

The seven basic principles of biomechanics can be applied or modified to describe the reaction of forces to any living organism.

- The lower the center of mass, the larger the base of support; the closer the center of mass to

the base of support, and the greater the mass, the more stability increases.
- The production of maximum force requires the use of all possible joint movements that contribute to the task's objective.
- The production of maximum velocity requires the use of joints in order—from largest to smallest.
- The greater the applied impulse, the greater increase in velocity.
- Movement usually occurs in the direction opposite that of the applied force.
- Angular motion is produced by the application of force acting at some distance from an axis, that is, by torque.
- Angular momentum is constant when a body or object is free in the air.

The forces studied can be combinations of internal, external, static, or dynamic, and all are important in the analysis of complex biochemical and biophysical processes. Even the mechanics of a single cell, including growth, cell division, active motion, and contractile mechanisms, can provide insight into mechanisms of stress, damage of structures, and disease processes at the microscopic level. Imaging and computer simulation allow precise measurements and observations to be made of the forces impacting the smallest cells.

APPLICATIONS AND PRODUCTS

Biomechanical engineering advances in modeling and simulation have tremendous potential research and application uses across many health care disciplines. Modeling has resulted in the development of designs for implantable devices to assist with organs or areas of the body that are malfunctioning. The biomechanical relationships between organs and supporting structures allow for improved device design and can assist with planning of surgical and treatment interventions. The materials used for medical and surgical procedures in humans and animals are being evaluated and some redesigned, as biomechanical science is showing that different materials, procedures, and techniques may be better for reducing complications and improving long-term patient health. Evaluating the physical relationship between the cells and structures of the body and foreign implements and interventions can quantify the stresses and forces on the system, which provides more accurate prediction of patient outcomes.

Biomechanical engineering professionals apply their knowledge to develop implantable medical devices that can diagnose, treat, or monitor disease and health conditions and improve the daily living of patients. Devices that are used within the human body are highly regulated by the U.S. Food and Drug Administration (FDA) and other agencies internationally. Pacemakers and defibrillators, also called cardiac resynchronization therapy (CRT) devices, can constantly evaluate a patient's heart and respond to changes in heart rate with electrical stimulation. These devices greatly improve therapeutic outcomes in patients afflicted with congestive heart failure. Patients with arrhythmias experience greater advantages with implantable devices than with pharmaceutical options. Cochlear implants have been designed to be attached to a patient's auditory nerve and can detect sound waves and process them in order to be interpreted by the brain as sound for deaf or hard-of-hearing patients. Patients who have had cataract surgery used to have to wear thick corrective lenses to restore any standard of vision but with the development of intraocular lenses that can be implanted into the eye, their vision can be restored, often to a better degree than before the cataract developed.

Artificial replacement joints comprise a large portion of medical-implant technology. Patients receive joint replacement when their existing joints no longer function properly or cause significant pain because of arthritis or degeneration. More than 220,000 total hip replacements were performed in the United States in 2003, and this number is expected to grow significantly as the baby boomer portion of the population ages. Artificial joints are normally fastened to the existing bone by cement, but advances in biomechanical engineering have lead to a new process called "bone ingrowth," in which the natural bone grows into the porous surface of the replacement joint. Biomechanical engineering contributes considerable knowledge to the design of the artificial joints, the materials from which they are made, the surgical procedure used, fixation techniques, failure mechanisms, and prediction of the lifetime of the replacement joints.

Computer-aided (CAD) design has allowed biomechanical engineers to create complex models of organs and systems that can provide advanced analysis and instant feedback. This information provides insight into the development of designs for artificial organs that align with or improve on the mechanical properties of biological organs.

Biomechanical engineering can provide predictive values to medical professionals, which can help them develop a profile that better forecasts patient outcomes and complications. An example of this is using finite element analysis in the evaluation of aortic-wall stress, which can remove some of the unpredictability of expansion and rupture of an abdominal aortic aneurysm. Biomechanical computational methodology and advances in imaging and processing technology have provided increased predictability for life-threatening events.

Nonmedical applications of biomechanical engineering also exist in any facet of industry that impacts human life. Corporations employ individuals or teams to use engineering principles to translate the scientifically proven principles into commercially viable products or new technological platforms. Biomechanical engineers also design and build experimental testing devices to evaluate a product's performance and safety before it reaches the marketplace, or they suggest more economically efficient design options. Biomechanical engineers also use ergonomic principles to develop new ideas and create new products, such as car seats, backpacks, or even equipment and clothing for elite athletes, military personnel, or astronauts.

SOCIAL CONTEXT AND FUTURE PROSPECTS

The diversity of studying the relationship between living structure and function has opened up vast opportunities in science, health care, and industry. In addition to conventional implant and replacement devices, the demand is growing for implantable tissues for cosmetic surgery, such as breast and tissue implants, as well as implantable devices to aid in weight loss, such as gastric banding.

Reports of biomechanical engineering triumphs and discoveries are appearing in the mainstream media, making the general public more aware of the scientific work being done and how it impacts daily life. Sports fans learn about the equipment, training, and rehabilitation techniques designed by biomechanical engineers that allow their favorite athletes to break performance records and return to work sooner after being injured or having surgery. The public is accessing more information about their own health options than ever before, and they are becoming knowledgeable about the range of treatments available to them and the pros and cons of each.

Biomechanical engineering and biotechnology is an area that is experiencing accelerated growth, and billions of dollars are being funneled into research and development annually. This growth is expected to continue.

——*April D. Ingram*

FURTHER READING

Ethier, C. Ross, and Craig A. Simmons. *Introductory Biomechanics: From Cells to Organisms.* Cambridge, England: Cambridge University Press, 2007. Provides an introduction to biomechanics and also discusses clinical specialties, such as cardiovascular, musculoskeletal, and ophthalmology.

Hall, Susan J. *Basic Biomechanics.* 5th ed. New York: McGraw-Hill, 2006. A good introduction to biomechanics, regardless of one's math skills.

Hamill, Joseph, and Kathleen Knutzen. *Biomechanical Basis of Human Movement.* 4th ed. Philadelphia: Lippincott, 2015. Integrates anatomy, physiology, calculus, and physics and provides the fundamental concepts of biomechanics.

Hay, James G., and J. Gavin Reid. *Anatomy, Mechanics, and Human Motion.* 2d ed. Englewood Cliffs, N.J.: Prentice Hall, 1988. A good resource for upper high school students, this text covers basic kinesiology.

Peterson, Donald R., and Joseph D. Bronzino, eds. *Biomechanics: Principles and Applications.* 2d ed. Boca Raton, Fla.: CRC Press, 2008. A collection of twenty articles on various aspects of research in biomechanics.

Prendergast, Patrick, ed. *Biomechanical Engineering: From Biosystems to Implant Technology.* London: Elsevier, 2007. One of the first comprehensive books for biomechanical engineers, written with the student in mind.

Biomechanics

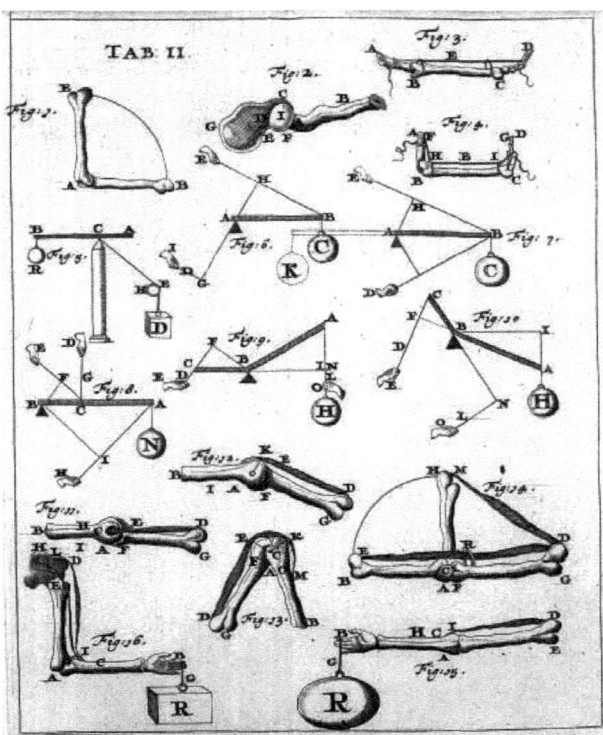

Page of one of the first works of biomechanics (*De Motu Animalium* by Giovanni Alfonso Borelli) [Public domain], via Wikimedia Commons

FIELDS OF STUDY

Biology; Biomechanical engineering; Bioengineering; Civil engineering; Computer engineering; Mechanics

ABSTRACT

Biomechanics is the study of the application of mechanical forces to a living organism. It investigates the effects of the relationship between the body and forces applied either from outside or within. In humans, biomechanists study the movements made by the body, how they are performed, and whether the forces produced by the muscles are optimal for the intended result or purpose. Biomechanics integrates the study of anatomy and physiology with physics, mathematics, and engineering principles. It may be considered a subdiscipline of kinesiology as well as a scientific branch of sports medicine.

HISTORY

Biomechanics has a long history even though the actual term and field of study concerned with mechanical analysis of living organisms was not internationally accepted and recognized until the early 1970s. Definitions provided by early biomechanics specialists James G. Hay in 1971 and Herbert Hatze in 1974 are still accepted. Hatze stated, "Biomechanics is the science which studies structures and functions of biological systems using the knowledge and methods of mechanics."

Highlights throughout history have provided insight into the development of this scientific discipline. The ancient Greek philosopher Aristotle was the first to introduce the term "mechanics," writing about the movement of living beings around 322 BCE. He developed a theory of running techniques and suggested that people could run faster by swinging their arms. In the 1500s, Leonardo da Vinci proposed that the human body is subject to the law of mechanics, and he contributed significantly to the development of anatomy as a modern science. Italian scientist Giovanni Alfonso Borelli, a student of Galileo, is often considered the father of biomechanics. In the mid-1600s, he developed mathematical models to describe anatomy and human movement mechanically. In the late 1600s, English physician and mathematician Sir Isaac Newton formulated mechanical principles and Newtonian laws of motion (inertia, acceleration, and reaction) that became the foundation of biomechanics.

British physiologist A. V. Hill, the 1923 winner of the Nobel Prize in Physiology or Medicine, conducted research to formulate mechanical and structural theories for muscle action. In the 1930s, American anatomy professor Herbert Elftman was able to quantify the internal forces in muscles and joints and developed the force plate to quantify ground reaction. A significant breakthrough in the understanding of muscle action was made by British physiologist Andrew F. Huxley in 1953, when he described his filament theory to explain muscle shortening. Russian physiologist Nicolas Bernstein published a paper in

1967 describing theories for motor coordination and control following his work studying locomotion patterns of children and adults in the Soviet Union.

HOW IT WORKS

The study of human movement is multifaceted, and biomechanics applies mechanical principles to the study of the structure and function of living things. Biomechanics is considered a relatively new field of applied science, and the research being done is of considerable interest to many other disciplines, including zoology, orthopedics, dentistry, physical education, forensics, cardiology, and a host of other medical specialties. Biomechanical analysis for each particular application is very specific; however, the basic principles are the same.

Newton's Laws of Motion. The development of scientific models reduces all things to their basic level to provide an understanding of how things work. This also allows scientists to predict how things will behave in response to forces and stimuli and ultimately to influence this behavior.

Newton's laws describe the conservation of energy and the state of equilibrium. Equilibrium results when the sum of forces is zero and no change occurs, and conservation of energy explains that energy cannot be created or destroyed, only converted from one form to another. Motion occurs in two ways, linear motion in a particular direction or rotational movement around an axis. Biomechanics explores and quantifies the movement and production of force used or required to produce a desired objective.

Seven Principles. Seven basic principles of biomechanics serve as the building blocks for analysis. These can be applied or modified to describe the reaction of forces to any living organism.

- The lower the center of mass, the larger the base of support; the closer the center of mass to the base of support and the greater the mass, the more stability increases.
- The production of maximum force requires the use of all possible joint movements that contribute to the task's objective.
- The production of maximum velocity requires the use of joints in order, from largest to smallest.
- The greater the applied impulse, the greater increase in velocity.
- Movement usually occurs in the direction opposite that of the applied force.
- Angular motion is produced by the application of force acting at some distance from an axis, that is, by torque.
- Angular momentum is constant when an athlete or object is free in the air.

Static and dynamic forces play key roles in the complex biochemical and biophysical processes that underlie cell function. The mechanical behavior of individual cells is of interest for many different biologic processes. Single-cell mechanics, including growth, cell division, active motion, and contractile mechanisms, can be quite dynamic and provide insight into mechanisms of stress and damage of structures. Cell mechanics can be involved in processes that lie at the root of many diseases and may provide opportunities as focal points for therapeutic interventions.

APPLICATIONS AND PRODUCTS

Biomechanics studies and quantifies the movement of all living things, from the cellular level to body systems and entire bodies, human and animal. There are many scientific and health disciplines, as well as industries that have applications developed from this knowledge. Research is ongoing in many areas to effectively develop treatment options for clinicians and better products and applications for industry.

Dentistry. Biomechanical principles are relevant in orthodontic and dental science to provide solutions to restore dental health, resolve jaw pain, and manage cosmetic and orthodontic issues. The design of dental implants must incorporate an analysis of load bearing and stress transfer while maintaining the integrity of surrounding tissue and comfortable function for the patient. This work has lead to the development of new materials in dental practices such as reinforced composites rather than metal frameworks.

Forensics. The field of forensic biomechanical analysis has been used to determine mechanisms of injury after traumatic events such as explosions in military situations. This understanding of how parts of the body behave in these events can be used to develop mitigation strategies that will reduce injuries. Accident and injury reconstruction using biomechanics is an emerging field with industrial and legal applications.

Biomechanical Modeling. Biomechanical modeling is a tremendous research field, and it has potential uses across many health care applications. Modeling has resulted in recommendations for prosthetic design and modifications of existing devices. Deformable breast models have demonstrated capabilities for breast cancer diagnosis and treatment. Tremendous growth is occurring in many medical fields that are exploring the biomechanical relationships between organs and supporting structures. These models can assist with planning surgical and treatment interventions and reconstruction and determining optimal loading and boundary constraints during clinical procedures.

Materials. Materials used for medical and surgical procedures in humans and animals are being evaluated and some are being changed as biomechanical science is demonstrating that different materials, procedures, and techniques may be better for reducing complications and improving long-term patient health. Evaluation of the physical relationship between the body and foreign implements can quantify the stresses and forces on the body, allowing for more accurate prediction of patient outcomes and determination of which treatments should be redesigned.

Predictability. Medical professionals are particularly interested in the predictive value that biomechanical profiling can provide for their patients. An example is the unpredictability of expansion and rupture of an abdominal aortic aneurysm. Major progress has been made in determining aortic wall stress using finite element analysis. Improvements in biomechanical computational methodology and advances in imaging and processing technology have provided increased predictive ability for this life-threatening event.

As the need for accurate and efficient evaluation grows, so does the research and development of effective biomechanical tools. Capturing real-time, real-world data, such as with gait analysis and range of motion features, provides immediate opportunities for applications. This real-time data can quantify an injury and over time provide information about the extent that the injury has improved. High-tech devices can translate real-world situations and two-dimensional images into a three-dimensional framework for analysis. Devices, imaging, and modeling tools and software are making tremendous strides and becoming the heart of a highly competitive industry aimed at simplifying the process of analysis and making it less invasive.

SOCIAL CONTEXT AND FUTURE PROSPECTS

Biomechanics has gone from a narrow focus on athletic performance to become a broad-based science, driving multibillion dollar industries to satisfy the needs of consumers who have become more knowledgeable about the relationship between science, health, and athletic performance. Funding for biomechanical research is increasingly available from national health promotion and injury prevention programs, governing bodies for sport, and business and industry. National athletic programs want to ensure that their athletes have the most advanced training methods, performance analysis methods, and equipment to maximize their athletes' performance at global competitions.

Much of the existing and developing technology is focused on increasingly automated and digitized systems to monitor and analyze movement and force. The physiological aspect of movement can be examined at a microscopic level, and instrumented athletic implements such as paddles or bicycle cranks allow real-time data to be collected during an event or performance. Force platforms are being reconfigured as starting blocks and diving platforms to measure reaction forces. These techniques for biomechanical performance analysis have led to revolutionary technique changes in many sports programs and rehabilitation methods.

Advances in biomechanical engineering have led to the development of innovations in equipment, playing surfaces, footwear, and clothing, allowing people to reduce injury and perform beyond previous expectations and records.

Computer modeling and virtual simulation training can provide athletes with realistic training opportunities, while their performance is analyzed and measured for improvement and injury prevention.

—*April D. Ingram*

FURTHER READING

Bronzino, Joseph D., and Donald R. Peterson. *Biomechanics: Principles and Practices*. Boca Raton: CRC, 2014. eBook Collection (EBSCOhost). Web. 25 Feb. 2015.

Hamill, Joseph, and Kathleen Knutzen. *Biomechanical Basis of Human Movement*. 4th ed. Philadelphia: Lippincott, 2015. Print. Integrates anatomy, physiology, calculus, and physics and provides the fundamental concepts of biomechanics.

Hatze, H. "The Meaning of the Term 'Biomechanics.'" *Journal of Biomechanics* 7.2 (1974): 89–90. Print.

Hay, James G. *The Biomechanics of Sports Techniques*. 4th ed. Englewood Cliffs: Prentice, 1993. Print.

Kerr, Andrew. *Introductory Biomechanics*. London: Elsevier, 2010. Print.

Peterson, Donald R., and Joseph D. Bronzino, eds. Biomechanics: Principles and Applications. 2d ed. Boca Raton, Fla.: CRC Press, 2008. A collection of twenty articles on various aspects of research in biomechanics.

Watkins, James. *Introduction to Biomechanics of Sport and Exercise*. London: Elsevier, 2007. Print.

Biometric eye scanners

U.S. Marine Corps Sergeant A.C. Wilson uses a retina scanner to positively identify a member of the Baghdaddi city council prior to a meeting with local tribal figureheads, sheiks. By Gunnery Sergeant Michael Q. Retana, U.S. Marine Corps [Public domain], via Wikimedia Commons

FIELDS OF STUDY

Bioinformatics; Biophysics; Biorobotics; Computer engineering

ABSTRACT

Biometric eye scanning can facilitate the automated control of access to areas where high levels of security must be maintained, such as correctional institutions and military and government installations that house sensitive materials.

BACKGROUND

The goal of biometric identification systems is to provide automated identity assurance—that is, the capability to recognize individuals accurately—with reliability, speed, and convenience. The complex nature of the human eye provides two of the most accurate biometric measures available. The iris and the retina, located on the front and back of the eye, respectively, are individually distinguishing structures. Retinal recognition became commercially available in the early 1980's, preceding iris recognition systems by about five years.

The iris is the round, pigmented membrane that surrounds the pupil of the eye. The intricate pattern of furrows and ridges in the iris is randomly formed prior to birth and remains stable from early childhood until death. In a typical iris scan, the person being identified aligns one eye close to a wall-mounted scanner for a few seconds. The scanner uses a near-infrared light to scan an image of the eye, and computer software then isolates the iris in the image and performs size and contrast corrections. Computer software then compares the final digital image with other iris images stored in a database; when a match is made, the person is identified.

APPLICATIONS

Prisons throughout the United States use iris-scanning technology to verify the identities of convicts before release. Correctional facilities also enroll visitors in their iris image databases and scan the irises of people leaving the facilities to be certain they are visitors, not inmates. Some organizations use small, semiportable iris scanners to control access to sensitive computer files and information.

Retina biometric identification is based on the individually distinguishing characteristics of blood vessel patterns on the back of the eye. These patterns are thought to be created by a random biological process and remain unchanged throughout life in a healthy individual. During retina scanning, the person being identified aligns one eye with a wall-mounted scanner for several seconds. The scanner illuminates the retina with a low-intensity infrared light and creates an image of the patterns formed by the major blood vessels. The image is then digitally encoded, stored, and compared using computer software.

Because the retina is located on the back of the eye, this type of scan requires a high degree of cooperation from the user to ensure proper illumination and alignment. Given that retina scanning is more complex than the iris-scanning process, retina-scanning technology is best deployed in high-security, controlled-access environments where user convenience is not a priority. Employees in military weapons facilities, power plants, and sensitive laboratory environments are commonly required to undergo retina scanning to gain access.

—Ruth N. Udey

FURTHER READING

Coats, William Sloan, et al. *The Practitioner's Guide to Biometrics*. Chicago: American Bar Association Publishing, 2007.

Nanavati, Samir, Michael Thieme, and Raj Nanavati. *Biometrics: Identity Verification in a Networked World*. New York: John Wiley & Sons, 2002.

Woodward, John D., Jr., Nicholas M. Orlans, and Peter T. Higgins. *Biometrics*. New York: McGraw-Hill, 2003.

Biometric identification systems

At Walt Disney World in Lake Buena Vista, Florida, biometric measurements are taken from the fingers of guests to ensure that a ticket is used by the same person from day to day. Taken by Raul654. CC BY-SA 3.0, https://commons.wikimedia.org/w/index.php?curid=363596

FIELDS OF STUDY

Bioinformatics; Biophysics; Biorobotics; Computer engineering

ABSTRACT

Biometric identification systems are becoming increasingly important given heightened concerns with security in many contexts. Compared with many other means of authorization and authentication, including password recognition, biometric technologies represent a significant advance in terms of ease of use, reliability, and validity.

BACKGROUND

The constantly evolving science of biometrics has produced a wide variety of systems capable of comparing hand, facial, eye, signature, vocal, brain, and genetic measures of given individuals against profiles of such measures stored in large databases. The applications of this technology for law-enforcement purposes are extensive. Biometric systems have been used to identify offenders who are using aliases, to fight illegal immigration, and to identify inmates as they are moved through various phases of the correctional system. Biometric data can be used to verify identity claims or to screen for persons who have been identified as potential security risks.

ACCURACY

Biometric identification systems represent a huge improvement over the traditional "token" (credit card or document) and password systems. Credit cards can be lost or stolen and then used as false identification. Similarly, passwords can be "cracked," forgotten, or stolen. Biometric characteristics, on the other hand, are much more stable and permanent. Their inherent complexity renders them difficult or impossible to replicate, and the person being identified usually needs to be physically present at the time of the verification attempt. In addition, biometric systems can couple identifying information with other important background data, such as health or employment records (a fact that has led some to criticize the use of these systems as infringing on civil liberties).

The components of the typical biometric system are relatively straightforward; they consist of a sensor and a computer. The sensor is the device that gathers the biometric data from the individual being evaluated. The computer then processes the data collected; in some cases, the computer may refine the data by removing irrelevant information and background "noise" that may interfere with the interpretation of the results. The computer captures the biometric features being measured and creates a template, which it then compares to a database of biometric information on known individuals, looking for an identification match, or "hit." The consequences of a successful identification are as varied as the systems themselves. At the point of identification, an individual might be allowed into a restricted area, picked up for further questioning in a specific investigation, or observed further for any suspicious behavior.

The accuracy of a biometric system is typically assessed using one or more of the following measures: the failure-to-acquire rate (a measure of the percentage of unsuccessful attempts by the system to obtain specific biometric information from subjects), the false-accept rate (also known as the false-positive rate, a measure of the percentage of incorrect matches of subjects' biometric profiles to profiles already included in the database), and the false-reject rate (also known as the false-negative rate, the percentage of failures to match subjects' biometric profiles with identical profiles already included in the database). Minimization of all these kinds of error rates reduces the numbers of suspects who are needlessly detained, restricted from air travel, or otherwise affected by law-enforcement "false alarms" while maximizing the appropriate identification of true security threats.

APPLICATIONS

Law-enforcement agencies employ biometric technologies in many ways, including for facial recognition, fingerprint identification, iris recognition, and voice recognition. Facial recognition systems use specific aspects of facial features from scanned photographs to make identifications. The features analyzed may include the physical distance between specific features, skin color, thermal patterns of blood flow, and facial lines. One application of facial recognition technology is the establishment by police departments of archives containing many thousands of offender photographs. These are matched with suspects' pictures or used to produce photo lineups that can be shown to crime victims or witnesses.

Numerous evaluations of facial recognition technology have produced mixed results. One Australian system, for example, tested in the Sydney airport, was found to have a false-reject rate of 2 percent. This rate was confirmed by tests sponsored by the U.S. government. Although this error rate seems low, major world airports typically service several million passengers annually, which means that the systems could potentially falsely reject many thousands of people. One meta-analysis of facial recognition systems produced accuracy rates ranging from 51 percent to 94 percent. Factors affecting the rates included lighting, the quality of the photographs taken, movements of the subjects, the angles of the poses in the photographs, and the presence of eyeglasses on subjects. In general, male subjects and older persons were more easily recognized than were female and younger subjects. An inverse relationship was also found between accuracy and the size of the database against which the subjects' facial features were compared.

Fingerprint identification is the oldest form of biometric identification, having been in use since the late nineteenth century. The Federal Bureau of Investigation (FBI) established a central database of fingerprints in 1924 against which law-enforcement agencies can seek to match the prints of crime suspects and victims. With modern electronic and laser technology, fingerprint images are often taken and transmitted "live" to the database. Efforts to automate the analysis and identification of fingerprints began in the 1960's.

Fingerprint identification systems use electronic fingerprint readers to locate where the ridges of fingerprints start, end, or split up. These areas, known as minutiae points, form the basis for the identification. Each fingerprint typically contains thirty to forty minutiae points, and no two people's prints will match on more than eight such points.

In terms of accuracy, the false-accept rates of fingerprint identification have generally been very close to zero, and false-reject rates have been 3 percent or less, according to a 2015 study by the Miami-Dade Police Department and the Office of Justice Programs. The accuracy of the analysis of fingerprints taken from crime scenes, however, is often reduced because of the poor quality of the prints themselves. In addition, although it is often assumed that fingerprints are stable over a lifetime, research has shown that they in fact can change in response to physiological growth, activity, or intentional alteration; it has also been shown that many fingerprint matching systems can be "spoofed." Despite some limitations, fingerprinting is less controversial and more highly developed than any other type of biometric identification system. This is reflected in court acceptance of fingerprinting evidence. Moreover, fingerprint identification is no longer the sole purview of law enforcement and military forces; by 2014, commercial fingerprint-scanning modules were available for the protection of business and consumer computers as well as smartphones.

In iris recognition identification systems, an image of the iris of the eye (the colored ring surrounding the pupil) of the person to be identified is recorded by a digital camera and then converted into a template, which is checked for matches against an existing database. False-positive rates for such systems have averaged 0.1 percent, and false-negative rates have averaged 1.5 percent. An advantage of using this biometric technique is that, unlike fingerprints, the structure of the iris is permanent by the age of one and is unique for each person (this includes comparisons between identical twins and even between the left and right eyes of the same person). Unlike with fingerprint identification, however, iris evidence is not left at crime scenes. In addition, failure rates as high as 15 percent have been found when iris-scanning technology is used in brightly lit settings. This technology has many potential applications, including security screening at airports and borders, passport and immigration control, and identification for banking and issuance of drivers' licenses. By 2015, the FBI had created a pilot program to launch a nationwide iris recognition database comparable to those used for fingerprints.

Voice recognition systems use physical and behavioral aspects of the voice to identify individuals; the voice features measured are based on the physiology of the windpipe, nasal cavity, and vocal cords. A digital "voice signature" is recorded, and a computer measures the features and compares them against known samples for identification and verification. One drawback to the use of voice biometrics is that voice patterns can vary with age, and they can also be affected by medical problems (including even a cold) and the emotional state of the examinee. Background noise can also be a problem with the use of this identification technology.

DNA identification relies on an individual's unique "variable number tandem repeats (VNTRs)" within that person's DNA. VNTR sequences are similar for relatives and the same for identical twins but unique among strangers. Thus, analysis of VNTRs can help scientists identify individuals and relatives.

Another biometric identification technology that has been investigated is hand geometry scanning, which involves more than ninety measurements of different parts of the hand. To detect forgery, dynamic signature identification has been developed; in this system, the specific dimensions of the pen strokes a person makes while writing his or her signature (including pressure, speed, and direction) are recorded and stored for later matching. This technology is prone to high false-negative rates, however, because even though signatures are ubiquitous in daily transactions, only specific parts of a person's signature remain constant across every signing. Gait analysis, which focuses on people's unique walking patterns, is another type of biometric technique. Limitations to gait analysis include the fact that making gait measurements may be invasive; also, gait can be affected by injury or by a change in shoes.

——*Eric Metchik*

Further Reading

Crompton, Malcolm. "Biometrics and Privacy: The End of the World as We Know It or the White Knight of Privacy?" In *Biometrics: Security and Authentication.* Sydney: Biometrics Institute, 2003.

Das, Ravindra. *Biometric Technology: Authentication, Biocryptography, and Cloud-Based Architecture.* CRC Press, 2015.

Federal Bureau of Investigation. "Next Generation Identification (NGI)." *Fingerprints & Other Biometrics,* 2016.

"Fingerprint Examiners Found to Have Very Low Error Rates." *PR Newswire US,* February 2, 2015.

Jain, Anil K., Arun Russ, and Sharath Pankanti. "Biometrics: A Tool for Information Security." *IEEE Transactions on Information Forensics and Security* 1, no. 2 (2006): 125–143.

Krishnan, K. N., with D. R. Berwick. *Developing a Police Perspective and Exploring the Use of Biometrics and Other Emerging Technologies as an Investigative Tool in Identity Crimes.* Payneham, S. Aust.: Australasian Centre for Policing Research, 2004.

Mansfield, A. J., and J. L. Wayman. *Best Practices in Testing and Reporting Performance of Biometric Devices: Version 2.01.* Teddington, Middlesex, England: National Physical Laboratory, 2002.

Mills, Kelly. "University Opts for Biometric Security." *Computerworld,* January 25, 2002, 3–4.

Vacca, John R. *Biometric Technologies and Verification Systems.* Burlington, Mass.: Elsevier, 2007.

Bionics and Biomedical Engineering

Velcro was inspired by the tiny hooks found on the surface of burs. By Zephyris. (Own work) [CC-BY-SA-3.0 (http://creativecommons.org/licenses/by-sa/3.0) or GFDL (http://www.gnu.org/copyleft/fdl.html)], via Wikimedia Commons

FIELDS OF STUDY

Bioengineering; Bioinformatics; Biomechanical engineering; Biophysics; Biorobotics; Computer engineering; Medical biotechnology (or Red biotechnology)

ABSTRACT

Bionics combines natural biologic systems with engineered devices and electrical mechanisms. An example of bionics is an artificial arm controlled by impulses from the human mind. Construction of bionic arms or similar devices requires the integrative use of medical equipment such as electroencephalograms (EEGs) and magnetic resonance imaging (MRI) machines with mechanically engineered prosthetic arms and legs. Biomedical engineering further melds biomedical and engineering sciences by producing medical equipment, tissue growth, and new pharmaceuticals. An example of biomedical engineering is human insulin production through genetic engineering to treat diabetes.

BACKGROUND

The fields of biomedical engineering and bionics focus on improving health, particularly after injury or illness, with better rehabilitation, medications, innovative treatments, enhanced diagnostic tools, and preventive medicine.

Bionics has moved nineteenth-century prostheses, such as the wooden leg, into the twenty-first century by using plastic polymers and levers. Bionics integrates circuit boards and wires connecting the nervous system to the modular prosthetic limb. Controlling artificial limb movements with thoughts provides more lifelike function and ability. This mind and prosthetic limb integration is the "bio" portion of bionics; the "nic" portion, taken from the word "electronic," concerns the mechanical engineering that makes it possible for the person using a bionic limb to increase the number and range of limb activity, approaching the function of a real limb.

Biomedical engineering encompasses many medical fields. The principle of adapting engineering techniques and knowledge to human structure and function is a key unifying concept of biomedical engineering. Advances in genetic engineering have produced remarkable bioengineered medications. Recombinant DNA techniques (genetic engineering) have produced synthetic hormones, such as insulin. Bacteria are used as a host for this process; once human-insulin-producing genes are implanted in the bacteria, the bacteria's DNA produce human insulin, and the human insulin is harvested to treat diabetics. Before this genetic technique was developed in 1982 to produce human insulin, insulin-dependent diabetics relied on insulin from pigs or cows. Although this insulin was life saving for diabetics, diabetics often developed problems from the pig or cow insulin because they would produce antibodies against the foreign insulin. This problem disappeared with the ability to engineer human insulin using recombinant DNA technology.

HISTORY

In the broad sense, biomedical engineering has existed for millennia. Human beings have always envisioned the integration of humans and technology to increase and enhance human abilities. Prosthetic devices go back many thousands of years: a three-thousand-year-old Egyptian mummy, for example, was found with a wooden big toe tied to its foot. In the fifteenth century, during the Italian Renaissance, Leonardo da Vinci's elegant drawings demonstrated some early ideas on bioengineering, including his helicopter and flying machines, which melded human and machine into one functional unit capable of flight. Other early examples of biomedical engineering include wooden teeth, crutches, and medical equipment, such as stethoscopes.

Electrophysiological studies in the early 1800s produced biomedical engineering information used to better understand human physiology. Engineering principles related to electricity combined with human physiology resulted in better knowledge of the electrical properties of nerves and muscles.

X rays, discovered by Wilhelm Conrad Röntgen in 1895, were an unknown type of radiation (thus the "X" name). When it was accidentally discovered that they could penetrate and destroy tissue, experiments were developed that led to a range of imaging technologies that evolved over the next century. The first formal biomedical engineering training program, established in 1921 at Germany's Oswalt Institute for Physics in Medicine, focused on three main areas: the effects of ionizing radiation, tissue electrical characteristics, and X-ray properties.

In 1948, the Institute of Radio Engineers (later the Institute of Electrical and Electronics Engineers), the American Institute for Electrical Engineering, and the Instrument Society of America held a conference on engineering in biology and medicine. The 1940s and 1950s saw the formation of professional societies related to biomedical engineering, such as the Biophysics Society, and of interest groups within engineering societies. However, research at the time focused on the study of radiation. Electronics and the budding computer era broadened interest and activities toward the end of the 1950s.

James D. Watson and Francis Crick identified the DNA double-helix structure in 1953. This important discovery fostered subsequent experimentation in molecular biology that yielded important information about how DNA and genes code for the expression of traits in all living organisms. The genetic code in DNA was deciphered in 1968, arming researchers with enough information to discover ways that DNA could be recombined to introduce genes from one organism into a different organism, thereby allowing the host to produce a variety of useful products. DNA recombination became one of the most important tools in the field of biomedical engineering, leading to tissue growth as well as new pharmaceuticals.

In 1962, the National Institutes of Health created the National Institute of General Medical Sciences, fostering the development of biomedical engineering programs. This institute funds research in the diagnosis, treatment, and prevention of disease.

Bionics and biomedical engineering span a wide variety of beneficial health-related fields. The common thread is the combination of technology with human applications. Dolly the sheep was cloned in 1996. Cloning produces a genetically identical copy of an existing life-form. Human embryonic cloning presents the potential of therapeutic reproduction of needed organs and tissues, such as kidney replacement for patients with renal failure.

In the twenty-first century, the linking of machines with the mind and sensory perception has provided hearing for deaf people, some sight for the blind, and willful control of prostheses for amputees.

HOW IT WORKS

Restorative bionics integrates prosthetic limbs with electrical connections to neurons, allowing an individual's thoughts to control the artificial limb. Tiny arrays of electrodes attached to the eye's retina connect to the optic nerve, enabling some visual perception for previously blind people. Deaf people hear with electric devices that send signals to auditory nerves, using antennas, magnets, receivers, and electrodes. Researchers are considering bionic skin development using nanotechnology to connect with nerves, enabling skin sensations for burn victims requiring extensive grafting.

Many biomedical devices work inside the human body. Pacemakers, artificial heart valves, stents, and even artificial hearts are some of the bionic devices correcting problems with the cardiovascular system. Pacemakers generate electric signals that improve abnormal heart rates and abnormal heart rhythms. When pulse generators located in the pacemakers sense an abnormal heart rate or rhythm, they produce shocks to restore the normal rate. Stents are inserted into an artery to widen it and open clogged blood vessels. Stents and pacemakers are examples of specialized bionic devices made up of bionic materials compatible with human structure and function.

Cloning. Cloning is a significant area of genetic engineering that allows the replication of a complete living organism by manipulating genes. Dolly the sheep, an all-white Finn Dorset ewe, was cloned from a surrogate mother blackface ewe, which was used as an egg donor and carried the cloned Dolly during gestation (pregnancy). An egg cell from the surrogate was removed and its nucleus (which contains DNA) was replaced with one from a Finn Dorset ewe; the resulting new egg was placed in the blackface ewe's uterus after stimulation with an electric pulse. The electrical pulse stimulated growth and cell duplication. The blackface ewe subsequently gave birth to the all-white Dolly. The newborn all-white Finn Dorset ewe was an identical genetic twin of the Finn Dorset that contributed the new nucleus.

Recombinant DNA. Another significant genetic engineering technique involves recombinant DNA. Human genes transferred to host organisms, such as bacteria, produce products coded for by the transferred genes. Human insulin and human growth hormone can be produced using this technique. Desired genes are removed from human cells and placed in circular bacterial DNA strips called plasmids. Scientists use enzymes to prepare these DNA formulations, ultimately splicing human genes into bacterial plasmids. These plasmids are used as vectors, taken up and reproduced by bacteria. This type of genetic adaptation results in insulin production if the spliced genes were taken from the part of the human genome producing insulin; other cells and substances, coded for by different human genes, can be produced this way. Many biologic medicines are produced using recombinant DNA technology.

APPLICATIONS AND PRODUCTS

Medical Devices. Biomedical engineers produce lifesaving medical equipment, including pacemakers, kidney dialysis machines, and artificial hearts. Synthetic limbs, artificial cochleas, and bionic sight chips are among the prosthetic devices that biomedical engineers have developed to enhance mobility, hearing, and vision. Medical monitoring devices, developed by biomedical engineers for use in intensive care units and surgery or by space and deep-sea explorers, monitor vital signs such as heart rate and rhythm, body temperature, and breathing rate.

Equipment and Machinery. Biomedical engineers produce a wide variety of other medical machinery, including laboratory equipment and therapeutic equipment. Therapeutic equipment includes laser devices for eye surgery and insulin pumps (sometimes called artificial pancreases) that both monitor blood sugar levels and deliver the appropriate amount of insulin when it is needed.

Imaging Systems. Medical imaging provides important machinery devised by biomedical engineers. This specialty incorporates sophisticated computers and imaging systems to produce computed tomography (CT), magnetic resonance imaging (MRI), and positron emission tomography (PET) scans. In naming its National Institute of Biomedical Imaging and Bioengineering (NIBIB), the U.S. Department of Health and Human Services emphasized the equal importance and close relatedness of these subspecialties by using both terms in the department's name.

Computer programming provides important circuitry for many biomedical engineering applications, including systems for differential disease diagnosis. Advances in bionics, moreover, rely heavily on computer systems to enhance vision, hearing, and body movements.

Biomaterials. Biomaterials, such as artificial skin and other genetically engineered body tissues, are areas promising dramatic improvements in the treatment of burn victims and individuals needing organ transplants. Bionanotechnology, another subfield of biomedical engineering, promises to enhance the surface of artificial skin by creating microscopic messengers that can create the sensations of touch and pain. Bioengineers interface with the fields of physical therapy, orthopedic surgery, and rehabilitative medicine in the fields of splint development, biomechanics, and wound healing.

Medications. Medicines have long been synthesized artificially in laboratories, but chemically synthesized medicines do not use human genes in their production. Medicines produced by using human genes in recombinant DNA procedures are called biologics and include antibodies, hormones, and cell receptor proteins. Some of these products include human insulin, the hepatitis B vaccine, and human growth hormone.

Bacteria and viruses invading a body are attacked and sometimes neutralized by antibodies produced by the immune system. Diseases such as Crohn's disease, an inflammatory bowel condition, and psoriatic arthritis are conditions exacerbated by inflammatory antibody responses mounted by the affected person's immune system. Genetic antibody production in the form of biologic medications interferes with or attacks mediators associated with Crohn's and arthritis and improves these illnesses by decreasing the severity of attacks or decreasing the frequency of flare-ups.

Cloning and Stem Cells. Cloned human embryos could provide embryonic stem cells. Embryonic stem cells have the potential to grow into a variety of cells, tissues, and organs, such as skin, kidneys, livers, or heart cells. Organ transplantation from genetically identical clones would not encounter the recipient's natural rejection process, which transplantations must overcome. As a result, recipients of genetically identical cells, tissues, and organs would enjoy more successful replacements of key organs and a better quality of life. Human cloning is subject to future research and development, but the promise of genetically identical replacement organs for people with failed hearts, kidneys, livers, or other organs provides hope for enhanced future treatments.

SOCIAL CONTEXT AND FUTURE PROSPECTS

Bionics technologies include artificial hearing, sight, and limbs that respond to nerve impulses. Bionics offers partial vision to the blind and prototype prosthetic arm devices that offer several movements through nerve impulses. The goal of bionics is to better integrate the materials in these artificial devices with human physiology to improve the lives of those with limb loss, blindness, or decreased hearing.

Cloned animals exist but cloning is not a yet a routine process. Technological advances offer rapid DNA analysis along with significantly lower cost genetic analysis. Genetic databases are filled with information on many life-forms, and new DNA sequencing information is added frequently. This basic information that has been collected is like a dictionary, full of words that can be used to form sentences, paragraphs, articles, and books, in that it can be used to create new or modified life-forms.

Biomedical engineering enables human genetic engineering. The stuff of life, genes, can be modified or manipulated with existing genetic techniques. The power to change life raises significant societal concerns and ethical issues. Beneficial results such as optimal organ transplantations and effective medications are the potential of human genetic engineering.

—*Richard P. Capriccioso, MD, and Christina Capriccioso*

FURTHER READING

"Biomedical Engineers." *Occupational Outlook Handbook.* Bureau of Labor Statistics, 17 Dec. 2015, www.bls.gov/ooh/architecture-and-engineering/biomedical-engineers.htm. Accessed 27 Oct. 2016.

Braga, Newton C. *Bionics for the Evil Genius: Twenty-five Build-It-Yourself Projects*. McGraw-Hill, 2006.

Fischman, Josh. "Merging Man and Machine: The Bionic Age." *National Geographic*, vol. 217, no. 1, 2010, pp. 34–53.

Hung, George K. *Biomedical Engineering: Principles of the Bionic Man*. World Scientific, 2010.

Richards-Kortum, Rebecca. *Biomedical Engineering for Global Health*. Cambridge University Press, 2010.

Smith, Marquard, and Joanne Morra, editors. *The Prosthetic Impulse: From a Posthuman Present to a Biocultural Future*. MIT, 2007.

BIOPESTICIDES AND THE ENVIRONMENT

Comparison of the Properties of *Bacillus Thuringiensis* and *Bacillus Popilliae* as Microbial Biocontrol Agents

	BACILLUS THURINGIENSIS	BACILLUS POPILLIAE
Pest controlled	Lepidoptera (many)	Coleoptera (few)
Pathogenicity	low	high
Response time	immediate	slow
Formulation	spores and toxin crystals	spores
Production	in vitro	in vivo
Persistence	low	high
Resistance in pests	developing	reported

Source: Data adapted from J. W. Deacon, *Microbial Control of Plant Pests and Diseases* (1983).

© EBSCO

FIELDS OF STUDY

Agricultural engineering; Agronomy; Animal Science; Biochemistry; Environmental biotechnology (or Green biotechnology)

ABSTRACT

Biopesticides have significant advantages over commercial pesticides in that they appear to be environmentally safer, given that they do not accumulate in the food chain and they have only slight effects on ecological balances.

APPLICATION OF BIOPESTICIDES

Pests are any unwanted animals, plants, or microorganisms. When the environment has no natural resistance to a pest and when no natural antagonists are present, pests can run rampant. For example, the fungus *Endothia parasitica*, which entered New York State in 1904, caused the nearly complete destruction of the American chestnut tree because no natural control was present.

Biopesticides represent the biological, rather than the chemical, control of pests. Many plants and animals are protected from pests by passive means. For example, plant rotation is a traditional method of insect and disease protection in which the host plant is removed for a period long enough to reduce pathogen and pest populations.

Biopesticides have several significant advantages over commercial pesticides. They appear to be ecologically safer than commercial pesticides because they do not accumulate in the food chain. Some biopesticides also provide persistent control, because pests require more than a single mutation to adapt to them and because they can become an integral part of a pest's life cycle. In addition, biopesticides have only slight effects on ecological balances because they do not affect nontarget species. Finally, biopesticides are compatible with other control agents. The major drawbacks to using biopesticides are that, in comparison with chemical pesticides, biopesticides work less efficiently and take more time to kill their targets.

Viruses, bacteria, fungi, protozoa, mites, and flowers have all been used as biopesticides. Viruses have been developed against insect pests such as *Lepidoptera*, *Hymenoptera*, and *Dipterans*. These viruses cause hyperparasitism. Gypsy moths and tent caterpillars, for example, periodically suffer from epidemic virus infestations.

Many saprophytic microorganisms that occur on plant roots and leaves can protect plants against microbial pests. *Bacillus cereus* has been used as an inoculum on soybean seeds to prevent infection by the fungal pathogen *Cercospora*. Some microorganisms used as biopesticides produce antibiotics, but the major mechanism for protection is probably competitive exclusion of a pest from sites on which the pest must grow. For example, *Agrobacterium radiobacter* antagonizes *Agrobacterium tumefaciens*, which causes crown gall disease. Two bacteria—*Bacillus* and *Streptomyces*—added as biopesticides to soil help control the damping off disease of cucumbers, peas, and lettuces caused by *Rhizoctonia solani*. *Bacillus subtilis* added to plant tissue also controls stem rot and wilt rot caused by the fungus *Fusarium*. Mycobacteria produce cellulose-degrading enzymes, and their addition to young seedlings helps control fungal infection by *Pythium*, *Rhizoctonia*, and *Fusarium*. *Bacillus* and *Pseudomonas* are bacteria that produce enzymes that dissolve fungal cell walls.

The best examples of microbial insecticides are *Bacillus thuringiensis* (*B.t.*) toxins, which were first used in 1901. They have had widespread commercial production and use since the 1960s and have been successfully tested on 140 insect species, including mosquitoes. *B.t.* produces insecticidal endotoxins during sporulation and also produces exotoxins contained in crystalline parasporal protein bodies. These protein crystals are insoluble in water but readily dissolve in an insect's gut. Once dissolved, the proteolytic enzymes paralyze the gut. *Bacillus* spores that have also been consumed germinate and kill the insect. *Bacillus popilliae* is a related bacterium that produces an insecticidal spore that has been used to control Japanese beetles, a pest of corn.

Saprophytic fungi can compete with pathogenic fungi. Among the fungi used as biopesticides are *Gliocladium virens*, *Trichoderma hamatum*, *Trichoderma harzianum*, *Trichoderma viride*, and *Talaromyces flavus*. For example, *Trichoderma* competes with the pathogens *Verticillium* and *Fusarium*. *Peniophora gigantea* antagonizes the pine pathogen *Heterobasidion annosum* through three mechanisms: It prevents the pathogen from colonizing stumps and traveling down into the root zone; it prevents the pathogen from traveling between infected and uninfected trees along interconnected roots; and it prevents the pathogen from growing up to stump surfaces and sporulating.

Nematodes are pests that interfere with commercial button mushroom (*Agaricus bisporus*) production. Several types of nematode-trapping fungi can be used as biopesticides to trap, kill, and digest the nematode pests. The fungi produce structures such as constricting and nonconstricting rings, sticky appendages, and spores, which attach to the nematodes. The most common nematode-trapping fungi are *Arthrobotrys oligospora*, *Arthrobotrys conoides*, *Dactylaria candida*, and *Meria coniospora*.

Protozoa have occasionally been used as biopesticide agents, but their use has suffered because of such difficulties as slow growth and complex culture conditions associated with their commercial production. Predaceous mites are used as a biopesticide to protect cotton from other insect pests such as the boll weevil.

Dalmatian and Persian insect powders contain pyrethrins, which are toxic insecticidal compounds produced in *Chrysanthemum* flowers. Synthetic versions of these naturally occurring compounds are found in products used to control head lice. Molecular genetics has also been used to insert the gene for the *B.t.* toxin into cotton and corn. *B.t.* cotton and *B.t.* corn both express the gene in their roots, which provides them with protection from root worms. Ecologists and environmentalists have expressed concern that constantly exposing pests to the toxin will cause insect resistance to develop rapidly and thus reduce the effectiveness of traditionally applied *B.t.*

—*Mark Coyne*

FURTHER READING

Churchill, B. W. *Biological Control of Weeds with Plant Pathogens.* Edited by R. Charudattan and H. Walker. New York: Wiley, 1982. Print.

Deacon, J. W. *Microbial Control of Plant Pests and Diseases.* Research Triangle Park: Instrumentation Systems & Automation, 1983. Print.

Metz, Matthew, ed. *Bacillus Thuringiensis: A Cornerstone of Modern Agriculture.* Binghamton: Haworth, 2003. Print.

Ohkawa, H., H. Miyagawa, and P. W. Lee, eds. *Pesticide Chemistry: Crop Protection, Public Health, Environmental Safety.* New York: Wiley, 2007. Print.

Bioprocess engineering

FIELDS OF STUDY

Biochemistry; Environmental biotechnology (or Green biotechnology); Industrial biotechnology (or White biotechnology); Industrial fermentation; Medical biotechnology (or Red biotechnology)

ABSTRACT

Bioprocess engineering is an interdisciplinary science that combines the disciplines of biology and engineering. It is associated primarily with the commercial exploitation of living things on a large scale. The objective of bioprocess engineering is to optimize either growth of organisms or the generation of target products. This is achieved mainly by the construction of controllable apparatuses. Both government agencies and private companies invest heavily in research within this area of applied science. Many traditional bioprocess engineering approaches (such as antibiotic production by microorganisms) have been advanced by techniques of genetic engineering and molecular biology.

HISTORY

People have been using bioprocessing for making bread, cheese, beer, and wine—all fermented foods—for thousands of years. Brewing was one of the first applications of bioprocess engineering. However, it was not until the nineteenth century that the scientific basis of fermentation was established, with the studies of French scientist Louis Pasteur, who discovered the microbial nature of beer brewing and wine making.

During the early part of the twentieth century, large-scale methods for treating wastewater were developed. Considerable growth in this field occurred toward the middle of the century, when the bioprocess for large-scale production of the antibiotic penicillin was developed. The World War II goal of industrial-scale production of penicillin led to the development of fermenters by engineers working together with biologists from the pharmaceutical company Pfizer. The fungus *Penicillium* grows and produces antibiotics much more effectively under controlled conditions inside a fermenter.

Later progress in bioprocess engineering has followed the development of genetic engineering, which raises the possibility of making new products from genetically modified microorganisms and plants grown in bioreactors. Just as past developments in bioprocess engineering have required contributions from a wide range of disciplines, including microbiology, genetics, biochemistry, chemistry, engineering, mathematics, and computer science, future developments are likely to require cooperation among scientists in multiple specialties.

HOW IT WORKS

Living cells may be used to generate a number of useful products: food and food ingredients (such as cheese, bread, and wine), antibiotics, biofuels, chemicals (enzymes), and human health care products such as insulin. Organisms are also used to destroy or break down harmful wastes, such as those created by the 2010 oil spill in the Gulf of Mexico, or to reduce pollution.

A good example of how bioprocess engineering works is the development of a bioprocess using bacteria for industrial production of the human hormone insulin. Without insulin, which regulates blood sugar levels, the body cannot use or store glucose properly. The inability of the body to make sufficient insulin causes diabetes. In the 1970's, the U.S. company Genentech developed a bioprocess for insulin production using genetically modified bacterial cells.

The initial stages involve genetic manipulation (in this case, transferring a human gene into bacterial DNA). Genetic manipulation is done in laboratories by scientists trained in molecular biology or biochemistry. After creating a genetically engineered bacterium, scientists grow it in a small tubes or flasks and study its growth characteristics and insulin production.

Once the bacterial growth and insulin production characteristics have been identified, scientists increase the scale of the bioprocess. They use or build small bioreactors (1-10 liters) that can monitor temperature, pH (acidity-alkalinity), oxygen concentration, and other process characteristics. The goal of this scale-up is to optimize bacterial growth and insulin production.

The next step is another scale-up, this time to a pilot-scale bioreactor. These bioreactors can be as large as 1,000 liters and are designed and built by engineers to study the response of bacterial cells to large-scale production. During a scale-up, decreased product yields are often experienced because the conditions in the large-scale bioreactors (temperature, pH, aeration, and nutrient supply) differ from those in small, laboratory-scale systems. If the pilot-scale bioreactors work efficiently, engineers will design industrial-scale bioreactors and supporting facilities (air supply, sterilization, and process-control equipment).

All these stages are part of upstream processing. An important part of bioprocess engineering is the product recovery process, or so-called downstream processing. Product recovery from cells often can be very difficult. It involves laboratory procedures such as mechanical breakage, centrifugation, filtration, chromatography, crystallization, and drying. The final step in bioprocess engineering is testing of the recovered product, in which animals are often used.

APPLICATIONS AND PRODUCTS

A wide range of products and applications of bioprocess engineering are familiar, everyday items.

Foods, Beverages, Food Additives, and Supplements. Living organisms play a major role in the production of food. Foods, beverages, additives, and supplements traditionally made by bioprocess engineering include dairy products (cheeses, sour cream, yogurt, and kefir), alcoholic beverages (beer, wines, and distilled spirits), plant products (soy sauce, tofu, sauerkraut), and food additives and supplements (flavors, proteins, vitamins, and carotenoids).

Traditional fermenters with microorganisms are used to obtain products in most of these applications. A typical industrial fermenter is constructed from stainless steel. Mixing of the microbial culture in fermenters is achieved by mechanical stirring, often with baffles. Airlift bioreactors have also been applied in the manufacturing of food products such as crude proteins synthesized by microorganisms. Mixing and liquid circulation in these bioreactors are induced by movement of an injected gas (such as air).

Biofuels. Bioprocess engineering is used in the production of biofuels, including ethanol (bioethanol), oil (biodiesel), butanol, biohydrogen, and biogas (methane). These biofuels are produced by the action of microorganisms in bioreactors, some of which use attached (immobilized) microorganisms. Cells, when immobilized in matrices such as agar, polyurethane, or glass beads, stabilize their growth and increase their physiological functions. Many microorganisms exist naturally in a state similar to immobilization, either on the surface of soil particles or in symbiosis with other organisms.

Environmental Applications. Bioprocess engineering plays an important role in removing pollution from the environment. It is used in treatment of wastewater and solid wastes, soil bioremediation, and mineral recovery. Environmental applications are based on the ability of organisms to use pollutants or other compounds as their food sources. One of the most important and widely used environmental applications is the treatment of wastewater by microorganisms. Microbes eat organic and inorganic compounds in wastewater and clean it at the same time. In this application, microorganisms are placed inside bioreactors (known as digesters) specifically designed by engineers. Engineers have also developed biofilters, bioreactors for removing pollutants from the air. Biofilters are used to remove pollutants, odors, and dust from air by the action of microorganisms. In addition, the mining industry uses bioprocess engineering for extracting minerals such as copper and uranium through the use of bacteria. Microbial leaching uses leaching dumps or tank bioreactors designed by engineers.

Enzymes. Enzymes are used in the health, food, laundry, pulp and paper, and textile industries. They are produced mainly from fungi and bacteria using bioprocess engineering. One of these enzymes is glucose isomerase, important in the production of fructose syrup. Genetic manipulation provides the means to produce many different enzymes, including those not normally synthesized by microorganisms. Fermenters for enzyme production are usually up to 100,000 liters in volume, although very expensive enzymes may be produced in smaller bioreactors, usually with immobilized cells.

Antibiotics and Other Health Care Products. Most antibiotics are produced by fungi and bacteria.

Industrial production of antibiotics usually occurs in fermenters (stirred tanks) of 40,000- to 200,000-liter capacity. The bioprocess for antibiotics was developed by engineers during World War II, although it has undergone some changes since the 1980's. Various food sources, including glucose and sucrose, have been adopted for antibiotic production by microorganisms. The modern bioprocess is highly efficient (90 percent). Process variables such as pH and aeration are controlled by computer, and nutrients are fed continuously to sustain maximum antibiotic production. Product recovery is also based on continuous extraction.

The other major health care products produced with the help of bioprocess engineering are steroids, bacterial vaccines, gene therapy vectors, and therapeutic proteins such as interferon, growth hormone, and insulin. Steroids are important hormones that are manufactured by the process of biotransformation, in which microorganisms are used to chemically modify an inexpensive material to create a desired product. Health care products are produced in traditional fermenters.

Biomass Production. Biomass is used as a fuel source, as a source of protein for human food or animal feed, and as a component in agricultural pesticides or fertilizer. Baker's yeast biomass is a major product of bioprocess engineering. It is required for making bread and other baked goods, beer, wine, and ethanol. Yeast is produced in large aerated fermenters of up to 200,000 liters. Molasses is used as a nutrient source for the cells. Yeast is recovered from the fermentation liquid by centrifugation and then is dried. People also use the biomass of algae. Algae are a source of animal feed, plant fertilizer, chemicals, and biofuels. Algal biomass is produced in open ponds, in tubular glass, or in plastic bioreactors.

Animal and Plant Cell Cultures. Bioprocess engineering incorporating animal cell culture is used primarily for the production of health care products such as viral vaccines or antibodies in traditional fermenters or bioreactors with immobilized cells. Antibodies, for example, are produced in bioreactors with hollow-fiber immobilized animal cells. Plant cell culture is also an important target of bioprocess engineering. However, only a few processes have been successfully developed. One successful process is the production of the pigment shikonin in Japan. Shikonin is used as a dye for coloring food and has applications as an anti-inflammatory agent.

Chemicals. There is an on-going trend in the chemical industry to use bioprocess engineering instead of pure chemistry for production of a variety of chemicals such as amino acids, polymers, and organic acids (citric, acetic, and lactic). Some of these chemicals (citric and lactic acids) are used as food preservatives. Many chemicals are produced in traditional fermenters by the action of microbes.

SOCIAL CONTEXT AND FUTURE PROSPECTS

The role of bioprocess engineering in industry is likely to expand because scientists are increasingly able to manipulate organisms to expand the range and yields of products and processes. Developments in this field continue rapidly.

Bioprocess engineering can potentially be the answer to several problems faced by humankind. One such problem is global warming, which is caused by rising levels of carbon dioxide and other greenhouse gases. A suggested method of addressing this issue is carbon dioxide removal, or sequestration, based on bioprocess engineering. This bioprocess uses microalgae (microscopic algae) in photobioreactors to capture the carbon dioxide that is discharged into the atmosphere by power plants and other industrial facilities. Photobioreactors are various types of closed systems made of an array of transparent tubes in which microalgae are cultivated and monitored under illumination.

The health care industry is another area where bioprocess engineers are likely to be active. For example, if pharmaceutical applications are found for stem cells, a bioprocess must be developed to produce a reliable, plentiful source of stem cells so that these drugs can be produced on a large scale. The process for growing and harvesting cells must be standardized so that the cells have the same characteristics and behave in a predictable manner. Bioprocess engineers must take these processes from laboratory procedures to industrial protocols.

In general, the future of bioprocess engineering is bright, although questions and concerns, primarily about using genetically modified organisms, have arisen. Public education in such a complex area of science is very important to avoid public mistrust of bioprocess engineering, which is very beneficial in most applications.

—Sergei A. Markov, PhD

FURTHER READING

Bailey, James E., and David F. Ollis. *Biochemical Engineering Fundamentals*. 2d ed. New York: McGraw-Hill, 2006. Covers all aspects of biochemical engineering in an understandable manner.

Bougaze, David, Thomas R. Jewell, and Rodolfo G. Buiser. *Biotechnology. Demystifying the Concepts.* San Francisco: Benjamin/Cummings, 2000. Classical book on biotechnology and bioprocessing.

Doran, Pauline M. *Bioprocess Engineering Principles*. London: Academic Press, 2009. A solid, basic textbook for students entering the field.

Glazer, Alexander N., and Hiroshi Nikaido. *Microbial Biotechnology: Fundamentals of Applied Microbiology.* New York: Cambridge University Press, 2007. In-depth analysis of the application of microorganisms in bioprocessing.

Heinzle, Elmar, Arno P. Biwer, and Charles L. Cooney. *Development of Sustainable Bioprocesses: Modeling and Assessment.* Hoboken, N.J.: John Wiley & Sons, 2007. Looks at making bioprocesses sustainable by improving them. Includes case studies on citric acid, biopolymers, antibiotics, and biopharmaceuticals.

Nebel, Bernard J., and Richard T. Wright. *Environmental Science: Towards a Sustainable Future.* 10th ed. Englewood Cliffs: Prentice Hall, 2008. Describes several bioprocesses used in waste treatment and pollution control.

Yang, Shang-Tian. *Bioprocessing for Value-Added Products from Renewable Resources: New Technologies and Applications.* Amsterdam: Elsevier, 2007. Reviews the techniques for producing products through bioprocesses and lists suitable organisms, including bacteria and algae, and describes their characteristics.

Bioremediation

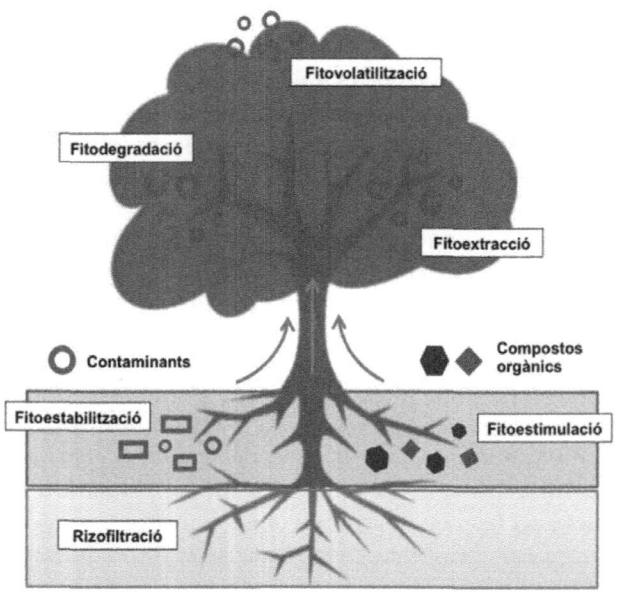
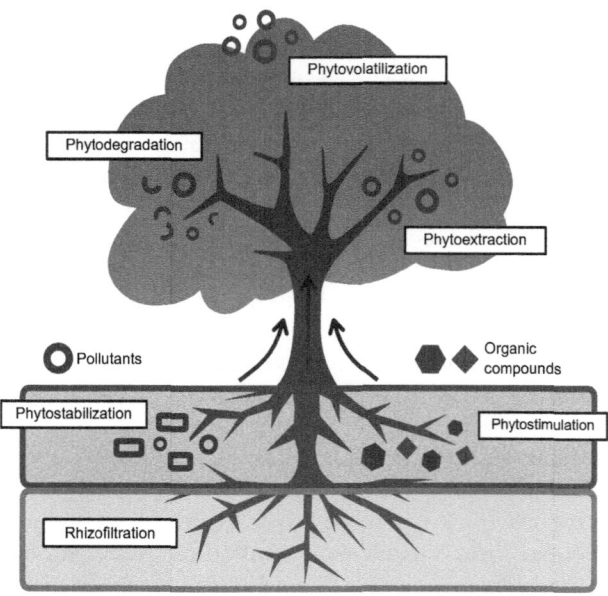

Phytoremediation process. By Townie (Arulnangai & Xavier Dengra. (Own work.) [CC BY-SA 4.0 (https://creativecommons.org/licenses/by-sa/4.0)], via Wikimedia Commons

FIELDS OF STUDY

Biochemistry; Environmental biotechnology (or Green biotechnology); Industrial biotechnology (or White biotechnology); Industrial fermentation

ABSTRACT

The environmentally beneficial and inexpensive waste management strategy of bioremediation enables the degradation of toxic organic and inorganic compounds into environmentally harmless products.

BACKGROUND

Bioremediation uses biological agents to degrade or decompose toxic environmental compounds into less toxic forms. It is a beneficial and inexpensive strategy for waste management that is environmentally friendly in comparison with other remediation technologies. The products of waste decomposition are usually simple inorganic nutrients or gases.

Bioremediation works because, as a general rule, all naturally occurring compounds in the environment are ultimately degraded by biological activity. Toxic and industrial wastes, and even some chemically synthesized compounds that do not naturally occur, can also be decomposed because parts of their structures resemble naturally occurring compounds that are sources of carbon and energy for biological systems. Wastes are either metabolized, in which case they are used as a source of carbon and energy, or cometabolized, and in which case they are simply modified so that they lose their toxicity or are bound to organic material in the environment and rendered unavailable.

Bioremediation can occur in situ (at the contaminated site) or ex situ, in which case contaminated soil or water is removed to a treatment facility where bioremediation takes place under controlled environmental conditions. Bioremediation can use organisms that naturally occur at a site, or it can be stimulated through the addition of organisms, sometimes genetically engineered organisms, to the contaminated site in a process known as "seeding." The first organism ever patented was a genetically engineered bacterium that had been designed to degrade the components of oil.

TECHNIQUES

Numerous approaches to bioremediation have been developed. One of the simplest is to fertilize a contaminated site to optimal nutrient levels and allow naturally occurring biodegrading populations to increase and become active. Organic contaminants have been mixed with decomposed and partially decomposed organic material and composted as a bioremediation process. In a method analogous to the activated sludge process in wastewater treatment, contaminants are mixed in slurries and aerated to promote their decomposition. It is possible to obtain biosolids that are specially adapted for slurry systems because they have previously been exposed to similar organic wastes.

In situ restoration of contaminated groundwater is often accomplished through the injection of nutrients and oxygen into the aquifers to promote the population and activity of indigenous microorganisms. Trichloroethylene (TCE), for example, is cometabolized by methane-oxidizing bacteria and can be bioremediated through the injection of oxygen and methane into contaminated aquifers to stimulate the activity of these bacteria. Nitrate-contaminated aquifers have been successfully treated through the pumping of readily available carbon-containing methanol or ethanol into the aquifers to stimulate denitrifying bacteria, which subsequently convert the nitrate to harmless nitrogen gas.

Bioreactors have been used in which the contaminant is mixed with a solid carrier, or the organisms are immobilized to a solid surface and continuously exposed to the contaminant. This has been used with both bacteria and fungi. For example, *Phanerochaete chysosporium*, which produces an extracellular peroxidase and hydrogen peroxide (H_2O_2), has been used to cleave various organic contaminants such as dichlorodiphenyl-trichloroethane (DDT) in bioreactors.

Highly chlorinated organic contaminants such as TCE and polychlorinated biphenyls (PCBs) resist degradation aerobically, but the contaminants can be dechlorinated by anaerobic bacteria, which decreases their toxicity and makes them easier to decompose. High concentrations of PCBs in the Hudson River in New York have been dechlorinated to less toxic forms by anaerobic bacteria. Methanogens—anaerobic bacteria that produce methane—have been observed to dechlorinate TCE in anaerobic bioreactors.

One of the problems with some wastes is that they are mixed with radioactive materials that are highly toxic to living organisms. One solution to this problem has been the genetic engineering of radiation-resistant bacteria so that they also have the ability to bioremediate. For example, *Deinococcus radiodurans*, a bacterium that can survive in nuclear reactors, has been genetically engineered to contain genes for the metabolism of toluene, which will enable it to be used in the bioremediation of radiation- and organic waste-contaminated sites.

PHYTOREMEDIATION

Phytoremediation is a special type of bioremediation in which plants—grasses, shrubs, trees, and algae—are used to biodegrade or immobilize environmental contaminants, usually metals. Types of phytoremediation include phytoextraction, in which the contaminant is extracted from soil by plant roots; phytostabilization, in which the contaminant is immobilized in the vicinity of plant roots; phytostimulation, in which the plant root exudates stimulate rhizosphere microorganisms that bioremediate the contaminant; phytovolatilization, in which the plant helps to volatilize the contaminant; and phytotransformation, in which the plant root and its enzymes actively transform the contaminant. For example, horseradish peroxidase is a plant enzyme that is used to oxidize and polymerize organic contaminants. The polymerized contaminants become insoluble and relatively unavailable.

Plants such as Indian mustard (*Brassica juncea*) and loco weed (*Astragalus*) are heavy metal accumulators and remove selenium and lead from soil. The aboveground plant parts are harvested to dispose of the metals. Algae are used to accumulate dissolved selenium in some treatments. Poplar trees have even been genetically engineered to contain a bacterial methyl reductase that lets them methylate and volatilize arsenic, mercury, and selenium absorbed by their roots.

EXAMPLES

A 1992 U.S. Environmental Protection Agency (EPA) survey indicated that of 132 well-documented bioremediation studies, 75 involved petroleum or related compounds, 13 involved wood preservatives such as creosote, 7 involved agricultural chemicals, 5 examined tars, 4 treated munitions such as trinitrotoluene (TNT), and the rest involved miscellaneous compounds. As this list suggests, bioremediation of oil spills has been the single best example of successful bioremediation in practice.

In March, 1989, the *Exxon Valdez* oil tanker spilled millions of gallons of crude oil in Prince William Sound, Alaska. On many beaches, the EPA authorized the use of simple bioremediation techniques, such as stimulating the growth of indigenous oil-degrading bacteria by adding common inorganic fertilizers. Beaches cleaned by this method did as well as beaches cleaned by mechanical methods. In another instance of successful bioremediation, selenium-contaminated soil in the Kesterson National Wildlife Refuge in California was partially decontaminated in the 1980s through the method of supplying indigenous fungi with organic substrates such as casein and waste orange peels. This promoted as much as 60 percent selenium volatilization in less than two months.

A great deal of research into biodegradation and bioremediation of oil spills was conducted following the British Petroleum (BP) Deepwater Horizon oil spill in the Gulf of Mexico. Because the oil flowed from an offshore wellhead, affecting deepwater, surface water, and coastal areas alike, the cleanup conditions differed from those experienced in previous spills. Moreover, oil-degrading bacteria were already prevalent at the spill location, feeding on naturally occurring oil seeps along the seafloor. Researchers proposed enhancing anaerobic degradation in affected marshes and introducing genetically modified bacteria. They also learned that the indigenous bacteria were not consuming polycyclic aromatic hydrocarbons (PAHs) and that certain microorganisms naturally inhabit water trapped within oil and feed off it, which may lead to improved methods of bioremediation in future.

——*Mark Coyne*

FURTHER READING

Alexander, Martin. *Biodegradation and Bioremediation.* 2d ed. San Diego, Calif.: Academic Press, 1999.

Atlas, Ronald M., and Jim Philp, eds. *Bioremediation: Applied Microbial Solutions for Real-World Environmental Cleanup.* Washington, D.C.: ASM Press, 2005.

Boopathy, Raj, Sara Shields, and Siva Nunna. "Biodegradation of Crude Oil from the BP Oil Spill in the Marsh Sediments of Southeast Louisiana,

USA." *Applied Biochemistry And Biotechnology* 167, no. 6 (July 2012): 1560–1568.

"A Citizen's Guide to Bioremediation." *EPA*, Sept. 2012. Accessed from https://clu-in.org/download/Citizens/a_citizens_guide_to_bioremediation.pdf.

Fingas, Merv. *The Basics of Oil Spill Cleanup*. 2d ed. Boca Raton, Fla.: CRC Press, 2001.

Frankenberger, William, and Sally Benson, eds. *Selenium in the Environment*. Boca Raton, Fla.: CRC Press, 1994.

Singh, V. P., and R. D. Stapleton, Jr., eds. *Biotransformations: Bioremediation Technology for Health and Environmental Protection*. New York: Elsevier Science, 2002.

Skipper, H. D., and R. F. Turco, eds. *Bioremediation: Science and Applications*. Madison, Wis.: American Society of Agronomy, 1995.

Stallard, Brian. "Oil Eaters: How Nature Cleans Up the Deepwater Horizon Spill." *Nature World News*, August 11, 2014.

Biosensors

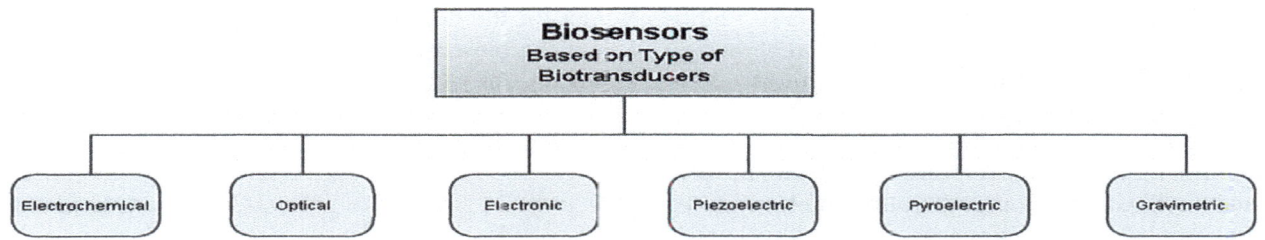

Classification of Biosensors based on type of biotransducer. By Ferbay. (Own work.) [CC-BY-SA-3.0 (http://creativecommons.org/licenses/by-sa/3.0)], via Wikimedia Commons

FIELDS OF STUDY

Biotechnology; Environmental biotechnology (or Green biotechnology); Forensics

ABSTRACT

Biosensors have attracted a lot of interest for their potential in countering the use of chemical and biological weapons by terrorists and for their applications as on-site forensic analytical devices at crime scenes. Biosensors potentially offer sensitive and rapid detection of harmful organisms and substances in food and water supplies. Such instruments have demonstrated usefulness for measuring many substances that are of interest to forensic science, such as toxins, drugs of abuse, poisonous chemicals, and DNA.

BACKGROUND

Biosensor devices differ in the biological components they use for sensing chemicals. Examples are enzymes, antibodies, receptors, and whole cells. The most common biological components used in biosensors are enzymes and antibodies. Different types of biological components result in different types of signals that must be converted into readouts.

Biosensors can be classified according to the ways in which the detection that is mediated by their biological components is converted into measurable signals. After the initial recognition of a chemical species by the biological component, a biosensor generates a readout signal in a process called transduction. At least five different kinds of transducers are used in biosensors: Amperometric transducers involve the movement of electrons resulting from a biorecognition event among three electrodes; potentiometric transducers exploit biological sensor-induced changes in the movement of ions, which results in the generation of an electric potential; thermal transducers utilize heat from biorecognition events that are endothermic reactions; optical transducers make use of the production or absorption of light resulting from biological recognition of detected chemicals or biological molecules; and piezoelectric transducers react to changes in mass produced by biological recognition of target chemicals or biological molecules.

The physical component of a biosensor's transducer, which is in contact with the biological sensor, may comprise electrodes, semiconductors, and optical constructions such as fiber optics and nanoparticles. Most biosensors use electrochemical types of transduction, such as amperometric and potentiometric methods, and enzymatic, antibody, or DNA biological recognition components.

WORKING AND ORGANIZATION

A biosensor contains an external and an internal interface. In the first step, at the external interface of the device, the substance being measured (analyte) binds with the biological recognition component of the biosensor. In the second step, at the internal interface, the biological recognition system interacts with the transducer component, and this produces a physical or chemical response. This response may involve the production of hydrogen ions, other ions, or electrons for amperometric, potentiometric, and conductimetric biosensors. A second type of transducer response may involve the biologically coupled production or absorption of light (fluorescence, chemiluminescence, or visible wavelength). A third type of transducer response would be a change in mass at the transducer such as occurs in piezoelectric (or microelectromechanical) systems. A fourth type of transducer response involves changes in temperature for thermal or calorimetric systems. The physical or chemical response produced by the transducer is processed and amplified to produce a readout signal that serves to indicate the presence and amount of a substance of interest.

APPLICATIONS

Nanotechnology—that is, the application and study of the structuring and behavior of materials at nanometer scale—has also been used in making biosensors. Gold, cadmium selenide, and zinc selenide nanoparticles and single-walled carbon nanotubes are among the nanoscale substances that are being used to make biosensors to detect metal ions, biological molecules, and even viruses such as those responsible for strains of influenza (such as influenza A and the avian flu virus H5N1).

Challenges in the uses of biosensors arise from the need for small, portable devices, the inherent instability of most biological molecules and cells, and the need for highly sensitive devices that can measure a wide range of substances simultaneously. Biosensors used at crime scenes by forensic investigators and in national defense applications must perform reliably and produce quick results under field conditions. In the United States, in addition to their uses by law-enforcement personnel and by national security agencies for the detection and prevention of bioterrorist attacks, biosensors are used for environmental monitoring, for quality control during food processing and the processing of pharmaceuticals, and for monitoring of agriculture.

——*Oluseyi A. Vanderpuye*

FURTHER READING

Cooper, Jon, and Tony Cass, eds. *Biosensors: A Practical Approach.* 2d ed. New York: Oxford University Press, 2004.

Eggins, Brian R. *Biosensors: An Introduction.* New York: John Wiley & Sons, 1996.

Hall, Elizabeth A.H. *Biosensors.* Englewood Cliffs, N.J.: Prentice Hall, 1991.

Karunakaran, Chandran, et al. *Biosensors and Bioelectronics.* Elsevier, 2015.

Kress-Rogers, Erika, ed. *Handbook of Biosensors and Electronic Noses: Medicine, Food, and the Environment.* Boca Raton, Fla.: CRC Press, 1997.

Vigneshvar, S., et al. "Recent Advances in Biosensor Technology for Potential Applications—An Overview." *Frontiers in Bioengineering and Biotechnology,* 16 Feb. 2016, journal.frontiersin.org/article/10.3389/fbioe.2016.00011/full.

BIOSTRATIGRAPHY

Amplexograptus, a graptolite index fossil, from the Ordovician near Caney Springs, Tennessee. By Wilson44691. (Own work.) [Public domain], via Wikimedia Commons

FIELDS OF STUDY

Animal Science; Archaeology; Botany; Environmental biotechnology (or Green biotechnology); Geology

ABSTRACT

Biostratigraphy is that branch of the study of layered rocks—stratigraphy—that focuses on fossils. Its goals are the identification and organization of strata based on their fossil content. Biostratigraphy thus investigates one of the principal bases of the geologic time scale of earth history.

BIOSTRATIGRAPHIC ZONES

Biostratigraphy is the method of identifying and differentiating layers of sedimentary rock (strata) by their fossil content. Strata with distinctive fossil content are termed biostratigraphic units, or zones. Zones vary greatly in thickness and in lateral extent. A zone may be a single layer that is a few centimeters thick and of very local extent, or it may encompass thousands of meters of rocks extending worldwide. The defining feature of a zone is its fossil content; the fossils of a given zone must differ in some specific way from the fossils of other zones.

Zones are usually recognized after fossils have been collected extensively over the lateral and vertical extent of a rock sequence or at many sequences over a broad region. The positions of the fossils in the strata are carefully recorded in the field. Fossils that co-occur in a single layer are noted, as are fossils found isolated in the strata. In the laboratory, the biostratigrapher, usually a paleontologist, then tabulates the vertical and lateral ranges of the fossils collected. It is from these ranges that the paleontologist recognizes zones. Different types of zones are recognized depending on the way in which the fossils in the strata prove to be distinctive. Assemblage zones are strata distinguished by an association (assemblage) of fossils. Thus, not one type but many types of fossils are used to define an assemblage zone. All dinosaur fossils, for example, can be thought of as defining an assemblage zone that encompasses earth history from about 220 to 66 million years ago.

Range zones are strata that encompass the vertical distribution, or range, of a particular type of fossil. Thus, one fossil type—not many—is used to define a range zone. In contrast to the example just given, one type of dinosaur, *Tyrannosaurus rex*, lived only between 68 and 66 million years ago. Its fossils thus define a range zone that corresponds temporally to this 2-million-year interval.

Acme zones are rock layers recognized by the abundance, or acme, of a type (or types) of fossil (or fossils) regardless of association or range. Horned dinosaurs (*Triceratops* and its allies) reached an acme between 70 and 66 million years ago. That is, during this period they were most diverse and most numerous. This acme zone thus overlaps the *Tyrannosaurus rex* range zone and represents a small portion of the dinosaur assemblage zone.

Finally, interval zones are recognized as strata between layers where a significant change in fossil content takes place. For example, the mass extinctions that took place 250 and 66 million years ago bound a 184-million-year-long interval zone that is popularly referred to as the "age of reptiles."

DEVELOPMENT OF BIOSTRATIGRAPHY

Biostratigraphy developed independently in England and France just after 1800. British civil engineer

William Smith worked in land surveying throughout the country. From his vast field experience, he recognized that a given stratum usually contains distinctive fossils and that the fossils (and the stratum) could often be recognized across a large area. Smith's work culminated in his geological map of England (1815), based on his tracing of rock-fossil layers across much of the country.

Meanwhile, in France, Georges Cuvier and Alexandre Brongniart studied the succession of rocks and fossils around Paris. They, too, discovered a definite relationship between strata and fossils and used it to interpret the geological history of the rocks exposed near Paris. In this history, Cuvier saw successive extinctions of many organisms coinciding with remarkable changes in the strata. To him, these represented vast "revolutions" in geological history, which Cuvier argued were of worldwide significance. It is now known that Cuvier was mistaken, but the discovery that a particular fossil type (or types) was confined to a particular stratum became the basis for biostratigraphy. This allowed geologists to identify strata from their fossil content and to trace these strata across broad regions of the earth's crust.

Almost simultaneous with the development of biostratigraphy was the development of biochronology. Biochronology is the recognition of intervals of geologic time by fossils. It stemmed from the realization that during earth's history, different types of organisms lived during different intervals of time. Thus, the fossils of any organism represent a particular interval of geologic time. (Such fossils are called index fossils because they act as an "index" to a geologic time interval.) Biochronology thus identifies intervals of geologic time based on fossils. These time-distinctive fossils are the fossils by which zones are defined, which is to say that each zone represents, or is equivalent to, some interval of geologic time.

The time value of zones made them more useful in tracing strata and deciphering local geological histories. Biostratigraphy now became one of the central methods of stratigraphic correlation. With the aid of fossils, it became possible to determine the ages of strata and thus demonstrate the synchrony or diachrony of these strata in different areas. Through its use in stratigraphic correlation, biostratigraphy became one of the bases for constructing what is called the relative geological time scale of earth history composed of eons, eras, periods, epochs, and ages. This time scale is the "calendar" by which all geologists temporally order their understanding of the history of the earth.

APPLICATION OF BIOSTRATIGRAPHY

Biostratigraphy is generally used as a method of stratigraphic correlation, the process of determining the equivalence of age or stratigraphic position of layered rocks in different areas. Stratigraphic correlation by biostratigraphy is extremely important in deciphering geological history. It reveals the sequence of geological events in one or more regions. Understanding geological history is of interest for its own sake to scientists and laypersons alike. It is crucial to the discovery of mineral deposits and energy resources within the earth's crust. In addition, it provides insight into the biological events that have taken place on this planet for the last 3.9 billion years.

A good example of the use of biostratigraphy in this last regard comes from the study of dinosaur extinction. When dinosaurs were first discovered in England in 1824, and when the term "dinosaur" was coined by the British anatomist Sir Richard Owen in 1841, no one realized that dinosaurs had lived on Earth for only 150 million years and that their extinction had taken place rather rapidly about 66 million years ago. By 1862, however, enough dinosaur fossils had been collected around the globe that a biostratigraphic pattern was beginning to emerge. In that year, the American geologist James Dwight Dana, in his classic *Manual of Geology*, noted that all dinosaurs disappeared before the end of the Mesozoic era, which is now considered as the interval of earth history between 250 and 66 million years ago. This biostratigraphic generalization was possible because geologists noticed that many Mesozoic rocks (but no older or younger rocks) were full of dinosaur fossils, and thus the Mesozoic came to be termed "the age of reptiles." It might just as well be referred to as the "dinosaur zone," except of course for the first 30 or so million years of the Mesozoic, during which dinosaurs apparently did not exist.

More than a century of research has confirmed Dana's biostratigraphic generalization and considerably refined it. Scientists now generally agree that the last dinosaurs disappeared worldwide approximately 66 million years ago. It is also known that dinosaurs first appeared about 220 million years ago. Thus, scientists are able to recognize a dinosaur zone and erect many types of zones based on the ranges and acmes of specific types of dinosaurs. This biostratigraphy of dinosaurs is the basis for informed discussion of the sequence and timing of events during the evolution of the dinosaurs. For example, scientists are now confident that *Stegosaurus* lived long before *Tyrannosaurus* and that stegosaurs as a group of dinosaurs became extinct long before the end of the Mesozoic.

Although discussion here has relied heavily on dinosaurs for examples of biostratigraphy at work, the fossils of these giant reptiles are not ideal for use in biostratigraphy because it is not easy to identify most dinosaur fossils precisely and because most dinosaurs were not animals with broad geographic ranges. Indeed, the fossils of most use in biostratigraphy, index fossils, are those that are easy to identify precisely and that represent organisms that had wide geographic ranges, enjoyed broad environmental tolerances, and lived only for a brief period of geologic time.

Usually an entire skull or skeleton is needed to identify a dinosaur fossil precisely. The isolated bones most often found are not enough, although they do indicate the fossil is that of a dinosaur. Most dinosaurs (there are some notable exceptions) seem to have lived in one portion of one continent; indeed, fossils of the horned dinosaur *Pentaceratops* (a cousin of *Triceratops*) have been found only in New Mexico. There is strong evidence that some dinosaurs preferred coastlines, whereas others preferred dry areas. Thus, many, if not most, dinosaurs did not live in a wide range of environments. Finally, although many dinosaurs apparently lived for only brief intervals of geologic time, the fossil record of most of these giant reptiles is not extensive enough to pin down their exact interval of existence.

The factors that argue against the use of most dinosaur fossils in biostratigraphy are quite different for microscopic fossils of pollen grains and the shelled protozoans known as foraminiferans. These microscopic fossils fit well the four criteria that identify fossils most useful in biostratigraphy. Indeed, such "microfossils" (studied by micropaleontologists) are some of the mainstays of biostratigraphy.

SIGNIFICANCE

Biostratigraphy—the recognition of strata by their fossil content—is a cornerstone of stratigraphic correlation. To identify bodies of rock, the presence of fossils can be traced over broad areas, and their sequence in distant areas can often be determined. Stratigraphic correlation by biostratigraphy is critical to deciphering geological history; without it, the search for mineral deposits and energy resources would be considerably more difficult. Furthermore, understanding the history of geological disasters—earthquakes, volcanic eruptions, meteorite impacts, and the like—and thereby being able to predict future disasters, relies on knowledge of the sequence and timing of geological events, knowledge often derived from biostratigraphy. Deciphering the history of life on this planet, including the myriad appearances, changes, and extinctions of Earth's biota during the last 3.9 billion years, largely depends on the sequence and timing established by biostratigraphy.

Biostratigraphy has also given rise to biochronology, the recognition of intervals of geologic time based on fossils. As a result, scientists have been able to construct a relative global geologic time scale, and it is within the context of this time scale that all geological and biological events in earth history have been placed.

—*Spencer G. Lucas*

FURTHER READING

Ager, Derek V. *The Nature of the Stratigraphical Record.* 3rd ed. Hoboken, N.J. John Wiley & Sons Inc., 1993. A witty and unabashedly curious look at stratigraphy; some of the discussion centers on biostratigraphy. An extensive bibliography, index, and a few well-chosen illustrations illuminate the text.

Barry, W. B. N. *Growth of a Prehistoric Time Scale.* Rev. ed. Palo Alto, Calif.: Blackwell Scientific Publications, 1987. Largely devoted to the history of how the global geologic time scale was formulated, much of this book is a history of biostratigraphy. Well illustrated, with a good bibliography and an index.

Brenner, R. L., and T. R. McHargue. *Integrative Stratigraphy Concepts and Applications.* Englewood Cliffs, N.J.: Prentice-Hall, 1988. Chapter 11 of this college-level textbook provides a detailed look at biostratigraphic concepts, methods, and applications. Well illustrated, with extensive reference lists and an index.

Brookfield, Michael E. *Principles of Stratigraphy.* Hoboken, N.J.: Wiley-Blackwell, 2004. Written for undergraduate students, this text provides an overview of the principles and applications of stratigraphy. Includes stratigraphic techniques and case studies. Organized into three sections, beginning with foundational material, followed by data collection and research topics, and completed with interpretations and analysis.

Dott, Robert H., Jr., and Donald R. Prothero. *Evolution of the Earth.* 8th ed. New York: McGraw-Hill, 2009. This basic textbook on historical geology is aimed at students of geology. However, it is very readable by anyone with a backgroundin science. Presents an up-to-date account of the earth's history from the viewpoint of plate tectonics. Includes a glossary.

Grotzinger, John, and Tom Jordan. *Understanding Earth,* 6th ed. New York: W. H. Freeman, 2009. This comprehensive physical geology text covers the formation and development of the earth. Readable by high school students as well as by general readers. Includes an index and a glossary of terms.

Hedberg, H. D., ed. *International Stratigraphic Guide.* New York: John Wiley & Sons, 1976. The international "rule book" for stratigraphy. It sets procedures and standards to be met when naming stratigraphic units. It also defines many terms used in stratigraphy and has an extensive bibliography. Chapter 6 is devoted to biostratigraphy.

McGowran, Brian. *Biostratigraphy: Microfossils and Geological Time.* New York: Cambridge University Press, 2008. This text expands the application of biostratigraphy from the original use in paleontology to the fields of petroleum exploration and deep-ocean drilling. Addresses the relatively new methodology of studying microfossils and absolute aging.

Ogg, James G., Gabi Ogg, and Felix M. Gradstein. *The Concise Geologic Time Scale.* New York: Cambridge University Press. 2008. This book is a complete overview of the geological time scale, including stratigraphy topics such as chronostratigraphy and magnetic stratigraphy. It is organized by geological time periods. Also contains a reference appendix, the geological time scale table, and indexing.

Prothero, Donald R. *Bringing Fossils to Life.* 2d ed. Boston: McGraw-Hill, 2004. This well-illustrated and entertaining text covers a broad range of paleontological topics, including biostratigraphy. Glossary, bibliography, and index.

Stanley, S. M. *Exploring Earth and Life Through Time.* 2d ed. New York: W. H. Freeman, 1989. An excellent introductory-level college textbook on historical geology. It reviews the history of life and the many fossil forms found in strata in the earth's crust. Chapter 5 includes a discussion of biostratigraphy. Lavishly illustrated, with extensive references, glossaries, appendices on fossil groups, and an index.

Winchester, Simon. *The Map That Changed the World.* London: Penguin, 2001. An entertaining study of William Smith, who developed the idea of biostratigraphy and used it to produce the first geologic map of Great Britain. Written for the interested reader but valuable to students and professionals also.

BIOSYNTHETICS

FIELDS OF STUDY

Biochemistry; Bioengineering; Biopharmaceutics; Environmental biotechnology (or Green biotechnology); Food science; Medical biotechnology (or Red biotechnology)

ABSTRACT

Biosynthesis is the process of using small, simple molecules to make larger, more complex molecules, either inside the body or in the laboratory. Numerous applications for drug development and medicine include the synthesis of proteins, hormones, dietary supplements, blood products, and surgical dressings for wounds. Additional techniques to facilitate the diagnosis and treatment of disease include protein biomarkers for immune assays, the development of proteomics to analyze changes in proteins in response to a drug, the development of polyclonal and monoclonal antibodies, immunizations, and various drug delivery systems.

BACKGROUND

In general, the term "biosynthetic" refers to any type of material produced via a biosynthetic process. A biosynthetic process uses enzymes and energetic molecules to transform small molecules into larger molecules within the cells of organisms. The two types of metabolites produced from cellular biosynthetic pathways include the primary metabolites of fatty acids and DNA needed by cells and the secondary metabolites of pheromones, antibiotics, and vitamins that assist the entire organism. Additional small molecules, such as adenosine triphosphate (ATP), provide the energetic driving force for the biosynthetic pathways, and other small molecules, including enzymes, further facilitate the reactions in these pathways. Thus, there have been many possibilities for numerous types of scientists, including chemists, biochemists, biologists, and geneticists, to create innovations.

The term "biosynthetic" differs from the term "chemosynthetic," because chemosynthetic indicates the production of materials that cannot take place within a living organism. Scientists generally begin the process of developing a new medical application or dietary supplement by first isolating and characterizing the DNA of the proteins or other small molecules directly involved in the biological process. They then try to duplicate this naturally occurring biological process to produce massive quantities of the desired material, and ultimately they combine these naturally occurring processes with chemicals that can mimic the process during laboratory manufacturing processes.

HISTORY

The biochemical pharmacologist Hermann Karl Felix "Hugh" Blaschko was a trailblazer whose discoveries in the 1930's initiated the field of biosynthetics. His work elucidated the biosynthetic pathway for adrenaline, which is often called the fight-and-flight hormone, and encompassed the study of the enzymes important for regulation of this hormone. This work led the way toward the development of syntheses using amino acids for therapeutic applications.

Another key development in biosynthetics was the discovery of the role of the amino acid L-arginine in the synthesis of creatine, an important biomolecule, by G. L. Foster, Rudolf Schoenheimer, and D. Rittenberg in 1939. Since that time, L-arginine has also been shown to be a precursor to nitrous oxide and nitric oxide, as well as a component of the urea cycle, which is important for ammonia regulation and thus influences the operation of the kidneys and other organs. Nitric oxide is important in the regulation of blood flow to muscles. These discoveries involving L-arginine have led to dietary supplements useful to bodybuilders who wish to enhance their weight-lifting performance.

Throughout the 1940's, 1950's, and 1960's, progress was made toward understanding the genetic composition of organisms, enzymes, and biosynthetic pathways. Researchers made contributions to understanding pyrimidine, galactosidase, *Escherichia coli*, and chlorophyll. Practical biosynthetic applications that were made possible by these fundamental discoveries began to manifest themselves throughout the 1970's, 1980's, and 1990's, with the development of surgical dressings, therapeutic hormones, and plant supplements for increased nutritional value.

HOW IT WORKS

Often the isolation and characterization of a specific gene responsible for producing an important enzyme or other small molecule is the first step in a lengthy process toward synthesis of a product that undergoes lengthy clinical trials before the final, approved product is ready for manufacture. Once the gene has been characterized, its DNA is further characterized to facilitate the process of peptide synthesis (the process of producing long peptides is known as protein biosynthesis).

The process of peptide synthesis involves the general concepts of antigenicity, hydrophilicity, and surface probability, as well as flexibility indexes. The process involves an analysis of the peptide's characteristics, the use of software and databases to determine hydrophilicity (affinity for water), study of the antigenicity (capacity to stimulate the production of antibodies) to assist with antibody production, the study of surface probability (which determines the likelihood of inducing the formation of antibodies), the determination of the protein sequence, phosphorylation (process that activates or deactivates many protein enzymes), and then selection of two to three peptides, followed by comparison of their homology (similarity of structure).

In a general process called screening, the efficacy of an antibiotic is first tested using bacterial cultures, followed by injection of the antibiotic into laboratory animals, such as rats, rabbits, or guinea pigs; then clinical trials are conducted according to protocols established by the Food and Drug Administration (FDA). Combinatorial chemistry, a faster screening method, is often used instead. FDA-approved products are then manufactured on a larger scale.

Antibody Production. The application of a binding assay is used for isolation of the purified protein that is to be the source of an antigen. This antigen is then used as a conjugate to a carrier protein, such as kehole limpet hemocyanin (KLH), to produce a target peptide with a length of thirteen to twenty amino acids to stimulate the immune system. A carrier protein is a membrane protein that can bind to a substance to facilitate the substance's passive transport into a cell. Injection into a laboratory animal occurs next, and then the animals undergo a series of four to six immunizations separated by about twenty days. Enzyme-linked immunosorbent assay (ELISA) is used to detect antibodies. ELISA is based on the antibody-antigen binding interaction and often uses color to visually indicate the concentration of antibodies. Purification of antibodies obtained from the antiserum for specific antigen binding completes the antibody production process.

Antigen Preparation. This process is facilitated through bioinformatics analysis to choose the appropriate two to three peptides based on the protein sequence provided by a customer. KLH conjugation used for immunization, and bovine serum albumin (BSA) conjugation is carried out for screening. After immunization protocols and specific antibodies have been selected during fusion and screening, a cell can be cryopreserved.

Combinatorial Chemistry. In combinatorial chemistry synthesis, a high-throughput screening method, the starting small molecule is attached to a type of polymeric resin, followed by different permutations of reagents, to produce large libraries containing hundreds of unique products that can be rapidly screened for enzymatic activity, specific antigen-binding, or protein-protein interactions. Often the process is controlled by a computer and completed through the application of robotics. A customer can specify antigen details, and a pharmaceutical company can design a protocol involving the general phases of preparation of antigen, immunization, fusion and screening of assays, and finally selection, purification, and production of antibodies.

APPLICATIONS AND PRODUCTS

Biosensors. Biosensors are microelectronic devices that use antibodies, enzymes, or other biological molecules to interact with an optical device or electrode to record data electronically. These devices can be operated by home health care providers to transmit data obtained from blood or urine samples, for example, to a clinical laboratory some distance away.

Therapeutic Proteins. Plasmids are used to transfer human genes that provide the code for proteins important for growth hormones, blood clotting, and insulin production to bacterial cells.

Disposable Micropumps for Drug Delivery. Disposable micropumps manufactured by Acuros in Germany are capable of delivering a preset amount of liquid hormones, proteins, antibodies, or other medications. An osmotic microactuator, based on osmotic pressure, is used to regulate the amount of drug delivered, and there are no moving parts or power supply components.

High-Throughput Screening. High-throughput screening can assay more than twenty thousand potentially useful drugs per week by using multiwell plates, standard binding assay methodologies, and robotics.

Protein Biomarker Assays. NextGen Sciences has developed a mass spectrometry method for protein biomarker assays that does not depend on antibodies but instead uses surrogate proteins to facilitate development of assays. The mass spectrometer measures the amount of surrogate peptides and applies statistical evaluation to assess each biomarker. This first stage requires that a protein be confirmed; then only these selected proteins are used for the second stage of validation of these protein biomarkers. The mass spectrometry data are used along with carbon-13 or nitrogen-15 isotopically labeled standards to calculate protein concentrations. Reporting the protein biomarkers in terms of concentration is important to allow batches containing hundreds of samples to be analyzed and validated. This technique uses proteomics (the quantitative analysis of proteins based on a physiological response) to allow for much faster development of assays than immunoassays. A wide range of at least 500 plasma proteins and 3,000 tissue proteins can be analyzed at once.

Gene Expression Databases. Gene Logic's BioExpress System is a comprehensive genome-wide gene expression database. The BioExpress System allows cells from a patient to be collected and analyzed to develop a useful biomarker profile for comparison with a database sample to indicate a therapeutic target. This process is made possible by the use of high-throughput gene expression profiling of the mononuclear cell fractions present in a blood sample. The software is capable of mining a database that has access to more than 18,000 samples containing biomarkers for the expression of the gene associated with ovarian cancer. This system is also capable of developing biomarker profiles to help diagnosis autoimmune diseases. Autoimmune diseases include rheumatoid arthritis, Crohn's disease, multiple sclerosis, systemic lupus erythematosus, and psoriasis, which affect about 20 million people in the United States.

Biosynthetic Temporary Skin Substitute. A biosynthetic skin substitute is a useful treatment for partial-thickness wounds, including skin tears, burns, and abrasions. After applying a gel to the surface of the wound, a semipermeable membrane of biosynthetic skin is used to cover the wound for protection from infection. Before the development of biosynthetic skin grafts, a physician had to choose between an allograft, which uses cadaver skin, and a xenograft, which uses tissue from another species. Biosynthetic dressings have also been developed. The dressing called Hydrofiber contains ionic silver and has been shown to prevent the spread of bacteria.

Needle-Free Drug Delivery Systems. The three types of needle-free drug delivery systems are liquid, powder, and depot injections. Each of these types uses some form of mechanical compression to create enough pressure to force the medication into the skin. Although these needle-free delivery systems cost more initially and require more technical expertise because of their complexity, they also have many advantages. In addition to eliminating pain from needle injections and reducing physician visits, these needle-free delivery systems decrease the frequency of incorrect doses. They are being used to deliver anesthetics, chemotherapy injections, vaccines, and hormones.

Nanoparticles. DNA nanotechnology uses discoveries involving nanoparticles and nanomaterials to manipulate DNA's molecular recognition abilities to build tiny medical robots that mimic bond parts or function within cells.

SOCIAL CONTEXT AND FUTURE PROSPECTS

The Human Genome Project has facilitated the mapping of genes, which has been instrumental to the development of vaccines to treat influenza, cervical cancer, and malaria, as well as the creation of new diagnostic tools for analysis. As a result, the pharmaceutical industry in the United States has become a multibillion-dollar industry. The generation of biosynthetic products has enhanced the lives of thousands of people through the development of treatments for many types of cancer, pneumonia, cardiovascular diseases, diabetes, tuberculosis, neurological disorders, strokes, blood disorders, and many other diseases.

Combinatorial chemistry has allowed for rapid screening of potentially successful medications that may enhance and extend the lives of many people. Normally, only one out of every 5,000 to 10,000 compounds screened makes it through the multiyear process of clinical trials to become an FDA-approved drug. However, the desire to recoup the money spent during the years of research required to bring a drug to market has caused some pharmaceutical companies to launch a product as early as possible, which has resulted in serious litigation because some drugs proved to have harmful side effects. The application of biosynthetic growth hormones for nonmedical applications, such as bodybuilding, has also caused ethical and medical controversy. However, as the global population continues to grow and the percentage of elderly persons increases, the need for the products of biosynthetic research will continue to grow.

—*Jeanne L. Kuhler, PhD*

FURTHER READING

Arya, Dev. *Aminoglycoside Antibiotics: From Chemical Biology to Drug Discovery.* New York: Wiley-Interscience, 2007. Describes the design and synthesis of antibiotics and the process of antibiotic resistance.

Dewick, Paul. *Medicinal Natural Products: A Biosynthetic Approach.* New York: John Wiley & Sons, 2009. Comprehensive textbook describing biosynthetic methods and processes, including new techniques in genetic engineering and isolation of genes.

Lazo, John, and Peter Wipf. "Combinatorial Chemistry and Contemporary Pharmacology." *The Journal of Pharmacology and Experimental Therapeutics* 293, no. 3 (February, 2000): 705-709. Describes the process of combinatorial chemistry. Includes experimental strategies and flow charts describing the screening of compounds.

Pettit, George. *Biosynthetic Products for Cancer Chemotherapy.* Vol. 5 London: Elsevier Science, 1985. A discussion of the fundamental processes involved with screening for antitumor agents.

Savageau, Michael. *Biochemical Systems Analysis: A Study of Function and Design in Molecular Biology.* New York: CreateSpace, 2010. Detailed textbook describing the immune system and gene regulation.

Spentzos, Dimitri. "Gene Expression Signature with Independent Prognostic Significance in Epithelial Ovarian Cancer." *Journal of Clinical Oncology* 22, no. 23 (December, 2004): 4648-4658. The research article describes the diagnosis of ovarian cancer and the use of biomarkers for detection.

Stanforth, Stephen. *Natural Product Chemistry at a Glance.* New York: Wiley-Blackwell, 2006. An introductory textbook that describes much of the organic chemistry involved in biosynthesis.

BIOTECHNOLOGY AND GENETIC ENGINEERING

Genetic engineering, the manipulation of genetic material, can be used to synthesize large quantities of drugs or hormones, such as insulin. © EBSCO. EBSCO

FIELDS OF STUDY

Biotechnology; Forensics; Genetics; Environmental biotechnology (or Green biotechnology); Industrial biotechnology (or White biotechnology); Marine biology (or Blue biotechnology); Medical biotechnology (or Red biotechnology)

ABSTRACT

Biotechnology has made tremendous advances possible in human and veterinary medicine, agriculture, food production, and other fields. However, debates continue regarding the potential of biotechnology, in particular genetic engineering, to produce organisms that may disrupt ecosystems, negatively affect human health, or be used in ethically inappropriate ways.

HISTORY

The term "biotechnology" is relatively new, but the practice of biotechnology is as old as civilization. Civilization did not evolve until humans learned to produce food crops and domestic livestock through the controlled breeding of selected plants and animals. The pace of modifying organisms accelerated during the twentieth century. Through carefully controlled breeding programs, plant architecture and fruit characteristics of crops were modified to facilitate mechanical harvesting. Plants were developed to produce specific drugs or spices, and microorganisms were selected to produce antibiotics such as penicillin and other useful medicinal and food products.

The ability to utilize artificial media to propagate plants led to the development of a technology called tissue culture. In some plant tissue culture, the tissue is treated with the proper plant hormones to produce masses of undifferentiated cells called callus tissue, which can also be separated into single cells to establish a cell suspension culture. Specific drugs or other chemicals can be produced with callus tissue and cell suspensions, or this tissue can be used to regenerate entire plants. Tissue culture technology is used as a propagation tool in commercial-scale plant production.

Numerous advances have also occurred in animal biotechnology. Artificial insemination, the process in which semen is collected from the male animal and deposited into the female reproductive tract through artificial techniques rather than natural mating, emerged as a practical procedure roughly a century ago, although as early as 1784 Italian biologist Lazzaro Spallanzani successfully inseminated a dog. Males in species such as cattle can sire hundreds of thousands of offspring through artificial insemination, whereas they could sire only fifty or fewer through natural means.

Embryo transfer is a technique used in humans to facilitate conception after in vitro fertilization, a procedure in which eggs are surgically removed from the ovaries and manually combined with sperm in a laboratory. Once fertilization and cell division are confirmed, the embryos are placed in the uterus. The eggs may be supplied by a woman who is unable to conceive naturally but who can carry a child to term. They may also be provided by an egg donor to a woman who cannot otherwise get pregnant; or a woman who cannot carry a child to term may supply eggs to be fertilized in vitro and implanted in a surrogate mother. Superovulation is the process in which females that are to provide eggs are injected with hormones to stimulate increased egg production. Embryo splitting is the mechanical division of an embryo into identical twins, quadruplets, sextuplets, and so on. Both superovulation and embryo splitting have made routine embryo transfers possible. In livestock, embryo transfer technology is used to combine the sperm from a superior male animal and several eggs, each of which can then be split into several offspring, from a superior female. The resulting embryos can then be transferred to the reproductive tracts of inferior surrogate females.

RECOMBINANT DNA TECHNOLOGY

Biotechnological advances have enabled scientists to tap into the world gene pool. This technology has great potential, and its full magnitude is far from being fully realized. Theoretically, it is possible to transfer one or more genes or gene segments from any organism in the world into any other organism. Because genes ultimately control how an organism functions, gene transfer can have a dramatic impact on agricultural resources and human health.

Research has provided the means by which genes can be identified and manipulated at the molecular and cellular levels. This identification and manipulation depend primarily on recombinant DNA technology. In concept, recombinant DNA methodology is fairly easy to comprehend, but in practice it is rather complex. The genes in all living cells are very similar in that they are all composed of the same chemical, deoxyribonucleic acid, or DNA. The DNA of all cells, whether from bacteria, plants, lower animals, or humans, is very similar, and when DNA from a foreign species is transferred into a different cell, it functions exactly as the native DNA functions; that is, it codes for protein.

The simplest protocol for this transfer involves the use of a vector, usually a piece of circular DNA called a plasmid, which is removed from a microorganism such as a bacterium and cut open by an enzyme called a restriction endonuclease or restriction enzyme. A section of DNA from the donor cell that contains a previously identified gene of interest is cut out from the donor cell DNA by the same restriction endonuclease. The section of donor cell DNA with the gene of interest is then combined with the open plasmid DNA, and the plasmid closes with the new gene as part of its structure. The recombinant plasmid (DNA from two sources) is placed back into the bacterium, where it will replicate and code for protein just as it did in the donor cell. The bacterium can be cultured and the gene product (protein) harvested, or the bacterium can be used as a vector to transfer the gene to another species, where it will also be expressed. This transfer of genes, and therefore of inherited traits, between different species has revolutionized biotechnology and provides the potential for genetic changes in plants and animals that have not yet been envisioned.

BIOTECHNOLOGY AND AGRICULTURE

Biotechnology has had a tremendous impact on agriculture. Traditional breeding programs may be too slow to keep pace with the needs of a rapidly expanding human population. Biotechnology provides a means of developing higher-yielding crops in one-third of the time it takes to develop them though traditional plant breeding programs because the genes for desired characteristics can be inserted directly into a plant without having to go through several generations to establish the trait. Also, there is often a need or desire to diversify agricultural production in a given area, but soil or climate conditions may severely limit the amount of diversification that can take place. Biotechnology can provide the tools to help solve this problem: Crops with high cash value can be developed to grow in areas that would not support unmodified versions of such crops. In addition, biotechnology can be used to increase the cash value of crops, as plants can be developed that can produce new and novel products such as antibiotics, hormones, and other pharmaceuticals.

As public pressure has grown for crop production to be friendlier to the environment, biotechnology has been touted as an important tool for the development of a long-term, sustainable, environmentally friendly agricultural system. Biotechnology is already being used to develop crops with improved resistance to pests. For example, a gene from the bacterium *Bacillus thuringiensis* (*B.t.*) codes for an insecticidal protein that kills insects but is harmless to other organisms. When this gene is transferred from the bacterium to a plant, insect larvae are killed if they eat from the leaves or roots of the plant. A number of *B.t.* plants have been developed, including cotton and potatoes. Crop varieties engineered for improved pest resistance have the potential to reduce reliance on pesticides; however, insect pests have developed resistance to some of these crops.

Biotechnology also plays an important role in the livestock industry. Bovine somatotropin, a hormone that stimulates growth in cattle, is harvested from recombinant bacteria and injected in dairy cattle to enhance milk production. However, questions have arisen as to whether overstimulating milk production is humane or healthy for cows, and fears regarding the health implications for humans consuming milk that contains bovine hormone residues have made many people seek organic dairy products free from artificial hormones. Some countries do not allow the use of these hormones in milk intended for human consumption.

Researchers are exploring the possibilities of genetically engineering animals that can resist disease or produce novel and interesting products such as pharmaceuticals. The cloning of Dolly the sheep in Scotland in 1996 opened a whole new avenue in the use of biotechnology for livestock production. The use of cloning technology in conjunction with surrogate mothers provides the means to produce a whole herd of genetically superior animals in a short period of time. However, reproductive cloning is expensive, its success rate is low, and many cloned animals have been found to be unhealthy and short-lived.

BIOTECHNOLOGY AND MEDICINE

DNA technology also has a direct impact on human health and is used to manufacture a variety of gene products that are utilized in the clinical treatment of diseases. Several human hormones produced by this methodology are already in use. The hormone insulin, for instance, which is used to treat insulin-dependent diabetics, was the first major success in using a product of recombinant technology. Recombinant DNA-produced insulin has been used to treat diabetic patients since 1982. Genetic engineering has also been used to synthesize protropin, a human growth hormone (HGH) employed in the treatment of growth failure conditions such as hyposomatotropism. Without treatment with HGH, people suffering from these conditions do not produce enough growth hormone to achieve a typical adult height.

Somatostatin, another pituitary hormone, has also been produced through recombinant DNA techniques. This hormone controls the release of insulin and HGH. Small proteins called interferons normally produced by cells to combat viral infections have been produced using recombinant DNA methodology, as have some vaccines against viral diseases. Recombivax HB, the first of these vaccines, is used in vaccinating against hepatitis B, an incurable and sometimes fatal liver disease.

The potential for the future application of gene therapy has also been enhanced by advances in biotechnology. Among the forms of gene therapy currently being considered are gene surgery, in which a mutant gene that may or may not be replaced by its normal counterpart is excised from the DNA; gene repair, in which defective DNA is repaired within the cell to restore the genetic code; and gene insertion, in which a normal gene complement is inserted in cells that carry a defective gene.

Gene surgery and gene repair techniques are extremely complex and remain in the experimental stages. Gene insertion can potentially be done in germ-line cells such as the egg or sperm, the fertilized ovum or zygote, the fetus, or the somatic cells (nonreproductive cells) of children or adults. Although zygote therapy holds the most promise, as this technique could eliminate genetic disease, gene insertion in zygotes also represents a means by which traits such as strength or intelligence might be enhanced and the genetic traits of future generations artificially selected, a possibility that raises a host of ethical questions. Germ-line genetic modification has been performed successfully in laboratory animals, but unwanted mutations with serious or lethal consequences have also sometimes resulted.

Gene insertion into somatic cells does not make changes that are passed on to subsequent generations, so it does not present the ethical dilemma that germ-line manipulation does. In this technique, a gene or gene segment is inserted into specific organs or tissues as a treatment for an existing condition. In human clinical trials, somatic gene therapy has shown success in treating advanced melanoma, myeloid disorders, inherited childhood blindness, and severe combined immunodeficiency. However, the carrier molecules used to deliver the therapeutic gene to the target cells have the potential to provoke a serious or fatal immune response in the patient.

ENVIRONMENTAL ISSUES

The potential benefits of biotechnology for human health, agriculture, and the environment are accompanied by potential drawbacks. Since the first recombinant DNA experiments in 1973, numerous social, ethical, and scientific questions have been raised about the possible detrimental effects of genetically engineered organisms on public health and the environment. The major environmental concerns are related to containment, or how to prevent genetically engineered organisms from escaping into the environment.

In the mid-1970's U.S. scientists invoked a self-imposed moratorium on genetic engineering experiments until the government could establish committees to develop safety guidelines that would apply to all recombinant DNA experimentation in the United States. This resulted in the formulation of guidelines specifying the degree of containment required for various types of genetic engineering experiments. Two types of containment, biological and physical, are addressed by the guidelines. "Physical containment" refers to the methods required to prevent an engineered organism from escaping from the laboratory; "biological containment" refers to the techniques used to ensure that an engineered organism cannot survive outside the laboratory. The guidelines associated with containment, particularly physical containment, are sometimes difficult to monitor and enforce.

Some observers have noted that despite the rigors of the containment guidelines, the possibility remains that an engineered organism will eventually escape into the environment. Should this occur, the organism could cause environmental damage as great as or greater than that caused in the past by the introduction of foreign species to new habitats. For example, the introduction of rabbits to Australia dramatically upset the ecological balance on that continent. Hence field experiments with genetically engineered organisms must be strenuously controlled and monitored.

Although numerous safe field trials have been conducted with genetically engineered organisms, such as *B.t.* plants, widespread opposition to such practices remains. There appear to be few risks that cannot be ascertained within the laboratory associated with the release of genetically engineered higher plants, but opponents have expressed the fear that engineered genes could possibly be transferred by cross-pollination to other species of plants. Such a transfer could, for example, produce a highly vigorous species of weed. In addition, such gene transfers could potentially result in a plant that produces a toxin that would be detrimental to other plants, animals, or humans.

Because viruses and bacteria are major components of numerous natural biochemical cycles and readily exchange genetic information in a variety of ways, it is even more difficult to envision all the ramifications associated with releasing these genetically altered organisms into the environment. Field testing of genetically engineered organisms will always involve some element of risk, and assessment of the risks of such testing is easier for some species, such as higher plants, than for other species, such as bacteria.

A clear need exists for rigid controls, and minimizing the risks also requires integral cooperation among industry, governments, and regulatory organizations. Under the Cartagena Protocol on Biosafety, which entered into force in 2003, before an importing nation may release living modified organisms (LMOs) into the environment, the country into which the LMO is to be imported must first give its informed consent. The importer must clearly identify the LMO, detail its traits and characteristics, and explain its proper handling, storage, transport, and use.

With advances in the cloning of plants and animals, environmentalists and others have expressed concerns about losses in genetic variability. In nature, species survival is dependent on the genetic variability, or diversity, of the population. Genetic variability obtained through normal sexual reproduction provides a species with the ability to adapt to changes in the environment; because the environment is continually changing, loss of genetic variability usually leads to extinction of the species. Because cloning results in genetically identical individuals, the cloning of large numbers of animals or plants of particular species at the expense of those produced through sexual reproduction can lead to the loss of genetic variability and thus to eventual extinction of those species.

—*D. R. Gossett, updated by Karen N. Kähler*

FURTHER READING

Chrispeels, Maarten J., and David E. Sadava, eds. *Plants, Genes, and Crop Biotechnology.* 2d ed. Sudbury, Mass.: Jones and Bartlett, 2003.

Drlica, Karl. *Understanding DNA and Gene Cloning: A Guide for the Curious.* 4th ed. Hoboken, N.J.: John Wiley & Sons, 2004.

Field, Thomas G., and Robert E. Taylor. *Scientific Farm Animal Production: An Introduction to Animal Science.* 10th ed. Upper Saddle River, N.J.: Prentice Hall, 2011.

Grace, Eric S. *Biotechnology Unzipped: Promises and Realities.* Rev. 2d ed. Washington, D.C.: National Academies Press, 2006.

Groves, M. J., ed. *Pharmaceutical Biotechnology.* 2d ed. Boca Raton, Fla.: Taylor & Francis, 2006.

Hill, Walter E. *Genetic Engineering: A Primer.* London: Taylor & Francis, 2002.

BIOTOXINS

FIELDS OF STUDY

Biology; Medical biotechnology (or Red biotechnology); Microbiology; Molecular biology; Molecular ecology

ABSTRACT

Biocrimes present law-enforcement agencies with serious challenges, as the perpetrators of such crimes can use numerous pathogens that exist naturally and do not require sophisticated expertise or technology to prepare. Further, because the effects of biotoxins are as diverse as the substances' multiple origins, it can be difficult for investigators to ascertain the types of biotoxins employed in particular crimes or terrorist attacks.

BACKGROUND

The use of biological agents and their toxins in criminal acts and as weapons of war has a long history. In the Far East, opium was the poison used for murder and suicide for several centuries. In the fourteenth century, Mongol warriors used plague-infected bodies as weapons of war, triggering an outbreak that killed thousands. During the French and Indian War (1754-1763), the British approved a plan to distribute to Native American tribes blankets contaminated with smallpox. These examples, however, pale

In the 1790s, British physician Edward Jenner developed a vaccine from material taken from infected cows that protected people against the biotoxin smallpox. This 1802 cartoon shows a public hospital in which Jenner is vaccinating a frightened woman as cows emerge from the bodies of people already vaccinated. (Library of Congress) [Public domain], via Wikimedia Commons

in comparison with the chilling prospects of modern bioterrorism aided by a rapidly expanding knowledge of biological agents, biotoxins, and their potential to wreak havoc in complex, interdependent societies.

COMMON MICROBIAL AGENTS

Various microbes can be the sources of biotoxins, including viruses, bacteria, and fungi. It is relatively easy to propagate bacteria and fungi with small samples, but the propagation of viruses for use in biocrimes requires certain training and access to specific technologies. Some of the common viruses that produce devastating effects include smallpox, Ebola, and Marburg. Smallpox, a highly contagious virus, is transmitted easily and carries a high mortality rate. By the 1970's, a worldwide vaccination program had eradicated smallpox. Three decades later, only two places in the world still officially maintained live cultures of the virus: a laboratory of the Centers for Disease Control and Prevention (CDC) in the United States and a lab in Russia. The Ebola and Marburg viruses are also extremely lethal; both cause hemorrhagic fever and profuse bleeding from bodily orifices. No cure or effective treatments for either virus have yet been found.

Bacterial biotoxins include anthrax, botulism, plague, and tularemia. *Bacillus anthracis*, the bacterium that causes anthrax, produces spores that are extremely resistant to the environment and are highly infectious when inhaled. Botulism is caused by a potent neurotoxin produced by the bacterium *Clostridium botulinum*. Once inhaled or ingested, the toxin causes respiratory failure and paralysis. Plague is also highly contagious; it causes a type of pneumonia and can be fatal if not treated early. *Francisella tularensis* causes tularemia, a generally nonlethal disease that is extremely incapacitating; symptoms include weight loss, fever, and headaches.

Many fungi produce remarkable amounts of toxic secondary metabolites, some of which are toxins. Fungal toxins are grouped into two categories: mycotoxins, which are produced by common molds, and mushroom toxins, which are formed in the fleshy fruiting bodies of sac or club fungi. Mycotoxins are major contributing factors to many cases of food poisoning. Some mycotoxins, such as aflatoxins, are believed to be among the most potent carcinogens. Ingestion of even minute amounts of aflatoxins over long periods of time through contaminated food can cause liver cancer. In 1974, hundreds of people were poisoned by aflatoxin-contaminated corn in India; more than one hundred died. Several members of the mushroom genus *Amanita* contain amanitin, one of the deadliest poisons found in nature. The poison contained in false morels, monomethyl hydrazine (MMH), can cause diarrhea, vomiting, and severe headaches; ingestion of this poison occasionally results in death.

MARINE AND PLANT BIOTOXINS

Many plants produce poisonous secondary metabolites that induce toxic effects when the plants or their extracts are consumed. Although sensitivity to plant toxins may vary among individuals, a good correlation generally exists between the amount of poison ingested and the severity of the clinical symptoms. Some highly toxic substances derived from plants include ricin (derived from castor beans), aconitine (from monkshood), strychnine (from the vomit nut), and huratoxin (from jimsonweed, also known as thorn apple). Ricin has been employed as a murder weapon in many cultures. In South America, native tribes have long used various plants to prepare curare, a common name for a deadly poison used on the tips of arrows or darts.

Harmful algal blooms represent a real threat to virtually all U.S. coastal and fresh waters. Potential impacts range from devastating economic effects to public health risks to ecosystem alterations. The phenomena commonly known as "red tides" produce extremely potent biotoxins. When such toxins accumulate in marine food chains, they cause mass mortalities of birds, fish, and marine mammals and often lead to closures of commercial and recreational fisheries. When humans accidentally consume seafood contaminated with algal toxins, illness develops and even death occurs in extreme cases. Two classes of algal toxins have been well studied: the paralytic shellfish poisoning (PSP) toxins and domoic acid, both of which act on nerve systems.

MICROBIAL FORENSICS

Criminal investigations involving biotoxins rely on forensic scientists who work in the cross-discipline known as microbial forensics. It can be challenging at times to distinguish symptoms and signs that may be caused by toxins from those that are just variants of normal health. Physicians and forensic scientists may not be able to recognize early symptoms associated with particular pathogens or biotoxins. Often, the identification of particular biotoxins requires the careful study of highly skilled professionals using sophisticated analytical instruments. Furthermore, confirmation of the presence of biological agents or toxins in evidence samples is generally not enough to guarantee conviction of a suspect without other supporting evidence.

—Ming Y. Zheng

FURTHER READING
Beasley, Val Richard, et al. "Diagnostic and Clinically Important Aspects of Cyanobacterial (Blue-Green Algae) Toxicoses." *Journal of Veterinary Diagnostic Investigation* 1 (October, 1989): 359-365. Scholarly article focuses on the diagnosis of biotoxins in animals.
Breeze, Roger G., Bruce Budowle, and Steven E. Schutzer, eds. *Microbial Forensics*. Burlington, Mass.: Elsevier Academic Press, 2005. Reviews the relationships between microbe physiology and forensics.
Cooper, Marion R., Anthony W. Johnson, and Elizabeth A. Dauncey. *Poisonous Plants and Fungi: An Illustrated Guide.* 2d ed. London: TSO, 2003. Comprehensive volume describes the many varieties of poisonous plants and fungi.
Garrett, Laurie. *The Coming Plague: Newly Emerging Diseases in a World Out of Balance.* New York: Farrar, Straus and Giroux, 1994. Discusses the increase in outbreaks of infectious diseases in the late twentieth century as well as ways to prevent such outbreaks.
Nelson, Lewis S., Richard D. Shih, and Michael J. Balick. *Handbook of Poisonous and Injurious Plants.* 2d ed. New York: Springer, 2007. Provides useful information on many different plant biotoxins.

BOTANY AND GENETIC ENGINEERING

FIELDS OF STUDY

Agricultural engineering; Agronomy; Biology; Botany; Genetics; Food science; Horticulture; Human nutrition; Plant pathology

ABSTRACT

Any topic dealing with plants, from the level of their cellular biology to the level of their economic production, is considered part of the field of botany.

BACKGROUND

The origins of botany, beginning around 5000 BCE, are rooted in human attempts to improve their lot by raising better food crops. This practical effort developed into intellectual curiosity about plants in general, and the science of botany was born. Some of the earliest botanical records are included with the writings of Greek philosophers, who were often physicians and who used plant materials as curative agents. In the second century BCE, Aristotle had a botanical garden and an associated library.

As more details became known about plants and their function, particularly after the discovery of the microscope, the growing body of knowledge became too great for general understanding, so a number of subdisciplines arose. Plant anatomy is concerned chiefly with the internal structure of plants. Plant physiology delves into the living functions of plants. Plant taxonomy has as its interest the discovery and

Picture of genetically modified corn in Yellow Springs, Ohio. By Lindsay Eyink from San Francisco, CA, USA. (Research field.) [CC-BY-2.0 (http://creativecommons.org/licenses/by/2.0)], via Wikimedia Commons

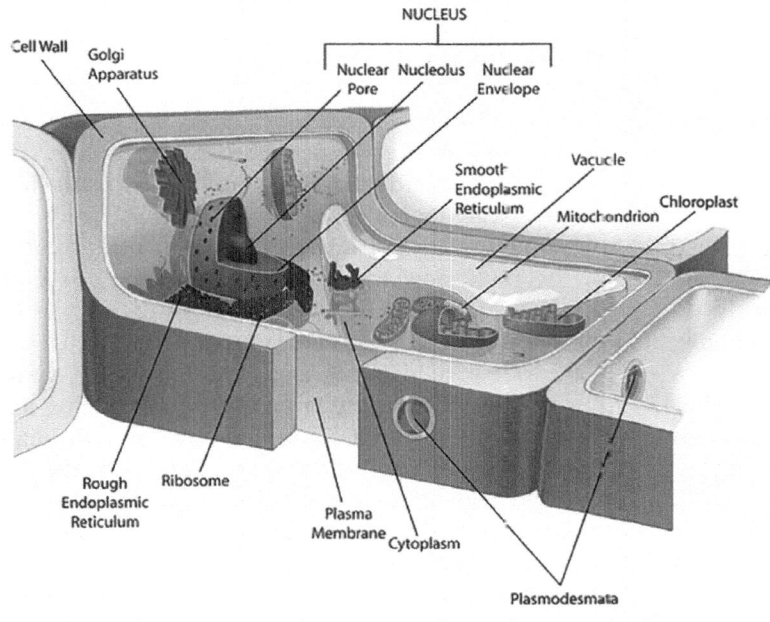

Plant Cell: The plant cell includes chloroplasts, several vacuoles, plasmodesmata, and cell wall, which the animal cell does not have. © EBSCO. EBSCO

systematic classification of plants. Plant geography deals with the global distribution of plants. Plant ecology studies the interactions between plants and their surroundings. Plant morphology studies the form and structure of plants. Plant genetics attempts to understand and work with the way that plant traits are inherited. Plant cytology, often called cell biology, is the science of cell structure and function. Economic botany, which traces its interest back to the origins of botany, studies those plants that play important economic roles (these include major crops such as wheat, rice, corn, and cotton). Ethnobotany is a rapidly developing subarea in which scientists communicate with indigenous peoples to explore the knowledge that exists as a part of their folk medicine. Several new drugs and the promise of others have developed from this search.

At the forefront of modern botany is the field of genetic engineering, including the cloning of organisms. New or better crops have long been developed by the technique of cross-breeding, but genetic engineering offers a much more direct course. Using its techniques scientists can introduce a gene carrying a desirable trait directly from one organism to another. In this way scientists hope to protect crops from frost damage, to inhibit the growth of weeds, to provide insect repulsion as a part of the plant's own system, and to increase the yield of food and fiber crops.

The role that plants play in the energy system of the Earth (and may someday play in space stations or other closed systems) is also a major area of study. Plants, through photosynthesis, convert sunlight into other useful forms of energy upon which humans have become dependent. During the same process carbon dioxide is removed from the air, and oxygen is delivered. Optimization of this process and discovering new applications for it are goals for botanists.

——*Kenneth H. Brown*

Botulinum toxin as a biological weapon

A lab technician with the Centers for Disease Control and Prevention grinds food with a mortar and pestle to enable the extraction of botulinum toxin. The CDC treats every case of food-borne botulism as a public health emergency. (Centers for Disease Control and Prevention.) [Public domain], via Wikimedia Commons

FIELDS OF STUDY

Biology; Medical biotechnology (or Red biotechnology); Microbiology; Molecular biology; Molecular ecology

ABSTRACT

Botulinum toxin is one of the most lethal known toxic substances; a few grams of the toxin introduced into the food supply could kill millions of people, making it an attractive agent for potential use as a biological weapon. In addition to that possibility, nonintentional poisonings sometimes occur through the consumption of food containing the toxin or through contamination of wounds with the toxin. Whenever botulinum toxin is suspected in cases of poisoning, law-enforcement agencies are concerned with identifying the toxin and its source.

BACKGROUND

Although the possibility that botulinum toxin could be used in biological warfare has been acknowledged for many years, no uses of the poison as a weapon have been reported in any major wars. Despite the Biological Weapons Convention of 1972, however, it is generally believed that many countries have stockpiles of the *Clostridium botulinum* bacterium and toxin as part of their biological warfare programs.

The most common form of botulinum poisoning occurs through the ingestion of foods containing the toxin. Food products contaminated with *C. botulinum* spores that are stored at room temperature can cause poisoning if they are consumed without first being adequately heated. Canned cheeses, ham, and sausage are common sources of the toxin. In a typical incident that took place in Italy in 1996, eight people contracted the poison by eating commercial cream cheese. One died, and the others had prolonged medical recoveries. In a 1995 incident in Canada, a sixteen-year-old girl was poisoned when she ate smoked fish. She died a few months later despite having received intensive medical treatment. In September 2006, four cases of botulism in the United States and two cases in Canada were traced to the consumption of contaminated carrot juice.

MECHANISM OF TOXICITY

The toxin, which was first isolated from *C. botulinum* in 1944 by Edward Schantz, must come into contact with nerve tissue to cause damage. The toxin attaches to the axon terminal of nerve endings, where it blocks the release of the principal neurotransmitter in the body, acetylcholine. This blockage prevents transmission of nerve impulses, resulting in loss of muscle contractility and flaccid paralysis.

In food-related poisoning, symptoms occur six to thirty-six hours after ingestion of food containing the toxin. Symptoms include excessive dry mouth, diarrhea, and vomiting. These may be followed by blurred vision, droopy eyelids, generalized muscle weakness, and progressive difficulty in breathing. Death may occur as a result of paralysis of the respiratory muscles. Symptoms of botulinum poisoning may occur more rapidly if the toxin is inhaled rather than ingested.

MEDICAL AND COSMETIC USES

Some medical treatments have been developed that take advantage of the botulinum toxin's neuromuscular blocking action; tiny concentrations of the toxin are used, for example, in the treatment of involuntary eye muscle contractions (blepharospasm). The toxin is also used in the treatment of migraine headaches and cervical dystonia, a neuromuscular condition involving the head and neck. Another important medical use of the toxin is in the treatment of excessive underarm perspiration (severe primary axillary hyperhidrosis). The toxin has also been employed at times in the treatment of the following ailments and symptoms, although it is not approved by the U.S. Food and Drug Administration (FDA) for these uses: overactive bladder, anal fissure, stroke, multiple sclerosis, Parkinson's disease, excessive salivation, neurological complications of diabetes mellitus, and muscle problems affecting the limbs, face, jaw, and vocal cords.

Commercial botulinum toxins, marketed under the names Botox and Dysport, among others, are used cosmetically to remove facial wrinkles and improve facial appearance. The toxin works on wrinkle lines that have been formed in the upper part of the face, particularly the forehead and around the eyes. Because very low concentrations of the toxin are used in these cosmetic preparations, treatment is usually safe. However, occasional adverse effects, such as allergic reactions and paralysis of the wrong muscles, have been reported. Four cases of poisoning caused by cosmetic use of a type of botulinum toxin that had not been approved by the FDA were reported in Florida in 2004.

INVESTIGATION OF BOTULINUM POISONING

When deaths or illnesses are suspected to be attributable to botulinum toxin poisoning, both forensic scientists and public health experts are usually involved in investigating the incidents. The immediate goal in any case is to identify the source of the toxin as quickly as possible to prevent any further harm. In the United States, law-enforcement agencies are required to report all cases of such poisoning to the Centers for Disease Control and Prevention (CDC).

Evidence at the suspected poisoning site must be preserved so that it can be analyzed for clues that may point to the source of the toxin. Apart from food, botulinum toxin and the toxin-producing *C. botulinum* bacterium may be found in the blood and feces of patients suffering from botulinum poisoning. In some fatal cases, forensic examination of tissue samples and suspensions of body fluids have been used to demonstrate the presence of the toxin even after advanced putrefaction.

———*Edward C. Nwanegbo*

FURTHER READING

Balkin, Karen F., ed. *Food-Borne Illnesses*. San Diego, Calif.: Greenhaven Press, 2004.

Breeze, Roger G., Bruce Budowle, and Steven E. Schutzer, eds. *Microbial Forensics*. Burlington, Mass.: Elsevier Academic Press, 2005.

Scott, Elizabeth, and Paul Sockett. *How to Prevent Food Poisoning: A Practical Guide to Safe Cooking, Eating, and Food Handling*. Hoboken, N.J.: John Wiley & Sons, 1998. Provides thorough information on food poisoning's causes and symptoms. Includes a chapter devoted to the science of food poisoning.

Smith, Louis D. S., and Hiroshi Sugiyama. *Botulism: The Organism, Its Toxins, the Disease*. 2d ed. Springfield, Ill.: Charles C Thomas, 1988.

Tucker, Jonathan B., ed. *Toxic Terror: Assessing Terrorist Use of Chemical and Biological Weapons*. Cambridge, Mass.: MIT Press, 2000.

BUBONIC PLAGUE AS A BIOLOGICAL WEAPON

FIELDS OF STUDY

Biology; Medical biotechnology (or Red biotechnology); Microbiology; Molecular biology; Molecular ecology

ABSTRACT

Natural outbreaks of bubonic plague still occur periodically, with an average of 18 cases in the United States and 1,666 cases worldwide per year. A larger cause for concern however, is the possibility that weaponized plague bacteria could be used in biological terrorism.

BACKGROUND

Bubonic plague is caused by a gram-negative, facultative anaerobe bacterial species, *Yersinia pestis*, acting as an intracellular parasite. The disease is transmitted

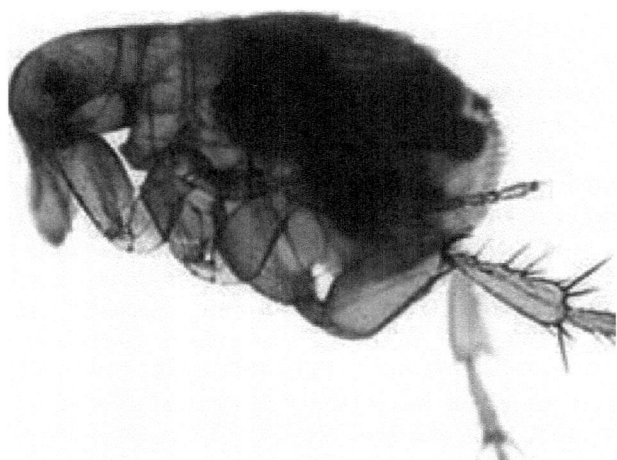

Scanning electron micrograph depicting a mass of *Yersinia pestis* bacteria (the cause of bubonic plague) in the foregut of the flea vector [Public domain], via Wikimedia Commons

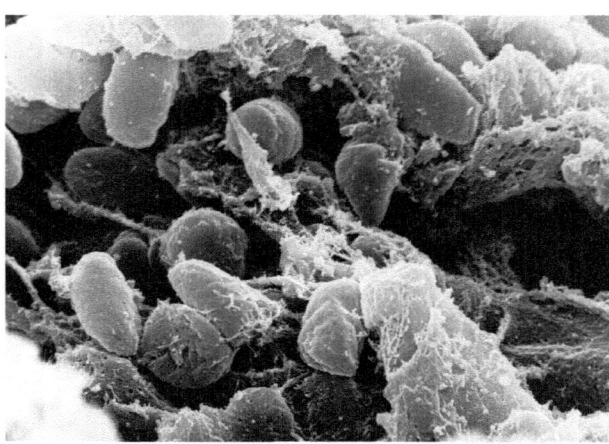

Oriental rat flea (*Xenopsylla cheopis*) infected with the *Yersinia pestis* bacterium which appears as a dark mass in the gut. The foregut of this flea is blocked by a *Y. pestis* biofilm; when the flea attempts to feed on an uninfected host, *Y. pestis* from the foregut is regurgitated into the wound, causing infection. By National Institute of Allergies and Infectious Diseases. [Public domain], via Wikimedia Commons

primarily by fleas from infected hosts, including more than two hundred species of rodents as well as domestic cats, dogs, rabbits, and even sheep or camels. Transmission may also occur through contact with infected bodily fluids or tissues as well as through aerosol exposure from a coughing patient. The bubonic plague is also known as the Black Death because it results in buboes, infected and inflamed lymph nodes that turn black as they become necrotic and hemorrhagic.

Three forms of plague are known. The skin form of the disease, bubonic plague, has a mortality rate of 50-90 percent if untreated and up to 15 percent if treated. A second form, pneumonic plague, results when the bacteria invade the lungs. Pneumonic plague is especially virulent, with mortality of 100 percent if not treated within twenty-four hours. Moreover, it causes bronchial pneumonia, which leads to coughing of highly infective aerosols of bacteria. The third form of plague is septicemic plague, in which blood-borne bacteria are widespread throughout the body, invading almost all organs. Septicemic plague is 100 percent fatal if untreated, and some 40 percent of those who contract it die even with treatment. Incubation time for plague before symptoms appear is one to six days.

Symptoms of bubonic plague include fever (as high at 105 degrees Fahrenheit), chills, muscular pain, sore throat, headache, severe weakness, extreme malaise, and enlarged, painful lymph nodes especially in the groin, armpits, and neck. In later stages, accelerated heart rate, accelerated breathing, and low blood pressure ensue. The normal course of treatment is antibiotics of the tetracycline or sulfonamide families. A vaccine does exist, but it is no longer available in the United States; it is used to contain local outbreaks in other parts of the world.

Because of the highly contagious nature of *Y. pestis*, this organism poses a grave danger as an agent in a biological terrorism attack. Aerosolized plague organisms as well as antibiotic-resistant strains of plague have been developed in former biological weapons facilities in Russia and the United States. Rapid identification of the agent is essential in any bioterrorism event.

—*Ralph R. Meyer*

FURTHER READING

Brubaker, Bob. "*Yersinia pestis* and the Bubonic Plague." In *The Prokaryotes*, edited by Martin Dworkin et al. 3d ed. Vol. 6. New York: Springer, 2006.

Orent, Wendy. *Plague: The Mysterious Past and Terrifying Future of the World's Most Dangerous Disease.* New York: Free Press, 2004.

Parker, Philip M., and James N. Parker. *Bubonic Plague: A Medical Dictionary, Bibliography, and Annotated Research Guide to Internet References.* San Diego, Calif.: ICON Health Publications, 2003.

Cell and tissue engineering

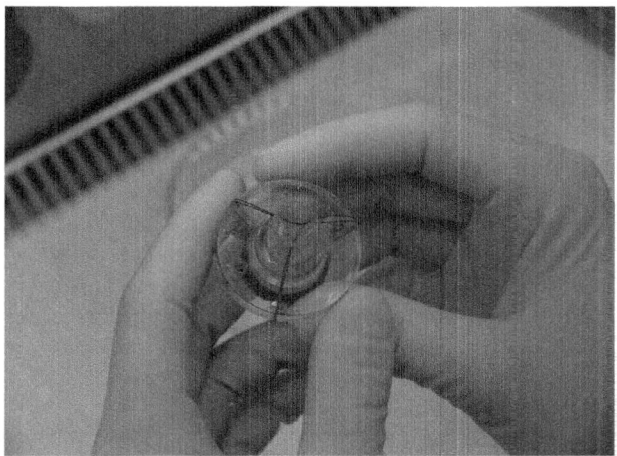

Tissue engineered heart valve. By HIA. (Own work.) [CC BY-SA 3.0 (https://creativecommons.org/licenses/by-sa/3.0)], via Wikimedia Commons

FIELDS OF STUDY

Biology; Genetics; Medical biotechnology (or Red biotechnology); Microbiology; Molecular biology

ABSTRACT

Cell and tissue engineering are fields dedicated to discovering the mechanisms that underlie cellular function and organization to develop biological or hybrid biological and nonbiological substitutes to restore or improve cellular tissues. The most immediate goal of cell and tissue engineering is to allow physicians to replace damaged or failing tissues within the body. The field was first recognized as a distinct branch of bioengineering in the 1980s and has since grown to attract participation from numerous medical and biological disciplines.

Engineered cellular materials may be used to grow new tissue within a patient's heart or to replace damaged bone, cartilage, or other tissues. In addition, research into the mechanisms affecting cellular organization and development may aid in the treatment of congenital and developmental disorders. Cell and tissue engineering has developed in conjunction with stem cell research and is therefore subject to debate over the ethics of stem cell research.

DEFINITION AND BASIC PRINCIPLES

cell and tissue engineering: A branch of bioengineering concerned with two basic goals: studying and understanding the processes that control and contribute to cell and tissue organization and developing substitutes to replace or improve existing tissues in an organism. Substitute tissues can be composed either of biological materials or of a blend of biological and nonbiological materials.

BACKGROUND

The basic goal of cell and tissue engineering is to create more effective treatments for tissue degeneration and damage resulting from congenital disorders, disease, and injury. Engineers may, for instance introduce foreign tissues that have been modified to stimulate healing within the patient's own tissues, or they may implant synthetic structures that help control and stimulate cellular development. Another goal in cell and tissue engineering is to create tissues that are resistant to rejection from the host organism's immune system. Rejection is one of the primary difficulties in organ transplantation and limb replacement surgery.

One of the basic principles of cell and tissue engineering is to use and enhance an organism's innate regenerative capacity. Engineers therefore examine the ways that tissues grow and change during development. Using cutting-edge development in genomics and gene therapy, engineers are working to develop ways to stimulate a patient's immune system and enhance healing.

Cell and tissue engineering have a wide variety of potential applications. In addition to creating new

A rose plant that began as cells grown in a tissue culture. By Scott Bauer. [Public domain], via Wikimedia Commons

therapies, engineering principles can be used to create new methods for delivering drugs and engineered cells to target locations within a patient. The potential applications of cell and tissue engineering depend on the capability to create cultures of cells and tissues to use for experimentation and transplantation. Research on cell growth is a major facet of the bioengineering field.

HISTORY

Cell and tissue engineering emerged from a field of study known as regenerative medicine, a branch concerned with developing and using methods to enhance the regenerative properties of tissues involved in the healing process. Ultimately, cell and tissue engineering became most closely associated with transplant medicine and surgery.

Medical historians have found documents from as early as 1825 recording the successful transplantation of skin. The first complete organ transplants occurred in the 1950s, and the first heart transplant was completed successfully in 1964.

The science of cell and tissue engineering arose from attempts to combat the problems that affect transplantation, including scarcity of organs and frequent issues involving rejection by the host's immune system. In the 1970s and 1980s, scientists began working on ways to build artificial or semi-artificial substitutes for organ transplants. Most early work in tissue engineering involved the search for a suitable artificial substitute for skin grafts.

By the mid-1980s, physicians were using semi-synthetic compounds to anchor and guide transplanted tissues. The first symposium for tissue engineering was held in 1988, by which time the field had adherents around the world. The rapid advance of research into the human genome and genetic medicine in the mid-1990s had a considerable effect on bioengineering. In the twenty-first century, cell and tissue engineers work closely with genetic engineers in an effort to create new and better tissue substitutes.

HOW IT WORKS

Broadly speaking, cell and tissue engineering involves creating cell cultures and tissues that are introduced to an organism to repair damaged or degenerated tissues. There are a wide variety of techniques and specific applications for cell and tissue engineering, ranging from cellular manipulation at the chemical or genetic level to the creation of artificial organs for transplant.

Most cell and tissue engineering methods share several common procedures. First, scientists must produce cells or tissues. Next, engineers must tell the cells what to do. This can be done in a variety of ways, from physically manipulating cellular development and tissue formation to altering the genes of cells in such a way as to direct their function. Finally, engineered tissues and cells must be integrated into the body of the host organism under controlled conditions to limit the potential for rejection. Cell and tissue engineering can be divided into two main categories, in vitro engineering and in vivo engineering.

In Vitro Engineering. In vitro engineering is the development of cell cultures and tissues outside of the body in a controlled laboratory environment. This method has several advantages. Producing tissues in a laboratory has the potential for growing large amounts of tissue and eventually entire organs. This could help solve a major issue with transplant surgery: the scarcity of viable organs for transplantation. Scientists can more precisely control the growing environment and can therefore exert greater control over developing cells and tissues. In vitro engineering allows engineers to modify and adjust cellular properties without the need for surgery or invasive techniques.

In vitro engineering is commonly used in the creation of skin tissues, cartilage, and some bone replacement tissues. Although in vitro techniques have certain advantages, they have serious drawbacks, including a higher rejection rate for cells and tissues created in vitro. In addition, there are physiological advantages to engineering within the host organism's body, including the presence of accessible cellular nutrients.

In Vivo Engineering. In vivo engineering is the family of techniques that involves creating engineered cellular cultures or tissues within the host's body. It involves the use of chemicals to alter cellular function and the use of synthetic materials that interact with the host's body to stimulate or direct cellular growth.

In vivo procedures typically involve introducing only minor changes to the host's internal environment, and therefore, these tissues are more likely to be resistant to rejection. In addition, working in vivo allows engineers to take full advantage of the host's existing cellular networks and the physiological environment of the body. The body provides the essential nutrients, exchange of materials, and disposal of waste, helping create healthy tissues.

The primary disadvantages of the in vivo approach are that engineers have less direct control over the development of the cells and tissues and cannot make exact changes to the microenvironment during development. In addition, in vivo engineering does not allow for the production of mass quantities of cells and is therefore not an avenue toward addressing the shortage of available tissues and organs for transplant.

APPLICATIONS AND PRODUCTS

Hundreds of bioengineers are working around the world, and they have created a wide variety of applications using cell and tissue engineering research. Among the most promising applications are cell matrices and bioartificial organ assistance devices.

Cell Matrices. In an effort to improve the success of tissue transplants, bioengineers have developed a method for using artificial matrices, also called "scaffolds," to control and direct the growth of new tissues. Using cutting-edge microengineering techniques and materials, engineers create three-dimensional structures that are implanted into an organism and thereafter serve as a "guide" for developing tissues.

The scaffold acts like an extracellular matrix that anchors growing cells. New cells anchor to the artificial matrix rather than to the organism's own extracellular material, allowing engineers to exert control over the eventual size, shape, and function of the new tissue. In addition, scaffolds can aid in the diffusion of resources within the growing tissue and can help engineers direct the placement of functional cells, as the scaffold can be installed directly at the site of an injury.

Matrices may be constructed from a variety of materials, including entirely synthetic combinations of polymers and other structures that are created from derivatives of the extracellular matrix. Many researchers have been designing scaffolds that dissolve as the tissues form and are then absorbed into the organism. These biodegradable scaffolds allow engineers to avoid further surgical procedures to remove implanted material

Cellular scaffolds represent a middle ground between in vivo and in vitro engineering. Engineers can create a scaffold in a laboratory environment and can allow tissue to anchor and grow around the matrix before implantation, or they can place a scaffold in their target area within the organism and allow the organism's own cells to populate the matrix.

Scaffolds have been used successfully in cardiac repair, especially in conjunction with stem cells. A scaffold seeded with stem cells may be implanted directly into a heart valve, roughly at the site where a cardiac infarction has occurred. The scaffold then directs the growing cells toward the injured area and facilitates regeneration of damaged tissue.

Artificial matrices have also been successful in treating disorders that affect the kidney, bone, and cartilage. Researchers are hopeful that cellular scaffolds could eventually allow the creation of entire organs by coaxing cells to develop around a scaffold designed as an organ template.

Bioartificial Organs. One of the major areas of research in tissue engineering is the creation of machines that assist organs damaged by disease or injury. Made from a combination of synthetic and organic materials, these machines are sometimes called bioartificial devices.

One of the most promising organ assistance devices is the bioartificial liver (BAL), which has been developed to help patients suffering from congenital liver disease, acute liver failure, and other metabolic disorders affecting the liver. The BAL consists of cells incorporated into a bioreactor, which is a small machine that provides an environment conducive to biological processes. Cells growing within the BAL receive optimal nutrients and are exposed to hormones and growth factors to stimulate development. The bioreactor is also designed to facilitate the delivery of any chemicals produced by the developing tissues to surrounding areas.

The BAL performs some of the functions usually performed by the liver: It processes blood, removes impurities, produces proteins, and aids in the synthesis of digestive enzymes. The BAL is not intended to permanently replace the liver but rather to supplement liver function or to allow a patient to survive until a liver transplant can be arranged. The bioartificial liver enables patients to forgo dialysis treatments, and some researchers hope to develop BAL devices that may function as a permanent replacement for patients in need of dialysis.

Researchers are working on bioartificial kidney devices that would aid patients with diabetes and other disorders leading to kidney failure. Again, the bioartificial kidney devices are bioreactors, using stem cells and kidney cells to perform some of the purification and detoxification functions of the kidney. Researchers are also developing bioartificial devices to treat disorders of the pancreas and the heart and to help patients suffering from nervous system or circulatory disorders. Taken as a whole, the development of organ assistance devices may be a step toward the development of bioartificial devices that can function to fully replace a patient's malfunctioning organ.

SOCIAL CONTEXT AND FUTURE PROSPECTS

Bioengineering is intended to improve daily life, both for those suffering from injury and illness and for the population at large. Cell and tissue engineers are focusing on ways to replace damaged tissues, providing, for instance, new skin where skin has been destroyed, and technology to supplement the function of essential organs. One of the ultimate goals of the industry is to create artificial organs that can fully and permanently replace damaged organs. Bioengineers are confident that in the future it will be possible to provide patients with a variety of organs including a heart, liver, or pancreas.

Although most cell and tissue engineers focus on combating physical illness and injury, bioengineering also has the potential to produce technology that will allow humans to improve their functional abilities. At some point, combinations of synthetic computer technology and biological components could be used to improve human visual capacity or to endow humans with more precise access to memory.

Humans are not the only targets for bioengineers, as other organisms may also be altered to improve their basic physiological functions. Take, for instance, a 2008 project from the Australian Center for Plant Functional Genomics in which researchers are attempting to bioengineer plants that can withstand higher levels of salt in the soil, a breakthrough that could turn into a major benefit for agriculture. Salt-resistant strains of important agricultural crops could grow where agriculture was previously impossible because of the soil's alkalinity.

As a distinct discipline, bioengineering is relatively new and scientists have only begun to investigate the potential applications and discoveries possible with further research. As the field has begun to expand, so too have opportunities for scientists, engineers, and physicians interested in exploring the future of medicine and science. The bioengineering field has already created billions in revenue and is still in a state of rapid growth. Universities, hospitals, and biomedical corporations are likely to increase their investment in these emerging technologies and techniques, creating a strong and growing industry for many years to come.

—*Micah L. Issitt*

FURTHER READING

Chien, Shu, Peter C. Y. Chen, and Y. C. Fung, eds. *An Introductory Text to Bioengineering.* Hackensack, N.J.: World Scientific Publishing, 2008.

De Gray, Aubrey, and Michael Rae. *Ending Aging: The Rejuvenation Breakthroughs That Could Reverse Aging in Our Lifetime.* New York: St. Martin's Griffin, 2008.

Mataigne, Fen. *Medicine by Design: The Practice and Promise of Biomedical Engineering.* Baltimore: The Johns Hopkins University Press, 2006.

Rose, Nickolas. *The Politics of Life Itself: Biomedicine, Power, and Subjectivity in the Twenty-first Century.* Princeton, N.J.: Princeton University Press, 2006.

Valentinuzzi, Max. *Understanding the Human Machine: A Primer for Bioengineering.* Hackensack, N.J.: World Scientific Publishing, 2004.

Zenios, Stefanos, Josh Makower, and Paul Yock, eds. *Biodesign: The Process of Innovating Medical Technologies.* New York: Cambridge University Press, 2010.

CLONING

A vineyard in the Napa Valley showing which particular clone of Cabernet Sauvignon is planted in this block. By star5112. (Flickr: JOH_6393.) [CC BY-SA 2.0 (https://creativecommons.org/licenses/by-sa/2.0)], via Wikimedia Commons

FIELDS OF STUDY

Biology; Genetics; Medical biotechnology (or Red biotechnology); Philosophy and Religious Studies; Political Science; Reproductive technology

ABSTRACT

Cloning is any type of biological reproduction that produces offspring that are genetically identical to their parents. Cloning occurs naturally, since many organisms routinely reproduce through natural cloning processes. Artificial cloning technologies include molecular cloning, which reproduces large quantities of discrete segments of DNA; reproductive cloning, which uses assisted reproductive technologies to produce animals that share the same desirable genetic characteristics as another living or previously existing organism; and therapeutic cloning, which uses the same techniques as reproductive cloning but instead derives useful cell lines from cloned embryos.

DEFINITION AND BASIC PRINCIPLES

Cloning is a means of producing biological organisms, cells, or DNA molecules that are genetically identical to their progenitors. There are natural forms of cloning and three main types of artificial cloning: molecular, reproductive, and therapeutic cloning.

Natural mechanisms of cloning occur in organisms such as bacteria that simply split or fragment into identical copies of themselves. In other organisms, reproductive cells, or gametes, undergo a process called parthenogenesis, in which they initiate development without the benefit of fertilization. Cloning is uncommon in mammals, but rarely, early mammalian embryos undergo a form of cloning called twinning, in which the embryo splits into two embryos, which develop into genetically identical twins.

Molecular cloning, also known as recombinant DNA technology or DNA cloning, involves the transfer of an isolated fragment of DNA from an organism of interest to a host cell that replicates it. Such isolated DNA fragments are known as cloned DNA or genes.

Reproductive cloning uses assisted reproductive technologies to generate animals with the same nuclear genome as another animal. The particular procedure used during reproductive cloning is called somatic cell nuclear transfer (SCNT). Cloned embryos are gestated in the womb of a surrogate mother until they come to term. Cloned organisms are not genetically modified organisms but are simply produced through a type of assisted reproduction.

Therapeutic cloning uses the same procedures as reproductive cloning; however, instead of transferring the cloned embryo into the womb of a surrogate mother, the embryo is further manipulated in the laboratory to make cell cultures of embryonic cells for basic or clinical research.

BACKGROUND

Sea urchins were the first animal cloned in the laboratory. In 1894, Hans Dreisch isolated sea urchin embryo cells and watched them develop into small, separate larvae. In 1902, Hans Spemann used the same procedure, embryo splitting, to isolate cells from salamander embryos, which also developed into identical adult salamanders. In 1903, U.S. Department of Agriculture employee Herbert Webber coined the word "clon" for asexually produced cells or organisms, which later evolved into "clone." This term comes from the Greek *klon*, which means "trunk" or "branch." Horticulturists have used this term for more than a century, since an entire new plant can grow from a cutting, resulting in a plant that is genetically identical to the plant from which the cutting was taken.

In 1928, Spemann cloned salamanders by transferring the nucleus, the subcellular compartment that houses the chromosomes, from one salamander embryo into the egg of another. Since Spemann's seminal experiments, scientists have adapted nuclear transfer technology to clone other organisms. In 1952, frogs were cloned, and in 1963, the Chinese embryologist Tong Dizhou cloned a carp to produce the first cloned fish. During the 1980's and 1990's, sheep, cows, and mice were cloned. However, all these animals were cloned by using nuclei from embryos. In 1996, Ian Wilmut and his team at the Roslin Institute in Edinburgh, Scotland, cloned a sheep from an adult cell, demonstrating that adult cells could serve as the source of genetic material for animal clones. This technological feat was followed by the cloning of goats, mules, gaurs (an endangered species), horses, pigs, mouflons (a wild sheep), mice, rats, dogs, cats, water buffalos, camels, rabbits, deer, wolves, and African wildcats, and even embryos from nonhuman primates and humans.

HOW IT WORKS

Molecular Cloning. To clone a gene, the DNA of the model organism is selectively fragmented by enzymes called restriction endonucleases (REs) and inserted into another piece of DNA called a cloning vector. Cloning vectors are either small circles of DNA called plasmids, bacterial viruses, or bacterial or yeast artificial chromosomes. They ferry the DNA fragments from the genome of the model organism into a host cell (either a bacterium or yeast). This population of host cells collectively carries the entire genome of the model organism in small fragments, and is called a gene library.

To isolate a gene from a gene library requires a probe, which is a fragment of DNA or RNA of any length that has a sequence that is complementary to the sequence of the gene that is to be isolated. Probes can be made synthetically or can come from the genes of closely related organisms. By screening the gene library with the probe, the gene of interest is cloned, which simply means to isolate it from all the other sequences found in the genome of the model organism.

Alternatively, scientists can synthesize small strands of DNA called primers, whose sequences are complementary to different locations in the gene. These primers can be used to specifically amplify the gene from the library by means of a polymerase chain reaction (PCR). A polymerase chain reaction makes large quantities of the gene of interest from a very small amount of starting material, and the amplified DNA can also be cloned into a cloning vector or analyzed directly.

Reproductive Cloning. To clone an animal, mature eggs are isolated from females of the animal species that is to be cloned. The egg is enucleated by piercing it with a microscopically narrow (0.0002-inch-wide) glass tube that is used to vacuum out the egg nucleus. The enucleated egg is fused with a cell from the body of the animal to be cloned and activated with either chemicals or an electric current. This procedure is called somatic cell nuclear transplantation (SCNT).

After activation, the egg divides and grows like a newly formed embryo. However, if the animal is a mammal, the embryo can survive only for a limited period of time before it must implant into the inner layer of the mother's womb. Therefore, a surrogate female from the same species of the animal to be cloned, or a closely related species, is made pseudopregnant by feeding her hormones, and the embryo is released into her receptive womb, where it implants. Barring any technical or biological mishap, the cloned embryo will develop, and the process will result in a live birth.

Therapeutic Cloning. To make embryonic cell cultures, cloned embryos are made by means of somatic cell nuclear transplantation. They are then either disassembled in the laboratory and used to establish embryonic cell cultures or gestated in a surrogate mother to the fetal stage, at which time the fetus is aborted, and cells from the fetus are used to establish fetal cell cultures.

By culturing specific cells from cloned embryos, scientists can make embryonic stem cell (ESC) cultures. During mammalian development, two distinct cell populations form after the first few days of embryonic development. The trophoblast, or the flattened, outer layer of cells, will eventually form the placenta and its associated structures. The inner cell mass (ICM) is the round, inner clump of cells that develop to form the embryo proper and a few structures associated with the placenta. If ICM cells are isolated and cultured on feeder cells, a layer of nondividing skin cells that secrete a cocktail of growth-promoting chemicals, the ICM cells will grow and spread over the surface of the culture dish. Such a culture is an embryonic stem cell culture, and these cells are pluripotent, which means that they can differentiate into any cell type in the adult body.

APPLICATIONS AND PRODUCTS

Molecular Cloning. Organisms that express cloned genes make many useful pharmaceuticals such as human insulin, growth hormone, clotting factors, fertility drugs, and vaccines. Cloned genes are also used to genetically screen individuals for genetic diseases. Pharmacologists even use cloned genes for pharmacogenetics, which screens patients for the presence of gene variants that can profoundly affect the efficacy and toxicity of particular drugs. This allows clinicians to tailor treatment to the exact genetic makeup of the patient to maximize treatment efficacy and minimize side effects. Such a strategy is called personalized medicine. Cloned genes are also used in gene therapy, which delivers cloned genes into the bodies of patients who suffer from genetic diseases in an attempt to cure them. Patients with cancer and inherited deficiencies of the immune system, blindness, and blood-based defects have been treated with gene therapy protocols.

In agriculture, the introduction of cloned genes into plants that are used as food crops has generated transgenic crops. These crops display several advantageous traits: reduced dependence on agrochemical applications (for example, Bt-corn and herbicide-resistant crops), increased nutritional value (for example, Golden Rice), increased resistance to environmental stresses, and reduced spoilage (for example, the Flavr Savr tomato).

Reproductive Cloning. When farmers identify food animals with desirable traits, they typically breed those animals as much as possible to improve the genetic quality of their herds and flocks. However, such prize animals inevitably die. Propagating these animals by reproductive cloning and mating them to as many animals as possible preserves the exceptional genetic content of a prize animal and allows it to produce far more offspring. This significantly raises the genetic quality of the flock or herd, and commercial dissemination of such cloned animals to other farmers raises the overall genetic quality of food animals. Reproductive cloning also eliminates the need for artificial insemination, which is often expensive and inconvenient.

Cloning effectively maintains high-quality animal stocks. Reproductive cloning of only the healthiest and most productive animals increases their numbers and improves the gene pool (sum total of genetic diversity) and overall health of food animals. This results in safer and healthier food and reduces the use of growth hormones, antibiotics, and other chemicals in the raising of animals.

In the field of conservation biology, the numbers of endangered species are often increased by captive breeding programs. However, not all endangered species can effectively breed in captivity. Reproductive cloning can aid in the preservation of those organisms that do not reproduce in captivity. Cloning can also resurrect genetic material from dead animals and potentially expand the gene pool of endangered species. In 2001, scientists at the University of Teramo, Italy, cloned the European mouflon, an endangered sheep, from cells sampled from a dead animal. When combined with other reproductive technologies, cloning can help save endangered species.

Cloned animals also serve as excellent research models. Because each cloned animal is genetically identical, experiments on cloned animals are devoid of differences caused by heterogeneous genetic backgrounds. Genetic manipulation of cloned animals allows researchers to modify genes of interest and more completely analyze their contribution to development and disease. Modifying particular genes of cloned animals also generates model systems for particular genetic diseases. Cloned, transgenic mice and cloned knockout mice, which have had a specific gene inactivated, are examples of the vast usefulness of such model systems.

Of enormous interest is modifying the genomes of cloned animals so that they can produce clinically and pharmaceutically significant products. By genetically modifying pigs, it is possible to make cloned pigs that contain organs that are fit for transplantation into humans (xenotransplantation). Also, producing antibodies, clotting factors, or even vaccines in the blood or milk of farm animals provides a means to mass-produce potentially expensive pharmaceutical agents at a fraction of the normal cost. This process is called pharming.

Therapeutic Cloning. Therapeutic cloning has tremendous potential for numerous clinical applications. Embryonic stem cells (ESCs) made from therapeutic cloning procedures are pluripotent. Therefore, injured, diseased, or failing tissues or organs could potentially be replaced by tissues or organs manufactured from embryonic stem cells in the laboratory or fetal cells from cloned fetuses. Furthermore, embryonic stem cells made from cloned embryos, or any tissues or organs fashioned from these cells, would not be regarded by the patient's body as foreign. Experiments in laboratory animals have shown that such scenarios are possible. Therapeutic cloning, coupled with embryonic stem cells technology, could christen a new era of regenerative medicine.

Embryonic stem cells from cloned embryos have toxicological applications. Toxicologists typically use laboratory animals or cultured cells to gauge the biological effects of natural or industrially produced molecules on human beings. Unfortunately, laboratory animals show limited utility as a model for human toxicology, and cultured cells do not represent the response of an organ or tissue to foreign molecules. Furthermore, neither of these model systems can assess the individual responses people will have to such molecules, because the genetic variation between individual humans causes differential responses to drugs, toxins, or environmental pollutants. However, cultured embryonic stem cells from cloned embryos can test the biological effects of drugs or environmental pollutants on cells made from a specific person. In addition, because these cells can be differentiated into various tissues and even organs, they can be used to evaluate the individual and tissue-specific responses people might have to particular drugs or pollutants.

IMPACT ON INDUSTRY

Biotechnology companies that use cloning technology in the United States, Europe, Canada, and Australia reported a combined net profit of $3.7 billion in 2009. The United States has been the leader in cloning research, but there are many high-quality laboratories that study cloning technology in the United Kingdom, continental Europe, South Korea, Australia, China, Canada, Iran, and Israel.

Governmental Regulatory Agencies. The U.S. governmental agencies that regulate cloning are the Food and Drug Administration (FDA), Environmental Protection Agency (EPA), and the Department of Agriculture (USDA). The FDA regulates any foods made by genetically modified organisms. This agency concerns itself with only the safety of foods and not the manner in which they are made. The EPA has regulatory authority over all pest-resistant plants to ensure that genetically engineered crops do not adversely affect the environment. Field testing of genetically modified organisms is overseen by a division of the USDA, the Animal and Plant Health Inspection Service.

Government and University Research. The largest funder of cloning research is the National Institutes of Health (NIH). Other governmental funding agencies include the National Science Foundation and the USDA. The NIH not only funds the research of other laboratories but also houses many of its own laboratories, some of which use investigate cloning technologies.

Most of the cloning on university campuses is basic research. Many universities house cloning research centers on their campuses. In other cases, the cloning centers are extensions of state universities. The Roslin Institute, for example, where Dolly the cloned sheep was made, is an extension of the University of Edinburgh. Some universities have even formed partnerships with biotechnology companies that allow the company to work on university property in exchange for funds and increased collaboration between the company and the university.

Industry and Business. Biotechnology companies from all over the world participate in cloning research. Many of these companies have even formed associations. Ausbiotech represents Australian biotechnology companies, the European Federation of Biotechnology represents institutions from European and non-European countries, and BIOTECanada represents more than 250 Canadian biotechnology companies. These trade associations represent the interests of biotechnology to governing bodies.

Pharmaceuticals made by transgenic organisms that express cloned genes constitute the largest proportion of products developed and manufactured by biotechnology companies. These products include diabetes treatments, vaccines, cytokines (special proteins that signal to white blood cells), and other medicines. The demand for new medicines drives research and development in this area, and the understanding of the entire human genome is ushering in many previously unknown medical treatment strategies.

Agricultural biotechnology companies focus largely on developing new crops with improved characteristics. Some of their work is focused on making crops that can grow in underdeveloped countries. For example, Monsanto has begun field trials of a genetically engineered cassava plant that is virus resistant, less poisonous, and much more nutritious than its native counterpart. Some 800 million people globally rely on cassava as their main food staple, but viral infections, poor processing that tends to generate poisonous cyanides, and a lack of nutritional content tend to limit the food potential of cassava.

A few animal cloning industries market techniques for cloning pets. Genetic Savings and Clone is one such company. A related company, Viagen, which is part of Exeter Life Sciences, offers commercial cloning services for farm animals.

Human cloning industries are working toward therapeutic cloning strategies. Advanced Cell Technologies (ACT) works on human cloning. Based in Worchester, Massachusetts, ACT seeks to produce patient-specific stem cells from cloned embryos and cloned fetuses that can cure degenerative diseases without the risk of rejection by the immune system.

SOCIAL CONTEXT AND FUTURE PROSPECTS

Despite the reservations of some people, cloning is a part of everyday life. Many of the foods Americans consume contain some genetically engineered products. Physicians prescribe medicines, give vaccines, and apply other biological products made by genetically engineered microorganisms on a quotidian basis. Given the inroads molecular cloning has already made into people's lives, it is unlikely that people would suffer any revulsion from eating meat from cloned cattle or sheep or having their lives saved by the transplantation of an organ that came from a cloned pig. People would also probably not protest seeing cloned versions of endangered species at their local zoos.

Nevertheless, many people have raised concerns over cloning technologies. First, conservation biologists have suggested that cloning endangered species does not address the habitat destruction and environmental degradation that pushed these species to near extinction in the first place. Second, cloning only makes one species and does not re-create an ecosystem. For example, cloning cannot recapitulate a coral reef or an old growth forest. Thus, it is the wrong solution for the problem.

Genetically modified organisms have become the focal point of concern for several environmental activism groups. Such groups oppose GMOs because they believe that the cloned genes inserted into them can spread to other species and cause severe environmental disruption and that genetically engineered foods have not been sufficiently tested and are potentially dangerous to human health.

The most contentious aspect of cloning technologies is human genetic engineering and reproductive cloning. Transhumanists are some of the most energetic proponents of human cloning and genetic enhancement. As a movement, Transhumanism regards infirmity, disease, aging, and death as undesirable and unnecessary and views science and technology as the means to defeat human limitations. Transhumanists' main argument for human cloning is that reproductive freedoms extend to everyone, and therefore, every human being has an inherent right to clone himself or herself.

Opponents of human cloning object to the manufacturing of human beings. Cloned children are made to be identical to someone else and therefore will always live in the shadow of the original person and never be completely the person they choose to be. These unreasonable expectations can psychologically damage them and violate their human dignity and individuality. Cloning would also alter the concept of human nature and therefore undermine the very foundation of liberal democracy.

In the future, the argument over cloning will not dissipate, but cloning research will certainly advance and provide more and more examples of the utility of this remarkable technology.

FURTHER READING

Alexander, Brian. *Rapture: A Raucous Tour of Cloning, Transhumanism, and the New Era of Immortality.* New York: Basic Books, 2004. A reporter examines the fringe groups that support human cloning and genetic enhancement and finds people who want to defeat the effect of entropy and live forever.

Fukuyama, Francis. *Our Posthuman Future: Consequences of the Biotechnology Revolution.* New York: Picador, 2003. A historian's admonition of the consequences of the biotechnology revolution and its potential to abolish human rights and erode the foundations of liberal democracy.

Mitchell, C. Ben, et al. *Biotechnology and the Human Good.* Washington D.C.: Georgetown University Press, 2007. A distinctly Christian assessment of the application of biotechnology to humans that remains optimistic but cautious and concerned.

Shanks, Pete. *Human Genetic Engineering: A Guide for Activists, Skeptics, and the Very Perplexed.* New York: Nation Books, 2005. A helpful explication of the science behind cloning, coupled with stern warnings against it, by a noted social activist.

Silver, Lee. *Challenging Nature: The Clash Between Biotechnology and Spirituality.* New York: Harper Perennial, 2006. A Princeton stem cell scientist explains the science behind biotechnology and stem cells. He offers some rather harsh critiques of more conservative thinkers who do not agree with his optimistic views of genetic enhancement and embryonic stem cells.

_____. *Remaking Eden: How Genetic Engineering and Cloning Will Transform the American Family.* New York: Harper Perennial, 2007. A very readable introduction to the science of cloning and genetic engineering by a noted mammalian embryologist, who believes that humans should be cloned and that people should welcome the profound changes that it will invoke within human societies.

Wilmut, Ian, Keith Campbell, and Colin Trudge. *The Second Creation: Dolly and the Age of Biological Control.* New York: Farrar, Straus and Giroux, 2000. The two researchers who made Dolly team up with a noted British science writer to give a personal but rigorous explanation and thoughtful examination of cloning. Contains a helpful glossary of terms.

——*Michael A. Buratovich, MA, PhD*

FASCINATING FACTS ABOUT CLONING

- Scientists at Advanced Cell Technology used fetal heart muscle cells from cloned cow fetuses to reverse the effects of heart attacks in adult cows.
- The first cloned cat, CC (CopyCat), made at Texas A & M University in 2001, has a completely different personality than the donor cat. Even though CC is genetically identical to her donor, she is shy and timid whereas the donor cat is outgoing and playful.
- In 2008, BioArts International held an essay contest that invited people to argue why their dog should be cloned. The winner was Trakr, a German Shepherd police dog, who discovered the last survivor of the September 11, 2001, terrorist attacks on the World Trade Center in New York City.
- By cloning vaccines into plants, scientists have made edible vaccines against digestive diseases such as cholera, the Norwalk virus, some food poisonings, and enterotoxigenic *Escherichia coli*. These vaccines are not injected but rather eaten.
- Ingo Potrykus and Peter Beyer invented Golden Rice in the 1990's. This genetically engineered strain of rice produces beta-carotene, a precursor for vitamin A biosynthesis, which is not found in normal rice in appreciable quantities. Children who live in countries where rice is the main food staple are at higher risk for vitamin A deficiency, and Golden Rice was developed to help prevent this deficiency. Subsequent development has increased the nutritional value of Golden Rice even further. Even though the makers of Golden Rice want to give it to farmers completely free of charge, opposition to genetically modified organisms has prevented it from ever being cultivated for food.

CLONING OF PLANTS

Softwood stemcuttings rooting in a controlled environment. By KVDP (Own work [Public domain], via Wikimedia Commons

FIELDS OF STUDY

Agricultural engineering; Agronomy; Biology; Botany; Genetics; Food science; Horticulture; Human nutrition; Plant pathology

ABSTRACT

Plant cloning is the production of a cell, cell component, or plant that is genetically identical to the unit or individual from which it was derived. The term "clone" is derived from the Greek word *klōn*, meaning a slip or twig. Hence, it is an appropriate choice. Plants have been "cloned" from stem cuttings or whole-plant divisions for many centuries, perhaps dating back as far as the beginnings of agriculture.

BACKGROUND

In 1838 German scientists Matthias Schleiden and Theodor Schwann presented their *cell theory*, which states, in part, that all life is composed of cells and that all cells arise from preexisting cells. This theory formed the basis for the concept of *totipotency*, which states that since cells must contain all of the genetic information necessary to create an entire, multicellular organism, all of the cells of a multicellular organism retain the potential to re-create, or regenerate, the entire organism. Thus was the basis for plant cell culture research.

The first attempt at culturing isolated plant tissues was by Austrian botanist Gottlieb Haberlandt at the beginning of the twentieth century, but it was unsuccessful. In 1939 Roger Jean Gautheret and his colleagues demonstrated the first successful culture of isolated plant tissues as a continuously dividing callus tissue. The term *callus* is defined as an unorganized mass of dividing cells, such as in a wound response. It was not until 1954, however, that the first whole plant was regenerated, or cloned, from a single adult plant cell by W. H. Muir, Albert Hildebrandt, and Albert Riker. Thereafter, an increased understanding of plant physiology, especially the role of plant hormones in plant growth and development, contributed to rapid advances in plant cell and tissue culture technologies in the 1970s and 1980s. Many plant species have been successfully cloned from single cells, thus demonstrating and affirming the concept of totipotency.

HORTICULTURE

By far, the greatest impact of cloning plants *in vitro* (Latin for "in glass," meaning in the laboratory or outside the plant) has been on the horticultural industry. In the 1980s plant tissue culture technologies propagated and produced many millions of plants. Since then, many economically important plants have been commonly propagated via tissue culture techniques, including vegetable crops (such as the potato), fruit crops (strawberries and dates), floriculture species (orchids, lilies, roses, Boston ferns), and even woody species (pines and grapes).

The advantages of plant cell, tissue, and organ culture technologies include a more rapid production of plants, taking weeks instead of months or years. Much less space is required (square feet instead of field plots). Plants can be produced year-round, and economic, political, and environmental considerations that hamper the propagation of regional or endangered plant species can be reduced. The disadvantages include the high start-up costs for facilities, the skilled labor required, and the need to maintain sterile conditions.

Two other significant considerations must be considered as a result of plant propagation technologies.

As illustrated by the Irish Potato Famine of the 1840s, the cultivation of whole fields of genetically identical plants (*monoculture*) leaves the entire crop vulnerable to pest and disease infestations. The second important consideration when generating entire populations of clones, especially using tissue culture technologies, is the potential for introducing genetic abnormalities, which then are present in the entire population of plants produced, a process termed *somaclonal variation*.

BIOTECHNOLOGY

An absolute requirement for the genetic engineering of plants is the ability to regenerate an entire plant from a single, genetically transformed cell, thus emphasizing the second major impact of plant cell culture technologies. In 1994 the U.S. Food and Drug Administration (FDA) approved the first genetically modified whole food crop, Calgene's Flavr Savr tomato. This plant was produced using what is termed anti-sense technology. One of the tomato's genes involved in fruit ripening was reversed, thus inactivating it and allowing tomatoes produced from it to have significantly delayed ripening. Although no longer commercially marketed, the Flavr Savr demonstrated the impact of genetic engineering in moving modern agriculture from the Green Revolution into what has been termed the Gene Revolution.

Other examples of agricultural engineering exist, such as Roundup Ready soybeans, engineered to resist the herbicide used on weeds where soybeans are grown, and Bt corn, which contains a bacterial gene conveying increased pest resistance. Since 1987, the U.S. Department of Agriculture (USDA) has required field testing of genetically modified crops to demonstrate that their use will not be disruptive to the natural ecosystem. Thousands of field trials have been completed or are in progress for genetically modified versions of various crop species, including potatoes, cotton, alfalfa, canola, and cucumbers.

These studies have not done enough to alleviate consumer concerns, however. Genetically modified foods (produced from genetically modified organisms, or GMOs) have caused major controversy not only over how they might affect nature—in terms of reduced biodiversity, new pests, and other environmental factors—but over how they could affect human health over the long term as well. There has been concern that GMOs could introduce a new allergen, for example. There has also been also concern that a gene transfer of harmful, antibiotic-resistant bacteria could occur, when animals consume GM plants and then humans consume the animal or the plant itself. Consumers have also shown concern over a lack of labeling; a person may unwittingly eat a GMO product when trying to avoid it.

——*Henry R. Owen*

FURTHER READING

Acquaah, George. *Principles of Plant Genetics and Breeding*. 2nd ed. Hoboken: Wiley, 2012. Print.

Barbosa-Cánovas, Gustavo, et al. *Global Issues in Food Science and Technology*. Burlington: Academic, 2009. Print.

Barnum, Susan R. *Biotechnology: An Introduction*. 2nd ed. Belmont: Wadsworth, 2005. Print.

Chrispeels, Maarten J., and David E. Sadava. *Plants, Genes, and Crop Biotechnology*. 2nd ed. Sudbury: Jones & Bartlett, 2003. Print.

Conger, B. V., ed. *Cloning Agricultural Plants via In Vitro Techniques*. Boca Raton: CRC, 1981. Print.

Dodds, John H., and Lorin W. Roberts. *Experiments in Plant Tissue Culture*. 3rd ed. New York: Cambridge University Press, 1995. Print.

Kyte, Lydiane, and John Kleyn. *Plants from Test Tubes: An Introduction to Micropropagation*. 3rd ed. Portland: Timber, 1996. Print.

Pierce, Benjamin A. *Genetics: A Conceptual Approach*. 4th ed. New York: W. H. Freeman, 2012. Print.

Pierik, R. L. M. *In Vitro Culture of Higher Plants*. Dordrecht: Martinus Nijhoff, 1987. Print.

Thieman, William J., and Michael A. Palladino. *Introduction to Biotechnology*. 3rd ed. Boston: Pearson, 2013. Print.

Weasel, Lisa H. *Food Fray: Inside the Controversy over Genetically Modified Food*. New York: Amacom-American, 2009. Print.

CRISPR-Cas9

FIELDS OF STUDY

Biology; Genetics; Medical biotechnology (or Red biotechnology); Philosophy and Religious Studies; Political Science

ABSTRACT

CRISPR-Cas9 is a fast, inexpensive, accurate way of editing DNA. It has a wide range of potential applications, but some scientists worry about the way this tool might be used. This unique technology allows scientists to remove, add, or alter sections of a person's DNA and could be used to cure genetic diseases. However, it can also be used to edit reproductive cells, which could affect generation after generation of humans.

BASIC PRINCIPLES

CRISPR stands for "Clustered Regularly Interspaced Short Palindromic Repeats." These are segments of prokaryotic DNA with short base sequences that are repeated over and over. Each repetition of the DNA is followed by spacer DNA, short segments of DNA that are left over from exposure to foreign DNA. Cas-9 is an enzyme that is produced by CRISPR that can bind with DNA and cut it, thereby editing a gene.

CRISPR-Cas9 works by causing a mutation in the DNA of a cell. Cas9 acts like a pair of scissors that cuts DNA at a specific chosen location in the cell's DNA so that it can be changed or removed. Guide RNA (gRNA) is used to bind to the DNA and guide Cas9 to the previously selected part of the gene to make sure that Cas9 is cutting the DNA at the correct place. After Cas9 cuts across both strands of DNA, the cell recognizes that its DNA is damaged and begins to repair it.

APPLICATIONS

Gene editing has great potential as a tool for helping people who have conditions that are genetically related, from conditions such as cancer to conditions like high cholesterol. However, gene editing of reproductive cells has caused many people to question it. Any change made in reproductive cells will be passed on from generation to generation, causing people to wonder about the long-term implications and consequences.

———*Marianne Moss Madsen, MS*

FURTHER READING

Doudna, Jennifer A. and Samuel H. Sternberg. *A Crack in Creation: Gene Editing and the Unthinkable Power to Control Evolution.* Houghton Mifflin Harcourt, 2017. Written by one of the scientists who discovered this earthshaking technology; focuses on whether or not to actually use this method to change our DNA and the promises and perils of this gene-editing tool.

Enriquez, Juan and Steve Gullans. *Evolving Ourselves: Redesigning the Future of Humanity—One Gene at a Time.* Current, 2016. Discusses the rapidly changing field of altering human evolution; shows the inner workings of innovative molecular biology and how this will affect who humans become in the future.

Kozubek, James. *Modern Prometheus: Editing the Human Genome with Crispr-Cas9.* Cambridge University Press, 2016. Discusses the potential for gene editing, including ethical and legal implications; tells the story across a 50-year timeline, including stories of the scientists involved in the process.

Lipkin, Steven Monroe and John Luoma. *The Age of Genomes: Tales from the Front Lines of Genetic Medicine.* Beacon Press, 2016. Focuses on the real-life stories of patients who may be helped by this type of gene editing in an easy-to-read and accessible way.

McGovern Institute for Brain Research at MIT, Genome Editing with CRISPR-Cas9, https://www.youtube.com/watch?v=2pp17E4E-O8.

New England Bio Labs, CRISPR/Cas9 and Targeted Genome Editing: A New Era in Molecular Biology, https://www.neb.com/tools-and-resources/feature-articles/crispr-cas9-and-targeted-genome-editing-a-new-era-in-molecular-biology.

Saboowala, Hakim. *CRISPR Cas 9: An Enzymatic Scissor for Specific Site Modification of Genome.* Amazon Digital Services, 2016. Discusses how the components of CRISPR-Cas9 can be combined in multiple ways to edit the genome.

Your Genome, Facts, What is CRISPR-Cas9? http://www.yourgenome.org/facts/what-is-crispr-cas9.

> **FASCINATING FACTS ABOUT THE CRISPR-CAS9**
>
> - Some types of bacteria have a built-in gene editing system that is similar to the CRISPR-Cas9 system. They use this system to respond to pathogens like viruses by snipping out a piece of the bacteria and keeping it so that they recognize the virus the next time it attacks.
> - Geneticists use gene mutation to study its effects and discover what the function of that particular gene is.
> - Scientists have used chemicals or radiation to alter genes, but CRISPR-Cas9 is faster, less expensive, and more reliable than these other methods.
> - Currently, gene editing of reproductive cells is illegal in many countries.

Cryogenics

Cryopreservation of plant shoots. Open tank of liquid nitrogen behind. Agricultural Research Service, the research agency of the United States Department of Agriculture, with the ID D055-16. [Public domain], via Wikimedia Commons

FIELDS OF STUDY

Biology; Genetics; Medical biotechnology (or Red biotechnology); Philosophy and Religious Studies

ABSTRACT

Cryogenics is the branch of physics concerned with creation of extremely low temperatures and the natural phenomena that result from subjecting various substances to those temperatures. At temperatures near absolute zero, the electric, magnetic, and thermal properties of most substances are greatly altered, allowing useful industrial, automotive, engineering, and medical applications.

DEFINITION AND BASIC PRINCIPLES

Cryogenics comes from two Greek words: *kryo*, meaning "frost," and *genic*, "to produce." This science studies the implications of producing extremely cold temperatures and how those temperatures affect substances such as gases and metal. Cryogenic temperature levels are not found naturally on Earth.

The usefulness of cryogenics is based on scientific principles. The three basic states of matter are gas, liquid, and solid. Matter moves from one state to another by the addition or subtraction of heat (energy). The molecules or atoms in matter move or vibrate at different rates depending on the level of heat. Extremely low temperatures, such as those achieved through cryogenics, slow the vibration of atoms and can change the state of matter. For example, cryogenic temperatures are used in the liquefaction of atmospheric gases such as oxygen, nitrogen, hydrogen, and methane for diverse industrial, engineering, automotive, and medical applications.

Sometimes cryogenics and cryonics are mistakenly linked, but the use of subzero temperatures is the only thing these practices share. Cryonics is the practice of freezing a body right after death to preserve it for a future time when a cure for fatal illness or remedy for fatal injury may be available. The practice of cryonics is based on the belief that technology from cryobiology can be applied to cryonics. Adherents believe that if cells, tissues, and organs can be preserved by cryogenic temperatures, then perhaps whole bodies can be preserved for future thawing and life restoration.

BACKGROUND

The history of cryogenics follows the evolution of low-temperature techniques and technology. Principles of cryogenics can be traced back to 2500 BCE, when Egyptians and Indians evaporated water through porous earthen containers to produce cooling. The ancient Chinese, Romans, and Greeks collected ice and snow from the mountains and stored it in cellars to preserve food. In the early 1800s, American inventor Jacob Perkins created a sulfuric-ether ice machine, a precursor to the refrigerator. In the mid-1800s, William Thomson, a British physicist better known as Lord Kelvin, theorized that extremely cold temperatures could stop the motion of atoms and molecules. This theoretical temperature became known as absolute zero, and the Kelvin scale of temperature measurement emerged.

Scientists of the time focused on liquefaction of permanent gases. By 1845, British physicist Michael Faraday had achieved liquefaction of permanent gases by cooling immersion baths of ether and dry ice followed by pressurization. Six permanent gases—oxygen, hydrogen, nitrogen, methane, nitric oxide, and carbon monoxide—still resisted liquefaction. In 1877, French physicist Louis-Paul Cailletet and Swiss physicist Raoul Pictet produced drops of liquid oxygen, working separately and using completely different methods. In 1883, S. F. von Wroblewski at the University of Krakow discovered that oxygen would liquefy at 90 kelvins (K) and nitrogen at 77 K. In 1898, Scottish chemist and physicist James Dewar discovered the boiling point of hydrogen to be 20 K and its freezing point to be 14 K.

Helium has the lowest boiling point of all known substances. It was liquefied in 1908 by Dutch physicist Heike Kamerlingh Onnes at the University of Leiden, who was also the first person to use the word "cryogenics." In 1892, Dewar invented the Dewar flask, a vacuum flask designed to maintain temperatures necessary for liquefying gases—the precursor to the thermos. The liquefaction of gases had many important commercial applications, and many industries use Dewar's concept in applying cryogenics to their processes and products.

The usefulness of cryogenics continued to evolve, and by 1934 the concept was well established. During World War II, scientists discovered that metals became resistant to wear when frozen. In the 1950s, the Dewar flask was improved with the multilayer insulation (MLI) technique for insulating cryogenic propellants used in rockets. Over the next thirty years, Dewar's concept led to the development of small cryocoolers, useful to the military in national defense. The National Aeronautics and Space Administration (NASA) space program applies cryogenics to its programs. Cryogenics can be used to preserve food for long periods—this is especially helpful during natural disasters. Cryogenics continues to grow globally and serve a wide variety of industries.

HOW IT WORKS

Cryogenics is an ever-expanding science. The basic principle of cryogenics that the creation of extremely low temperatures will affect the properties of matter so the changed matter can be used for a number of applications. Four techniques can create the conditions necessary for cryogenics: heat conduction, evaporative cooling, rapid-expansion cooling (Joule-Thomson effect), and adiabatic demagnetization.

Creating Low Temperatures. With heat conduction, heat flows from matter of higher temperature to matter of lower temperature in what amounts, basically, to a transfer of thermal energy. As the process is repeated, the matter cools. This principle is used in cryogenics by allowing substances to be immersed in liquids with cryogenic temperatures or in an environment such as a cryogenic refrigerator for cooling.

Evaporative cooling is demonstrated in the human body when heat is lost through liquid (perspiration) to cool the body via the skin. Perspiration absorbs heat from the body, which evaporates after it is expelled. In the early 1920s in Arizona during the summers, people hung wet sheets inside screened sleeping porches. Electric fans pulled air through the sheets to cool the sleeping space. In the same way, a container of liquid can evaporate so that the heat is removed as gas; repeating the process drops the temperature of the liquid. An example is reducing the temperature of liquid nitrogen to its freezing point.

The Joule-Thomson effect occurs without the transfer of heat. Temperature is affected by the relationship between volume, mass, pressure, and temperature. Rapid expansion of a gas from high to low pressure results in a temperature drop. This principle was employed by Onnes to liquefy helium, and it is useful in home refrigerators and air conditioners.

Adiabatic demagnetization uses paramagnetic salts to absorb energy from liquid, resulting in a temperature drop. The principle of adiabatic demagnetization is the removal of the isothermal magnetized field from matter to lower the temperature. This principle is useful in application to refrigeration systems, which may include a superconducting magnet.

Cryogenic Refrigeration. Cryogenic refrigeration, used by the military, laboratories, and commercial businesses, employs gases such as helium (valued for its low boiling point), nitrogen, and hydrogen to cool equipment and related components to temperatures lower than 150 K. The selected gas is cooled through pressurization to liquid or solid forms (dry ice used in the food industry is solidified carbon dioxide). The cold liquid may be stored in insulated containers until used in a cold station to cool equipment in an immersion bath or with sprayer.

Cryogenic Processing and Tempering. Cryogenic processing or treatment increases the length of wear of many metals and some plastics using a deep-freezing process. Metal objects are introduced to cooled liquid gases such as liquid nitrogen. The computer-controlled process takes about seventy-two hours to affect the molecular structure of the metal. The next step is cryogenic or heat tempering to improve the strength and durability of the metal object.

APPLICATIONS AND PRODUCTS

Early applications of cryogenics targeted the need to liquefy gases. The success of this process in the late 1800s paved the way for more study and research to apply cryogenics to developing life needs and products. Examples include applications in the automobile and health-care industries, the development of rocket fuels, and food preservation. Cryogenic engineering has applications related to commercial, industrial, aerospace, medical, domestic, and defense ventures.

Superconductivity Applications. One property of cryogenics is superconductivity. This occurs when the temperature is dropped so low that the electrical current experiences no resistance. Superconductivity is important in magnetic resonance imaging (MRI), which uses a powerful magnetic field generated by superconducting electromagnets to diagnose certain medical conditions. The superconducting coils are cooled by liquid helium, which becomes a free-flowing superfluid, and liquid nitrogen cools the superconducting compounds, making cryogenics an integral part of this process. Another application is the use of liquefied gases to spray on buried electrical cables to minimize wasted power and energy and to maintain cool cables with decreased electrical resistance.

Health Care Applications. The health care industry recognizes the value of cryogenics. Medical applications using cryogenics include preservation of cells or tissues, blood products, semen, corneas, embryos, vaccines, and skin for grafting. Cryotubes with liquid nitrogen are useful in storing strains of bacteria at low temperatures. Chemical reactions needed to release active ingredients in statin drugs, used for cholesterol control, must be completed at very low temperatures (−100 degrees Celsius). High-resolution imaging, like MRI, depends on cryogenic principles for the diagnosis of disease and medical conditions. Dermatologists uses cryotherapy to treat warts or skin lesions.

Food and Beverage Applications. The food industry uses cryogenic gases to preserve and transport mass amounts of food without spoilage. This is particularly useful when supplying food to war zones or natural-disaster areas. Deep-frozen food retains its color, taste, and nutrient content, while its shelf life is significantly increased. Certain fruits and vegetables can be deep frozen for consumption out of season. Freeze-dried foods and beverages, such as coffee, soups, and military rations, can be safely stored for long periods without spoilage. Restaurants and bars use liquid gases to store beverages while maintaining the taste and look of the drink.

Automotive Applications. The automotive industry employs cryogenics in diverse ways. One technique makes use of the property of thermal contraction. Because materials will contract when cooled, the valve seals of automobiles are treated with liquid nitrogen, which causes them to shrink to allow insertion and then expand as they warm up, resulting in a tight fit. The automotive industry also uses cryogenics to increase strength and minimize wear of metal engine parts, pistons, cranks, rods, spark plugs, gears, axles, brake rotors and pads, valves, rings, rockers, and clutches. Cryogenic-treated spark plugs can increase an automobile's horsepower as well as its gasoline mileage. The use of cryogenics allows a race car to race as many as thirty times without a major rebuild on the motor, in contrast to an untreated car, which could only race twice.

Aerospace Industry Applications. NASA's space program uses cryogenic liquids to propel rockets. Rockets carry liquid hydrogen for fuel and liquid oxygen for combustion. Cryogenic hydrogen fuel is what enabled NASA's workhorse space shuttle to get into orbit. In addition, liquid helium is used to cool the infrared telescopes on rockets.

Tools, Equipment, and Instrument Applications. Metal tools can be cryogenically treated to increase their wear resistance. Surgery or dentistry tools can be expensive, and cryogenic treatment can prolong their usage. Sports equipment, such as golf clubs, also benefits from cryogenics, as it provides increased wear resistance and better performance. Scuba divers are able to stay submerged for hours using an insulated Dewar flask of cryogenically cooled nitrogen and oxygen. Some people claim musical instruments receive benefits from cryogenic treatment; in brass instruments, a crisper and cleaner sound is allegedly produced with cryogenic enhancement.

Other Applications. Other applications are evolving as industries recognize the benefits of cryogenics to their products and programs. The military have used cryogenics in various ways, including infrared tracking systems, unmanned vehicles, and missile-warning receivers. Companies can immerse discarded recyclables in liquid nitrogen to make them brittle, making them easier to pulverize or grind down to a more eco-friendly form.

SOCIAL CONTEXT AND FUTURE PROSPECTS

The economic and ecological impact of cryogenic research and applications holds global promise for the future. In 2009, Dutch company Stirling Cryogenics built a liquid-argon cooling system for the ICARUS experiment, a neutrino study being carried out by Italy's Istituto Nazionale di Fisica Nucleare (National Institute of Nuclear Physics); the system, which maintains four hundred liters of liquid argon at a temperature of 94 kelvins, was designed to run nonstop for ten years. The Cryogenic and Refrigeration Engineering Research Centre (CRERC) at China's Technical Institute of Physics and Chemistry was created to explore new innovations and technology in cryogenic engineering. Both government agencies and private industries in the United States are pursuing innovative ways to use existing applications and define future implications of cryogenics. Although cryogenics has proved useful to many industries, its full potential as a science has not yet been realized.

——*Marylane Wade Koch, MSN, RN*

FURTHER READING

Hayes, Allyson E., ed. *Cryogenics: Theory, Processes and Applications.* Hauppauge: Nova, 2010. Print.

Jha, A. R. *Cryogenic Technology and Applications.* Burlington: Elsevier, 2006. Print.

Maytal, Ben-Zion, and John M. Pfotenhauer. *Miniature Joule-Thomson Cryocooling: Principles and Practice.* New York: Springer, 2013. Print.

Pavese, Franco, and Gianfranco Molinar Min Beciet. *Modern Gas-Based Temperature and Pressure Measurements.* 2nd ed. New York: Springer, 2013. Print.

Van Sciver, Steven W. *Helium Cryogenics.* 2nd ed. New York: Springer, 2012. Print.

Ventura, Guglielmo, and Lara Risegari. *The Art of Cryogenics: Low-Temperature Experimental Techniques.* Burlington: Elsevier, 2008. Print.

D

DESALINATION PLANTS AND TECHNOLOGY

Reverse osmosis desalination plant in Barcelona, Spain. By James Grellier. (Own work.) [CC BY-SA 3.0 (https://creativecommons.org/licenses/by-sa/3.0) or GFDL (http://www.gnu.org/copyleft/fdl.html)], via Wikimedia Commons

FIELDS OF STUDY

Agricultural engineering; Agronomy; Biology; Chemistry; Civil engineering; Genetics; Horticulture

ABSTRACT

Seawater and other salt-containing waters are converted into potable water by distillation, reverse osmosis, and other processes experimentally, and increasingly practically, in regions where water resources are limited or expensive.

HISTORY

For many years, large ships at sea have used distillation processes to convert seawater into usable water for passengers and crews because it is more economical than carrying enormous quantities of fresh water for drinking, cooking, and cleaning. In desert regions and some areas that have limited suitable fresh water available, distillation and, more recently, membrane processes have been introduced for the conversion of brackish water, industrial effluents, wastewater, and seawater. Large-scale pilot processes have been rare. One notable example is a plant that was built in San Diego in the 1950's and later shipped to the U.S. naval base at Guantánamo Bay in Cuba. It can produce 13 million liters of distilled water per day.

Because brackish water and various wastewaters contain between 500 and 5,000 parts per million of dissolved solids, and seawater and geothermally produced brines contain up to 50,000 or more, a number of different processing methods have been developed. In addition, the end use of the water may dictate the superiority of one method above the others. For many agricultural purposes, water containing a few thousand parts per million can be used, whereas U.S. drinking water standards are set at a maximum of 500 (in actuality, many U.S. cities' water supplies exceed this standard).

DISTILLATION METHODS

Distillation methods were first described by Aristotle, but they had their first practical use aboard English naval vessels in the 1600's. Since then they have become much more complex, but they still involve a high-cost, energy-intensive boiling process, and subsequently a cooling process for liquefaction of the steam generated. The original processes required submerged tubes, which became encrusted with chemical deposits. Multistage flash process plants are currently used in which the latent heat of evaporation of the water is captured and reused, and the scaling is diminished by adding chemicals or removing the ions causing the deposits. Newer variations of these processes are being investigated. Some attempts have been made to couple power generation plants with distillation units, which may provide more desirable economy of operation.

Various versions of the multistage flash process are used in many parts of the Middle East and in more than three-quarters of the currently operating systems. Other designs for distillation plants have been proposed, and some have been built. Most of these have used horizontal tube processes with a design that permits multiple stages with vacuum distillation and a gradual reduction of saline content by incorporating steam with the brine. Large installations are currently incorporating this design. Smaller plants have employed a vapor compression procedure for industrial plants and resort hotels, but these are gradually being replaced by reverse osmosis facilities.

Solar distillation procedures would appear to offer great future alternatives in the very regions where water is in short supply. If solar energy could be more cheaply and efficiently obtained, and the land area needed made available, the saline water conversion problem would be solved relatively easily.

MEMBRANE METHODS

Although reverse osmosis has been most heavily promoted, there is actually a large group of related procedures that utilize membrane separations to purify water. In ordinary osmosis, such as occurs through cell walls, a semipermeable membrane (one through which only the solvent can flow) allows water to flow from a less concentrated solution into a more concentrated one (thus exhibiting an "osmotic pressure"). In reverse osmosis, pressure is exerted on the more concentrated solution, overcoming the osmotic pressure and reversing the flow. After the brine (saline water) has been concentrated in this manner, the process is repeated with fresh brine.

Among the membranes that have been utilized, most are polyamides and polyimides, which closely resemble protein structures. Reverse osmosis has been most effective with brackish waters, which do not have the high osmotic pressure of seawater to overcome. However, improved membrane systems have permitted construction of larger seawater charged reverse osmosis plants in the 13-million-liters-a-day range. A procedure known as electrodialysis permits an electric field to assist in directing ion flow through membranes, which are permeable to either cations or anions; some success in using this method with brackish water has been achieved. Pressurization cycles with ion exchange resins or membranes have been successful with low energy requirements, but experiments have failed to find the high-strength materials required to survive the high pressures needed.

ION EXCHANGE METHODS

Utilizing ion exchange resins in a normal flow-by mode is very reasonable for purifying slightly brackish water. In fact, it is used to soften water in many communities with hard-water supplies. Resins that replace metallic ions with positive hydrogen ions, and nonmetal ions with negative hydroxide ions, can readily accomplish that limited task, but they are not adequate for seawater conversion. The necessity of regenerating the exhausted resins with acid or base make designing a continuous process more difficult.

FREEZING AND SOLVENT EXTRACTION METHODS

When a solution freezes under equilibrium conditions, the solid formed is pure solvent. Therefore, when an iceberg forms, it contains very pure water. It has been proposed that icebergs could be towed to water-short regions. However, mechanical problems, such as providing appropriate freezing chambers and removing brine from the ice surface have prevented these methods from being seriously explored. Solvent extraction procedures have been tried experimentally, but solvent use and removal are costly.

——*William J. Wasserman*

FURTHER READING

Khan, Arshad Hassan. *Desalination Processes and Multistage Flash Distillation Practice*. New York: Elsevier, 1986.

Lauer, William C., ed. *Desalination of Seawater and Brackish Water*. Denver, Colo.: American Water Works Association, 2006.

National Research Council of the National Academies. *Desalination: A National Perspective*. Washington, D.C.: National Academies Press, 2008.

Simon, Paul. *Tapped Out: The Coming World Crisis in Water and What We Can Do About It*. New York: Welcome Rain, 1998.

Spiegler, K. S., and A. D. K. Laird, eds. *Principles of Desalination*. 2d ed. New York: Academic Press, 1980.

National Academies Press *Desalination: A National Perspective*. http://books.nap.edu/openbook.php?record_id=12184&page=R1

Detection and prevention of food poisoning

Food Poisoning Pathogens and Their Sources

Pathogen / Substance	Sources
Aeromonas	Untreated spring or well water
Arsenic	Pesticides and industrial chemicals
Bacillus cereus	Contaminated fried rice, meatballs
Clostridium botulinum	Improperly canned foods
Campylobacter jejuni	Contaminated meats (including beef, pork, lamb, goat, venison), poultry
Clostridium perfringens	Undercooked meats, poultry, legumes
Ciguatera	Carnivorous reef fish in Hawaii, Florida, the Caribbean
Entamoeba histolytica	Contaminated food, water
Enterohemorrhagic *Escherichia coli* (*E. coli* O157:H7)	Undercooked hamburger meat, contaminated leafy greens
Enteroinvasive *E. coli*	Contaminated imported cheeses
Enterotoxic *E. coli*	Contaminated water, foods
Giardia lamblia	Contaminated groundwater
Hepatitis A	Contaminated water, foods; person-to-person transmission
Listeria monocytogenes	Unpasteurized, raw milk; soft cheeses; contaminated raw vegetables, shrimp
Mercury	Inorganic mercuric salts
Mold	Fruits, grains, nuts, other foods past prime
Neurotoxic shellfish poison	Mollusks in coastal Florida
Norwalk virus	Contaminated water, shellfish; person-to-person transmission
Paralytic shellfish poison	Bivalve mollusks from temperate, coastal areas
Salmonella	Contaminated beef, poultry, eggs, dairy
Scombroid	Tuna, mahimahi, kingfish
Shigella	Contaminated potato salad, egg salad, lettuce, vegetables, ice cream, water, milk
Staphylococci	Improperly stored foods high in salt or sugar
Trichinella spiralis	Contaminated water, meats (pork, wild game)
Tetrodotoxin	Puffer fish from Japan
Vibrio cholerae	Contaminated water, foods
Vibrio parahaemolyticus	Raw, undercooked seafood
Vibrio vulnificus	Raw oysters
Yersinia	Contaminated milk, ice cream

© EBSCO.

FIELDS OF STUDY

Agricultural engineering; Agronomy; Biology; Chemical engineering; Chemistry; Food science; Forensics

ABSTRACT

Food-borne illnesses affect an estimated seventy-six million Americans every year. Although most victims recover quickly, some five thousand deaths from food poisoning occur in the United States annually.

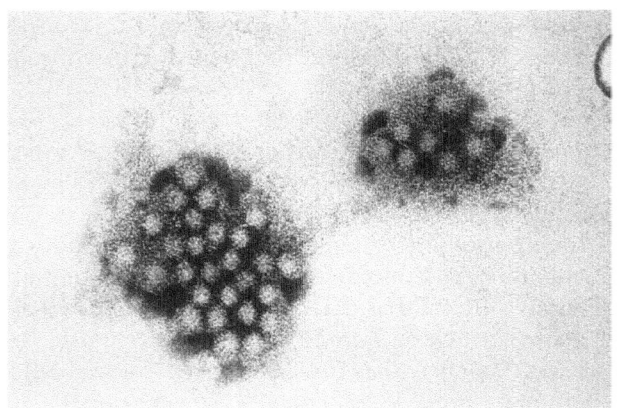

Norwalk viruses (and related caliciviruses) are important causes of nonbacterial gastroenteritis in the United States. An estimated 181,000 cases of this type of food poisoning occur annually. [Public domain], via Wikimedia Commons

Approximately 75 percent of all food poisoning cases are caused by known pathogens; only a small percentage of food poisoning cases are caused by unknown sources or substances. Viruses, bacteria, parasites, toxins, metals, and prions that are consumed through contaminated foods and liquids, including water, cause food poisoning.

More than 250 different food-borne diseases have been identified, and the Centers for Disease Control and Prevention (CDC) estimates that 97 percent of food poisoning illnesses result from improper food handling; of these, 79 percent are caused by foods prepared in commercial kitchens and 21 percent are caused by foods prepared in home kitchens. Prepared foods left at unsafe temperatures, inadequate cooking or reheating, cross-contamination of foods, and infections in food handlers cause most food poisoning cases.

When contaminated food or water is consumed, the digestive system is usually able to destroy any harmful pathogens, but some pathogens can survive and cause illness. Usual symptoms include nausea, abdominal pain, vomiting, diarrhea, and headache; these occur normally within one to three days of ingestion of the contaminated food. Sometimes, however, a food-borne pathogen begins to multiply in the victim's stomach and intestine, producing toxins. These toxins then travel into the bloodstream, where they are transported to vital body organs and muscles, often with lethal results. Rare pathogens such as *Clostridium botulinum* (the bacterium that causes botulism), which grows in improperly canned foods, and harmful substances such as arsenic or pesticides cause severe symptoms that usually result in death.

DETECTION

When a serious case of food poisoning is diagnosed by a physician—after other illnesses that mimic food poisoning have been ruled out—it is confirmed by laboratory analysis of fecal and blood samples and oral history from the victim. The local health department and public health scientists are usually the first responders to a suspected case of food poisoning. Preservation of evidence is critical, and scientists carefully collect suspected food materials and samples of hair and stomach contents from the victim. Microbial forensic analysts must identify, collect, and preserve samples properly for transportation to the lab to avoid contamination of the samples.

A cluster of cases may be detected if several people connected by a common experience or event are suspected of having food poisoning. Scientists must determine whether the cases constitute more than the usual expected number of cases of a given food-borne illness or whether the reports constitute a false cluster (for example, backlogged cases reported all at one time).

Key components of a food poisoning investigation include the selection of investigatory method, analysis of samples, interpretation and validation of results, and quality assurance. If a food poisoning outbreak is identified in a particular area or in relation to a particular food, food distributors and vendors are notified immediately, and public recall notices are issued to prevent an epidemic.

Epidemiologists, forensic toxicologists and pathologists, microbiologists, and food safety and public health officials may all be called in to determine the source of a food-related contamination. Prior to 1996, many food poisoning cases went unreported to health authorities. The CDC established the Foodborne Diseases Active Surveillance Network (FoodNet) in 1996 to monitor food-borne illnesses. PulseNet, a branch of FoodNet, networks public health and food regulatory laboratories nationally and helps identify cases that are spread out over large geographic locations. PulseNet allows rapid analysis of suspected pathogens and identifies them through DNA (deoxyribonucleic acid) fingerprinting, enabling quick detection of specific contamination sources. By combining local and regional surveillance reports, PulseNet quickly identifies suspected cases and reduces epidemic outbreaks.

—Alice C. Richer

FURTHER READING

Balkin, Karen F., ed. *Food-Borne Illnesses.* San Diego, Calif.: Greenhaven Press, 2004. Collection of essays presents a variety of perspectives on food safety issues.

National Center for Food Protection and Defense. *Food Defense Education: Post 9/11.* Minneapolis: Author, 2007. Report on a three-year study explores food safety education programs in the United States, with emphasis on the work of criminal justice professionals.

Scott, Elizabeth, and Paul Sockett. *How to Prevent Food Poisoning: A Practical Guide to Safe Cooking, Eating, and Food Handling.* Hoboken, N.J.: John Wiley & Sons, 1998. Provides thorough information on food poisoning's causes and symptoms. Includes a chapter devoted to the science of food poisoning.

Trestrail, John Harris, III. *Criminal Poisoning: Investigational Guide for Law Enforcement, Toxicologists, Forensic Scientists, and Attorneys.* 2d ed. Totowa, N.J.: Humana Press, 2007. Focuses on intentional poisonings and the techniques used to investigate poisoning crimes.

DIAMOND V. CHAKRABARTY

FIELDS OF STUDY

Chemical engineering; Chemistry; Environmental biotechnology (or Green biotechnology); Microbiology; Molecular biology

ABSTRACT

The Supreme Court's ruling on genetic engineering in the case of *Diamond v. Chakrabarty*, decided on June 16, 1980, was pivotal, as the Court determined that genetically engineered microorganisms are patentable products of human ingenuity.

In 1972 Ananda Chakrabarty, a microbiologist at the General Electric Research and Development Center in Schenectady, New York, attempted to patent a genetically engineered bacterium that could decompose compounds such as camphor and octane in crude oil. Chakrabarty's patent application was initially rejected because the patent office had a long history of excluding living organisms from patent protection. Chakrabarty, through General Electric, successfully appealed this decision. In 1979 the acting commissioner of patents and trademarks appealed the reversal. The case was argued before the U.S. Supreme Court on March 17, 1980.

In a five-to-four decision, the Supreme Court ruled that living things are patentable if they represent novel, genetically altered variants of naturally occurring organisms. The majority decision held that Chakrabarty's organism is manufactured since he had inserted new genetic information into it and that the organism is new because a similar organism is unlikely to occur in nature without human intervention. The organism thus falls within the meaning of the patent statute: It is a product of human ingenuity with a distinctive name, character, and use. The minority opinion held that previous congressional acts that specifically excluded living organisms from patent protection were clearly intended to apply in this case.

This decision let emerging biotechnology companies get patent protection for their living products and allowed them potentially to capitalize on the revolution in genetic engineering. The justices of the Supreme Court realized the ramifications of their decision in terms of its impact on the ethics of patenting living things and its potential to accelerate the release of possibly harmful genetically engineered organisms. However, the basis of their decision was fundamentally narrow: Did Chakrabarty's work constitute patentable material? The Court held that the further development of biotechnology or its restrictions is a congressional and executive concern, not a judicial one.

Diamond v. Chakrabarty did not greatly influence the extent to which genetically altered organisms have been released into the environment; rather, public opposition to the release of genetically engineered organisms has played the dominant role in this area. Instead, the lasting impact of the Court's decision in *Diamond v. Chakrabarty* lies in its extension of the definition of patentable products to compounds or organisms that exist in nature but can be further manipulated by biotechnological means. This had been true only for certain hybrid plants developed through the use of conventional breeding techniques. Furthermore, the decision became the judicial basis for later decisions regarding attempts to patent genetic sequences that may be common to living organisms but require human ingenuity if they are to be extracted, sequenced, replicated, and reinserted into new organisms with their properties intact.

—*Mark Coyne*

FURTHER READING

Resnik, David B. *Owning the Genome: A Moral Analysis of DNA Patenting.* Albany: State University of New York Press, 2004.

Rimmer, Matthew. *Intellectual Property and Biotechnology: Biological Inventions.* Northampton, Mass.: Edward Elgar, 2008.

DNA ANALYSIS

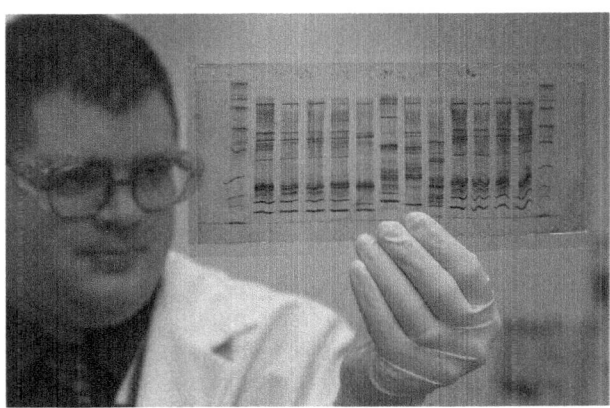

C3P chemist reads a DNA profile to determine the origin of a commodity. By James Tourtellotte, photo editor of *CBP Today*, [Public domain], via Wikimedia Commons

FIELDS OF STUDY

Bioinformatics; Biophysics; Biotechnology; Forensics; Genetics; Political Science

ABSTRACT

DNA analysis involves the use of scientific tools to access the information found in DNA to identify its source, whether some infectious agent, another organism of interest, or a particular individual, such as in forensic applications. Medical applications of this technology include the search for mutations associated with genetic disorders and the design of probes that are able to diagnose these disorders in a timely fashion.

DEFINITION AND BASIC PRINCIPLES

DNA analysis is, in the strictest sense of the term, an actual observation of the length or sequence of a portion of DNA. The length of a fragment of DNA can be determined using gel electrophoresis. This technique involves placing DNA onto a semisolid support, or gel, and applying an electric current to the gel so that DNA migrates toward the positive pole. The migration of DNA in gel electrophoresis is proportional to its mass, which is, in turn, proportional to its length. Determining the actual sequence of bases in a strand of DNA is much more complicated.

Although DNA analysis has at times been equated with genetic testing, the two are not always the same. Certain types of genetic testing developed in the 1960s did not technically involve DNA analysis. Amniocentesis, which allowed for Down syndrome testing, actually involved chromosomal analysis following the creation of a karyotype (organized profile of a person's chromosomes). Similarly, genetic testing for phenylketonuria (PKU) and Tay-Sachs disease originally involved enzyme assays, not an analysis of the defective genes themselves. Thus, actual DNA analysis did not begin in earnest until the mid-1970s.

BACKGROUND AND HISTORY

Although the double helical structure of DNA was first described in 1953 by American molecular biologist James D. Watson and British biophysicist Francis

Crick, more than twenty years passed before scientists developed methods of comparing DNA for the purpose of identification. In 1974, British molecular biologist Joseph Sambrook described the differentiation of human tumor viruses following cleavage by a restriction endonuclease (an enzyme that cleaves DNA at a specific nucleotide sequence). He noticed that different-sized bands of DNA were visible following their separation on a gel. This discovery formed the basis for what has become known as Restriction fragment length polymorphism (RFLP). Although the DNA of viruses and even bacteria could be analyzed directly by RFLP, the restriction enzyme cleavage patterns of higher organisms were of sufficient complexity that only a subset of the bands that were produced could be analyzed. DNA analysis of higher organisms was made possible in 1975 by the development of the Southern blotting technique by British biochemist Edwin Southern.

The following decade saw two ideas that would revolutionize the field of DNA analysis. In 1983, American biochemist Kary Mullis developed the Polymerase chain reaction (PCR), a process that enabled the amplification across several orders of magnitude of small amounts of starting sample DNA in the laboratory. Then, in 1985, British geneticist Alec Jeffreys realized that human DNA was peppered with regions of repeating sequences, or variable number tandem repeats (VNTRs), and that comparisons of these regions could create a unique DNA fingerprint for any given individual.

HOW IT WORKS

Probes and Primers. Most of the methods of DNA analysis take advantage of DNA's natural tendency to form a double helix. RFLP analysis has long been paired with Southern blotting. This procedure involves binding a small synthetic fragment of DNA to a region of interest that is contained within one or more of the bands of DNA that have been separated by gel electrophoresis and then blotted onto some type of membrane. This binding is made possible by the complementary nature of the DNA bases, the fact that adenine forms hydrogen bonds to pair with thymine and that cytosine pairs with guanine, a concept called Watson-Crick base pairing after the discoverers of DNA structure. The synthetic DNA, called a probe, is designed to contain about twenty complementary nucleotides to the sequence of interest. Binding of this probe to the blotted DNA is called hybridization. Originally, DNA probes were labeled with a radioactive marker to enable the detection of their position on a membrane, but later probe labeling has included nonradioactive alternatives such as fluorescent dyes.

The polymerase chain reaction (PCR) also involves the binding of small synthetic fragments of DNA to a region of interest, but in this process, two such fragments bind to opposite strands of the target DNA. These fragments, although identical in structure to the probes described earlier, are called primers because they are used to prime a DNA synthesis reaction. Also, the binding reaction, which involves the same process of complementary bases coming together to form hydrogen bonds, is referred to as annealing. It had been previously discovered that DNA could be made in the laboratory by taking a given single strand of DNA and adding a specific primer, the four types of nucleotides, and purified DNA polymerase (the enzyme normally involved in the polymerization process), but Mullis's insight in the early 1980s was that this process could be converted into a chain reaction that produced large amounts of DNA. By adding two primers instead of one and by using double-stranded DNA as a target, twice as many molecules of DNA could be created, but this necessitated a step in which the DNA had to be heated to near-boiling temperatures to separate, or melt, the two strands of the double helix. Mullis reasoned that a DNA polymerase that had been purified from a thermophilic microbe would be able to survive this heating step. This would allow the stringing together of a number of cycles with three different temperatures—one each for annealing, polymerization, and DNA melting—without having to add more DNA polymerase enzyme. Thus, after each cycle, the amount of DNA double helix would be doubled, resulting in more than a billion molecules of DNA after thirty cycles, even if the cycle started with only one strand of DNA.

DNA Polymorphism. DNA analysis takes advantage of the intrinsic variability that exists among organisms as well as among members of the same species. Polymorphism, a word derived from the Greek for "many forms," is used to describe this variability, the simplest form of which is a single nucleotide polymorphism (SNP). Single nucleotide polymorphisms, which are also referred to as point mutations in genetics, are detectable by RFLP analysis only when they occur within the recognition sequence for a particular restriction enzyme because the enzyme fails to cleave the altered sequence. RFLP analysis also readily detects deletions or insertions of DNA sequences that have occurred between restriction enzyme cleavage sites. What Jeffreys realized in the 1980s was that most restriction fragment length variation in humans was not caused by large insertions or deletions of unique DNA sequences but by a variation in the number of repetitive DNA elements that were found in tandem with one another. He did not, however, use the PCR method that was being developed at the time because a practitioner of PCR must know the precise sequences that flank a site of interest to design the primers used in this procedure. Instead, Jeffreys performed Southern blotting using a probe designed to hybridize with the about fifteen-nucleotide-long sequence that he was studying. This probe specifically labeled the regions of the membrane that contained these variable number tandem repeats (VNTRs). For this contribution, Jeffreys has been called the father of DNA fingerprinting.

Subsequent analysis of various regions in human DNA has taken advantage of PCR to produce results, focusing on even smaller tandem repeats with repeating units that are only one to six nucleotides in length. Discovered in 1989, these short tandem repeats (STRs) were eventually found to outnumber variable number tandem repeats by nearly one hundredfold, being found at more than 100,000 sites in human DNA. As more and more of these STR sites were characterized over time, primers that annealed to their flanking sequences were designed to amplify the repeat area in question.

APPLICATIONS AND PRODUCTS

Of Microbes and Man. Although the tools involved in DNA analysis are often used in basic research such as determining the evolutionary relationships between organisms, much of the application of this technology involves analysis for the purposes of identification. Although identification could potentially include any organism of interest, the primary focus of DNA analysis has been disease-causing viruses and microorganisms along with humans. Ever since Sambrook and colleagues first applied RFLP analysis to differentiate between two strains of viruses, viral epidemiology has remained an important application for tools such as PCR. For example, around the beginning of the twenty-first century, nucleic acid amplification testing (NAAT) was developed to detect the viral load of the human immunodeficiency virus (HIV). The procedure is a faster and more effective way to test for the presence of HIV in a person. NAAT has also been applied as a diagnostic test for certain bacterial infections. Other PCR-based methods have been adapted to test for bacterial contamination of foods as well as of hospital areas and supplies. In most cases, the identification of the precise strain of virus or microbe present is unnecessary because the physical presence of an infectious agent, not its detailed classification, is of interest. Tandem-repeat-based methods of identification are largely useless when analyzing such infectious agents because these agents tend to lack such repetitive DNA sequences. Because the DNA sequence of the entire genome (the complete set of DNA found in a particular organism) of most known infectious agents has been determined, it is possible to design primers that will specifically amplify DNA from a given target species.

In some cases, as in life-threatening illnesses, potential epidemics, and acts of bioterrorism, the speed at which an infectious agent is identified is critical to saving lives. For such applications, a type of PCR called real-time PCR has been developed. Rather than waiting to run gel electrophoresis after the full thirty or so cycles of a traditional PCR reaction have been completed, real-time PCR measures the production of a fluorescent-tagged product in real time, during the early phases of the reaction. This allows for an agent to be detected in minutes rather than hours.

Crime Scenes and Beyond. The best-known use of DNA analysis is probably in the area of forensics. The first case in which Jeffreys applied DNA fingerprinting was an immigration dispute. In 1983, British authorities had denied a thirteen-year-old boy entry into the country, claiming that his passport was forged and that his stated mother, a British subject, was not his biological mother. The dispute continued until 1985, when Jeffreys was able to apply his new technique to prove that the maternal relationship stated on the passport was indeed correct. Since that time, maternity tests have been vastly outnumbered by paternity tests, but the principle used in both types of parental testing remains the same.

The first use of DNA fingerprinting in a criminal case occurred in 1986, when it was used to exonerate a suspect accused of the rape and murder of a teenage girl near Leicester, England. Later, the same technique was used to identify the real killer. Since this early case, evidence from DNA fingerprinting has helped convict thousands of criminals. The source of DNA is blood in about half of all cases; other common sources are semen and hair. DNA analysis also plays an important role in the identification of human remains following disasters, acts of terrorism, and war.

Limitations of PCR. Following the advent of PCR, the amount of forensic sample required for analysis was reduced significantly. The original DNA fingerprinting procedure developed by Jeffreys required a blood sample about the size of a quarter, but later methods needed only a few cells swabbed from a person's cheeks to perform an analysis. Although PCR requires much less starting material than RFLP analysis and is also a more rapid procedure to perform, it does have a number of limitations. The first limitation, that flanking DNA sequences must be known ahead of time, was largely overcome as more and more human short tandem repeats were characterized along with the DNA that surrounded them. A second limitation is that the method is so sensitive that it is prone to contamination by outside sources. Because even a single fragment of DNA can be amplified into large amounts on a gel, care must be taken not to introduce foreign DNA from an investigator's hair or fingertips. A third limitation is that only a single area, or locus, of DNA can be analyzed at one time. To overcome this limitation, a procedure called multiplex PCR has been developed. This method simultaneously employs a number of primers that have been labeled with fluorescent tags. These can be identified during the subsequent gel electrophoresis step based on their specific labels.

Medical Applications. Besides using DNA analysis to identify infectious agents, the medical community has begun to use this technique to study genetic disorders. However, common methods of DNA analysis cannot identify most genetic disorders, with the exception of a class of disorders called trinucleotide repeat expansion disorders. This rare class of disorders, which includes Huntington's disease as well as fragile X syndrome, is readily detectable using PCR amplification of the short tandem repeats that contribute to the disorders in question. A more common class of genetic disorders results from point mutations in genes and can therefore be linked to particular single nucleotide polymorphisms in the human genome. Unfortunately, single nucleotide polymorphisms are not detectable by PCR and will show up in RFLP analysis only if they occur in the restriction enzyme recognition site itself, which is a rare occurrence. The identification of genetic disorders is therefore largely dependent on determining the actual sequence of the DNA, still a technically challenging and expensive undertaking despite progress that has been made since the inception of the Human Genome Project DNA sequencing program in the 1990s.

Methods involved in DNA sequencing include many of the same principles as other forms of DNA analysis. A single primer is labeled with a fluorescent dye and mixed with a target sequence in the presence of a thermostable DNA polymerase. This procedure does not amplify the DNA as in PCR but results in primer extension for a certain length along the target sequence. Another difference from PCR is that modified nucleotides are added to this mixture so that the primer extension is halted whenever these particular nucleotides are incorporated into a growing DNA strand. Four separate tubes are used in this method, one for each of the four DNA bases. Once these four reactions are separated by electrophoresis, the order of bases can be determined using computer software that monitors the relative migration of the bands that occurs from each of the four reaction tubes.

SOCIAL CONTEXT AND FUTURE PROSPECTS

Single nucleotide polymorphisms (SNPs), although not used extensively in forensic applications, potentially contain valuable information that can be of use to crime scene investigators. For example, the presence of particular SNPs may indicate a perpetrator's race, while others could indicate hair color. One disadvantage of SNPs, besides the relative difficulty of identifying them, is that many more of them are needed to provide a unique identification (compared to the number of short tandem repeats needed for PCR). Because most SNPs are biallelic, they contain one base or another but generally not all four possible bases, and it is estimated that as many as fifty would have to be analyzed to obtain the same level of confidence as provided by the thirteen STR loci contained in CODIS. This may not prove as difficult as it sounds because it is estimated that there are probably about 10 million SNP sites scattered throughout the human genome. If accurate, that would mean that SNPs outnumber short tandem repeats to the same degree that short tandem repeats outnumber variable number tandem repeats.

DNA sequencing in some form or another is likely to continue to play an increasing role in DNA analysis. The cost of DNA sequencing is beginning to drop as it becomes more prevalent and increasingly automated. Although the first human genome sequence was produced at a cost of billions of dollars, scientists have set a goal of reducing the cost of DNA sequencing to about one thousand dollars; the cost will likely dip below that in the not-too-distant-future. At the same time, scientists are developing a number of methods that allow SNPs to be determined without first finding the sequence of the 99.7 percent of DNA bases that do not exist as SNPs. These methods include directed hybridizations, ligations, primer extensions, or nuclease cleavages that specifically involve SNPs while leaving the rest of the DNA alone.

With any increase in the involvement of DNA sequencing in forensics comes the likelihood that debate will intensify concerning privacy issues regarding the use of sequence information. Unlike commonly used methods of PCR analysis, SNP determination will reveal certain details about suspects that could be open to abuse. Ethical issues involving the use and dissemination of DNA data will have to be resolved as the methods of DNA analysis continue to evolve.

——*James S. Godde, PhD*

FURTHER READING

McClintock, J. Thomas. *Forensic DNA Analysis: A Laboratory Manual.* CRC Press, 2008.

Nakamura, Yusuke. "DNA Variations in Human and Medical Genetics: Twenty-Five Years of My Experience." *Journal of Human Genetics*, vol. 54, 2009, pp. 1–8.

Pereira, Filipe, et al. "Identification of Species with DNA-Based Technology: Current Progress and Challenges." *Recent Patents on DNA and Gene Sequence*, vol. 2, 2008, pp. 187–200.

Roper, Stephan M., and Owatha L. Tatum. "Forensic Aspects of DNA-Based Human Identity Testing." *Journal of Forensic Nursing*, vol. 4, 2008, pp. 150–56.

Rudin, Norah, and Keith Inman. *An Introduction to Forensic DNA Analysis.* 2d ed. Boca Raton, Fla.: CRC Press, 2002. Provides a good introduction to the use of biological evidence in forensics as well as the history and application of DNA fingerprinting in

Watson, James D., and Andrew Berry. *DNA: The Secret of Life.* Alfred A. Knopf, 2006.

DNA BANKS FOR ENDANGERED ANIMALS

Florida panther may avoid extinction with the help of DNA banks. Photo by Rodney Cammauf. By National Park Service, [Public domain], via Wikimedia Commons

FIELDS OF STUDY

Animal science; Bioinformatics; Biophysics; Biotechnology; Reproductive technology

ABSTRACT

Building on the advances being made in DNA-related technologies in forensic science and the knowledge being accumulated in the related field of wildlife forensics, some organizations have undertaken to collect and store the genetic materials of endangered animal species in the hope that someday technological advancements will enable scientists to use these materials to restore the species.

BACKGROUND

Building on the advances being made in DNA-related technologies in forensic science and the knowledge being accumulated in the related field of wildlife forensics, some organizations have undertaken to collect and store the genetic materials of endangered animal species in the hope that someday technological advancements will enable scientists to use these materials to restore the species, potentially through cloning.

Organizations that are concerned with the loss of animal species to extinction have established banks to preserve the DNA (deoxyribonucleic acid) of endangered animals. In addition to collecting and storing biological samples from endangered species (sperm, embryos, and body tissues), preserving them in liquid nitrogen at nearly −400 degrees Fahrenheit, these organizations store information on the species' natural habitats and maintain databases to keep track of the materials that have been collected. Organizations devoted to preserving animal DNA have been established in the United States, Great Britain, China, India, and Australia; among the most widely known animal DNA banks are those maintained by the Frozen Ark Project in England and the Frozen Zoo Project in San Diego, California.

Several factors have come together to fuel the animal DNA bank movement, including advances in DNA technology and growing environmental activism. The banking of animal DNA is a conscience-driven effort by people who also want to increase awareness of the threats posed to existing species by human advancement. Estimates of potential extinctions in the twenty-first century are ominous, with some researchers asserting that the world is in the midst of a mass extinction period. Many species considered to be endangered or threatened in the early twenty-first century may someday benefit from DNA banks, including the California condor, the Florida panther, the polar bear, the killer whale, the black rhino, the panda, and the yellow seahorse.

Critics of the organizations that maintain DNA banks for endangered animals have asserted that these organizations may inadvertently create an underground market for the animals they mean to protect; the stored genetic materials could potentially have high monetary value. Moreover, some scientists believe that some species extinction is a natural part of the planet's life cycle, one that humans should not tamper with, at least until they have had much more time to observe the interactions between species and Earth's environment. The animal DNA banks, however, enjoy widespread support among scientists, if only for the value they provide as historical databases.

———*Brion Sever*

FURTHER READING

Frozen Ark. Frozen Ark Project, 2011. Web. 12 Mar. 2015.

McGavin, George. *Endangered: Wildlife on the Brink of Extinction.* Richmond Hill: Firefly, 2006. Print.

Ryder, Oliver A., et al. "DNA Banks for Endangered Animal Species." *Science* 288.5464 (2000): 127–77. Print.

Stone, Richard. *Mammoth: The Resurrection of an Ice Age Giant.* New York: Basic, 2002. Print.

DNA DATABASE CONTROVERSIES

Banner - International DNA Bank. By INTERDIPCO. (Own work.) [CC-BY-SA-3.0 (http://creativecommons.org/licenses/by-sa/3.0)], via Wikimedia Commons

FIELDS OF STUDY

Biology; Forensics; Genetics; Medical biotechnology (or Red biotechnology); Philosophy and Religious Studies; Political Science

ABSTRACT

DNA databases constitute extremely important tools for law enforcement, but the existence of such repositories raises social, ethical, and legal issues, particularly concerning privacy rights and confidentiality. Because of the potential for misuse of the information stored in DNA databases, issues of public safety need to be balanced with the protection of civil liberties.

BACKGROUND

Each person has a unique DNA (deoxyribonucleic acid) profile that may be used to identify that individual. DNA samples, however, may also be used for other purposes; a person's DNA can reveal susceptibility to certain diseases, for example, or predisposition to certain behaviors. The establishment of DNA databases has helped law-enforcement agencies greatly in identifying suspects, but many observers have expressed concerns regarding the potential for misuse of the information stored in these databases.

DNA Dragnets and Fourth Amendment Issues

After serious crimes have been committed, law-enforcement agencies sometimes conduct so-called DNA dragnets to attempt to identify suspects; that is, they obtain DNA samples from all persons in selected groups of individuals to compare with DNA found at the crime scenes. In these cases, the people involved are generally pressured to "volunteer" DNA samples or the samples are taken without due process. In one case, all men in Truro, Massachusetts (a town with a population of approximately eighteen hundred), were requested to volunteer DNA samples after a murder. In Baton Rouge, Louisiana, samples were collected from about twelve hundred men during a hunt for a serial killer. The Fourth Amendment to the U.S. Constitution protects against unreasonable searches and seizures, and it has been argued that such collection of DNA samples without probable cause (a reasonable belief that the individuals have some involvement with the crimes) may constitute "unreasonable search" under the Fourth Amendment. It is likely that this issue will ultimately be decided by the U.S. Supreme Court.

The DNA profiles produced as a result of DNA dragnets are usually entered into state criminal databases. In most cases, removal of a profile from such a database requires a court order. In some states, including Virginia and Louisiana, the DNA profiles of any individuals who are arrested, even if they are not convicted, are also retained in databases. Some critics have argued that the inclusion of unconvicted individuals in DNA databases may be viewed as a violation of the concept of presumption of innocence.

The national DNA database maintained by the Federal Bureau of Investigation (FBI), known as CODIS (Combined DNA Index System), does not permit the entering of "volunteer" DNA profiles. The U.S. Congress, however, has made it legal for federal agencies to collect DNA samples from suspected illegal immigrants, and those profiles are entered into CODIS.

Familial Issues

People who are related by blood have similar, but not identical, DNA profiles. Given this fact, law-enforcement agencies sometimes search DNA databases for less-than-perfect matches to their suspects' DNA profiles; this is known as familial searching. By finding first-degree relatives, the police may identify suspects. Such searches have been criticized as violating the privacy rights of the parties involved; they have yet to be tested in the courts. Familial searching has also been criticized as racially discriminatory. For instance, because African Americans are disproportionately represented in CODIS, they are approximately four times more likely than Caucasians to be "findable" through familial searching.

In addition to the use of DNA databases by law-enforcement agencies, the storage of DNA profiles in such repositories raises many other issues, particularly in the areas of privacy and confidentiality. Information on adoption and sperm and egg donation, for example, can be very sensitive, with many parties wishing to remain anonymous. Such anonymity is threatened by the placement of DNA profiles in searchable databases.

Universal Databases and Discrimination

By 2007, nine countries had established population-based DNA databases as resources for the study of genealogy and gene-disease relationships. In some nations, initiatives have been undertaken to include DNA profiles of all newborns in these databases. The storage of this information may be useful to scientists who seek to understand the genetic components of disease, but it raises issues of privacy and confidentiality. Of particular importance is the question of who has the right to access the data. In the United States, the courts have yet to decide whether DNA database information falls under the security and privacy provisions of the Health Insurance Portability and Accountability Act (HIPAA) of 1996.

It has been argued that universal DNA databases and disease gene databases have great potential for abuse. Disease genes are being identified at an increasing rate, and screening for many such genes is becoming easier and cheaper. In addition to government databases, private DNA databases are being established, and these pose additional issues of confidentiality because they contain more detailed information than that found in the government databases and so will be less likely to protect individuals' anonymity. As critics have pointed out, the availability of information on individuals' genetic characteristics has the potential to lead to discrimination by insurance carriers, employers, educational institutions, and government agencies.

Retention of DNA samples

A DNA profile only identifies the individual. The genetic markers used are short tandem repeat (STR) regions of the genome where there are variable numbers of certain DNA segments. These regions do not encode functional genes, so a person's DNA profile does not contain information on the individual's genetic makeup. In contrast, the DNA sample from which the profile was created contains all of that person's genetic information. Most U.S. states do not have laws that require the destruction of DNA samples after DNA profiling is complete.

Moreover, hospitals, clinics, and doctors' offices store tissue samples taken for biopsies or retained for research from which DNA may be extracted and analyzed. No set policy exists among forensic laboratories regarding destruction of DNA samples, and many preserve such samples in case additional testing is deemed necessary or the results of previous tests need to be confirmed. It has been argued that government agencies have not adequately regulated the retention and future potential uses of such DNA samples.

—*Ralph R. Meyer*

Further Reading

Greeley, Henry T. "The Uneasy Ethical and Legal Underpinnings of Large-Scale Genomic Biobanks." *Annual Review of Genomics and Human Genetics* 7 (2007): 343-364. Discusses the various issues related to the existence of private databases that contain genetic information on individuals.

Lazer, David, ed. *DNA and the Criminal Justice System: The Technology of Justice.* Cambridge, Mass.: MIT Press, 2004. Collection of essays explores the ethical and procedural issues related to DNA evidence. Includes a chapter by Associate Justice Stephen G. Breyer of the U.S. Supreme Court.

Moulton, Benjamin W. "DNA Fingerprinting and Civil Liberties." *Journal of Law, Medicine and Ethics* 34 (Summer, 2006): 147-148. Presents an overview of the issues addressed at a symposium on the topic of DNA fingerprinting and civil liberties. Contributions to the symposium appear as articles in the same issue of the journal.

Swede, Helen, Carol L. Stone, and Alyssa R. Norwood. "National Population-Based Biobanks for Genetic Research." *Genetics in Medicine* 9, no. 3 (2007): 141-149. Focuses on the ethics issues related to the establishment of national genetic databases.

Weiss, Marcia J. "Beware! Uncle Sam Has Your DNA: Legal Fallout from Its Use and Misuse in the U.S." *Ethics and Information Technology* 6, no. 1 (2004): 55-63. Discusses the constitutionality of DNA profiling.

DNA Extraction from Hair, Bodily Fluids, and Tissues

FIELDS OF STUDY

Biology; Forensics; Genetics; Medical biotechnology (or Red biotechnology); Political Science

ABSTRACT

DNA comparison has become a critical tool for identifying victims and suspects in a variety of crimes, and biological evidence such as hairs, bodily fluids, and tissues can provide the DNA needed for comparisons.

A forensic scientist at the U.S. Army Criminal Investigation Laboratory at Fort Gillem, Ga., processes evidence in one of the DNA extraction rooms. Thirty-two more forensic examiners and specialists were added to support the projected increased workload. By CID Command Public Affairs (United States Army). [Public domain], via Wikimedia Commons

BACKGROUND

DNA comparison has become a critical tool for identifying victims and suspects in a variety of crimes, and biological evidence such as hairs, bodily fluids, and tissues can provide the DNA needed for comparisons.

When biological materials that have been found at crime scenes—such as hairs, bodily fluids, and tissues—are submitted for DNA (deoxyribonucleic acid) analysis, the samples must first undergo DNA extraction procedures. The most common methods for extracting DNA from such materials are organic extraction, Chelex extraction, extraction using preservation paper, and extraction using silica-based columns. The common organic extraction uses detergent and proteinase to break open (lyse) cells, followed by introduction of an organic solvent to separate proteins and other cellular debris away from the DNA. Chelex extraction is a quick and easy procedure, but the purity of the DNA extracted is low. Chelex binds metal ions that could otherwise lead to poor DNA typing results, but little other purification is done. Preservation papers provide another quick method for extracting DNA. Bodily fluids are applied directly to the paper, where the cells are immediately lysed. Once the sample is dried, a small portion can be punched out, washed briefly, and then moved directly to DNA amplification. Silica-based columns bind DNA following cell lysis, allowing cellular debris to be washed through. The DNA is then eluted in relatively pure form.

HAIR

Given their ubiquity, hairs are often found at crime scenes. Hairs with root tags attached are excellent sources of DNA, whereas hairs without their roots are not good sources of nuclear DNA for short tandem repeat (STR) testing. Shed hairs, which are the kind most often found at crime scenes, generally do not have root tags and so require mitochondrial DNA testing.

Hair roots are processed like other tissues. A questioned shed hair is first cleaned to remove any exogenous DNA; an enzymatic detergent such as Terg-A-Zyme is often used for this purpose. The keratin (protein) of the hair is broken down with proteinase, detergent, and dithiothreitol, releasing any trapped DNA. Prior to this, the hair may be ground or homogenized, which further helps to free DNA.

An alternative method for extracting DNA from hair is alkaline extraction. The hair is washed and then is exposed to a strong basic solution (such as sodium hydroxide), which destroys the protein without harming the DNA. The solution is neutralized and filtered, leaving the DNA ready for analysis.

BODILY FLUIDS

The bodily fluids most commonly processed in crime laboratories (generally in dried form) are semen and blood; these are followed in frequency by saliva and urine. Other fluids, including vaginal secretions and perspiration, are also sources of DNA. In blood, the abundant red blood cells do not have nuclei and therefore do not contain DNA. In contrast, white blood cells, which make up less than 1 percent of cells present in blood, harbor a full DNA component. This means that blood is a valuable source of DNA, although not an ideal one.

Semen, which contains huge numbers of spermatozoa, each containing its complement of DNA, is considered one of the richest sources of genetic material. Owing to the strength of sperm cell walls, isolation of DNA from semen requires treatment similar to that of hair shafts. The other bodily fluids contain DNA somewhat by chance, in that epithelial cells are shed into them, such as from the mouth (saliva) or urinary tract (urine). DNA from these sources can be isolated from bodily fluids using several of the procedures noted above.

Skin, Bone, and Other Tissues

Shed skin cells are a viable source of DNA and may potentially be collected from any item that has come into contact with skin, such as clothing, keys, or backpacks. Generally, organic extraction of such DNA is performed after a cutting or swab is collected from the item that has had skin contact. It should be noted that individuals can shed cells at very different rates; thus two pieces of evidence that seem similar may produce variable levels of DNA typing success.

Bones are frequently included in forensic investigations, in general owing to their relative longevity. Fresh skeletal material is a rich source of DNA, but the longer bones are in contact with the environment, the more degraded the DNA becomes. Likewise, inhibitory chemicals such humic acids can leach from soil into bone, making DNA analysis difficult. This is particularly true of spongy bones, where rain, soil, and microorganisms easily enter and destroy DNA. In contrast, skeletal materials with more cortical (compact) bone (such as the femur, or thighbone) resist DNA degradation. The DNA is extracted as described above after the bone is cleaned and then ground or drilled to create a powder. In some instances, bone may be decalcified using EDTA (ethylenediaminetetraacetic acid), which helps to augment its breakdown.

Other tissues are also potential sources of DNA. Generally, tissues that do not harbor degradative enzymes (as does the digestive tract, for instance) are favored. Small pieces of tissue may be homogenized to disrupt cells before the tissue is incubated in detergent and proteinase. Organic extraction is the most common method of DNA purification, but DNA has been extracted successfully using the Chelex and silica-based column methods.

———*Brianne M. Kiley and David R. Foran*

Further Reading

Belgrader, P., et al. "Automated DNA Purification and Amplification from Blood-Stained Cards Using a Robotic Workstation." *BioTechniques* 19 (September, 1995): 426–32.

Butler, John M. *Forensic DNA Typing: Biology, Technology, and Genetics of STR Markers*. 2d ed. Burlington, Mass.: Elsevier Academic Press, 2005.

Deedrick, Douglas W. "Hairs, Fibers, Crime, and Evidence: Part 1—Hair Evidence." *Forensic Science Communications* 2 (July, 2000).

Graffy, Elizabeth A., and David R. Foran. "A Simplified Method for Mitochondrial DNA Extraction from Head Hair Shafts." *Journal of Forensic Sciences* 50 (September, 2005): 1119–22.

Nagy, M., et al. "Optimization and Validation of a Fully Automated Silica-Coated Magnetic Beads Purification Technology in Forensics." *Forensic Science International* 152, no. 1 (2005): 13–22.

Walsh, P. S., D. A. Metzger, and R. Higuchi. "Chelex 100 as a Medium for Simple Extraction of DNA for PCR-Based Typing from Forensic Material." *BioTechniques* 10 (April, 1991): 506–13.

DNA Fingerprinting as Evidence

FIELDS OF STUDY

Biology; Forensics; Genetics; Medical biotechnology (or Red biotechnology); Political Science

ABSTRACT

The development of DNA fingerprinting represents one of the major breakthroughs in forensic science. Using techniques from molecular biology and increasingly detailed databases, investigators are able to examine the DNA contained within biological evidence obtained from crime scenes and compare their findings with known samples, resulting in a high degree of probability that particular pieces of evidence can be associated with individual suspects or victims.

Definition

DNA fingerprinting: Laboratory procedure for analyzing patterns of sequence variation in DNA samples for the purpose of identifying evidence in forensic investigations.

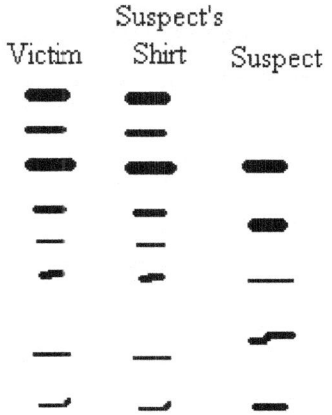

Forensic DNA evidence. By Nkonopli. (Own work.) [Public domain], via Wikimedia Commons

BACKGROUND

The development of DNA fingerprinting represents one of the major breakthroughs in forensic science. Using techniques from molecular biology and increasingly detailed databases, investigators are able to examine the DNA contained within biological evidence obtained from crime scenes and compare their findings with known samples, resulting in a high degree of probability that particular pieces of evidence can be associated with individual suspects or victims.

In 1985, the English geneticist Alec Jeffreys proposed that newly discovered repetitive sequences in DNA (deoxyribonucleic acid) could be used as a form of genetic fingerprint to identify individuals. DNA fingerprinting represents a combination of both molecular biology and population genetics in that in this process, pieces of DNA are examined for the presence of specific markers and the findings are compared against known samples in a database to establish the prevalence of those markers in the general population.

The human genome comprises more than 3.2 billion nucleotides, the letters that are responsible for coding for the proteins that make up and carry out bodily functions. The sequences of the human genome that code for these proteins are called genes. Humans are believed to have approximately twenty-five thousand genes. Between these genes are vast stretches of DNA that do not code for proteins.

Within these areas are repetitive sequences called variable number of tandem repeats, or VNTRs. One class of VNTRs is made up of the minisatellites, sequences of up to one hundred nucleotides that may be repeated in tandem up to one thousand times. In addition to the minisatellites, microsatellites have been identified. These are also known as short tandem repeats (STRs) or simple sequence repeats (SSRs). Microsatellites are sequences of two to seven nucleotides that may be repeated hundreds of times. An example is the CA repeat (CACACACA) that occurs on average every thirty-thousand base pairs in the human genome.

Every human being has two copies of genetic information in each cell. This represents the genetic information contributed by each of the individual's parents. Each STR thus exists in two copies. Often, the number of repeats within a specific STR differs in the parents. These differences are called alleles. In a population, a given STR may be polymorphic, meaning that many different forms (or alleles) of the STR exist in the population. For each allele used in forensic analysis, population geneticists have determined the percentage of the population at large that contains that given allele. This information forms the basis of DNA fingerprinting.

ANALYSIS OF DNA FINGERPRINTS

When DNA fingerprinting was initially developed, analysts examined DNA patterns using a procedure called the Southern blot. In a Southern blot, DNA is extracted from cells and then cut with a special enzyme called a restriction endonuclease. Restriction endonucleases recognize specific sequences of nucleotides in the DNA. When the sequence has been identified, the enzyme makes a cut in the DNA, generating short fragments that may be separated by size through gel electrophoresis. A radioactive probe is then used to identify specific fragments that contain sequences of nucleotides of interest. Initially, analysts accomplished this task by using minisatellites and restriction fragment length polymorphisms (RFLPs). When exposed to photographic film, the radioactive probes revealed patterns in the DNA that could be used to identify evidence.

Soon after DNA fingerprinting began, the entire process was greatly simplified by the invention of the polymerase chain reaction (PCR). Instead of cutting the DNA with restriction enzymes, the analyst copies specific sections of the genome, in this case the area containing the microsatellite repeats, millions of times. The amplified sections are then separated by gel electrophoresis, stained, and photographed. Because the fragments are much smaller than those generated during a Southern blot, the results may be ready in just a few hours of time. As with a Southern blot, the length of the fragment is determined by the number of repeats. The larger the number of repeats in the amplified section of DNA, the slower its movement during gel electrophoresis. This allows the analyst to discriminate among STRs that differ in the numbers of repeats within the amplified sequence.

In the United States, thirteen STRs have been identified for use during forensic and criminal investigations. Most investigative laboratories also include an additional test for amelogenin, a gene associated with dental pulp, which allows an investigator to determine whether a sample comes from a male or a female subject. For each of the STRs, researchers have determined the prevalence of that allele in the general population. Although the allele for one STR may be shared by a large percentage of the population, the power of discrimination becomes much greater as the number of STRs being analyzed is increased. For example, when three STRs (A, B, and C) are used, the probability of a certain combination (A^1, B^3, C^2) is equal to the product of the frequency of that allele in the population (A B C). As the number of STRs increases, the chances that two individuals will share the same identical pattern decrease.

When coupled with additional evidence found at a crime scene, DNA fingerprinting can be a powerful tool for proving the guilt or innocence of a suspect. DNA fingerprinting is also used to identify human remains that have degraded over time or that have been badly damaged by exposure to chemicals or fire.

——*Michael Windelspecht*

FURTHER READING

Butler, John M. *Forensic DNA Typing: Biology, Technology, and Genetics of STR Markers*. 2d ed. Burlington, Mass.: Elsevier Academic Press, 2005. Provides a detailed examination of DNA fingerprinting analysis using STR markers. Intended for readers with background in the sciences.

Jeffreys, A. J., V. Wilson, and S. L. Thein. "Individual-Specific 'Fingerprints' of Human DNA." *Nature* 316 (1985): 76-79. Landmark paper that introduced the concept of DNA fingerprinting as a method of identification.

Kobilinsky, Lawrence F., Louis Levine, and Henrietta Margolis-Nunno. *Forensic DNA Analysis*. New York: Chelsea House, 2007. Presents a comprehensive introduction to the use of STRs in DNA fingerprinting. Includes discussion of future directions, including mitochondrial and Y chromosome analyses.

Rudin, Norah, and Keith Inman. *An Introduction to Forensic DNA Analysis*. 2d ed. Boca Raton, Fla.: CRC Press, 2002. Provides a good introduction to the use of biological evidence in forensics as well as the history and application of DNA fingerprinting in forensic investigations.

DNA ISOLATION METHODS

FIELDS OF STUDY

Biology; Biotechnology; Forensics; Chemical engineering; Genetics; Medical biotechnology (or Red biotechnology)

ABSTRACT

DNA obtained from forensic samples can be used to link suspects to crime scenes, associate suspects and victims, identify the remains of missing individuals, or determine parentage. Numerous techniques are available for isolating DNA from cellular material; the choice of the most appropriate helps to ensure that a DNA profile will be obtained successfully.

HOW IT WORKS

The isolation of DNA (deoxyribonucleic acid) from biological material can be relatively straightforward, but forensic scientists must consider several factors before commencing. The first of these is the planned

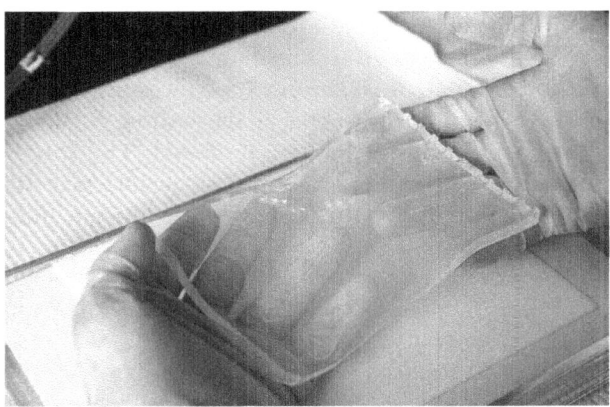

Agarose gel ready for use in the Southern method of DNA isolation. By Linda Bartlett. [Public domain], via Wikimedia Commons

subsequent DNA analysis, including whether the DNA can be single-stranded, as for polymerase chain reaction (PCR) analyses, or must be double-stranded, as for restriction fragment length polymorphism (RFLP) testing. The source of the sample (whether blood, semen, hair, or bone) will influence processing choices and may mean that a tissue must be processed in advance (for example, skeletal material may need to be ground). Other considerations include the desired level of DNA cleanliness, the maximization of yield, the minimization of processing steps, the number of samples, the presence of mixtures, and the presence of potential PCR inhibitors (such as iron in blood or humic acid in soil).

Commonly used methods of DNA isolation in forensic laboratories include organic extraction, differential extraction, Chelex, the use of preservative papers, the use of isolation kits, and the use of robotics. All of these involve breaking open cells (lysis) to release DNA, purification to remove unwanted material, and harvesting the DNA for analysis.

Once an extraction method is chosen, the analyst must take steps to prevent contamination. Proper training of laboratory personnel is imperative; disposable gloves and protective clothing should be worn, and equipment must be cleaned regularly. A reagent blank control, containing no tissue but undergoing the same extraction process, should be included to ensure that reagents are not contaminated.

Organic Extraction

Organic extractions are widely used in forensic laboratories owing to their general applicability and the purity of the resultant DNA. Following cell lysis (with a proteinase and detergent), undesired materials (such as fats and proteins) are solubilized into an organic solvent such as phenol or chloroform.

Organic extractions can be time-consuming, but the DNA collected is relatively clean, can be used for any type of subsequent analysis, and is amenable to any tissue type. Disadvantages of these methods include lengthy time expenditure and exposure to hazardous chemicals.

Differential Extraction

An expanded organic method is the differential extraction, which is used on samples from sexual assaults, particularly vaginal swabs, which can contain epithelial cells from the victim and sperm from the perpetrator. Differential extractions take advantage of the dissimilar nature of sperm and epithelial cell walls. The sample is first placed in a mild lysis buffer that releases epithelial cell DNA while sperm remain intact. The sperm are pelleted by centrifugation, and the liquid containing epithelial DNA (the nonsperm fraction) is removed and purified organically. The sperm are then lysed under stronger conditions, and this male/sperm fraction is purified. Differential extraction allows enrichment of each fraction by upward of 90 percent, helping to clarify mixture results.

Chelex Extraction

DNA preparation using Chelex (iminodiacetic acid bound to polystyrene beads) has two positive attributes: It is fast and it is easy. The entire procedure is carried out in a single tube and is generally performed on blood and saliva, although other tissues may be considered. The major objective of a Chelex extraction is to bind (chelate) unwanted metals that can inhibit PCR, notably polyvalent cations (such as iron and calcium). The sample is boiled, and upon centrifugation the beads are forced to the bottom of the tube, leaving the DNA in solution ready for quantification and amplification. Because minimal purification steps are involved in this method, the DNA is not pristine (hence it may not amplify or store well). Also, owing to the boiling step, the DNA is single-stranded.

Preservative Papers

One of the simplest methods for extracting DNA is through the use of special papers chemically treated to lyse cells and denature proteins on contact. A liquid sample (blood, saliva) is applied to the paper and dried, then stored at room temperature or used immediately. A small punch of the stained paper is collected, washed, and subjected directly to PCR. Extended sample stability is the major advantage of the use of preservative papers; disadvantages include possible difficulty in manipulating the papers and the fact that this method produces no DNA quantification.

COMMERCIAL DNA ISOLATION KITS

Several companies have developed commercial kits for DNA purification. These tend to be quick (as little as thirty minutes) and easy to use, but they are often expensive. Kits can allow for a large number of samples to be processed simultaneously, and manufacturers provide necessary solutions as well as other materials, such as tubes and columns. Generally, cells are lysed and the DNA is bound in place (for example, to silica on a column), followed by washing and DNA release. The DNA isolated tends to be pure, but sample digestion is short, and thus yield may be sacrificed; yield can particularly suffer when limited amounts of DNA exist.

AUTOMATION

The use of robotic means of DNA preparation allows the processing of large numbers of samples in short amounts of time while eliminating the human factor. Automated DNA isolation is most desirable for work with high-quality material, such as database samples. Material involved in law-enforcement investigations is less often processed in this manner. In automated methods, the DNA isolation procedures are similar to those detailed above (particularly those for kits), with reagent transfer being automated. The robots involved are expensive, as are the proprietary reagents required, but the savings in technician time can be substantial.

—*Amy L. Barber and David R. Foran*

FURTHER READING

Belgrader, P., et al. "Automated DNA Purification and Amplification from Blood-Stained Cards Using a Robotic Workstation." *BioTechniques* 19 (September, 1995): 426-432.

Butler, John M. *Forensic DNA Typing: Biology, Technology, and Genetics of STR Markers*. 2nd ed. Burlington, Mass.: Elsevier Academic Press, 2005.

Greenspoon, S. A., et al. "Application of the BioMek 2000 Laboratory Automation Workstation and the DNA IQ System to the Extraction of Forensic Casework Samples." *Journal of Forensic Sciences* 49 (2004): 29-39.

Nagy, M., et al. "Optimization and Validation of a Fully Automated Silica-Coated Magnetic Beads Purification Technology in Forensics." *Forensic Science International* 152, no. 1 (2005): 13-22.

Walsh, P. S., D. A. Metzger, and R. Higuchi. "Chelex 100 as a Medium for Simple Extraction of DNA for PCR-Based Typing from Forensic Material." *BioTechniques* 10 (April, 1991): 506-513.

DNA PROFILING

FIELDS OF STUDY

Biology; Biotechnology; Chemical engineering; Forensics; Genetics; Political Science

ABSTRACT

When DNA is analyzed and typed, a graph (called an electropherogram) containing peaks representing the different alleles (different forms of a gene) inherent to that particular sample is obtained. These data must be statistically interpreted by a forensic scientist, who calculates the frequency of that "fingerprint" combination in the general population, before they are presented and accepted in a court of law. Without statistical value, DNA processing serves no practical purpose in criminal investigations.

HOW IT WORKS

In legal cases involving DNA (deoxyribonucleic acid) evidence, statistical analysis provides investigators with a tool that can potentially exclude innocent individuals from suspect lists. In forensic science, DNA samples are used not only to establish similarities between evidence and suspects or victims but also to identify

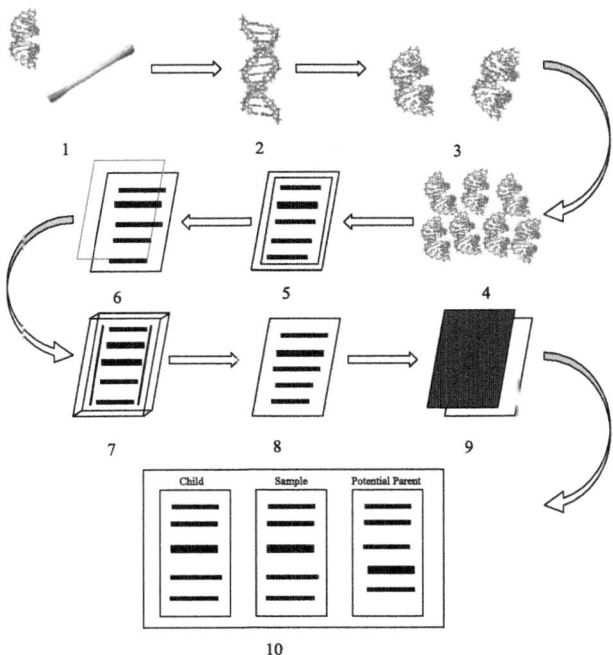

1: A cell sample is taken- usually a cheek swab or blood test 2: DNA is extracted from sample 3: Cleavage of DNA by restriction enzyme- the DNA is broken into small fragments 4: Small fragments are amplified by the polymerase chain reaction- results in many more fragments 5: DNA fragments are separated by electrophoresis 6: The fragments are transferred to an agar plate 7: On the agar plate specific DNA fragments are bound to a radioactive DNA probe 8: The agar plate is washed free of excess probe 9: An x-ray film is used to detect a radioactive pattern 10: The DNA is compared to other DNA samples. By Sneptunebear16. (Own work.), [CC BY-SA 4.0 (https://creativecommons.org/licenses/by-sa/4.0)], via Wikimedia Commons

victims of mass murders and catastrophes and to determine the parentage of children. The statistical analyses employed vary depending on the situation.

Because 99 percent of the bases that form the human genome are identical among all individuals, analysts need to use several different DNA markers to encounter differences in the remaining 1 percent. Most crime laboratories in the United States use the thirteen short tandem repeat (STR) markers used by the national DNA database, the Combined DNA Index System (CODIS), to compare DNA samples. Of these thirteen markers, only one needs to differ to exclude a particular individual (except identical twins) from being the source of an evidence sample. In that case, no statistical analysis is necessary. The inclusion of individuals, however, is somewhat more complicated.

Allele frequencies for each STR marker (the number of occurrences of a particular allele) have been determined by scientists working for the Federal Bureau of Investigation (FBI), a project funded by the National Science Foundation (NSF), and the National Institute of Science and Technology for various geographic regions and ethnic populations. These databases were created through the collection of DNA results from many (at least two hundred) individuals and the calculation of a percentage frequency of allele X in the total population surveyed for a specific marker. Individual crime laboratories determine which of the databases they use for frequency calculations, but the end results are similar in that they have either a very low or very high probability.

STATISTICAL APPROACH

Each human being is made up of thousands of genes, each of which is inherited from the parents. Most individuals possess two copies (alleles) of a gene, one donated by the mother and one donated by the father. These copies can have either the same form (for example, two alleles for black hair) or different forms (one allele for black hair, one allele for blond hair). If the occurrence of inheriting one marker has no effect on the occurrence of the other, statisticians and analysts are able to multiply the frequencies of the individual alleles to establish the overall frequency of the DNA profile.

In criminal investigations, the races or geographic origins of the perpetrators are often not known; thus, a forensic scientist cannot place a suspect into an ethnic category (with any degree of certainty high enough to stand up in court) when determining which allele frequencies to use. Therefore, the most conservative approach is commonly used. For example, if given an allelic frequency database that has African American, Caucasian, and Asian populations, and the frequency of allele A of marker B is to be determined, the analyst will probably select the population in which allele A is more common or has the highest percentage. In doing this for each allele in question, the analyst will obtain a final number that is the highest possible for that specific profile. This approach ultimately favors the suspect and removes any biases that might compromise the investigation. Once all the allelic frequencies corresponding to the sample are obtained, the Hardy-Weinberg probabilistic principle is used to calculate the occurrence of that DNA fingerprint.

The Hardy-Weinberg principle is based on the assumption that the probability of two independent events occurring at the same time is the product of the probability of them occurring separately. The equilibrium formula for the Hardy-Weinberg principle is $p^2 + 2pq + q^2$, in which p and q represent the two possible forms of a particular gene. Thus, when an individual is homozygous (has two identical copies of the allele) for a particular allele, the frequency of that allele is squared (p^2 or q^2) to obtain the overall frequency for that marker in that particular individual. If the individual is heterozygous (has two different copies of the allele), then the frequency of the first allele is multiplied by the frequency of the second allele and the product is multiplied by two ($2pq$). Once all marker frequencies are obtained, the individual frequencies are multiplied to obtain the rarity of the overall DNA fingerprint.

INTERPRETATION

When attempting to establish biological relationships, such as paternity, analysts do not attempt to find the probabilistic nature of randomly finding other individuals in the general population with matching profiles. Instead, they take into account the increased probability of similarity between samples (because there is a chance that they are related) to prevent under- or overestimation of the likelihood ratio. Take, for example, a paternity dispute in which the DNA of a child's alleged father is analyzed. It is known that the child has a father, and because there is only one alleged father, a fifty-fifty chance exists that he is the father. When the calculation is done, a paternity likelihood probability will be obtained, and the known 50 percent default probability will increase this likelihood by a certain factor, thus making the results stronger.

Forensic scientists performing DNA analysis must be cautious, however, when profiles are incomplete or the DNA is degraded, given that markers that are actually heterozygous could be interpreted as homozygous because only one allele was amplified from the damaged DNA. Additionally, scientists need to perform more thorough and complicated analyses when they are dealing with mixed profiles or when there is suspicion of contamination of the DNA evidence.

—*Lilliana I. Moreno*

FURTHER READING

Barbaro, Anna, Patrizia Cormaci, and Aldo Barbaro. "DNA Analysis from Mixed Biological Materials." *Forensic Science International* 146, supp. 1 (Fall, 2004): S123-S125.

Buckleton, John, Christopher M. Triggs, and Simon J. Walsh, eds. *Forensic DNA Evidence Interpretation*. Boca Raton, Fla.: CRC Press, 2005.

Butler, John M., et al. "Allele Frequencies for Fifteen Autosomal STR Loci on U.S. Caucasian, African American, and Hispanic Populations." *Journal of Forensic Sciences* 48 (Summer, 2003): 908-911.

Fung, Wing K. "User-Friendly Programs for Easy Calculations in Paternity Testing and Kinship Determinations." *Forensic Science International* 136 (Fall, 2003): 22-34.

Lucy, David. *Introduction to Statistics for Forensic Scientists*. Hoboken, N.J.: John Wiley & Sons, 2005.

DNA RECOGNITION INSTRUMENTS

FIELDS OF STUDY

Biology; Biotechnology; Chemical engineering; Forensics; Genetics

ABSTRACT

The use of biological evidence, specifically DNA, to prove the guilt or innocence of suspects is an important component in modern criminal investigations. By using DNA recognition instruments at crime scenes to identify the presence of materials from which DNA evidence may be isolated—such as hair, saliva, or semen from suspects or victims—investigators increase the efficiency of evidence gathering.

BACKGROUND

The detection of DNA (deoxyribonucleic acid) at a crime scene begins with the identification and isolation of biological evidence such as blood, semen, saliva, or hair. Historically, such evidence has been

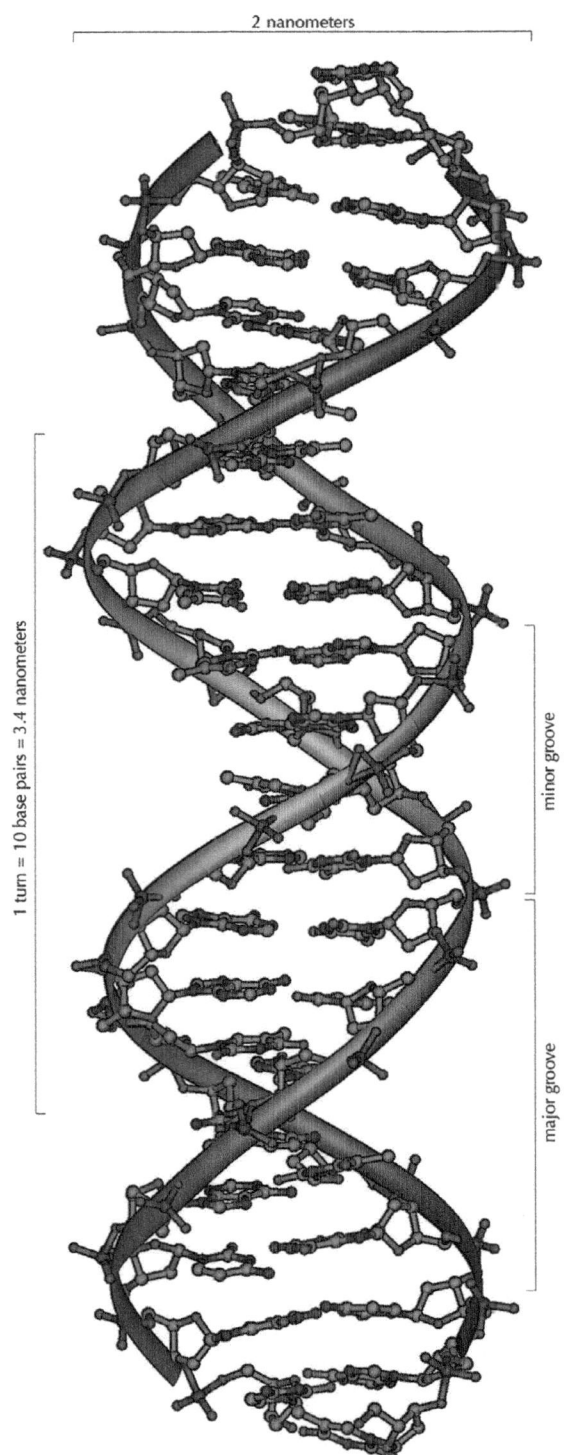

An overview of the structure of DNA. By Michael Ströck. (mstroeck.) [GFDL (www.gnu.org/copyleft/fdl.html) or CC-BY-SA-3.0 (http://creativecommons.org/licenses/by-sa/3.0/)], via Wikimedia Commons

found through physical searches of crime scenes, but the search process has been expedited by the development of specialized light sources and chemical tests.

HOW IT WORKS

Most commonly, the detection of biological samples has been aided by the use of ultraviolet (UV) light sources. UV lights belong to a class of instruments known as alternate light source (ALS) instruments. Unlike light sources that emit wavelengths of light across a broad spectrum, ALS lights use filters or special bulbs to emit a much narrower range of wavelengths. For example, UV lights emit wavelengths in the 400-200 nanometer range. The most common UV light is called a "black light," which emits wavelengths in the 400-320 nanometer range, also known as UVA. These wavelengths are invisible to the human eye but may cause certain chemicals, mainly proteins, to fluoresce. UV lights are useful in the detection of blood, semen, and saliva. Other ALS instruments include copper and argon lasers and modified arc lamps. The use of lasers in DNA detection is becoming increasingly popular because lasers do not damage DNA strands and can emit very specific wavelengths of light.

ALS instruments can indicate fluorescing molecules, but upon finding such samples, investigators must confirm the presence of specific forms of biological evidence by using certain chemicals to perform presumptive tests. These are usually colometric tests, meaning that the reagents used in the tests change colors when exposed to specific compounds. For example, the presumptive test for semen is specific for acid phosphatase, an enzyme that is more abundant in semen than in other body fluids. The presumptive tests for blood use human-specific antibodies that allow investigators to distinguish between human blood sources and the blood of other animals.

In addition to light sources and chemical tests, several commercially produced detection kits are available to forensic investigators. Among these are kits that can detect the proteins present in semen and kits that can detect the presence of DNA. Kits are also available that can disregard female DNA and indicate the presence of male DNA only, making them especially useful in rape and sexual assault cases.

——*Michael Windelspecht*

FURTHER READING

Butler, John M. *Forensic DNA Typing: Biology, Technology, and Genetics of STR Markers.* 2d ed. Burlington, Mass.: Elsevier Academic Press, 2005

James, Stuart H., and Jon J. Nordby, eds. *Forensic Science: An Introduction to Scientific and Investigative Techniques.* 2d ed. Boca Raton, Fla.: CRC Press, 2005.

DNA SEQUENCING AND CRIME SCENES

FIELDS OF STUDY

Biology; Biotechnology; Chemical engineering; Forensics; Genetics; Political Science

ABSTRACT

Biological specimens obtained at scenes of crimes are not always intact; some have been exposed to harsh environmental conditions and others do not possess the necessary amount of nuclear DNA for standard analysis. When analysts encounter these types of specimens, they can use the alternative method of DNA sequencing to discriminate between samples.

DNA Sequencing. By Linda Bartlett, [Public domain], via Wikimedia Commons

BACKGROUND

In some instances, biological samples are not suitable for performing the relatively fast and standard nuclear analysis and typing of DNA (deoxyribonucleic acid). On occasion, the copy number of nucleated cells is too low, and the analyst cannot obtain enough nuclear DNA to acquire a strong enough signal to interpret after typing the DNA. DNA sequencing, although more time-consuming and complicated than the standard analysis, may provide the necessary clues when other methods to obtain a good profile using conventional typing methods are exhausted.

HOW IT WORKS

The first step in sequencing a DNA sample is to determine what fraction of the genome is to be read (usually between three hundred and one thousand base pairs) and perform an amplification reaction using unlabeled primers that target the region of choice. After the DNA is amplified, the Polymerase chain reaction (PCR) product is cleaned to remove any unbound nucleotides and residual, nonincorporated primers. This purified product is subsequently quantified and diluted to a predetermined concentration that is dependent on the length of the fragment of DNA being sequenced. The sequencing template is amplified with a single primer (forward or reverse) so that the end products are single-stranded DNA molecules. Unlike in the more common PCR reactions, a pair of primers cannot be used because the end product would result in a double-stranded molecule that could not be interpreted. This would be analogous to attempting to read two lines of text when one is superimposed on the other, a task that is difficult to perform.

In the sequencing reaction, fluorescently tagged modified nucleotides called dideoxyribonucleic acids (ddNTPs) are present. Each is labeled with a different color fluor and is randomly incorporated into the newly synthesized DNA strand. Because ddNTPs lack the hydroxyl group normally involved in the elongation step, the synthesis of the new strand terminates with the labeled ddNTP. The cycle-sequencing product is then cleaned to remove any unbound nucleotides and dried.

The samples are now ready to be "read," or separated. The dried samples are resuspended in formamide, a denaturant, before they are loaded onto high-throughput genetic analyzers. The sample is electrokinetically injected into a capillary filled with a gel polymer that is able to provide the one-base-pair resolution needed to determine the order of nucleotides in a particular DNA fragment. Because each nucleotide is labeled with fluors that will emit at different wavelengths, color distinction can be used to determine the sequence of the sample. The fluorescence is captured by a camera and transferred to the computer's software, which makes the data available to the analyst for further interpretation.

—Lilliana I. Moreno

FURTHER READING

Kieleczawa, Jan. *DNA Sequencing: Optimizing the Process and Analysis.* Sudbury, Mass.: Jones & Bartlett, 2005.

Nunnally, Brian K. *Analytical Techniques in DNA Sequencing.* New York: Taylor & Francis, 2005.

DNA TYPING

FIELDS OF STUDY

Biology; Biotechnology; Chemical engineering; Forensics; Genetics; Political Science

ABSTRACT

In forensics, the collection and processing of evidence are steps in crime scene investigations that are expected ultimately to aid in the process of conviction or exoneration of potential suspects. DNA typing can provide a unique "picture" that can identify an individual. If the DNA found in a crime scene sample is a perfect match for that of a known sample, this constitutes a powerful piece of evidence that can often help in the conviction of the guilty; lack of an exact match can potentially exonerate an individual.

BACKGROUND

The human genome comprises approximately three billion nitrogenous bases, of which 99 percent are identical across the human population, leaving only 1 percent that makes each person's DNA (deoxyribonucleic acid) unique; the only exception is identical twins, who have identical DNA. When DNA is amplified, an analyst uses a set of primers to target locations in the molecule that are known to vary between individuals. Within these varying regions, however, possible similarities still exist; for example, at locus A, person X and person Y are both heterozygous with alleles named 13, 15. In locus C, these same individuals are homozygous for allele 12, but the fact that they share the same alleles for these two markers does not indicate that they are the same person. The power of DNA discrimination is evident when a combination of markers (usually thirteen) gives at least one nonmatching allele between the samples being typed, indicating that the samples did not come from the same individual.

HOW IT WORKS

Among the different human DNA typing techniques are restriction fragment length polymorphisms (RFLPs; often referred to as variable number of tandem repeats, or VNTRs), single nucleotide polymorphisms (SNPs), short tandem repeats (STRs), mitochondrial (mtDNA), and Y chromosome. The technique selected depends on the source of the sample and the degree of separation needed. Some of the products of the techniques noted above are often visualized on agarose or polyacrylamide slab gels, whereas others can be loaded into genetic analyzers. Although gels separate DNA, the resolution obtained is much less exact than that required for DNA typing in criminal cases and often more complicated to analyze. Genetic analyzers are instruments based on capillary electrophoresis technology. They use a capillary loaded with a gel polymer that acts like a sieve and is able to separate amplified DNA fragments based on size and charge. The DNA is electrokinetically injected into the capillary and kept at a constant heat to keep the DNA traveling through it in a denatured (single-stranded) form. The shorter fragments travel faster and elute first out of the capillary; the longer fragments move more slowly and are retained longer in the capillary. As the fragments are eluted, the fluorescent tag associated with each fragment is detected and recorded by a camera, and the data are transferred to a computer. When all the fragments present have traveled through the capillary, a unique DNA fingerprint is obtained. This DNA fingerprint is now ready to be profiled.

—*Lilliana I. Moreno*

FURTHER READING

Butler, John M. *Forensic DNA Typing: Biology, Technology, and Genetics of STR Markers*. 2d ed. Burlington, Mass.: Elsevier Academic Press, 2005

Carracedo, Angel. *Forensic DNA Typing Protocols*. Totowa, N.J.: Humana Press, 2005.

Moreno, Lilliana I., and Bruce McCord. "Separation of DNA for forensic Applications Using Capillary Electrophoresis." In *Handbook of Capillary and Microchip Electrophoresis and Associated Microtechniques*, edited by James P. Landers. 3d ed. Boca Raton, Fla.: CRC Press, 2008.

DNA: RECOMBINANT TECHNOLOGY

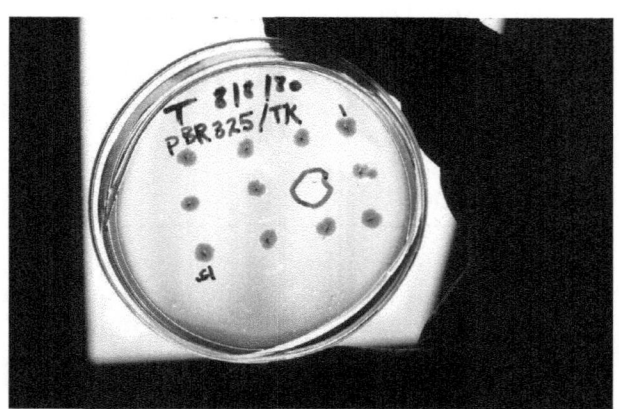

A culture medium and the hands of a technician. In this recombinant DNA technology, the thymidine kinase gene of *herpes simplex* virus is being cloned in bacteria. Those bacteria that have incorporated the gene are no longer resistant to the antibiotic tet. By Linda Bartlett. [Public domain], via Wikimedia Commons

Recombinant DNA technology has been essential for understanding DNA sequences. Because of their large, complex genomes, it was difficult to study one gene in eukaryotes, but recombinant DNA technology has allowed the isolation and amplification of specific DNA fragments facilitating the molecular analysis of genes. In addition, the tools of recombinant DNA technology have been used to create genetically modified plants. Such modifications include the introduction of resistance to insects, herbicides, viruses, and bacterial and fungal diseases into plants. Plants have also been made to produce antibodies so that plants can serve as edible vaccines.

DNA STRUCTURE

Organisms contain two kinds of nucleic acids: ribonucleic acid, or RNA, and deoxyribonucleic acid, or DNA. DNA is made of a double chain, or helix. The structure of one chain, or strand, is a backbone made up of repetitions of the same basic unit. That unit is a five-carbon sugar molecule called 2'-deoxyribose attached to a phosphate residue. RNA contains a ribose sugar instead. Also attached to the sugar part of the backbone are other molecules called bases. The four bases are adenine (A), guanine (G), cytosine (C), and thymine (T). DNA molecules are double strands that are held together because each base in one strand is paired to (hydrogen-bonds with) a base in the other strand. Adenine always pairs with the base thymine, and guanine always pairs with cytosine. A and T are called complementary bases. Likewise, G and C are complementary bases.

FIELDS OF STUDY

Biology; Biotechnology; Chemical engineering; Medical biotechnology (or Red biotechnology)

ABSTRACT

Recombinant DNA technology makes use of science's understanding of the molecular structure of DNA, the nucleic acid that encodes genetic information, to alter DNA in order to manipulate genetic traits. Such technology has immense implications for agriculture, horticulture, and the generation of medicinal compounds from plants.

DNA is shaped much like a helical ladder, with the sugar and phosphate backbones being the sides of a ladder and the base pairs that hold the two strands together being the rungs of a ladder. DNA is often represented as a string of letters, with each letter representing a base. The order of A's, T's, G's, and C's (the rungs of the ladder) along a DNA double helix is the sequence of that DNA which contains the genetic information.

RESTRICTION ENZYMES

In recombinant DNA technology, scientists are able to use molecules called *restriction enzymes* to make cuts at specific sequences. Some restriction enzymes make cuts straight across the two strands of DNA in the double helix, creating blunt ends. Other restriction enzymes cut the two strands in a staggered pattern, leaving short, specific single strands at the cut sites. These single-stranded regions, called *sticky ends* or cohesive ends, can base-pair (hydrogen-bond) with complementary base sequences from other, similarly cut DNAs. These sticky ends allow joining of DNA from any source cut with restriction enzymes that create the same ends.

Cutting with restriction enzymes creates fragments of DNA with sequence-specific ends that can be spliced into small, self-replicating vehicle, or vector, molecules and introduced into a host cell where the vector molecules with the added DNA fragments replicate to produce a large amount of specific DNA for analyses. This process is *recombinant DNA cloning*.

DNA CLONING

One way to clone a specific gene is to clone all the DNA fragments generated from cutting with a restriction enzyme and then screen for the clone containing the desired gene. This method of cloning random DNA segments into a vector is called *shotgunning*. The entire collection of such cloned fragments, which together represent the entire genome of the organism, is called a gene library. Genomic DNA libraries are made by cloning the total genomic DNA of an organism.

Another way to clone a specific gene is to begin with messenger RNA (mRNA) from the organism. (Messenger RNA is a molecule that functions to create complementary copies of DNA strands. At a ribosome the messenger RNA then determine the order of amino acids that are joined to make a protein.) Using *reverse transcriptase* (an enzyme encoded by some RNA viruses that uses RNA as a template for DNA synthesis), scientists can make a DNA copy of the mRNA. The complementary strand is also synthesized to create a double-stranded DNA called cDNA (complementary DNA) that is complementary to the mRNA. These cDNAs are then cloned to create a complementary DNA library. Individual cloned cDNAs can be used to trap the corresponding mRNA on a nitrocellulose filter. At this point, the mRNA can be used in a cell-free protein synthesis system to allow identification of the protein encoded by that cDNA clone. Alternatively, the cDNA can be used to find sequences complementary to it in a genomic library to obtain a clone of the specific gene.

NUCLEIC ACID HYBRIDIZATION

The ability to hybridize nucleic acids to find sequences complementary to a particular DNA is another essential tool that offers another way to identify cloned genes. This method is called *nucleic acid hybridization* or Southern blotting (named after E. M. Southern, who developed the method). In this procedure, DNA is cut with restriction endonucleases, and the resulting DNA fragments are separated by size using agarose gel electrophoresis. The DNA in the gel is denatured (made single-stranded) by high pH and transferred to a nitrocellulose filter. The DNAs are immobilized on the nitrocellulose in the same pattern as on the gel (a Southern blot). A probe—a specific DNA or RNA—is hybridized to the nitrocellulose. The probe is "labeled" with a radioactive or fluorescent (non-radioactive) tag so it can be detected. The probe is denatured by heat so it is single-stranded and able to anneal (hybridize) with its complementary sequence among the single-stranded DNAs tethered to the nitrocellulose. The probe is then detected to reveal the position of the DNAs that hybridized with the probe.

POLYMERASE CHAIN REACTIONS

Another tool of molecular biology is to use a *polymerase chain reaction* (PCR) to amplify specific segments of DNA in vitro (in the test tube). PCR requires a pair of sequences, called primers, about twenty base pairs long that are complementary to the ends of the region of DNA to be amplified. High temperature is used to denature the double-stranded DNA. At a lower temperature, the primers anneal (base-pair) to their complementary sequences, and a thermal-stable DNA polymerase copies the single-stranded templates. After the replication of the segment between the two primers (one cycle), the newly synthesized double-stranded DNA molecules are denatured by high temperature, the temperature is lowered, primers anneal, and a second cycle of replication occurs. The number of DNA molecules produced doubles with each cycle of replication. As a result, a million copies of a single DNA molecule can be produced in only a few hours, if the appropriate sequences for the two primers are known. PCR is a very sensitive method: Even a single DNA molecule can be amplified. PCR is much faster than recombinant DNA cloning and can produce a large amount of a specific piece of DNA.

Developing ways to determine DNA sequences (the sequences of adenine, thymine, guanine, and cytosine base pairs on the DNA "rungs") has led to the identification of the complete DNA sequences of the genomes of a number of organisms, including the model plant *Arabidopsis* as well as the much more complex human genome.

———*Susan J. Karcher*

DOLLY THE SHEEP

FIELDS OF STUDY

Biology; Genetics; Medical biotechnology (or Red biotechnology); Philosophy and Religious Studies; Political Science; Reproductive technology

ABSTRACT

The birth of Dolly and subsequently cloned animals raised a host of ethical issues and opened the door to possible means of improving human health and the environment. Dolly's comparatively short life span, about half that of the typical sheep of her breed, may have been related to her clone origins.

On February 5, 1997, Ian Wilmut of the Roslin Institute in Edinburgh, Scotland, announced the birth of Dolly the sheep, the first clone produced from a cell taken from an adult mammal. Scientists and the general public were shocked at this announcement, because it was believed that cloning a mammal from an adult cell was, at the time, technically impossible. Dolly was seven months old before her birth was made

FURTHER READING

Hill, Walter E. *Genetic Engineering: A Primer.* The Netherlands: Harwood Academic Publishers, 2000. Provides a concise overview. Illustrations, glossary, appendix, and index.

Kreuzer, Helen, and Adrianne Massey. *Recombinant DNA and Biotechnology: A Guide for Students.* Oxford: Blackwell Science, 2000. Introductory overview. Illustrations and index.

Old, R. W., and S. B. Primrose. *Principles of Gene Manipulation: An Introduction to Genetic Engineering.* Boston: Blackwell Scientific, 1994. Focuses on methods of cloning. Illustrations, appendices, references, and index.

Reece, Jane B., and Neil A. Campbell. *Biology.* 6th ed. Menlo Park, Calif.: Benjamin Cummings, 2002. Introductory textbook. Illustrations, problems sets, glossary, index.

Watson, James, Michael Gilman, Jan Witkowski, and Mark Zoller. *Recombinant DNA.* 2d ed. New York: W. H. Freeman, 1992. Emphasizes applications. Illustrations, reading lists, and index.

public. Because she was cloned from mammary cells, the research team named her after well-endowed country-western singer Dolly Parton.

HOW DOLLY WAS CREATED

A clone is an organism developed from a single cell isolated from another organism. The cell donor and the clone are genetically identical. Prior to the creation of Dolly, no attempts at cloning a mammal from adult cells had been unequivocally successful. In the early 1980's scientists had created clones of mammals by using donor cells from young embryos. The research team that cloned Dolly first cloned a pair of sheep, Megan and Morag, from embryonic cells grown in the laboratory. Adult cells and embryonic cells have identical genetic material, or deoxyribonucleic acid (DNA); however, adult cells produce proteins specific to the type of cell they become. For example, brain cells produce neurotransmitters and do not produce hemoglobin, even though they possess the hemoglobin gene. Scientists believed that the structure of the DNA in an

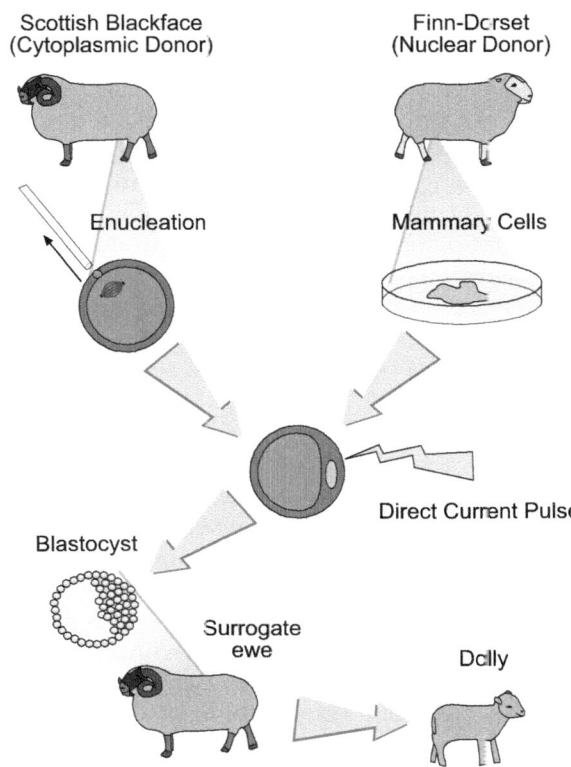

This is diagram of how Dolly the sheep was made. By Squidonius (Own work (Original text: self-made)) [Public domain], via Wikimedia Commons

adult cell was irreversibly altered during the process of maturation to gain this specificity and therefore could not be used to produce a clone.

Wilmut and his colleagues at the Roslin Institute used a novel approach to clone Dolly. The donor cells were sheep udder cells from a six-year-old pregnant ewe. The cells had been frozen for about three years, and the donor was long deceased. The researchers believed that prior attempts at cloning mammals from adult cells had failed because the cells were too active or in the wrong phase of their life. To make the cells quiescent, the research team starved them for several days. Meanwhile, the researchers removed genetic material of eggs from a different breed of sheep, a process called enucleation. They then fused the starved cells with the enucleated donor eggs and implanted them into surrogate mother sheep of a third breed. Of 277 attempts, only a single egg went full-term, resulting in the birth of Dolly. Dolly looked strikingly different from the breed of the egg donor or the surrogate mother but identical to the breed that donated the adult DNA, an observation that provides suggestive evidence that she developed from the donor DNA.

Initially many questions surrounded the validity of the experiment that produced Dolly. Some scientists believed that she could have been the result of a contaminating fetal cell, and the results of the experiment were not easily replicated by other researchers. However, in July, 1998, Japanese scientists announced the birth of two calves cloned from adult cow uterus cells, and researchers in Hawaii successfully produced more than fifty mouse clones from adult mouse ovarian cells. Also, DNA analysis of Dolly confirmed that she was indeed the first clone of an adult mammal. After reaching sexual maturity, Dolly mated naturally and on April 13, 1998, gave birth to a lamb, showing that she was a healthy young adult whose ability to reproduce was not compromised by her unusual origin. She would have two more pregnancies and bear five more lambs during her life.

IMPLICATIONS

In May, 1999, researchers found that the telomeres in Dolly's cells were shorter than those in other mammals of similar age, a finding borne out in other cloning research. Telomeres, which are sequences at the ends of chromosomes, become progressively shorter as an organism ages, but it was not clear whether Dolly's life span would be shorter than usual. By January, 2002, Dolly was showing signs of arthritis; whether the condition was the result of premature aging or the amount of time she had spent indoors on a hard floor was unclear. She subsequently developed a progressive lung disease.

The ailing Dolly was euthanized in February, 2003, at the age of six and a half. Sheep of her breed typically live to be about twelve. A postmortem examination confirmed that she had arthritis in her hind legs and that her lung ailment was sheep pulmonary adenomatosis (a lung tumor associated with a retrovirus), a fatal disease common in older sheep, particularly those living indoors. The examination revealed no other abnormalities. During her lifetime, she had tended toward stoutness, but her keepers attributed Dolly's weight problem not to her clone origins but to her living mostly indoors and being fed treats by her many visitors. Dolly's remains were donated to the National Museum of Scotland in Edinburgh, where her preserved body was placed on exhibit.

There are numerous potential applications for the cloning technology that produced Dolly. Scientists envision using cloning in tandem with genetic engineering to create animals with organs suitable for transplant into humans, or ones that produce human proteins for use in pharmaceuticals. In fact, Wilmut's research was sponsored by the Scottish pharmaceutical company PPL Therapeutics, Ltd. His subsequent research involved sheep cloned from fetal cells that had been genetically altered to carry a human gene that caused the animals' milk to contain a blood-clotting protein with the potential for use in treating human hemophilia.

Some researchers envision entire herds of genetically identical cattle. Because it is very difficult to produce prize milk- or meat-producing animals consistently with traditional breeding methods, repeated cloning of one prize breeding animal would greatly speed the process. A potentially serious problem with genetically identical herds, however, is that genetic diversity allows species to survive changes in their environment and attacks by disease. Diseases affecting only a few individuals of a genetically diverse species may become rampant in a genetically identical one. If genetic diversity is lost, it could lead to the extinction of that species.

Cloning has long been regarded as a possible means of bringing endangered animal species back from the brink of extinction, particularly those species that do not breed well in captivity. However, the expense and failure rate of the technique may be too great for it to be practicable. The year 2001 saw the birth of the world's first cloned endangered wild animals. A cloned gaur (a type of Southeast Asian ox) was successfully brought to term by a domestic cow that served as its surrogate mother, but the newborn succumbed to common dysentery two days after its birth. A cloned European mouflon, one of the world's smallest wild sheep species, fared better. Born to a domestic sheep, the mouflon was raised at a wildlife center in Sardinia. Other endangered species that have since been successfully cloned include the Javan banteng (a type of wild cattle) and the African wildcat. The first clone of an extinct animal was born in 2009. Cloned from the frozen skin of a bucardo, or Pyrenean ibex, a subspecies of wild goat that went extinct in 2000, the animal died minutes after birth.

—*Karen E. Kalumuck, updated by Karen N. Kähler*

FURTHER READING

Einsiedel, Edna. "Brave New Sheep: The Clone Named Dolly." In *Biotechnology: The Making of a Global Controversy*, edited by Martin W. Bauer and George Gaskell. New York: Cambridge University Press, 2002.

Franklin, Sarah. *Dolly Mixtures: The Remaking of Genealogy*. Durham, N.C.: Duke University Press, 2007.

Kolata, Gina Bari. *Clone: The Road to Dolly, and the Path Ahead*. New York: HarperCollins, 1998.

Morgan, Rose M. *The Genetics Revolution: History, Fears, and Future of a Life-Altering Science*. Westport, Conn.: Greenwood Press, 2006.

Wilmut, Ian, and Roger Highfield. *After Dolly: The Uses and Misuses of Human Cloning*. New York: W. W. Norton & Company, 2006.

———, Keith Campbell, and Colin Trudge. *The Second Creation: Dolly and the Age of Biological Control*. New York: Farrar, Straus and Giroux, 2000. The two researchers who made Dolly team up with a noted British science writer to give a personal but rigorous explanation and thoughtful examination of cloning. Contains a helpful glossary of terms.

Drug testing

FIELDS OF STUDY

Biology; Biotechnology; Forensics; Medical biotechnology (or Red biotechnology); Political Science

ABSTRACT

Drug testing is done to ensure the safety of the general public, to maintain standards at schools and places of employment, and to make sure that athletes do not gain unfair advantage through the use of performance-enhancing drugs. The goal of these tests is detect whether a person has used drugs such as alcohol, marijuana, cocaine, amphetamines, barbiturates, benzodiazepines, lysergic acid diethylamide (LSD), opiates, phencyclidine (PCP), synthetic hormones, and steroids. Commonly used drug tests analyze a person's breath, urine, saliva, sweat, blood, or hair.

BACKGROUND

Drug testing in the workplace and schools has become commonplace. A variety of tests are used to detect elevated levels of the most common drugs that can impair job performance or are illegal to use. The importance of drug testing has continued to increase since the Controlled Substances Act of 1970 placed all regulated drugs into five classifications based on their medicinal value, their potential to harm people, and their likelihood of being abused or causing addiction. Schedule I drugs have no known medical value and are most likely to be abused, while Schedule V drugs have little potential for abuse. These scheduled drugs are called controlled substances because their use, manufacture, sale, and distribution are subject to control by the federal government.

There are two general types of drug testing. Federally regulated drug testing, according to the National Institute on Drug Abuse (NIDA), requires testing for cannabinoids (THC, marijuana, hashish), cocaine, amphetamines, opiates (morphine, heroin, and codeine), and PCP. Nonfederally regulated drug testing is often used to test athletes in various sports for the use of creatine, hormones, steroids, and other performance-enhancing drugs. Additional tests are used to detect barbiturates and alcohol.

Urinalysis is typically used as a preliminary test because it is less expensive and more convenient than the other tests. Saliva tests and breathalyzers are commonly used. Blood tests, although less frequently employed because they are generally more expensive and invasive, are more dependable, as are hair strand tests. A preliminary positive test using urinalysis must be confirmed by diagnostic tests completed in an analytical laboratory setting, which can take several days to complete. These diagnostic tests include the analytical instruments of gas chromatography (GC), mass spectrometry (MS), ion scanning, high-pressure liquid chromatography (HPLC), immunoassay (IA), and inductively coupled plasma spectrometry (ICP-MS).

HISTORY

The detection of ingested drugs in various body fluids first sparked the interest of the ancient alchemists. In 1936, Rolla N. Harger of Indiana University patented the Drunkometer, a breath test to measure a person's level of alcohol intoxication. In 1954, Robert F. Borkenstein of Indiana University invented the breathalyzer, which had the benefit of greater portability, to measure blood-alcohol content. However, it was not until the widespread use of recreational drugs in the 1960's that the National Institute of Drug Abuse was established to monitor drug use. With its creation, federal funding became available to researchers to develop drug testing methods, which led to rapid advances. In 1973, physician Robert L. DuPont was appointed director of the National Institute of Drug Abuse. As director, DuPont implemented the use of the urine test and further developed immunoassays to test for several controlled substances.

In 1981, an airplane crashed on the USS *Nimitz*, and the investigation into the incident revealed drug use to be a contributing factor. As a result, the United States Navy began random drug testing of all active-duty personnel in 1982. In the 1980's, the U.S. Department of Transportation began to test all of its employees. In September, 1986, President Ronald Reagan signed Executive Order 12564, making drug testing mandatory for federal employees and all employees in safety-sensitive positions, such as employees in the nuclear power industry. The National Institute of Drug Abuse extended this mandatory testing to include truck drivers working in the petroleum industry. This testing has come to be regulated by the Substance Abuse and Mental Health Services Administration (SAMHSA), which is part of the U.S. Department of Health and Human Services.

HOW IT WORKS

Because of the commercial availability of so many masking agents, the most effective drug testing occurs when the subject has had no previous notification. Thus, random drug testing is very effective and has become common in the workplace, schools, and for athletes. The National Collegiate Athletic Association and the National Football League provide only one to two days notice before drug testing, and the United States Olympic Committee has a no-notice policy for drug testing.

The first commonly available drug testing method was the breathalyzer, followed by urinalysis. The usage of saliva tests continues to increase, while sweat tests remain the least-used testing method. Blood tests require additional medical staff and are also the most invasive testing method; therefore, although they are very accurate, they are not as commonly used as urinalysis. Hair tests are very accurate but do not detect drug use in the last four to five days. In terms of validity for legal purposes, any of these preliminary, or screening, tests must be confirmed by an analytical technique, most often gas chromatography/mass spectrometry, performed by trained personnel within a diagnostic laboratory.

Breath. Borkenstein received a patent in 1958 for the breathalyzer, which determines an individual's blood-alcohol content (BAC) from a breath sample. The ethanol in the breath of an individual reacts with the dichromate ion, which has a yellow-orange color, in the presence of acid to form the green chromate ion. This color change from pale orange to green can easily be observed. All fifty states and the District of Columbia have laws that forbid a person to drive with a BAC of 0.08 percent or greater, a level at which the individual is judged to be legally impaired.

Urine. Urinalysis became a common method of detecting drugs in the 1980's and has continued to be widely used. The urine sample is collected and sealed to ensure that it remains tamper-free. It is generally subjected to an immunoassay test first because this test is very fast.

Saliva. Testing of oral fluids is becoming increasingly common because of its convenience for random testing, and it is more resistant to adulteration than urine samples. Saliva testing can detect cocaine, amphetamine, methamphetamine, marijuana, bezodiazepines, PCP, opiates, and alcohol if the substance was ingested between six hours and three days before the test was administered.

Sweat. Although traditionally not considered to be as useful as the other methods because of the dilute sample obtained, patches that can be worn on the skin and collect samples over several hours are increasingly popular. This method of drug testing is preferred by government agencies such as parole departments and child protective services in which urine testing is not the method of choice.

Blood. Because blood tests are the most invasive and expensive method, requiring additional medical personnel, they are not as widely used as the other tests as a screening method. However, blood testing is very accurate and reliable, so it is often used to confirm a positive result from another type of drug test.

Hair. Hair samples from any part of the body can be used and are extremely resistant to any type of tampering or adulteration. Special fatty esters are permanently formed in the hair as a result of alcohol and drug metabolism, and therefore, this method is very reliable.

Gas Chromatography/Mass Spectrometry (GC/MS). This tandem analytical instrumentation must be done by trained personnel within a laboratory setting and therefore is not as convenient as the other testing methods. However, it is much more accurate and is used to confirm more rapid, preliminary tests. The gas chromatograph is able to separate molecules based on their attractive interactions with the material that packs a column. Molecules take varying amounts of time to travel, or elute, from the column, resulting in different retention times, or amounts of time retained on the packing material of the column. These separated molecules are then ionized in the mass spectrometer, which is able to produce molecular weight information.

APPLICATIONS AND PRODUCTS

Drug Test Dips. Test strips known as drug test dips, dip strips, drug test cards, or drug panels use a single immunoassay panel to test for several common drugs at once. Specific reactions between antibodies and antigens allow marijuana, cocaine, amphetamine, opiates, and methamphetamine to be detected in urine samples. These assay strips are so easy to use that the staff of many schools, sports clubs, and offices can use them. However, they must be used only as a preliminary test. Positive results should be confirmed using gas chromatography/mass spectrometry conducted by an independent diagnostic laboratory.

The test strip is removed from its protective pouch and allowed to equilibrate to room temperature. Meanwhile, a urine sample is obtained in a small cup and also allowed to equilibrate to room temperature. The test strip is dipped vertically, with the arrow on the test strip pointing down into the sample, and remains immersed in the urine sample for ten to fifteen seconds. Then the test strip is removed from the urine sample and placed on a flat, nonabsorbent surface. After five minutes, the test strip is checked for the appearance of any horizontal lines. The appearance of a colored line in the control region of the dip strip and a faded color line in the test region of the dip strip indicates a negative test and that the concentration of a drug is too low to be detected. If only one line appears in the control region, with no line visible in the test region, then the test is considered to be positive for the presence of drugs. The test is considered to be invalid if only one line appears in the test region or no line appears at all. An invalid test is usually the result of either not following the procedure correctly or not using a large enough urine sample.

The test strips for marijuana use a monoclonal antibody to detect levels of THC, the active ingredient in marijuana, in excess of 50 nanograms per milliliter (ng/ml), the level recommended by SAMHSA. Methamphetamine can be detected in urine samples for three to five days after usage by using a test strip equipped with a monoclonal antibody. A positive test indicates a level in excess of 1,000 ng/mL. A test strip detects the major metabolite of cocaine, benzoylecgonine, for up to twenty-four to forty-eight hours after use. Morphine in excess of 2,000 ng/ml can also be detected by using a test strip containing a specific antibody. Morphine is the primary metabolite product of heroin and codeine.

Kits. Drug tests used for fast, preliminary screening include easy-to-use kits that can test samples of urine, saliva, breath, hair, or sweat. Of these tests, the hair test for drugs is considered to be the most accurate but is still considered to be a preliminary test. To confirm preliminary results or to obtain results for legal purposes, a sample of saliva or urine must be sent to a laboratory when a more reliable test such as GC/MS must be performed. It can take three to seven days to obtain the results. Urine drug test kits are less expensive than other tests, provide instantaneous results, and are easy to store. However, because of variations in metabolism rate, there can be a three-day to one-month detection window, making these tests easier to adulterate than the saliva, hair, sweat, or blood tests.

Adulteration of Specimens. Adulteration generally refers to intentional tampering with a urine sample, and certain substances can be added to urine to create a false-negative test result. Adulteration of urine samples is a common problem because four types of masking products—dilution substances, synthetic urine, cleaning substances, and adulterants—are readily available. More than four hundred readily available commercial products can mask urine samples. Dilution substances, including diuretics, lower the concentration of drug in a sample. An individual can either ingest one of these substances before submitting a urine sample or add the substance directly to the urine sample. Synthetic and dehydrated human urine can be bought and submitted for testing. An individual can also purchase a cleaning substance such as an herbal supplement for $30 to $70 and ingest the substance before submitting a sample. The herbal supplement reacts chemically with the drug to essentially nullify its active ingredient. Adulterants are chemicals that can actually react with the drugs, but these are actually added to the sample rather than ingested by the individual.

Methods to Detect Adulteration. Several methods can be used to detect the use of some type of adulter-

> **FASCINATING FACTS ABOUT DRUG TESTING**
>
> - Each year in the United States, drug and alcohol use costs more than $100 billion in lost productivity.
> - A single urine testing kit can detect up to twelve drugs.
> - The drug detection window for testing urine samples is wider than for other methods for drug testing because it depends on metabolic rate, age, amount and frequency of use, urine pH, drug tolerance, body mass, and overall health. Therefore, urine testing is considered less accurate than other testing methods.
> - Sweat patches for drug testing are about the size of a playing card and have a tamper-proof feature. They can be worn for up to seven hours to collect samples.
> - The annual cost of conducting more than 100,000 drug tests to detect performance-enhancing drugs in athletes in the United States is about $30 million.
> - There are more than four hundred commercial products readily available to mask urine samples.
> - An aerosol detecting agent called DrugAlert can be sprayed on a paper towel that has been used to wipe any household surface to detect a variety of drugs on that surface.

ant. If the specific gravity of urine is outside the normal range of 1.003 and 1.030, then the sample may have been diluted. Another indication of adulteration via dilution is to test the level of creatine. If the level is too low (less than 5 milligrams, or mg, per deciliter) then dilution took place. Oxidants, such as pyridinium chlorochromate (PCC), bleach, or hydrogen peroxide, can react chemically with the drug to essentially nullify it. One commonly sold PCC adulterant is called Urine Luck. Tests can also detect the presence of any type of additional oxidant. If the pH (acidity-alkalinity) value of the urine is outside the normal range of 4.0 to 9.0, then an adulterant was added. Two common adulterants are sold under the names of Whizzies or Klear, and these react chemically with drugs in the urine by oxidizing the active ingredient in marijuana. Another chemical reaction occurs when an adulterant called Clear Choice or Urine Aid prevents enzyme activity in the test, which results in the presence of glutaraldehyde, which causes a false negative. Among the states that have passed laws to prevent the sale of masking agents are Florida, South Carolina, North Carolina, New Jersey, Maryland, Virginia, Kentucky, Oklahoma, Nebraska, Illinois, Pennsylvania, and Arkansas.

IMPACT ON INDUSTRY

Drug testing was initially developed for military use and then mandated for federal employees. However, the widespread use of recreational drugs beginning in the 1960's caused private sector employers and then academic institutions to routinely administer drug tests. As a result, a huge market has developed not only for rapid, on-site testing methods but also for at-home testing methods, which many parents use to monitor their children. This demand produced explosive growth by companies that manufacture on-site or at-home drug testing kits, and the kits are sold by mass merchandisers and in drugstores such as Target, Walmart, Walgreens, and CVS. In addition, many of these inexpensive kits can be purchased on the Internet.

SOCIAL CONTEXT AND FUTURE PROSPECTS

Mandatory testing is regulated by SAMHSA, part of the U.S. Department of Health and Human Services. This mandatory testing does not yet test for semisynthetic opioids, such as oxycodone, oxymorphone, and hydrocodone, which are often used to relieve pain but have the potential to be abused. However, many employers, athletic organizations, and schools test for these drugs, and ongoing research is directed toward increasing the convenience and reliability of methods of detecting these drugs. Because so many masking agents are readily available, random testing without prior notification is the most effective method, although it is not without controversy. Schools are increasingly performing random drug testing, often leading to protests that the tests are an invasion of privacy and a violation of Fourth Amendment rights.

Organizations such as the International Olympic Committee, National Collegiate Athletic Association, National Basketball Association, and National Football

League monitor athletes for the use of more than one hundred anabolic-androgenic steroids. Efforts are being made to eliminate the use of performance-enhancing drugs in all sports. The International Olympic Committee led a collective initiative in creating the World Anti-Doping Agency (WADA) in Switzerland in 1999. WADA created a code in an attempt to standardize regulation and procedures in all sporting countries and keeps a list of prohibited substances. Banned substances include anabolic steroids, hormones, masking agents, stimulants, narcotics, cannabinoids, glucocorticosteroids, and for some sports, alcohol and beta-blockers during competition. Also forbidden are methods of enhancing oxygen transfer (such as blood doping) and gene doping. The UNESCO International Convention Against Doping in Sport, which came into force in 2007, is a global treaty designed to help governments align their policies with the WADA code.

—Jeanne L. Kuhler, B.S., M.S., PhD

FURTHER READING

Jenkins, Amanda J., and Bruce A. Goldberger, eds. *On-Site Drug Testing*. Totowa, N.J.: Humana Press, 2002. Discusses on-site methods of testing for drugs in hospital, criminal, workplace, and school settings. Looks at many specific tests, discussing their efficacy and their underlying principles.

Karch, Stephen B., ed. *Workplace Drug Testing*. Boca Raton, Fla.: CRC Press, 2008. Examines regulations and mandatory guidelines for federal workplace drug testing and describes techniques. Provides sample protocols from the nuclear power and transportation industries.

Liska, Ken. *Drugs and the Human Body with Implications for Society*. Upper Saddle River, N.J.: Pearson/Prentice Hall, 2004. Simply describes the various classes of drugs and drug testing methods.

Mur, Cindy, ed. *Drug Testing*. Farmington Hills, Mich.:Greenhaven Press/Thomson Gale, 2006. A collection of essays on drug testing in schools and the workplace, discussing efficacy and ethical issues such as privacy.

Pascal, Kintz. *Analytical and Practical Aspects of Drug Testing in Hair*. Boca Raton, Fla.: CRC Press, 2006. Looks at advances in the use of strands of hair for drug testing in the workplace and in forensic crime laboratories and techniques for detecting specific drugs.

Thieme, Detlef, and Peter Hemmersbach. *Doping in Sports*. Berlin: Springer, 2010. Examines sports doping from its beginning, covering the use of anabolic steroids, erthyropoietin, human growth hormone, and gene doping in humans and the doping of race horses. Effects of the drugs, detection methods, and regulations are also discussed.

Drug and Alcohol Testing Industry Association http://www.datia.org

Substance Abuse and Mental Health Services Administration http://www.samhsa.gov

Substance Abuse Program Administrators Association http://www.sapaa.com

U.S. Department of Labor Drug-Free Workplace Adviser http://www.dol.gov/elaws/drugfree.htm

World Anti-Doping Agency http://www.wada-ama.org

E

Engineering

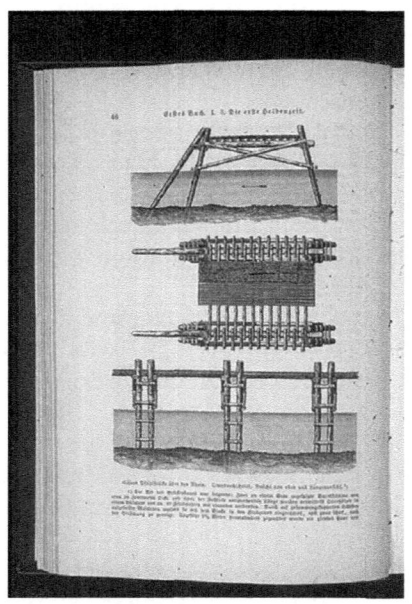

Caesar's bridge of pilings over the Rhine, a feat of early military engineering. (Library of Congress) [Public domain], via Wikimedia Commons

FIELDS OF STUDY

Agricultural engineering; Bioengineering; Biomechanical engineering; Bioprocess engineering; Biosystems engineering; Chemical engineering

ABSTRACT

Engineering is the application of scientific and mathematical principles for practical purposes. Engineering is subdivided into many disciplines; all create new products and make existing products or systems work more efficiently, faster, safer, or at less cost. The products of engineering are ubiquitous and range from the familiar, such as microwave ovens and sound systems in movie theaters, to the complex, such as rocket propulsion systems and genetic engineering.

BACKGROUND

Engineering is a broad field in which practitioners attempt to solve problems. Engineers work within strict parameters set by the physical universe. Engineers first observe and experiment with various phenomena, then express their findings in mathematical and chemical formulas. The generalizations that describe the physical universe are called laws or principles and include gravity, the speed of light, the speed of sound, the basic building or subatomic particles of matter, the chemical construction of compounds, and the thermodynamic relationship that to produce energy requires energy. The fundamental composition of the universe is divided into matter and energy. The potential exists to convert matter into energy and vice versa. The physical universe sets the rules for engineers, whether the project is designing a booster rocket to lift thousands of tons into outer space or creating a probe for surgery on an infant's heart.

Engineering is a rigorous, demanding discipline because all work must be done with regard to the laws of the physical universe. Products and systems must withstand rigorous independent trials. A team in Utah, for example, must be able to replicate the work of a team in Ukraine. Engineers develop projects using the scientific method, which has four parts: observing, generalizing, theorizing, and testing.

HISTORY

The first prehistoric humans to use a branch as a lever might be called engineers, although they never knew about fulcrums. The people who designed and built the pyramids of Giza (2500 BCE) were engineers. The term "engineer" derives from the medieval Latin word *ingeniator*, a person with "ingenium," connoting curiosity and brilliance. Leonardo da Vinci, who used mathematics and scientific principles in everything from his paintings to his designs for military fortifications, was called the Ingegnere Generale (general

Genetically engineered mice expressing green fluorescent protein, which glows green under blue light. The central mouse is wild-type. [CC BY 2.0 (http://creativecommons.org/licenses/by/2.0)], via Wikimedia Commons

engineer). Galileo is credited with seeking a systematic explanation for phenomena and adopting a scientific approach to problem solving. In 1600, William Gilbert, considered the first electrical engineer, published *De magnete, magneticisque corporibus et de magno magnete tellure* (*A New Natural Philosophy of the Magnet, Magnetic Bodies, and the Great Terrestrial Magnet*, 1893; better known as *De magnete*) and coined the term "electricity."

Until the Industrial Revolution of the eighteenth and nineteenth centuries, engineering was done using trial and error. The British are credited with developing mechanical engineering, including the first steam engine prototype developed by Thomas Savery in 1698 and first practical steam engine developed by James Watt in the 1760s.

Military situations often propel civilian advancements, as illustrated by World War II. The need for advances in flight, transportation, communication, mass production, and distribution fostered growth in the fields of aerospace, telecommunication, computers, automation, artificial intelligence, and robotics. In the twenty-first century, biomedical engineering spurred advances in medicine with developments such as synthetic body parts and genetic testing.

HOW IT WORKS

Engineering is made up of specialties and subspecialties. Scientific discoveries and new problems constantly create opportunities for additional subspecialties. Nevertheless, all engineers work the same way. When presented with a problem to solve, they research the issue, design and develop a solution, and test and evaluate it.

For example, to create tiles for the underbelly of the space shuttle, engineers begin by researching the conditions under which the tiles must function. They examine the total area covered by the tiles, their individual size and weight, and temperature and frictional variations that affect the stability and longevity of the tiles. They decide how the tiles will be secured and interact with the materials adjacent to them. They also must consider budgets and deadlines.

Collaboration. Engineering is collaborative. For example, if a laboratory requires a better centrifuge, the laboratory needs designers with knowledge in materials, wiring, and metal casting. If the metal used is unusual or scarce, mining engineers need to determine the feasibility of providing the metal. At the assembly factory, an industrial engineer alters the assembly line to create the centrifuge. Through this collaborative process, the improved centrifuge enables a biomedical engineer to produce a life-saving drug.

Communication. The collaborative nature of engineering means everyone relies on proven scientific knowledge and symbols clearly communicated among engineers and customers. The increasingly complex group activity of engineering and the need to communicate it to a variety of audiences has resulted in the emergence of the field of technical communications, which specializes in the creation of written, spoken, and graphic materials that are clear, unambiguous, and technically accurate.

Design and Development. Design and development are often initially at odds with each other. For example, in an architectural team assigned with creating the tallest building in the world, the design engineer is likely to be very concerned with the aesthetics of the building in a desire to please the client and the city's urban planners. However, the development engineer may not approve the design, no matter how beautiful, because the forces of nature (such as wind shear on a mile-high building) might not allow for facets of the design. The aesthetics of design and the practical concerns of development typically generate a certain level of tension. The ultimate engineering challenge is to develop materials or methods that withstand these forces of nature or otherwise circumvent them, allowing designs, products, and processes that previously were impossible.

Testing. With computers and tools such as computer-aided design (CAD), designs that at one time took days to draw can be created in hours. Similarly, computers allow a prototype (or trial product) to be quickly produced. Advances in computer simulation make it easier to conduct tests. Testing can be done multiple times and under a broad range of harsh conditions. For example, computer simulation is used to test the composite materials that are increasingly used in place of wood in building infrastructures. These composites are useful for a variety of reasons, including fire retardation. If used as beams in a multistory building, they must be able to withstand tremendous bending and heat forces. Testing also examines the materials' compatibility with the ground conditions at the building site, including the potential for earthquakes or other disasters.

Financial Considerations. Financial parameters often vie with human cost, as in biomedical advancements. If a new drug or stent material is rushed into production without proper testing to maximize the profit of the developing company, patients may suffer. Experimenting with new concrete materials without determining the proper drying time might lower the cost of their development, but buildings or bridges could collapse. Dollars and humanity are always in the forefront of any engineering project.

APPLICATIONS AND PRODUCTS

The collaborative nature of engineering requires the cooperation of engineers with various types of knowledge to solve any single problem. Each branch of engineering has specialized knowledge and expertise.

Aerospace. The field of aerospace engineering is divided into aeronautical engineering, which deals with aircraft that remain in the Earth's atmosphere, and astronautical engineering, which deals with spacecraft. Aircraft and spacecraft must endure extreme changes in temperature and atmospheric pressure and withstand massive structural loads. Weight and cost considerations are paramount, as is reliability. Engineers have developed new composite materials to reduce the weight of aircraft and enhance fuel efficiency and have altered spacecraft design to help control the friction generated when spacecraft leave and reenter the Earth's atmosphere. These developments have influenced earthbound transportation from cars to bullet trains.

Architectural. The field of architectural engineering applies the principles of engineering to the design and construction of buildings. Architectural engineers address the electrical, mechanical, and structural aspects of a building's design as well as its appearance and how it fits in its environment. Areas of concern to architectural engineers include plumbing, lighting, acoustics, energy conservation, and heating, ventilation, and air conditioning (HVAC). Architectural engineers must also make sure that buildings they design meet all regulations regarding accessibility and safety in addition to being fully functional.

Bioengineering. The field of bioengineering involves using the principles of engineering in biology, medicine, environmental studies, and agriculture. Bioengineering is often used to refer to biomedical engineering, which involves the development of artificial limbs and organs, including ceramic knees and hips, pacemakers, stents, artificial eye lenses, skin grafts, cochlear implants, and artificial hands. However, bioengineering also has many other applications, including the creation of genetically modified foods that are resistant to pests, drugs that prevent organ rejection after a transplant operation, and chemical coatings for a stent placed in a heart blood vessel that will make the implantation less stressful for the body. Bioengineers must concern themselves with not only the biological and mechanical functionality of their creations but also financial and social issues such as ethical concerns.

Chemical. Everything in the universe is made up of chemicals. Engineers in the field of chemical engineering develop a wide range of materials, including fertilizers to increase crop production, the building materials for a submarine, and fabric for everything from clothing to tents. They may also be involved in finding, mining, processing, and distributing fuels and other materials. Chemical engineers also work on processes, such as improving water quality or developing less-polluting, readily available, inexpensive fuels.

Civil. Some of the largest engineering projects are in the field of civil engineering, which involves the design, construction, and maintenance of infrastructure such as roads, tunnels, bridges, canals, dams, airports, and sewage and water systems. Examples include the interstate highway system, the Hoover Dam, and the Brooklyn Bridge. Completion of civil engineering projects often results in major shifts in population distribution and changes in how people live. For example, the highway system allowed fresh produce to be shipped to northern states in the wintertime, improving the diets of those who lived there. Originally, the term "civil engineer" was used to distinguish between engineers who worked on public projects and military engineers who worked on military projects such as topographical maps and the building of forts. The subspecialties of civil engineering include construction engineering, irrigation engineering, transportation engineering, soils and foundation engineering, geodetic engineering, hydraulic engineering, and coastal and ocean engineering

Computer. The field of computer engineering has two main focuses, the design and development of hardware and of the accompanying software. Computer hardware refers to the circuits and architecture of the computer, and software refers to the computer programs that run the computer. The hardware does only what the software instructs it to do, and the software is limited by the hardware. Computer engineers may research, design, develop, test, and install hardware such as computer chips, circuit boards, systems, modems, keyboards, printers, or computers embedded in various electronic products, such as the tracking devices used to monitor parolees. They may also create, maintain, test, and install software for mainframes, personal computers, electronic devices, and smartphones. Computer programs range from simple to complex and from familiar to unfamiliar. Smartphone applications are extremely numerous, as are applications for personal computers. Software is used to track airplanes and other transportation, to browse the Web, to provide security for financial transactions and corporations, and to direct unmanned missiles to a precisely defined target. Computers can operate from a remote location. For example, anaerobic manure digesters are used to convert cattle manure to biogas that can be converted to energy, a biosolid that can be used as bedding or soil amendment, and a nonodorous liquid stream that can be used as fertilizer. These digesters can be placed on numerous cattle farms in different states and operated and controlled by computers miles away.

Electrical. Electrical engineering studies the uses of electricity and the equipment to generate and distribute electricity to homes and businesses. Without electrical engineering, smartphones, televisions, home appliances, and many life-saving medical devices would not exist. Computers could not turn on. The Global Positioning System (GPS) would be useless, and starting a car would require using a hand crank. This field of engineering is increasingly involved in investigating different ways to produce electricity, including alternative fuels such as biomass and solar and wind power.

Environmental. The growth in the population of the world has been accompanied by increases in consumption and the production of waste. Environmental engineering is concerned with the reduction of existing pollution in the air, in the water, and on land, and the prevention of future harm to the environment. Issues addressed include pollution from manufacturing and other sources, the transportation of clean water, and the disposal of nonbiodegradable materials and hazardous and nuclear waste. Because pollution of the air, land, and water crosses national borders, environmental engineers need a broad, global perspective.

Industrial. Managing production and delivery of any product is the expertise of industrial engineers. They observe the people, machines, information, and technology involved in the process from start to finish, looking for any areas that can be improved. Increasingly, they use computer simulations and robotics. Their goals are to increase efficiency, reduce costs, and ensure worker safety. For example, worker safety can be improved through ergonomics and the use of less-stressful, easier-to-manipulate tools. The expertise of industrial engineers can have a major impact on the profitability of companies.

Manufacturing. Manufacturing engineering examines the equipment, tools, machines, and processes involved in manufacturing. It also examines how manufacturing systems are integrated. Its goals are to increase product quality, safety, output, and profitability by making sure that materials and labor are used optimally and waste—whether of time, labor, or materials—is minimized. For example, engineers may improve machinery that folds disposable diapers or that machines the gears for a truck, or they may reconfigure the product's packaging to better protect it or facilitate shipping. Increasingly, robots are used to do hazardous, messy, or highly repetitive work, such as painting or capping bottles.

Mechanical. The field of mechanical engineering is the oldest and largest specialty. Mechanical engineers create the machines that drive technology and industry and design tools used by other engineers. These machines and tools must be built to specifications regarding usage, maintenance, cost, and delivery. Mechanical engineers create both power-generating machinery such as turbines and power-using machinery such as elevators by taking advantage of the compressibility properties of fluids and gases.

Nuclear. Nuclear engineering requires expertise in the production, handling, utilization, and disposal of nuclear materials, which have inherent dangers as well as extensive potential. Nuclear materials are used in medicine for radiation treatments and diagnostic testing. They also function as a source of energy in nuclear power plants. Because of the danger of nuclear materials being used for weapons, nuclear engineering is subject to many governmental regulations designed to improve security.

SOCIAL CONTEXT AND FUTURE PROSPECTS

Engineering can both prolong life through biomedical advances such as neonatal machinery and destroy life through unmanned military equipment and nuclear weaponry. An ever-increasing number of people and their concentration in urban areas means that ways must be sought to provide more food safely and to ensure an adequate supply of clean, safe drinking water. These needs will create projects involving genetically engineered crops, urban agriculture, desalination facilities, and the restoration of contaminated rivers and streams. The never-ending quest for energy will remain a fertile area for research and development. Heated political debates about taxing certain fuels and subsidizing others are part of the impetus behind solar, wind, biomass, and other alternative fuel development.

Indians, in U.S. engineering programs is being addressed through education initiatives. Women's enrollment in engineering schools has hovered around 20 percent since about 2000, though the percentage has increased steadily over that time. African Americans make up about 13 percent of the U.S. population, yet only about 4 percent of the total number of students earning bachelor's degrees in engineering each year are black; in fact, from 2003 to 2011, the percentage of African Americans earning bachelor's degrees in engineering dropped from 5.1 percent to 4.2 percent (peaking at 5.3 percent in 2005). However, Hispanics, who represent about 17 percent of the U.S. population, have experienced an upward trend in earning bachelor's degrees in engineering. In 2003, 5.4 percent of students earning BAs in engineering were Hispanic; that percentage rose steadily to 8.5 in 2011. In 2013 9 percent of all U.S. postsecondary degrees in science, technology, engineering, and mathematics (STEM) fields were earned by people of Hispanic descent.

—*Judith L. Steininger, BA, MA*

FURTHER READING

Addis, Bill. *Building: Three Thousand Years of Design Engineering and Construction.* New York: Phaidon, 2007. Print.

Baura, Gail D. *Engineering Ethics: An Industrial Perspective.* Boston: Elsevier, 2006. Print.

Carmichael, D. G. *Infrastructure Investment: An Engineering Perspective.* Boca Raton: Taylor, 2015. Print.

Dieiter, George E., and Linda C. Schmidt. *Engineering Design.* 5th ed. Boston: McGraw, 2013. Print.

Nemerow, Nelson Leonard, et al., eds. *Environmental Engineering: Environmental Health and Safety for Municipal Infrastructure, Land Use and Planning, and Industry.* 6th ed., John Wiley & Sons, 2009. 3 vols.

Petroski, Henry. *Success through Failure: The Paradox of Design.* 2006. Rpt. Princeton: Princeton UP, 2008. Print.

Yount, Lisa. *Biotechnology and Genetic Engineering.* 3rd ed. New York: Facts On File, 2008. Print.

Environmental biotechnology

FIELDS OF STUDY

Agricultural engineering; Bioengineering; Biophysics; Bioremediation; Environmental biotechnology (or Green biotechnology); Food science; Horticulture; Marine biology (or Blue biotechnology)

ABSTRACT

Environmental biotechnology, also known as biotechnical pollution control, is a rapidly developing science that uses biological resources to protect and restore the environment. It has significant implications and applications in both the prevention of air, soil, and water pollution and the restoration of contaminated environments.

BACKGROUND

Fundamentally, environmental biotechnology is the study of how microorganisms, plants, and their enzymes can assist in the restoration, remediation, preservation, and sustainable use of the world's natural environment. The primary role of environmental biotechnologists is to create a better balance between human development and the natural environment. All fields of biotechnology have seen rapid growth and progress in the 1990's and 2000's. The field of environmental biotechnology, in particular, has benefited from advancements in genetic engineering and modern microbiological concepts, which offer both traditional and innovative solutions to different forms of contamination occurring in different mediums (soil, water, and air).

HISTORY

The term "biotechnology" was being used as early as 1919 but did not occur frequently in scientific literature until the 1960's and 1970's, with the publication of the *Journal of Biotechnology*. Although the use of biotechnology within the medical and agricultural industry can be traced back many years, the purposeful use of biotechnology to mitigate environmental issues has a much shorter history.

Although people were familiar with the concept of biotechnology, for the most part, efforts were focused on the medical and agricultural industries. The arrival of the Industrial Revolution, however, rapidly altered and influenced the environment through the release of toxic pollutants into the waterways and soil. As people became wealthier, the demand for goods grew, and with the rise in industrial and agricultural production came the increase in the impact on the environment. Although the concept of using natural degradation processes was not new—for many years, communities had composted and relied on natural processes and microbes in the breakdown and treatment of sewage—environmental biotechnology was not yet considered a science.

In the 1960's, however, chemical pollution and its adverse effects came under significant public scrutiny. One of the landmark cases involved the chemical Dichloro-diphenyl-trichloroethane (DDT), which was widely used as a pesticide but later was found to seriously affect bird populations and to cause cancer in humans. Other cases of chemical pollution and its adverse effects on people and the environment were becoming known, including mercury poisoning in Japan, Agent Orange in Vietnam, industrial sludge in the United States, and the devastating effects of oil spills. By the 1970's, it had become clear that the environment was becoming sullied and that contaminants were adversely affecting people. Concern regarding the environment led to the development of many laws and regulations in both developed and developing countries regarding the proper management of waste and pollution control. It was in the light of this political and social awareness that the field of environmental biotechnology emerged. Although environmental biotechnology originally focused on the treatment of wastewater, the field expanded to include areas of study such as soil contamination, solid-waste treatment, and air purification methods.

In 1992, during the United Nations Conference on Environment and Development in Rio de Janeiro, environmental biotechnology was recognized and embraced as a crucial tool for both repairing and preventing environmental and health issues caused by humans. Since this conference, the field of environmental biotechnology has advanced at a rapid rate and has grown to provide innovative approaches to the sustainable development and protection of the world's ecosystems.

HOW IT WORKS

Traditional methods of waste removal, such as landfill and incineration, cannot cope with the sheer volume of waste created by human populations. This situation has increased the need to develop alternative environmentally sound treatments and techniques. Environmental biotechnology seeks to positively affect pollution control and waste management.

A rise in consumption since the 1990's has been accompanied by a corresponding increase in the release of pollutants into the environment. Some of these contaminants, particularly those that also are naturally occurring, can be digested, degraded, or removed from the soil and water through the action of microorganisms. However, some of these human-created pollutants rarely occur naturally, and the accumulation of such substances can have a serious ecological impact.

The use of biotechnology in the treatment of waste and pollution is not a new idea. For more than a century, many communities have relied on natural processes and microbes to break down and treat sewage. The fundamental aim of environmental biotechnology is to use organisms to control contamination and treat waste. In the process called bioremediation, microorganisms, including fungi and bacteria and their enzymes, are used to return a contaminated environment to its original condition. Naturally occurring biological degradation processes are purposely employed to remove contaminants from areas where they have been released. The use of such processes requires a solid scientific understanding of the contaminant, its impact, and the affected ecosystem.

The concept of environmental biotechnology depends on the notion that all living organisms, such as flora, fauna, bacteria, and fungi, consume nutrients for their survival and, in doing so, produce waste by-products. Not all organisms require the same nutrients nor react in the same manner, however. Some organisms, such as certain bacteria and microorganisms, flourish on chemicals and toxins that are actually poisonous or harmful to other organisms or ecosystems. The fact that some microorganisms and various strains of microbial species react differently to chemical toxins and environmental pollutants has advanced the concept of using genetic manipulation techniques.

Environmental biotechnology aims to provide a natural approach to tackling environmental issues, from identification of biohazards to restoration of industrial, agricultural, and natural areas affected by contamination. Central to the concept of environmental biotechnology is the ability to determine which contaminants are present, for how long, and in what quantity, and what recovery method is applicable. There are four basic concepts and approaches in the field of environmental biotechnology: bioremediation, prevention, detection and monitoring, and genetic engineering.

APPLICATIONS AND PRODUCTS

Environmental pollution can be a legacy of former industrial practices or a product of unsustainable present-day practices. One of the most serious environmental issues facing the world in the twenty-first century is the production of very large quantities of waste, the majority of which becomes landfill. Industrialized nations also produce significant quantities of chemicals that often end up in the soil and water. Environmental biotechnology seeks ways to combat the escalating ecological problems associated with such pollution and waste. Technologies that have been developed and implemented include bioremediation of water and soil, biomonitoring using biosensors and bioassays, and bioprocessing.

Bioremediation. Bioremediation is usually classified as either in situ or ex situ. In situ bioremediation entails treating the contamination in place and relies on the ability of the microorganisms to metabolize or remove the contaminants inside the naturally occurring system; ex situ remediation entails the polluted material being removed from the contaminated site and treated elsewhere and relies on some form of artificial engineering and input. The process of bioremediation can occur naturally or be encouraged through artificial stimulus. The process of attenuation occurs under natural conditions and incorporates the normal chemical, biological, and physical processes, such as aerobic and anaerobic degradation, that eliminate or reduce soil and water contaminants.

Biostimulation and Bioaugmentation. Biostimulation is a form of bioremediation in which the natural processes of degradation are encouraged through the introduction of certain stimuli, such as nutrients and additional substrates. Bioaugmentation is a form of bioremediation that involves increasing the activity of the microorganisms that assist in pollutant reduction or augmenting existing yet insufficient populations of microorganisms. Bioenrichment is another form of bioremediation that involves adding nutrients or oxygen to environments to increase the breakdown of contaminants.

Biodetectors. Biodetectors such as bioassays and biosensors are used to monitor, assess, and analyze biological material, and provide important information and data about the effects and concentration of pollutants in the natural environment. Bioassays are procedures or experiments in which the quantity of a contaminant is estimated by measuring its effect on living organisms. Although they can be relatively slow and expensive, they are essential for the assessment and prediction of real and potential effects of pollution on the natural environment.

Biosensors are devices that detect and measure minute amounts of or changes in concentration of chemical substances within an environmental area and translate that information into data. Because of their ability to detect even tiny quantities of targeted chemicals with greater speed and at less cost than bioassays, biosensors have become important tools for the monitoring and control of pollution levels, both before and after bioremediation measures are implemented.

Bioprocessing. Bioprocessing is a process that uses living cells or organisms to produce specific outcomes. Communities of microbial organisms perform a comprehensive range of bioprocesses within the natural environment, which can be exploited to benefit both the environment and industry. Bioprocesses used in environmental biotechnology include microbial enhanced oil recovery (MEOR), biological treatments of polluted air, biodesulfurization, conversion of pollutants into useful products such as fertilizers and green energy, and microbial exploration technology.

Biofilm Control. Biofilms, often referred to as slime, occur on the surface of aqueous environments and are caused by a complex accumulation of microorganisms. Biofilms are usually considered undesirable, as they are frequently associated with odors, infections, fouling, and corrosion, but they can be beneficial under some circumstances such as the treatment of wastewater.

Genetic Engineering and Manipulation. The advancement of genetics and genetic manipulation has had significant impact on environmental biotechnology. Research in molecular genetics has provided novel techniques for the detection and degradation of contaminants through the manipulation and enhancement of the microbes' ability to adapt themselves genetically to different pollutants. The ecologically useful and improved organisms are classified as genetically engineered microbes (GEM).

SOCIAL CONTEXT AND FUTURE PROSPECTS

In the early twenty-first century, the population of the world edged toward 7 billion, increasing the amount of pollution that reaches the land, water, and atmosphere. Many natural ecosystems are struggling to cope with the remnants of old toxic contamination and the influx of new contamination. The ecological, social, and economic costs of pollution are immeasurable, and environmental recovery is one of the most important problems facing the global community.

Conventional biotechnology processes and techniques have relied on end-of-pipe technologies, that is treatment of waste and pollution that has already contaminated the air, soil, or water. Although such methods are necessary, particularly in remediation of existing pollution, many environmental biotechnologists think that end-of-pipe methods should be regarded as last-resort efforts, rather than preferred methods. As such, environmental biotechnology is moving from first-generation technology based on naturally occurring processes and microorganisms to second-generation technology based on high-tech anthropomorphic enhancement and manipulation of natural processes and microorganisms.

The future of environmental biotechnology lies in following an integrated environmental protection approach, with the fundamental goal being to control pollution before it enters the natural ecosystem and to recover already polluted areas. An essential step in this goal is to create controls and pass legislation to reduce the incidence of contamination. However, many researchers believe that the future of environmental engineering also will be closely aligned with the advancement and application of molecular and genetic methods. Decreasing or mitigating greenhouse gas pollution in the Earth's atmosphere is of vital importance to global health, so research in environmental biotechnology is focusing on biological organisms and processes that may help.

—*Christine Watts, PhD*

FURTHER READING

Cummings, Stephen, ed. *Bioremediation: Methods and Protocols*. New York: Humana Press, 2010. Experts in the field of environmental biotechnology present innovative and imaginative bioremediation techniques in pollution removal.

Evans, Gareth, and Judith Furlong. *Environmental Biotechnology: Theory and Application*. New York: John Wiley & Sons, 2003. A detailed examination of environmental biotechnology, focusing on present-day practices, the potential for biotechnological interventions, and microbial techniques and methods.

Illman, Walter, and Pedro Alvarez. "Performance Assessment of Bioremediation and Natural Attenuation." *Critical Reviews in Environmental Science and Technology*, 39, no. 4 (April, 2009): 209-270. A critical review of the state-of-the-art in performance assessment methods. Discusses future research directions in bioremediation, natural attenuation, chemical fingerprinting, and molecular biological tools.

Jördening, Hans-Joachim, and Josef Winter. *Environmental Biotechnology: Concepts and Applications*. Weinheim, Germany: Wiley-VCH, 2005. A solid foundation for students wishing to study environmental biotechnology. Examines in detail the microbiological treatment of waste and pollution in water, soil, and air.

Scragg, Alan. *Environmental Biotechnology*. 2d ed. New York: Oxford University Press, 2005. Examines the multitude of ways in which environmental biotechnology is applied in pollution control, environmental management, and removal of oil and minerals.

Thakur, Indu Shekhar. *Environmental Biotechnology: Basic Concepts and Applications*. New Delhi: I. K. International, 2006. A comprehensive examination of environmental processes and the many possible applications of environmental biotechnology, such as bioremediation, bioprocessing, and bioleaching.

ENZYME ENGINEERING

FIELDS OF STUDY

Biology; Biomechanical engineering; Biotechnology; Chemical engineering; Medical biotechnology (or Red biotechnology)

ABSTRACT

Catalysts accelerate the rate of chemical reactions without being essentially changed, and enzymes are biological catalysts that accelerate the rate of reactions that occur in living systems. Enzyme engineering identifies enzymes that have potentially useful catalytic activities and chemically or structurally modifies them to increase their activity, change their substrate specificity, may change the types of reactions they catalyze, or change the properties of enzymes and the manner in which they are regulated. Engineered enzymes can generate completely novel molecules or new, improved ways to synthesize useful molecules.

BACKGROUND

Enzyme engineers use catalytic antibodies or abzymes. Antibodies are Y-shaped proteins made by vertebrate immune systems that bind to specific chemicals. Abzymes bind to chemicals and force them into the transition state of a chemical reaction, which accelerates the formation of the product from the reactants.

HISTORY

Enzyme engineering arose only after advances in several other fields made it possible to determine the primary amino acid sequences and three-dimensional structure of enzymes and directly manipulate them at the molecular level. Swedish biochemist Pehr Victor Edman gave birth to protein sequencing in 1950, when he designed the Edman degradation reactions that can determine the primary amino acid sequence of proteins. In 1958, English biochemist

John Cowdery Kendrew used X-ray crystallography to solve the three-dimensional structure of the muscle oxygen-storing protein myoglobin. In the 1970's, American biochemist Herbert Wayne Boyer and American geneticist Stanley Norman Cohen pioneered molecular cloning techniques that gave scientists the means to clone genes and insert them into bacteria for propagation.

The first studies in enzyme engineering examined the effects of mutations on enzyme active sites. Beta-lactamase, the enzyme used by bacteria to degrade beta-lactam antibiotic (penicillin, ampicillin, and amoxicillin) was one of the first enzymes examined by enzyme engineering. In 1978, Canadian chemist Michael Smith and his colleagues invented site-directed mutagenesis, which gave biochemists a much better way to place targeted mutations into the genes that encode enzymes and thereby change their primary amino acid sequence. In 1986, the laboratories of Peter Schultz (University of California, Berkeley) and Richard Lerner (Research Institute of Scripps Clinic) made the first catalytic antibodies that could split ester bonds.

HOW IT WORKS

Semisynthetic Enzymes. Enzymes that are modified by chemical means are known as semisynthetic enzymes. There are two main ways to produce semisynthetic enzymes: atom replacement or group attachment.

Atom replacement exchanges one atom within an enzyme for a different atom. Such replacements can modify enzyme activity or change the substrate specificity of the enzyme. Group attachment involves the use of particular chemical reagents to attach particular molecules to enzymes. Attaching additional molecules to enzymes can also markedly change enzyme activity and substrate specificity.

Directed Evolution. Directed evolution randomly changes amino acids in a protein without prior knowledge of the exact function of each amino acid. The first step, diversification, takes the gene that encodes the enzyme of interest and replicates it many times while using a copying machinery that is inherently error-prone. This introduces random mutations into the gene and creates a large collect of gene variants that are usually grown in bacteria.

The second step, selection, tests or screens these enzyme variants for a desired property. Once the desired variants are identified, they undergo the third step, amplification, which replicates the identified variants and sequences them in order to determine which mutations produced the desired properties. Collectively, these three steps constitute one round of directed evolution, and the vast majority of such experiments require multiple rounds. The goal is to find those variant enzymes that show the most desired characteristics to the greatest extent. Directed-evolution studies suffer from the need to make huge numbers of mutants that produce no discernable effect, since up to 90 percent of all mutants made are uninformative.

Semirational Design. This enzyme engineering strategy employs sophisticated computer programs that assemble all the available structural information of the enzyme under study and predict how the mutations introduced into different locations within the enzyme might affect its activity. The enzyme engineer then notes the predicted changes that will potentially generate the desired property changes and uses this information to conduct targeted mutagenesis experiments. Targeted mutagenesis experiments introduce mutations into specific locations of a protein. Once these mutations are made, the variant enzyme with the engineered changes is tested to determine if it has the specific properties the enzyme engineer was hoping to produce in the enzyme. These approaches combine structural information with rational design. Two computer programs that make such predictions include Protein Sequence-Activity Relationship or ProSAR and Combinatorial Active-Site Saturation Test, otherwise known as CASTing.

Rational Design. If a great deal of structural information about the enzyme in question is available, then that structural information informs which amino acids should be changed. Many rational design attempts have not succeeded because of uncertainties regarding protein structure.

De Novo Design. A computer builds an enzyme around the transition state of a reaction from scratch. The computer begins by designing the active site by placing specific amino acids in strategic positions so that they efficiently bind the transition state of the chemical reaction and stabilize it. The program then constructs a protein backbone that supports and properly positions the active-site amino acids but still provides a coherent protein structure that is predictably stable under the desired conditions.

This particular strategy suffers from gaps in the ability to predict protein structure accurately and correlate this ideal structure with enzymatic activity. For example, two enzymes (retro-aldol enzyme and a Kemp elimination catalyst) were built completely from scratch by using computer programs. However, both enzymes required further optimization by directed evolution to achieve maximum activity.

Catalytic Antibodies. The immune system of some vertebrates makes Y-shaped proteins that specifically bind to and neutralize foreign substances that invade the body. Immunizing laboratory animals with stable analogues of the transition states of various reactions directs the immune systems of those animals to synthesize antibodies that cannot only bind particular chemical reactants but force them into the transition state of the reaction, which subsequently forms the product.

APPLICATIONS AND PRODUCTS

Pharmaceutical Production. Beta-lactam and cephalosporin antibiotics are commonly prescribed to combat various illnesses. Both of these drugs kill bacteria by inhibiting the synthesis of the bacterial cell wall. Beta-lactam antibiotics include such widely recognized drugs as penicillin, ampicillin, and amoxicillin, whereas cephalosporin antibiotics include such popularly used antibiotics as Ceftin (cefuroxime), Kephlex (cephalexin), and Ceclor (cefaclor). Unfortunately, with repeated use, bacteria can become resistant to commonly used antibiotics, and making new, improved antibiotics is essential to treat some of the more recent and aggressive infectious diseases. To make new cephalosporin antibiotics, enzyme engineers have used enzymes called acylases to convert simple starting chemicals into various versions of these drugs. By engineering these acylase enzymes, pharmaceutical companies have been able to make new cephalosporin and beta-lactam antibiotics that have novel properties and can kill bacteria that are resistant to older drugs.

Enzymes as Medicines. When a person is cut, blood oozes from the damaged tissue. Fortunately, blood clotting (also known as coagulation) eventually stanches this blood flow. Blood clotting is an essential part of wound healing, but it is also a very highly regulated event. The formation of blood clots inside undamaged blood vessels clogs those vessels and leads to heart attacks if clots form inside the vessels that surround the heart, or a stroke, if they occur within vessels that surround the brain. The human body has ways to destroy unnecessary clots. An enzyme called tissue plasminogen activating factor (TPA) activates other enzymes in the body that degrade harmful clots. Commercially available, native TPA is called Alteplase, which has a half-life in the bloodstream of four to six minutes. Engineered forms of TPA are also clinically available. Reteplase, a shortened version of TPA (consists of 357 of the 527 amino acids of Alteplase), has a longer half-life (thirteen to sixteen minutes). Tenecteplase, which has two amino acid changes (substitutes asparagine[103] with a threonine and asparagine[114] with glutamine), has an even longer half-life of twenty to twenty-four minutes.

Engineered enzymes are also used in enzyme-replacement therapies. Several genetic diseases, known as lysosomal storage diseases, result from the inability to make functional versions of enzymes that degrade various biological molecules. The accumulation of these molecules kills brain cells and causes the death of the patient. Engineered enzymes used in enzyme-replacement therapies include Cerezyme (imiglucerase, used to treat Gaucher's disease), Naglazyme (galsulfase, used to treat mucopolysaccharidosis VI), Myozyme (alglucosidase alfa, used to treat Pompe disease), and Aldurazyme (laronidase, used to treat mucopolysaccharidosis I).

Enzyme Immobilization. By attaching enzymes to surfaces, embedding them in gel matrices, hollow fibers, or cross-linking them to each other, enzymes are immobilized on insoluble surfaces. This increases their stability, simplifies their recycling, and increases the tolerance of enzymes to high levels of substrate and products. Detergent enzyme preparations, such as Alcalase, immobilize the protease subtilisin by attaching it to insoluble particles. Attaching the enzyme to inert material increases its reuse as it degrades proteinaceous matter.

Making Enzymes Soluble in Organic Solvents. Enzymes usually work in water, but many reactions between organic chemicals occur in organic solvents. Although Russian chemist Alexander Klibanov showed that several enzymes are active in organic solvents, many enzymes are neither soluble in organic solvents nor work properly in such environments. Attaching a molecule called polyethylene glycol (PEG) to some enzymes makes them soluble and active in organic solvents and allows them to make things such as polyester, peptides (small proteins), esters (sweet-smelling things found in foods), and amides (nitrogen-containing compounds). Such modified enzymes also have clinical uses. For example, the enzyme asparaginase can kill cancer cells but is toxic, unstable, and some patients have severe allergies to it. PEG-treated asparaginase is not as toxic as the native enzyme, is much more stable, and does not cause allergy. PEG-asparaginase is used to treat tumors in humans.

Abzymes. A notable variety of reactions are catalyzed by catalytic antibodies that range from forming or breaking carbon-carbon bonds, rearrangements, hydrolysis of various bonds, transfer of chemical groups, and even an industrial reaction called the Diels-Alder reaction. However, abzymes are very expensive and tedious to make, and their catalytic activity is well below that of enzymes. Yet they do provide tailor-made catalysts when no other such reagent exists.

SOCIAL CONTEXT AND FUTURE PROSPECTS

Because modified enzymes can make certain products more cheaply, the public response to modified enzymes is generally positive. However, the genetically modified organisms (GMOs) that are used to produce these enzymes give many people pause, since the introduction of GMOs into the environment may have long-term consequences that are presently unrecognized. Strict government regulation that forbids the release of GMOs into the environment without approval allays most of these concerns, but some people are still troubled by the use of GMOs to make products that they eventually end up eating or using in some other manner.

Enzyme engineering is one of the up-and-coming fields in chemistry and biochemistry. Since the 1990's, the use of enzymes in industrial and academic chemistry has greatly increased. There are many advantages to using enzymes in that they can act outside cells and under mild conditions that minimize troublesome side effects, are environmentally innocuous, compatible with other enzymes, and are very efficient, though highly selective catalysts. The largest drawback of using enzymes is that the right enzyme is sometimes not available to catalyze the desired reaction. Enzyme engineering can eliminate this significant drawback.

Furthermore, as biochemists achieve a more profound understanding of protein structure, cheaper and faster ways of doing enzyme engineering, such as rational design, become more successful and practical. This will shorten the time required for enzyme engineering experiments and reduce its cost. Companies are already looking intently at enzyme engineering as a significant investment for their research and development departments.

——*Michael A. Buratovich, PhD*

FURTHER READING

Arnold, Frances H., and George Georgiou, eds. *Directed Enzyme Evolution: Screening and Selection Methods.* Totowa, N.J.: Humana Press, 2010. Laboratory protocol book that describes, in great detail with figures and graphs, some rather ingenious techniques for screening mutant clones of enzyme genes.

_____. *Directed Evolution Library Creation: Methods and Protocols.* Totowa, N.J.: Humana Press, 2010. Encyclopedic collection of protocols for generating libraries of randomly mutagenic enzyme genes in bacteria, with tables, graphs, and some figures.

Faber, Kurt. *Biotransformations in Organic Chemistry: A Textbook.* 5th ed. New York: Springer-Verlag, 2004. A very clear, useful textbook on the uses of enzymes in chemistry that includes a chapter on engineered enzymes.

Park, Sheldon J., and Jennifer R. Cochran, eds. *Protein Engineering and Design.* Boca Raton, Fla.: CRC Press, 2010. Covers the broader field of protein engineering—methods of developing altered proteins for novel applications—in two sections: one on experimental protein engineering and the other on computational design. Includes discussion of enzyme engineering using both rational and combinatorial approaches.

Scheindlin, Stanley. "Clinical Enzymology: Enzymes As Medicine." *Molecular Interventions* 7, no. 1 (February, 2007): 4-8. An absorbing and readable summary of the use of engineered enzymes in clinical diagnoses and treatments.

ESTROGENS FROM PLANTS

Chemical structures of the most common phytoestrogens found in plants (top and middle) compared with estrogen (bottom) found in animals. By Boghog2. [Public domain], via Wikimedia Commons

FIELDS OF STUDY

Biology; Biomechanical engineering; Biotechnology; Chemical engineering; Food science; Human nutrition; Medical biotechnology (or Red biotechnology); Pharmacogenomics; Plant pathology

ABSTRACT

While the female hormones called estrogens are common in mammals, only a few plants contain estrogens. Others synthesize compounds which are chemically unrelated to estrogens but resemble them in their molecular size and shape. These compounds are called phytoestrogens (plant estrogens) and may, when ingested by animals or humans, have properties similar to those of mammalian estrogens.

BACKGROUND

The precursor of estrogens in plants and animals is the linear (straight-chain) triterpene known as squalene. Cyclization of squalene, via the intermediate cycloartenol in plants and via the intermediate lanosterol in animals, forms a group of very important compounds known as the *steroids*. Steroids include cholesterol, mammalian sex hormones (including the estrogens and androgens), corticosteroids, insects' molting hormones, and plant brassinosteroid hormones. All steroids have a tetracyclic (four-ringed) structure; the rings are named A, B, C, and D. Differences in the functional groups attached to the tetracyclic skeleton, differences in the side-chain attached to ring D, and differences in the overall shape of the molecule determine a steroid's biological activity.

Estrogens have an aromatic A ring (a ring of six carbon atoms joined by alternating single and double bonds). This constrains the junction between the A and B rings, resulting in a "flat," or planar, molecule. This shape is essential for the potent biological activity of estrogens, and chemical modifications that alter the planar nature of an estrogen molecule reduce its biological activity.

Estrogens in Plants Versus Animals

In animals, one of the most potent estrogens is *estradiol*. It triggers the production of gonadotropins leading to ovulation. It is metabolized to the less active estrogens, *estrone* and *estriol*. Estrone and estriol are produced by the placentas of pregnant mammals, and both compounds accumulate in the urine during pregnancy.

In plants, estrogens are *secondary metabolites*. Although many thousands of secondary metabolites occur in plants, the distribution of particular secondary metabolites is often limited to just a few genera. This appears to be the case for estrogens. Estrone has been isolated from the seeds of pomegranate and from the date palm, in which it is a component of the kernel oil. Estriol has been isolated from the pussy willow. It is not known what function these estrogens have in plants. It is possible that, like other secondary metabolites, they may function in plant defense.

Isoflavonoids as Phytoestrogens

Isoflavonoids are a type of secondary metabolite and are found almost exclusively in the legume (pea) family of plants. They are known to function in plant defense. They have been shown to deter herbivores and also to facilitate a plant's defense response to pathogen attack. Interestingly, some isoflavonoids have chemical structures that, in overall size, shape, and polarity, resemble estrogens. The resemblance includes the flatness, or planarity, of the molecules and the positions and orientation of oxygen atoms. Isoflavonoids that have these molecular characteristics can mimic the biological activity of estrogens and are called *phytoestrogens*.

In terms of biosynthetic origins and chemical structure, phytoestrogens and estrogens are quite different. Phytoestrogens, being isoflavonoids, are phenolic compounds, formed from phenylalanine (an amino acid) by the shikimate pathway. In contrast, estrogens are triterpenoids, formed from acetyl coenzyme A by the isoprenoid pathway.

Isoflavonoids that are considered to be phytoestrogens exhibit only weak estrogenic activity in animals and humans. Examples of phytoestrogens are *coumestrol, daidzein,* and *genistein*. Coumestrol and daidzein are found in alfalfa (known as lucerne in Europe) and clover. Both of these plants belong to the legume family and are important forage crops for animals. If the content of phytoestrogens in alfalfa or clover is high, the reproductive cycles of grazing animals may be adversely affected. This can pose a problem for farmers wanting to breed livestock in the normal way. For this reason, the amount of grazing in fields of alfalfa or clover has to be restricted. Alternatively, varieties of alfalfa or clover that have been bred to contain lower levels of isoflavonoids can be grown. Unfortunately, plant varieties with lower isoflavonoid content are often more susceptible to both pathogen attack and attack by herbivorous pests.

Phytoestrogens in the Human Diet

One of the major sources of phytoestrogens in the human diet is the soybean. Genistein is the major phytoestrogen in soybeans. It is present in some soybean products such as tofu, although it is not present in soy sauce. Genistein, extracted from soybean plants, can also be obtained as a dietary supplement. Dietary supplements, which are often pills, powders, or tinctures containing plant-derived products, can be purchased over the counter. In the United States, the manufacture and sale of such products, classified as "dietary supplements," is far less closely regulated and standardized than the manufacture and sale of food and drugs.

Genistein has been promoted as a possible preventive treatment or therapy for several diseases and conditions. There are claims that it reduces hot flashes associated with menopause, that it can prevent or delay the onset of osteoporosis in postmenopausal women, and that it can lower blood cholesterol levels. In each instance the potential effectiveness of genistein would be attributable to its acting as an estrogen replacement in older women, in whom the level of estradiol is naturally low. Genistein may also be effective in the treatment of certain breast cancers that require estrogen in order to grow. In this case it is theorized that the genistein, with weak estrogen activity, acts to reduce cancer growth by competing with the more potent estradiol for the estrogen receptor.

Some of the evidence for the role of phytoestrogens in women's health is circumstantial. It is based, in part, on observations that women who live in countries such as Japan and China, where soy products are widely consumed, have a lower incidence of diseases such as osteoporosis and breast cancer. Clearly, other factors, genetic and environmental, may be contributory. Health claims attributed to phytoestrogens, including genistein, need further evaluation in well-designed clinical trials before such claims can be accepted by the scientific and medical communities or relied upon by those using dietary supplements.

—*Valerie M. Sponsel*

FURTHER READING

Dolby, Victoria. *All About Soy Isoflavones and Women's Health*. New York: Avery, 1999. Frequently asked questions, and answers to them, written for the nonscientist.

Levetin, Estelle, and Karen McMahon. *Plants and Society*. Boston: WCB/McGraw-Hill, 1999. Chapter 13, "Legumes," and the unit on plants and human health are relevant.

F

FIBER TECHNOLOGIES

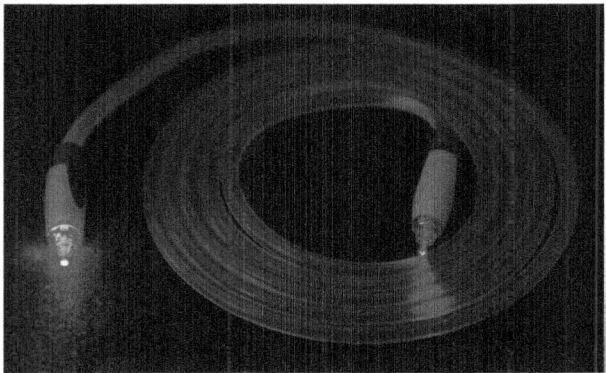

A TOSLINK fiber optic audio cable being illuminated at one end. By Hustvedt (Template:One), [CC-BY-SA-3.0 (http://creativecommons.org/licenses/by-sa/3.0) or GFDL (http://www.gnu.org/copyleft/fdl.html)], via Wikimedia Commons

FIELDS OF STUDY

Biology; Biomechanical engineering; Biotechnology; Chemical engineering; Industrial biotechnology (or White biotechnology)

ABSTRACT

Fibers have been used for thousands of years, but not until the nineteenth and twentieth centuries did chemically modified natural fibers (cellulose) and synthetic plastic or polymer fibers become extremely important, opening new fields of application. Advanced composite materials rely exclusively on synthetic fibers. Research has also produced new applications of natural materials such as glass and basalt in the form of fibers. The current "king" among fibers is carbon, and new forms of carbon, such as carbon nanotubes, promise to advance fiber technology even further.

DEFINITION AND BASIC PRINCIPLES

A fiber is a long, thin filament of a material. Fiber technologies are used to produce fibers from different materials that are either obtained from natural sources or produced synthetically. Natural fibers are either cellulose-based or protein-based, depending on their source. All cellulosic fibers come from plant sources, while protein-based fibers such as silk and wool are exclusively from animal sources; both fiber types are referred to as biopolymers. Synthetic fibers are manufactured from synthetic polymers, such as nylon, rayon, polyaramides, and polyesters. An infinite variety of synthetic materials can be used for the production of synthetic fibers.

Production typically consists of drawing a melted material through an orifice in such a way that it solidifies as it leaves the orifice, producing a single long strand or fiber. Any material that can be made to melt can be used in this way to produce fibers. There are also other ways in which specialty fibers also can be produced through chemical vapor deposition. Fibers are subsequently used in different ways, according to the characteristics of the material.

BACKGROUND AND HISTORY

Some of the earliest known applications of fibers date back to the ancient Egyptian and Babylonian civilizations. Papyrus was formed from the fibers of the papyrus reed. Linen fabrics were woven from flax fibers. Cotton fibers were used to make sail fabric. Ancient China produced the first paper from cellulose fiber and perfected the use of silk fiber.

Until the nineteenth century, all fibers came from natural sources. In the late nineteenth century, nitrocellulose was first used to develop smokeless gunpowder; it also became the first commercially successful plastic: celluloid.

As polymer science developed in the twentieth century, new and entirely synthetic materials were discovered that could be formed into fine fibers. Nylon-66 was invented in 1935 and Teflon in 1938. Following World War II, the plastics industry grew rapidly as new materials and uses were invented. The immense variety of polymer formulations provides an almost limitless array of materials, each with its own unique characteristics. The principal fibers used today are varieties of nylons, polyesters, polyamides, and epoxies that are capable of being produced in fiber form. In addition, large quantities of carbon and glass fibers are used in an ever-growing variety of functions.

HOW IT WORKS

The formation of fibers from natural or synthetic materials depends on some specific factors. A material must have the correct plastic characteristics that allow it to be formed into fibers. Without exception, all natural plant fibers are cellulose-based, and all fibers from animal sources are protein-based. In some cases, the fibers can be used just as they are taken from their source, but the vast majority of natural fibers must be subjected to chemical and physical treatment processes to improve their properties.

Cellulose Fibers. Cellulose fibers provide the greatest natural variety of fiber forms and types. Cellulose is a biopolymer; its individual molecules are constructed of thousands of molecules of glucose chemically bonded in a head-to-tail manner. Polymers in general are mixtures of many similar compounds that differ only in the number of monomer units from which they are constructed. The processes used to make natural and synthetic polymers produce similar molecules having a range of molecular weights. Physical and chemical manipulation of the bulk cellulose material, as in the production of rayon, is designed to provide a consistent form of the material that can then be formed into long filaments, or fibers.

Synthetic Polymers. Synthetic polymers have greatly expanded the range of fiber materials that are available, and the range of uses to which they can be applied. Synthetic polymers come in two varieties: thermoplastic and thermosetting. Thermoplastic polymers are those whose material becomes softer and eventually melts when heated. Thermosetting polymers are those whose the material sets and becomes hard or brittle through heating. It is possible to use both types of polymers to produce fibers, although thermoplastics are most commonly used for fiber production.

The process for both synthetic fibers is essentially the same, but with reversed logic. Fibers from thermoplastic polymers are produced by drawing the liquefied material through dies with orifices of the desired size. The material enters the die as a viscous liquid that is cooled and solidifies as it exits the die. The now-solid filament is then pulled from the die, drawing more molten material along as a continuous fiber. This is a simpler and more easily controlled method than forcing the liquid material through the die using pressure, and it produces highly consistent fibers with predictable properties.

Fibers from thermosetting polymers are formed in a similar manner, as the unpolymerized material is forced through the die. Rather than cooling, however, the material is heated as it exits the die to drive the polymerization to completion and to set the polymer.

Other materials are used to produce fibers in the manner used to produce fibers from thermoplastic polymers. Metal fibers were the first of these materials. The processes used for their production provided the basic technology for the production of fibers from polymers and other nonmetals. The best-known of these fibers is glass fiber, which is used with polymer resins to form composite materials. A somewhat more high-tech variety of glass fiber is used in fiber optics for high-speed communications networks. Basalt fiber has also been developed for use in composite materials. Both are available commercially in a variety of dimensions and forms.

Production of carbon fiber begins with fibers already formed from a carbon-based material, referred to as either pitch or PAN. Pitch is a blend of polymeric substances from tars, while PAN indicates that the carbon-based starting material is polyacrylonitrile. These starting fibers are then heat-treated in such a way that essentially all other atoms in the material are driven off, leaving the carbon skeletons of the original polymeric material as the end-product fiber.

Boron fiber is produced by passing a very thin filament of tungsten through a sealed chamber, during which the element boron is deposited onto the tungsten fiber by the process of chemical vapor deposition.

APPLICATIONS AND PRODUCTS

All fiber applications derive from the intrinsic nature of the material from which the fibers are formed. Each material, and each molecular variation of a material, produces fibers with unique characteristics and properties, even though the basic molecular formulas of different materials are very similar. As well, the physical structure of the fibers and the manner in which they were processed work to determine the properties of those fibers. The diameter of the fibers is a very important consideration. Other considerations are the temperature of the melt from which fibers of a material were drawn; whether the fibers were stretched or not, and the degree by which they were stretched; whether the fibers are hollow, filled, or solid; and the resistance of the fiber material to such environmental influences as exposure to light and other materials.

Structural Fibers. Loosely defined, all fibers are structural fibers in that they are used to form various structures, from plain weave cloth for clothing to advanced composite materials for high-tech applications. That they must resist physical loading is the common feature identifying them as structural fibers. In a stricter sense, structural fibers are fibers (materials such as glass, carbon, aramid, basalt, and boron) that are ordinarily used for construction purposes. They are used in normal and advanced composite materials to provide the fundamental load-bearing strength of the structure.

A typical application involves "laying-up" a structure of several layers of the fiber material, each with its own orientation, and encasing it within a rigid matrix of polymeric resin or other solidifying material. The solid matrix maintains the proper orientation of the encased fibers to maintain the intrinsic strength of the structure.

Materials so formed have many structural applications. Glass fiber, for example, is commonly used to construct different fiberglass shapes, from flower pots to boat hulls, and is the most familiar of composite fiber materials. Glass fiber is also used in the construction of modern aircraft, such as the Airbus A-380, whose fuselage panels are composite structures of glass fibers embedded in a matrix of aluminum metal.

Carbon and aramid fibers such as Kevlar are used for high-strength structures. Their strength is such that the application of a layer of carbon fiber composite is frequently used to prolong the usable lifetime of weakened concrete structures, such as bridge pillars and structural joists, by several years. While very light, Kevlar is so strong that high-performance automotive drive trains can be constructed from it. It is the material of choice for the construction of modern high-performance military and civilian aircraft, and for the remote manipulators that were used aboard the space shuttles of the National Aeronautics and Space Administration. Kevlar is recognizable as the high stretch-resistance cord used to reinforce vehicle tires of all kinds and as the material that provides the impact-resistance of bulletproof vests.

In fiber structural applications, as with all material applications, it is important to understand the manner in which one material can interact with another. Allowing carbon fiber to form a galvanic connection to another structural component such as aluminum, for example, can result in damage to the overall structure caused by the electrical current that naturally results.

Fabrics and Textiles. The single most recognized application of fiber technologies is in the manufacture of textiles and fabrics. Textiles and fabrics are produced by interweaving strands of fibers consisting of single long fibers or of a number of fibers that have been spun together to form a single strand. There is no limit to the number of types of fibers that can be combined to form strands, or on the number of types of strands that can be combined in a weave.

The fiber manufacturing processes used with any individual material can be adjusted or altered to produce a range of fiber textures, including those that are soft and spongy or hard and resilient. The range of chemical compositions for any individual polymeric material, natural or synthetic, and the range of available processing options, provides a variety of properties that affect the application of fabrics and textiles produced.

Clothing and clothing design consume great quantities of fabrics and textiles. Also, clothing designers seek to find and utilize basic differences in fabric and textile properties that derive from variations in chemical composition and fiber processing methods.

Fibers for fabrics and textiles are quantified in units of deniers. Because the diameter of the fiber can be produced on a continuous diameter scale, it is therefore possible to have an essentially infinite range of denier weights. The effective weight of a fiber may also be adjusted by the use of sizing materials added to fibers during processing to augment or improve their stiffness, strength, smoothness, or weight. The gradual loss of sizing from the fibers accounts for cotton denim jeans and other clothing items becoming suppler, less weighty, and more comfortable over time.

The high resistance of woven fabrics and textiles to physical loading makes them extremely valuable in many applications that do not relate to clothing. Sailcloth, whether from heavy cotton canvas or light nylon fabric, is more than sufficiently strong to move the entire mass of a large ship through water by resisting the force of wind pressing against the sails. Utility covers made from woven polypropylene strands are also a common consumer item, though used more for their water-repellent properties than for their strength. Sacks made from woven materials are used worldwide to carry quantities of goods ranging from coffee beans to gold coins and bullion. One reason for this latter use is that the fiber fabric can at some point be completely burned away to permit recovery of miniscule flakes of gold that chip off during handling.

Cordage. Ropes, cords, and strings in many weights and winds traditionally have been made from natural fibers such as cotton, hemp, sisal, and manila. These require little processing for rough cordage, but the suppleness of the cordage product increases with additional processing. Typically, many small fibers are combined to produce strands of the desired size, and these larger strands can then be entwined or plaited to produce cordage of larger sizes. The accumulated strength of the small fibers produces cordage that is stronger than cordage of the same size consisting of a single strand. The same concept is applied to cordage made from synthetic fibers.

Ropes and cords made from polypropylene can be produced as a single strand. However, the properties of such cordage would reflect the properties of the bulk material rather than the properties of combined small fibers. It would become brittle when cold, overly stretchy when warm, and subject to failure by impact shock. Combined fibers, although still subject to the effects of heat, cold, and impact shock, overcome many of these properties as the individual fibers act to support each other and provide superior resistance.

SOCIAL CONTEXT AND FUTURE PROSPECTS

One could argue that the fiber industry is the principal industry of modern society, solely on the basis that everyone wears clothes of some kind that have been made from natural or synthetic fibers. As this is unlikely ever to change, given the climatic conditions that prevail on this planet and given the need for protective outerwear in any environment, there is every likelihood that there will always be a need for specialists who are proficient in both fiber manufacturing and fiber utilization.

—Richard M. Renneboog, MSc

FURTHER READING

Fenichell, Stephen. *Plastic: The Making of a Synthetic Century.* New York: HarperCollins, 1996. A well-researched account of the plastics industry, focusing on the social and historical contexts of plastics, their technical development, and the many uses for synthetic fibers.

Morrison, Robert Thornton, and Robert Nielson Boyd. *Organic Chemistry.* 5th ed. Newton, Mass.: Allyn & Bacon, 1987. Provides one of the best and most readable introductions to organic chemistry and polymerization.

Selinger, Ben. *Chemistry in the Marketplace.* 5th ed. Sydney: Allen & Unwin, 2002. The seventh chapter of this book provides a concise overview of many fiber materials and their common uses and properties.

Weinberger, Charles B. *"Instructional Module on Synthetic Fiber Manufacturing."* Gateway Engineering Education Coalition: 30 Aug. 1996. This article presents an introduction to the chemical engineering of synthetic fiber production, giving an idea of the sort of training and specialization required for careers in this field.

G

GENETIC ENGINEERING

FIELDS OF STUDY

Biotechnology; Genetics; Medical biotechnology (or Red biotechnology); Philosophy and Religious Studies

ABSTRACT

Genetic engineering, also known as genetic modification, is an interdisciplinary scientific technique using molecular techniques to directly alter the basic genetic blueprint (DNA) of bacteria, plants, animals, humans, and other living organisms to achieve or enhance a specific trait or useful characteristic. Genetic engineering, though often controversial, is used in diverse areas, including medicine and agriculture, to diagnose and treat diseases, produce industrial products, neutralize pollutants, create higher-yielding crops, and perform scientific research. The genetic engineering process uses the tools of molecular genetics to explore and change living systems on a fundamental level and has revolutionized scientists' ability to understand, modify, and enhance the natural world.

Bt-toxins present in peanut leaves (bottom image) protect it from extensive damage caused by European corn borer larvae (top image). By Herb Pilcher, USDA ARS. [Public domain], via Wikimedia Commons

BACKGROUND

Although genetic engineering is most often discussed in the controversial arenas of crop production or theoretical human genetic manipulation, genetic engineering is used in diverse areas such as medicine, industry, and agriculture to treat disease, diagnose problems, produce industrial products, convert industrial waste, create hardier crops, and perform better scientific research.

The focus of genetic engineering is the gene. Genes are the basic units of inheritance that contain information and instructions for the creation, maintenance, and reproduction of living organisms. Genes are composed of DNA, a highly organized molecule located in almost every cell of an organism's

193

body. In genetic engineering, scientists add very specific pieces of useful genetic material to another organism's genes to change an organism's natural characteristics.

Genetic engineering was made possible by the development of new molecular genetic procedures, often called recombinant DNA technology, that can identify the particular DNA sequence of a gene or an entire genome, allow scientists to find the genetic material that codes for useful or desired features, and then insert the new material into the correct place in another organism's genetic code.

HISTORY

Before modern genetic engineering was possible, farmers had long selected for desired traits by breeding plants and animals with the desirable traits, a process known as selective breeding. Brewers and bakers also changed grains and flour into preferred products such as beer and bread through the use of small organisms called yeast and the process of fermentation.

By the early twentieth century, plant scientists had begun to use the work done by Gregor Mendel in the nineteenth century on the inheritance patterns of specific plant features to more formally introduce improvements in a plant species in a process called classic selection. However, the features of the basic unit of inheritance were not known until James D. Watson and Francis Crick identified the structure of DNA in 1953 with the double-helix model.

The nature of DNA and the technology to manipulate and modify the genetics of an organism was not available until twenty years later, when the first successful recombinant organism was created by Herbert Boyer and Stanley Cohen. Boyer and his laboratory had isolated an enzyme that could precisely cut segments of DNA in an organism, and Cohen found a way to introduce antibiotic-carrying plasmids into bacteria and a way to isolate and clone the genes in the plasmids. They combined their knowledge to create a way to clone genetically engineered molecules in foreign cells. Their discoveries led to the creation of a quick and easy way to make chemicals such as human growth hormone and synthetic insulin.

After Boyer and Cohen, many other scientists worked with recombinant DNA techniques to improve the procedures and develop a variety of genetically modified organisms designed to meet specific scientific, agricultural, industrial, and medical needs. Over time, these techniques and applications in genetic engineering spawned the multibillion-dollar biotechnology industry.

As the biotechnology industry grew and genetically modified organisms became more widespread, it became important to define which organisms were genetically modified organisms and which were products of classic selection. It also became necessary to determine if living organisms produced through genetic engineering could be patented by the companies and universities designing them. In 1980, the U.S. Supreme Court ruled in the case *Diamond v. Chakrabarty* that genetically altered life-forms can be patented.

In 1982, the U.S. Food and Drug Administration (FDA) approved the first consumer product developed through modern genetic engineering: a biosynthetic human insulin, sold under the trademark Humulin. The bacterially produced insulin created by Genentech and marketed by Eli Lilly revolutionized the treatment of diabetes, as it produced fewer immune reactions and its supply no longer depended on the availability of animals.

In 1996, Genzyme Transgenics (which in 2002 became GTC Biotherapeutics) created a transgenic goat that produced milk containing a cancer-fighting protein. It soon created additional transgenic animals that could produce specific human proteins to treat human disease. The ability to produce human hormones, enzymes, and other therapeutic products has decreased the risk of disease transmittal from donors to recipients of human products, increased supply, decreased immune reactions, and decreased the variability between medication batches that had been seen in the past.

In the 1990s, scientists sought to develop genetically modified plants and crops, including genetically engineered foods. By 1992, the first plant designed for human consumption (the Flavr Savr tomato) was approved for commercial production by the U.S. Department of Agriculture. In 1994, the European Union approved genetically modified tobacco in France. After these genetically modified crops gained approval, genetically engineered plants and other organisms became more widespread in the United States and reached supermarket shelves. It has been estimated that more than 75 percent of food products on store shelves may contain at least a small quantity of genetically engineered crops.

The Human Genome Project (1990–2003), a collaborative international scientific research initiative spearheaded by the National Institutes of Health, advanced genetic engineering by its publication of human DNA sequencing data. These data allowed scientists to learn more about the physical and functional aspects of genes and DNA. By its completion, the Human Genome Project had fully sequenced the human genome and provided a basic genetic road map for scientists to find human DNA segments of interest.

During the 1900s, scientists also used genetic engineering technology to develop numerous varieties of investigational organisms with very specific characteristics for use in research. Some genetically modified organisms were used to learn more about the natural progression of particular diseases. Others were created to test experimental therapies before moving to humans. These genetically modified organisms have helped scientists learn more about genetic disease, cancer, aging, and other chronic diseases.

In academic and industry laboratories, modern genetic engineering continues to solve problems related to health, disease, industry, and agriculture. Additional applications in humans and human disease have been assisted by government-funded initiatives such as the Human Genome Project. Although controversial at times, genetic engineering is a modern tool to be used in addressing a wide range of problems.

HOW IT WORKS

Although the types of organisms modified vary substantially in genome size and structure, all genetic engineering involves several general steps: identifying the desired feature or end application, isolating the gene segment that codes for the feature, inserting the gene segment into a vector, and adding it to the target organism, a process called transformation.

Identification. To create genetically modified organisms that will meet a specific need or solve a particular issue, the best way to engineer a solution must be determined. For example, if a large oil spill required cleanup, the first step would be to determine the type of organism and the desired features that would be most effective at removing the spilled oil. Issues to consider in solving a problem through genetic engineering include the desired size and type of organism to be modified, the availability of desired characteristics or features with a known DNA segment, and the possible positive and negative environmental impact resulting from a modified organism.

Isolation of the Proper Gene Segment. To modify an organism through insertion of a gene or DNA segment, the specific DNA segment in the donor organism must be known and be able to be effectively removed from its host genome. In some cases, the DNA segment coding for the desired characteristic is known and available because of previous scientific work. In other cases, this step can be very labor intensive and require long-term research.

After the DNA segment is identified, a particular recombinant technique, often a restriction enzyme, is used to cut the desired gene or DNA segment out of the donor organism and move it into a vector. Depending on the genetic engineering requirements, other techniques such as polymerase chain reaction or agarose gel electrophoresis can be used to isolate a gene or gene segment.

Insertion. Insertion is the genetic engineering step during which the desired gene or DNA segment is integrated into the vector. In this step, restriction enzymes are used to cut the vector open in a particular place so that the desired gene or DNA segment can attach itself to the vector. Then a special enzyme glue called ligase is used to attach the DNA segment to the vector. The most commonly used vectors in genetic engineering are circular form of DNA called bacterial plasmids; however, the type of vector used for a particular application depends on the size of the gene or segment being moved and the organism being modified.

Transformation. The next step of genetic engineering is called transformation. During this step, the desired gene or DNA segment is introduced and successfully added to the organism being modified. A variety of methods can be used to send the vector containing the desired DNA segment into the organism being modified in such a way that the new genetic information is added to the organism's standard genes. These methods include microinjection, use of a gene gun, electroporation, or use of viruses. In each of these methods, the vector carrying the desired genetic segment is forced into the new cells of the organism being modified. Completion of the transformation step relies on testing that determines whether the inserted DNA segment is producing the desired effect or trait in the newly modified organism.

CRISPR. A specific method of genetic engineering, known as CRISPR (clustered regularly interspaced short palindromic repeats) or CRISPR-Cas9, was developed in the first decades of the twenty-first century and was soon viewed as a revolutionary technique. The process is based on segments of DNA, also called CRISPR, from microorganisms including certain bacteria that are able to manipulate the DNA of invading viruses. The enzyme Cas9 and a segment of guide RNA (gRNA) are used as scissors on the molecular level to cut DNA at precise points, allowing fine control of gene editing. The process allowed for much faster and successful genetic engineering than previous methods, opening up the possibility of a much wider application of genetic modification.

APPLICATIONS AND PRODUCTS

Genetic engineering has far-reaching applications in food production, industry, medicine, and research.

Crops. One of the most widespread but also most controversial uses of genetic engineering is in the creation of genetically modified crops and food. The goal of genetic modification varies from crop to crop. Soybeans have been modified with a DNA segment conveying resistance to herbicides sprayed over fields to kill weeds growing amid the soybeans. The Flavr Savr tomato was engineered to decrease ripening time and increase shelf life. Varieties of rice and corn have been engineered using DNA segments for other plant genomes to have increased levels of vitamins.

From the first commercially grown genetically engineered product for human consumption (the Flavr Savr tomato), adoption of genetically engineered crops in the United States has increased quickly. According to data from the U.S. Department of Agriculture's Economic Research Service, fields planted with genetically engineered cotton (herbicide-tolerant and insect-resistant cotton) reached 96 percent of the total cotton acreage in 2014. That same year, herbicide-tolerant soybeans and biotech corn accounted for 94 percent and 93 percent, respectively, of their crop populations.

Supporters of genetically modified crops feel that the plants can increase food production to meet the world's needs using lower amounts of pesticides and increasing farmer profits. Opponents of genetically engineered crops are concerned about perceived safety issues regarding food produced from these crops, ecological issues around increased use of herbicides, contamination between genetically modified and naturally grown crops because of cross-pollination of fields, and economic difficulties dealing with patents on genetically modified crops.

Livestock. Although farmers have bred particular varieties of livestock such as cows, goats, chickens, and sheep for thousands of years to maximize desirable qualities, genetic engineering allows a more rapid introduction of specific qualities that may or not occur naturally in the animals. The benefits of genetically engineered livestock are numerous and affect the producers, environment, and consumers. Producers benefit by having disease-resistant, increasingly productive, or fast-growing animals. For example, the gene responsible for regulating milk production in cows can be modified to increase milk production. Also, if animals are engineered to have milder waste, the environment will benefit. The FDA has reviewed genetically modified pigs that are better able to digest and process phosphorus in ways that release up to 70 percent less phosphorus in their waste. Consumers benefit from more nutritious, vitamin-enriched meat, as in the case of pigs that are engineered to produce omega-3 fatty acids through the expression of a roundworm gene.

Many of the concerns that apply to crops also apply to livestock. In 2015, the FDA stated that a type of salmon called AquAdvantage, genetically modified to grow twice as fast as conventional Atlantic salmon, was safe for human consumption, the first time a genetically engineered meat was approved by the organization. The FDA has placed several genetically modified animals under review in a category called food-drug. The FDA considers a modified DNA segment to be like a drug and is regulating transgenic organisms in the same way it oversees animals that receive growth hormones or antibiotics.

Diagnosis. Genetic engineering has allowed the development of faster, cheaper, and more accurate diagnostic tests for certain diseases to be used both in the laboratory and in the body. The tests based on genetic engineering are used to identify infectious diseases, hormonal changes, pregnancies, cancer, and other diseases and conditions. For example, a series of faster and more accurate tests for the presence of the human immunodeficiency virus (HIV) have been developed based on genetically modified HIV antigens. Other tests can diagnose diseases by detecting particular substances in specific locations in the body. These exams rely on genetically modified antibodies with markers that can be injected into the body.

Medications. The use of genetically modified organisms to produce human hormones, enzymes, vaccines, and medications has revolutionized the pharmaceutical industry. Since 1982, when Genentech's biosynthetic human insulin was introduced, the ability to manufacture new products in a controlled environment instead of collecting similar substances from the limited supply of human and animal sources has led to more readily available, effective, and reliable medications. Products include human growth hormone to treat children with insufficient growth, plasminogen activator to dissolve blood clots, and erythropoietin to treat low blood iron (anemia). In 1994, genetic engineering also led to innovative treatments for rare genetic disorders such as Gaucher disease with the production of specific human enzymes in genetically modified Chinese hamster ovary cells. The point of this enzyme replacement therapy is to replace the enzyme that the affected individuals are missing through intravenous injections of genetically engineered human enzymes. In multiple situations, the use of genetically engineered organisms to create medications has saved lives and decreased the burden of disease in ways that could not be imagined before. Future applications of genetic engineering in medicine are likely to focus on the creation of better medications for life-threatening indications.

Disease Cures. Genetic engineering made it possible to develop gene therapy. Genetic diseases are inherited conditions that occur because of one or more genetic changes or mutations that prevent the correct functioning of a particular gene. Most genetic diseases do not have a treatment or cure. However, with genetic engineering techniques, scientists hope that they will be able to transform an affected individual's mutated gene into a working gene by replacing it with a functional copy of the gene. Gene therapy has shown some success in helping individuals with severe combined immunodeficiency (SCID), hemophilia type B, and several other genetic diseases, and even has been applied to cancer. However, this type of treatment is still under investigation to determine if it can safely and permanently cure genetic conditions, and other concerns also remain, including ethoeconomic issues over the potential high cost and limited availability of effective treatments.

Research. Genetic engineering and genome sequencing have been used to improve investigative techniques through the ability to manipulate organisms on a basic, genetic level. In genetic research, genetic engineering techniques have allowed scientists to create mice and other organisms affected by a specific gene change for detailed study of a specific genetic disease. For example, a genetically modified mouse that lacks the gene to produce amyloids has been used to study Alzheimer's disease. On a broader scale, genetic engineering allows detailed analysis of an organism's structure, function, and development. Through the insertion of a marker in or near a gene coding for a product of interest, scientists can track the location of that gene's product over time.

Industrial Applications. Genetically engineered organisms are used in several manufacturing arenas in production, processing, and waste removal. Most industrial applications of biotechnology are based on naturally occurring processes using modified bacteria, yeast, and other small organisms to digest, transform, and synthesize natural materials from one form into another. More specifically, genetically modified microorganisms have been used to produce industrial chemicals such as ethylene oxide (for making plastics), ethylene glycol (antifreeze), and alcohol. Bacteria have also been engineered to remove toxic wastes from the environment, for example, the varieties of genetically engineered bacteria that consume oil by chemically transforming its compounds into usable basic molecules. Future directions in industry include production of textile fibers, fuels, plastics, and other industrial chemicals out of industrial wastes or raw materials.

SOCIAL CONTEXT AND FUTURE PROSPECTS

Genetic engineering has already altered the course of agriculture, industry, and medicine with its life-changing applications. Crops have been modified so that they are more nutritious and naturally produce pesticides. Life-saving medications made of human hormones integrated into bacteria are widely available in consistent and purified forms. Bacteria that convert toxic chemicals into harmless basic elements have been developed. Great strides have been made in using gene therapy to cure genetic diseases. Much of the future of genetic engineering will be marked by further refinement of these applications and processes, as well as by unforeseeable breakthroughs made possible by future technological development. However, these significant scientific strides also come with important ethical questions and safety concerns.

Environmentalists are concerned about the impact of genetically modified crops on ecosystems, in particular whether the genes introduced into genetically modified crops will be transferred to conventional crops through cross-pollination. Some researchers express concern that the true impact of GMOs is untested, and that widespread availability of such organisms should be delayed until more data is collected.

Despite general scientific consensus that GMOs as currently produced pose no health risks to humans, some individuals and organizations strongly protest otherwise. While many such opponents cross into the realm of conspiracy theories, some legitimate organizations continue to pose questions. Advocacy groups such as Greenpeace and the World Wildlife Fund are concerned about the safety of genetically modified food and feel that the available data do not prove that there are no risks to human health from consumption. Despite statements from the Royal Society of Medicine and the U.S. National Academy of Sciences in support of the safety of such foods, these groups have called for additional and more rigorous testing before genetically engineered foods are marketed.

Other countries have significant concerns about the safety of genetically modified foods. The European Union regulates genetically modified food imported from other nations, including the United States. Venezuela has banned the growing of genetically modified crops, and India has issued a moratorium on the cultivation of genetically modified foods pending an investigation into safety concerns. In December 2014, however, following an investigation into potential threats to health and environment presented by genetically engineered crops, India's Genetic Engineering Appraisal Committee approved twelve genetically modified crops for experimental trials to gather more information on their safety and utility. Other counties such as Japan and Zambia have also registered concerns over the safety of genetically modified foods. In November 2014, a Russian NGO announced what it claimed to be the most comprehensive and largest study yet examining the health effects of genetically modified foods. The three year study will involve feeding rats a diet of genetically modified maize and an extremely common herbicide to gauge what effect, if any, the products have on the rats' health.

There is significantly less controversy over the use of genetically modified organisms in industrial production and medicines. However, the use of genetic engineering techniques for human gene therapy and related applications has touched off a firestorm of ethical debate. Controversy continues to flare up on issues including the ethoeconomic issues of genetic testing, the potential for designer babies, and the possibility of body enhancement through genetics, as well as over the general ethical and religious implications of altering the human genome at any level. These debates are complicated by a lack of cohesive regulation regarding human genetic engineering.

—*Dawn A. Laney, M.S., C.G.C., C.C.R.C.*

FURTHER READING

Avise, John C. *The Hope, Hype, and Reality of Genetic Engineering: Remarkable Stories from Agriculture, Industry, Medicine, and the Environment.* New York: Oxford UP, 2004. Print.

Cathomen, Toni, Matthew Hirsch, and Matthew H. Porteus. *Genome Editing: The Next Step in Gene Therapy.* Springer, 2016.

Hodge, Russ. *Genetic Engineering: Manipulating the Mechanisms of Life.* New York: Facts On File, 2009. Print.

Jabr, Ferris. "Building Tastier Fruits & Veggies." *Scientific American* 311.1 (2014): 56–61. Print.

Khan, Muhammad Sarwar, Iqrar A. Khan, and Debmalya Barh. *Applied Molecular Biotechnology: The*

Next Generation of Genetic Engineering. CRC Press, 2016.

Nicholl, Desmond S. T. *An Introduction to Genetic Engineering.* New York: Cambridge University Press, 2008. Print.

Powell, Russell. "The Evolutionary Biological Implications of Human Genetic Engineering." *Journal of Medicine & Philosophy* 37.3 (2012): 204–25. Print.

Shanks, Pete. *Human Genetic Engineering: A Guide for Activists, Skeptics, and the Very Perplexed.* New York: Nation Books, 2005. A helpful explication of the science behind cloning, coupled with stern warnings against it, by a noted social activist.

Yount, Lisa. *Biotechnology and Genetic Engineering.* 3d ed. New York: Facts On File, 2008. Print.

Genetic resources

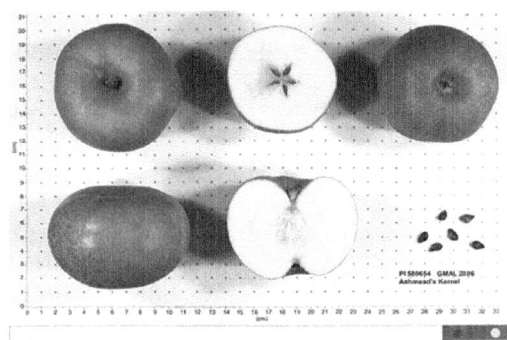

Size and cross sections of Apples of the cultivar Ashmead's Kernel. By USDA, ARS, National Genetic Resources Program. Germplasm Resources Information Network - (GRIN), [Public domain], via Wikimedia Commons

FIELDS OF STUDY

Biotechnology; Genetics; Horticulture; Medical biotechnology (or Red biotechnology); Microbiology; Pharmacogenomics; Philosophy and Religious Studies; Plant pathology

ABSTRACT

The raw material used in biotechnology is the genetic code found within the DNA of living organisms. While not viewed as a natural resource historically, genetic material has, with the advent of modern biotechnology, become a commodity that not only can be manipulated to improve agricultural yield but also can be used as a source by which to produce novel pharmaceutical or chemical products.

BACKGROUND

Biotechnology can be defined as the use of living organisms to achieve human goals; in this sense, humans have used biotechnology throughout history to provide themselves with such things as food, clothing, shelter, cosmetics, and medicine. Starting around 10,000 BCE, humans began to alter the genetic makeup of the plants and animals that they used by artificially selecting certain traits in the crops and livestock that they were breeding. Because farmers lived in different areas around the world with varying environmental conditions, the varieties of domesticated organisms that eventually developed initially preserved what is known as "genetic diversity." Every organism on Earth has a particular genome, its entire set of DNA, which is specific to that particular living thing. Therefore, genetic diversity is at its greatest when the widest variety of organisms available are in existence in a particular area. The term "biodiversity" refers to the number of different species (or other taxonomical units) that inhabit a given ecosystem or geographical area.

GENETIC EROSION OF PLANTS AND ANIMALS

A process known as "genetic erosion" decreases the biodiversity of cultivated areas. In this process, local species are lost from an area as they are replaced by less diverse, domesticated varieties. Human activities such as urbanization, the replacement of traditional agriculture with more modern techniques, and the introduction of high-yield varieties of crops have been blamed for such erosion of genetic resources. One example can be seen in the crops that are utilized for food in modern society. Among the 300,000 or so flowering plants that have been characterized to date, estimates indicate that humankind has used around 7,000 of these throughout history to satisfy basic human needs. However, only 30 of these account for 95 percent of the world's dietary calories, less than 10 account for 75 percent, and a mere 3 (corn, wheat, and rice) make up nearly 50 percent of the caloric intake of humankind. Not only has the number of different crops decreased over time, but also the variety of crop species has declined. Such a narrow genetic base of crops puts the food supply at risk from pests or diseases that affect a specific type of crop. Apparently, humans, in their eagerness to improve crop varieties, have somehow robbed the Earth of a portion of the genetic diversity that has taken millions of years to develop. Traditional subsistence agriculture, although it may have lacked the productivity of modern methods, actually increased the likelihood of a reasonable level of production by preserving the genetic diversity of the crops that were being grown.

This is not to say that using plants as a food source is their only economically viable use. Terrestrial plants have long been used as medicine or for other chemical applications. The recent loss of plant biodiversity is alarming for this reason also. Possibly, some undiscovered cure for a particular disease is at risk from disappearing permanently from the Earth, if it has not done so already. Of the remaining flowering plants on Earth, estimates indicate that one in four could become extinct by 2050. Compounding the problem, most of the world's biodiversity is located in geographically or politically unstable areas, namely in tropical or subtropical developing countries. A full two-thirds of all plant species known to humankind are located in the tropics; about 60,000 species are found in Latin America alone.

Animals have also served as an important source of food and medicine throughout history and have also been submitted to artificial genetic selection, along with accompanying genetic erosion. Marine invertebrates, in particular, have been investigated as a source of molecular compounds with medicinal properties. It has also been during more recent times that the importance of microorganisms in producing therapeutically relevant products has become known. While the full extent of microbial diversity remains unclear and bacterial diversity in particular appears to follow different biogeographical patterns from those found for plants and animals, it is evident that many natural habitats that may harbor medicinally relevant microbes are disappearing rapidly. The worldwide loss of biodiversity comprises all types of living organisms, including plants, animals, fungi, protists, and bacteria.

PARADIGM SHIFT

In the 1990's, a significant shift occurred in the way that genetic resources were viewed as well as how their ownership was determined. Before this time, genetic resources were considered to be a common heritage of humankind and were to be treated so that they were used to the benefit of all. The only problem with this notion is that it gave host countries little economic incentive for conservation. Another reason for this paradigm shift was a revolution in biology which had taken place in the decade that preceded the change. Biological tools that allowed for genetic engineering had been developed during this time, thereby expanding the number of organisms amenable to biotechnology. These included those that could be artificially selected for particular traits and bred with one another to any organism from which DNA could be extracted. This extracted DNA could then be introduced into a number of living vectors that may have been completely unrelated to the original source of genetic material. This new technology not only ensured that virtually any organism could be used as a source of genetic innovation with a potential for practical application but also decreased the cost of working with genetic material to a level at which many more laboratories could afford to participate in genetic engineering efforts.

The United Nations Convention on Biological Diversity (CBD) was signed at a meeting in Rio de Janeiro, Brazil, in 1992 and went into effect the following year. The CBD affirmed the sovereign right of individual nations to their biodiversity and gave them a means by which to regulate access to their genetic resources, creating the stipulation that entities such as bioengineering firms secure informed written consent before collecting genetic material from any particular country. The export of a seed, microbe, or other plant- or animal-derived sample has been compared to exporting a very small chemical factory, complete with blueprints and its own source of venture capital. While most countries would not allow this to happen using conventional technology, prior to 1992 this had been the norm for biological goods.

Despite the establishment of the CBD, a number of potential problems concerning genetic resources remain. Developing countries are leery of corporations from developed nations that may, given the opportunity, engage in biopiracy. This includes taking advantage of indigenous knowledge and local technologies without providing adequate compensation. Genetic material is similar to electronic media in that it can be reproduced easily and relatively inexpensively, a fact that makes enforcing antipiracy legislation difficult. Conditions of contracts and changes in patent legislation must be followed closely by developing nations to ensure that undue control is not handed to foreign investors.

In addition, just because a country as a whole receives compensation for a particular genetic resource does not mean that a given region of that country will see any economic benefit. Two examples from the United States (which predate the CBD) include the cancer drug Taxol and Taq polymerase, an enzyme used in genetic engineering. These products were discovered respectively in Pacific yew bark from the Pacific Northwest and from hot springs in Yellowstone National Park. Despite the fact that both of these products have produced millions of dollars of profits, the regions of the country where they were first discovered ended up receiving little or no financial benefit. This somewhat flawed system of compensation and financial incentive is not expected to work much better in developing countries. Historically, even when indigenous knowledge was used to develop a specific product, indigenous peoples often received little or no benefit from sharing their knowledge.

CONSERVATION EFFORTS

The same year the CBD was signed, an interdepartmental effort in the U.S. government created the International Conservation of Biodiversity Groups (ICBG) initiative. The objectives of the ICBG were to establish an inventory of species that have been used in traditional medicine, identify lead compounds for the treatment of human disease from this group, conduct economic assessments of species in the host country, establish study plots in developing countries to study changes in rain-forest ecology, and train local scientists in the principles of drug-development and biodiversity conservation. Conservation of genetic resources typically fits into one of two categories, in situ or ex situ: The former is Latin for "in the place," and the latter means "out of the place." In situ conservation takes place on farms, for agricultural crops, or in natural reserves, for wild plants. This type of conservation preserves the evolutionary dynamics of the species in question. Ex situ conservation usually involves storing samples, called accessions, of seeds or vegetative material for plants in what are known as gene banks. This type of conservation can also be applied to animals, where embryos or germ cells are stored frozen. This latter conservation technique has the disadvantage of being able to preserve only a small amount of the genetic diversity present in a given population but often plays a critical role in the preservation of many varieties of organisms, particularly those which are endangered or have already become extinct.

SCREENING FOR COMPOUNDS

Biological organisms of interest to the pharmaceutical or chemical industries are typically those which produce small organic compounds known as secondary metabolites. Some hypothesize that these compounds serve either defensive or signaling roles in the cell: Plants and animals use these compounds to defend themselves from potential predators, and microbes use these to defend themselves from and signal to the other organisms that surround them. Overall, more than one-half of the best-selling pharmaceuticals in use are derived from such natural products. Bioprospecting is the act of systematically searching through given genetic resources for compounds that may have a commercial application. Scientists are thus screening large numbers of extracts from plants, microbes, and marine organisms for secondary metabolites containing antifungal, antiviral, or antitumor activities.

There are a number of hurdles that must be overcome before a specific activity can be gleaned from a particular natural product. Because most natural products consist of mixtures of crude extracts, a certain degree of purification must take place before a lead compound can be tested for a desired application. "Time-to-lead" is a term that refers to the degree of purification and structural characterization that is necessary before a sample can be effectively assayed for a given activity. Another issue is the continued supply of a given natural product. In the past several decades, techniques for the extraction, fractionation, and chemical identification of secondary metabolites have become more routine and less expensive to perform. Before this was the case, it was often necessary to re-collect samples of particular natural products for use in large-scale purifications. Frequently, developers would then discover that it was impossible to reproduce the originally detected activity. Advances in genetic engineering as well as cell culture techniques have largely eliminated the need to re-collect an original sample. These advances actually make it more challenging for a supplier country to adequately charge for the use of a natural resource, because they can no longer rely on the need for recollection of biological material to take place. This leaves two basic strategies for institutions seeking to benefit from international biotrade: becoming a low-cost supplier or becoming a value-added supplier.

This latter strategy relies on the fact that selection of natural products for testing purposes does not have to occur randomly: Both chemotaxonomic and ethnomedical techniques can be applied to create a value-added product. Chemotaxonomic strategies rely on the selection of organisms from a related taxonomic group that are expected to produce a similar chemical category of substances as the original sample. An example of this can be seen in the soil-derived filamentous fungi as well as in the Actinobacteria. Since the antibiotics penicillin and streptomycin were isolated from the former group in the 1930's and from the latter group a decade later, taxonomically related groups have been successfully screened for secondary metabolites. In contemporary society, such compounds are used to treat cancer, arteriosclerosis, and infectious disease and are even used as immunosuppressive agents. In ethnomedical selection, knowledge of the use of a natural product in traditional medicine is expected to increase the chance of getting positive results with a particular extract. This approach involves sending experts into the field to conduct interviews with traditional healers. While this type of value-added product is more likely to generate a positive "hit," it is time-consuming and therefore often slow to generate high numbers of potential compounds. Another disadvantage of this type of approach is that it has proven difficult to select with efficacy for agents against complex diseases like cancer, because indigenous traditional healers may be unfamiliar with such maladies.

The most recent approach to the isolation of bioactive natural products eliminates the supply and subsequent screening of living organisms altogether. Because it is actually the genetic data that are of interest to most researchers and not the isolated organism, collecting DNA from environmental samples and directly cloning it into a host vector is becoming more commonplace. While the nature of the organism which contributed its genetic material to any metagenome, the collection of a large number of genomes, may not be determined with any certainty, the end result of having a gene that produces a particular compound of interest has been achieved. This approach is especially adaptable to microorganisms that inhabit soil and water samples in high numbers, the DNA of which can be extracted with relative ease. This approach gained favor when it became evident that a minority of microbial diversity exists in those microbes amenable to being grown under laboratory conditions, and that vast amounts of biodiversity are present in the microorganisms, which resist culturing in the lab for some reason. While activity-based screening of cloned metagenomic libraries is, by definition, a random process, it is believed that new classes of useful compounds are bound to be discovered using this technique.

———*James S. Godde*

FURTHER READING

Esquinas-Alcazar, José. "Protecting Crop Genetic Diversity for Food Security: Political, Ethical, and Technical Challenges." *Nature Reviews: Genetics* 6, no. 12 (December, 2005): 946-953.

Ferrer, Manuel, et al. "Metagenomics for Mining New Genetic Resources of Microbial Communities." *Journal of Molecular Microbiology and Biotechnology* 16, nos. 1/2 (2009): 109-123.

Reid, Walter V. "Gene Co-ops and the Biotrade: Translating Genetic Resource Rights into Sustainable Development." *Journal of Ethnopharmacology* 51, nos. 1-3 (April, 1996): 75-92.

Schuster, Brian G. "A New Integrated Program for Natural Product Development and the Value of an Ethnomedical Approach." *Journal of Alternative and Complementary Medicine* 7, no. 1 (2001): S61-S72.

Singh, Sheo B., and Fernando Pelaez. "Biodiversity, Chemical Diversity, and Drug Discovery." *Progress in Drug Research* 65, no. 141 (2008): 142-174.

Genetically engineered pharmaceuticals

FIELDS OF STUDY

Biotechnology; Genetics; Medical biotechnology (or Red biotechnology); Microbiology; Pharmacogenomics

ABSTRACT

Genetic technology is being used to develop state-of-the-art drugs to treat human diseases and health conditions. Genetic engineering can create carefully targeted drugs that offer greater effectiveness and potency than conventional pharmaceuticals while causing fewer side effects.

BACKGROUND

Genetics is the scientific study of heredity—the biological factors that determine the characteristics of all living things. All reproductive life-forms develop under the laws of genetics. The basis of genetics is the gene, a tiny unit of matter that determines some identifiable characteristic of an individual. Genes are located in fixed positions on chromosomes (molecular chains) that reside in the center of cells. A major part of the chromosome is deoxyribonucleic acid (DNA), which is responsible for transmitting genetic information, in the form of genes, when new life is created. This transfer of traits applies to organisms of all sizes, from microscopic to larger and more complex systems, such as humans. Inherited traits include color of hair, eyes, and skin, as well as susceptibility to various ailments.

As the complex chemical-biological process of reproduction takes place at the gene-chromosome level, it is possible for a random processing error called a mutation to be introduced. Mutant cells may cause many human defects and diseases. Genetic abnormalities, also known as birth defects, include such ailments as hemophilia (resistance to blood clotting), color blindness, anatomical defects, speech disorders, hormonal disorders, brain disorders, and psychiatric illness. Aside from birth defects, genetic cellular mutations can occur anytime during a lifetime. Normally the body contains certain controlling genes that destroy mutant genes that spontaneously appear. If these controlling genes become defective, the mutant genes can take over the body, as occurs with cancerous tumors.

GENETIC TECHNOLOGY IN MEDICINE

Pharmaceuticals are drugs used to treat human diseases and conditions. The term "engineered drug" implies that scientific principles and manufacturing processes are applied in creating the drug. A genetically engineered pharmaceutical is a specialized drug made by the application of specific genetic principles. Gene-based technology is used to investigate, test, and apply state-of-the-art pharmaceuticals to invasive and widespread diseases. Its potential for fighting illnesses such as cancer and acquired immunodeficiency syndrome (AIDS) is of particular interest to medical science.

For example, scientists have developed ways to control cancer cells using genetic medicines instead of killing the cancer with radiation, conventional drugs, or surgical removal. Herceptin, a genetically engineered drug approved by the U.S. Food and Drug Administration (FDA), is used in treating certain breast cancers. Herceptin is an antibody engineered to attack specific cancer cells, helping to reduce the cancer tumor by keeping a particular protein from reproducing. Some tumor cells are inherently resistant to the drug or become resistant over time.

In the past the production of natural body chemicals required the harvesting of the needed chemicals from human or animal materials. Supplies of such sources are sometimes minimal, and concentrating chemicals from human and animal tissue can also multiply the chances of carrying diseases from those sources. For instance, during the 1960's through the mid-1980's some children suffering from growth failure were treated with human growth hormone (HGH) extracted from the pituitary glands of human cadavers. In 1985 three adults in the United States who had been treated with this HGH during childhood died from Creutzfeldt-Jakob disease (CJD), a rare, incurable, and fatal brain disease with a long incubation period. Other recipients of HGH, both in the United States and abroad, also contracted CJD, apparently from HGH that was contaminated with the infectious agent that causes CJD.

Genetic engineering of substances such as growth hormone circumvents many traditional problems. The genes for producing the desired chemicals can be implanted in the genetic code of plants or microorganisms, especially the benign bacterium *Escherichia coli*. These sources can enable high-volume production at high levels of concentration. Because plants and microorganisms are very different from people, the chance of spreading disease through this production method is minimal.

GENETICALLY ENGINEERED VACCINES

Genetically engineered pharmaceuticals can increase the body's production of naturally occurring chemical substances and supply toxins to attack targeted pathogens (disease-causing viruses, bacteria, fungi, or parasites). Vaccines work by triggering the body's immune system, which then defends itself. Compared with traditional methods of creating vaccines, genetic engineering enables faster development of safer vaccines.

A live attenuated vaccine is one that contains the living pathogen, but in a weakened form that cannot induce illness. Recombinant DNA technology can be used to remove key genes from microbes to render them harmless. This method has been used to engineer a vaccine against *Vibrio cholerae*, the bacterium that causes cholera. Another live attenuated vaccine, one used against the rotavirus pathogen responsible for serious diarrheal illness in infants and young children, was created through a technique that combines genes from different strains of the pathogen in a way that makes a harmless simulator virus. At the cellular or molecular level, the simulator virus appears to the immune system to be a pathogenic invader, causing antibodies to develop that attack the active pathogen.

A subunit vaccine does not employ the entire pathogenic organism; rather, it relies on its antigens, substances that trigger the body's immune system. For this technique, select antigens or portions thereof are used to provoke an immune response. While the antigens could be harvested from laboratory-grown microbes, recombinant DNA technology makes it possible to manufacture the antigen molecules. These parts of the pathogen's genetic code are inserted into common baker's yeast, a harmless microbe. Hepatitis B subunit vaccine uses a portion of the protein coat surrounding the virus's DNA. Because the rest of the microbe is not included, the possibility of an adverse reaction is greatly reduced.

In the creation of conjugate vaccines, the bacterial antigens selected for the immune response they provoke are not easily recognized by the body. These are joined with easily recognizable antigens located on a harmless bacterial shell and injected into the body to trigger the immune system. Conjugate vaccines for children have been developed against middle-ear infections and other diseases caused by pneumococci, a group of common bacteria. In 2000 the FDA licensed Prevnar, a conjugate vaccine that targets the seven most common types of pneumococci causing invasive disease in infants and toddlers. Ten years later, with other types of pneumococci becoming increasingly common in young children, the FDA licensed a conjugate vaccine to replace Prevnar, one that targets thirteen pneumococcal strains.

Naked DNA vaccines have been tested on diseases such as AIDS and some cancers. This experimental method involves injecting a person with some of a pathogen's DNA—specifically, with the genes that code for antigens. Some of the body's cells accept this added DNA as instructions to produce antigens, thereby triggering immune response. A similar experimental technique, recombinant vector vaccination, employs attenuated microorganisms that act as carriers for the DNA while further stimulating immune response.

The use of genetically engineered pharmaceuticals has raised concerns about their possible effects on the environment. In 1989 the Virginia Department of Health approved a field test of baits spiked with genetically engineered oral vaccine to control the spread of rabies in raccoons. Health officials were worried, however, about the possible danger to humans posed by the vaccinia virus used as the vaccine; thus an island location was chosen to prevent the possible spread of vaccinated animals to larger, mainland populations. Several researchers expressed concerns about the long-term effects of releasing a nonnative virus into the environment; although vaccinia had been used for many years to prevent smallpox, little was known about its host range or its ability to cause disease. The U.S. Department of Agriculture concluded in a 1991 report, however, that laboratory and field tests had shown the genetically engineered rabies vaccine to have had no adverse effects on any species. In the same report, the department approved further field tests on the grounds that such tests were safe and posed no significant environmental risk. This rabies-control method has since become common practice in North America and Europe.

—*Robert J. Wells, updated by Karen N. Kähler*

FURTHER READING

Aldridge, Susan. *The Thread of Life: The Story of Genes and Genetic Engineering*. 1996. Reprint. New York: Cambridge University Press, 2000.

Castilho, Leda R., et al., eds. *Animal Cell Technology: From Biopharmaceuticals to Gene Therapy*. New York: Taylor & Francis, 2008.

Crommelin, Daan J. A., Robert D. Sindelar, and Bernd Meibohm, eds. *Pharmaceutical Biotechnology: Fundamentals and Applications*. 3d ed. New York: Informa Healthcare, 2008.

Gad, Shayne Cox, ed. *Handbook of Pharmaceutical Biotechnology*. Hoboken, N.J.: John Wiley & Sons, 2007.

Groves, M. J., ed. *Pharmaceutical Biotechnology*. 2d ed. Boca Raton, Fla.: Taylor & Francis, 2006.

Guzman, Carlos Alberto, and Giora Z. Feuerstein, eds. *Pharmaceutical Biotechnology*. New York: Springer, 2009.

Rehbinder, E., et al. *Pharming: Promises and Risks of Biopharmaceuticals Derived from Genetically Modified Plants and Animals*. Berlin: Springer, 2009.

Walsh, Gary. *Biopharmaceuticals: Biochemistry and Biotechnology*. 2d ed. New York: John Wiley & Sons, 2003.

GENETICALLY MODIFIED FOOD PRODUCTION

Plums genetically engineered for resistance to plum pox, a disease carried by aphids. By Scott Bauer, USDA ARS, [Public domain], via Wikimedia Commons

FIELDS OF STUDY

Agricultural engineering; Agronomy; Animal science; Biotechnology; Environmental biotechnology (or Green biotechnology); Food science; Genetics; Human nutrition; Microbiology

ABSTRACT

Genetically modified food production is a subset of biotechnology and genetic engineering. This developing field offers both hope and concern for global food production. Genetically modified food production is the direct result of the development of genetically modified organisms. Conventional plant breeding is a slow process, and it takes several years to develop plants with desirable traits. Advances in genetic engineering have allowed scientists to speed up the process of developing plants with the most desirable traits and with greater predictability. These plants help farmers increase production and obtain higher yields. However, the use of genetically modified organisms in food production is controversial in some countries because the effect of these organisms on humans has not been assessed.

BASIC PRINCIPLES

Genetically modified food production is the creation of food products using genetically modified organisms. Some of the food production problems that can be addressed using genetically modified organisms are limiting or eliminating the damage caused by pests, weeds, and diseases, and providing tolerance to specific herbicides, extreme temperatures, and drought, as well as other production-related issues.

One of the initial goals of developing genetically modified plants was to achieve higher crop yields by creating versions of plants such as corn and soybeans that offered greater resistance to diseases caused by pests and viruses. Genetically modified foods, also known as genetically engineered food, are developed, produced, and marketed mainly because they present an advantage over traditionally produced food to either the farmer/producer or the consumer. The benefit to the producers comes from increased productivity and the reduction of lost crops. The consumer benefits from genetically engineered food by having access to food with better nutritional value and, in some cases, lower prices.

BACKGROUND

People began harvesting plants and domesticating animals around 10,000 BCE, and they soon were selecting the best seed and breeding the best animals through a process of trial and error. In the latter part of the nineteenth century, the monk Gregor Mendel used peas to demonstrate heredity in plants, thereby laying the foundation for modern plant breeding and genetics. The early work of Mendel and others led to the development of commercial crops by the 1930s. The field of molecular biology, which emerged in the twentieth century, has led to a better understanding of the cells and molecular processes of living organisms. This understanding has allowed researchers to develop genetically engineered plants and animals to address and solve many of the problems faced by farmers in crop and livestock production.

The first field trials of crops developed using genetically modified organisms took place in 1990, and by 1992, the first genetically modified corn was approved for use by the U.S. Food and Drug Administration (FDA). In the 1990s, additional advances led to genetically engineered vaccines and hormones as well as to the cloning of animals. Since their introduction in the United States, genetically modified crop plants have been widely used by American farmers, and the percentage of acreage planted with genetically modified crops has been steadily increasing, reaching more than 90 percent for some crops. Soybeans and cotton genetically modified to tolerate herbicides are the two most widely adopted genetically modified crops. Cotton and corn with insect resistance are the next most common crops grown by American farmers.

HOW IT WORKS

Genetic engineering allows scientists to insert a gene from one organism into another, resulting in a genetically modified organism. Therefore, genetic engineering begins with the identification and isolation of a gene that expresses a desirable trait. This gene can be found in a relative of the target species or in a completely unrelated species. A recipient plant or animal is selected, and the gene is inserted and incorporated into the genome of the recipient. The desired gene is inserted by various techniques such as using *Agrobacterium* as a vector or using a gene gun. The newly inserted gene becomes part of the genome of the recipient and is regulated in the same way as its other genes. Genetic engineering confers a new ability on the organism that has received the new gene. One advantage of genetic engineering is that genes can be introduced in a plant even if they do not occur in the genome of the target plant.

The use of genetically modified food production, a new and valuable tool in agriculture, fisheries, and forestry production, has allowed significant improvements in food production to meet the needs of the ever-expanding world population. With conventional plant breeding techniques, researchers crossbred plants, taking five to seven years to generate a plant with the desired traits. In conventional breeding, half of an individual's genes come from each parent, but in genetically modified organisms, one or more genetically desirable traits has been added to the genetic material of the desired plant. One of the main differences between conventional breeding and genetic engineering is that in the conventional process, crosses are possible only between close relatives, whereas with genetic engineering, scientists can transfer genes between plants that are not related and might not be able to crossbreed in nature.

For example, in the case of Bt-corn, a gene from a naturally occurring soil bacterium, *Bacillus thuringiensis*, was inserted into corn to provide resistance to the corn borer. The gene from the bacterium produces a protein, Bt delta endotoxin, which kills the European and southwestern corn borer larvae. Bt-corn eliminates the need to spray insecticides to control corn borers. Although planting these crops reduces the amount of pesticides released into the environment, the long-term effects of Bt-corn on human health and the environment are not known.

Companies involved in the research and development of plant and animals derived from genetic engineering patent these new products and processes. The patent allows them to protect their investment; however, it costs farmers who use the seed. A contract between the farmers and these companies prohibits the farmers from saving seed for use the following year, reselling seed to a third party, or exchanging seed with other farmers.

Scientists envisage numerous future applications of genetically modified food. For example, food could be used to produce drugs to address human health problems including infectious diseases. One of the crops that is being considered for such an application is bananas, which could be used to produce a human vaccine.

APPLICATIONS AND PRODUCTS

Over the years, many biotechnology firms and well-established companies have become involved in research and development using genetically modified organisms. The genetic engineering technique and biotechnology are likely to have many potential commercial uses. Genetically modified organisms have many applications in food production and livestock. Many products, including plants and animals, have been developed using genetic engineering techniques.

Insect Resistance. The use of biotechnology to achieve insect resistance in food plants destined for human or animal consumption is accomplished by incorporating a gene into a particular food plant such as corn. In the case of Bt-corn, the gene for toxin production from the bacterium *B. thuringiensis* was incorporated into the corn plant. This toxin is an insecticide commonly used in agriculture and is safe for human consumption. Genetically modified plants that permanently produce this toxin have been shown to require lower quantities of insecticides in specific situations, for example, where pest pressure is high.

Virus Resistance. Virus resistance is achieved through the introduction of a gene from certain viruses that cause disease in plants. Virus resistance makes plants less susceptible to diseases caused by such viruses, resulting in higher crop yields.

Herbicide Tolerance. Herbicide tolerance is achieved through the introduction of a gene from a bacterium conveying resistance to some herbicides. In situations where weed pressure is high, the use of such crops has resulted in a reduction in the quantity of herbicides used. For example, corn resistant to the popular Monsanto weedkiller Roundup (glyphosate) has been developed so that farmers can treat their crops with Roundup without damaging their corn crop. This means that farmers can reduce the number and amount of herbicides used in any given year and situation.

Genetically Modified Animals. Genetic engineering of animals has taken place for different purposes and using a number of different species such as cattle, sheep, goats, rabbits, pigs, chickens, and fish. For example, well-performing bulls have been cloned to create better breeding stock, and animals have been used to produce useful human proteins. Research using genetic engineering in cattle production is trying to produce cows that are resistant to mad cow disease or that have the capacity to produce milk with higher levels of protein.

Scientists have been able to genetically engineer a variety of salmon that grows at twice the rate of Atlantic salmon. The fish is no bigger than the Atlantic salmon, but it reaches that size in half the time. Scientists inserted a Chinook salmon growth hormone gene into a fertilized egg of an Atlantic salmon. To ensure that the gene remains active all year round, scientists added a "switch" from the ocean pout to the Atlantic salmon. This genetically engineered fish reduces the production cost of salmon. AquaBounty Technologies, which developed the AquAdvantage salmon applied to the FDA for permission to market the fish. In September 2010, the FDA concluded that the fish was safe to eat but felt that more scientific research was needed, particularly on the possible environmental impact of the modified salmon. By late 2015, AquaBounty Technologies had received approval from the FDA to sell the genetically modified salmon in the United States.

Possible Risks. Most scientists and regulators agree that the use of genetically modified organisms in food production may pose some risks. These potential risks usually fall into two basic categories: the effects on human and animal health and the impact on the environment where the genetically modified organisms are grown. Scientists and regulators advise that care must be exercised to reduce these risks, especially the possibility of transferring toxins or allergenic compounds used in genetically modified organisms to ordinary plants or animals. This cross-contamination could result in unexpected allergic reactions in humans and animals. One of the major risks to natural resources and the environment is the possibility of outcrossing, the transfer of genes from genetically modified organisms to regular crops or related wild species. For example, the use of herbicide-resistant corn and soybeans raises the possibility of outcrossing, which could lead to the development of more aggressive weeds or herbicide-resistant wild relatives of these cultivated plants. This outcrossing could upset the balance of the natural ecosystem. The introduction of genetically modified plants could also lead to a loss of biodiversity as traditional varieties of plants are displaced by a smaller number of genetically modified varieties.

Genetically modified organisms are not always advantageous: the cost of research and development can be prohibitively high, much cheaper ways to control the undesired pests or diseases may exist, and the still unknown effects on humans and the environment may potentially result in lawsuits against the developers of the plant or animal.

SOCIAL CONTEXT AND FUTURE PROSPECTS

The social debate about the use of genetically modified food centers on the level of risk to health and the environment that they present and whether these types of food are necessary. The possible risks to human health presented by genetically modified food fall into two categories, direct and indirect. Direct effects include toxicity, an allergic reaction (also called allergenicity), and negative nutritional effects, and indirect effects include the stability of the inserted gene, outcrossing, and unintended effects as a result of the gene insertion. Therefore, genetically modified foods must be tested very carefully and thoroughly to ensure that the benefits of such plants outweigh their risks and the hidden costs of developing them.

The debate about whether genetically modified food needs to be created to deal with hunger among people in the developing world is taking place in many venues, both political and scientific. Some scientists opposed to genetically engineered food argue that there is more than enough food in the world and that the hunger crisis in some countries is the result of problems with food distribution and the politics in those countries rather than production levels and systems. They also argue that offering food with unknown levels of risk to those in need is unethical. This argument assumes that genetically modified foods have risks not present in traditional foods; however, the proponents of genetically modified food argue that traditional foods are not devoid of risk. However, as production of genetically modified foods has increased and no major adverse effects have emerged, some of the earlier criticism has disappeared. At the same time, a consensus does not exist among U.S. scientists on the risks and the safety of genetically modified plants and animals.

———*Lakhdar Boukerrou, PhD*

FURTHER READING

Fernandez-Cornejo, Jorge, et al. "Genetically Engineered Crops in the United States." U.S. Department of Agriculture/Economic Research Service, Feb. 2014.

Fedoroff, Nina, and Nancy Marie Brown. *Mendel in the Kitchen: A Scientist's View of Genetically Modified Food.* Joseph Henry Press, 2004.

King, Robert C., et al. *A Dictionary of Genetics.* 8th ed., Oxford UP, 2013.

Lurquin, Paul F. *High Tech Harvest: Understanding Genetically Modified Food Plants.* Westview, 2002.

Weasel, Lisa. *Food Fray: Inside the Controversy over Genetically Modified Food.* American Management Association, 2009.

Genetically modified organisms

GloFish, the first genetically modified animal to be sold as a pet. By www.glofish.com. (http://www.glofish.com/images/glofish_005.jpg, [Public domain], via Wikimedia Commons

FIELDS OF STUDY

Agricultural engineering; Agronomy; Animal science; Biotechnology; Environmental biotechnology (or Green biotechnology); Food science; Genetics; Human nutrition; Microbiology

ABSTRACT

Genetically modified organisms are produced through biotechnology and genetic engineering and basically involve genetic modifications in which genetic material is added or removed to alter the genetic structure of the organism. Many organisms have undergone genetic modification, including bacteria and viruses, plants and animals, and even human beings. The majority of genetically modified organisms are created for therapeutic reasons, such as medicine and food for human consumption. Such organisms have the potential to affect all members of human society and their surrounding environment and have therefore become one of the most controversial ethical and ecological issues of the twenty-first century.

BACKGROUND

Humans have selectively bred and crossbred plants and animals for desired traits since almost the dawn of agriculture, but advances in genetic technology have given people novel ways in which to manipulate plants and animals. These advances are motivated by the desires to develop new medical treatments for genetic diseases and disorders and to increase food production to satisfy the world's growing population. Most advances involve recombinant DNA technology, in which an organism's genes are altered by removing a specific gene from the cell of one organism and inserting it into the cell of another. This splicing together of gene fragments from different species produces a new organism that would not be produced through natural reproduction processes or would not be feasible because of the impossibility of interspecies breeding. This new organism is defined as a genetically modified organism (GMO).

The advancement of genetic technology and the introduction of GMOs into the human food chain has prompted controversy over the ethics of manipulating nature and the potential for GMOs in worldwide agricultural production and medicine. Although many experts state that GMOs are safe for human consumption and offer myriad benefits to humankind, others claim that the production and consumption of GMOs is unethical and untested, which means that GMOs involve unknown consequences, which potentially could be dangerous.

HISTORY

The process of natural selection, first described by Charles Darwin in 1859 in his seminal *On the Origin of Species by Means of Natural Selection*, states that species evolve over time. Individual organisms that possess the most desired and useful characteristics survive, reproduce, and give birth to offspring; these offspring, in turn, possess the same positive characteristics. For many hundreds of years, people have manipulated the process of natural selection though traditional agricultural selection and crossbreeding to create or eliminate specific characteristics in plant and animal species, producing a wide variety of cereal crops, livestock animals, and pets.

Human interference has altered many plants and animals through crossbreeding or selection, but the desirable traits initially appeared through naturally occurring genetic variation. Because the desired traits were already in existence, human interference in the breeding process was often viewed as relatively benign and within natural bounds. Although humans have manipulated the breeding of plants and animals based on phenotypic characteristics for a long time, the ability to directly manipulate the genotype developed much later. Specifically, to feed a growing and hungry world population and develop medical treatments, medical and agricultural scientists have researched and advanced genetic modification technology.

Although genetic engineering is a phenomenon of the late twentieth century, the building blocks for such technology began with the first isolation of DNA in 1869 and the subsequent awareness of its relevance to heredity in 1928. The first accurate double-helix model of DNA was developed in 1953 by James D. Watson and Francis Crick, and the first gene sequence and recombinant DNA was created in 1972 by researchers from Stanford University. The latter discovery truly heralded the beginning of the biotechnological industry and the development of GMOs.

Genetic engineering research continued during the 1970s, and the first publicly and commercially available GMO, a form of human insulin produced by bacteria, was developed in the United States in 1982. However, for the most part, the majority of commercial GMOs sold and used in the twenty-first century are found in agriculture and food production. GMO research scientists believe that genetic engineering is the only method that will guarantee global food production, particularly as predications regarding global climates have indicated that traditional agriculture practices will fail to meet demand.

HOW IT WORKS

Fundamentally, the development and manufacture of GMOs is the replacement of natural selection processes with artificial genetic manipulation. At its most basic, GMO technology relies on a sound understanding of DNA and involves the subtraction of specific genetic material or substitution of material from one species with that from another. Genetic engineering, a complex endeavor, deals with the most fundamental building blocks of an organism. Within a cell are tiny strand like structures called chromosomes, which contain a nucleic acid called DNA. This molecule contains all the genetic material required for inheritance and thus is the basis for genetic manipulation technology.

Initially, the term genetic manipulation referred to a vast array of techniques for the modification of organisms through reproduction and gene inheritance. Later, however, the definition became more restricted and refers specifically to recombinant DNA technology, a form of genetic engineering in which the genome of a cell or organism is artificially modified. The fundamental concept of this technology is that genetic material from different species is combined to create a new species or organism. That is, molecules of DNA from more than one source are united together inside a cell, which is then inserted into a new organism or host, where it is able to reproduce. Because the genome is passed on to offspring, the modification is considered to be self-perpetuating. An organism's biological activities and physical characteristics are controlled by its genome, so modification of the genome can significantly influence the organism's biological functions and traits. The objective behind such technology is to advance the fields of medicine and agriculture to develop more effective medical treatments and to improve crop yield and disease resistance.

APPLICATIONS AND PRODUCTS

The possible applications and products of genetic engineering are vast, perhaps limited only by the imagination. For the most part, however, the major function of genetic engineering, and hence the development of GMOs, is related to their potential in agricultural, medical, and environmental applications.

Agricultural Applications. The continuing rapid expansion of the human population is necessitating an increase in the supply of food. Providing adequate food for a hungry world has become a significant issue for science. The need for food has been instrumental in promoting and advancing genetic modification techniques to produce new and improved organisms, particularly those that, for example, have higher yields or are drought and disease resistant. Through recombinant gene technology, it has become possible to create plant species that are capable of surviving in extreme temperatures and with low rainfall, that can convert atmospheric nitrogen into a useable form (thereby eliminating the need for nitrogen fertilizer), and that have the ability to produce their own resistance to pests and pathogens (thereby eliminating the need for chemical pesticides).

Versions of soybeans, canola, corn, potatoes, sugar beets, and cotton that have been genetically modified to increase herbicide tolerance and resist insects are all available for purchase.

With some specific exceptions, research and the development of genetically engineered animals has proven to be less straightforward than the genetic modification of plants and certainly more ethically problematic. In addition, although the public definitely shows some resistance to the idea of introducing genetically modified plants into the human food chain, most people express much greater resistance to the idea of directly consuming genetically modified animals. Therefore, research on genetically modified animals for use in agriculture has stayed in a relatively early stage of development. However, there has been some research into and experimentation with the genetic manipulation of animals to increase production and meat yield; such experimentation is most promising in fish species rather than in hoofed farm animals.

Medical Applications. Although agricultural applications are very important, the potential medical applications of GMOs are perhaps even more significant. The world's first commercial applications of genetic engineering were, in fact, medically oriented and included synthetic human insulin, approved for public sale in the United States by the Food and Drug Administration (FDA) in 1982, and a human hepatitis-B vaccine, approved by the FDA in 1987. Before the 1980s, synthetic human insulin (produced from animals) was available only in relatively limited quantities. Since the 1980s, research into medical applications for GMOs has rapidly advanced. Of particular benefit is the ability of genetic engineering to produce GMOs on a previously unavailable scale.

Perhaps the most significant potential application of genetic engineering and GMOs is in the treatment and possible cure of genetic diseases. Human society is plagued with both serious diseases and mild disorders, more than three-thousand of which are genetic in origin and therefore difficult to cure using conventional medicine. Although this technology is still in its infancy, gene therapy is perhaps limitless in its potential to assist people with genetic disease.

Environmental Applications. Increasing human populations are important not only in relation to agricultural food production but also in terms of their impact on the environment. Genetic engineering and GMOs could potentially solve some of the world's most serious ecological problems. Research has produced genetically modified viruses that can be used to create ecologically friendly lithium batteries, modified bacteria that can produce biodegradable plastic, and genetically manipulated bacteria that have been encoded for use in bioremediation. Genetic modification technology may even be of use in the fight for survival of some of the world's most vulnerable and endangered species.

SOCIAL CONTEXT AND FUTURE PROSPECTS

Genetically modified organisms potentially could be very advantageous to people. Specifically, scientists claim that GMOs will be vital to the future of food production and therapeutic medicine. Given the possibility of climate change due to global warming, crops that can produce higher yields, resist pests and pathogens, and better tolerate drought are very attractive. The use of GMOs in the treatment of genetic disorders makes them potentially life-saving. Supporters of GMO technology argue that genetic engineering has become an economic and environmental necessity in regard to agriculture, environmental bioremediation, and medicine.

Despite their obvious benefits, GMOs also hold many potential dangers. In addition, GMOs are not well received by the public. Surveys in some countries have revealed that the majority of people are actually against the creation and production of genetically modified foods, animals in particular. This opinion is shared by many environmental organizations, which claim that the undeniable benefits of GMOs are far outweighed by their possible effects on ecosystems, native flora and fauna, and human health. Of particular concern is that many of the potential risks of GMOs are as yet unknown. Opponents of GMO technology have stated that imposing GMOs onto the public without long-term rigorous testing is irresponsible and that more research is required.

Critics of GMOs have also argued that the Food and Drug Administration (FDA) should play a greater role in testing the safety of these products and regulating their sale. However, in 2015, the FDA instituted a new rule stating that foods containing genetically modified ingredients from sources approved by the FDA do not necessarily need to have a label identifying these genetically modified ingredients, leaving that choice up to the producer. Additionally, the organization released a statement stipulating that it preferred the use of terms such as "not bioengineered" over "non-GMO" on any labeling. Many people expressed disagreement with these decisions. That same year, the FDA reiterated that it did not have any substantive evidence proving that GMOs differ in any significant way from their natural counterparts or that they presented any greater safety risk. Around the same time, the FDA also officially approved the first genetically modified animal for sale in the U.S. food market, salmon. After analyzing data submitted by the manufacturer of the fish, which was modified to grow at faster speeds, the FDA found that the salmon was safe to eat by both humans and animals and that its genes remained stable throughout generations.

——Christine Watts, PhD, B. App.Sc., B.Sc.

FURTHER READING

Bertheau, Yves. *Genetically Modified and Non-Genetically Modified Food Supply Chains: Coexistence and Traceability.* Chichester: Wiley-Blackwell, 2013. Print.

Howe, Christopher. *Gene Cloning and Manipulation.* 2nd ed., Cambridge: Cambridge University Press, 2007. Print.

Nelson, Gerald C., ed. *Genetically Modified Organisms in Agriculture: Economics and Politics.* San Diego, Calif : Academic Press, 2008.

Newton, David E. *GMO Food: A Reference Handbook.* Santa Barbara, California: ABC-CLIO, 2014.

Nicholl, Desmond S. T. *An Introduction to Genetic Engineering.* 3rd ed., Cambridge University Press, 2008.

Primrose, Sandy B., et al. *Principles of Gene Manipulation.* 6th ed., Blackwell, 2003.

Strom, Stephanie. "F.D.A. Takes Issue with the Term 'Non-G.M.O.'" *The New York Times*, 20 Nov. 2015, www.nytimes.com/2015/11/21/business/fda-takes-issue-with-the-term-non-gmo.html. Accessed 28 Oct. 2016.

Tutelyan, Victor. *Genetically Modified Food Sources: Safety Assessment and Control.* Elsevier, 2013.

Watson, James D., et al. *Recombinant DNA: Genes and Genomes—A Short Course.* 3rd ed., Freeman, 2007.

Young, Tomme R. *Genetically Modified Organisms and Biosafety: A Background Paper for Decision-Makers and Others to Assist in Consideration of GMO Issues.* International Union for Conservation of Nature, 2004.

GENETICALLY ALTERED BACTERIA

FIELDS OF STUDY

Agricultural engineering; Agronomy; Animal science; Biotechnology; Environmental biotechnology (or Green biotechnology); Medical biotechnology (or Red biotechnology); Microbiology

ABSTRACT

The genetic modification of bacteria has made possible the manufacture of medically important human proteins such as insulin and growth hormone. Genetically altered bacteria have also been used as a means to introduce into plants genetic material that increases the plants' resistance to disease, pests, or freezing.

BACKGROUND

The ability to alter bacteria genetically is the outcome of several independent discoveries. In 1944 Oswald Avery and his coworkers demonstrated gene transfer among bacteria using purified deoxyribonucleic acid (DNA), a process called known as transformation. In the 1960's the discovery of restriction enzymes permitted the creation of hybrid molecules of DNA. Such enzymes cut DNA molecules at specific sites, allowing fragments from different sources to be joined within the same piece of genetic machinery. Restriction enzymes are not species-specific in choosing their targets. Therefore, DNA from any source, when treated with the same restriction enzyme, will generate identical cuts. The treated DNA molecules are allowed to bind with each other, while a second set of enzymes called ligases are used to fuse the hybrids. The recombinant molecules may then be introduced into bacteria cells through transformation. In this manner, the cell acquires whatever genetic information is found in the DNA. Descendants of the transformed cells will be genetically identical, forming clones of the original.

The most common forms of genetically altered DNA are bacterial plasmids, small circular molecules separate from the cell chromosome. Plasmids may be altered to serve as appropriate vectors for genetic engineering, usually containing an antibiotic resistance gene for selection of only those cells that have incorporated the DNA. Once the cell has incorporated the plasmid, it acquires the ability to produce any gene product encoded on the molecule. The resulting artificially produced DNA is called recombinant DNA. The first bacterium to be genetically altered through recombinant DNA technology for medical purposes, *Escherichia coli*, contained the gene for the production of human insulin. Prior to creation of the insulin-producing bacterium in the 1970's, diabetics were dependent on insulin purified from animals. In addition to being relatively expensive, insulin obtained from animals produced allergic reactions among some diabetes patients. Insulin obtained from genetically altered bacteria, by contrast, is identical to human insulin. Subsequent recombinant DNA research has led to the manufacture of a variety of human proteins, including human growth hormone, parathyroid hormone, several kinds of interferons, many monoclonal antibodies, hepatitis B surface antigen, clotting factors, and granulocyte colony-stimulating factor.

Genetically altered bacteria may also serve as vectors for the introduction of genes into plants. The bacterium *Agrobacterium tumefaciens*, the etiological agent for a plant disease called crown gall, contains a plasmid known as Ti. Following infection of the plant cell by the bacterium, the plasmid is integrated into the host chromosome, becoming part of the plant's genetic material. Any genes that were part of the plasmid are integrated as well. Desired genes can be introduced into the plasmid, promoting pest or disease resistance within plants infected by the bacterium.

In April, 1987, scientists in California sprayed strawberry plants with genetically altered bacteria to improve the plants' freeze resistance; this event marked the first deliberate release of genetically altered organisms in the United States to be sanctioned by the Environmental Protection Agency (EPA). The release of the bacteria represented the climax of more than a decade of public debate over what would happen when the first products of biotechnology became commercially available. Fears centered on the creation of bacteria that might radically alter the environment through elaboration of gene products not normally found in such cells. Other concerns included the creation of super bacteria with unusual resistance to conventional medical treatment.

Despite these fears, approval for further releases of genetically altered bacteria soon followed, and the restrictions on release were greatly relaxed. By 1991 permits for field tests of more than 180 genetically altered plants and microorganisms had been granted. Between 1987 and 2004 more than 10,000 trials were conducted at more than 39,000 sites, and more than sixty biotechnology products entered the market. Among future planned tests are clinical trials of the use of a modified *Streptococcus mutans* in fighting dental cavities. Scientists have modified this bacterium, responsible for tooth decay in its unaltered form, so that it does not produce the lactic acid that ordinarily erodes tooth enamel. In animal tests, it has been found that the modified bacterium eventually replaces the *S. mutans* naturally occurring in the mouth.

In general, "red biotechnology" (the application of biotechnology in medicine) tends to generate less controversy than "green biotechnology" (use of biotechnology in food production). In the United States, anyone intending to produce or import genetically altered microorganisms for commercial purposes must submit a notice to the EPA, which assesses whether the organism constitutes an unreasonable risk to human health or the environment.

——*Richard Adler, updated by Karen N. Kähler*

FURTHER READING

Drlica, Karl. *Understanding DNA and Gene Cloning: A Guide for the Curious.* 4th ed. Hoboken, N.J.: John Wiley & Sons, 2004.

Food and Agriculture Organization and World Health Organization. *Safety Assessment of Foods Derived from Genetically Modified Microorganisms.* Geneva: World Health Organization, 2001.

Han, Lei. "Genetically Modified Microorganisms: Development and Applications." In *The GMO Handbook: Genetically Modified Animals, Microbes, and Plants in Biotechnology*, edited by Sarad R. Parekh. Totowa, N.J.: Humana Press, 2004.

Stemke, Douglas J. "Genetically Modified Organisms: Biosafety and Ethical Issues." In *The GMO*

Handbook: Genetically Modified Animals, Microbes, and Plants in Biotechnology, edited by Sarad R. Parekh. Totowa, N.J.: Humana Press, 2004.

Watson, James D., et al. *Recombinant DNA: Genes and Genomes—A Short Course.* 3d ed. New York: W. H. Freeman, 2007.

Genomics

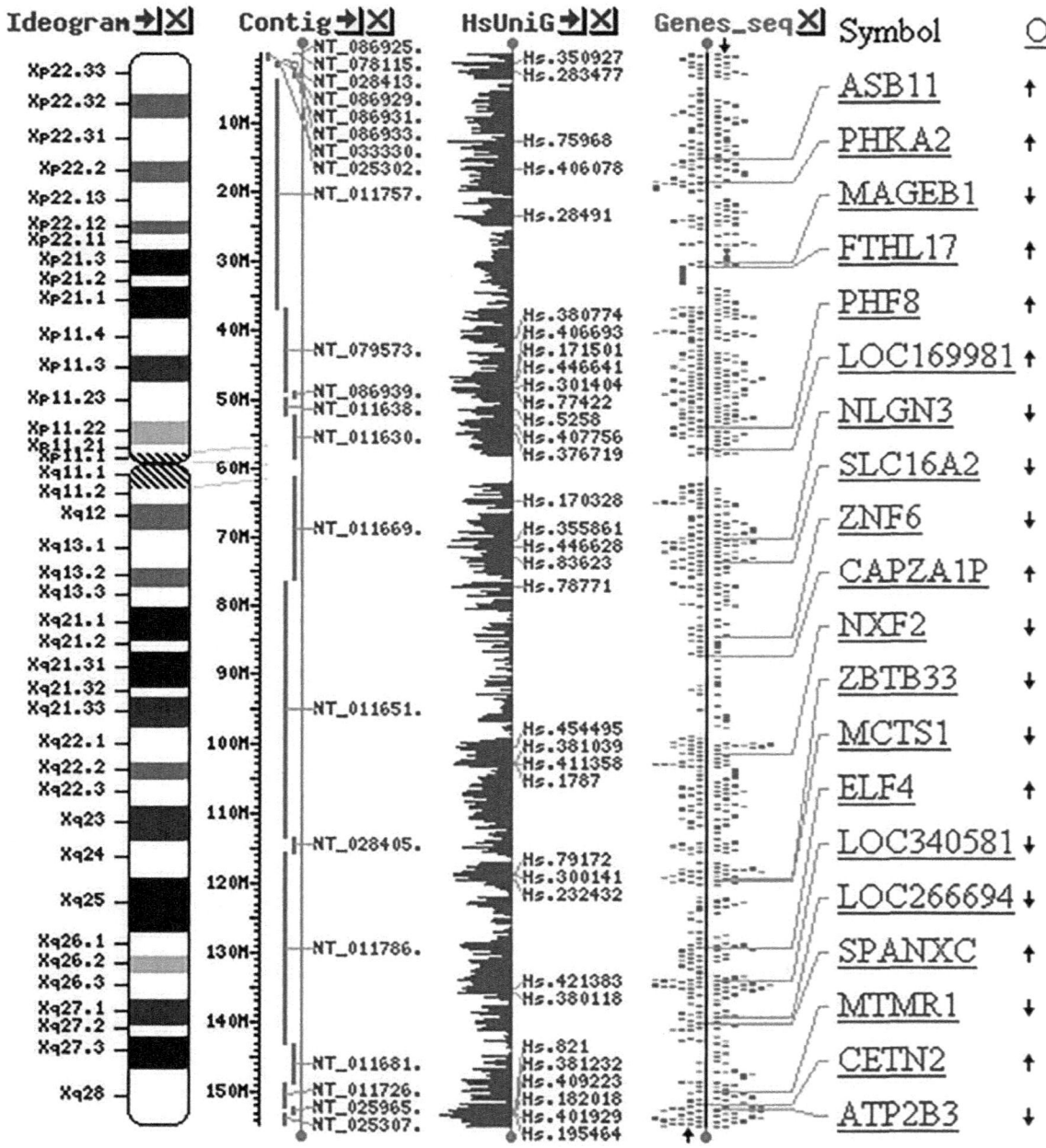

Map of the human X chromosome (from the National Center for Biotechnology Information website), [Public domain], via Wikimedia Commons

FIELDS OF STUDY

Biotechnology; Environmental biotechnology (or Green biotechnology); Genetics; Medical biotechnology (or Red biotechnology)

ABSTRACT

Genomics is the branch of biotechnology that focuses on the genome, the entire set of genes in an organism, as well as the interaction of individual genes with one another and the organism's environment. Broad applications include comparative and functional genomics.

BACKGROUND

Genomics studies organisms' entire genomes, focusing on structure, function, and inheritance. It examines complex diseases, including cancer, heart disease, and diabetes, in which both genetic and environmental factors play a part. In contrast, genetics focuses on individual genes and their role in inheritance and heritable genetic diseases such as phenylketonuria and cystic fibrosis. Of central importance in genomics is determining the sequences of DNA that make up an organism. DNA sequencing can be used to find mutations or variations that may be involved in causing a disease. Functional genomics examines how individual genes function and interact with each other within a given genome at the functional level, and comparative genetics examines the relationships between different species. Comparative genomics may be used to study evolutionary relationships between species by comparing their chromosomes.

HISTORY

Scientists became aware of the existence of genomes in the late nineteenth century when they first viewed chromosomes under a microscope, although the word "genome" was not coined until 1920. In the twentieth century, scientists studied the frequency at which chromosomes exchanged parts during meiosis to map the genes on chromosomes. However, this technique was useful for mapping primarily genes that had mutant phenotypes, which accounted for only a small part of the total genome.

Numerous scientists discovered the base-pair sequences of many human genes, but mapping the entire human genome remained a challenge. Determining the entire human genome was generally regarded as a worthwhile endeavor, but the process would take billions of dollars, taking funding away from more established, mainstream biomedical research. Ethicists, scientists, and economists debated the merits of the research versus its exorbitant cost. Nevertheless, the Human Genome Project was launched in 1990, supported by the U.S. Department of Energy (Human Genome Program, directed by Ari Patrinos) and the National Institutes of Health (National Human Genome Research Institute, directed by Francis S. Collins). The Human Genome Project became an international effort, with scientists from around the globe joining the team. Newly developed computer software helped facilitate the process. In 1998, Celera Genomics, a private company headed by J. Craig Venter, began an independent effort to complete the mapping of the human genome using an unorthodox shotgun, or whole-genome, method of sequencing. In 2000, Collins, Venter, and Patrinos jointly announced the completion of a rough draft of the human genome. In 2003, fifty years after Francis Crick and James D. Watson discovered the double-helix structure of DNA, the human genome had been sequenced, and the project was declared complete, although final papers were published in 2006.

HOW IT WORKS

Determining an organism's genome requires the sequencing of its DNA, which involves finding the order of the bases in a given stretch of DNA. Several sequencing methods were developed in the 1970's, but sequencing was a very slow process.

One problem was that the total amount of DNA necessary for sequencing and analyzing a genome of interest could be several times the total amount of available DNA. To increase the amount of DNA, scientists turned to cloning. The DNA fragments are replicated inside a bacterial cell, then the cloned DNA is extracted and placed in a machine for sequencing. The polymerase chain reaction, developed in 1983, allows ancient DNA or degraded DNA to be sequenced using special protocols to amplify the sample before sequencing.

Genomic sequences are usually elucidated with automatic sequencers; shotgun, (whole-genome) sequencing, developed by Venter, is the most popular DNA sequencing technique. The DNA is randomly

215

cut into fragments, and then the ends of each fragment are sequenced, resulting in two reads per fragment. The reads are used to reconstruct the original DNA sequence. Newer sequencing techniques generally follow this model but may use different strategies along the way. As the fragments are sequenced, the data are set aside. When enough sequences have been gathered, they are linked through sequence overlaps. The resulting genomic sequence is then deposited into a publicly accessible database.

The genomic sequence is analyzed to discover the individual genes and how they are regulated. Bioinformatics is the discipline that relies on computer programs to search for genes in DNA sequences, using what is known about the gene of interst. After scientists find the sequences that make up a gene, the structure and function can be compared with similar gene sequences in other organisms.

APPLICATIONS AND PRODUCTS

Genomics has many applications. Among the most important applications, because of their impact on society, are those that deal with evolution and the environment. Other areas, such as synthetic biology and personal genomics, may lead to innovations that can be of great benefit to individuals and populations and to public health.

Evolutionary Genomics. Until about 30,000 years ago, Neanderthals, the closest relatives of modern humans, inhabited Europe and Asia. Since the early twentieth century, anthropologists and paleontologists have attempted to demonstrate an evolutionary relationship between Neanderthals and modern humans, who emerged about 400,000 years ago. In 1997, Svante Pääbo, director of the Department of Genetics at the Max Planck Institute for Evolutionary Anthropology, furthered the understanding of the genetic relationship of Neanderthals and modern humans when he sequenced Neanderthal mtDNA. In 2009, scientists from the Max Planck Institute announced that they had generated a draft sequence equivalent to more than 60 percent of the complete Neanderthal genome and had begun comparing it to that of humans. In May, 2010, these researchers revealed that the human genome contains some of the same genes as were found in the Neanderthal genome. Their continued efforts may shed light on the origin of humankind as well as the evolutionary process.

Environmental Genomics. In 2009, the International Union for Conservation of Nature estimated that Earth is home to 8 million to 14 million animals and plants, of which only 1.8 million have been identified and classified. Naturalists such as Charles Darwin classified plants and animals using taxonomic systems. Later scientists compared the genetic differences among species; a species was defined as a group of organisms that could breed and produce fertile offspring. Determining where one species ends and another begins is difficult, as in the case of wolves and coyotes. Modern technology allows species to be identified and classified by comparing modern and ancient samples of genomic DNA and tracking how a species descended from an ancestor. The question of what constitutes a species is important in determining which species can be considered endangered. Global climate change has led to the extinction of many plant and animal species, and many scientists hope to be able to preserve the world's biodiversity using knowledge gleaned from genomics.

In 2004, Venter and his colleagues at the Institute for Biological Energy Alternatives (later part of the J. Craig Venter Institute), of Rockville, Maryland, applied whole-genome sequencing to microbial populations collected from the Sargasso Sea, close to Bermuda, where researchers expected a low diversity of species. They analyzed more than 1 billion base pairs and found 1.2 million new genes and 1,800 species of microbes (including 150 new species of bacteria). According to Venter and his colleagues, the number of species suggests that microbial life in the ocean is more plentiful and diverse than previously thought.

Synthetic Biology. One of the goals of synthetic biology, which creates artificial biological systems (not existing in nature) and redesigns existing biological systems, is intelligent design. Synthetic biology applies the principles of large-scale engineering to biology and is built on the premise that organisms can be divided into discrete parts. In 2010, researchers at the J. Craig Venter Institute created a self-replicating bacterial cell containing totally synthetic DNA. The cell's genome was designed in the computer. No natural DNA was used, and chemical synthesis was used to make a viable cell.

Synthetic biologists believe that someday it may be possible to program bamboo to grow into preformed chairs or to program trees to spew oil from their stems. Reprogrammed bacteria may be able to heal, not harm, humans and animals. Some experts, such as David Rejeski of the Woodrow Wilson International Center for Scholars in Washington, D.C., believe that synthetic biology may fundamentally change the way things are made within the next one hundred years, creating manufacturing shift as significant as the Industrial Revolution.

IMPACT ON INDUSTRY

Genome Decoding for Individuals. The cost of sequencing the first human genome was about $3 billion. However, advances in technology have caused the price of obtaining a complete genome to drop rapidly. In 2010, Ilumina, the market leader in DNA-sequencing machines, offered to provide people with their genome for $9,500 if they might benefit from the information (as is the case with people with rare cancers). Life Technologies, Ilumina's closest competitor, is working with a group of cancer research centers to see how genetic information might help in the treatment of cancer. It is also offering a sequencer that sells for around $100,000, in contrast to the typical $700,000. The DNA-sequencing industry saw the entrance of Pacific Biosciences, which offers sequencing in a matter of minutes rather than days, and Complete Genomic, which has created a sequencing facility for drug companies to use.

Although genome decoding had focused on healthy individuals, in March of 2010, two research teams independently decoded the entire genome of patients who had genetic diseases to find the exact genes that caused their diseases. The lower cost of DNA sequencing may revive the largely unsuccessful efforts to identify the genetic causes of major killers such as heart disease, diabetes, and Alzheimer's disease. For example, James R. Lupski, a medical geneticist who has a Charcot-Marie-Tooth, a neurological disease, used whole-genome sequencing on his own DNA in an effort to better understand his illness. He was able to find the genetic variation causing his condition, although earlier research had not been able to pinpoint it.

Genomic Ventures. Many genomics companies formed in the 1990's and 2000's, but many are still in the research and development process and have yet to market products. Most of these companies apply genomics to the creation of pharmaceuticals, although some use genomic studies for environmental purposes or in synthetic biology. Many are involved in cooperative ventures with pharmaceutical companies, and mergers and acquisitions are common.

In October, 2006, Venter (former head of Celera Genomics) brought together several affiliated organizations, The Institute for Genomic Research (TIGR), The Center for the Advancement of Genomics (TCAG), The J. Craig Venter Science Foundation, The Joint Technology Center, and the Institute for Biological Energy Alternatives to create the J. Craig Venter Institute, with offices in Rockville, Maryland, and San Diego, California. The institute, known for high-throughput DNA sequencing, has research groups working on human genomic medicine, microbial and environmental genomics, plant genomics, infectious diseases, and synthetic biology and bioenergy. Its approach toward infectious disease is to examine the genome of the microbes and viruses that cause disease and the microbial flora in various cavities of the human body. Another focus is its bioinformatic group, which works on the software, databases, and mathematics to analyze the data created by DNA sequencing.

Other smaller companies are also active in various aspects of genomics. Millenium: The Takeda Oncology Company, based in Cambridge, Massachusetts, was established in 1993 as a genomics company and has evolved into a biopharmacological company focused on cancer. Its leading product is injectable VELCADE (bortezomib) for the treatment of people with multiple myeloma and a type of lymphoma. Rockville, Maryland-based Human Genome Sciences (founded 1992) is developing drugs to treat hepatitis C and systemic lupus erythematosus. Incyte, based in Wilmington, Delaware, is developing a drugs to treat rheumatoid arthritis, psoriasis, and cancer.

Affymetrix, based in Santa Clara, California, developed the first microarray (for genome-wide analysis) in 1989 and the first commercial microarray in 1996. Its genomic analysis tools are invaluable for researchers and pharmaceutical companies. San Diego, California-based Verdezyne, switched from being a provider of synthetic genes to drug and industrial enzyme companies to focus on a fermentation process to produce renewable energy fuels and chemicals, such as ethanol and adipic acid. It is using

> **FASCINATING FACTS ABOUT GENOMICS**
>
> - In 2005, the dog genome sequence was published in *Nature* magazine. The DNA sequenced came from Tasha, a purebred female boxer. Scientists think that the genetic contribution to disease may be easier to determine in dogs than in humans. About 5 percent of the human genome is also present in dogs.
> - Humans and chimpanzees share 96 percent of their genomes.
> - The average difference in the genomes of two people is 0.2 percent, or one in five hundred bases.
> - About 97 percent of the DNA in the human genome consists of so-called junk DNA—sequences with no known functions.
> - Scientists at the J. Craig Venter Institute are studying the genome of the SARS coronavirus, which causes severe acute respiratory syndrome, in humans and animals in an effort to determine how the virus crosses the species barrier.
> - In 2010, the complete draft genome sequence of the soybean was published in *Nature* magazine. It was expected to increase understanding of the nitrogen-fixing process and to help scientists develop better soybeans, including a more digestible version.

genomics technology to create proprietary metabolic pathways in yeast for enhanced conversion of hexose and pentose sugars to ethanol.

Government Agencies. One of the main governmental agencies involved in genomics is the National Human Genome Research Institute, part of the National Institutes of Health. It began as part of the Human Genome Project and continues to conduct research into the genetic basis of human disease. It is divided into seven branches: Cancer Genetics, Genetic Disease Research, Genetics and Molecular Biology, Genome Technology, Inherited Disease Research, Medical Genetics, and Social and Behavioral Research.

The U.S. Department of Energy is also involved in genomics research. Its Genomic Science Program focuses on using genomics to deal with issues regarding the environment, energy, and the climate. The agency is involved in the genome sequencing of microbes that can help cycle carbon from the atmosphere, clean up toxic waste, and create biofuels. The DOE Joint Genome Institute, operated by the University of California, Berkeley, concentrates on providing clean energy and environmental solutions.

SOCIAL CONTEXT AND FUTURE PROSPECTS

The U.S. Department of Energy and the National Institutes of Health have devoted between 3 and 5 percent of their annual genome project budgets toward the study of ethical, legal, and social issues. Societal concerns include privacy and confidentiality issues, possible social stigma and discrimination, and how genetic information will be used by insurance companies, the legal system, and academia. Various government agencies have taken the lead in addressing the existing and potential social and ethical issues that may arise from genome-centered biology.

The Genetic Information Nondiscrimination Act of 2008 prohibits insurance companies and employers from discriminating against people based on the results of genetic testing. In addition, under the law, employers and insurers are prohibited from requesting or demanding that individuals take genetic tests.

In an interview in 2010, National Institutes of Health director and former Human Genome Project member Francis S. Collins expressed support for genomic research because it is likely to enable people to prevent and treat disease. He also noted that the patenting of human genes, once very controversial, has become more common, with more than 20 percent of all know genes having been patented. The patents, he argued, help private biotech companies fund costly research into genetic diseases. He also made a point of stating that his religious beliefs do not interfere with his work in evolutionary genetics.

Aware of the legal and social issues and of most people's limited understanding of genetics, the J. Craig Venter Institute has created a division to promote understanding of genomics among policy makers and the general public and to foster a positive image for the biotechnology industry.

The possible benefits of genomic research are immense. Some experts feel that better understanding of the genomics may lead to radical innovations in disease treatment and prevention, and environmental and synthetic genetics may produce ways to create renewable fuels. The decreasing cost of obtaining a person's genome may mean that physicians can take an individualized approach to medicine. Therefore, although controversy over genetics remains, this field is likely to remain an active area of research, yielding numerous applications, and providing numerous work opportunities to those interested in the field.

———*Cynthia F. Racer, M.A., M.P.H.*

FURTHER READING

Davies, Kevin. *The $1,000 Genome: The Revolution in DNA Sequencing and the New Era of Personalized Medicine.* New York: Free Press, 2010. Looks at how less expensive, faster means of obtaining a person's genome will change medicine and make it more tailored to the individual.

DeSalle, Michael, and Michael Yudell. *Welcome to the Genome: A User's Guide to the Genetic Past, Present, and Future.* Hoboken, N.J.: Wiley-Liss, 2005. Starts with a brief history of genetics and description of the science before examining how the genome was sequenced. Analyzes the likely medical and agricultural applications.

DOE Joint Genome Institute. http://www.jgi.doe.gov

Fairbanks, Daniel J. *Relics of Eden: The Powerful Evidence of Evolution in Human DNA.* Amherst, N.Y.: Prometheus Books, 2007. An examination of the field of evolutionary genomics that asserts that there is no dichotomy between religion and science.

Gee, Henry. *Jacob's Ladder: The History of the Human Genome.* New York: W. W. Norton, 2004. Examines what human genome sequencing reveals and how this information may be used in the future.

Genomic Science Program, U.S. Department of Energy http://genomicscience.energy.gov/index.shtml

Gibson, D. G., et al. "Reation of a Bacterial Cell Controlled by a Chemically Synthesized Genome." *Science Express* (May 20, 2010). Announces the creation of a self-replicating bacterial cell governed by a synthetic genome.

J. Craig Venter Institute. http://www.jcvi.org

National Human Genome Research Institute. http://www.genome.gov

Shreeve, James. *The Genome War: How Craig Venter Tried to Capture the Code of Life and Save the World.* New York: Alfred A. Knopf, 2004. Describes the competition between Venter and Collins to decode the human genome.

The Human Genome Project. http://www.ornl.gov/sci/techresources/Human_Genome/home.shtml

H

HUMAN GENETIC ENGINEERING

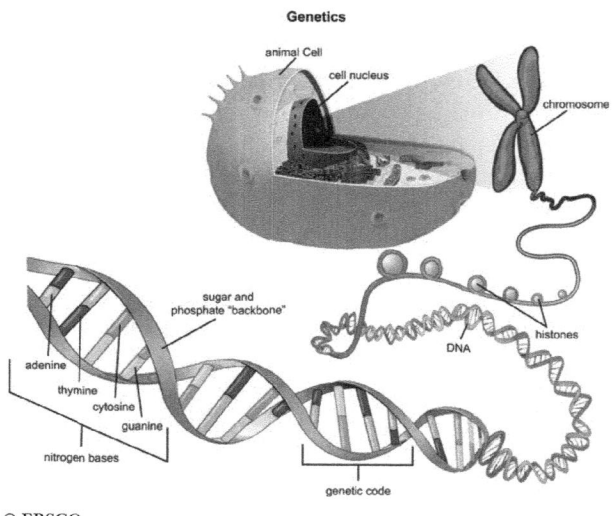

© EBSCO

FIELDS OF STUDY

Biotechnology; Genetics; Medical biotechnology (or Red biotechnology); Philosophy and Religious Studies; Reproductive technology

ABSTRACT

Human genetic engineering is a branch of genetic engineering that focuses on understanding human genes in order to produce applications that can improve human life. Genes, formulated by DNA (deoxyribonucleic acid), determine an individual's genotype—the complete genetic information carried by that individual, even the aspects that are not expressed. By contrast, an individual's phenotype—their visible and other physical characteristics—is formed as the result of human genes interacting with the environment. Human genetic engineering aims to alter genotypes to cause changes in phenotypes. Also, and more often, the knowledge of human genetics is used to engineer products, such as medications, that can cure or improve the quality of human life by addressing genetic disorders. Many of the applications of what is now known as human genetic engineering arose out of the mapping of the human genome during the Human Genome Project, completed in 2003.

BACKGROUND

An individual's full complement of genes is formulated by their DNA (deoxyribonucleic acid) and is called the genotype. That individual's observable traits or characteristics—that is, how those genes are expressed—are formed from the interaction of genes with the environment and are called the phenotype. Human genetic engineering aims to alter genotypes to cause changes in phenotypes. To understand human genetic engineering capabilities, it is important to understand basic genetic principles.

Human genetic engineering is a scientific endeavor, and as such this field builds on the information and knowledge gained from the decades of experimentation accomplished in years past. Without this foundation, human genetic engineering would not exist. This foundation brings the prospect of human cloning and the use of human genetics for therapeutic purposes.

A keystone event in modern genetics occurred in 1953, when American biologist James D. Watson and English physicist Francis Crick deduced the double-helix structure of DNA. This structural information enabled effective study of how genetic material codes for life. In the 1960s, the genetic code of DNA was deciphered. Armed with this important information, scientists Stanley Norman Cohen and Herbert Wayne Boyer undertook the first recombinant DNA experiments on bacteria in 1973. The ambitious Human Genome Project was started in 1990 with the goal of mapping out the entire human genetic sequence. In June 2000, the first working draft of the human genetic sequence was produced from the efforts of this project. April 2003 saw the announcement of the first complete human genetic sequence—breakthrough information for human genetic engineering.

Cloning, a subdiscipline of bioengineering, is the reproduction of genetically identical living organisms. In 1996, the first mammal was cloned, a sheep named Dolly. Other animals have been cloned since this pioneering event, including a bull in 1999 and a pig in 2000. The year 2003 saw the cloning of a mule, a horse, and a rat, followed by the cloning of a dog in 2005. Attempts at pet cloning have occurred: John Sperling, a wealthy and influential American educator, has funded pet-cloning projects, and researchers at Texas A&M University successfully cloned a cat in 2002. Commercial attempts at pet cloning started in April 2004 with a company called Genetics Savings and Clone (now defunct) offering pet gene banking and cloning. Korean researchers published claims of successful human embryonic cloning in 2004, but these claims were later retracted because of fabricated data and other problems with the research. In May 2010, the journal *Science* reported that scientists J. Craig Venter, Clyde Hutchison III, and Hamilton Smith had created a living creature in the laboratory. This new life-form, a bacterium, was artificially produced using genetic engineering techniques. It had no ancestor and it reproduced, a key ability of living organisms.

Further milestones in human genetic engineering occurred in the areas of medicine, pharmaceuticals, forensics, and even psychology, with more practicable and practical results. Many, if not most, of these blossomed shortly after the mapping of the human genome was completed in 2003. Many more will be developed as scientists and researchers continue to investigate the data that were gathered through that monumental accomplishment.

HOW IT WORKS

Until the middle of the eighteenth century, when the theory was disproved by Italian biologist Lazzaro Spallanzani, most biologists believed in spontaneous generation—that life arises from combinations of decaying matter, as if flies arose from garbage. It is now known that DNA genetically codes for many physical characteristics.

The story of genetics starts with DNA and ends with protein. DNA, the genetic material found in the nucleus of every cell, codes for (that is, creates instructions for the building of) various proteins by means of nucleotides, which form the building blocks of DNA. The nucleotides establish the code. Proteins make up the structural elements of the body, including collagen, ligaments, tendons, and muscles; some hormones, such as insulin, are made of protein as well. Perhaps most important, however, are the protein enzymes.

All the enzymes in the body are made up of protein and protein alone. Enzymes are key because they accelerate chemical reactions. Thousands of chemical reactions occur in human bodies all the time. Protein enzymes catalyze all these reactions. DNA, by dictating the production of enzymes, controls these chemical reactions.

DNA dictates which proteins are produced by living and staying in the cell nucleus during the entire protein-making process. Much like a general in a command center, the DNA sends out orders but does not leave the nucleus. DNA is made up of nucleotide bases, and the first step in protein production involves the reading of the nucleotide code in the DNA. When the nucleotides are read, a strand of messenger ribonucleic acid (mRNA) is produced and sent from the nucleus into the cytoplasm of the cell. The DNA stays in the nucleus, while the mRNA leaves the nucleus. This process of reading the DNA nucleotide code and producing mRNA is called transcription.

The next step involves reading the nucleotide code on the mRNA in the cytoplasm of the cell. The cytoplasm is the liquid environment inside the cell where all the cellular organelles float. Transfer RNA (tRNA) reads the code on the mRNA in a process called translation. Transfer RNA is so called because it transfers a specific amino acid when it reads the appropriate code on the mRNA. Amino acids are the basic building blocks of protein. About twenty-two different amino acids build all the various proteins the body uses. It is like the English alphabet: twenty-six letters and vowels comprise the English alphabet, and combining these various letters and vowels results in tens of thousands of different combinations and all the various words in the English language. Likewise, the body uses the different amino acids to form the tens of thousands of different proteins in the body.

During translation, a specific amino acid is coded for and carried by the transfer RNA to a ribosome.

Ribosomes (along with the rough endoplasmic reticulum) are where the cell's proteins are produced. Along ribosomes, different amino acids are transported by tRNA and linked, forming a protein molecule. Some proteins may be only eighty or ninety amino acids long, whereas others, such as hemoglobin, may have more than 300 amino acids as their amino acid backbone.

The way DNA codes for all this involves the nucleotide bases that make up DNA. Four nucleotide bases make up DNA: adenine, cytosine, guanine, and thymine. Adenine will chemically bind with thymine, and cytosine always chemically binds with guanine. When DNA is transcribed to form mRNA, if the nucleotide sequence in the DNA reads cytosine-cytosine-guanine, these nucleotide bases will code for guanine-guanine-cytosine in the mRNA. Then when mRNA is translated by tRNA, the code goes back to the original DNA code, cytosine-cytosine-guanine. Cytosine-cytosine-guanine can code for a specific amino acid, and in that fashion DNA codes for the amino acid sequence of all protein molecules. The nucleotide base sequence in the mRNA is called the codon and the complimentary base sequence found in tRNA is called the anticodon.

The example of how Dolly the sheep was cloned demonstrates how genetic engineering in mammals works, and, hence, how human cloning could work. Cloning is an ultimate example of genetic engineering because cloning produces an entire living organism via genetic engineering. Dolly was a Finn Dorset sheep, which is all white. A Blackface ewe, named because of the distinctive black face these sheep have, was used as an egg donor and as a surrogate mother.

Cells taken from a Finn Dorset ewe were grown in a tissue culture. An egg cell, from the Blackface ewe, had the nucleus removed. The nucleus contained the genes and DNA. The nucleus and genetic information from the Finn Dorset ewe were placed in the enucleated Blackface ewe egg cell. The Blackface ewe egg cell, now containing genetic information from the Finn Dorset ewe, was placed in the uterus of the Blackface ewe after an electric pulse is applied to stimulate growth and duplication of the cells. The Blackface ewe gave birth to Dolly, the all-white Finn Dorset ewe. The newborn Finn Dorset ewe was an identical genetic copy of the Finn Dorset ewe originally used to harvest the genetic information found in the nucleus.

Recombinant DNA refers to DNA transfer from one cell to another. In human genetic engineering, genes transfer from a human chromosome to another cell, usually bacteria. If the transferred human genes code for insulin, the bacteria accepting the transferred genes will now produce human insulin.

In this process, the desired genes are isolated and removed from the human cell. Bacterial cells have small, circular strips of DNA called plasmids. These circular plasmids are removed from the bacterial cells and opened up. Various enzymes are used to cut the human DNA and bacterial DNA sequences at specific points. Restriction enzymes cut the original DNA in specific locations. DNA ligase pastes strips of DNA together. Scientists mix isolated human genes with the opened bacterial plasmids, along with DNA ligase. The human genes are spliced into the bacterial plasmid and the circle of genetic information in the bacterial plasmid closes.

The bacterial plasmid with the spliced human genes is now called a vector. The plasmid vectors are taken up by the bacterial cells. Once inside the bacterial cells, the bacteria multiply and reproduce the spliced human genes. Whatever specific human genes were selected for splicing, for example, human insulin genes, are now functioning in the reproduced bacteria, and human insulin is harvested from the bacterial clones.

APPLICATIONS AND PRODUCTS

Early Medical Applications. Recombinant DNA techniques are remarkable biological life adaptations, and many medicines based on this technology are used. Medications generated through human genetic engineering techniques have been in use since 1982, with the production of human insulin using recombinant DNA techniques. Human genes are inserted into a bacterial host that then makes human insulin. Prior to this type of genetic manipulation, diabetics needing insulin had to rely on insulin harvested from pigs or cows. Genetic techniques produce human growth hormone, previously only available from human cadavers. A genetically engineered hepatitis B vaccine has been in use since 1987.

Since these first human genetic medicines and vaccines, many types of biological products have been introduced or are under current investigation and development. These new medicinal products are called biologics to distinguish them from chemically synthesized medicines. Genes and genetic manipulations produce biologics. Major types of biologics include hormones, antibodies, and cell-receptor proteins.

Insulin and human growth hormone, discussed above, are classic protein hormones produced with recombinant DNA technology. The immune system produces protein antibodies that attack disease causing-agents such bacteria and viruses. Genetic antibody production interferes with or attacks entities associated with diseases such as psoriatic arthritis and Crohn's disease. Recombinant DNA technology produces proteins binding with specialized white blood cells to reduce inflammation associated with rheumatoid arthritis.

Bioinformatics. The purpose of bioinformatics is to help organize, store, and analyze genetic biological information in a rapid and precise manner, dictated by the need to be able to access genetic information quickly. In the United States the online database that provides access to these gene sequences is called GenBank, which is under the purview of the National Center for Biotechnology Information (NCBI) and has been made available on the Internet. In addition to human genome sequence records, GenBank provides genome information about plants, bacteria, and animals other than humans.

Proteomics. Bioinformatics provides the basis for all modern studies of human genetics, including analyzing genes and gene sequences, determining gene functions, and detecting faulty genes. The study of genes and their functions is called proteomics, which involves the comparative study of protein expression. That is, it studies the metabolic and morphological relationship between the protein encoded within the genome and how that protein works. Geneticists are now classifying proteins into families, superfamilies, and folds according to their configuration, enzymatic activity, and sequence. Ultimately proteomics will complete the picture of the genetic structure and functioning of all human genes.

Toxicogenomics. Another newly developing field that relies on bioinformatics is the study of toxicogenomics, which is concerned with how human genes respond to toxins. One particular area of study in this field is specifically concerned with evaluating how environmental factors negatively interact with mRNA translation, resulting in disease or dysfunction.

Gene Testing. In a gene-testing protocol, a sample of blood or body fluids is examined to detect a genetic anomaly such as the transposition of part of a chromosome or an altered sequence of the bases that comprise a specific gene, either of which can lead to a genetically based disorder or disease. There are thousands of genetic tests available to detect malfunctioning or nonfunctioning genes. Most early gene tests focused on various types of human cancers, but others have since been developed to detect a wide range of conditions, from serious and life-threatening diseases to such conditions as genetic susceptibility to tobacco addiction. As of 2016, the National Center for Biotechnology Information's Genetic Testing Registry listed more than forty-eight thousand tests available to detect more than ten thousand conditions.

The emphasis on the relationship between genetics and cancer lies in the fact that all human cancers are genetically triggered or have a genetic basis. Some cancers are inherited as mutations, but most result from random genetic mutations that occur in specific cells, often precipitated by viral infections or environmental factors not yet well understood.

At least four types of genetic problems have been identified in human cancers. The normal function of oncogenes, for example, is to signal the start of cell division. However, when mutations occur or oncogenes are overexpressed, the cells keep on dividing, leading to rapid growth of cell masses. The genetic inheritance of certain kinds of breast and ovarian cancers results from the nonfunctioning tumor-suppressor genes that normally stop cell division. When genetically altered tumor-suppressor genes are unable to stop cell division, cancer results. Conversely, the genes that cause inheritance of colon cancer result from the failure of DNA repair genes to correct mutations properly. The accumulation of mutations in these "proofreading" genes makes them inefficient or less efficient, and cells continue to replicate, producing a tumor mass.

If a gene screening reveals a genetic problem several options may be available, including gene therapy and genetic counseling. If the detected genetic anomaly results in disease, then pharmacogenomics holds promise of patient-specific drug treatment

Gene Therapy. The science of gene therapy uses recombinant DNA technology to cure diseases or disorders that have a genetic basis. Still in its experimental stages, gene therapy may include procedures to replace a defective gene, repair a defective gene, or introduce healthy genes to supplement, complement, or augment the function of nonfunctional or malfunctioning genes. Several hundred protocols are being used in gene-therapy trials, and many more are under development. As of 2011, trials are focusing on two major types of gene therapy, somatic cell gene therapy and germ-line gene therapy.

Somatic cell gene therapy concentrates on altering a defective gene or genes in human body cells in an attempt to prevent or lessen the debilitating impact of a disease or other genetic disorder. Some examples of somatic cell gene therapy protocols now being tested include ones for adenosine deaminase (ADA) deficiency, cystic fibrosis, lung cancer, brain tumors, ovarian cancer, and AIDS.

In somatic cell gene therapy a sample of the patient's cells may be removed and treated and then reintegrated into body tissue carrying the corrected gene. An alternative somatic cell therapy is called gene replacement, which typically involves insertion of a normally functioning gene. Some experimental delivery methods for gene insertion include use of retroviral vectors and adenovirus vectors. These viral vectors are used because they are readily able to insert their genomes into host cells. Hence, adding the needed (or corrective) gene segment to the viral genome guarantees delivery into the cell's nuclear interior. Nonviral delivery vectors that are being investigated for gene replacement include liposome fat bodies, human artificial chromosomes, and naked DNA (free DNA, or DNA that is not enclosed in a viral particle or any other "package").

Another type of somatic gene therapy involves blocking gene activity, whereby potentially harmful genes such as those that cause Marfan syndrome and Huntington's disease are disabled or destroyed. Two types of gene-blocking therapies being investigated include the use of antisense molecules that target and bind to the mRNA produced by the gene, thereby preventing its translation, and the use of specially developed ribozymes that can target and cleave gene sequences that contain the unwanted mutation.

Germ-line therapy is concerned with altering the genetics of male and female reproductive cells (gametes) as well as other body cells. Because germ-line therapy will alter the individual's genes as well as those of his or her offspring, both concepts and protocols are still very controversial. Some aspects of germ-line therapy now being explored include human cloning and genetic enhancement.

Clinical Genetics. Clinical genetics is that branch of medical genetics involved in the direct clinical care of people afflicted with diseases caused by genetic disorders. Clinical genetics involves diagnosis, counseling, management, and support. Genetic counseling is a part of clinical genetics directly concerned with medical management, risk determination and options, and decisions regarding reproduction of afflicted individuals. Support services are an integral feature of all genetic counseling themes.

Clinical genetics begins with an accurate diagnosis that recognizes a specific, underlying genetic cause of a physical or biochemical defect following guidelines outlined by the National Institutes of Health (NIH) Counseling Development Conference. Clinical practice includes several hundred genetic tests that are able to detect mutations such as those associated with breast and colon cancers, muscular dystrophy, cystic fibrosis, sickle-cell disease, and Huntington's disease.

Genetic counseling follows clinical diagnosis and focuses initially on explaining the risk factors and human problems associated with the genetic disorder. Both the afflicted individual and family members are involved in all counseling procedures. Important components include a frank discussion of risks, of options such as preventive operations, and of options involved with regard to reproduction. All reproductive options are described along with their potential consequences, but genetic counseling is a support service rather than a directive mode. That is, it does not include recommendations. Instead, its ultimate mission is to help both the afflicted individuals and their families recognize and cope with the immediate and future implications of the genetic disorder.

Pharmacogenomics. That branch of human medical genetics dealing with the correlation of specific drugs to fit specific diseases in individuals is called pharmacogenomics. This field recognizes that individuals may metabolically respond differentially to therapeutic medicines based on their genetic makeup. It is anticipated that testing human genome data will greatly speed the development of new drugs that not only target specific diseases but also will be tailored to the specific genetics of patients.

Forensic Genetics. Forensic genetics is the use of human genetics in criminal or paternity cases. For example, DNA testing on blood, saliva, or other tissue can be used to determine the source of evidence, such as blood stains or semen, left at a crime scene. Forensic DNA analysis is also used to determine paternity and other kinship. Finally, with the increasing use of forensic genetics since the 1990's, some incarcerated prisoners have been released after it was clearly determined that they could not possibly have been guilty of crimes they were convicted of, as DNA evidence eliminated them from suspicion.

Potential for Human Cloning. Human therapeutic cloning involves the production of cloned human embryos, with the idea of harvesting embryonic stem cells. The hope is that the stem cells can be grown into a wide variety of cells to replace or repair organs, such as liver, kidney, or heart cells. Although human cloning has not yet reached this potential, future applications could offer identically matched kidneys for people with failing kidneys or even a genetically duplicate heart for someone in severe heart failure.

SOCIAL CONTEXT AND FUTURE PROSPECTS

The Human Genome Project painstakingly mapped out the human DNA sequence in 2003, after a decade and a half of meticulous multicenter collaboration. Genetic databases are now rapidly filling with genetic detail because of technological advances in the speed of analyzing DNA sequences. While the speed of this analysis has increased considerably, the price of such investigations has dropped significantly. Genetic databases currently hold information on a wide variety of life-forms, and significant amounts of new generic information is added frequently.

The DNA sequencing found in genetic databases provides the burgeoning field of synthetic biology with important basic information needed for human genetic engineering. This information can be used for modeling and as supply depots for the mixing and matching of genes. As the speed of genetic analysis has increased significantly and the price of genetic investigations has dropped considerably, the process of DNA synthesis is much less expensive and faster than it was in the beginning of the twenty-first century.

More genetic information, faster artificial DNA synthesis, and significant technological cost savings result in more feasible human genetic engineering projects. Genes are the stuff of life, and the field is on the verge of changing life and even making new life-forms, via genetic engineering. How and what changes are made will present significant bioethical and societal challenges, along with potentially fantastic and beneficial results.

———*Richard P. Capriccioso, M.D.*

FURTHER READING

Andrews, Lori B. *The Clone Age: Adventures in the New World of Reproductive Technology.* Henry Holt, 1999.

Baudrillard, Jean. *The Vital Illusion.* Edited by Julia Witwer. Columbia UP, 2000.

Capriccioso, Richard P. "Genetic Testing." *Cancer,* edited by Jeffrey A. Knight, Salem Press, 2009. Salem Health.

Genetic Testing Registry. National Center for Biotechnology Information, National Institutes of Health, 2016, www.ncbi.nlm.nih.gov/gtr/. Accessed 28 Oct. 2016.

Hartwell, Leland H., et al. *Genetics: From Genes to Genomes.* 5th ed., McGraw-Hill Education, 2015.

Hekimi, Siegfried, editor. *The Molecular Genetics of Aging:.* Springer, 2000. Results and Problems in Cell Differentiation.

Jorde, Lynn B., et al. *Medical Genetics.* 5th ed., Elsevier, 2016.

Lewis, Ricki. *Human Genetics: Concepts and Applications.* 11th ed., McGraw-Hill Education, 2015.

Pasternak, Jack J. *An Introduction to Human Molecular Genetics: Mechanisms of Inherited Diseases.* 2nd ed., Wiley-Liss, 2005.

Rudin, Norah, and Keith Inman. *An Introduction to Forensic DNA Analysis.* 2d ed. Boca Raton, Fla.: CRC Press, 2002. Provides a good introduction to the use of biological evidence in forensics as well as the history and application of DNA fingerprinting in

Shostak, Stanley. *Becoming Immortal: Combining Cloning and Stem-Cell Therapy.* State U of New York P, 2002.

Wilson, Edward O. *On Human Nature.* 1978. Harvard UP, 2004.

Human-computer interaction

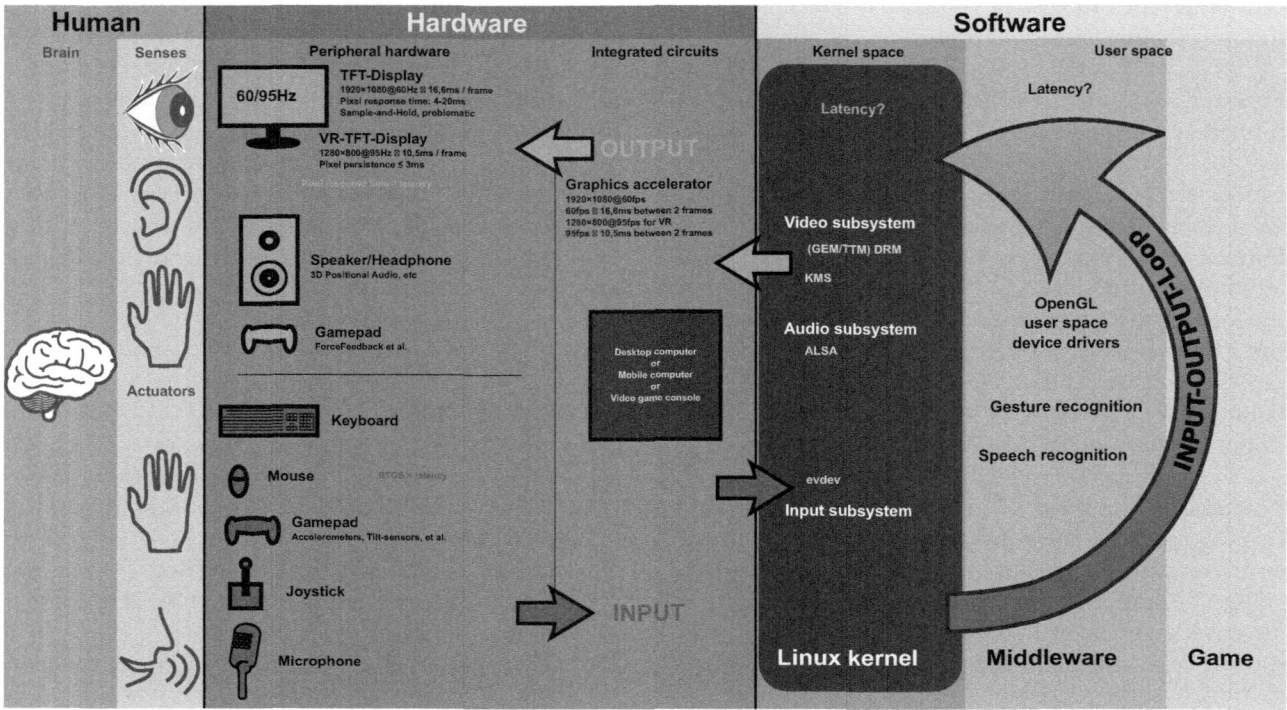

The input/output loop: Human Machine Interface (HMI) peripherals. By Shmuel Csaba Otto Traian, [CC BY-SA 3.0 (https://creativecommons.org/licenses/by-sa/3.0) or GFDL (http://www.gnu.org/copyleft/fdl.html)], via Wikimedia Commons

FIELDS OF STUDY

Bioinformatics; Biomathematics; Biomechanical engineering; Biorobotics; Computer engineering

ABSTRACT

Human-computer interaction (HCI) is a field concerned with the study, design, implementation, evaluation, and improvement of the ways in which human beings use or interact with computer systems. The importance of human-computer interaction within the field of computer science has grown in tandem with technology's potential to help people accomplish an increasing number and variety of personal, professional, and social goals. For example, the development of user-friendly interactive computer interfaces, websites, games, home appliances, office equipment, art installations, and information distribution systems such as advertising and public awareness campaigns are all applications that fall within the realm of HCI.

BACKGROUND

The fundamental philosophy that guides HCI is the principle of user-centered design. This philosophy proposes that the development of any product or interface should be driven by the needs of the person or people who will ultimately use it, rather than by any design considerations that center around the object itself. A key element of usability is affordance, the notion that the appearance of any interactive element should suggest the ways in which it can be manipulated. For example, the use of shadowing around a button on a website might help make it look three-dimensional, thus suggesting that it can be pushed or clicked. Visibility is closely related to affordance; it is the notion that the function of all the controls with which a user interacts should be clearly mapped to their effects. For example, a label such as "Volume Up" beneath a button might indicate exactly what it does. Various protocols facilitate the creation of highly usable applications. A cornerstone of HCI is iterative design, a method of development that uses repeated cycles of feedback and analysis to improve each prototype version of a product, instead of simply creating a single design and launching it immediately. To learn more about the people who will eventually use a product and how they will use it, designers also make use of ethnographic field studies and usability tests.

HISTORY

Before the advent of the personal computer, those who interacted with computers were largely technology specialists. In the 1980s, however, more and more individual users began making use of software such as word-processing programs, computer games, and spreadsheets. HCI as a field emerged from the growing need to redesign such tools to make them practical and useful to ordinary people with no technical training. The first HCI researchers came from a variety of related fields: cognitive science, psychology, computer graphics, human factors (the study of how human capabilities affect the design of mechanical systems), and technology. Among the thinkers and researchers whose ideas have shaped the formation of HCI as a science are John M. Carroll, best known for his theory of minimalism (an approach to instruction that emphasizes real-life applications and the chunking of new material into logical parts), and Adele Goldberg, whose work on early software interfaces at the Palo Alto Research Center (PARC) was instrumental in the development of the modern graphical user interface.

In the early days of HCI, the notion of usability was simply defined as the degree to which a computer system was easy and effective to use. However, usability has come to encompass a number of other qualities, including whether an interface is enjoyable, encourages creativity, relieves tension, anticipates points of confusion, and facilitates the combined efforts of multiple users. In addition, there has been a shift in HCI away from a reliance on theoretical findings from cognitive science and toward a more hands-on approach that prioritizes field studies and usability testing by real participants.

HOW IT WORKS

Input and Output Devices. The essential goal of HCI is to improve the ways in which information is transferred between a user and the machine he or she is using. Input and output devices are the basic tools HCI researchers and professionals use for this purpose. The more sophisticated the interaction between input and output devices—the more complex the feedback loop between the two directions of information flow—the more the human user will be able to accomplish with the machine.

An input device is any tool that delivers data of some kind from a human to a machine. The most familiar input devices are the ones associated with personal computers: keyboards and mice. Other commonly used devices include joysticks, trackballs, pen styluses, and tablets. Still more unconventional or elaborate input devices might take the shape of head gear designed to track the movements of a user's head and neck, video cameras that track the movements of a user's eyes, skin sensors that detect changes in body temperature or heart rate, wearable gloves that precisely track hand gestures, or automatic speech recognition devices that translate spoken commands into instructions that a machine can understand. Some input devices, such as the sensors that open automatic doors at the fronts of banks or supermarkets, are designed to record information passively, without the user having to take any action.

An output device is any tool that delivers information from a machine to a human. Again, the most familiar output devices are those associated with personal computers: monitors, flat-panel displays, and audio speakers. Other output devices include wearable head-mounted displays or goggles that provide visual feedback directly in front of the user's field of vision and full-body suits that provide tactile feedback to the user in the form of pressure.

Perceptual-Motor Interaction. When HCI theorists speak about perceptual-motor interaction, what they are referring to is the notion that users' perceptions—the information they gather from the machine—are inextricably linked to their physical actions, or how they relate to the machine. Computer systems can take advantage of this by using both input and output devices to provide feedback about the user's actions that will help him or her make the next move. For example, a word on a website may change in color when a user hovers the mouse over it, indicating that it is a functional link. A joystick being used in a racing game may exert what feels like muscular tension or pressure against the user's hand in response to the device being steered to the left or right. Ideally, any feedback a system gives a user should be aligned to the physical direction in which he or she is moving an input device. For example, the direction in which a cursor moves on screen should be the same as the direction in which the user is moving the mouse. This is known as kinesthetic correspondence.

Another technique HCI researchers have devised to facilitate the feedback loop between a user's perceptions and actions is known as augmented reality. With this approach, rather than providing the user with data from a single source, the output device projects digital information, such as labels, descriptions, charts, and outlines, on the physical world. When an engineer is looking at a complex mechanical system, for example, the display might show what each part in the system is called and enable him or her to call up additional troubleshooting or repair information.

APPLICATIONS AND PRODUCTS

Computers. At one time, interacting with a personal computer required knowing how to use a command-line interface in which the user typed in instructions—often worded in abstract technical language—for a computer to execute. A graphical user interface, based on HCI principles, supplements or replaces text-based commands with visual elements such as icons, labels, windows, widgets, menus, and control buttons. These elements are controlled using a physical pointing device such as a mouse. For instance, a user may use a mouse to open, close, or resize a window or to pull down a list of options in a menu in order to select one. The major advantage graphical user interfaces have over text-based interfaces is that they make completing tasks far simpler and more intuitive. Using graphic images rather than text reduces the amount of time it takes to interpret and use a control, even for a novice user. This enables users to focus on the task at hand rather than to spend time figuring out how to manipulate the technology itself. For instance, rather than having to recall and then correctly type in a complicated command, a user can print a particular file by selecting its name in a window, opening it, and clicking on an icon designed to look like a printer. Similarly, rather than choosing options from a menu in order to open a certain file within an application, a user might drag and drop the icon for the file onto the icon for the application. Besides helping individuals navigate through and execute commands in operating systems, software engineers also use HCI principles to increase the usability of specific computer programs. One example is the way pop-up windows appear in the word-processing program Microsoft Word when a user types in the salutation in a letter or the beginning item in a list. The program is designed to recognize the user's task, anticipate the needs of that task, and offer assistance with formatting customized to that particular kind of writing.

Consumer Appliances. Besides computers, a host of consumer appliances use aspects of HCI design to improve usability. Graphic icons are ubiquitous parts of the interfaces commonly found on cameras, stereos, microwave ovens, refrigerators, and televisions. Smartphones such as Apple's iPhone rely on the same graphic displays and direct manipulation techniques as used in full-sized computers. Many also add extra tactile, or haptic, dimensions of usability such as touchscreen keyboards and the ability to rotate windows on the device by physically rotating the device itself in space. Entertainment products such as video game consoles have moved away from keyboard and joystick interfaces, which may not have kinesthetic correspondence, toward far more sophisticated controls. The hand-held device that accompanies the Nintendo Wii, for instance, allows players to control the motions of avatars within a game through the natural movements of their own bodies. Finally, HCI research influences the physical design of many household devices. For example, a plug for an appliance designed with the user in mind might be deliberately shaped so that it can be inserted into an outlet in any orientation, based on the understanding that a user may have to fit several plugs into a limited amount of space, and many appliances have bulky plugs that take up a lot of room.

Increasingly, HCI research is helping appliance designers move toward multimodal user interfaces. These are systems that engage the whole array of human senses and physical capabilities, match particular tasks to the modalities that are the easiest and most effective for people to use, and respond in tangible ways to the actions and behaviors of users. Multimodal interfaces combine input devices for collecting data from the human user (such as video cameras, sound recording devices, and pressure sensors) with software tools that use statistical analysis or artificial intelligence to interpret these data (such as natural language processing programs and computer vision applications). For example, a multimodal interface for a GPS system installed in an automobile might allow the user to simply speak the name of a destination aloud rather than having to type it in while driving. The system might use auditory processing of the user's voice as well as visual processing of his or her lip movements to more accurately interpret speech. It might also use a camera to closely follow the movements of the user's eyes, tracking his or her gaze from one part of the screen to another and using this information to helpfully zoom in on particular parts of the map or automatically select a particular item in a menu.

Similarly, in 2015, Amazon took the technology of the Bluetooth speaker one step further with its release of the Echo device. This speaker has a built-in program that allows the user to give voice commands to instruct the device to play certain music or to sync up with other applications and devices.

Workplace Information Systems. HCI research plays an important role in many products that enable people to perform workplace tasks more effectively. For example, experimental computer systems are being designed for air traffic control that will increase safety and efficiency. Such systems work by collecting data about the operator's pupil size, facial expression, heart rate, and the forward momentum and intensity of his or her mouse movements and clicks. This information helps the computer interpret the operator's behavior and state of mind and respond accordingly. When an airplane drifts slightly off its course, the system analyzes the operator's physical modalities. If his or her gaze travels quickly over the relevant area of the screen, with no change in pupil size or mouse click intensity, the computer might conclude that the operator has missed the anomaly and attempt to draw attention to it by using a flashing light or an alarm.

Other common workplace applications of HCI include products that are designed to facilitate communication and collaboration between team members, such as instant messaging programs, wikis (collaboratively edited Web sites), and videoconferencing tools. In addition, HCI principles have contributed to many project management tools that enable groups to schedule and track the progress they are making on a shared task or to make changes to common documents without overriding someone else's work.

Education and Training. Schools, museums, and businesses all make use of HCI principles when designing educational and training curricula for students, visitors, and staff. For example, many school districts are moving away from printed textbooks and toward interactive electronic programs that target a variety of information-processing modalities through multimedia. Unlike paper and pencil worksheets, such programs also provide instant feedback, making it easier for students to learn and understand new concepts. Businesses use similar programs to train employees in such areas as the use of new software and the company's policies on issues of workplace ethics. Many art and science museums have installed electronic kiosks with touchscreens that visitors can use to learn more about a particular exhibit. HCI principles underlie the design of such kiosks. For example, rather than using a text-heavy interface, the screen on an interactive kiosk at a science museum might display video of a museum staff member talking to the visitor about each available option.

SOCIAL CONTEXT AND FUTURE PROSPECTS

As HCI moves forward with research into multimodal interfaces and ubiquitous computing, notion of the computer as an object separate from the user may eventually be relegated to the archives of technological history, to be replaced by wearable machine interfaces that can be worn like clothing on the user's head, arm, or torso. Apple released its second version of a "smartwatch" in 2016, which is designed to have all of the features of smartphones in a wearable, theoretically more convenient format. Much like other wearable gadgets such as the Fitbit, playing into society's increased concern with exercise and overall health, the watch has the ability to track human components such as heart rate and serve as a GPS that can map running, walking, and biking routes. Virtual reality interfaces have been developed that are capable of immersing the user in a 360-degree space that looks, sounds, feels, and perhaps even smells like a real environment—and with which they can interact naturally and intuitively, using their whole bodies. As the capacity to measure the physical properties of human beings becomes ever more sophisticated, input devices may grow more and more sensitive; it is possible to envision a future, for instance, in which a machine might "listen in" to the synaptic firings of the neurons in a user's brain and respond accordingly. Indeed, it is not beyond the realm of possibility that a means could be found of stimulating a user's neurons to produce direct visual or auditory sensations. The future of HCI research may be wide open, but its essential place in the workplace, home, recreational spaces, and the broader human culture is assured. As further evidence of the significance of human-computer interactions and its place in modern technology, the organization Advancing Technology for Humanity held its second annual International Conference on Human-Computer Interactions in 2016.

——M. Lee, M.A.

FURTHER READING

Bainbridge, William Sims, editors. *Berkshire Encyclopedia of Human-Computer Interaction: When Science Fiction Becomes Fact.* Berkshire, 2004. 2 vols.

Gibbs, Samuel, and Alex Hern. "Apple Watch 2 Brings GPS, Waterproofing and Faster Processing." *The Guardian,* 8 Sept. 2016, www.theguardian.com/technology/2016/sep/07/iphone-7-launch-apple-watch-2-gains-gps-longer-battery-life. Accessed 28 Oct. 2016.

Helander, Martin. *A Guide to Human Factors and Ergonomics.* 2nd ed., CRC Press, 2006.

Hughes, Brian. "Technology Becomes Us: The Age of Human-Computer Interaction." *The Huffington Post,* 20 Apr. 2016, www.huffingtonpost.com/brian-hughes/technology-becomes-us-the_b_9732166.html. Accessed 28 Oct. 2016.

Jokinen, Jussi P. P. "Emotional User Experience: Traits, Events, and States." *International Journal of Human-Computer Studies,* vol. 76, 2015, pp. 67–77.

Purchase, Helen C. *Experimental Human-Computer Interaction: A Practical Guide with Visual Examples.* Cambridge University Press, 2012.

Sears, Andrew, and Julie A. Jacko, editors. *The Human-Computer Interaction Handbook: Fundamentals, Evolving Technologies, and Emerging Applications.* 2nd ed., Erlbaum, 2008.

Sharp, Heken, et al. *Interaction Design: Beyond Human-Computer Interaction.* 2nd ed., John Wiley & Sons, 2007.

Soegaard, Mads, and Rikke Friis Dam, editors. *The Encyclopedia of Human-Computer Interaction.* 2nd ed., Interaction Design Foundation, 2014.

Thatcher, Jim, et al. *Web Accessibility: Web Standards and Regulatory Compliance.* Springer, 2006.

Tufte, Edward R. *The Visual Display of Quantitative Information.* 2nd ed., Graphics, 2007.

Hybridization (botany)

FIELDS OF STUDY

Agricultural engineering; Botany; Biotechnology; Environmental biotechnology (or Green biotechnology); Forestry; Medical biotechnology (or Red biotechnology); Microbiology; Plant pathology

ABSTRACT

Hybridization is the process of crossing two genetically different individuals to create new genotypes. For example, a cross between a parent 1, with the genetic makeup (genotype) *BB*, and parent 2, with *bb*, produces progeny with the genetic makeup *Bb*, which is a hybrid (the first filial generation or F_1). Hybridization was the basis of Gregor Mendel's historic experiments with garden peas. Inheritance studies require crossing plants with contrasting or complementary traits.

BACKGROUND

Hybridization of plants occurs in nature through various mechanisms. Some plants (such as the oil palm) are insect-pollinated, and others (such as maize, or corn) are wind-pollinated. Such plants are referred to as *cross-pollinated plants*. Natural hybridization has played a significant role in producing new genetic

Sample leaves of common garden dahlia, [Public domain], via Wikimedia Commons

combinations and is the norm in cross-pollinated plants. It is a common way of generating genetic variability.

In plants with perfect flowers (*autogamous*, having flowers with both stamens and pistils), cross-pollination rarely occurs. Such plants (such as wheat and rice) are called *self-pollinated plants*. Flowers bearing only pistils or stamens are said to be imperfect flowers. Plants that have separate pistillate and staminate flowers on the same plant (such as maize) are called *monoecious*. Plants that have male and female flowers on separate plants (such as asparagus) are called *dioecious*.

Through artificial means (controlled pollination), hybridization of both cross-pollinated and self-pollinated plants can be accomplished. Artificial hybridization is an important aspect of improving both cross-pollinated and self-pollinated plants. The breeder must know the time of development of reproductive structures of the species, treatments to promote and synchronize flowering, and pollinating techniques.

APPLICATIONS TO AGRICULTURE

The concept of *hybrid vigor*, or *heterosis*, resulted from hybridization. Heterosis (or heterozygosis) occurs when the hybrid outperforms its parents for a certain trait. Around 1761 Joseph Gottlieb Kölreuter was the first to report on hybrid vigor in interspecific crosses of various species of *Nicotiana*. He concluded that cross-fertilization was generally beneficial and self-fertilization was not. In 1799 T. A. Knight conjectured that because of widespread existence of cross-pollination in nature, it must be the norm. Charles Darwin reported the results of his experiments with maize. He indicated that in twenty-four crosses, there was an increase in plant height, which was attributed to hybridization, and that decrease in plant height was associated with self-pollination (or selfing). He also noted that crossing of inbred plants could reverse the deleterious effects of selfing or *inbreeding*. In 1862 Darwin wrote, "Nature tells us, in the most emphatic manner, that she abhors perpetual self-fertilization." In the late 1800's William J. Beal evaluated hybrids between maize varieties. He observed that some hybrids yielded 50 percent more than the mean of their parents. S. W. Johnson provided an explanation for hybrid vigor in 1891. G. W. McClure reported in 1892 that hybrids between maize varieties were superior to the mean of the two parents.

EXPLOITATION OF HETEROSIS

The phenomenon of heterosis has been exploited in crop plants, such as maize, sorghum, sunflower, onion, and tomato. Maize (corn) was the first crop in the United States in which hybrids were produced from *inbred lines*. It was George Shull who, following the rediscovery of Mendel's laws of inheritance in 1900, conducted the first experiments on inbreeding and crossing, or hybridizing, of inbred lines. Shull suggested that inbreeding within a maize variety resulted in pure (*homozygous*) lines and that hybrid vigor resulted from crossing of pure lines because *heterozygosity* was created at many allelic sites. Hybrid maize was introduced in the United States in the late 1920's and early 1930's, after which U.S. maize production increased dramatically from the use of hybrids.

Heterosis now drives a multibillion-dollar business in agriculture. Yield improvement made in various crops in which heterosis was detected has been tremendous. In 1932 in the United States, 44.8 million hectares (111 million acres) were required to produce 51 million metric tons of maize grain, with a mean yield of 1.66 metric tons per hectare. In 1994 it took only 32 million hectares (79 million acres) to produce 280 million metric tons of grain, with a mean yield of 8.69 metric tons per hectare. In the United States in 1996, twenty-one vegetable crops occupied 1,576,494 hectares (3.9 million acres), with a mean of 63 percent of the crop in hybrids. Heterosis saved an estimated 220,337 hectares (544,459 acres) of agricultural land per year, feeding 18 percent more people without an increase in land use. From 1986 to 1995, the best rice hybrids showed a 17 percent yield advantage over the best inbred-rice varieties at the International Rice Research Institute.

Despite the impact that heterosis has had on crop production, its molecular genetic basis is still not clear. It is hoped that with the progress being made in the genetic sequencing of various plant species, a better understanding of heterosis will emerge. Plant breeding entails hybridization within a species as well as hybridization between species or even genera, called *wide crosses*. The latter are important for generating genetic variability or for incorporating a desirable gene not available within a species. There are barriers, however, for accomplishing *interspecific* and *intergeneric* crosses. Plants of the same species cross easily and produce fertile progeny. Wide crosses are difficult to make and generally produce sterile progeny because of chromosome-pairing difficulties during meiosis.

Triticale is the only human-made cereal crop, which is a cross between the genus *Triticum* (wheat) and the genus *Secale* (rye). The first fertile triticale was produced in 1891. Some of the interspecific and intergeneric barriers should be overcome via the newer techniques of gene transfer. It is expected that genes from wild relatives of cultivated plants will continue to be sought to correct defects in otherwise high-yielding varieties.

—*Manjit S. Kang*

FURTHER READING

Basra, A. S., ed. *Heterosis and Hybrid Seed Production in Agronomic Crops*. Binghamton, N.Y.: Food Products Press, 1999. This book discusses current research in some of the most important crops of the world.

Coors, J. G., and S. Pandey. *Genetics and Exploitation of Heterosis in Crops*. Madison, Wis.: American Society of Agronomy and Crop Science Society of America, and Soil Science Society of America, 1999. Provides an account of the various issues related to hybrid vigor, or heterosis.

Fehr, W. R., and H. H. Hadley. *Hybridization of Crop Plants*. Madison, Wis.: American Society of Agronomy and Crop Science Society of America, 1980. Brings together the experience of plant breeders and scientists in a form that can be used by the layperson.

INDUSTRIAL FERMENTATION

Beer brewing was an early application of biotechnology. By Jost Amman [Public domain], via Wikimedia Commons

FIELDS OF STUDY

Biochemistry; Bioengineering; Biology; Bioprocess engineering; Biosystems engineering; Industrial biotechnology (or White biotechnology); Industrial fermentation

ABSTRACT

Industrial fermentation is an interdisciplinary science that applies principles associated with biology and engineering. The biological aspect focuses on microbiology and biochemistry. The engineering aspect applies fluid dynamics and materials engineering. Industrial fermentation is associated primarily with the commercial exploitation of microorganisms on a large scale. The microbes used may be natural species, mutants, or microorganisms that have been genetically engineered. Many products of considerable economic value are derived from industrial fermentation processes. Common products such as antibiotics, cheese, pickles, wine, beer, biofuels, vitamins, amino acids, solvents, and biological insecticides and pesticides are produced via industrial fermentation.

BACKGROUND

The goal of industrial fermentation is to improve biochemical or physiological processes that microbes are capable of performing while yielding the highest quality and quantity of a particular product. The development of fermentation processes requires knowledge from disciplines such as microbiology, biochemistry, genetics, chemistry, chemical and bioprocess engineering, mathematics, and computer science. The major microorganisms used in industrial fermentation are fungi (such as yeast) and bacteria. Fermentation is performed in large fermenters or other bioreactors often of several thousand liters in volume. Industrial fermentation is a part of many industries, including microbiology, food, pharmaceutical, biotechnology, and chemical.

HISTORY

Traditional fermentations such as those for making bread, cheese, yogurt, vinegar, beer, and wine had been used by people for thousands of years before its microbial nature was understood. Brewing beer was one of the first applications of fermentation in ancient Egypt as long as 10,000 years ago. The exact origins of dairy products are unknown—it may have been as early as 8000 BCE. It was probably nomadic Turkish tribes in Central Asia who invented cheese and yogurt making. Traditionally, dairy fermentation

Fermentation tanks for red wine making. By Cjp24. (Own work.) [GFDL (http://www.gnu.org/copyleft/fdl.html) or CC-BY-SA-3.0-2.5-2.0-1.0 (http://creativecommons.org/licenses/by-sa/3.0)], via Wikimedia Commons

was a means of milk preservation. The scientific understanding of fermentation began only in the nineteenth century after French scientist Louis Pasteur published the results of his studies on the microbial nature of wine making.

The first industrial fermentation bioprocesses based on knowledge of microbes appeared in the early twentieth century. Russian biochemist Chaim Weizmann is considered to be the father of industrial fermentation. Weizmann used the bacterium *Clostridium acetobutylicum* for the production of acetone from starch in 1916. Acetone was used to make explosives during World War I.

Significant growth of this field began in the middle of the twentieth century, when the fermentation process for the large-scale production of antibiotic penicillin was developed. The goal of industrial-scale production of penicillin during World War II led to development of fermenters by engineers that were working together with biologists from the pharmaceutical company Pfizer. The fungus *Penicillium* grows and produces an antibiotic much more effectively under controlled conditions inside the fermenter. Continuous progress in industrial fermentation technology in the twentieth century has followed the development of genetic engineering. Genetic engineering allows gene transfer between species and creates possibilities to generate new products from genetically modified microorganisms that are grown in fermenters.

The twenty-first century has been characterized by the introduction of biofuels, which are made by industrial fermentation processes. Once again, past and future developments in fermentation technology require contributions from a wide range of disciplines, including microbiology, genetics, biochemistry, chemistry, engineering, mathematics, and computer science.

HOW IT WORKS

Industrial fermentation is based on microbial metabolism. Microbes produce different kinds of substances that they used for growth and maintenance of their cells. These substances can be useful for humans. The goal of industrial fermentation technology is to enhance the microbial production of useful substances.

Process of Fermentation. In biology, fermentation is a process of harvesting energy of organic molecules in oxygen-free conditions. Sugars are a prime example of what can be fermented, although, there are many other organic molecules that can be used. Different fermentations are known and are categorized by the substrate metabolized or the type of the product.

In industry, any large microbiological process is called fermentation. Thus, the term fermentation has a different meaning than in biology. Most industrial fermentations require oxygen.

Industrial Fermentation Organisms. Different organisms, such as bacteria, fungi, and plant and animal cells, are used in industrial fermentation processes. An industrial fermentation organism must produce the product of interest in high yield, grow rapidly on inexpensive culture media available in bulk quantities, be open to genetic manipulation, and be nonpathogenic (does not cause any diseases).

Fermentation Media. To make a desired product by fermentation, microorganisms need nutrients (substrates). Nutrients for microbial growth are known as media. Most fermentation requires liquid media or broth. General media components include carbon, nitrogen, oxygen, and hydrogen in the form of organic or inorganic compounds. Other minor or trace elements must also be supplied, for example, iron, phosphorus, or sulfur.

Fermentation Systems. Industrial fermentation takes place in fermenters, which are also called bioreactors. Fermenters are closed vessels (to avoid microbial contamination) that reach vast volumes, as many as several hundred thousand liters. Designed by engineers, the main purpose of a fermenter is to provide controllable conditions for growth of microbial cells or other cells. Parameters such as pH, temperature, nutrients, fluid flow, and other variables are controlled. There are two kinds of fermenters, those for anaerobic processes (oxygen-free) and aerobic processes. Aerobic fermentation is the most common in industry. Anaerobic fermenters can be as simple as stainless-steel tanks or barrels. Aerobic fermenters are more complicated. The most critical part in these systems is aeration. In a large-scale fermenter transfer of oxygen is very important. Oxygen transfer and dispersion are provided by stirring with impellers or oxygen (air) sparging.

Fermentation Control and Monitoring. Industrial fermentation control is very important to ensure that organisms behave properly. In most cases computers are used for controlling and monitoring the fermentation process. Computers control temperature, pH, cell density, oxygen concentration, level of nutrients, and product concentration.

APPLICATIONS AND PRODUCTS

There are a wide range of industrial fermentation products and applications.

Food, Beverages, Food Additives, and Supplements. Industrial fermentation plays a major role in the production of food. Food products traditionally made by fermentation include dairy products (cheeses, sour cream, yogurt, and kefir); food additives and supplements (flavors, proteins, vitamins, and carotenoids); alcoholic beverages (beer, wines, and distilled spirits); plant products (bread, coffee, soy sauce, tofu, sauerkraut); and fermented meat and fish (pepperoni and salami).

Industrial Fermentation. The primary and largest industry revolves around food products. Milk from cows, sheep, goats, and horses have traditionally been used for the production of fermented dairy products. These products include cheese, sour cream, kefir, and yogurt. More recently so-called probiotics appeared and have been marketed as health-food drinks. Dairy products are produced via fermentation using lactic bacteria such as *Lactobacillus acidophilus* and *Bifidobacterium*. Fungi are also involved in making some cheeses. Fermentation produces lactic acid and other flavors and aroma compounds that make dairy products taste good.

Many products of industrial fermentation are added into food as flavors, vitamins, colors, preservatives, and antioxidants. These products are more desirable than food additives produced chemically. Many of the vitamins are made by microbial fermentations including thiamine (vitamin B_1), riboflavin (vitamin B_2), cobalamin (vitamin B_{12}), and vitamin C (ascorbic acid). Vitamin C is not only a vitamin but is also an important antioxidant that helps to prevent heart diseases. Carotenoids are another effective antioxidant. They are also used as a natural food color for butter and ice cream. Carotenoids are red, orange, and yellow pigments produced by bacteria, algae, and plants.

Food preservatives are yet another product of industrial fermentation. Organic acids, particularly lactic and citric acids, are extensively used as food preservatives. Some of these preservatives (such as citric acid) are used as flavoring agents. A mixture of two bacterial species (*Lactobacillus* and *Streptococcus*) is usually used for industrial production of lactic acid. The mold *Aspergillus niger* is used for citric acid manufacturing. Another common preservative is the protein nisin. Nisin is produced via fermentation by the bacterium *Lactococcus lactis*. It is employed in the dairy industry especially for production of processed cheese.

Antibiotics and Other Health Care Products. Antibiotics are chemicals that are produced by fungi and bacteria that kill or inhibit the growth of other microbes. They are the second most significant product of industrial fermentation. Most antibiotics are generated by molds or bacteria called actinomycetes. More than 4,000 antibiotics have been isolated from microorganisms, but only about 50 are produced regularly. Among them, beta-lactams, such as penicillins and tetracyclines, are most common. Penicillin is produced by the mold *Penicillium chrysogenum* via corn fermentation in bioreactors of up to 200,000 liters.

The other major health care products produced with the help of industrial fermentation are bacterial vaccines, therapeutic proteins, steroids, and gene therapy vectors. There are two categories of bacterial vaccines: living and inactivated vaccines. Living vaccines consist of weakened, also known as attenuated, bacteria. Examples of living vaccines include those for diseases such as anthrax, which is caused by *Bacillus anthracis*, and typhoid fever, which is caused by *Salmonella typhi*. Inactivated vaccines are composed of bacterial cells or their parts that have been inactivated by heat or formaldehyde. Examples of these vaccines are those for meningitis, whooping cough, and cholera. Vaccine production takes place in fermenters no bigger than 1,000 liters in volume. It requires highly controlled operations to avoid the release of bacteria into the environment. All exhaust gases pass through sterilization processes.

Therapeutic proteins include growth hormone, insulin, wound-healing factors, and interferon. Previously, such compounds were made from animal tissues and were very expensive to manufacture. Genetic engineering now allows their production by fermentation from bacteria. Human growth hormone is synthesized in the human brain and controls growth. Too little growth hormone can cause some cases of dwarfism. The American company Genentech started production of human growth hormone from genetically modified *Escherichia coli* by fermentation in 1985. Insulin is an animal and human hormone that is involved in the regulation of blood sugar. The body's inability to make sufficient insulin causes diabetes. Insulin extracted from pigs had been used to treat diabetes, but it has been replaced by insulin produced by industrial fermentation from genetically modified bacteria.

Chemicals. Numerous chemicals, such as amino acids, polymers, organic acids (citric, acetic, and lactic), and bioinsecticides are produced by industrial fermentation. Amino acids are used as a food and animal feed, as well as in the pharmaceutical, cosmetic, and chemical industries. Bacteria such as *Micrococcus luteus* and *Corynebacterium glutamicum* are used for industrial fermentation to produce chemicals. Bacterial toxins are effective against different insects. Since the 1960's, preparations of the bacteria *Bacillus thuringiensis* have been produced by fermentation as a biological insecticide.

Enzymes. Enzymes are used in many industries as catalysts. Microorganisms are the favored source for industrial enzymes. Seventy percent of these enzymes are made from *Bacillus* bacteria via fermentation. Most commercial microbial enzymes are hydrolases, which break down different organic molecules such as proteins and lipids. The enzyme glucose isomerase is important in the production of fructose syrups from corn and is widely used in the food industry.

Biomass Production. During biomass production by fermentation, the cells produced are the products. Biomass is used for four purposes: as a source of protein for human food or animal feed, in industry as fermentation starter cultures, in agriculture as a pesticide or fertilizer, and as a fuel source.

One major product of this application of industrial fermentation is baker's yeast biomass. Baker's yeast is required for making bread, bakery products, beer, wine, ethanol, microbial media, vitamins, animal feed, and biochemicals for research. Yeast is produced in large aerated fermenters of up to 200,000 liters. Molasses is used as a nutrient source for the cells. Yeast is recovered from fermentation liquid by centrifugation and then dried. It can then be sold as compressed yeast cakes or dry yeast.

Many bacteria have been considered as potential sources of protein to fulfill the food needs in some countries of the world. As of 2011, only a few species are cultivated around the world as a source of food and feed. Among them, cyanobacteria are the most popular. The protein level of *Cyanobacterium Spirulina* can be as high as those found in meat, nuts, and soybeans, from 50 to 70 percent. This cyanobacterium has been used as a human food for millennia in Asia, parts of Africa, and in Mexico.

Apart from yeast and bacteria, people are also using the biomass of algae. Algae are a source of animal feed, plant fertilizer, chemicals, and biodiesel. Because light is necessary to grow algae, the biomass is produced in open ponds or in transparent tubular glass or plastic bioreactors, called photobioreactors.

Biofuels. Industrial fermentation is used in the production of biofuels, mainly ethanol and biogas. These two biofuels are produced by the action of microorganisms in bioreactors. Fermentation can also be used for generation of biodiesel, butanol, and biohydrogen. Biofuels are considered, by many, as a future substitute for fossil fuels. Pollution from fossil fuels affects public health and causes global climate change due to the release of carbon dioxide (CO_2). Using biofuels as an energy source generates less pollutants and little or no CO_2.

Production of ethanol is a process based on fungal or bacterial fermentation of a variety of materials. In the United States, most of the ethanol is produced by yeast (fungal) fermentation of sugar from cornstarch. Sugar is extracted using enzymes, and then yeast cells convert the sugar into ethanol and CO_2. Ethanol is separated from the fermentation broth by distillation. Brazil, the second largest ethanol producer after the United States, uses sugarcane fermentation to generate ethanol. The Brazilian production of ethanol from sugarcane is more efficient than the American corn-based ethanol.

Biogas is produced during the anaerobic (non-oxygen) fermentation of organic matter by communities of microorganisms (bacteria and *Archaea*). There are different types of biogas. One type contains a mixture of methane (50 to 75 percent) and CO_2. Another type is composed primarily of nitrogen, hydrogen, and carbon monoxide (CO) with trace amounts of methane. Methane is generated by microorganisms called *Archaea* and is an integral part of their metabolism. For practical use, biogas is generated from wastewater, animal waste, and "gas wells" in landfills.

SOCIAL CONTEXT AND FUTURE PROSPECTS

Industrial fermentation plays a major role in providing food, chemicals, and fuels. End users are consumers, farmers, medical doctors, and industrialists. Industrial fermentation is changing the course of history. People have made food by fermentation for centuries. In the twentieth century, the development of antibiotics and their production by industrial fermentation had the most significant impact on the practice of medicine than any other development. The growth of the industrial fermentation field is continuing rapidly. Since the beginning of the twenty-first century, industrial fermentation underwent an unprecedented growth and expansion due to biofuel introduction. This record growth is particularly visible in the U.S. ethanol industry. In 1980, the U.S. ethanol industry produced 175 million gallons of ethanol by fermentation, and in 2009, 10.6 billion gallons.

The role of industrial fermentation in human society is likely to expand in the future because of increasing requirements for resources.

—*Sergei A. Markov, PhD*

FURTHER READING

Bailey, James E., and David F. Ollis. *Biochemical Engineering Fundamentals.* 2d ed. New York: McGraw-Hill, 1986. Classic textbook on biochemical engineering.

Bourgaize, David, Thomas R. Jewell, and Rodolfo G. Buiser. *Biotechnology: Demystifying the Concepts.* San Francisco: Benjamin/Cummings, 2000. Classic text on biotechnology.

Doran, Pauline M. *Bioprocess Engineering Principles.* San Diego: Academic Press, 1995. Describes various bioreactors and fermenters used for industrial fermentation.

Glazer, Alexander N., and Hiroshi Nikaido. *Microbial Biotechnology: Fundamentals of Applied Microbiology.* 2d ed. New York: Cambridge University Press, 2007. In-depth analysis of application of microorganisms in industrial fermentation.

Lydersen, Bjorn K., Nancy A. D'Elia, and Kim L. Nelson, eds. *Bioprocess Engineering: Systems, Equipment and Facilities.* New York: John Wiley & Sons, 1994. Describes equipment and facilities for industrial fermentation.

Wright, Richard T., and Dorothy F. Boorse. *Environmental Science: Toward a Sustainable Future.* 11th ed. Boston: Benjamin/Cummings, 2011. This textbook describes several bioprocesses used in waste treatment and pollution control.

INTELLIGENCE

FIELDS OF STUDY

Bioinformatics; Biomathematics; Biomechanical engineering; Biorobotics; Computer engineering

ABSTRACT

Animals are guided by more than instinct when interacting with their environment, yet the exact measurement of intelligence in various species remains problematic. While scientists devise numerous problem-solving tasks to assess intelligence, anthropomorphism leads to exaggerated claims of intelligence through anecdotal evidence.

BACKGROUND

Both the general public and scientific community have long been intrigued with questions about how animals think and what are they thinking. Published reports of animal cognition increased dramatically in the last half of the twentieth century. Chimpanzees in the Ivory Coast have demonstrated extensive use of rocks as tools in cracking nuts. These primates have also been reported to hide undesirable expressions from their faces and act as if blind or deaf. Vervets have been found to use an elaborate system of alarm calls that seem to function as words. Parrots can demonstrate the ability to count, and birds exhibit the capacity to make and use tools to gather food. Dolphins apparently understand and follow simple commands. Primates have been trained to use signs in a symbolic fashion, communicating their needs, desires, and thoughts.

THEORIES OF COGNITIVE ETHOLOGY

Cognitive ethology is a relatively new discipline that studies animal intelligence. Donald Griffin is considered to have founded this branch of study through the publication of *Animal Thinking* (1984) and *Animal Minds* (1992). Since the appearance of his books, numerous instances of animal intelligence have been gathered from observation and experimentation.

Traditionally, attitudes about animal intelligence can be sorted into those that place animals on a continuum with humans and those that see animals as distinct from humans. From the former perspective, animal behavior is readily interpreted as a definite sign of various cognitive skills and special abilities along a continuum of development. From a discontinuity perspective, only humans are considered to possess the higher cognitive skill of reasoning. The higher cognitive abilities are considered to be a uniquely human capacity that sets them apart from the lower animals, who are controlled by instinct.

Charles Darwin, in *The Descent of Man* (1871), defended the idea of the intelligence of animals existing on a continuum with humans. Since animals and humans have a common ancestry, animals would have the fundamental capacities for rational choice, reflection, and insight. Darwin concluded that the differences between the minds of humans and animals were of degree rather than of kind. Following Darwin's proclamation, a number of anecdotal studies concerning animal intelligence appeared that suggested extensive cognitive ability in animals. Unfortunately, many of the examples illustrated anthropomorphism. This is the process whereby humanlike characteristics are attributed to animal behavior.

Some interpretations of Darwin's statement created a distorted view about evolution that persisted long into the twentieth century. The idea that life on earth represents a chain of progress from inferior to superior forms began to influence the view of animal intelligence. The theory that ontogeny recapitulates phylogeny also became popular in the early years of the twentieth century. This theory, which does not have any scientific support, suggested that the advancement of life forms corresponded to the stages of development for humans. This stepladder approach to animal intelligence led to a ranking of animals compared to the developmental stages of human infants and children. This approach to animal intelligence is flawed because it relies on the notion that some animals are more highly evolved than others are. Evolution does not have a single point of greatest evolution. The branches of the evolutionary tree have culminated with many different species occupying special niches. Thus, the "degree" of a species' evolution depends on the extent to which it successfully occupies its niche.

ANIMALS WHO MIGHT THINK

In addition to this tendency to attribute states of mind to animals that are found in humans, there were a number of cases of labeling trained behavior in animals as signs of reasoning skills. One of the most famous examples was the case of the horse, Clever Hans, in the early 1900's. Wilhelm von Osten owned a horse that demonstrated extensive arithmetic skills. When von Osten presented a written arithmetic problem to Hans, the horse would tap out the answer with his forefoot. Clever Hans also appeared adept at telling time, and answered questions about sociopolitical events by nodding or shaking his head yes or no. The horse's abilities suggested to

> **MEASURING ARITHMETIC SKILLS IN PRIMATES**
>
> The chimpanzee Sheba was involved in a number of widely reported experiments that purported to demonstrate arithmetic ability in primates. Sheba was taught to associate a tray containing one, two, or three pieces of candy with cards containing the corresponding number of marks. If two candies were present on the tray, Sheba learned to pick the card with two marks. If the correct card was selected, Sheba would be rewarded by being allowed to eat the candy. In the next stage of the experiment, the marks on the cards were replaced by the numerals 1, 2, and 3. Then Sheba had to match the number of candy pieces with the correct number. After Sheba showed success in this phase of the experiment, the candy pieces were replaced with other inedible objects. If Sheba chose the number that matched the number of objects, she was rewarded with the corresponding number of candies. Eventually the numbers were expanded to include 0 and 4. Next Sheba was given the challenge of counting the total number of candies presented on a series of trays. Sheba demonstrated a rudimentary ability for addition showing correct responses 75 percent of the time. Initially the sums were restricted to one, two, or three. Sheba was eventually able to exhibit the skill to add numbers when the candies were replaced with the numerals 1, 2, and 3. Although Sheba was able to show mathematical ability, even the experimenter acknowledged some important limitations. Sheba required extensive training and "heroic" effort on the part of the trainer to accomplish the counting performances. A true mathematical ability should generalize to other situations, yet animals such as Sheba do not demonstrate easy or automatic transfer of numerical performance from one realm to another.

many individuals the similarity between animal and human minds. Eventually, the Prussian Academy of Sciences discovered that Hans was not answering the questions by means of any reasoning skills, but was an astute observer of the behavior of his owner and those around him. When questions were posed to Hans, cues were provided unconsciously to the horse about the correct answer. Since horses have evolved to ascertain subtle visual cues from others in their herd, Hans was able to form a number of cued associations which led to a reward. The owner of Clever Hans was not attempting to perpetrate fraud. He believed in the possibility that a horse could have reasoning ability, but von Osten was not sophisticated in how he tested for the skills. The inadvertent cueing of an animal to respond in a certain fashion is one of the major confounding factors found in the investigation of animal intelligence.

The case of Clever Hans illustrates two other problems that confound reports concerning the level of intelligence in animals. First is the problem of anthropomorphism. People develop an emotional bond with animals and interpret behavior in order to enhance the closeness they feel to them. The second problem concerns the methods used to measure intelligence. The classic case of Köhler's chimpanzees illustrates this problem.

In the early part of the twentieth century, Wolfgang Köhler assessed the reasoning ability of chimpanzees to obtain food outside of an enclosure. After a rake was left in the enclosure, food was placed out of reach of the caged chimpanzees. The chimpanzees were able to use the rake to bring food to the cage. Köhler concluded that the animals had insight into the nature of the problem and used reasoning to achieve a solution. A further study, requiring the fitting together of two sticks in order to reach the food, also supported Köhler's conclusions. However, later experimentation has revealed that chimpanzees without a history of playing with sticks could not solve the problem. Apparently, in order to solve the problem, the chimpanzees needed an extensive history of playing with sticks, which enabled them to learn how sticks could be used at a later time. In solving the problem, they were using an instinctual tendency to play with sticks and scraping them over the ground.

PRIMATES AND SIGN LANGUAGE

A contemporary example of the problem of measurement can be provided with the case of Washoe, the first chimpanzee to be taught sign language. Because of physical inability to vocalize human speech, chimpanzees were taught sign language as a mode of communication with humans. Soon Washoe and another signing chimpanzee, Nim Chimsky, were reported to have spontaneously created novel sentences through their signing. For example, Washoe was reported to have signed the combination water and bird after seeing a swan. Being a novel combination of signs, the trainers of Washoe explained the behavior as creative insight. Unfortunately, Washoe had also shown repeated signing of meaningless combinations, leading to the conclusion that a significant pairing of signs would eventually appear not because of the primate's cognitive reasoning but as a result of chance. Inevitably, these early attempts to demonstrate animal intelligence were widely discredited as exaggeration or self-delusion on the part of the animal's trainers, and this animal language research from the late 1970's fell into disrepute.

In order avoid the ambiguities of sign language, later researchers used keyboards that related symbols to a variety of objects, people, and places. Much of this research has taken place at the Language Research Center at Georgia State University in Atlanta under the guidance of Dr. Sue Savage-Rumbaugh. In the first experiments, two chimpanzees, Austin and Sherman, were familiarized with a system of symbols or lexigrams. Each was abstract and arbitrarily associated with an object, person, place, or situation. Eventually Austin and Sherman learned to communicate with symbols illustrated on a keyboard. For example, an experiment was devised where one chimpanzee was shown where food was being deposited in a certain container while the other had control of a tool to open the container. With the keyboard present, the chimpanzees were able to communicate with one another to use the tool on the correct container.

Soon a bonobo chimpanzee, Kanzi, became the star pupil of this technique and learned a vocabulary of two hundred symbols. Kanzi eventually showed the capacity to construct rudimentary sentences that were generated spontaneously. The chimpanzees trained using the keyboards appear to be exhibiting a protogrammar. This is a term to indicate the beginnings of grammar, roughly equivalent to the verbal skills seen in a human child about two to three years old.

In the late 1990's, another bonobo chimpanzee, Panbanisha, surpassed the capacities evidenced by Kanzi. Panbanisha has been reported to understand complex sentences and use the keyboard to communicate spontaneously with the outside world. Although the results have been impressive, critics of the Center's activities remain. The question remains whether the chimpanzees are demonstrating extremely effective training or some level of abstract reasoning.

—*Frank J. Prerost*

FURTHER READING

Budiansky, Stephen. *If a Lion Could Talk*. New York: Free Press, 1998. A good collection of contemporary and historical cases of animal intelligence. The stories cover a wide range of examples seen in various animal species.

Moss, Cynthia. *Elephant Memories*. New York: William Morrow, 1988. An interesting account of thirteen years of field observations concerning the behavior of elephants in the Amboseli National Park in Kenya.

Page, George. *Inside the Animal Mind*. New York: Doubleday, 1999. The author begins with a historical account of the popular and scientific views about animal intelligence. He provides good details about the various attempts to communicate with primates by teaching them sign language or through the use of keyboards.

Savage-Rumbaugh, Sue. *Kanzi: The Ape at the Brink of the Human Mind*. New York: John Wiley & Sons, 1994. This book presents an apparent breakthrough in the communication with chimpanzees using symbols and a keyboard. It includes a number of incidences of the spontaneous construction of sentences by the primates.

Savage-Rumbaugh, Sue, Stuart G. Shanker, and Talbot J. Taylor. *Apes, Language, and the Human Mind*. New York: Oxford University Press, 1998. A book written from an academic and scientific perspective about the ability of chimpanzees to communicate by means of symbols and a keyboard display.

Intensive farming

Terrace rice fields in Yunnan Province, China. By Jialiang Gac, www.peace-on-earth.org. (Original Photograph.), [GFDL (http://www.gnu.org/copyleft/fdl.html), CC-BY-SA-3.0 (http://creativecommons.crg/licenses/by-sa/3.0/) or CC BY-SA 2.5 (https://creativecommons.org/licenses/by-sa/2.5)], via Wikimedia Commons

FIELDS OF STUDY

Agricultural engineering; Botany; Biotechnology; Environmental biotechnology (or Green biotechnology); Forestry; Medical biotechnology (or Red biotechnology); Microbiology; Plant pathology

ABSTRACT

The world's growing population requires ever-increasing amounts of food, and intensive farming provides high levels of crop production on the land available. The greatest challenge to such farming is its long-term sustainability; intensive farming's dependence on fossil fuel–burning heavy equipment and its reliance on chemicals to eliminate pests and replace nutrients in soil are damaging to the environment.

BACKGROUND

During the nineteenth century, the economy of the United States was largely dependent on the agricultural production of thousands of small farms. As the country industrialized, and particularly after World War II, small farmers began to consolidate their lands, with the result that there were fewer farm operators and much larger farms. Mechanization, in the form of tractors, seed application implements, and other heavy equipment, as well as technological advancements in chemical pesticides, herbicides, and fertilizers contributed to supporting the infrastructure of these large-scale farming operations and lowered the farmers' labor costs.

As world population continues to rise, demand for food also rises, and food crops require land for production. With the spread of urbanization and suburbanization, productive cropland has become increasingly scarce. Intensive farming, also known as industrial farming, maximizes crop production on the land available by relying heavily on the mechanization of many processes, by treating even marginal soils with chemical fertilizers to increase yield, and by using chemical pesticides and herbicides. Such large-scale commercial agriculture is very capital-intensive in nature, and most intensive farming operations are thus found in the world's economically developed regions.

Industrial farming operations make extensive use of technology to run systems and monitor resources, both to keep labor costs low and to increase productivity. For example, many such vast farms have switched from labor-intensive manually operated irrigation systems to computer-controlled center-pivot irrigation systems. Some farms use geographic information system (GIS) technology to facilitate the monitoring and control of their land resources.

BENEFITS OF LARGE-SCALE FARMING

From the 1940's into the 1980's, the so-called Green Revolution in agriculture did much to transform regions that had been dependent on more traditional forms of farming into areas that could produce greater amounts of food for their populations through more intensive agriculture. The impact of the Green Revolution was especially strong in some developing nations, such as India, where the shift to larger farms and greater mechanization was highly successful.

Increased food production is the largest benefit of intensive commercial farming. In the United States, the level of crop production on farms doubled between the nineteenth and twenty-first centuries. It has been estimated that in the 1960's one U.S. farmworker produced enough fiber and food for about thirty people; by 2010, the number of people whose food needs could be met by a single U.S. farmworker had increased to more than one hundred.

ENVIRONMENTAL IMPACTS

The practices associated with industrial farming have a number of negative impacts on the environment. Running heavy machinery, producing fertilizers and pesticides, and processing, storing, marketing, selling, and transporting massive amounts of crops require a great deal of energy, most of which comes from natural gas and oil. For example, grain grown in the midwestern United States might be shipped to Michigan for milling and processing into cereal; the finished product, a breakfast cereal, thus might travel thousands of miles before it reaches a consumer's table. This intense dependence on petroleum for agricultural production has long-term negative impacts for the environment, as fossil-fuel supplies are finite and the burning of fossil fuels contributes to carbon dioxide emissions, which have been linked to global warming.

The pesticides used in intensive farming have come under scrutiny for health reasons, with many critics pointing out the danger of pesticide exposure to farmworkers and questioning how residues that may remain on crops could affect consumers of the resulting food product. In addition, pesticides and fertilizers used across vast fields are carried into water sources by irrigation and rainfall runoff. Furthermore, the soil of industrial farms is degraded over time and requires large amounts of fertilizers to rebuild nutrients; yet another impact on the soil is the erosion that can be caused by the use of heavy machinery and intensive irrigation.

The water requirements of huge industrial farms are another factor with impacts on the environment. Many dams have been built specifically to direct water resources to enable crop irrigation, and other sources have also been affected. For example, it has been projected that the deep wells being used as water sources for industrial farms in the American Midwest could eventually run low given the demands of these farms and the continual encroachment of cities and suburbs into formerly rural areas.

In recognition of the negative impacts that traditional industrial farming practices can have on the environment, some farm operators have instituted conservation methods such as minimal tillage, which reduces erosion, and integrated pest management, which reduces pesticide use. Additionally, some farms have invested in fuel-efficient heavy equipment and in technologies, such as GIS, that can increase efficiency of land management while keeping crop production high.

Another way in which industrial farms have tried to improve yields while reducing the use of pesticides is by planting genetically modified (GM) seeds. Such genetically modified organisms are often engineered for greater disease and pest resistance and for tolerance to herbicides. This method is not without controversy, however. Opponents fear that modified genes could be transmitted by pollination to nearby conventional crops and weaken the genetic diversity of the particular plant species. Others are concerned that not enough research has been conducted on the health effects of consuming GM foods.

——*M. Marian Mustoe*

FURTHER READING

Avery, Alex, and Dennis Avery. "High-Yield Conservation: More Food and Environmental Quality through Intensive Agriculture." In *Agricultural Policy and the Environment*, edited by Roger E. Meiners and Bruce Yandle. Lanham, Md.: Rowman & Littlefield, 2003.

Barrows, Geoffrey, Steven Sexton, and David Zilberman. "Agricultural Biotechnology: The Promise and Prospects of Genetically Modified Crops." *Journal of Economic Perspectives* 28, no. 1 (2014): 99–120.

Conkin, Paul K. *A Revolution Down on the Farm: The Transformation of American Agriculture Since 1929*. Lexington, Ky.: University Press of Kentucky, 2008.

Filson, Glen C., ed. *Intensive Agriculture and Sustainability: A Farming Systems Analysis*. Vancouver: University of British Columbia Press, 2004.

Gliessman, Stephen R., and Martha Rosemeyer, eds. *The Conversion to Sustainable Agriculture: Principles, Processes, and Practices*. Boca Raton, Fla.: CRC Press, 2010.

Laidlaw, Stuart. *Secret Ingredients: The Brave New World of Industrial Farming*. Toronto: McClelland & Stewart, 2004.

M

Medicinal plants

A dozen species growing in the shade. Carex (Sparkler Sedge), Fern, Viola tricolor, Coral Bells, and Strawberries. [Public domain], via Wikimedia Commons

FIELDS OF STUDY

Agricultural engineering; Botany; Biotechnology; Environmental biotechnology (or Green biotechnology); Forestry; Medical biotechnology (or Red biotechnology); Microbiology; Plant pathology

ABSTRACT

Because plants are so biochemically diverse, they produce thousands of substances commonly referred to as secondary metabolites. Many of these secondary metabolites have medicinal properties that have proven to be beneficial to humankind.

BACKGROUND

The use of plants for medicinal purposes predates recorded history. Primitive people's use of trial and error in their constant search for edible plants led them to discover plants containing substances that cause appetite suppression, stimulation, hallucinations, or other effects. Written records show that drugs such as opium have been in use for more than five thousand years.

From antiquity until fairly recent times, most physicians were also botanists or at least herbalists. Because modern commercial medicines are marketed in neat packages, most people do not realize that many of these drugs were first extracted from plants. Chemists have learned how to synthesize many natural products that were initially identified in a plant. However, in many cases a plant is still the only economically feasible source of the drug.

ANTIBACTERIAL AND ANTI-INFLAMMATORY AGENTS

The first effective antibacterial substance was carbolic acid, but the first truly plant-derived antibacterial drug was *penicillin*, which was extracted from a very primitive plant, the fungus *Penicillium*, in 1928. The success of penicillin led to the discovery of other fungal and bacterial compounds that have antibacterial activity. The most notable of these are cephalosporin and griseofulvin.

Inflammation can be caused by mechanical or chemical damage, radiation, or foreign organisms. For centuries poultices of leaves from coriander (*Coriandrum sativum*), thornapple (*Datura stramonium*), wintergreen (*Gaultheria procumbens*), witch hazel (*Hamamelis virginiana*), and willow (*Salix niger*) were used to treat localized inflammation. In the seventeenth and eighteenth centuries, cinchona bark was used as a source of quinine, which could be taken internally. In 1876 salicylic acid was obtained from the salicin produced by willow (*Salix*) leaves. Today, salicylic acid, also known as aspirin, and its derivatives, such as ibuprofen, are the most widely used anti-inflammatory drugs in the world.

DRUGS AFFECTING THE REPRODUCTIVE SYSTEM

A home remedy for preventing pregnancy was a tea made from the leaves of the Mexican plant zoapatle

(*Montana tomentosa*). The drug zoapatanol and its derivatives were extracted from this plant to produce the first effective birth control substance. It has not been used in human trials, however, because of potential harmful side effects. Other plant compounds that affect the reproductive system include diosgenin, extracted from *Dioscorea* species and used as a precursor for the progesterone used in birth control pills; gossypol from cotton (*Gossypium* species), which has been shown to be an effective birth control agent for males; ergometrine, extracted from the ergot fungus (*Claviceps*) and used to control postpartum bleeding; and yohimbine, from the African tree *Corynanthe yohimbe*, which apparently has some effect as an aphrodisiac.

CIRCULATORY, ANALGESIC, AND CANCER-FIGHTING DRUGS

Through the ages, dogbane (*Apocynum cannabinum*) and milkweed (*Asclepias*) have been prized for their effects on the circulatory system. These plants contain compounds called cardiac glycosides. Foxglove (*Digitalis*) has produced the most useful cardiac glycosides, digitalis and digoxin.

Opiate alkaloids such as opium, extracted from a poppy (*Papaver sonniferum*), and its derivatives, such as morphine as well as cocaine, from *Erythroxylum coca* and *Erythroxylum truxillense*, have long been known for their *analgesic* (pain-relieving) properties through their extremely dangerous and addictive effects on the central nervous system.

The primary plant-derived anticancer agents are vincristine and vinblastine, extracted from *Catheranthus roseus*, maytansinoids from *Maytentus serrata*, ellipticine and related compounds from *Ochrosia elliptica*, and paclitaxel (commonly known as taxol) from the yew tree *Taxus baccata*.

FIGHTING ASTHMA, GASTROINTESTINAL DISORDERS, AND PARASITES

The major anti-asthma drugs come from ephedrine, extracted from the ma huang plant (*Ephedra sinaica*), and its structural derivatives. Plant-derived drugs that affect the gastrointestinal tract include castor oil, senna, and aloes as laxatives, opiate alkaloids as antidiarrheals, and ipecac from *Cephaelis acuminata* as an emetic. The most useful plant-derived antiparasitic agent is quinine, derived from the bark of the chincona plant (*Chincona succirubra*). Quinine has been used to control malaria, a disease that has plagued humankind for centuries.

COMPLEMENTARY AND ALTERNATIVE MEDICINES

There are medicinal philosophies outside of Western medicine that rely heavily on plants and herbal comounds to treat illness. Examples are Traditional Chinese Medicine (TCM) and Ayurvedic Medicine, both of which are thousands of years old and tied, at least historically, to religious practice. Plants and plant roots such as ginger, turmeric, and ginseng have been used as supplements to maintain health or to treat ailments, such as nausea.

THE FUTURE

More plant-derived medicines await discovery, many from tropical rain-forest vegetation. Biotechnology has provided methods by which plants can be genetically modified to produce novel pharmaceuticals. Progress toward the production of specific proteins in transgenic plants provides opportunities to produce large quantities of complex pharmaceuticals and other valuable products in traditional farm environments rather than in laboratories. These novel strategies open up routes for production of a broad array of natural or nature-based products, ranging from foodstuffs with enhanced nutritive value to biopharmaceuticals.

—*D. R. Gossett*

FURTHER READING

Cutler, Stephen J., and Horace G. Cutler, eds. *Biologically Active Natural Products: Pharmaceuticals*. Boca Raton, Fla.: CRC Press, 2000. Demonstrates the connections between agrochemicals and pharmaceuticals and explores the uses of plants and plant products in the formulation and development of pharmaceuticals.

Fetrow, Charles W. *The Complete Guide to Herbal Medicines*. Springhouse, Pa.: Springhouse, 2000. Accessible information available on more than three hundred herbal medicines.

Herrick, James W., and Dean R. Snow, eds. *Iroquois Medical Botany*. Syracuse, N.Y.: Syracuse University Press, 1997. A fascinating look at one Native American body of knowledge of herbal medicines. Important not only for those interested in herbal medicine but also for those studying American Indian cosmology as it relates to material culture. Illustrated, with references, index.

Lewis, Walter H., and Memory P. F. Elvin-Lewis. *Medical Botany: Plants Affecting Human Health*. 2d ed. Hoboken, N.J.: J. Wiley, 2003. An excellent in-depth study of plants and the medicines they produce, examining plants' effects on human health as injurious, remedial, or psychoactive. Includes bacteria, fungi, and seaweeds as well as flowering plants.

Mann, J. *Murder, Magic, and Medicine*. Rev. ed. New York: Oxford University Press, 2000. An interesting and readable book on the use of natural plant products for medicinal purposes.

Sneader, Walter. *The Evolution of Modern Medicines*. New York: Wiley, 1986. Provides excellent coverage of how plants contributed to the development of many pharmaceuticals.

Stannard, Jerry, Katherine E. Stannard, and Richard Kay, eds. *Pristina Medicamenta: Ancient and Medieval Medical Botany*. Brookfield, Vt.: Ashgate, 1999. Articles on premodern texts on plants.

Stockwell, Christine. *Nature's Pharmacy: A History of Plants and Healing*. London: Century, 1989. An excellent discussion of medicinal products from plants.

Sumner, Judith, and Mark J. Plotkin. *The Natural History of Medicinal Plants*. Portland, Oreg.: Timber Press, 2000. An accessible introduction to the world of medicinal plants by a Harvard University botanist, from Europe in the Middle Ages to the modern pharmacopeia.

Trease, G. E., and W. C. Evans. *Trease and Evans' Pharmacognosy*. 15th ed. W. B. Saunders, 2002. One of the most complete treatises on the production of drugs from plants. At nearly 600 pages, covers all scientific aspects of the topic, from taxonomy, cellular biology, and phytochemistry through genetics. Drugs are examined in chapters that group them by chemical class. The scope is broad, including vitamins and hormones and even alternative therapies such as homeopathic medicine and aromatherapy. Professionals will appreciate the chapters on investigative methodologies. Appendices, index.

Walter, Lynne Paige, and Ellen Hodgson Brown. *Nature's Pharmacy: Break the Drug Cycle with Safe, Natural Treatments for Two Hundred Everyday Ailments*. Upper Saddle River, N.J.: Prentice Hall, 1999. Typical of a wave of similar publications that began to appear after the 1996 law that removed "food supplements" from FDA regulatory responsibility, this 400-plus-page reference catalogs common ailments, from acne to whooping cough, offering signs, symptoms, and suggestions for alternative treatments in addition to traditional Western medicine.

METABOLIC ENGINEERING

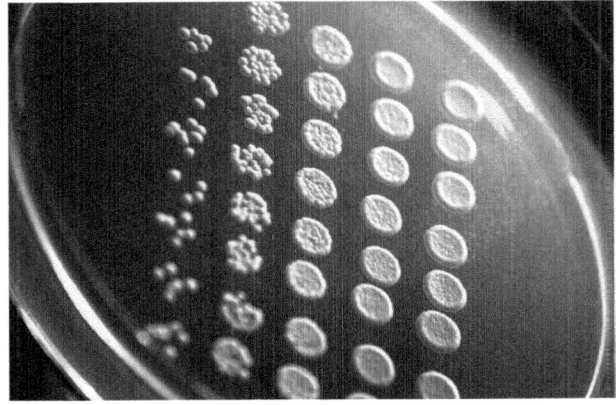

Yeast colonies on an agar plate. By Rainis Venta. (Own work) [CC BY-SA 3.0 (https://creativecommons.org/licenses/by-sa/3.0)], via Wikimedia Commons

FIELDS OF STUDY

Biochemistry; Biotechnology; Food science; Genetics; Human nutrition; Microbiology

ABSTRACT

Metabolic engineering is a new science that appeared in the 1990's. It is associated with biology and chemistry. Metabolic engineering allows the designing of biochemical pathways that do not exist in the natural world, as well as the redesign of existing biochemical pathways often with the use of genetic engineering. Metabolic engineers often modify biochemical pathways by reducing cellular energy use or

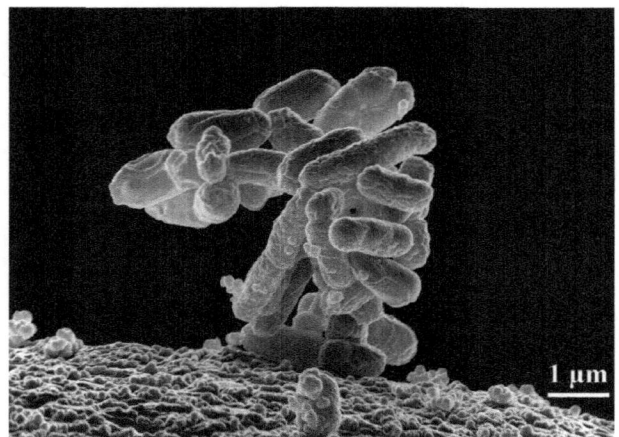

Low-temperature electron micrograph of a cluster of *E. coli* bacteria, magnified 10,000 times. Each individual bacterium is oblong shaped. By Photo by Eric Erbe, digital colorization by Christopher Pooley, both of USDA, ARS, EMU. [Public domain], via Wikimedia Commons

waste production, by changing the nutrient flow to the cells, or improving the productivity and yield of a particular pathway. In addition, metabolic engineers may potentially design new organisms that are tailor-made for the desired chemicals and production processes. Many novel compounds of industrial and medical interest can be produced by metabolic engineering. In the twenty-first century, the main efforts of metabolic engineers are concentrated on biofuels and pharmaceuticals.

DEFINITION AND BASIC PRINCIPLES

Metabolic engineering is a relatively new field that deals with the modification and optimization of metabolic pathways, mainly in microorganisms, by altering genes, nutrient uptake, or metabolic flow to allow production of novel compounds that are of industrial and medical interest. Metabolic pathways of living organisms are not optimal for specific practical applications, but they can be modified using the tools of modern biotechnology such as genetic engineering. The redesign of existing, natural metabolic pathways for useful purposes is a main objective of metabolic engineering. Metabolic engineering usually includes two phases: careful analysis of the metabolic pathway and genes involved in the pathway (analytical phase) and its modification (synthesis phase). Pathway analysis often includes the metabolic control analysis: determining which compounds can control the productivity and yield of particular pathway. Different tasks of metabolic engineering are as follows: improvements of productivity and yield of particular pathway; expansion of substrate range; elimination of waste; improvement of process performance; improvements of cellular activities; and extension of product array. Metabolic engineering is becoming one of the principal fields of biotechnology.

Production of many chemicals and fuels uses non-renewable resources or limited natural resources. Metabolic engineering creates many alternatives to replace dangerous chemicals and petroleum-based transportation fuels with clean, green, and renewable chemicals and biofuels.

BACKGROUND

The term "metabolic engineering" first appeared in the early 1990's. Since that time, the range of products that can be generated has increased significantly, partly because of remarkable advances in other fields related to metabolic engineering, such as DNA sequencing and genetic engineering. With DNA sequencing, scientists were able to identify the majority of metabolic genes and enzymes in many organisms. In the post-sequencing era, the obtained information is used for practical construction of biochemical pathways or whole organisms with optimized functions through metabolic engineering.

In the 1990's, scientists developed new genetic tools that gave metabolic engineers more precise control over metabolic pathways. They also created analytical tools that allowed the metabolic engineer to track metabolites in a cell to identify new biochemical pathways more precisely.

Earlier in the twenty-first century, metabolic engineers joined other scientists in their quest for alternative fuels, which are in high demand because of increasing oil prices and concern about climate change.

HOW IT WORKS

Metabolic engineering is based mainly on microbial metabolism. Microbes produce different kinds of substances that they use for the growth and maintenance of their cells. These substances can be useful for humans. The goal of metabolic engineering is to enhance the microbial production of useful substances. To achieve this goal, metabolic engineers must follow a particular route. They need to choose a friendly organism (host) for their metabolic manipulations. They need to find cheap and available substrates to use for modified metabolic pathways. Finally, metabolic engineers must be able to perform genetic manipulations of metabolic routes. Metabolic engineers can also alter nutrient uptake or metabolic flow. All these steps are dependent on each other. For example, genes cannot be manipulated in every organism; products or metabolic intermediates may be toxic to its host.

Host and Host Design. Generation of products by metabolic engineering has been achieved by transferring product-specific enzymes or entire metabolic pathways into so-called user-friendly microorganism hosts, which were used traditionally in industry. These industrial microorganisms grow rapidly on inexpensive culture media available in bulk quantities, are open to genetic manipulation (and genetic manipulation tools are available), and are nonpathogenic (do not cause disease). In addition, it is important that the host can survive (and thrive) under the desired process conditions (ambient versus extremes of temperature, pH). It is essential that the host is genetically stable (with the introduced pathway) and not susceptible to virus or another microbe's attack. Among the host microorganisms most widely used are *Saccharomyces cerevisiae* and *Escherichia coli*. *Saccharomyces cerevisiae*, or baker's yeast, has been used for making bread and alcohol for thousands of years. It is one of the earliest domesticated organisms. This organism has come to be used in a large number of different processes within the biotechnological and pharmaceutical industries. Comprehensive knowledge of *S. cerevisiae* has been accumulated over a long period of time. In addition, the complete genome sequence of yeast is available, and yeast is nonpathogenic. The well-established fermentation and process technology for large-scale production with *S. cerevisiae* in bioreactors makes this organism very attractive for several industrial purposes.

Escherichia coli, commonly known is *E. coli*, is a bacterium that is widely used as a research (model) organism. It is easy to grow and genetically manipulate this bacterium, and its genome sequence is available. Several important products such as interferon (flu-fighting drug), insulin, and growth hormone are manufactured by genetically modified *E. coli*.

In addition to *E. coli* and *S. cerevisiae*, several other microorganisms are widely used as hosts for metabolic engineering manipulations, including bacteria *Bacillus subtilis* and *Streptomyces coelicolor*.

Finally, in addition to redesigning particular metabolic processes, metabolic engineers may also design de novo artificial cells that will produce desired products.

Substrates. To make metabolically engineered products, chemical substrates are needed. To make these products economically viable, inexpensive sources of substrates are required. Substrates must contain different chemical components, such as carbon, nitrogen, oxygen, and hydrogen. For example, metabolic engineers are looking at sugars from cellulosic biomass as potential substrates for biofuel production. Cellulosic biomass is a very attractive biofuel feedstock because of its abundant supply. On a global scale, plants produce almost 100 billion tons of cellulose per year, making it the most abundant organic compound on Earth.

Genetic Manipulation of Metabolic Routes. Genetic manipulation of metabolic pathways by adding or deleting genes or modifying the expression of existing genes in the host can serve several useful purposes. It can extend the existing pathways or shifting metabolic route into a desired pathway or increase the rate-determined step of the particular metabolic route. Adding genes into the host consists of the following steps.

- The gene the for desired pathway is obtained from the non-host organism.
- The gene is inserted into the host cell.
- Host cells are induced to express (to cause the gene to manifest its effects) this "foreign" gene in order to produce the desired product.

One example of how gene manipulation is used in areas relevant to metabolic engineering is as follows: In the mold *Aspergillus terreus*, the producer of cholesterol-lowering drug lovastatin, genes were modified to increase their expression levels in order to change its metabolism in terms of drug production.

Another example is the introduction of bovine lactic acid pathway into *S. cerevisiae*. As a part of this, a gene responsible for speeding up removal of hydrogen, which participates in lactic acid production, was expressed in *S. cerevisiae*, and lactic acid was produced at rate of eleven grams per liter per hour. Because it tolerates acid, yeast may serve as an alternative to bacteria, which is usually used in industry for lactic acid production. Lactic acid is widely used as a food preservative.

Altering Nutrient Uptake or Metabolic Flow. Alteration of nutrient uptake or metabolic flow can be done not only by genetic manipulation but also by using inhibitors—simple chemicals or physical factors such as light or temperature.

The alteration of molecular hydrogen (H_2) production in green algae using high-intensity light is an example of metabolic flow modification by physical factors. H_2 is one of the possible energy carriers of the future. Microscopic green algae produce H_2 in photosynthetic reactions from water using sunlight as an energy source, usually in anoxic (without oxygen) conditions. Oxygen (O_2) produced by photosynthesis in green algae is an inhibitor of H_2 production. Brief illumination of algal cells by high-intensity light was accompanied by rapid suppression of photosynthetic O_2 evolution. The decline in the rate of O_2 evolution was accompanied by stimulation of H_2 production in algal cells.

Production Systems. All of the above-mentioned considerations are very important in metabolic engineering, although it is also important to ensure that the production of desired compounds by modified cells can be reproduced. This can be achieved by using bioreactors, in which the important parameters such as pH, temperature, substrate supply, and other variables are controlled. It is even possible to modify cell metabolism by using bioreactors.

APPLICATIONS AND PRODUCTS

There are a wide range of metabolic engineering products and applications. Undoubtedly, a number of novel applications and products will arise in the future.

Pharmaceuticals. Metabolic engineering is most promising in the production of pharmaceuticals. These include pharmaceuticals from different classes of natural products: alkaloids, isoprenoids, and flavonoids. Biosynthesis of natural products is an emerging area of metabolic engineering that offers significant advantages over conventional chemical methods. Some pharmaceutical compounds are too complex to be chemically synthesized or extracted from biomass organisms inexpensively.

Alkaloids are mainly plant-derived compounds that have been used as drugs such as morphine. Alkaloids are produced by simple extraction from plants. Studies show that alkaloids can be synthesized from amino acids by metabolic engineering in *E. coli* and *S. cerevisiae*.

Isoprenoids, organic compounds composed of two or more hydrocarbons, have a range of functions: pigments, fragrances, and vitamins. Isoprenoids are also the precursors to sex hormones. Many isoprenoids have been produced using microorganisms, including carotenoids and various plant-derived terpenes. Metabolic engineers are using *S. cerevisiae* as a cell factory for the biosynthesis of isoprenoids. One metabolic-engineering success is the production of Taxol, which is used to treat breast cancer. It is an isoprenoid that was first isolated in the bark of the Pacific yew (*Taxus brevifolia*). The demand for Taxol greatly exceeds the supply that can be obtained from its natural source. A partial Taxol biosynthetic pathway has been engineered in *S. cerevisiae*.

Another metabolic engineering success is the production of isoprenoids-carotenoids. Carotenoids are naturally occurring yellow, orange, and red pigments commonly found in plants such as carrots as well as in bacteria, algae, and fungi and play an important role in fighting disease. Metabolic engineers have successfully introduced carotenoid biochemical pathways into nonproducing carotenoid microbes such as *E. coli* and *S. cerevisiae*.

Flavonoids are a group of secondary plant metabolites. These compounds can be used as antioxidants or antiviral, antibacterial, and anticancer drugs. Many flavonoid biosynthetic pathways are known, and a wide array of flavonoid compounds from *S. cerevisiae* are expected to be produced by metabolic engineering in the near future.

Chemicals. Numerous chemicals, such as amino acids, organic acids, vitamins, flavors, fragrances, and nutraceuticals can be manufactured by metabolic engineering.

Glycerol (or glycerin) is a chemical produced by metabolic engineering. Glycerol is used to synthesize many products, ranging from cosmetics to lubricants. It is a by-product of soap or biodiesel manufacturing and its production is 1.2 billion liters annually. It can be also used a fuel. Metabolically engineered *S. cerevisiae* strain produced more than 200 grams of glycerol per liter of liquid medium.

Another example of chemicals produced with help of metabolic engineering are sterols. The most well-known sterol is cholesterol. Sterols are important for living organisms as they are a part of the cellular membrane, participate in the synthesis of several hormones, and are also nutrient supplements. Several sterols are being produced from metabolically engineered *S. cerevisiae*.

Fuels. Metabolic engineering can be used in the production of biofuels. Several scientific laboratories have demonstrated the feasibility of manipulating microorganisms to produce molecules similar to oil-derived products, although the yield is very low. Adjusting metabolic pathways of microbes to produce fuels similar to gasoline has the potential to save an enormous amount of money. These fuels can be used in existing engines, unlike other biofuels that require modified engines or fueling stations.

Several research groups are trying to metabolically engineer microorganisms to produce ethanol fuel using cellulose as substrate. Another example of the work of metabolic engineers is biodiesel production. Biodiesel is a diesel substitute primarily obtained from vegetable oils such as soybean. However, the production of this fuel is limited by the absence of sufficient vegetable oil feedstocks. Another problem is that in order to produce biodiesel, oils should be modified by transesterification, a chemical reaction with methanol, catalyzed by acids or bases (such as sodium hydroxide). *E. coli* has been metabolically engineered to produce biodiesel directly, using low-cost materials.

SOCIAL CONTEXT AND FUTURE PROSPECTS

Though the redesign of life forms for the benefit of mankind is definitely an exciting career, metabolic engineers are paying particular attention to ethical, legal, and political issues. To continue in this work, the field as a whole will need sustained support from the public and government.

At present, metabolic engineering is more a collection of successful experiments than an established science. In the future, metabolic engineering may play a significant role in production of chemicals and fuels from inexpensive and renewable starting materials. Continued development of the techniques of metabolic engineering will be necessary to expand the range of products. The role of metabolic engineering in science is likely to expand in the future as a result of increasing needs for pharmaceuticals and biofuels.

—*Sergei A. Markov, PhD*

FURTHER READING

Bailey, James E., and David F. Ollis. *Biochemical Engineering Fundamentals.* 2d ed. New York: McGraw-Hill, 1986. Classic textbook on biochemical engineering.

Bourgaize, David, Thomas R. Jewell, and Rodolfo G. Buiser. *Biotechnology: Demystifying the Concepts.* San Francisco: Benjamin Cummings, 2000. Excellent introduction to biotechnology.

Lewin, Benjamin. *Genes VIII.* San Francisco: Benjamin Cummings, 2003. In-depth look at genes and molecular biology.

Madigan, Michael T., et al. *Brock Biology of Microorganisms.* 12th ed. San Francisco: Benjamin Cummings, 2008. Several chapters of this popular textbook describe microbial metabolism and the application of microorganisms in industry.

Marguet, Philippe, et al. "Biology by Design: Reduction and Synthesis of Cellular Components and Behavior." *Journal of the Royal Society Interface* 4, no. 15 (2007): 607–623. Review on metabolic engineering and synthetic biology written for the general public.

Ostergaard, Simon, Lisbeth Olsson, and Jens Nielsen. "Metabolic Engineering of *Saccharomyces cerevisiae*." *Microbiology and Molecular Biology Reviews* 64, no. 1 (2000): 34–50. Describes metabolic engineering techniques using *S. cerevisiae* as an example.

Stephanopoulos, Gregory N., Aristos A. Aristidou, and Jens Nielsen. *Metabolic Engineering: Principles and Methodologies.* San Diego: Academic Press, 1998. Classic text on metabolic engineering.

Microscopy

FIELDS OF STUDY

Bioengineering; Biophysics; Chemical engineering; Molecular biology

ABSTRACT

Microscopy is the science of creating, observing, analyzing, and capturing visible images of objects and their components that are too small to be seen by the naked eye. It also refers to research conducted with the aid of microscopes, or instruments used for visual magnification. Microscopy is an essential tool for conducting research in a large number of scientific disciplines, including chemistry, biology, and medicine. For example, microscopy enables biologists to examine, in fine detail, the structure and function of individual components of a cell. The field also has a variety of industrial, materials science, and other practical applications. Powerful microscopes are used, for instance, to inspect the composition of the tiny silicon crystals used to manufacture semiconductors and integrated circuits and to detect minute defects in glass.

DEFINITION AND BASIC PRINCIPLES

Microscopes are scientific instruments whose purpose is to create enlarged visual images of objects so tiny they cannot be seen by the unaided human eye. Microscopy is an applied science concerned with ways of developing and improving microscope technology and relies heavily on knowledge gained from physics, mathematics, and engineering. Different varieties of microscopes function in various ways, but it is useful to understand two basic principles that apply to how well a microscope performs.

First, it might seem that the fundamental purpose of a microscope is to magnify an object. In reality, however, photographic enlargements can always be used to further enlarge the image any given microscope creates. Therefore, for a microscope to be truly useful to scientists and other researchers, it must not only magnify the specimen being observed but also properly separate (or "resolve") the details of individual components within the image. In effect, the more resolving power a microscope has, the crisper and clearer the magnified image it can produce and the more information it can provide about the object in question. If the resolution of a microscope is not high enough, for instance, two tiny dots next to each other might be perceived as a single element, no matter how much the image of the specimen was magnified.

Second, the image a microscope creates must possess a high enough degree of contrast to allow the viewer to clearly distinguish the object from its background and to differentiate various details within the object. A mostly translucent specimen, for example, might be impossible to make out against a bright background. Special techniques are used to increase the contrast in microscopic images. Phase contrast microscopy takes advantage of differences in the refractive indexes (the extent to which a material bends light) of various components of the specimen. Microscopes using this technology translate these variations into differences in the amplitude of the light waves reflected from each component. This results in light and dark areas that can be seen by the viewer. Interferometry is another important technique used to increase contrast. It does so by creating two images of a single specimen, superimposed on top of each other.

BACKGROUND

The first ground glass lenses that had the ability to magnify objects were created in the late Middle Ages by monks who used them for reading. In the sixteenth century, the first microscopes were created in the Netherlands by inventors Hans Janssen and Zacharias Janssen, who placed two lenses into a series of tubes to create a primitive compound microscope. The device was focused by drawing one of the tubes in and out of the other, and it had a magnifying power of about 10 times (10x). The seventeenth century saw a flurry of interest in microscope technology. The Dutch amateur scientist Antoni van Leeuwenhoek designed hundreds of microscopes in which a bi-convex lens was placed between two glass plates. He was the first person to ever observe bacteria and other single-celled organisms under a microscope. In the eighteenth century, cuff-style microscopes were created, whose design prefigured the modern laboratory optical microscope. These were instruments with two brass tubes, one fixed and one sliding. By sliding the assembly up and down and turning a small thumbscrew, the object under observation could be brought into fine focus. The whole mechanism was mounted on a solid wooden base and had a magnifying power of up to 100x.

In the nineteenth century, advances in optical science and lens production pushed microscope technology forward by leaps and bounds. Two people whose work was prominent in this era were the German physicists and engineers Ernst Abbe and Carl Zeiss; together they designed sophisticated lenses that cut down drastically on spherical aberration (which causes points of lights to look like discs) and chromatic aberration (which distorts the colors of objects). The twentieth and twenty-first centuries have seen major changes in the field of microscopy, with the development of new technologies such as electron and scanning probe microscopes and huge improvements in the design of optical microscopy. Digital imaging—the transformation of an image created in an optical microscope to digital form—is making it easier than ever to analyze magnified specimens.

HOW IT WORKS

Since the 1800's, hundreds of different varieties of microscopic technologies have been developed, each useful for performing certain types of observations on specific kinds of materials. Three broad categories of microscopy exist into which the majority of these technologies can be categorized.

Optical Microscopy. Optical microscopes, also known as light microscopes, create a magnified image by using a series of glass lenses to manipulate visible light, or light from the portion of the electromagnetic spectrum that can be seen by the naked eye. A condenser focuses light onto the specimen to be observed. As this light passes through or is reflected off the specimen, it is collected by one or more objectives. (Most microscopes have multiple objectives contained in a long tube with magnification powers ranging from 4x to 100x.) The objectives focus the light they have gathered into parallel rays; the result is a magnified image of the specimen. This image, however, is projected to a distance of infinity. To focus the image at a distance comfortable for the human eye, an ocular lens is required, through which the viewer looks. Most ocular lenses further magnify the image by another 10x.

In a conventional bright-field optical microscope, the source of this light is usually an incandescent or halogen lightbulb positioned directly below the specimen to be examined. Bright-field microscopy is useful for observing specimens that are either naturally dark or can be stained a dark color—such as cells or thin cross sections of biological material, usually placed on a glass slide. Images produced by a bright-field microscope appear dark against a bright white background.

Some specimens, such as living organisms, are difficult to see under bright-field microscopes. Other forms of optical microscopy, such as dark-field microscopy, have been developed to combat this problem. In dark-field microscopy, opaque material inside the condenser blocks the most central source of light, causing light to hit the specimen at oblique angles. When this angled light hits even the tiniest particle, the light scatters and makes the particle visible—like dust motes catching the angled light coming in from a window. Dark-field microscopes show specimens as bright points of light against a dark background.

Fluorescence microscopy is a special form of optical microscopy in which the specimen itself acts as the source of light. A fluorescence microscope irradiates the specimen with light of a certain wavelength, causing its atoms to become excited and emit energy as visible light. Some specimens, like chlorophyll, fluoresce naturally; others can be made to fluoresce through the use of chemicals. Other special forms of optical microscopy include phase contrast microscopy, in which small changes in the wavelength of light as it passes through transparent regions of the specimen are intensified so that they show up as areas of greater brightness, and confocal microscopy, in which light coming from out-of-focus regions of the specimen is filtered out of the final image through a pinhole aperture, eliminating blurry regions in the image.

Electron Microscopy. Electron microscopes operate using the same basic principles as optical microscopes—with one important difference. Where optical microscopy manipulates focused beams of visible light, electron microscopy manipulates focused beams of highly excited electrons, which have wavelengths much shorter than those of visible light. This technique enables objects to be magnified at far higher levels and resolved in far finer detail than optical microscopy. In addition, electron microscopes have larger depths of field, allowing a larger area of an object to be in focus at one time. The source of the electrons in an electron microscope is most often a thermionic electron gun—a device that shoots out a stream of electrons produced by heating a charged electrode. The path these electrons take is shaped by a series of lenses, just as in an optical microscope. However, rather than being made out of glass, the lenses in an electron microscope consist of coils of wire (solenoids). An electric current passed through a solenoid creates an electromagnetic field that can direct the flow of electrons and focus it into a thin beam that can be directed toward the object under study.

In a transmission electron microscope (TEM), the beam of electrons enters the specimen and passes through it. When electrons hit dense regions of the specimen, they bounce off and are not included in the resulting image. The remaining electrons travel through the object and then pass through more electromagnetic lenses that create a final, magnified image on a fluorescent screen. In a scanning electron microscope (SEM), the beam of electrons sweeps across the surface of the specimen in a back-and-forth pattern of parallel lines known as a raster. As the beam scans over the object, it causes atoms within it to become excited and emit electrons that escape from the object. These electrons, known as deflected secondary electrons, are collected, counted, and measured. The information from this analysis is then used to create a magnified pixelized image on a computer screen. Because the electron beam sweeps over the entire surface of the object under study, electron microscopes are able to produce an image of the specimen's structure in three dimensions. Electron microscopes generally require samples to be placed in a vacuum in order to operate because molecules of air might disturb the movement of the electrons used to form images.

Scanning Probe Microscopy. Scanning probe microscopy abandons lenses altogether and makes use of very fine mechanical tips, or probes, attached to a cantilever. The probes delicately scan back and forth over the surface of the specimen being studied in order to inspect it. Scanning probe microscopes can deliver information about not only the topography of an object but also its internal properties. Some can even map a specimen's properties on a nanoscale.

Scanning tunneling microscopes (STM) rely on a phenomenon discovered by quantum mechanics, in which electrons—which have wavelike properties—are able to "tunnel" outside of the surface of a solid object into surrounding space. Scanning tunneling microscopes have incredibly sharp metallic probes, often made of tungsten or an alloy of platinum and iridium, with tips that are a mere one or two atoms in size. The tip of the probe does not touch the surface of the specimen but is held very close to it. An electric current of low voltage is applied to the gap between the two. In response to the current, electrons from the object tunnel across the gap. Changes in the intensity of the tunneling electrons are analyzed to produce an image of the object that can then be magnified. Scanning tunneling microscopes can be used only to examine specimens that conduct electricity.

Atomic force microscopes (AFM), whose tips are typically made of silicon or diamond, can probe surfaces made of practically any material. As the tip of an atomic force microscope is dragged across the surface of a specimen, it is either deflected by or drawn toward the object, depending on whether the atoms in the object are repelled by or attracted to the microscope's tip. By measuring these forces, a magnified representation of the physical structure of the sample can be created. By changing the modes in which these microscopes operate, different properties such as magnetism, friction, and electrical conductivity can be assessed.

Near-field scanning optical microscopes (NSOM) use a probe that emits an incredibly fine beam of laser light very close to a specimen. These microscopes use the intensity of the reflected light to produce a magnified topographical image of the object. Scanning probe acoustic microscopes (SPAM) direct a focused, ultrasonic (high-pitched) sound wave toward the specimen being observed and form a magnified image of it based on how and how much the wave is reflected by the object's surface.

APPLICATIONS AND PRODUCTS

Scientific Research. At heart, all scientific research rests on the power of observation. Microscopes make it possible for scientists from fields such as biology, chemistry, metallurgy, mineralogy, and countless other disciplines to make more accurate and more complete observations of microscopic structures. Biologists, for example, use fluorescent microscopes to analyze the structure and function of minute intracellular organelles such as ribosomes, mitochondria, and even single strands of DNA. Using the microscope, researchers have been able to watch as individual cells undergo mitosis and viruses invade healthy cells to spread their own genetic material and also to identify the precise manner in which different kinds of proteins are folded. Analytical chemists and physicists use microscopes to conduct research at the scale of the atom or even on a nanoscale. For instance, researchers use scanning tunneling microscopy and atomic force microscopy to observe how peptides—organic compounds composed of two or more amino acids—interact with carbon and graphite nanotubes.

Electron microscopes are being used by botanical researchers to examine how leaves protect themselves from insects by forming crystals inside themselves, by ornithologists to figure out how minute structures on the surface of certain bird feathers create an iridescent effect, and by geoscientists to identify the weather-induced changes to geological features such as rocks. The field is even enabling complex scientific research to take place on other planets. Robotic space vehicles such as the Mars Exploration Rover are often equipped with autonomously operated scanning probe microscopes capable of studying the properties associated with the surfaces encountered by the vehicle. In all these applications, microscopy allows investigators to transcend the limitations of the human sense of sight and expand people's scientific understanding of the world.

Microscopy has a place in many applied sciences as well. It is a useful tool in food science, where it has provided a better understanding of the chemical properties of foods and how processing them alters their natural properties and also has helped isolate food contaminants. Microscopes enable materials scientists to analyze the three-dimensional structure of the plastics and polymers they are developing. Mechanical engineers use microscopes to develop sharper and more sophisticated edges on tools used for cutting.

Microscopy is indispensable in forensic science, where it is used to examine crime scene evidence such as blood, hair, dust, fingerprints, tiny shards of glass, and threads of fiber. Criminologists use high-powered microscopes to help them study the minute hand motions that were used to construct signatures on suspicious documents, looking for frequent stops and starts or other signs of possible forgery. Counterfeit currency makers are often foiled by microscopes, which help scientists detect very subtle discrepancies in the color and texture of the paper fibers used to manufacture counterfeit bills.

Medical Applications. Virtually all biomedical and bioengineering research projects make use of microscopy at some point. Scientists in the pharmaceutical industry, for example, need to closely examine the physical structure and dispersion characteristics of the active components (chemicals) used in the development of drugs, as well as the materials used to coat medical devices such as pacemakers or other implants. Fluorescence microscopy and confocal microscopy are commonly used for these purposes.

Besides their use in preclinical biomedical research, microscopes play a role in at least two other important areas of medical practice: diagnosis and surgery. In diagnosis, microscopes help physicians and laboratory technicians detect whether cells in a patient's tissue samples show signs of disease. When a female patient undergoes a Pap smear, for example, cells are taken from her cervix and analyzed under an optical or electron microscope. If the sample is cancerous or precancerous, a microscopic examination will show changes in the cervical cells that make them look flat or scaly. Microscopes are also used to detect the presence of pathogens in tissue samples. For example, blood cells from patients who have been infected by the malaria virus may appear enlarged or stippled; the malaria parasites themselves will also be visible under magnification of about 100x.

Surgery with the use of an operating microscope (sometimes called microsurgery) has become common in nearly all surgical fields, but it is vital for performing many brain, eye, and ear surgeries. Magnifying the sometimes minute biological structures involved in a procedure can help a surgeon perform delicate tasks that were practically impossible before the age of microscopes. One particularly significant microsurgery application, for example, is the ability to reattach limbs, fingers, or toes that have been severed from a patient's body. By magnifying the individual nerve fibers, blood vessels, and tendons both in the severed part and at the site of separation, a surgeon can connect them one by one.

The typical microsurgery setup involves a surgeon looking at an operating site through a microscope (or sometimes a television screen connected to the microscope) rather than facing the site directly. Many microsurgery procedures are minimally invasive, making use of instruments inserted through small cuts, or ports, in a patient's skin. Often, robotic instrumentation is used in conjunction with microsurgical tools to track a surgeon's hand motions and correct for tiny tremors in his or her movements. Using a combination of microscopes, remotely controlled tools, and large video screens, surgeons can conduct coronary artery bypasses without ever opening up a patient's chest.

Nanotechnology. Nanotechnology is an example of a scientific discipline whose very existence simply would not be possible without the use of extremely powerful microscopes. This emerging field takes advantage of the special ways in which molecules behave at the nanoscale to create nanoscale machinery such as tiny sensors that can detect and tally the number of specific types of molecules in a sample of chemicals or nanoparticles that systematically seek out and destroy cancerous cells within a patient's body. Optical microscopes do not have the magnifying power necessary to clearly resolve objects at the nanoscale (about 1^{-100} nanometers, or about one-thousandth the width of a strand of human hair), so electron microscopes and scanning probe microscopes serve as the foundational tools of nanoscientists.

Atomic force microscopes are particularly important in nanotechnology for several reasons. Like electron microscopes, they are capable of imaging structures that are incredibly small (including single atoms or molecules). Unlike electron microscopes, they do not have to operate in a vacuum, giving scientists the ability to work with a greater variety of samples, such as living biological cells. Most significantly, researchers can use the probes attached to atomic force microscopes not just for observing specimens but also for actually manipulating them.

Although most nanotechnology applications are still in the research and development stage, nanotechnology is already causing a transformation in manufacturing. With the help of microscopes that can characterize the behavior of nanoscale structures, scientists have created, for example, grease- and mildew-resistant paints, bacteria-killing storage containers, and nanoscale drug-delivery systems that introduce drugs directly into cells affected by disease.

Microscopy and Art. The applications of microscopy stretch far beyond the boundaries of science. Art historians make extensive use of microscopes to study paintings, sculptures, and other works of art in minute detail, using them to uncover insights about materials, artistic techniques, and what a piece has been through over the course of its history. For example, they often conduct microscopic analyses of the chemical and structural properties of the specific pigments used to create a painting. Museums also use microscopic inspections to authenticate artworks and accurately date them. Using an optical microscope, art historians in Belgium were able to detect minute quantities of a cobalt blue pigment mixed in with an ultramarine pigment in a painting that had been attributed to the Dutch master painter Jan Vermeer. However, Vermeer lived and worked in the seventeenth century, and cobalt blue pigment was not developed until the nineteenth century. Without the ability to scrutinize the precise morphology and crystalline structure of the pigments involved, the historians would never have been able to determine that the painting was, in fact, a forgery.

IMPACT ON INDUSTRY

Manufacturing. In many manufacturing industries, microscopes are essential for various stages of production and inspection. By examining materials and finished products on a microscopic scale, manufacturers can catch flaws, remove contaminants, and ensure the quality and safety of their products. They can also easily assemble products whose individual components are too small to be seen by the naked eye. For example, high-powered microscopes are essential for the manufacture of the silicon wafer microchips—whose circuits can be as small as 0.001 millimeter—used in computers and other electronic devices. Microscopes are particularly important in metallurgical industries because incredibly tiny discrepancies in the crystal structure of a metal alloy can cause significant differences in its physical properties, including hardness, toughness (how likely it is that small fissures in a metal will expand and cause it to break), and tensile strength (how much the metal can stretch before it is unable to return to its original shape). Both transmission and scanning electron microscopes are commonly used in the metal industry to examine the microscopic crystals, or grains, in samples of metal under production as a quality-control measure.

Product Development. Microscopy is an important tool in the research and development stages of many different products. Optical, electron, and atomic force microscopy enable manufacturers of contact lenses and intraocular implants, for example to characterize features such as the topography, adhesive quality, hardness, elasticity, and viscosity of a lens, as well as to determine how uniformly it has been made. Cosmetics firms use microscope technology to help them evaluate the effectiveness of products ranging from face creams to skin whiteners. Electron microscopes, for example, help biochemists determine how well shampoos smooth down the rough edges of hair, making it feel softer and look glossier.

Microscope Market. The sale of electron microscopes of all kinds is responsible for generating the biggest revenues in the market for microscopy products, largely because these machines are relatively large and expensive—the most sophisticated can cost up to $1 million—in comparison with other types of microscopes. Optical microscopes tend to dominate in contexts in which high-powered microscopes are not necessarily required, such as schools and smaller research laboratories. Scanning probe microscopes are less common than both electron and optical microscopes because they tend to be used in more specialized fields. The two biggest markets for microscopy applications are the life sciences research sector and the biomedical sector, including pharmaceutical companies and medical facilities. Other major consumers of microscope technology include manufacturers of semiconductors and textiles. Among the most important global corporations involved in the development, manufacture, and sale of microscopic equipment are Carl Zeiss, Nikon, Olympus, Leica Microsystems, JEOL, and Hitachi.

SOCIAL CONTEXT AND FUTURE PROSPECTS

Microscopy breaks down the barrier to knowledge created by the limitations of the human sense of sight. If knowledge is power, then microscopy represents one of science and technology's most powerful contributions to society. It enables researchers to discover more about the precise structure and behavior of healthy and diseased cells, pinpoint the mechanisms by which pathogens such as bacteria and viruses act in the body, and explore the chemical properties of potential pharmaceutical therapies. It even assists surgeons in performing difficult operations more safely and accurately, thereby saving countless lives. By providing scientists with an intimate knowledge of the way molecules, atoms, and subatomic particles interact, microscopy has propelled the formation of theories about the fundamental nature of the universe.

The growing needs of nanotechnology are inspiring further developments in microscopy. A scanning probe microscope built for the Argonne National Laboratory in Illinois allows researchers to "see" into an individual atom and observe its magnetic spin. The microscope, which cost $2 million, is itself very small—but it must be placed inside a machine 16 feet high and located in a soundproof room, so as to prevent even the tiniest vibration to throw off its focus.

On the other hand, microscopes intended for use in clinical settings in the developing world point the way toward ever smaller and cheaper instruments. A dime-sized microscope that sells for about $10 has no lenses but instead is made of a layer of metal set on top of an array of charge-coupled device (CCD) sensors arranged in a grid. The CCDs translate light into an electric signal; then, a great number of tiny channels are pierced into the metal. As a sample of blood, water, or other liquid flows over the channels, the particles in it block light from passing through to the CCDs in certain areas. The information about which channels are blocked and which remain open to light is used to create an image of the specimen.

—*M. Lee, MA*

FURTHER READING

Cardell, Carolina, Isabel Guerra, and Antonio Sánchez-Navas. "SEM-EDX at the Service of Archaeology to Unravel Historical Technology." *Microscopy Today* 17, no. 14 (August, 2009): 28-33. An overview of the use of scanning electron microscopy to analyze archaeological materials. Includes diagrams and full-color photomicrographs.

Dykstra, Michael J., and Laura E. Reuss. *Biological Electron Microscopy: Theory, Techniques, and Troubleshooting.* 2d ed. New York: Kluwer Academic, 2003. A guide to using microscopic instrumentation in cytological research. Covers conventional light microscopy, transmission electron microscopy, scanning electron microscopy, and photomicroscopy.

Reitdorf, Jens, et al., eds. *Microscopy Techniques.* New York: Springer, 2005. A technical reference book designed for those with a biomedical background, including numerous tables and diagrams, plus appendixes detailing mathematical formulas.

Sluder, Greenfield, and D. E. Wolf, eds. *Digital Microscopy.* 3d ed. Boston: Elsevier Academic Press, 2007. A guide to coordinating microscopes with digital cameras to capture and analyze microscopic images. Includes detailed laboratory exercises to demonstrate principles in action.

Yao, Nan, and Zhong Lin Wang, eds. *Handbook of Microscopy for Nanotechnology.* New York: Kluwer Academic, 2005. An overview of microscopy applications in nanotechnology. Each of the twenty-two chapters contains a discussion of a specific microscopic instrument or technique by nanotechnology specialists working in different fields.

FASCINATING FACTS ABOUT MICROSCOPY

- Microscopes need not be large themselves in order to enlarge other things. One of the world's smallest microscopes, the Cellvizio microscope, is less than one-tenth of an inch in diameter and can be inserted down a patient's throat to observe live cells.
- In 1986, no fewer than three recipients of the Nobel Prize in Physics were awarded their honors based on their work in improving microscopy technology. Ernst Ruska won for the invention of the electron microscope, and Gerd Binnig and Heinrich Rohrer were recognized for the invention of the scanning tunneling microscope.
- Microscopes have helped chemists at the University of California, Irvine, detect the presence of fat in strands of human hair. Their experiment seeks to determine whether fat is a natural component of hair or is deposited in hair by hair-care products.
- Atomic force microscopes, which probe the forces between atoms to produce an image of incredibly tiny particles, enable scientists to look at—and even pick up and move around—single strands of DNA or individual atoms.
- By examining either a rough or a cut-and-polished gem beneath a powerful microscope, a gemologist can easily tell whether it is an authentic natural stone or one that has been synthetically manufactured.
- Swiss inventor George de Mestral first got the idea for Velcro hook-and-loop fasteners when he used a microscope to examine the intricate hook-and-loop structure of the tiny burrs that had gotten firmly caught on his pant legs while he was walking through the forest.
- In 2008, a microscope attached to the National Aeronautics and Space Administration's Phoenix Mars Lander took a photograph of a single particle of the incredibly fine red dust that swirls around Mars and forms its soil. Dust particles on Mars are about 100 nanometers, about one-thousandth the width of a human hair—or even smaller.

Mitochondrial DNA analysis and typing

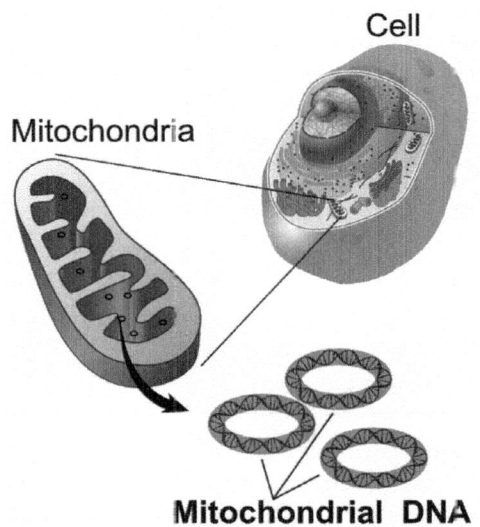

Mitochondrial DNA is the small circular chromosome found inside mitochondria. The mitochondria are organelles found in cells that are the sites of energy production. The mitochondria, and thus mitochondrial DNA, are passed from mother to offspring. By National Human Genome Research Institute. Public domain], via Wikimedia Commons

FIELDS OF STUDY

Biotechnology; Forensics; Genetics

ABSTRACT

Mitochondrial DNA analysis is often used when traditional DNA typing methods are unsuccessful because of biological degradation. The technique is employed in cases involving hair, teeth, skeletal remains, and other difficult forensic samples. It has also been used to identify missing persons and the victims of mass disasters.

HOW IT WORKS

Mitochondria are cellular organelles responsible for bodily energy production. Human mitochondria contain a circular genome of 16,569 bases, the bulk of which encodes thirty-seven RNA (ribonucleic acid)/proteins. The remaining segment, termed the control region, regulates DNA (deoxyribonucleic acid) replication and transcription.

The noncoding nature of the control region has allowed mutations (or polymorphisms) to accumulate over time, most of which are located in two hypervariable regions: HV1 and HV2. Scientists conduct mitochondrial DNA (mtDNA) typing by obtaining the HV DNA sequences and comparing them with the reference sequence known as the Anderson sequence or the Cambridge Reference Sequence. Differences between the sample mtDNA and the reference sequence are reported based on the type of polymorphism (base change, insertion/deletion) and its nucleotide position. For instance, if mtDNA from a hair has a C at position 152 while the reference sequence has a T, the DNA type (termed a haplotype) for that individual would be reported as 152C. Any other polymorphisms are reported as well, and the frequency of that haplotype in humans can be determined.

MtDNA typing does not have the discriminatory power of nuclear; however, there are instances in which mtDNA is the only DNA recoverable, particularly from materials of forensic interest. This can be the case for shed hair, aged bone or teeth, nails, and mummified tissue, among others. Nuclear DNA contained in such samples may be degraded, whereas mtDNA, owing to its high copy number (hundreds or thousands of copies per cell) and protection afforded by the mitochondrion, is still analyzable.

Another key feature of mtDNA is that it is maternally inherited, with all mitochondria stemming from the egg. As a result, siblings and other maternal relatives share the same mtDNA haplotype. This feature has made mtDNA typing an invaluable tool in forensic science. The Armed Forces DNA Identification Laboratory uses mtDNA to identify skeletal remains recovered from war casualties by comparing mtDNA samples from potential relatives. The Federal Bureau of Investigation (FBI) operates laboratories that focus solely on mtDNA analysis.

Several historical mysteries have been resolved using mtDNA. In 1998, mtDNA analysis was used to identify the remains of the Vietnam War service person interred in the Tomb of the Unknowns at Arlington National Cemetery as U.S. Air Force First Lieutenant Michael Blassie. Similarly, mtDNA aided in the identification of the members of the

Romanov family, the last Russian royal family, who were murdered during the Bolshevik Revolution in 1918.

—*Michael J. Mutolo and David R. Foran*

FURTHER READING

Butler, John M. *Forensic DNA Typing: Biology, Technology, and Genetics of STR Markers.* 2d ed. Burlington, Mass.: Elsevier Academic Press, 2005.

Cox, Margaret, et al. *The Scientific Investigation of Mass Graves.* New York: Cambridge University Press, 2008.

Hummel, Susanne. *Ancient DNA Typing: Methods, Strategies, and Applications.* New York: Springer, 2002.

MODEL ORGANISMS

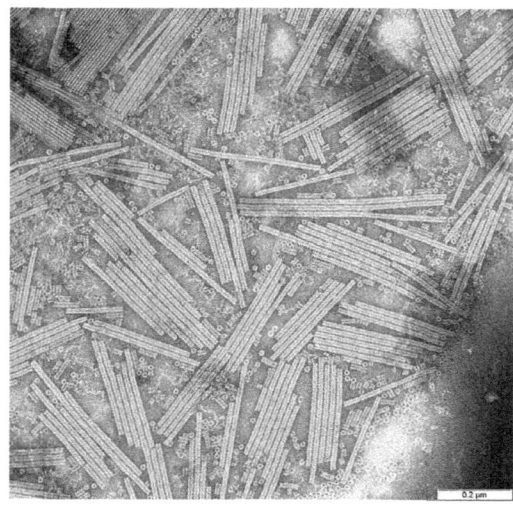

Low-temperature electron micrograph of a cluster of *E. coli* bacteria, magnified 10,000 times. Each individual bacterium is oblong shaped. By Photo by Eric Erbe, digital colorization by Christopher Pooley, both of USDA, ARS, EMU. [Public domain], via Wikimedia Commons

FIELDS OF STUDY

Biotechnology; Genetics

ABSTRACT

Practitioners in various areas of experimental biology recognize particular organisms as standard objects of study within their specialties. These model organisms are species that are used to exemplify a given category of organisms. For example, the laboratory rat is a common model used to study mammals, and *Arabidopsis thaliana* is commonly used as a model to study plants.

HOW IT WORKS

The characteristics of well-chosen model organisms, such as their genetic makeup or their development, make them suited to be the subjects of biological research. They are often developed specifically for laboratory study of pure-breeding strains that can be relied upon to provide a consistent medium for experimentation or examination. Biologists planning an experiment might specify a particular breed of rat, for example, whose traits are well defined and that are familiar to their peers. When a given model organism has become the standard in a particular field, it can become a kind of common currency that facilitates exchange among scientists.

Under these circumstances, a great deal of knowledge about individual model organisms may be accumulated rapidly. The crucial assumption, based on the theory of evolution, that underlies the use of specific organisms as models is that species sharing a common ancestor will have fundamental similarities of physiology and biochemistry. Among ubiquitous model organisms are the bacterium *Escherichia coli*, the yeast *Saccharomyces cerevisiae*, the roundworm *Caenorhabditis elegans*, the fruit fly *Drosophila melanogaster*, and the plant *Arabidopsis thaliana*.

COMMON FEATURES

Above all else, model organisms must be practical to observe and to use in experiments. They must be easy to breed or propagate and resilient enough to withstand manipulation. For knowledge gained in the study of model organisms to be applicable on a larger scale, the organisms must be representative of the taxonomic group in question. Clearly, the applicability of studies performed on a particular model organism varies, depending on the nature of the inquiry. For example, the yeast *S. cerevisiae* is broadly representative of the fungi as a whole, but its study may also provide insights into specific molecular processes common to all eukaryotes, including humans. Model organisms are often chosen because they are among the simplest examples of the group being studied. They may have a particularly small genome, a short life cycle, or even a small size that makes them convenient organisms with which to work. They may also lend themselves very well to the study of specific features. For example, fruit flies are commonly used in the study of genetics because they have a small genome from which it is easy to induce and detect mutations.

Well-chosen model organisms, those that possess some or all of the aforementioned characteristics, have been valuable tools for scientific research. Given these characteristics, it is easy to identify two types of scientific inquiry that are well served by the use of model organisms. There are studies in which a category of organisms is investigated by studying one of its simplest members and those in which a particular feature or biological process is illuminated by examining an organism in which it is especially accessible. Thus, the mouse is frequently used as a model for all mammals, and the green alga *Chlamydomonas* is used to study photosynthesis.

If a specific organism becomes the consensus model for a given category, the situation lends itself well to a speedy advancement of knowledge. The fact that many clusters of researchers choose to focus on the same model promotes collaboration and the more rapid accumulation of a body of knowledge about the organism, enhancing the likelihood of broader insights or theoretical advances. Having a research subject organism in common facilitates communication among researchers and leads to the formation of standard terminology. The widespread study of a single organism promotes the development and propagation of effective techniques for its use and allows for the introduction of standard experimental practices. Many observers have argued that the very success of science as a collaborative activity relies on scientists having some consensus about the tools and objects of their research and the terminology with which they describe it.

Although there are many advantages of a model organism becoming widespread in a particular field, there are some limitations to what can be achieved by the study of model organisms. There must always be a question of the applicability to other species of knowledge gained from the study of a model organism. A poor choice of an organism for a model can hinder the production of scientific knowledge just as much as research on a valid model can be beneficial. There is also the risk that focusing a discipline on one or a few models may inhibit our understanding of diversity. As the botanist Dina Mandoli said, "flowering plants have an estimated 300,000 species . . . no one plant, not even *Arabidopsis thaliana*, can encompass this enormous diversity at the whole plant, physiologic, chemical, genetic, or molecular level." It is important, therefore, that research be carried out on enough model organisms to produce an adequate breadth of knowledge. To that end, there are dozens of model plants in use representing a cross section of the kingdom, of which *Arabidopsis* has been the most widely and successfully employed.

ARABIDOPSIS

Arabidopsis is a genus of the mustard family that is closely related to food plants such as canola, cabbage, cauliflower, broccoli, radish, and turnip. Furthermore, although *Arabidopsis* is not used in agriculture, it is assumed that its study can lead to better knowledge of crop plants such as corn and soybeans because of evolutionary similarities among the genomes of all angiosperms. In the 1980s *Arabidopsis thaliana* (thale cress) became the primary model organism used in botany. Many characteristics lend it to such use, including its small size—the plants are a few inches tall when mature—and its short life cycle of less than six weeks. The short life cycle allows researchers to see the effects of experimentation across successive generations in a relatively short span of time. *A. thaliana* also has a small genome and the least amount of DNA (deoxyribonucleic acid) per haploid cell of any known flowering plant. As a result, it is comparatively easy to trace effects of experimentation to specific genes. It is valuable in the laboratory because of its prolific seed production and the availability of numerous mutations. It may be efficiently transformed with the bacterium *Agrobacterium tumefaciens*, which is used as a vector for the introduction of foreign DNA to the plant genome.

A. thaliana was publicly recognized for its potential as a model organism in the 1960s. In 1985 it was first promoted as a model for molecular genetic research, and the first molecular map of one of the five *A. thaliana* chromosomes was published in 1988. In 1996 the Arabidopsis Genome Initiative was begun. Thanks to a multinational effort, by the year 2000 the *A. thaliana* gene sequence was fully decoded. The sequencing project was itself acclaimed as a model, because the researchers strove to be systematic and comprehensive in their investigation of the genome.

Prior to the widespread use of *A. thaliana*, many prominent scientists claimed that progress in botanical research was hindered by the study of too many organisms at once. Since *A. thaliana* became a principal subject of research, botanical knowledge has advanced markedly. Researchers concentrating on *A. thaliana* have helped unify the studies of classical and molecular genetics, plant development, plant physiology, and plant pathology. These advances have led to a more fundamental understanding of many processes of plant growth and development at a molecular level.

Some specific areas in which *A. thaliana* research has produced important advances are light perception, floral induction, flower development, and response to pathogenic and environmental stresses. For example, the functions of individual phytochromes, which are photoreceptors involved in many aspects of plant growth and development, were elucidated in *A. thaliana*. Likewise, the first hormone receptor isolated in plants, that for ethylene, was discovered as a result of using *A. thaliana* mutants.

CHLORELLA AND *CHLAMYDOMONAS*

Chlorella pyrenoidosa and *Chlamydomonas reinhardtii* are unicellular green algae that have been used extensively as model organisms. They have many features in common with other model organisms, including short and simple life cycles and easily isolated mutants. Although there is debate as to whether green algae should be included in the plant kingdom, they have been important tools for botanically related research because they are photosynthetic eukaryotic organisms. They therefore offer less complex subjects through which to study many processes that are central to plant life. There are no other unicellular members of the plant kingdom, so study of many important botanical processes may be more easily undertaken on *Chlorella* or *Chlamydomonas* than on any plant.

In the mid-twentieth century Melvin Calvin used *Chlorella* in his Nobel Prize-winning research, which elucidated the cycle involved in photosynthetic carbon fixation that now bears his name, the Calvin cycle. This is a perfect example of model organisms' value in research. It is often easier to work out a mechanism in a simple organism and see whether it operates the same way in complex organisms—the understanding of which may be the ultimate purpose of the research—than to attempt the investigation on a complex organism in the first place. Once the Calvin cycle had been explained in *Chlorella*, it was shown to be ubiquitous in the chloroplasts of higher plants.

C. reinhardtii is the green alga most commonly used as a model organism in contemporary research; its genome has also been sequenced. Among the topics of research in which *C. reinhardtii* is the model organism of choice, one of the most compelling is that of chloroplast biogenesis and inheritance. *C. reinhardtii* is often referred to as the "green yeast," and like the yeast *S. cerevisiae*, it is an important eukaryotic model system. For studying certain aspects of cell biology to which yeast is not applicable, *C. reinhardtii* is chosen in preference. Such areas include cell motility caused by flagella, phototaxis (phototaxy), photosynthesis, and the study of centrioles, basal bodies, and chloroplasts.

ANIMAL MODEL ORGANISMS

Caenorhabditis elegans is a type of nematode, or roundworm, that is transparent and only about one millimeter in length. This model animal has been used by scientists to test the concepts of gene therapy and to develop methods for sequencing large amounts of DNA. *C. elegans* has also provided information about the biology of human diseases such as Alzheimer's disease and cancer. Additionally, research on this worm has enabled scientists to develop effective control measures for plant and animal parasitic roundworms.

Another model animal is the fruit fly, *Drosophila melanogaster*. Studies of *D. melanogaster* have allowed scientists to determine that genes reside on chromosomes and have given insight into the nature of mutations. Studies of the development of complex structures such as the eye have provided insight into some of the ways cell specialization is regulated and directed by DNA.

The transparent embryos of the zebra fish, *Danio rerio,* haved provided an excellent system with which to study the genes that regulate vetebrate development. Additionally, zebra fish have been used in studies investigating the bioaccumulation of organic compounds in the environment.

The common house mouse, *Mus musculus,* is well known for its use as a model for all mammal life. Mice are useful for genetic study because of the availability of hundreds of single gene mutations. Studies of mice demonstrated that Gregor Mendel's laws of inheritance are as applicable to mammals as to plants. Transgenic genetic analysis of mice has allowed for the creation of mouse strains that mimic human genetic diseases.

OTHER PROMINENT MODEL ORGANISMS

For areas other than those just mentioned, yeast is the most commonly used simple eukaryotic model organism. In 1996 *S. cerevisiae* became the first eukaryote to have its genome fully sequenced; its size is approximately one-tenth of that of *A. thaliana.* Around the same time, the genome project was completed for the preeminent prokaryotic model organism, *E. coli.* This bacterium has become crucial not only as a focus of experiment but also as a biotechnological workhorse. Genes can be cloned by their insertion into *E. coli,* and gene products can therefore be mass-produced in large-scale fermentations of the bacteria.

—*Alistair Sponsel*

FURTHER READING

Bowman, John L., ed. *Arabidopsis: An Atlas of Morphology and Development.* New York: Springer, 1994. Print.

Carroll, Pamela M., et al., eds. *Model Organisms in Drug Discovery.* Chichester: Wiley, 2003. Print.

Creager, Angela N. H. *The Life of a Virus: Tobacco Mosaic Virus as an Experimental Model, 1930–1965.* Chicago: Chicago UP, 2002. Print.

Davis, Rowland H. *Neurospora: Contributions of a Model Organism.* New York: Oxford UP, 2000. Print.

Graham, Linda E., James M. Graham, and Lee W. Wilcox. *Algae.* 2nd ed. San Francisco: Pearson, 2009. Print.

Kohler, Robert E. *Lords of the Fly: Drosophila Genetics and the Experimental Life.* Chicago: Chicago UP, 1994. Print.

Lamoreux, M. Lynn, et al. *The Colors of Mice: A Model Genetic Network.* Hoboken: Wiley, 2010. Print.

Meyerowitz, Elliot M., ed. *Arabidopsis.* Plainview: Cold Spring Harbor, 1994. Print.

Treuting, Piper M., and Suzanne M. Dintzis, eds. *Comparative Anatomy and Histology: A Mouse and Human Atlas.* Boston: Elsevier, 2012. Print.

MOLECULAR SYSTEMATICS

FIELDS OF STUDY

Biotechnology; Forensics; Genetics

ABSTRACT

Molecular systematics is the discipline of classifying organisms based on variations in protein and DNA in order to make fine taxonomic categorizations not solely dependent on morphology.

HOW IT WORKS

Taxonomy, sometimes called systematics, is the study of categorizing organisms into logically related groupings. Historically, the way to perform taxonomy was to examine physical characteristics of organisms and classify species according to the most commonly held traits. Unfortunately, this method of systematizing plants and animals assumed that because they have common physical traits, they have common ancestry. A gross form of this miscategorization might take place, for example, if one suggested that since both mushrooms and ivy can grow on the sides of trees, they are closely related. The two species certainly have common physical traits but only vaguely resemble each other.

It is such a realization that motivated systematists to begin using molecular differences to compare species and populations. Molecular systematics uses variations in protein and deoxyribonucleic acid (DNA) molecules to determine how similar, or dissimilar, sets of organisms are. These molecular differences provide a much more accurate taxonomic picture.

SYSTEMATICS AND EVOLUTION

The real power of molecular systematics is that it allows the examination of how species have changed over evolutionary time, as well as of the relationships between species that have no common physical characteristics. Molecular changes can be used to explore phylogenetics (how populations are related evolutionarily and genetically). It has been suggested that the amount of change that takes place in DNA over time can act as a molecular clock, gauging how much evolutionary time has passed. The clock is set by first examining geological and historical records to determine how long two species have been physically separated. By examination of the number of molecular changes that have occurred between those species over that known time, a time frame of change can be established. Genes are thought to evolve and mutate at a constant, predictable rate, giving rise to this evolutionary clock hypothesis.

There are three major domains of life: prokaryotes (modern bacteria), Archaebacteria (descendants of ancient bacteria), and eukaryotes (cellular organisms with nuclei and organelles). All these organisms share a common ancestry of hundreds of millions of years. All species over time are connected to one another through a web of interlacing DNA as they reproduce, separate to become new species, and reproduce again. All organisms carry their ancestors' genetic information with them as a bundle in each cell, and the more closely related organisms are to one another, the more similar the contents of that bundle will stay over time. Humans share common genes, unchanged over millennia, with all other organisms—from the bacterium *Escherichia coli* to barley to gophers. The more important the job of a gene, the less it changes over time; this concept is called conservation. Conservation is the force that keeps a biological or genetic link between every species on earth.

PROTEIN-LEVEL ANALYSIS

Proteins were the earliest biomolecules used to study phylogenetics. Initially, protein differences could be studied only at the grossest levels. It was found that populations of organisms could be distinguished based on possessing different alleles (genetic sites) that made proteins possessing the same function but with different chemical structures. These enzymes were called isozymes. Isozymes can be separated and compared for size by employing a technique called gel electrophoresis. Gel electrophoresis uses a slab of gelatin-like medium and an electric field to separate molecules on the basis of size and electric charge. The genetic similarity of two different species can be determined based on common molecular weight of the isozymes.

Proteins are composed of strings of the twenty amino acids common to all life on earth. It is possible to ascertain the amino acid sequence of a protein. If the amino acid sequence of the same protein is ascertained among several different species, that sequence should be more similar between closely related species than more distantly related species. These differences allow taxonomists to gauge similarity of populations.

Antibodies are biomolecules that are able to recognize and bind very specifically to other molecules. Biologists employ antibodies that specifically recognize molecules at the surface of cells to test relationships between species. Antibodies that recognize cell-surface molecules on one species should recognize those same molecules in closely related species, but not from distantly related species, allowing a researcher to gauge similarity between species.

DNA-LEVEL ANALYSIS

The most common method used to establish taxonomic relationships is to compare DNA sequences between species. DNA is the double-stranded, polymeric molecule that encodes the proteins that direct the inner workings of all cells. The DNA molecule is structure like a ladder, with rungs formed by pairings of of four molecules, the bases guanine (G), adenine (A), thymine (T), and and cytosine (C). These bases, arranged in unique order, are read by special enzymes and encode messages that are translated into proteins. Sequences encoding for the same protein can change between species. In taxonomy, DNA sequences are obtained from several populations of organisms. Analysis of these sequences allows one to obtain a picture of how different populations have changed over time. This DNA sequencing may be used to compare many different types of DNA: regions that encode for genes, do not encode for genes, reside in chloroplast DNA, or reside in mitochondrial DNA.

Another common method of DNA phylogenetic analysis is called restriction mapping. In this method, DNA from different species is subjected to enzymatic treatment from proteins called endonucleases. These endonucleases have the ability to cleave DNA into fragments. Where the enzymes cleave the DNA is determined by the DNA sequence itself. The size and pattern of the fragments created by this treatment should be more similar in related species than in unrelated species.

A fairly new method of DNA analysis examines repetitive DNAs, called microsatellite sequences, that are found in all eukaryotic organisms. Microsatellite sequences are short arrangements of bases, such as GATC, repeated over and over. The number of repeats at a particular genetic location is usually more similar in related species than in unrelated ones. The differences in these repeated sequences are called "simple sequence polymorphisms" and are detected by a special enzymatic reaction called the polymerase chain reaction. Once detected, the fragments are separated and compared for size by means of gel electrophoresis.

—*James J. Campanella*

FURTHER READING

Avise, John. *Molecular Markers, Natural History, and Evolution.* New York: Chapman and Hall, 1994. Examines relationships between genetic changes, taxonomy, and evolution.

Freeman, Scott, and Jon Herron. *Evolutionary Analysis.* New York: Prentice Hall, 2000. Covers the connections between evolution and taxonomy.

Nei, Masatoshi, and Suhir Kumar. *Molecular Evolution and Phylogenetics.* New York: Oxford University Press, 2000. Covers the mathematics of phylogenetic analysis.

Page, Roderic, and Edward Holmes. *Molecular Evolution: A Phylogenetic Approach.* Malden, Mass.: Blackwell Science, 1998. Concentrates on molecular evolution in populations and how species arise.

N

NANOTECHNOLOGY

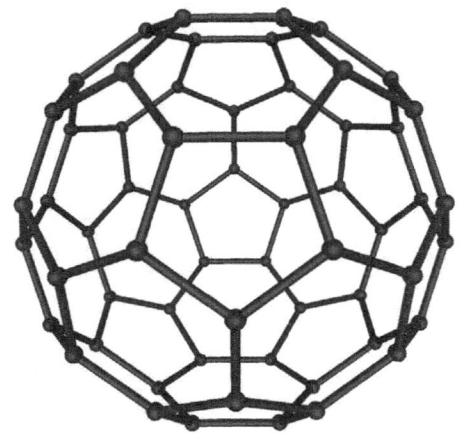

Buckminsterfullerene C60, also known as the buckyball, is a representative member of the carbon structures known as fullerenes. Members of the fullerene family are a major subject of research falling under the nanotechnology umbrella. By Mstroeck at en.wikipedia Later versions were uploaded by Bryn C at en.wikipedia. [GFDL (www.gnu.org/copyleft/fdl.html) or CC-BY-SA-3.0 (http://creativecommons.org/licenses/by-sa/3.0/)], from Wikimedia Commons

FIELDS OF STUDY

Biochemistry; Bioengineering; Biotechnology; Genetics; Medical biotechnology (or Red biotechnology); Molecular biology

ABSTRACT

Nanotechnology is dedicated to the study and manipulation of structures at the extremely small nano level. The technology focuses on how particles of a substance at a nanoscale behave differently than particles at a larger scale. Nanotechnology explores how those differences can benefit applications in a variety of fields. In medicine, nanomaterials can be used to deliver drugs to targeted areas of the body needing treatment. Environmental scientists can use nanoparticles to target and eliminate pollutants in the water and air. Microprocessors and consumer products also benefit from increased use of nanotechnology, as components and associated products become exponentially smaller.

DEFINITION AND BASIC PRINCIPLES

Nanotechnology is the science that deals with the study and manipulation of structures at the nano level. At the nano level, things are measured in nanometers (nm), or one billionth of a meter (10^{-9}). Nanoparticles can be produced using various techniques known as top-down nanofabrication, which starts with a larger quantity of material and removes portions to create the nanoscale material. Another method is bottom-up nanofabrication, in which individual atoms or molecules are assembled to create nanoparticles. One area of research involves developing bottom-up self-assembly techniques that would allow nanoparticles to create themselves when the necessary materials are placed in contact with one another.

Nanotechnology is based on the discovery that materials behave differently at the nanoscale, less than 100 nm in size, than they do at slightly larger scales. For instance, gold is classified as an inert material because it neither corrodes nor tarnishes; however, at the nano level, gold will oxidize in carbon monoxide. It will also appear as colors other than the yellow for which it is known.

Nanotechnology is not simply about working with materials such as gold at the nanoscale. It also involves taking advantage of the differences at this scale to create markers and other new structures that are of use in a wide variety of medical and other applications.

BACKGROUND

In 1931, German scientists Ernst Ruska and Max Knoll built the first transmission electron microscope (TEM). Capable of magnifying objects by a factor of up to one million, the TEM made it possible to see things at the molecular level. The TEM was used to study the proteins that make up the human body.

It was also used to study metals. The TEM made it possible to view particles smaller than 200 nm by focusing a beam of electrons to pass through an object, rather than focusing light on an object, as is the case with traditional microscopes.

In 1959 the noted American theoretical physicist Richard Feynman brought nanoscale possibilities to the forefront with his talk "There's Plenty of Room at the Bottom," presented at the California Institute of Technology in 1959. In this talk, he asked the audience to consider what would happen if they could arrange individual atoms, and he included a discussion of the scaling issues that would arise. It is generally agreed that Feynman's reputation and influence brought increased attention to the possible uses of structures at the atomic level.

In the 1970s scientists worked with nanoscale materials to create technology for space colonies. In 1974 Tokyo Science University professor Norio Taniguchi coined the term "nano-technology." As he defined it, nanotechnology would be a manufacturing process for materials built by atoms or molecules.

In the 1980s the invention of the scanning tunneling microscope (STM) led to the discovery of fullerenes, or hollow carbon molecules, in 1986. The carbon nanotube was discovered a few years later. In 1986, K. Eric Drexler's seminal work on nanotechnology, *Engines of Creation*, was published. In this work, Drexler used the term "nanotechnology" to describe a process that is now understood to be molecular nanotechnology. Drexler's book explores the positive and negative consequences of being able to manipulate the structure of matter. Included in his book are ruminations on a time when all the works in the Library of Congress would fit on a sugar cube and when nanoscale robots and scrubbers could clear capillaries or whisk pollutants from the air. Debate continues as to whether Drexler's vision of a world with such nanotechnology is even attainable

In 2000 the U.S. National Nanotechnology Initiative was founded. Its mandate is to coordinate federal nanotechnology research and development. Great growth in the creation of improved products using nanoparticles has taken place since that time. The creation of smaller and smaller components—which reduces all aspects of manufacture, from the amount of materials needed to the cost of shipping the finished product—is driving the use of nanoscale materials in the manufacturing sector. Furthermore, the ability to target delivery of treatments to areas of the body needing those treatments is spurring research in the medical field.

The true promise of nanotechnology is not yet known, but this multidisciplinary science is widely viewed as one that will alter the landscape of fields from manufacturing to medicine.

HOW IT WORKS

Basic Tools. Nanoscale materials can be created for specific purposes, but there exists also natural nanoscale material, like smoke from fire. To create nanoscale material and to be able to work with it requires specialized tools and technology. One essential piece of equipment is an electron microscope. Electron microscopy makes use of electrons, rather than light, to view objects. Because these microscopes have to get the electrons moving, and because they need several thousand volts of electricity, they are often quite large.

One type of electron microscope, the scanning electron microscope (SEM), requires a metallic sample. If the sample is not metallic, it is coated with gold. The SEM can give an accurate image with good resolution at sizes as small as a few nanometers.

For smaller objects or closer viewing, a TEM is more appropriate. With a TEM, the electrons pass through the object. To accomplish this, the sample has to be very thin, and preparing the sample is time consuming. The TEM also has greater power needs than the SEM, so SEM is used in most cases, and the TEM is reserved for times when a resolution of a few tenths of a nanometer is absolutely necessary.

The atomic force microscope (AFM) is a third type of electron microscope. Designed to give a clear image of the surface of a sample, this microscope uses a laser to scan across the surface. The result is an image that shows the surface of the object, making visible the object's "peaks and valleys."

Moving the actual atoms around is an important part of creating nanoscale materials for specific purposes. Another type of electron microscope, the scanning tunneling microscope (STM), images the surface of a material in the same way as the AFM. The tip of the probe, which is typically made up of a single atom, can also be used to pass an electrical current to the sample, which lessens the space between the probe and the sample. As the probe moves across the sample, the atoms nearest the charged atom move with it. In this way, individual atoms can be moved to a desired location in a process known as quantum mechanical tunneling.

Molecular assemblers and nanorobots are two other potential tools. The assemblers would use specialized tips to form bonds with materials that would make specific types of materials easier to move. Nanorobots might someday move through a person's blood stream or through the atmosphere, equipped with nanoscale processors and other materials that enable them to perform specific functions.

Bottom-Up Nanofabrication. Bottom-up nanofabrication is one approach to nanomanufacturing. This process builds a specific nanostructure or material by combining components of atomic and molecular scale. Creating a structure this way is time consuming, so scientists are working to create nanoscale materials that will spontaneously join to assemble a desired structure without physical manipulation.

Top-Down Nanofabrication. Top-down nanofabrication is a process in which a larger amount of material is used at the start. The desired nanomaterial is created by removing, or carving away, the material that is not needed. This is less time consuming than bottom-up nanofabrication, but it produces considerable waste.

Specialized Processes. To facilitate the manufacture of nanoscale materials, a number of specialized processes are used. These include nanoimprint lithography, in which nanoscale features are stamped or printed onto a surface; atomic layer epitaxy, in which a layer that is only one atom thick is deposited on a surface; and dip-pen lithography, in which the tip of an atomic force microscope writes on a surface after being dipped into a chemical.

APPLICATIONS AND PRODUCTS

Smart Materials. Smart materials are materials that react in ways appropriate to the stimulus or situation they encounter. Combining smart materials with nanoscale materials would, for example, enable scientists to create drugs that would respond when encountering specific viruses or diseases. They could also be used to signal problems with other systems, such as nuclear power generators or pollution levels.

Sensors. The difference between a smart material and a sensor is that the smart material will generate a response to the situation encountered, while the sensor will generate an alarm or signal that there is something that requires attention. The capacity to incorporate sensors at a nanoscale greatly enhances the ability of engineers and manufacturers to create structures and products with a feedback loop that is not cumbersome. Nanoscale materials can easily be incorporated into the product.

Medical Uses. The potential uses of nanoscale materials in the field of medicine are of particular interest to researchers. Theoretically, nanorobots could be programmed to perform functions that would eliminate the possibility of infection at a wound site. They could also speed healing. Smart materials could be designed to dispense medication in appropriate doses when a virus or bacteria is encountered. Sensors could be used to alert physicians to the first stages of malignancy. There is great potential for nanomaterials to meet the needs of aging populations without intrusive surgeries requiring lengthy recovery and rehabilitation.

Energy. Nanomaterials also hold promise for energy applications. With nanostructures, components of heating and cooling systems could be tailored to control temperatures with greater efficiency. This could be accomplished by engineering the materials so that some types of atoms, such as oxygen, can pass through, while others, such as mold or moisture, cannot. With this level of control, living conditions could be designed to meet the specific needs of different categories of residents.

Extending the life of batteries and prolonging their charge has been the subject of decades of research. With nanoparticles, researchers at Rutgers University and Bell Labs have been able to better separate the chemical components of batteries, resulting in longer battery life. With further nanoscale research, it may be possible to alter the internal composition of batteries to achieve even greater performance.

Light-emitting diode (LED) technology uses 90 percent less energy than conventional, non-LED lighting. It also generates less heat than traditional metal-filament light bulbs. Nanomanufacture would make it possible to create a new generation of efficient LED lighting products.

Electronics. Moore's law states that transistor density on integrated circuits doubles about every two years. With the advent of nanotechnology, the rate of miniaturization has the potential to double at a much greater rate. This miniaturization will profoundly affect the computer industry. Computers will become lighter and smaller as nanoparticles are used to increase everything from screen resolution to battery life while reducing the size of essential internal components, such as capacitors.

SOCIAL CONTEXT AND FUTURE PROSPECTS

Whether nanotechnology will ultimately be good or bad for the human race remains to be seen, as it continues to be incorporated into more and more products and processes, both common and highly specialized. There is tremendous potential associated with the ability to manipulate individual atoms and molecules, to deliver medications to a disease site, and to build products such as cars that are lighter yet stronger than ever. Much research is devoted to using nanotechnology to improve fields such as pollution mitigation, energy efficiency, and cell and tissue engineering. However, there also exists the persistent worry that humans will lose control of this technology and face what Drexler called a "gray goo" scenario, in which self-replicating nanorobots run out of control and ultimately destroy the world.

Despite fears linked to cutting-edge technology, many experts, including nanotechnology pioneers, consider such doomsday scenarios involving robots to be highly unlikely or even impossible outside of science fiction. More worrisome, many argue, is the potential for nanotechnology to have other unintended negative consequences, including health impacts and ethical challenges. Some studies have shown that the extremely small nature of nanoparticles makes them susceptible to being breathed in or ingested by humans and other animals, potentially causing significant damage. Structures including carbon nanotubes of graphene have been linked to cancer. Furthermore, the range of possible applications for nanotechnology raises various ethical questions about how, when, and by whom such technology can and should be used, including issues of economic inequality and notions of "playing God." These risks, and the potential for other unknown negative impacts, have led to calls for careful regulation and oversight of nanotechnology, as there has been with nuclear technology, genetic engineering, and other powerful technologies.

——*Gina Hagler, MBA*

FURTHER READING

Berlatsky, Noah. *Nanotechnology*. Greenhaven, 2014.

Binns, Chris. *Introduction to Nanoscience and Nanotechnology*. Hoboken: Wiley, 2010. Print.

Biswas, Abhijit, et al. "Advances in Top-Down and Bottom-Up Surface Nanofabrication: Techniques, Applications & Future Prospects." *Advances in Colloid and Interface Science* 170.1–2 (2012): 2–27. Print.

Demetzos, Costas. *Pharmaceutical Nanotechnology: Fundamentals and Practical Applications*. Adis, 2016.

Drexler, K. Eric. *Engines of Creation: The Coming Era of Nanotechnology*. New York: Anchor, 1986. Print.

Drexler, K. Eric. *Radical Abundance: How a Revolution in Nanotechnology Will Change Civilization*. New York: PublicAffairs, 2013. Print.

Khudyakov, Yury E., and Paul Pumpens. *Viral Nanotechnology*. CRC Press, 2016.

Ramsden, Jeremy. *Nanotechnology*. Elsevier, 2016.

Ratner, Daniel, and Mark A. Ratner. *Nanotechnology and Homeland Security: New Weapons for New Wars*. Upper Saddle River: Prentice, 2004. Print.

Ratner, Mark A., and Daniel Ratner. *Nanotechnology: A Gentle Introduction to the Next Big Idea*. Upper Saddle River: Prentice, 2003. Print.

Rogers, Ben, Jesse Adams, and Sumita Pennathur. *Nanotechnology: The Whole Story*. Boca Raton: CRC, 2013. Print.

Rogers, Ben, Sumita Pennathur, and Jesse Adams. *Nanotechnology: Understanding Small Systems*. 2nd ed. Boca Raton: CRC, 2011. Print.

Stine, Keith J. *Carbohydrate Nanotechnology*. Wiley, 2016.

Nanotechnology and the Environment

FIELDS OF STUDY

Biochemistry; Bioengineering; Biotechnology; Environmental biotechnology (or Green biotechnology); Genetics; Medical biotechnology (or Red biotechnology); Molecular biology

ABSTRACT

Nanotechnology receives large-scale public and private investment around the world. The field holds the promise of revolutionizing much of science and technology, but the potential impacts of the products of nanotechnology on the environment, health, and safety are poorly understood. Preliminary research findings suggest that much further study is needed.

Nanotechnology is a relatively new field concerned with a wide range of materials and processes that seeks to understand and control matter at dimensions of 1 to 100 nanometers. The increasing speeds and decreasing sizes of personal electronic devices owe much to nanotechnology, which enables the production of lighter and stronger materials that reduce energy usage and prolong the life spans of the devices made with them. Battery and lighting systems have been developed that use nanotechnology to be more fuel-efficient. Nanotechnology has also contributed to the development of improved medical diagnostic devices and drug-delivery systems. Research into the use of nanotechnology to detect pollutants and build more effective water-purification systems is ongoing. The promise of nanotechnology is immense, but the field is new, and much remains uncertain about its overall impacts on the environment and human health.

BACKGROUND

The term "nano" is a prefix for one-billionth, or 1×10^{-9}, of a unit. Nanotechnology is concerned with the study, development, and control of materials at nanoscale—that is, in the range of 1 to 100 nanometers (or 1 to 100 billionths of 1 meter). Nanotechnology includes processes and instruments that engineer materials at nanoscale, and nanoparticles have at least one dimension in the nanoscale range.

Many naturally occurring materials have nanoscale dimensions, including proteins, the genetic molecule deoxyribonucleic acid (DNA), and viruses. Volcanoes and forest fires produce nanoparticles, and many soils contain organic and inorganic nanoparticles. Nanoparticles are also produced by human activities, such as cigarette smoking and the burning of fuels in combustion engines. The polymer and plastics industry makes chemical molecules with nanoscale dimensions. Environmental issues associated with nanotechnology thus overlap other areas.

A significant segment of nanotechnology is concerned with newly discovered nanoparticles. These can have completely different electrical, magnetic, or biological properties compared to larger particles of the same substance. For example, gold nanoparticles can be red, blue, or gold, depending on their precise size. Titanium dioxide and zinc oxide are used in sunscreens to block the sun's ultraviolet (UV) rays, but large particles of these substances leave a white coating on the skin; as nanoparticles, the same substances are transparent and more appealing to sunbathers. Questions have arisen, however, about what else might be different about these nanoparticles and what effects they might have when they get into the body or are washed into the environment.

Richard E. Smalley has been called the grandfather of nanotechnology. In 1986 he and others discovered a completely new form of carbon. They named it buckminsterfullerene, but as the molecules look like soccer balls, they are more commonly called buckyballs. Several sizes and shapes have been identified and collectively are called fullerenes. They are being investigated as possible devices to transport drug molecules to specific tissues and cells.

Many different kinds of nanoparticles have been developed and given names such as dendrimers, nanowires, and quantum dots. Carbon nanotubes are particularly interesting. Identified in 1991, they are highly organized carbon atoms that form sheets that roll into long tubes. Having different forms, they usually are a few nanometers wide and can be millimeters long. They are extremely strong for their weight, can transport other materials, and have unique electrical properties. They may have uses in reinforcing car and airplane bodies and in making comfortable bulletproof clothing; they may also have applications in medicine and in new battery technology. Carbon nanotubes have been at the center of early debate over the potential environmental impacts of nanotechnology.

PRELIMINARY NANOTOXICOLOGY

Nanoparticles can enter living cells, making them useful as drugs but also raising concerns. Some nanoparticles enter the nuclei of cells, where genetic material is stored. This may have beneficial uses, but it could also lead to genetic damage. Nanoparticles smaller than 35 nanometers can penetrate the blood-brain barrier, which prevents most chemicals from reaching the brain. This property could help deliver drugs for brain disorders, but it might also cause harmful side effects.

The field of science that investigates such concerns is known as nanotoxicology. Some early nanotoxicological studies have provided worrisome results. Nanosilver has antibacterial properties and has been used in special clothing and on surfaces where bacteria might grow. Bulk silver is normally safe, but laboratory experiments have shown that nanosilver might interfere with the human immune system.

Carbon nanotubes account for much of nanoparticle manufacturing, and their production is predicted to increase dramatically. However, a 2009 review of research into the toxicity of carbon nanotubes found that only twenty-one studies had yet been conducted, and all had shown some damage to tissues and animals. None of the research studies had examined the effects of carbon nanotubes on humans.

Hardly any research has examined nanoparticles after they enter the environment from normal wear and tear or when products are discarded. One of the first studies exposed largemouth bass to buckyballs for forty-eight hours. Most of the organs of the fish were unharmed, but their brains showed evidence of oxidative damage. Because buckyballs are highly fat-soluble and thus prone to cause environmental damage, the researchers urged the widespread application of the precautionary principle to avoid the sort of environmental damage seen from earlier chemicals with similar solubility profiles.

NANOBOTS AND GRAY GOO

The vision for nanotechnology originated in a 1959 talk by American physicist Richard Feynman. He predicted the development of very fast and small computers, as well as tiny machines that could circulate through the body. Such hypothetical devices have come to be called nanobots or nanites and are commonly included in science-fiction scenarios. They also come up in discussions about the potential environmental impacts of nanotechnology.

Another early proponent of nanotechnology, K. Eric Drexler, published *Engines of Creation: The Coming Era of Nanotechnology* in 1986. This book had a significant impact on the development and popular understanding of nanotechnology. Drexler's proposal, called molecular manufacturing, involved building nanomachines (or assemblers) that would make things with atomic precision. Drexler's proposal remains scientifically controversial, with critics such as Richard Smalley asserting that the approach is physically impossible.

Others have used the idea of such assemblers to portray nanotechnology as extremely dangerous. Drexler suggested that nanomachines could be programmed to assemble copies of themselves. This would increase production but raised concerns about how to control them. Nanotechnology molecular assemblers have become associated with self-replication, although the ideas are not necessarily linked. Science-fiction authors have invented scenarios in which marauding nanobots wreak environmental havoc.

Drexler eventually distanced himself from the idea of assembler self-replication, but he coined the term "gray goo" to describe a situation in which self-replicating nanomachines get out of control and consume everything around them. In a later edition of his book, Drexler lamented how nanotechnology had become associated with so-called gray goo scenarios. He used the term in only one passage of the book, while repeatedly stressing nanotechnology's potential either to destroy the world or to remake it, curing illness and restoring the environment.

Such grand claims make it more difficult to evaluate the real potential and actual threat of nanotechnology. A distinction must be made between "normal" nanotechnology and "futuristic" nanotechnology. Molecular assemblers and gray goo scenarios belong well into the future, although the steps that current scientists take may determine whether or when these might develop. Normal nanotechnology in the early twenty-first century focuses on recent discoveries about nanoparticles and their properties. Because of the newness of the field, much remains unknown and uncertain, either positive or negative, and practical environmental concerns are being raised.

EXAMINING ENVIRONMENTAL CONCERNS

Given the scientific uncertainty about the impacts of nanoparticles, many scientists and policy makers have called for caution in the study and use of such particles. The European Union has adopted a precautionary approach in its chemical regulatory agency and in its voluntary code of conduct for nanotechnologists. Worldwide, concerted efforts have been undertaken to understand and regulate nanoparticles. The U.S. Environmental Protection Agency launched a major research initiative in 2009 into the health and environmental concerns raised by nanotechnology.

While nanotoxicology is starting to be addressed, many remain concerned about the funding of research into environmental concerns. In the United States, the National Nanotechnology Initiative (NNI) coordinates federal investment in nanotechnology. Between 2005 and 2010, less than 4 percent of the $9 billion invested went directly to the examination of environmental, health, and safety issues. In 2008 the NNI published its strategy for such research, but the National Research Council was critical of the approach, concluding that it overestimated how much research had already been conducted and thereby underestimated the funding necessary to address environmental issues adequately.

The potential benefits of nanotechnology are enormous, but they will be realized only if concerns about possible negative health and environmental impacts are addressed. A gray goo scenario is not necessary for nanoparticles to damage the environment, and much research has yet to be done in this young field before the risks of nanotechnology are fully understood and steps can be taken to ensure that any potential harms are minimized.

—*Dónal P. O'Mathúna*

FURTHER READING

Allhoff, Fritz, et al., eds. *Nanoethics: The Ethical and Social Implications of Nanotechnology*. Hoboken, N.J.: Wiley-Interscience, 2007.

Drexler, K. Eric. *Engines of Creation: The Coming Era of Nanotechnology*. New York: Anchor Books, 1986.

Edwards, Steven A. *The Nanotech Pioneers: Where Are They Taking Us?* Weinheim, Germany: Wiley-VCH, 2006.

O'Mathúna, Dónal P. *Nanoethics: Big Ethical Issues with Small Technology*. London: Continuum, 2009.

Ray, Paresh Chandra, Hongtao Yu, and Peter P. Fu. "Toxicity and Environmental Risks of Nanomaterials: Challenges and Future Needs." *Journal of Environmental Science and Health*, Part C, Environmental Carcinogenesis and Ecotoxicology Reviews 27, no. 1 (2009): 1-35.

NEURAL ENGINEERING

FIELDS OF STUDY

Biochemistry; Bioengineering; Biotechnology; Genetics; Medical biotechnology (or Red biotechnology)

ABSTRACT

Neural engineering is an emerging discipline that translates research discoveries into neurotechnologies. These technologies provide new tools for neuroscience research, while leading to enhanced care for patients with nervous-system disorders. Neural engineers aim to understand, represent, repair, replace, and augment nervous-system function. They accomplish this by incorporating principles and solutions derived from neuroscience, computer science, electrochemistry, materials science, robotics, and other fields. Much of the work focuses on the delicate interface between living neural tissue and nonliving constructs. Efforts focus on elucidating the coding and processing of information in the sensory and motor systems, understanding disease states, and manipulating neural function through interactions with artificial devices such as brain-computer interfaces and neuroprosthetics.

Neuron Types

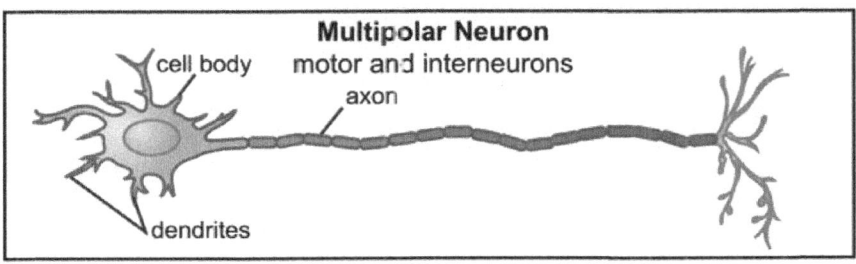

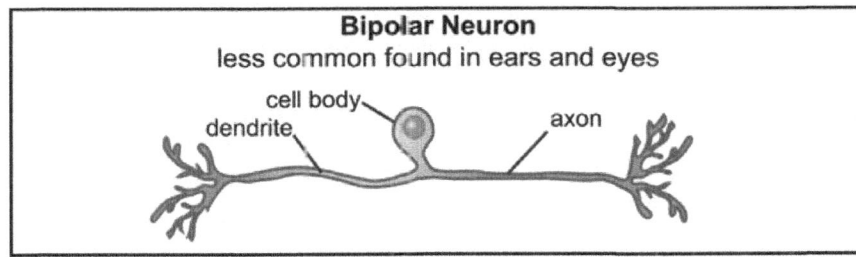

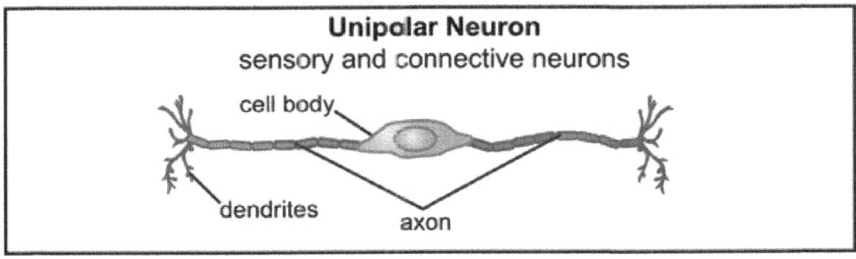

© EBSCO

DEFINITION AND BASIC PRINCIPLES

Neural engineering (or neuroengineering, NE) is an emerging interdisciplinary research area within biomedical engineering that employs neuroscientific and engineering methods to elucidate neuronal function and design solutions for neurological dysfunction. Restoring sensory, motor, and cognitive function in the nervous system is a priority. The strong emphasis on engineering and quantitative methods separates NE from the "traditional" fields of neuroscience and neurophysiology. The strong neuroscientific approach distinguishes NE from other engineering disciplines such as artificial neural networks. Despite being a distinct discipline, NE draws heavily from basic neuroscience and neurology and brings together engineers, physicians, biologists, psychologists, physicists, and mathematicians.

At present, neural engineering can be viewed as the driving technology behind several overlapping fields: functional electrical stimulation, stereotactic and functional neurosurgery, neuroprosthetics and neuromodulation. The broad scope of NE also encompasses neurodiagnostics, neuroimaging, neural tissue regeneration, and computational approaches. By using mathematical models of neural function (computational neuroscience), researchers can perform robust testing of therapeutic strategies before they are used on patients.

The human brain, arguably the most complex system known to humankind, contains about 10^{11} neurons and several times more glial cells. Understanding the functional neuroanatomy of this exquisite device is a sine qua non for anyone aiming to manipulate and repair it. The "neuron doctrine," pioneered by

Spanish neuroscientist Santiago Ramón y Cajal, considers the neuron to be a distinct anatomical and functional unit. The extension introduced by American neuroscientist Warren S. McCullogh and American logician Walter Pitts asserts that the neuron is the basic information-processing unit of the brain. For neuroengineers, this means that a particular goal can be reached just by manipulating a cell or group of cells. One argument in favor of this view is that stimulating groups of neurons produces a regular effect. Motor activity, for example, can be induced by stimulating the motor cortex with electrodes. In addition, lesions to specific brain areas due to neurodegenerative disorders or stroke lead to more or less predictable clinical manifestation patterns.

BACKGROUND

Electricity (in the form of electric fish) was used by ancient Egyptians and Romans for therapeutic purposes. In the eighteenth century, the work of Swiss anatomist Albrecht von Haller, Italian physician Luigi Galvani, and Benjamin Franklin set the stage for the use of electrical stimulation to restore movement to paralyzed limbs. The basis of modern NE is early neuroscience research demonstrating that neural function can be recorded, manipulated, and mathematically modeled. In the mid-twentieth century, electrical recordings became popular as a window into neuronal function. Metal wire electrodes recorded extracellularly, while glass pipettes probed individual cells. Functional electrical stimulation (FES) emerged with a distinct engineering orientation and the aim to use controlled electrical stimulation to restore function. Modern neuromodulation has developed since the 1970's, driven mainly by clinical professionals. The first peripheral nerve, then spinal cord and deep brain stimulators were introduced in the 1960's. In 1997, the Food and Drug Administration (FDA) approved deep brain stimulation (DBS) for the treatment of Parkinson's disease. An FES-based device that restored grasp was approved the same year.

In the 1970's, researchers developed primitive systems controlled by electrical activity recorded from the head. The U.S. Pentagon's Advanced Research Projects Agency (ARPA) supported research aimed at developing bionic systems for soldiers. Scientists demonstrated that recorded brain signals can communicate a user's intent in a reliable manner and found cells in the motor cortex the firing rates of which correlate with hand movements in two-dimensional space.

Since the 1960's, engineers, neuroscientists, and physicists have constructed mathematical models of the retina that describe various aspects of its function, including light-stimulus processing and transduction. In addition, scientists have made attempts to treat blindness using engineering solutions, such as nonbiological "visual prostheses." In 1975, the first multichannel cochlear implant (CI) was developed and implanted two years later.

HOW IT WORKS

Neuromodulation and Neuroaugmentation. Neural engineering applications have two broad (and sometimes overlapping) goals: neuromodulation and neuroaugmentation. Neuromodulation (altering nervous system function) employs stimulators and infusion devices, among other techniques. It can be applied at multiple levels: cortical, subcortical, spinal, or peripheral. Neural augmentation aims to amplify neural function and uses sensory (auditory, visual) and motor prostheses.

Neuromuscular Stimulation. Based on a method that has remained unchanged for decades, electrodes are placed within the excitable tissue that provide current to activate certain pathways. This supplements or replaces lost motor or autonomic functions in patients with paralysis. An example is application of electrical pulses to peripheral motor nerves in patients with spinal cord injuries. These pulses lead to action potentials that propagate across neuromuscular junctions and lead to muscle contraction. Coordinating the elicited muscle contractions ultimately reconstitutes function.

Neural Prosthetics. Neural prostheses (NP) aim to restore sensory or motor function—lost because of disease or trauma—by linking machines to the nervous system. By artificially manipulating the biological system using external electrical currents, neuroengineers try to mimic normal sensorimotor function. Electrodes act as transducers that excite neurons through electrical stimulation, or record (read) neural signals. In the first approach, stimulation is used for its therapeutic efficacy, for example, to alleviate the symptoms of Parkinson's disease, or to provide input to the nervous system, such as converting sound to neural input with a cochlear implant. The second paradigm uses recordings of neural activity to detect motor intention and provide input signal to an external device. This forms the basis of a subset of neural prosthetics called brain-controlled interfaces (BCI).

Microsystems. Miniaturization is a crucial part of designing instruments that interface efficiently with neural tissue and provide adequate resolution with minimal invasiveness. Microsystems technology integrates devices and systems at the microscopic and submicroscopic levels. It is derived from microelectronic batch-processing fabrication techniques. A "neural microsystem" is a hybrid system consisting of a microsystem and its interfacing neurons (be they cultured, part of brain slices, or in the intact nervous system). Technologies such as microelectrodes, microdialysis probes, fiber optic, and advanced magnetic materials are used. The properties of these systems render them suitable for simultaneous measurements of neuronal signals in different locations (to analyze neural network properties) as well as for implantation within the body.

APPLICATIONS AND PRODUCTS

Some of the most common applications of NE methods are described below.

Cochlear Implants. Cochlear implants (CI), by far the most successful sensory neural prostheses to date, have penetrated the mainstream therapeutic arsenal. Their popularity is rivaled only by the cardiac pacemakers and deep brain stimulation (DBS) systems. Implanted in patients with sensorineural deafness, these devices process sounds electronically and transmit stimuli to the cochlea. A CI includes several components: a microphone, a small speech processor that transforms sounds into a signal suitable for auditory neurons, a transmitter to relay the signal to the cochlea, a receiver that picks up the transmitted signal, and an electrode array implanted in the cochlea. Individual results vary, but achieving a high degree of accuracy in speech perception is possible, as is the development of language skills.

Retinal Bioengineering. Retinal photoreceptor cells contain visual pigment, which absorbs light and initiates the process of transducing it into electrical signals. They synapse onto other types of cells, which in turn carry the signals forward, eventually through the optic nerve and into the brain, where they are interpreted. Every neuron in the visual system has a "receptive field," a particular portion of the visual space within which light will influence that neuron's behavior. This is directly related to (and represented by) a specific region of the retina. Inherited retinal degenerations such as retinitis pigmentosa (RP) or age-related macular degeneration (AMD) are responsible for the compromised or nonexistent vision of millions of people. In these disorders, the retinal photoreceptor cells lose function and die, but the secondary neurons are spared.

Using an electronic prosthetic device, a signal is sent to these secondary neurons that ultimately causes an external visual image. A miniature video camera is mounted on the patient's eyeglasses that captures images and feeds them to a microprocessor, which converts them to an electronic signal. Then the signal is sent to an array of electrodes located on the retina's surface. The electrodes transmit the signal to the viable secondary neurons. The neurons process the signal and pass it down the optic nerve to the brain to establish the visual image.

Several different versions of this device exist and are implanted either into the retina or brain. Cortical visual prostheses could entirely bypass the retina, especially when this structure is damaged from diseases such as diabetes or glaucoma. Retinal prostheses, or artificial retinas (AR), could take advantage of any remaining functional cells and would target photoreceptor disorders such as RP. Two distinct retinal placements are used for AR. The first type slides under the retina (subretinal implant) and consists of small silicon-based disks bearing microphotodiodes. The second type would be an epiretinal system, which involves placing the camera or sensor outside the eye, sending signals to an intraocular receiver. In addition to challenges related to miniaturization and power supply, developing these systems faces obstacles pertaining to biocompatibility, such as retinal health and implant damage, and vascularization.

Functional Electrical Stimulation (FES). Some FES devices are commercialized, and others belong to clinical research settings. A typical unit includes an electronic stimulator, a feedback or control unit, leads, and electrodes. Electrical stimulators bear one or multiple channels (outputs) that are activated simultaneously or in sequence to produce the desired movement. Applications of FES include standing, ambulation, cycling, grasping, bowel and bladder control, male sexual assistance, and respiratory control. Although not curative, the method has numerous benefits, such as improved cardiovascular health, muscle-mass retention, and enhanced psychological well-being through increased functionality and independence.

Brain-Controlled Interfaces. A two-electrode device was implanted into a 1998 stroke victim who could communicate only by blinking his eyes. The device read from only a few neurons and allowed him to select letters and icons with his brain. A team of researchers helped a young patient with a spinal cord injury by implanting electrodes into his motor cortex that were connected to an interface. The patient was able to use the system to control a computer cursor and move objects using a robotic arm.

Brain-controlled interfaces (BCIs), a subset of NP, represent a new method of communication based on brain-generated neural activity. Still in an experimental phase, they offer hope to patients with severe motor dysfunction. These interfaces capture neural activity mediating a subject's intention to act and translate it into command signals transmitted to a computer (brain-computer interface) or robotic limb. Independent of peripheral nerves and muscles, BCI have the ability to restore communication and movement. This exciting technological advance is not only poised to help patients, but it also provides insight into the way neurons interact.

Every BCI has four main components: recording of electrical activity, extraction of the planned action from this activity, execution of the desired action using the prosthetic effector (actuator), and delivery of feedback (via sensation or prosthetic device).

Brain-controlled interfaces rely on four main recording modalities: electroencephalography, electrocorticography, local field potentials, and singe-neuron action potentials. The methods are noninvasive, semi-invasive, or invasive, depending on where the transducer is placed: scalp, brain surface, or cortical tissue.

The field is still in its infancy; however, several basic principles have emerged from these and other early experiments. A crucial requirement in BCI function, for example, is for the reading device to obtain sufficient information for a particular task. Another observation refers to the "transparency of action" in brain-machine interface (BMI) systems: Upon reaching proficiency, the action follows the thought, with no awareness of intermediate neural events.

Deep Brain Stimulation (DBS) and Other Modulation Methods. Deep brain stimulation of thalamic nuclei decreases tremors in patients with Parkinson's disease. It may alleviate depression, epilepsy, and other brain disorders. One or more thin electrodes, about 1 millimeter in diameter, are placed in the brain. An external signal generator with a power supply is also implanted somewhere in the body, typically in the chest cavity. An external remote control sends signals to the generator, varying the parameters of the stimulation, including the amount and frequency of the current and the duration and frequency of the pulses. The exact mechanism by which this method works is still unclear. It appears to exert its effect on axons and act in an inhibitory manner, by inducing an effect akin to ablation of target area, much like early Parkinson's treatment. One major advantage of DBS over other previously employed methods is its reversibility and absence of structural damage. Another valuable neuromodulatory approach, the electrical stimulation of the vagus nerve, can reduce seizure frequency in patients with epilepsy and alleviate treatment-resistant depression. Transcutaneous electrical nerve stimulation (TENS) represents the most common form of electrotherapy and is still in use for pain relief. Cranial electrotherapy stimulation involves passing small currents across the skull. The approach shows good results in depression, anxiety, and sleep disorders.

Transcranial magnetic stimulation uses the magnetic field produced by a current passing through a coil and can be applied for diagnostic (multiple sclerosis, stroke), therapeutic (depression), or research purposes.

SOCIAL CONTEXT AND FUTURE PROSPECTS

Bioelectrodes for neural recording and neurostimulation are an essential part of neuroprosthetic devices. Designing an optimal, stable electrode that records long-term and interacts adequately with neural tissue remains a priority for neural engineers. The implementation of microsystem technology opens new perspectives in the field.

More than 200 million people around the world suffer from hearing loss, mainly because sensory hair cells in the cochlea have degenerated. The only efficient therapy for patients with profound hearing loss is the CI. Improvements in CI performance have increased the average sentence recognition with multichannel devices. An exciting new development, auditory brainstem implants, show improved performance in patients with impaired cochlear nerves.

Millions of Americans have vision loss. The need for a reliable prosthetic retina is significant, and rivals the one for CI. Technological progress makes it quite likely that a functioning implant with a more sophisticated design and higher number of electrodes will be on the market soon. The epiretinal approach is promising, but providing interpretable visual information to the brain represents a challenge. In addition, even if they prove to be successful, retinal prostheses under development address only a limited number of visual disorders. Much is left to be discovered and tested in this field.

The coming years will also see rapid gains in the area of BCI. Whether they achieve widespread use will depend on several factors, including performance, safety, cost, and improved quality of life.

The advent of gene therapy, stem cell therapy, and other regenerative approaches offers new hope for patients and may complement prosthetic devices. However, many ethical and scientific issues still have to be solved.

Implanted devices are changing the way neurological disorders are treated. An unprecedented transition of NE discoveries from the research to the commercial realm is taking place. At the same time, new discoveries constantly challenge the basic tenets of neuroscience and may alter the face of NE in the coming decades. People's understanding of the nervous system, especially of the brain, changes, and so do the strategies designed to enhance and restore its function.

——*Mihaela Avramut, MD, PhD*

FURTHER READING

Blume, Stuart. *The Artificial Ear: Cochlear Implants and the Culture of Deafness.* New Brunswick, N.J.: Rutgers University Press, 2010. Historical study of implant development and implementation.

DiLorenzo, Daniel J., and Joseph D. Bronzino, eds. *Neuroengineering.* Boca Raton, Fla.: CRC Press, 2008. Essential review of neuroengineering developments written by leaders in the field.

Durand, Dominique M. "What Is Neural Engineering?" *Journal of Neural Engineering* 4, no. 4 (September, 2005). Written by the editor in chief of the journal, who defines NE and its scope.

He, Bin, ed. *Neural Engineering.* New York: Kluwer Academic/Plenum Publishers, 2005. Introductory overview of research in neural engineering.

Katz, Bruce F. *Neuroengineering the Future: Virtual Minds and the Creation of Immortality.* Hingham, Mass.: Infinity Science Press, 2008. Fascinating introduction to this field, describing the state of the art and speculating on long-term developments.

Montaigne, Fen. *Medicine By Design: The Practice and Promise of Biomedical Engineering.* Baltimore: The Johns Hopkins University Press, 2006. Bioengineering (including neuroengineering) applications made accessible to the nonspecialist through vignettes and portraits of researchers.

NIGHT VISION TECHNOLOGY

FIELDS OF STUDY

Biochemistry; Bioengineering; Biotechnology; Medical biotechnology (or Red biotechnology)

ABSTRACT

Night vision technology is used to allow for better night vision than is possible with the human eye alone. Night vision technology uses light amplification and thermal-imaging components incorporated into goggles, cameras, binoculars, and other devices to improve vision under low-light conditions.

DEFINITION AND BASIC PRINCIPLES

Night vision technology is the use of light-amplifying and thermal-imaging devices to enhance human vision performance in low light. These devices can take the form of cameras, goggles, binoculars, and spotting scopes. This technology takes ambient light and amplifies it through photoelectric techniques or thermal imaging that takes advantage of the energy released in the infrared spectrum in the form of heat.

Night vision devices use a photocathode that collects photons, which are light particles present even in dim light. These photons strike a photocathode, which then emits electrons. Photocathodes can be made of a variety of coated metallic materials. These electrons are multiplied by a microchannel plate and then transformed back into green light using a phosphor screen. Green light works well because of the sensitivity of the human eye to these wavelengths. There are variations on this technology, including early night vision systems that project infrared light and then amplify the reflected light.

BACKGROUND AND HISTORY

The groundwork for the development of night vision technology was laid by early scientists such as Heinrich Hertz who described the photoelectric effect in 1887. The discovery that electrons are emitted when light strikes metal was further developed by German physicists Max Planck and Albert Einstein in the early twentieth century. Their work confirmed the particle nature of light and provided the foundation for future applications, which included night vision technology.

William E. Spicer was a cofounder of the Stanford Synchotron Radiation Lightsource and was instrumental in the development of light amplification. His work paved the way for the first generation of night vision goggles and had applications in medical-imaging technology. Spicer's work provided the basis by which light in the infrared spectrum, which is not visible to the human eye, can be detected, amplified, and transformed into visible green light. All of the night vision devices rely on this basic technology.

As a result of the research done by Spicer night vision goggles were developed for use by the military in World War II in the 1940's. England, Germany, and the United States all developed sniper scopes using infrared cathodes. These devices used an infrared beam to generate reflected light from the surroundings that were then amplified by the scope. These devices had the disadvantages of low range and the ability of the enemy to detect the infrared beam. Early devices using an infrared beam to create reflected light are called active night vision devices and are referred to as generation zero.

Militaries around the world continued to work on improved night vision technology. Generation one devices, the next iteration, improved on the light amplification so that ambient light could be used without the need to use an infrared beam. These systems did not work well on very dark or cloudy nights. Early night vision devices were large and created distortion of images. The Starlight scope used in Vietnam is an example of this generation of devices. Generation zero and generation one night vision devices are now available to the general public.

As technology advanced, the next generation of night vision devices became more sensitive by the addition of microchannel plates, which further amplified the signal. A microchannel plate is manufactured from lead oxide cladding glass. Generation two devices have less distortion and increased brightness. Generation three night vision technology incorporates gallium arsenide cathodes, which further increases sensitivity. Generation four devices, which are typically used for military applications, incorporated changes to the microchannel and added gating. Gating is a system that switches on and off to allow for rapid response to changes in light. For example, if night vision goggles are on and then a light is suddenly switched on, the user will be then able to see under the lighted conditions.

Thermal imaging has been made possible with improved sensitivity and light amplification and also creates images using infrared wavelengths that are emitted as heat. Not all night vision devices are able to detect thermal energy.

HOW IT WORKS

To understand how night vision technology works, it is important to have a basic understanding of light and of how the human eye responds to light. Before the twentieth century, there was an ongoing debate as to whether light was a wave or a particle. Sir Isaac Newton favored a particle theory, which was later substantiated by Henrich Hertz, Max Planck, and Albert Einstein. However, modern understanding of light is that it behaves like both a wave and a particle.

For the purpose of understanding night vision technology, it is the photoelectric effect that forms the basis for these devices. When particles of light called photons strike metal, electrons are emitted. Specialized photocathodes are coated with various materials to make them more sensitive. The technical specifications of the photocathodes have improved over the generations since the 1940's, in part because of the use of different materials and coatings. The function of the photocathode in a night vision device is to convert the light into electrons. In low-light conditions the night vision devices are able to detect infrared light that is not detectable by the human eye.

The electrons are then converted into visible light by a phosphor screen, which then converts the electrons back into green light visible to the human eye. Later devices added a microchannel plate, which serves to amplify the electron energy while preserving the pattern or image. The microchannel plate is an array of tiny glass tubes. The electrons enter and are confined in each tube as they travel through, which results in the preservation of their entering pattern. While traveling through the microchannel plate, the electrons are further amplified by the application of voltage across the microchannel plate. This allows for more energy entering the phosphor screen and a subsequently brighter image. Infrared light travels from the environment to the photocathode, where it is translated into electrons, which in turn enter the microchannel plate. The amplified signal then strikes the phosphor screen, which turns the energy into green light that the viewer can see.

The human eye is most sensitive to visible light with wavelengths of about 400 to 700 nanometers (nm). Infrared light is in the 700 nm to 1 millimeter (mm) range. Infrared is further divided into near-infrared IR-A with 750 to 1400 nm wavelength range, medium wavelength IR-B of 1,400 to 3,000 nm range, and long wavelength or far IR-C with wavelengths of 3,000 nm to 1 mm. The long wavelengths are used in thermal-imaging devices. Infrared light is not detected by the human eye, so night vision devices are used to transcribe this light into visible green light. The human eye is particularly sensitive to green light. For example, 0.001 watt of green light will appear bright, while 0.001 watt of blue light will appear dim.

APPLICATIONS AND PRODUCTS

Military Applications. Military organizations used night vision technology in World War II and continue to be at the forefront of new developments. Military applications include night vision goggles for military personnel, sniper scopes, reconnaissance, and vehicle navigation. The advances that led to thermal imaging were on display in the media during the Gulf War in 1991. Those who may have watched the coverage of this war on television will remember the pictures with greenish images and periodic flashes of bright green corresponding to tracers and explosions.

Thermal forward-looking imaging (FLIR) devices are installed on vehicles and helicopters. Night vision devices are available to personnel for survival purposes even in a downed aircraft. This technology has continued to be employed in weapons-aiming devices. Data collection and communications technology have been added to some night vision devices in order to improve military communication and reconnaissance.

Law Enforcement. Law-enforcement applications are similar to military applications and include surveillance, weapons aiming, recording, and identification of suspects in situations of low light. Thermal imaging is used to identify illegal marijuana-growing operations, which are sometimes located in ordinary urban neighborhoods. The heat lamps used in growing the plants make it possible for law enforcement to identify these operations by air. A helicopter equipped with thermal-imaging equipment can detect an increased heat signature coming from the roof of the house that contains the growing operations. FLIR is also used on law-enforcement vehicles. Night vision technology is also used in search-and-rescue operations by law enforcement and other agencies.

Photography. Some photographers are using night vision cameras to create artistic images. To address the green images created by this technology, the photographers employ digital-editing techniques. The resulting images are unique works of art.

Recreational Use. Recreational use of night vision technology has expanded as the older generation of devices has become less expensive. Newer generations are still mostly used by the military and law enforcement because of the higher costs of these advanced devices. Spotting scopes, binoculars, and cameras are used by hunters, campers, hikers, and fisherman. Night vision devices are used for wildlife viewing and photography.

A unique activity that makes use of night vision goggles is dining in the dark. The servers use night vision goggles to provide a meal for diners who do not have the night vision goggles. The idea is to make the meal more of an adventure and to enhance the dining experience. Some companies use this as a team-building activity.

Scientific Research. Scientists use night vision devices to study nocturnal animals and other phenomena that might not otherwise be visible to the human eye. This has opened up a new area of study for wildlife biologists. In some parks, night vision technology is used to study wildlife and vehicle collisions in order to determine ways to reduce these incidents, which are dangerous to both humans and animals.

Astronomical research has also benefited from the use of night vision technology. The National Aeronautics and Space Administration (NASA) has used night vision technology to acquire images with the Hubble Space Telescope and the Mars Rovers. This technology is also being offered to amateur astronomers to enhance the images that can be acquired.

SOCIAL CONTEXT AND FUTURE PROSPECTS

The development of night vision technology has changed the way wars are fought. Before this technology was available most militaries avoided night operations. Militaries have competed to stay on the forefront of night vision technology research in order to maintain a tactical advantage. Night vision technology has been credited with the success of Desert Storm in 1991, giving the U.S. military an advantage in the conflict.

As this technology advances into solid-state formats, additional communications and analysis features will be added to allow real-time communication between soldiers. Remote surveillance and reconnaissance using thermal imaging is becoming more widely used. Night vision technology is already being used in the acquisition of astronomic images. NASA is already using thermal and infrared imaging in their missions to Mars.

Thermal-imaging systems are now being marketed for night driving, heavy equipment operators, maritime applications, and pilots. As the costs of these systems decline they will be more widely available for the general public and possibly may eventually become a standard option in passenger vehicles.

—*Ellen E. Anderson Penno,*
MD, MS, FRCSC, Dip ABO

FURTHER READING

American Academy of Ophthalmology. *Clinical Optics.* San Francisco: American Academy of Ophthalmology, 2006. This volume covers the fundamental concepts of optics as it relates to lenses, refraction, and reflection. It also covers the basic optics of the human eye and the fundamental principles of lasers.

Hobson, Art. *Physics: Concepts and Connections.* 5th ed. Boston: Pearson Addison-Wesley, 2010. Includes chapters on light, geometric optics, wave nature of light, and a section on night vision imaging.

Kakalios, James. *The Physics of Superheroes.* 2d ed. New York: Gotham Books, 2009. Uses comic-book references to cover basic physics theory. Includes chapters on mechanics, energy (heat and light), and modern physics.

Newell, Frank W. *Ophthalmology: Principles and Concepts.* 5th ed. St. Louis: Mosby, 1982. Covers basic eye anatomy, optics, and retinal physiology and biochemistry.

Tipler, Paul A., and Gene Mosca. *Physics for Scientists and Engineers.* 6th ed. New York: W. H. Freeman, 2008. Paul Tipler's physics text has been a staple for introductory university physics courses for many years. Chapters cover basic physics concepts including the basic physics of optics and the dual wave and particle nature of light.

P

PASTEURIZATION AND IRRADIATION

Cream pasteurizing and cooling coils at Murgon Butter Factory, 1939. [Public domain], via Wikimedia Commons

FIELDS OF STUDY

Biochemistry; Biotechnology; Environmental biotechnology (or Green biotechnology); Food science; Human nutrition

ABSTRACT

Pasteurization and irradiation are processes that partially sterilize food in order to make it safe to eat, without substantially altering its nutritional content, structure, and taste. Pasteurization uses mild heat treatment, whereas irradiation makes use of ionizing radiation. Both reduce the levels of pathogenic (disease-causing) and spoilage microorganisms to a level that renders the food safe to eat provided that it is stored appropriately for no longer than the prescribed time. Irradiation can also be used on fresh fruits and vegetables to kill insects and to retard biological processes, such as ripening. At high levels, irradiation will fully sterilize food, packing material, and disposable medical items.

DEFINITION AND BASIC PRINCIPLES

Pasteurization is the process of using mild heat to treat food, whereas irradiation, sometimes called radiation pasteurization or cold pasteurization, uses ionizing radiation. The primary purpose of each is to destroy microorganisms that would be pathogenic to human consumers, without significantly changing the food's attributes. In addition, these processes can be used to destroy microorganisms or enzymes that spoil food, leading to a longer shelf life and less waste. Irradiation can also be used on fresh fruits and vegetables to kill insects and to delay germination, ripening, or sprouting.

Pasteurization, used primarily with liquid foods, such as milk, fruit juices, and beer, refers to heat treatments that do not exceed 100 degrees Celsius (C), whereas heat sterilization (such as canning) uses temperatures of 100 degrees C or higher. In general, high-temperature short-time (HTST) pasteurization destroys undesirable organisms while minimizing deleterious effects on the food. In milk, comparable killing of microorganisms can be achieved by conventional pasteurization at 63 degrees C for thirty minutes or by HTST at 72 degrees C for fifteen seconds or 88 degrees C for one second. Sterilization of milk at ultrahigh temperature (UHT) typically requires 138 degrees C for two seconds.

For irradiation, the ionizing radiation used are gamma rays (generated from the decay of radioisotopes cobalt 60 or cesium 137), X rays, and electrons, the latter two generated by machines for such purposes. The operators of equipment involving ionizing radiation need to be protected from its effects. The ability subsequently to pasteurize or irradiate food should not compensate for best practices to minimize contamination of food before treatment. Moreover, treated food also needs to be protected from subsequent contamination.

The Radura symbol, as required by U.S. Food and Drug Administration regulations to show a food has been treated with ionizing radiation. [Public domain], via Wikimedia Commons

BACKGROUND AND HISTORY

Thermal and nonthermal processes have long been used to ensure the safety and storage of food. Cooking and smoking food were practiced in prehistoric times and likely permitted a survival advantage. Subsequently, drying, salting, and pickling were also used.

Pasteurization was developed in 1862 by French scientists Louis Pasteur and Claude Bernard, initially for the preservation of beer and wine. The pasteurization of milk became widespread in the early 1900s.

X-rays were discovered by German physicist Wilhelm Röntgen in 1895, and radiation emitted from uranium and other radioactive elements was discovered by French physicists Henri Becquerel, Marie Curie, and Pierre Curie shortly thereafter. Patents were issued for food preservation using ionizing radiation in 1905 and for the use of X rays to destroy *Trichinella* in pork in 1921.

By 2005, food irradiation was widely used in Asia, less often used in America and Eastern Europe, and rarely used in Western Europe, where consumer resistance is high. The primary use of irradiation is for herbs, spices, and dry vegetables, followed by root crops (such as potatoes and garlic, to inhibit sprouting), grains and fruits, meat, and seafood. Food irradiation is strictly regulated within each country, and all irradiated foods must be labeled with words such as "Treated by irradiation" and the international Radura logo. Global trade would be enhanced by harmonization of national regulations.

HOW IT WORKS

Pasteurization. Heat kills pathogenic and spoilage microorganisms by disrupting their cellular structure and metabolism. In pasteurization, sufficient heat is applied to the food being treated to kill undesirable organisms, but without damaging the food itself. In this balance, not all undesirable organisms are destroyed, but they are reduced to such a level that the product, if stored appropriately, will be safe for consumption until its use-by date.

Although some liquid foods, such as beer and fruit juices, may be pasteurized after filling containers (with warm water or steam applied to raise the temperature appropriately), most are pasteurized in a vat process or a continuous-flow process and then packaged. The vat (or batch) process involves heating in a well-agitated tank for the required time and temperature. The vat process is suitable for relatively small-scale operations.

The continuous flow process became possible when plate heat exchangers were developed in the late 1920s and has been enhanced by the development of concentric-tube heat exchangers. It is particularly well-suited to HTST and large-scale operations. In a typical system, the liquid to be pasteurized flows in a continuous tube from a cooled holding tank, through a preheater, through the heater (which heats the fluid to the required temperature), through a holding tube (whose size coupled with the flow rate determines the length of time that the liquid is held at the specified temperature), and then cooled down for storage. In practice, the preheater acts as a precooler, extracting heat from the heated liquid, before it is further cooled, permitting 85 to 90 percent of the heat to be reclaimed. The temperature in the holding tube must be monitored to ensure that the desired temperature has been maintained and, if not, the flow must be automatically diverted back to the starting tank.

In both vat and continuous-flow processes, the system must be thoroughly cleaned between uses, but a single use of the latter may last for many hours.

Irradiation. Radiation kills pathogenic and spoilage organisms, as well as insects. It also retards germination, ripening, and sprouting. It does so by disrupting cell structure, cell metabolism, and, most importantly, DNA molecules, preventing further growth and reproduction. Irradiation exerts its effects by direct action of the radiation or indirectly, principally via the radiolysis of water, which leads to the generation of highly reactive chemical species, such as hydroxyl radicals and hydrogen peroxide. Smaller organisms are more resistant than larger ones, for instance, viruses compared with bacteria. Spores of species such as *Clostridia* that cause botulism are more resistant than vegetative cells. Gram-negative bacteria, including primary food pathogens *Escherichia coli* and *Salmonella*, are more sensitive than gram-positive bacteria. Gamma rays are more effective than x-rays, which in turn are more effective than electrons. These differences relate to the penetrating power of the radiation, with electron irradiation only suitable for treatment of surfaces or thin packages.

In the process of irradiation, the product is brought in line with the source of radiation for the requisite period of time. Electrons or x-rays are generated by machines for these purposes and can be turned on and off as required. Because gamma rays result from radioactive decay, they cannot be turned on or off; when not needed, the sources of radioactivity are stored in a large water tank that absorbs the radiation. Irradiation with electrons and x-rays is well suited to a conveyor-belt system that brings the product into the radiation beam. With gamma rays, the use of an overhead rail system is preferred. The packages of product to be irradiated are suspended from that system and moved so that the package can be bombarded from various sides and angles to ensure uniformity of treatment. In all cases, a dosimeter (or dose meter) must be periodically included to ensure that the material has received the required dosage.

In food irradiation, sufficient radiation is applied to destroy undesirable organisms (including insects) or to inhibit a biological process, without adversely affecting the nutritive value and sensory characteristics of the food. As with pasteurization, organisms may not be completely eliminated but are reduced to a safe level provided the food is stored appropriately for no longer than the permitted time. Radiation is able to penetrate packaging materials, which reduces the risk of contamination after treatment. On the other hand, packaging materials can be affected by the ionizing radiation generating radiolysis that may migrate to the food and affect its taste. Careful choice of packaging material, as well as adhesives and printed material, must be made to avoid such problems. The fats in foods are susceptible to breakdown, forming products with unacceptable taste, but this effect is minimized by irradiating foods high in fat while frozen. Irradiation in the absence of oxygen minimizes the generation of byproducts that can affect the color and taste of the food.

APPLICATIONS AND PRODUCTS

Pasteurized Products. Pasteurization is typically applied to liquids—milk is the best-known example. Most milk consumed around the world is pasteurized or heat sterilized. Before the development of milk pasteurization, more than 25 percent of food-borne diseases were attributed to milk and milk products. Many microorganisms, including pathogenic ones, survive well in milk. Combined with aseptic packaging technology, pasteurization makes milk less prone to spreading disease and less perishable. Fruit juices and beers may be flash pasteurized (HTST) to minimize spoilage. A few wines are pasteurized; wines with less than 14 percent alcohol are sometimes pasteurized (or ultra-filtered) to stop any further fermentation. Flash pasteurization is also used to make wines acceptable to strict Orthodox Jews. Liquid eggs can be similarly pasteurized; those in shell can also be pasteurized in a series of warm-water baths.

Nonliquid pasteurized products include cheese, almonds, smokeless tobacco, crabmeat, bread, and ready-to-eat meals. Pasteurized cheese is made from pasteurized milk, so the liquid is treated in this process, although the cheese is subjected to heat treatments as well. Almonds can be pasteurized with a steam treatment, designed to kill any microorganisms on the outside of the nut; pasteurization of almonds can also refer to their treatment with propylene oxide, but that treatment is more appropriately called chemical fumigation. Smokeless (or chewing) tobacco is pasteurized by heating to 85 degrees C. Crabmeat labeled as pasteurized is heated to 113 degrees C for one minute in sealed cans or plastic containers. Because this temperature is higher than 100 degrees C, it is not properly termed pasteurization. Nevertheless, this treatment does not kill all pathogens present, it merely reduces them to a safe level, and the product must be stored at refrigerated temperatures for no longer than prescribed to ensure its safety. Bread and ready-to-eat meals are usually pasteurized by microwaves, which generate heat in the product being treated.

Irradiated Products. The extent of irradiation of a food will vary according to the desired end point and nature of the substance being irradiated. The absorbed doses are expressed in units of gray (Gy) and kilogray (kGy). Low doses (less than one kGy) will inhibit sprouting of potatoes, garlic, onions and other root foods, disinfect insects on fruits, grains and dry foods, delay ripening of fresh fruits and vegetables, and inactivate parasites on pork and fresh fish. Medium doses (one to ten kGy) will extend shelf life of strawberries, mushrooms, fresh fish, and meat (if stored at between 0 and 4 degrees C), destroy parasites in meats, control molds on fresh fruit, and destroy pathogenic and spoilage microorganisms in spices, raw or frozen poultry, meat, and shrimp. High doses (greater than ten kGy) will sterilize herbs, spices, meat, poultry, seafood, food additives (such as enzymes and natural gums), packaging materials (such as wine corks), disposable medical items (such as syringes, tubing, and gloves), and hospital food (especially for immune-compromised patients). In the United States, the only exception to a maximum of thirty kGy is for sterilizing frozen packaged meats for National Aeronautics and Space Administration (NASA) space flights. When in space, American astronauts have been eating irradiated foods, such as beef, pork, smoked turkey, and corned beef, since the beginning of the space program. Interestingly, milk and milk products are not good prospects for irradiation because it generates undesirable flavors.

In irradiation, gamma rays are used more often than electrons or x-rays, largely because of their greater penetrating power. Irradiation of fresh fruits and vegetables to destroy any mature or immature insects obviates any need to quarantine these products. In the past, such disinfection was done with methyl bromide, but its use is being phased out because it is an ozone-depleting chemical. The costs of food irradiation include high capital costs and modest operating costs. Irradiation facilities must protect workers from the radiation used; this involves thick walls and fail-safe design that prevents accidental radiation exposure to employees when in operation. Several methods are available to determine if a food has been irradiated, but biological methods that enumerate dead and live microorganisms of the species of concern are particularly useful because they not only show how many survived, but the total burden before irradiation, providing a check on good handling practices before treatment.

SOCIAL CONTEXT AND FUTURE PROSPECTS

Pasteurization is a well-accepted technology. Nevertheless, proponents of raw milk contend that pasteurization is unnecessary if milk is kept clean from the udder of the cow to the consumer and that it destroys some desirable components in milk. Replacing hand milking with milking machines and direct transfer of raw milk to a cooling tank have made contamination from microorganisms from the environment less likely but do not eliminate the risk. Recall that milk is a good medium for growing many microorganisms, including pathogens. With regard to components of milk that are destroyed in pasteurization, the only substantial loss is some enzymes, such as alkaline phosphatase. But, on eating, enzymes are inactivated and digested in the stomach and intestines of consumers and do not provide any demonstrable health benefits aside from being a source of amino acids. Public health and medical organizations promote the pasteurization of milk in the interest of food safety.

Irradiation is not as accepted, largely because of public concerns about nuclear radiation. Gamma irradiation does not increase radioactivity in food over what occurs naturally. Although it does produce radiolytic products in foods, animal testing indicates that irradiated food is safe and high doses of radiolytic products have no adverse effects. Public health and medical organizations, such as the American Medical Association, attest to the safety of irradiation in protecting the food supply. Public education at the point of sale has been demonstrated to be effective in overcoming resistance to irradiated foods.

—*James L. Robinson, PhD*

FURTHER READING
Berk, Zeki. *Food Process Engineering and Technology.* Academic Press, 2009.
Fellows, P. J. *Food Processing Technology: Principles and Practice.* 3rd ed., CRC Press, 2009.

Komolprasert, Vanee, and Kim M. Morehouse, editors. *Irradiation of Food and Packaging: Recent Developments*. American Chemical Society, 2004.

Ortega-Rivas, Enrique, editor. *Processing Effects on Safety and Quality of Foods*. CRC Press, 2010.

Stewart, Eileen. "Food Irradiation: More Pros than Cons? Part 1" *The Biologist*, vol. 51, no. 2, 2004, pp. 91–96.

———. "Food Irradiation: More Pros than Cons? Part 2." *The Biologist*, vol. 51, no. 3, 2004, pp. 141–44.

PATHOGEN GENOMIC SEQUENCING

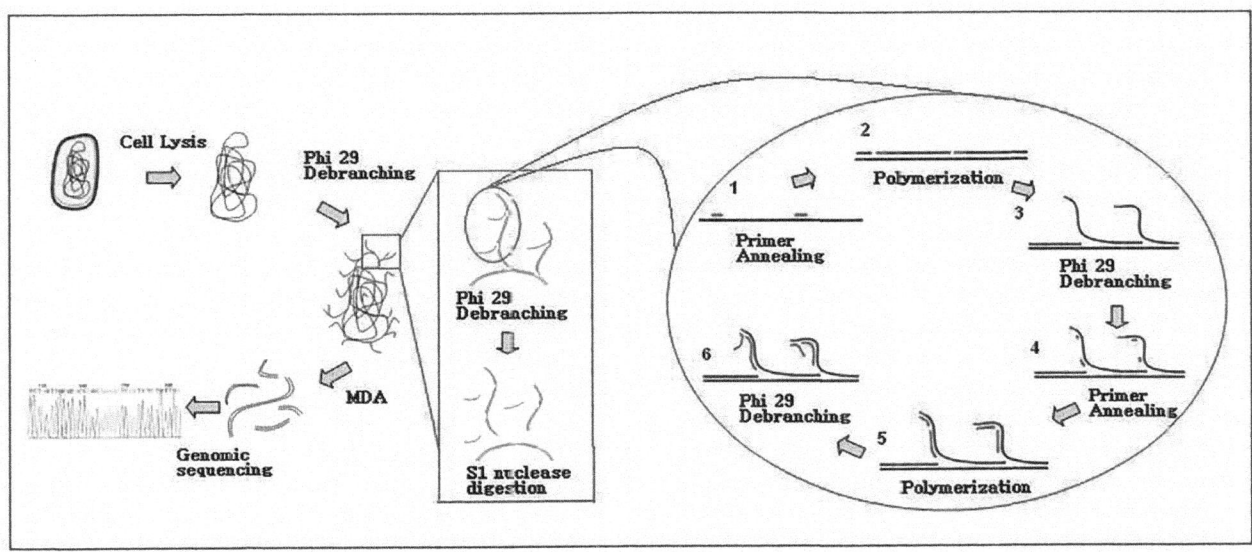

MDA mediated single cell genome sequencing. By Qianli Ma. (I made it myself.) [Public domain], via Wikimedia Commons

FIELDS OF STUDY

Biochemistry; Biotechnology; Environmental biotechnology (or Green biotechnology); Genetics; Medical biotechnology (or Red biotechnology)

ABSTRACT

Microbial forensics, the branch of forensic science that deals with microorganisms, extends the scope of epidemiology by going into greater detail to characterize pathogens for use as possible evidence in legal proceedings. The sequencing of pathogens is an especially important technique in forensic science as it pertains to increasingly dangerous global scourges of infectious emerging diseases.

DEFINITION

Techniques used to determine the linear order of monomers (small molecules such as nucleotides) that can be linked together to form polymers, mainly nucleic acids, found in the genomes of disease-causing microbes such as certain viruses, bacteria, and fungi.

HOW IT WORKS

Using a triad of techniques—Polymerase chain reaction (PCR) and genomic sequencing followed by phylogenetic studies, which infer relationships among microbial strains—forensic scientists have the ability to determine the origin of a disease-causing agent, or pathogen, used in a biocrime and to determine who or what organization was responsible for its dissemination. In 2000, the human genome was characterized through sequencing techniques; this accomplishment revolutionized molecular biology. The comparatively small viral genomes of human immunodeficiency virus (HIV), hantaviruses, and *Haemophilus influenzae* have been elucidated through sequencing, as have the larger genomes of bacteria including *Mycobacterium tuberculosis* (TB) and its drug-resistant strains, *Yersinia pestis* (plague), *Mycobacterium leprae* (leprosy or Hansen's disease), *Salmonella typhi* (typhus), *Bacillus anthracis* (anthrax), and *Neisseria meningitidis* (meningitis). All of these pathogens are considered threats to global health.

The techniques used by forensic scientists in the collection, handling, shipping, and preservation of potential pathogens are different from those employed for nonpathogenic samples. Forensic microbial analysis is based on a technique that identifies tandemly repeated sequences within the pathogen's genome. (Tandem repeats are repetitive sequences consisting of two or more nucleotides that serve as genetic markers and are frequently used to establish attribution—the source of a pathogen—in legal proceedings.)

After an unknown attacker sent letters containing *B. anthracis* to addresses in New York City, Washington, D.C., and Boca Raton, Florida, in the fall of 2001, careful preservation of initial and follow-up samples of the contents of the envelopes allowed forensic scientists to identify the Ames strain of *B. anthracis* as the specific pathogen used in the attacks. Before the entire genome was sequenced, the sample in question was compared with natural strains of *B. anthracis*; this comparison narrowed the possible source to a human-made strain of anthrax as opposed to one that might be found in nature. Further comparison to attribute the Ames strain to this biocrime was possible because the first strain, isolated from a victim in Florida (the index case), and the strains isolated from other victims as well as those found in anthrax spores recovered in the letters had previously been collected and preserved according to microbial forensic guidelines.

Specialized facilities, such as the J. Craig Venter Institute (JCVI), have been established to enable microbial forensic analysts to ascertain the entire genomic sequences of about five million bases of *B. anthracis* so as to identify polymorphisms (variations in DNA sequences) that serve as genomic signatures. In late 2007, the National Institute of Allergy and Infectious Diseases announced that twenty-eight hundred human and avian isolates had been completely sequenced and were publicly accessible.

———*Cynthia Racer*

FURTHER READING

Binnewies, Tim T., et al. "Ten Years of Bacterial Genome Sequencing: Comparative-Genomics-Based Discoveries." *Functional and Integrative Genomics* 6 (July, 2006): 165-185.

Breeze, Roger G., Bruce Budowle, and Steven E. Schutzer, eds. *Microbial Forensics*. Burlington, Mass.: Elsevier Academic Press, 2005.

Budowle, Bruce, et al. "Genetic Analysis and Attribution of Forensic Evidence." *Critical Review of Microbiology* 31 (October, 2005): 233-254.

_____. "Quality Sample Collection, Handling, and Preservation for an Effective Microbial Forensics Program." *Applied and Environmental Microbiology* 72 (October, 2006): 6431-6438.

Cole, Leonard A. *The Anthrax Letters: A Medical Detective Story*. Washington, D.C.: Joseph Henry Press, 2003.

PERFORMANCE-ENHANCING DRUGS

FIELDS OF STUDY

Biochemistry; Biotechnology; Medical biotechnology (or Red biotechnology); Philosophy and Religious Studies

ABSTRACT

As the level of athletic competition continues to rise throughout the world, athletes are always looking for ways to get a competitive edge. Superior nutrition and training programs are not enough for some athletes; many choose to use performance-enhancing drugs even though such substances are banned by sports federations and illegal unless prescribed for medical purposes. Law-enforcement agencies expend significant resources in efforts to address the illegal sale and use of such drugs.

BACKGROUND

Athletes' use of particular substances to improve their physical performance dates back to ancient Greece. Competitors in the ancient Olympics took

Ilona Slupianek failed a test along with three Finnish athletes at the 1977 European Cup, becoming the only East German athlete ever to be convicted of doping. From Bundesarchiv, Bild 183-Z0802-015. [CC BY-SA 3.0 de (https://creativecommons.org/licenses/by-sa/3.0/de/deed.en)], via Wikimedia Commons

stimulants such as strychnine and extracts from cola plants, cacti, and fungi to improve their performance. Although the beneficial effects of these substances are questionable, many believe that their widespread use was one of the elements that led to the termination of the Olympic Games and other sporting competitions in about 400 CE.

HOW IT WORKS

Competitive sports did not gain great popularity again until the time of the second phase of the Industrial Revolution, around 1850. With competition, the use of performance-enhancing drugs also returned. In particular, competitive swimmers, runners, and cyclists used caffeine, strychnine, codeine, cocaine, heroin, and nitroglycerin to stimulate their bodies to perform. Numerous athletes died from taking these drugs, but their deaths did deter many others from using such drugs. After World War II and throughout the Cold War period, the use of performance-enhancing drugs escalated, particularly the use of anabolic steroids.

Bodybuilding Supplements. The most widely known class of performance-enhancing drugs used since the 1930s is that of anabolic steroids (also known as anabolic-androgenic steroids). These are testosterone-like substances that augment male sex characteristics and the building of muscle. Anabolic steroids were first developed in Nazi Germany, where they were used to increase the aggressiveness of troops in battle.

Many studies have shown that anabolic steroids increase muscle mass and strength, which has made them popular among many different kinds of athletes, from football players to track-and-field athletes who participate in throwing events. Two strategies that athletes use to maximize strength and muscle mass with anabolic steroids are known as stacking and pyramiding. Stacking is the blending of different types of the drug in oral and injectable forms to maximize effect. Pyramiding is the continual increase in dosage over time to maximize benefit. Taking large doses of anabolic steroids comes with many dangerous side effects, however. Because the liver is responsible for breaking down and removing excess chemicals from the body, anabolic steroids can cause severe liver damage. These substances have also been linked to high blood pressure, adult-onset diabetes mellitus, increased blood clotting factors, and decreased high-density lipoproteins (good cholesterol) in the blood, all of which increase the risk of cardiovascular disease.

Another substance that has been reported to increase muscle mass and strength is human growth hormone (HGH). With the advances made in the field of genetic engineering during the 1980s, HGH became increasingly widely available and hit the black market, where athletes could get access to it. Research on the effects of HGH has been very limited, however, and any actual benefits of the substance for athletes have not been identified conclusively. Athletes at the 2012 Summer Olympics were tested for HGH, but because of the unreliability of one the test's constituent assays was taken off the market, rendering the tests invalid.

Two other substances that have been promoted as useful in increasing muscle mass and strength are dehydroepiandrosterone (DHEA) and androstenedione. Both are precursors to testosterone that are converted to testosterone by the body. Research has not found either substance to be effective for the enhancement of athletic abilities, and both decrease the high-density lipoproteins in the blood, which increases the risk of cardiovascular disease.

One supplement that has been shown to be effective in improving performance in high-intensity exercise is creatine. Research indicates that the ingestion of creatine in high doses helps the muscles to work harder and increases the body's ability to gain muscle and strength. Although creatine is not regulated by the U.S. Food and Drug Administration (FDA), it is banned by most sports federations. The long-term side effects of this substance have not been clearly identified.

Stimulants. The primary stimulant substances used by athletes for much of recent history are amphetamines. Athletes take these drugs to decrease fatigue, increase alertness, and decrease reaction time. Most research has found, however, that these drugs do not improve athletes' quickness; rather, under influence of the drugs, the athletes only perceive themselves as being quicker. In addition, amphetamines offer only short-term reduction of fatigue; thus in using amphetamines athletes gain no real performance benefits while exposing themselves to dangerous side effects: amphetamines are highly addictive, and those who take them experience increased metabolism, loss of appetite, and weight loss.

Athletes also use two stimulants that are not regulated drugs: caffeine and ephedrine. Endurance athletes use caffeine to increase their bodies' use of fat for energy and to conserve carbohydrates for later stages of their competitive events. Research has found such use to be effective and to have limited side effects, which include increased urine output and blood vessel spasms. The sports federations have not banned caffeine, but most place limits on the amount allowed in a competing athlete's body. Ephedrine is a naturally occurring stimulant similar to amphetamines. Like amphetamines, it has not been shown to have performance-enhancing benefits, and it has similar side effects. Ephedrine is banned by most sports federations.

Blood Doping. Many athletes in the past have improved their performance through blood doping—that is, by increasing the amount of red blood cells, which carry oxygen, in their blood. This increases the oxygen available to muscles and improves athletic performance in endurance events.

Historically, athletes who practiced blood doping would have several units of their own blood drawn and placed in storage six to eight weeks before competition. Their bodies would produce more red blood cells in the intervening time, and then, prior to the competition, the athletes would reinfuse their stored blood to increase their red blood cells. This process has generally been replaced by the use of the hormone erythropoietin, which causes the body to increase the production of red blood cells. All methods of blood doping are banned by sports federations.

IMPORTANT ISSUES

A major concern related to performance-enhancing drugs is the lack of information available about them. New drugs and variations on older drugs are continually being developed, in large part because manufacturers and users are interested in staying ahead of the technology available to test athletes for the use of banned and illegal drugs. When new drugs or new forms of older drugs are developed, several years of research are required to determine if they are effective, what the proper dosages are, and what their side effects are, as well as to develop new tests to detect their use by athletes.

Typically, a new drug formulation is available for more than a year before awareness of it becomes widespread enough that research on the drug is undertaken. After the research begins, more than another year might elapse before scientists are able to determine whether the drug is effective at all, and many years might pass before the negative side effects of the drug can be identified. Developing an effective method of testing for a new drug can also take months or even years. Given this lengthy process, athletes who use performance-enhancing drugs have a wide window of opportunity for cheating. Nonetheless, major professional sports associations have implemented stringent antidoping policies, and most, including that of Major League Baseball—an organization that has had well-documented problems with its athletes taking performance-enhancing drugs—have been successful in deterring athletes from using illegal substances.

Another serious concern raised by the use of performance-enhancing drugs is the effect of such substances on athletes' health. Many athletes, whether taking FDA-approved drugs or nontested substances, take very high doses. In fact, many take higher doses than what researchers can ethically test, and thus the true benefits and side effects of these substances are not known. Given that many of the known side effects have negative health implications, athletes who use illegal and banned substances to enhance their performance are not only breaking the rules but also risking serious health problems.

—Bradley R. A. Wilson

FURTHER READING

Aretha, David. *Steroids and Other Performance-Enhancing Drugs.* Berkeley Heights: Enslow, 2005. Print.

Bahrke, Michael S., and Charles E. Yesalis, eds. *Performance-Enhancing Substances in Sport and Exercise.* Champaign: Human Kinetics, 2002. Print

Espejo, Roman. *Performance-Enhancing Drugs.* Farmington Hills: Greenhaven, 2015. Print.

Haley, James, and Tamara Roleff. *Performance-Enhancing Drugs.* San Diego: Greenhaven, 2003. Print.

Monroe, Judy. *Steroids, Sports, and Body Image: The Risks of Performance-Enhancing Drugs.* Berkeley Heights: Enslow, 2004. Print.

Scott, Celicia. *Doping: Human Growth Hormone, Steroids, and Other Performance-Enhancing Drugs.* Broomall: Mason Crest, 2015. Print.

Yesalis, Charles E. *Anabolic Steroids in Sport and Exercise.* 2d ed. Champaign: Human Kinetics, 2000. Print.

PLANT BIOTECHNOLOGY

FIELDS OF STUDY

Biochemistry; Biology; Biotechnology; Botany; Environmental biotechnology (or Green biotechnology); Food science; Horticulture; Human nutrition

BACKGROUND

Plant biotechnology may be defined as the application of knowledge obtained from study of the life sciences to create technological improvements in plant species. By this very broad definition, plant biotechnology has been conducted for more than ten thousand years.

The roots of plant biotechnology can be traced back to the time when humans started collecting seeds from their favorite wild plants and began cultivating them in tended fields. It appears that when the plants were harvested, the seeds of the most desirable plants were retained and replanted the next growing season. While these primitive agriculturists did not have extensive knowledge of the life sciences, they evidently did understand the basic principles of collecting and replanting the seeds of any naturally occurring variant plants with improved qualities, such as those with the largest fruits or the highest yield, in a process that we call *artificial selection*. This domestication and controlled improvement of plant species was the beginning of plant biotechnology. This very simple process of selectively breeding naturally occurring variants with observably improved qualities served as the basis of agriculture for thousands of years and resulted in thousands of domesticated plant cultivars that no longer resembled the wild plants from which they descended.

The second era of plant biotechnology began in the late 1800s as the base of knowledge derived from the study of the life sciences increased dramatically. In the 1860s Gregor Mendel, using data obtained from controlled pea breeding experiments, deduced some basic principles of genetics and presented these in a short monograph modestly titled *Versuch über pflanzen-hybriden* in 1865 (*Experiments in Plant-Hybridisation*, 1910). In this publication, Mendel proposed that heritable genetic factors segregate during sexual reproduction of plants and that factors for different traits assort independently of each other. Mendel's work suggests a mechanism of heritable factors that could be manipulated by controlled breeding of plants through selective fertilization; the work also suggests that the pattern of inheritance for these factors could be analyzed or, in some cases, predicted by the use of mathematical statistics.

These findings complemented the work of Charles Darwin, whose 1859 book *On the Origin of Species by Means of Natural Selection* expounds the principles of descent with modification and selection as the chief factor of evolutionary change. The application of these principles to agriculture resulted in deliberately produced hybrid varieties for a large number of cultivated plants via selective fertilization. These artificially selected hybrids soon began to benefit humankind with tremendous increases in both the productivity and the quality of food crops.

GENETIC ENGINEERING

The third era of plant biotechnology involves a drastic change in the way crop improvement may be accomplished, by direct manipulation of genetic elements (genes). This process is known as *genetic engineering* and results in plants that are called genetically modified organisms (GMOs), to distinguish them from plants that are produced by conventional plant-breeding methods. Genetically modified plants can contribute desirable genes from outside traditional breeding boundaries. Even genes from outside the plant kingdom can be brought into plants. For example, animal genes, including human genes, have been transferred into plants, a feat not replicated in nature.

PUBLIC CONCERN

It is perhaps this lack of natural boundaries for genetic exchange that seems so foreign to conventional scientific thought and that makes plant genetic engineering controversial. The thought of taking genes from animals, bacteria, viruses, or any other organism and putting them into plants, especially plants consumed for food, has raised a host of questions among concerned scientists and the public alike. Negative public perception of genetically modified crops has affected the development and commercialization of many plant biotechnology products, especially food plants. While there are dozens of genetically engineered plants ready for field production, public pressure has delayed the release of some of these plants and has caused the withdrawal of others from the marketplace.

This public concern also appears to have driven increased government review of products and decreased government funding for plant biotechnology projects in Europe. Negative public perceptions have not seemed to be as strong in Asia, since the pressures of feeding large populations tend to outweigh the perceived risks. The social climate of the United States toward biotechnology, although guarded, has appeared to be less apprehensive than that of most European countries. Therefore, many agricultural biotechnology projects have moved from European countries to U.S. laboratories.

ECONOMIC GOALS

To what end are humans genetically engineering plants? This is an essential question for researchers, executives of biotechnology companies, and consumers at large. Before addressing technical questions about how to apply biotechnology, the desired goals must be clearly defined. The general goals of plant biotechnology appear to be (1) economic improvement of existing products, (2) improvement of human nutrition, and (3) development of novel products from plants.

Economic improvements include increases in yield, quality, pest resistance, nutritional value, harvestability, or any other change that adds value to an established agricultural product. Examples of this category include insect-protected tomatoes, potatoes, cotton, and corn; herbicide-resistant canola, corn, cotton, flax, and soybeans; canola and soybeans with genetically altered oil compositions; virus-resistant squash and papayas; and improved-ripening tomatoes. All of these examples were introduced to agriculture in the later half of the 1990s, though some have been removed from the market.

NUTRITIONAL GOALS

Additionally, some products appearing in the scientific literature but awaiting commercialization have the potential to dramatically improve human nutritional deficiencies, which are especially prevalent in developing countries. These products include "golden rice," genetically modified rice that produces carotenoids, a dietary source of vitamin A. Golden rice has the potential to prevent vitamin A deficiency in developing countries, where this vitamin deficiency is a leading cause of blindness. The product has been met with resistance by anti-GMO activists and with caution by the World Health Organization.

Researchers are also using genetic engineering to increase the amount of the iron-storing protein ferritin in seed crops such as legumes. Iron deficiency, which affects 30 percent of the human population, can impair cognitive development and cause other other health problems. This proposed enhancement of iron content in consumable plant products could help more than a billion people who suffer from chronic iron deficiency.

NOVEL PRODUCTS

Novel products include those not traditionally associated with plants and are limited only by imagination and available techniques. These include the production of plastics, vaccines, antibodies, human blood proteins, and new pharmaceuticals. One project has involved the production of hepatitis B vaccine in transgenic tomatoes. This project, which underwent clinical trials in the late 1990s, has the potential to provide a simple and inexpensive means of vaccinating people against hepatitis B. By oral administration of tomato juice containing the vaccine protein, humans are thought to develop an immune response that may protect them from infection by the hepatitis B virus. Hepatitis B has been epidemic in Asia and has increased at an alarming rate in the rest of the world. The disease ultimately causes liver disease, cancer, and death in nearly a million infected people worldwide.

PLANT TISSUE CULTURES

Central to plant biotechnology is the use of in vitro methods. Researchers use *plant tissue cultures,* for example, to grow plant cells on sterile nutrient media. Countless recipes for these nutrient media exist. The choice of which one to use is based on the plant species and the tissue type to be grown. All such media contain at least some of the important nutitional elements—such as nitrogen, potassium, calcium, magnesium, sulfur, phosphorus, iron, boron, manganese, zinc, iodine, molybdenum, copper, and cobalt, usually in the form of inorganic salts or as metal chelates—and an organic energy source, such as sucrose. The media may also contain vitamins, hormones, and other ingredients, depending on the intended use.

To initiate plant tissue culture, a piece of a living plant is excised and disinfected using a chemical disinfectant. This piece of plant tissue, called an *explant,* is placed on a sterile plant tissue culture medium to grow. Many plant tissues may be used to obtain explants for plant tissue culture, including those from leaves, petioles, shoots, tubers, roots, and meristematic regions. When an explant is placed in the sterile tissue culture medium, cells that are not terminally differentiated will grow and divide. If plant hormones are included in the recipe, the plant cells can be coaxed to develop into different types of tissues or organs. By using a succession of media containing different hormones, it is possible to regenerate whole plants from single cells. The choice of tissue used for the explant and the choice of hormones included in the tissue culture medium depend on the desired result.

MICROPROPAGATION

Micropropagation, another biotechnology technique, is the production of many clonal plants using tissue culture methods. By means of micropropagation, it is possible to generate many thousands of plant clones using tissue explants obtained from a single parent plant. The main advantage to micropropagation is the potential of producing thousands of exact copies of a plant with desirable traits. Micropropagation is especially important for rare plants, genetically engineered plants, and plants that have sexual reproductive problems. Many plant species are routinely propagated by micropropagation methods, including orchids, ferns, many flowering ornamentals, and vegetable plants.

STEPS IN GENETIC ENGINEERING

The first genetically engineered plants, tobacco plants, were reported in the scientific literature in 1984. Since 1984 there have been thousands of genetically engineered plants produced in laboratories worldwide. The process of genetically engineering a plant involves several key steps:
- isolating the genetic sequence (gene) to be placed from its biological source
- placing the gene in an appropriate vehicle to facilitate insertion into plant cells
- inserting the gene into the plant in a process known as *plant transformation*
- selecting the few plant cells that contain the new gene (transformed cells) out of all the plant cells in the explant
- multiplying the transformed cells in sterile tissue culture
- regenerating the transformed cells into a whole plant that can grow outside the tissue culture vessel

The gene or genes to be placed in the plant may be obtained from virtually any biological source: animals, bacteria, fungi, viruses, or other plants. Placing genes into an appropriate vehicle for transfer into a plant involves using various molecular biology techniques, such as restriction enzymes and ligation, to essentially "cut and paste" the gene or genes of interest into another DNA molecule, which serves as the transfer vehicle (vector).

PLANT TRANSFORMATION METHODS

Plant transformation with foreign genes may be accomplished by several proven methods, including bacteria-mediated transfer, microparticle bombardment, electroporation, microinjection, sonication, and chemical treatment.

By far, the most often utilized method of plant transformation involves the use of naturally occuring plant pathogenic bacteria from the genus *Agrobacterium*. In nature, this bacterium infects plants and transfers some of its own bacterial DNA into the plant. Through the action of proteins produced by the bacteria, bacterial DNA is made to integrate permanently into the plant's own genomic DNA. Expression of the bacterial DNA in the plant causes the plant to produce unusual quantities of plant hormones and other compounds, called opines, which provide food for the bacteria. The unusual quantities of plant hormones around the infection site cause the plant cells to grow abnormally, producing characteristic tumors. Scientists have harnessed this pathogenic bacterium to insert genes into plants by deleting the bacterial genes that cause tumors in the plant and then inserting desirable genes in their place. When the modified *Agrobacterium* infects a plant, it transfers the desirable genes into the plant genome instead of causing tumors. The desirable genes become a permanent part of the plant genome, and expression of these genes in plant cells produces desirable products.

One major drawback of the *Agrobacterium* method is that insertion of bacterial DNA into the plant genome is essentially random. The gene may not be efficiently transcribed at its location, or the insertion of bacterial DNA may knock out an important plant gene by inserting in the middle of it—or both may occur. Therefore, the fact that a cell is genetically transformed does not guarantee that it will perform as desired.

Microparticle bombardment is the introduction of foreign DNA constructs into plant cells by attaching the DNA to small metal particles and blasting the particles into plant cells using either a compressed air gun or a gun powered by a 0.22 caliber gun cartridge. This is truly a "brute force" method of introducing DNA into a cell that inadvertently causes many lethal casualties among the bombarded plant cells. However, some plant cells blasted with the DNA-containing metal particles will recover and survive. The plant cells may express the DNA for only a short time (transient expression), because the DNA does not readily integrate into the plant genome, but occasionally the foreign DNA may spontaneously recombine into the plant genome and become permanent.

Other ways of introducing foreign DNA into plant cells include electroporation, microinjection, sonication, and chemical treatment. These methods are not used extensively, because they generally require the production of *protoplasts* (plant cells that lack their cell walls) from plant cells before transformation. To create protoplasts, the plant cell wall is removed by digestion with the enzymes cellulase and pectinase. Protoplasts are fragile structures, but the absence of a cell wall is desirable because it leaves only the plasma membrane as a barrier to foreign DNA entering a plant cell.

Electroporation uses very brief pulses of high-voltage electrical energy to create temporary holes in the plasma membrane through which the foreign DNA can pass. *Microinjection* involves physically injecting a small amount of DNA into a plant cell using a microscope and an extremely fine needle. *Sonication* uses ultrasonic waves to punch temporary holes in the plasma membrane; this method is therefore similar to electroporation. *Chemical treatment* involves the use of polyethylene glycol to render the plasma membrane permeable to foreign DNA.

All of the transformation procedures produce only a few transformed cells out of the millions of cells in an explant, so selection of transformed cells is essential.

SELECTION OF TRANSFORMED PLANT CELLS

Selecting the few transformed plant cells out of all the plant cells in an explant requires some advance planning. Most foreign DNA constructs introduced into a plant are designed and built to contain additional genes that function as selectable markers or reporter genes. *Selectable markers* include genes for resistance to antibiotics or herbicides. Plant cells containing and expressing these genes will be tolerant of antibiotics or herbicides added to the plant tissue culture media, while the nontransformed plant cells will be killed off. The surviving cells in the tissue culture media are mostly transformed.

Instead of selectable markers, *reporter genes* may be used. Reporter genes induce an easily observable trait to transformed plant cells that facilitates the physical isolation of these cells. Reporter genes include beta-glucuronidase, luciferase, and plant pigment genes. Beta-glucuronidase (commonly known as GUS) allows the plant cells expressing this gene to metabolize colorigenic substrates, while nontransformed plant cells cannot. To use this test, researchers treat a small amount of plant tissue with the colorigenic chemical substrate. If the cell turns color (blue) it is known to be transformed and expressing the GUS gene. If the cell does not turn color, it probably is not transformed. Another reporter gene is luciferase, an enzyme isolated from fireflies. Luciferase makes plant cells glow in the presence of certain chemicals if the gene is present; hence, transformed cells glow, whereas nontransformed cells do not glow. Plant pigment genes, such as anthocyanin pigment genes, occur naturally in plants and produce pigments that impart color to flowers. Inclusion of these pigment genes as reporter genes will allow transformed plant cells to be selected by their color. Transformed cells have color, while nontransformed cells remain colorless. Both selectable markers and reporter genes allow selection of cells into which genes have been successfully inserted and are operating properly.

REGENERATING WHOLE TRANSFORMED PLANTS

After successfully getting a gene construct into a plant cell and selecting the transformed cells, it is possible to get the plant cells to multiply in tissue culture. Also, by treating the plant cells with combinations of plant hormones, the cells are made to differentiate into various plant organs or whole plants.

For example, treating transformed plant cells with a high concentration of the plant hormone cytokinin causes shoots to develop. Transferring these shoots to another medium, one that is high in the plant hormone auxin, will cause roots to develop on the shoots. In this way a whole *transgenic plant* may be regenerated from transformed plant cells. Once a transformed plant is regenerated in tissue culture, the plant may be transferred to a climate-controlled greenhouse, where it can grow to maturity.

Future generations of transgenic plants may then be propagated sexually via seeds or asexually via vegetative propagation methods. Often transgenic plants must be grown in containment greenhouses to prevent accidental release into the environment. In such high-tech greenhouses, all factors contributing to optimal plant growth—lighting, temperature, humidity, nutrients, and other environmental conditions—are tightly controlled. Often hydroponic systems, which use a solution of plant nutrients as a growth medium in place of soil, are employed to control all aspects of plant nutrition.

—*Robert A. Sinnott*

FURTHER READING

Acquaah, George. *Principles of Plant Genetics and Breeding*. 2nd ed. Hoboken: Wiley, 2012. Print.

Barbosa-Cánovas, Gustavo, et al. *Global Issues in Food Science and Technology*. Burlington: Academic, 2009. Print.

Barnum, Susan R. *Biotechnology: An Introduction*. 2nd ed. Belmont: Wadsworth, 2005. Print.

Chrispeels, Maarten J., and David E. Sadava. *Plants, Genes, and Crop Biotechnology*. 2nd ed. Sudbury: Jones & Bartlett, 2003. Print.

Dixon, Richard A., and Robert A. Gonzales, eds. *Plant Cell Culture: A Practical Approach*. 2nd ed. New York: Oxford UP, 1994. Print.

Gamborg Oluf L., and Gregory C. Phillips, eds. *Plant Cell, Tissue and Organ Culture: Fundamental Methods*. New York: Springer, 1995. Print.

International Food Information Council Foundation. "Food Biotechnology Timeline." *Food Biotechnology: A Communicator's Guide to Improving Understanding*. Food Insight, n.d. PDF. Web. 24 Sept. 2013.

Kyte, Lydiane, and John G. Kleyn. *Plants from Test Tubes: An Introduction to Micropropagation*. 3rd ed. Portland: Timber, 1996. Print.

Owen, Meran R. L., and Jan Pen, eds. *Transgenic Plants: A Production System for Industrial and Pharmaceutical Proteins*. New York: Wiley, 1996. Print.

Pierce, Benjamin A. *Genetics: A Conceptual Approach*. 4th ed. New York: W. H. Freeman, 2012. Print.

"Plant Biotechnology Timeline." *National Institute of Food and Agriculture*. USDA, 18 Mar. 2009. Web. 24 Sept. 2013.

Thieman, William J., and Michael A. Palladino. *Introduction to Biotechnology*. 3rd ed. Boston: Pearson, 2013. Print.

Weasel, Lisa H. *Food Fray: Inside the Controversy over Genetically Modified Food*. New York: Amacom-American, 2009. Print.

TIME LINE OF PLANT BIOTECHNOLOGY

Year	Event
1838	German scientists Matthias Schleiden and Theodor Schwann present their cell theory: that all life-forms are made up of cells.
1858	Biologist Rudolf Virchow adds to the cell theory, proposing that "where a cell exists, there must have been a preexisting cell."
1902	Austrian botanist Gottlieb Haberlandt completes the cell theory with his idea of totipotency: cells must contain all the genetic information necessary to create an entire, multicellular organism. Therefore, every plant cell is capable of developing into an entire plant.
1939	Roger Jean Gautheret demonstrates the first successful culture of isolated plant tissues as a continuously dividing callus tissue.
1953	James Watson and Francis Crick make their landmark proposal for the double-helical structure of deoxyribonucleic acid (DNA), the molecule that carries genetic material.
1954	The first whole plant is regenerated, or cloned, from a single adult plant cell by W. H. Muir, Albert Hildebrandt, and Albert Riker.
1967	DNA ligase, the enzyme that joins DNA molecules, is discovered.
1970–1971	Daniel Nathans, Hamilton Smith, and Werner Arber discover restriction endonucleases.
1972	Researchers at Stanford University construct the first recombinant DNA molecules, and the following year DNA is inserted into *Escherichia coli* cells.
1980	The U.S. Supreme Court rules that genetically altered life-forms can be patented, allowing Exxon to patent a microorganism that eats oil.
1983	The National Institutes of Health permit scientists at the University of California at Berkeley to release genetically engineered bacteria designed to retard frost formation on crop plants.
1983	The U.S. Patent Office begins to issue a series of patents for genetically modified plants.
1984	The Plant Gene Expression Center, a collaborative effort between academia and the U.S. Department of Agriculture, is established to research plant molecular biology, sequence plant genomes, and develop genetically modified plants.
1985	Field testing of plants genetically modified to resist plant pathogens and disease vectors begins.
1986	The first release of a genetically modified crop, genetically engineered tobacco plants, is approved by the Environmental Protection Agency.
1987	Calgene receives a patent for a DNA sequence that extends the shelf life of tomatoes.
1987	Advanced Genetic Sciences Inc. field-tests a recombinant organism designed to inhibit frost in strawberries.
1990	Calgene conducts field experiments with herbicide-resistant cotton plants.
1990	At the Plant Gene Expression Center, biologist Michael Fromm announces the use of a high-speed "gene gun" to transform corn. Gene guns are used to shoot genetic material directly into cells via DNA-coated microparticles.

1993	Kary Mullis wins the Nobel Prize in Chemistry for the development of polymerase chain reaction technology, a technique he invented in 1981 for quickly multiplying DNA sequences in vitro.
1993	The U.S. Food and Drug Administration (FDA) announces its finding that genetically modified foods are not "inherently dangerous" and not in need of regulation. The following year, the FDA approves the first genetically modified whole food crop, Calgene's Flavr Savr tomato (later removed from the market).
1996	The genomes of *Saccharomyces cerevisiae* (baker's yeast), with twelve million base pairs, and of ancient archaea cells, which live near thermal vents at the ocean bottom, advance biologists' understanding of the evolution of life.
1996	GM seeds of soybean, corn, canola, tomato, cotton, and potato are planted in the United States, Canada, Mexico, Australia, Argentina, and China.
1998	Virus-resistant papaya is planted in Hawaii.
1999	Swiss and German scientists develop golden rice, a grain fortified with betacarotene.
2000	Researchers complete the full genomic sequence for the model flowering plant *Arabidopsis thaliana*.
2001	A transgenic tomato that can thrive in salty conditions is developed by American and Canadian scientists.
2002	Researchers complete the genomic sequence for rice, *Oryza sativa*.
2008	Biotech sugar beets are commercialized.
2011	"High-oleic" soybean varieties, high in monounsaturated fats, become available in the United States. More GM foods are submitted for government review, including low-acrylamide potatoes and non-browning apples.
2012	A reported 420.8 million acres of biotech crops are grown in twenty-eight countries. More than 90 percent of the 17.3 million farmers growing these crops live in developing countries.

Plant Breeding and Propagation

Gentian seedlings in a plant nursery. By peganum (*Gentiana punctata* seedlings.) [CC BY-SA 2.0 (https://creativecommons.org/licenses/by-sa/2.0)], via Wikimedia Commons

FIELDS OF STUDY

Agricultural engineering; Biochemistry; Biology; Biotechnology; Botany; Environmental biotechnology (or Green biotechnology); Food science; Horticulture; Human nutrition

ABSTRACT

The science of plant breeding and propagation is the controlled, systematic identification and multiplication of useful plant varieties. It is critical to human survival. Of all discoveries furthering the advance of civilization, improvements in food production have arguably been the most significant. Breeding food crops for higher productivity, improved nutritional content, and greater resistance to stress and disease; preserving rare species and reintroducing them into the natural environment; and creating plant varieties that act as factories for complex pharmaceuticals are among the most active fields for the plant breeder. Genetic engineering techniques are revolutionizing the industry.

DEFINITION AND BASIC PRINCIPLES

Plant breeding and propagation science encompasses any systematic attempt to create or identify and select useful varieties of plants and multiply them for commercial production. Historically, the raw materials came from naturally occurring variations. Using increased knowledge of genetics and the mechanisms of inheritance, modern plant breeders have been able to crossbreed strains to produce hybrids with specific suites of characters. Breeders may use radiation or chemical mutagens to increase variability. Beginning in the 1980's, genetic engineering has enabled plant scientists to insert specific genes into genomes of cultivated plants.

Having produced and selected a desirable strain, the breeder must then produce sufficient numbers of plants to test for trait stability and determine optimal conditions for commercial production. Unlike annuals that produce abundant seed, perennials with long generation times, such as forest trees, are a challenge to the plant breeder. For species such as conifers, advances in tissue-culturing techniques have transformed production processes in both the development and the marketing phases.

The advent of genetic engineering in agriculture has produced a number of hotly debated issues concerning the safety, long-term environmental impact, and wisdom of introducing nonplant genes into human dietary staples on a global scale. This debate has slowed the spread of this technology.

BACKGROUND

Humans have been selecting and propagating useful varieties of plants for more than 10,000 years. The domestication of wheat, rice, and barley in the Old World and maize and potatoes in the New World was critical to the development of the earliest civilizations. When plant breeding became systematic enough to constitute science is difficult to pinpoint. Roman agricultural science, which drew heavily on Greek and Mesopotamian antecedents, included the deliberate introduction and propagation of novel plant varieties. Contemporary Chinese were at least as advanced as the Romans in agronomy and horticulture, and the sophisticated agricultural techniques and variety of cultivars in pre-Columbian Mexico and Peru argues for the existence of a scientific approach in those cultures.

Agricultural science came into renewed prominence in the eighteenth century as educated landowners brought the tools of the Enlightenment to bear. Private botanical gardens became establishments for introducing, testing, and propagating exotic species and varieties of food and ornamental plants. By midcentury, breeders recognized that flowering plants did have a sexual cycle and that the principles of animal breeding applied to them.

Beginning with the concept of genes pioneered by Gregor Mendel in the nineteenth century, the science of genetics progressed steadily through the twentieth century, giving plant breeders increasing understanding of the processes underlying their work. Plant breeders have contributed a great deal to genetics in general.

HOW IT WORKS

Conventional Plant Breeding. The plant breeder seeks to combine the desirable traits from different strains into a single variety. Sources for the genes producing these traits include existing commercial cultivars, field collections or germ plasm banks, closely related wild plants, and spontaneous or induced mutations. Under carefully controlled conditions, the researcher cross-pollinates two varieties, plants the resulting seed, and evaluates the progeny for the desired combination of characteristics. To produce uniform seed of the resulting hybrid for sale, a company will develop breeding stocks of the two parent strains, crossing them in production plots. A gene for male sterility is often incorporated into one of the parents, preventing self-pollination.

Tremendous improvements in methods of vegetative propagation have enabled it to be used for commercial production of perennials with a generation time of many years. There have also been significant advances in methods for evaluating desirable characteristics at an earlier stage in a plant's life cycle. If the biochemical basis for a mature trait is known, precursors can often be detected in very young seedlings. Alternatively, if a mature trait is closely linked genetically to a trait that is expressed earlier in the life cycle, the presence of that marker can be used for screening.

Many plants form viable interspecific crosses. Often these are sterile, which results in seedless fruits. Seedless varieties must either be hybrids of seed-producing parents or propagated vegetatively. Sometimes breeders use colchicine to induce polyploidy, resulting in a sexually reproducing interspecific hybrid with a double chromosome complement.

Genetic Engineering. Certain viruses and viruslike plasmids invade cells and attach their DNA to that of the host. Any DNA attached to the virus or plasmid will also be incorporated. Plant genetic engineering uses a plasmid from a phytopathogenic bacterium, *Agrobacterium tumefaciens*, to transform plant cells. Because transformation rates are very low, geneticists incorporate an antibiotic-resistance gene to facilitate screening. Plant cells in undifferentiated tissue culture are exposed to transformed *Agrobacterium* plasmids and transferred to an antibiotic-containing medium. The surviving cells are then grown out as plants.

In theory, the ability to synthesize any biologically produced compound can be transferred to any plant species by this method. In practice, the absence of activators or presence of inhibitory genes stymies many attempts.

Once engineered into a plant variety, a gene propagates normally from generation to generation and, under field conditions, into other populations of the same species. A gene for herbicide resistance, introduced into a crop to facilitate management by chemical means, can backfire if it spreads to closely related weeds. Of serious concern is so-called terminator technology, a sterility gene introduced into genetically engineered seed by seed companies to prevent patent infringement.

Modern Methods of Plant Propagation. A key feature of modern plant propagation is the use of explants and their proliferation under sterile conditions to produce large numbers of genetically identical, pathogen-free plant starts. Undifferentiated plant cells are grown in liquid or solid media promoting rapid growth. To produce a mature plant, subcultures are then subjected to growth regimes promoting differentiation and maturation. For biopharmaceutical production, plant cells may be grown indefinitely in liquid-culture bioreactors and harvested in the same manner as microbial populations, often heterotrophically.

APPLICATIONS AND PRODUCTS

Seed. The primary use of plant breeding is to improve the productivity, cultivation characteristics, stress tolerance, and nutritional content of food crops, particularly cereal grains. The green revolution stressed productivity under modern agricultural methods.

Later, more attention was paid to tailoring crops to specific environmental and social conditions. An example is yellow rice, genetically engineered for high vitamin A content, to combat vitamin A deficiency in rice-dependent populations.

Preservation of Biodiversity. Tissue culture methods for propagation have been a great boon for the proliferation of medicinal plants, heritage varieties, and stock for habitat restoration. Orchids and cycads, which are difficult to propagate from seed, are under tremendous pressure from collectors who encourage poaching in nature preserves. Modern nursery propagation helps protect these vulnerable species.

Molecular Farms. The use of genetically engineered plants to produce exotic organic compounds is still in the development phase but shows promise. Researchers have successfully produced a strain of *Arabis* (a mustard) that synthesizes hirudin, an anticlotting agent, from leeches. Molecular-farmed pharmaceuticals of plant origin are safer than those extracted from animals or human plasma. Another promising line of development is edible vaccines—food plants synthesizing proteins that provoke an antigen response to a particular disease-causing bacterium.

Reforestation. With tissue culture and rapid mass multiplication of stocks of woody plants, reforestation following disturbance has become more targeted. The forester can readily access seedlings adapted to the site, with built-in insect and disease resistance. Species such as the American chestnut, eliminated from most of its original range by blight in the early 1900's, are being successfully reintroduced as highly selected resistant strains.

Virus-Indexed Plants. Production of disease-free stock of potatoes and bananas has always been a challenge. The ability to produce commercial quantities of plant starts from a small amount of sterilized tissue greatly reduces the spread of viruses and other pathogens in these vegetatively propagated crops.

SOCIAL CONTEXT AND FUTURE PROSPECTS

Advances in plant breeding have provoked many controversies, the most pressing of which are the contribution of improved crop varieties to unsustainable population growth, ownership of rights to germ plasm and to the products of genetic engineering, the safety of transgenic plants as human food, the risks of the unplanned spread of modified genes, and the adverse effects on traditional agriculture in developing countries. Concern over safety has led to a patchwork of national laws that inhibit but fail to effectively regulate global commerce in the products of genetic engineering.

The relationship of increased crop productivity to exponential population growth, already noted by William Malthus in 1798, was a feature of the green revolution in the twentieth century. Gains from improved technology or new crops are quickly canceled out unless population growth slows. Rural populations may actually end up worse off because improved technology favors large-scale operations and displaces farmers. Institutes in developing countries, such as the International Maize and Wheat Improvement Center (Centro Internacional de Mejoramiento de Maíz y Trigo, or CIMMYT), are working to ensure that the new round of agricultural technology will have a more beneficial human impact.

New varieties of plants under patent can be propagated only under license from the originator, which prevents farmers from saving seed. This increases costs and forces small operations to use older, less productive strains. When genetically engineered strains cross with crops in adjacent fields, the farmer of the traditional crop may be unable to sell it for human consumption. There are conflicting claims over the ownership of rights to improved varieties whose base stock came from field collections in developing nations.

If these objections can be overcome, the potential for the new technology is tremendous. The combination of greater productivity and improved nutritional content in staple crops could be of immense benefit. Incorporating resistance to disease and stress should lower the need for pesticides and herbicides and make farming on marginal lands more environmentally friendly. Using genetically engineered green plants to produce pharmaceutical and other complex organic chemicals ought to lower their cost and increase availability. The ability to propagate a species from vegetative portions of a few individuals and reintroduce it into its native habitat will undoubtedly be a boon in preserving endangered species, restoring native vegetation, and promoting sustainable ecologically friendly landscaping. As long as science and the public do not lose sight of potential social costs and maintain adequate safeguards, plant breeding will continue to serve humankind well.

—Martha A. Sherwood, PhD

Further Reading

Hrazdina, Geza, ed. *The Use of Agriculturally Important Genes in Biotechnology*. Washington, D.C.: NATO Scientific Affairs Division, 2000.

Kalloo, G., and J. B Chowdhury, eds. *Distant Hybridization of Crop Plants*. Berlin: Springer-Verlag, 1992.

Liang, George H., and Daniel Z. Skinner. *Genetically Modified Crops: Their Development, Uses, and Risks*. New York: Food Products Press, 2004.

Parekh, Sarad R. *The GMO Handbook: Genetically Modified Animals, Microbes, and Plants in Biotechnology*. Totowa, N.J.: Humana Press, 2004.

Slater, Adrian, Nigel W. Scott, and Mark Fowler. *Plant Biotechnology: The Genetic Manipulation of Plants*. New York: Oxford University Press, 2008.

Plant cells: molecular level

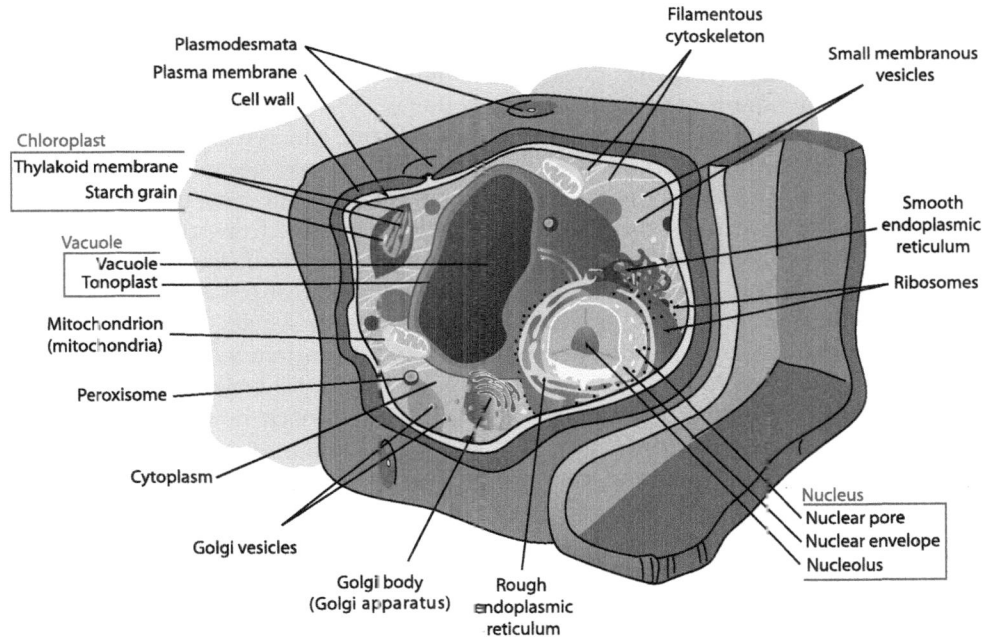

Gentian seedlings in a plant nursery. By peganum (Gentiana punctata seedlings.) [CC BY-SA 2.0 (https://creativecommons.org/licenses/by-sa/2.0)], via Wikimedia Commons

FIELDS OF STUDY

Biochemistry; Biology; Biotechnology; Botany; Environmental biotechnology (or Green biotechnology); Food science; Horticulture; Microbiology; Medical biotechnology (or Red biotechnology)

ABSTRACT

Water, ions, salts, and gases all are types of inorganic molecules that are essential to cellular function. The chemical properties of water make it an ideal solvent and buffer for the chemistry that occurs inside cells. The capillary action that helps water travel up plant tissues from the roots is a direct consequence of the polarity of the water molecule.

The chemistry of life on earth is carbon and water chemistry. Water is the most abundant compound in living cells and makes up as much as 90 percent of the weight of most plant tissues. Many of the molecules that are part of larger macromolecules in cells are linked together chemically by *dehydration synthesis*, or the loss of water. These macromolecules are broken up into their component units by the addition of a water molecule between the units, a process known as *hydrolysis*. The chemical properties of water make it an ideal solvent and buffer for the chemistry that occurs inside cells.

Because the electrons of the covalent bonds within the water molecule are more often orbiting the oxygen atom, the oxygen atom gains a slightly negative charge. The hydrogen atoms are slightly positive. This separation of charge across the water molecule is said to make it polar. Because of its polar nature, water is able to dissolve, or ionize, a variety of molecules. This gives water its buffering capacity.

Water molecules are attracted to one another because of this polarity. This weak attraction, which occurs in the form of hydrogen bonds, has great chemical consequences when many molecules of water are involved. Hydrogen bonding allows water to have surface tension. The capillary action that helps water travel up plant tissues from the roots is a direct consequence of the polarity of the water molecule. Water is also able to absorb heat without vaporizing (changing from a liquid to a gas state) quickly. Therefore, physiological temperatures can be maintained as water molecules absorb the heat from metabolic reactions.

Water, ions, salts, and gases all are types of *inorganic molecules* that are essential to cellular function. Inorganic molecules are chemical molecules that do not contain carbon. The remainder of the molecules within cells are built around the unique properties of the carbon atom and are called organic molecules.

ORGANIC MACROMOLECULES

There are four major classes of organic molecules in cells: *carbohydrates*, *lipids*, *nucleic acids*, and *proteins*. All of these molecules contain carbon backbones, and almost all of them contain oxygen and hydrogen as well as other elements. Some or all of the members of each class of organic molecules occur as very large molecules, called macromolecules, that are polymers of smaller molecules joined together by covalent bonds. For example, starch and cellulose are carbohydrate polymers of simpler carbohydrates called sugars. Likewise, fats and oils are lipid polymers composed of smaller lipids called fatty acids and the sugar alcohol called glycerol.

CARBOHYDRATES

Carbohydrates are molecules that consist of primarily carbon, hydrogen, and oxygen atoms. Carbohydrates are the primary source of stored energy in most living organisms. They can also serve as structural molecules in cell walls and as markers on some cell membranes, identifying different types of cells.

Simple sugars, or *monosaccharides*, are sugars that are small molecules composed of a chain of covalently bonded carbon atoms with associated hydrogen and oxygen atoms. These molecules always have a ratio of one carbon atom to two hydrogen atoms to one oxygen atom (CH_2O). The monosaccharide glucose is the primary sugar produced from simpler sugars made in photosynthesis.

When two simple sugars are covalently linked together, they form a *disaccharide*. In plants, the disaccharide sucrose, which is composed of one fructose molecule and one glucose molecule, is the most common sugar. Sucrose is the same thing as so-called table sugar, which is harvested from sugar cane or sugar beets.

Many sugars can be linked together to form a carbohydrate polymer, or *polysaccharide*. Starch is composed of many glucose molecules linked together and is the major form of carbohydrate storage in plants. When energy is required, the individual sugars of the polysaccharides are hydrolyzed (broken down to simpler molecules), and the glucose that is released is used by the mitochondria to generate energy. Polysaccharides are also important structural molecules in plants. The most abundant polysaccharide in nature is cellulose, another polymer of glucose and a major component of plant cell walls.

LIPIDS

Lipids are diverse group of unrelated molecules which includes fats, oils, steroids and sterols, waxes, and other water-insoluble molecules. Lipids are characterized by their hydrophobic, or "water-fearing," chemical behavior, which is what makes them insoluble in water. Unlike other molecules that ionize and are dissolved by water, lipid molecules are nonpolar. They are repelled by the polar nature of water and tend to aggregate in aqueous solutions. Lipids also are used to store energy and are especially abundant in seeds because lipids contain more energy by weight than carbohydrates.

Examples of lipids commonly found in biological systems include fats and oils that are storage molecules known as *triglycerides*. A triglyceride consists of glycerol (a three-carbon molecule) and three fatty acid molecules, long-chain hydrocarbon molecules that are attached to each of the three glycerol-carbon atoms by ester linkages.

The long chain of carbon atoms of the fatty acid can be saturated or unsaturated with respect to hydrogen content. *Saturated fatty acids* contain as many hydrogen atoms as allowed bonded to each carbon atom. Saturated fats tend to be solid at room temperature and include substances such as butter and lard. *Unsaturated fatty acids* do not have the maximum number of hydrogen atoms because some of the carbon atoms form double bonds with adjacent carbon atoms in the chain. Unsaturated fats tend to be liquid at room temperature and include substances such as corn oil and olive oil.

Plants have many lipids that are unique to them. For instance, cutin and suberin are two lipid polymers that form structural components of many plant cell walls. These two molecules form a meshwork that secures another type of lipid polymer found in plants, wax. *Waxes* are long-chain lipid compounds that are integrated into the cutin and suberin meshwork and are important in preventing water loss for plants. Waxes give apple peels their characteristic shiny appearance.

Phospholipids are a type of lipid molecule that is found in all living organisms. They are structurally similar to triglycerides, except instead of having three fatty acids attached to glycerol, they have only two. Replacing the third fatty acid is a charged phosphate group. This unique structure results in one end of the molecule being hydrophilic (the phosphate end, often called the head) and the other being hydrophobic (the end with the two fatty acids, often called the tail). Consequently, phospholipids will spontaneously form an oily layer at the water surface, orienting their charged phosphate heads toward the water and their fatty acid tails away from the water and toward the air. This is the basis for the phospholipid bilayer structure that underlies the formation of all cellular membranes. In the case of a lipid bilayer, because there is water on both sides, the two layers are tail to tail, with their heads oriented to the inside and outside of the membrane, where they come into contact with water.

NUCLEIC ACIDS

The information that directs all cellular activity is contained within the chemical structure of the *nucleic acids*. Nucleic acids are polymers of smaller molecules called *nucleotides*. Nucleotides, in turn, are composed of three types of covalently linked molecules: a ribose sugar, a phosphate group, and a nitrogen-containing base. The two major nucleotides that are found in cells are *deoxyribonucleic acid* (DNA) and *ribonucleic acid* (RNA).

DNA contains the genetic information that directs the development and activity of the organism. In eukaryotic cells DNA resides in the nucleus in linear molecules of repeating nucleotide units, although there are circular molecules of DNA found in the mitochondria and chloroplasts of eukaryotic cells. DNA nucleotides are composed of a five-carbon deoxyribose sugar, a phosphate group, and one of four possible bases: adenine (A), thymine (T), cytosine (C), and guanine (G). The information of the DNA molecule is found in the sequence of the nitrogenous bases along its length. Any region of DNA that directs a cellular function or encodes another molecule is called a *gene*. Not all DNA regions encode proteins. Some regions encode the instructions for RNA molecules that are used as catalysts and for protein synthesis reactions. Some genes are regulatory, controlling the time and place where certain genes are expressed. In many eukaryotes, genes only account for 10 percent of the DNA. Although some of the remaining 90 percent carries various structural functions, most of it is of uncertain function.

In 1953 Francis Crick and James Watson constructed a molecular structure for the DNA molecule, relying heavily on the experimental data generated by Rosalind Franklin. The structure they proposed, which has since been supported by additional experimental data, was that of a *double helix*. The DNA molecule can be envisioned as a ladder. The sugars and phosphates of the nucleotides alternate with each other to form the backbone, the outside vertical support, and the bases form the individual rungs of the ladder. The ladder is twisted to create a helical structure. DNA can exist as single strands and in other confirmations in the cell, but the "B-form" of the DNA double helix is the most common form in the cell.

RNA molecules are also polymers of nucleotides, but the nucleotides of the RNA molecule differ slightly from those of the DNA molecule. RNA nucleotides contain a five-carbon ribose sugar, a phosphate group, and one of four bases. Three of the four bases are the same as found in DNA: adenine, guanine, and cytosine. Instead of thymine, RNA uses the base uracil. RNA bases can pair in essentially the same way as DNA bases, but most often RNA exists as single-stranded molecules in cells. These long strands of RNA can often pair with other bases in short regions, causing the RNA to fold up into highly complex, three-dimensional structures important for RNA function.

RNA is found throughout cells. messenger RNA (mRNA) is made by the cell using the DNA sequence in genes as a template for making a complementary strand of RNA in a process called transcription. In After being transcribed and modified in certain complex ways, most mRNA is transported to the cytoplasm where it is used to direct the synthesis of proteins. Ribosomal RNA (rRNA) is a major component of ribosomes, which are responsible for coordinating protein synthesis, along with transfer RNA (tRNA). Some RNA molecules, like protein molecules, can also catalyze chemical reactions. Catalytic RNA molecules are called *ribozymes*, and they play roles in gene expression and protein synthesis.

Single nucleotides and compounds that are made from them are involved in many cellular processes. The universal unit of "energy currency" in the cell is *adenosine triphosphate* (ATP). Guanosine triphosphate (GTP) is a molecule that is involved in relaying signals received at the cell membrane to the nucleus of the cell. Compounds, such as NADH and NADPH, that are involved in metabolic reactions in the mitochondria and in energy capture reactions in the chloroplasts also contain nucleotides.

PROTEINS

Protein molecules are large, complex molecules with a huge variety of structures and functions within cells. Most chemical reactions in cells are catalyzed by proteins called *enzymes*. Proteins form the basis of the cytoskeleton of cells, providing structure and motility. Proteins are also essential for the communication between cells and within cells. In plants, the largest concentration of proteins can be found in some seeds.

Proteins are polymers of nitrogen-containing molecules called *amino acids*. The amino acids are much simpler molecules than the nitrogenous bases found in nucleic acids. The same twenty amino acids is are used in the manufacture of proteins in the cells of all living organisms. An amino acid is built around a single carbon atom called the *alpha carbon*. Bonded to the alpha carbon are a hydrogen atom (H), a carboxyl group (COOH), and an amino group that contains nitrogen (NH_2). A specialized "R" group is attached at the last site. The R-groups are different for each of the twenty amino acids, and their chemical properties, such as charge, hydrophilic or hydrophobic nature, and size, dictate protein function and shape.

The order and number of amino acids that are linked together to form a protein are determined by the order of the codons in the DNA that encode that protein. The order and number of the amino acids in a protein is called the *primary structure*, and it ultimately determines the shape of the protein. Proteins can have *secondary structures* formed by hydrogen bonding between the peptide bonds that link the amino acids together. The two common secondary structures in proteins are the *alpha helix* and the *beta pleated sheet*. The amino acid chain (also called a peptide chain) can fold up on itself to form globular structures. This is known as *tertiary structure*. Tertiary structure is determined by the number and order of amino acids in the protein and is formed when molecules in the R-groups of the amino acids interact with one another. When two or more peptide chains interact to form a single functional molecule, the protein is said to have *quaternary structure*.

——*Michele Arduengo*

FURTHER READING

American Society for Cell Biology. *Exploring the Cell.* Available on the World Wide Web at http://www.ascb.org/pubs/exploring.pdf. Features excellent images of cells and cellular processes.

Lodish, Harvey, et al. *Molecular Cell Biology.* 4th ed. New York: W. H. Freeman, 2000. One of the most authoritative cell biology reference texts available. Provides excellent graphics, CD-ROM animations of cellular processes, and primary literature references.

Raven, Peter H., Ray F. Evert, and Susan E. Eichhorn. *Biology of Plants.* 6th ed. New York: W. H. Freeman/Worth, 1999. Introductory botany text provides a chapter introducing eukaryotic cells in general and plant cells specifically.

Plants as a Medical Resource

Matthaeus Silvaticus teaching his students about medicinal plants in his physic garden in Salerno, from the frontispiece to a 1526 edition of *Opus Pandectarum Medicinae* by Giaros. [Public domain], via Wikimedia Commons

FIELDS OF STUDY

Biochemistry; Biology; Biotechnology; Botany; Environmental biotechnology (or Green biotechnology); Horticulture; Microbiology; Medical biotechnology (or Red biotechnology)

ABSTRACT

Because plants are so biochemically diverse, they produce thousands of natural products commonly referred to as secondary metabolites, and many of these secondary metabolites have medicinal properties that have proven to be beneficial to humankind.

BACKGROUND

The use of plants for medicinal purposes predates the recorded history of humankind. Primitive people's use of trial and error in the constant search for edible plants inevitably led them to the discovery of plants that contained substances that caused appetite suppression, stimulation, hallucination, or other side effects. Written records show that drugs such as opium have been in use for more than five thousand years. From antiquity until fairly recent times, most practicing physicians were also botanists or at least herbalists. In contemporary society medicinal plants are perhaps one of the most overlooked natural resources. Because modern commercial medicines are obtained in neat packages in the form of pills, capsules, or bottled liquids, most people do not realize that many of these drugs were first extracted from plants. In some cases, chemists have learned how to duplicate synthetically the natural product that was initially identified in a plant, but in many cases, a plant may still be the only economically feasible source of the drug.

PLANT-DERIVED MEDICINES

There are numerous ways to categorize medicinal compounds from plants. For this discussion, medicinal drugs will be categorized as antibacterial substances, anti-inflammatory agents, drugs affecting the reproductive system, drugs affecting the heart and circulation, drugs affecting the central nervous system, antiasthma drugs, drugs affecting the gastrointestinal tract, antiparasitic agents, and anticancer agents.

The first effective antibacterial substance was carbolic acid, but the first truly plant-derived antibacterial drug was penicillin, which was extracted from an extremely primitive plant, the fungus *Penicillium*, in 1928. The work with penicillin led to the discovery of other fungal and bacterial compounds that have antibacterial activity. The most notable of these are cephalosporin and griseofulvin.

Inflammation can be caused by mechanical or chemical damage, radiation, or foreign organisms. For centuries poultices of leaves from coriander (*Coriandrum sativum*), thornapple (*Datura stramonium*), wintergreen (*Gaultheria procumbens*), witchhazel (*Hamamelis virginiana*), and willow (*Salix niger*) were used to treat localized inflammation. In the seventeenth and eighteenth centuries, cinchona bark was used as a source of quinine, which could be taken internally. In 1876, salicylic acid was obtained from the salicin produced by the willow leaves. Today, salicylic acid, also known as aspirin, and derivatives such as ibuprofen, are the most widely used anti-inflammatory drugs in the world.

The most effective home remedy for preventing pregnancy was a tea made from the leaves of the Mexican plant zoapatle (*Montana tomentosa*). The drug zoapatanol and its derivatives were extracted from this plant to produce the first effective birth control substance—which has not been used in human trials, however, because of potential harmful side effects. Other plant compounds that affect the reproductive system include diosgenin, extracted from *Dioscorea* species and used as a precursor for the progesterone used in birth control pills; gossypol from cotton (*Gossypium* species.), which has been shown to be an effective birth control agent for males; ergometrine, extracted from the ergot fungus (*Claviceps* species.) and used to control postpartum bleeding; and yohimbine from the African tree (*Corynanthe yohimbe*), which apparently has some effect as an aphrodisiac.

Through the ages, dogbane (*Apocynum cannabinum*) and milkweeds (*Asclepias spp*) have been prized for their effects on the circulatory system because of the presence of a group of compounds called cardiac glycosides, but foxglove (*Digitalis* species) has produced the most useful cardiac glycosides, digitalis and digoxin. Opiate alkaloids such as opium extracted from the poppy (*Papaver sonniferum*) and its derivatives such as morphine, as well as cocaine from *Erythroxylum coca* and *Erythroxylum truxillense*, have long been known for their analgesic (pain-relieving) properties through their effects on the central nervous system. Both these drugs can also produce harmful side effects, however, and both have addictive properties.

The major antiasthma drugs come from ephedrine, extracted from the ma huang plant (*Ephedra sinaica*), and its structural derivatives. Plant-derived drugs that affect the gastrointestinal tract include castor oil, senna, and aloes as laxatives, opiate alkaloids as antidiarrheals, and ipecac from *Cephaelis acuminata* as an emetic. The most useful plant-derived antiparasitic agent is quinine, derived from the bark of the chinchona plant (*Chinchona succirubra*). Quinine has been used to control malaria, a disease that has plagued humankind for centuries. The primary plant-derived anticancer agents are vincristine and vinblastine, extracted from *Catheranthus roseus*, maytansinoids from *Maytentus serrata*, ellipticine and related compounds from *Ochrosia elliptica*, and taxol from the yew tree (*Taxus baccata*).

THE FUTURE

Many as-yet-unknown plant-derived medicinal drugs await discovery, particularly in the tropical rain forests. The threats to many plant species, and biodiversity in general, from development and industrialization may compromise the ability of humankind to take advantage of the unique compounds offered by these plants.

Modern biotechnology has provided the methods by which plants can be bioengineered to produce new and novel pharmaceuticals. Progress toward the production of specific proteins in transgenic plants provides opportunities to produce large quantities of complex pharmaceuticals and other valuable products in traditional farm environments rather than in laboratories. These novel strategies promise a broad array of natural or nature-based products, ranging from foodstuffs with enhanced nutritive value to the production of biopharmaceuticals.

——*D. R. Gossett*

FURTHER READING

Evans, William Charles. *Trease and Evans Pharmacognosy*. 15th ed. New York: W. B. Saunders, 2002.

Foster, Steven, and Rebecca L. Johnson. *Desk Reference to Nature's Medicine*. Washington, D.C.: National Geographic Society, 2006.

Hanson, Bryan. *Understanding Medicinal Plants: Their Chemistry and Therapeutic Action*. New York: Haworth Herbal Press, 2005.

Kar, Ashutosh. *Pharmacognosy and Pharmacobiotechnology*. 2d ed. Tunbridge Wells, England: Anshan, 2008.

Lewis, Walter H., and Memory P. F. Elvin-Lewis. *Medical Botany: Plants Affecting Human Health*. 2d ed. Hoboken, N.J.: J. Wiley, 2003.

Mann, John. *Murder, Magic, and Medicine*. Rev. ed. New York: Oxford University Press, 2000.

Plotkin, Mark J. *Medicine Quest: In Search of Nature's Healing Secrets*. New York: Viking, 2000.

Sneader, Walter. *Drug Discovery: A History*. Hoboken, N.J.: Wiley, 2005.

Stockwell, Christine. *Nature's Pharmacy: A History of Plants and Healing*. London: Century, 1989. An excellent discussion of medicinal products from plants.

World Health Organization *WHO Guidelines on Good Agricultural and Collection Practices (GACP) for Medicinal Plants*. http://whqlibdoc.who.int/publications/2003/9241546271.pdf

Polymerase chain reaction

FIELDS OF STUDY

Biotechnology; Environmental biotechnology (or Green biotechnology); Forensics; Microbiology; Medical biotechnology (or Red biotechnology)

ABSTRACT

The technique of polymerase chain reaction allows forensic scientists to use extremely small amounts of sample DNA to identify individual humans, animals, and other organisms. Such analyses are used for many purposes, including to link suspects to crime scenes, to identify the victims of mass disasters, to establish paternity, and to identify pathogens.

CONCEPTS

The molecule of life, DNA (deoxyribonucleic acid), is the unique genetic blueprint of an individual. DNA is present in virtually every cell in the body. The uniqueness of DNA, which is composed of two long strands, stems from the specific order (sequence) of its different nucleotide building blocks, which are linked together to form each strand. Because the nucleotide sequence of each person's DNA is unique (the only exception being identical twins, who have identical DNA), DNA analysis is an integral part of forensic investigations. DNA may be isolated from biological samples (such as blood, skin, semen, or hair) found at the scenes of mass disasters or crimes, from cheek cells swabbed from the insides of mouths (as may be taken from the parties in paternity determinations or from criminal suspects), or from the environment (as in the case of pathogens).

The specific sequence of nucleotides in several regions of the isolated DNA (short tandem repeats, known as STRs, or other specific sequences) is compared with the sequence in identical regions of a DNA standard. This standard comes from a known source; it may be from a predeath biological sample recovered from a disaster victim's personal effects, from a crime suspect, from a party involved in a paternity case, or, in the case of a pathogen, from a previously experimentally derived sequence. When a crime is being investigated, the standard may be on file in the Federal Bureau of Investigation's Combined DNA Index System (CODIS). Even if there is no match in a criminal case, such information may help exonerate a suspect.

Frequently, the amounts of DNA obtained at crime or disaster scenes and in other situations are insufficient to allow such comparisons. When this is the case, scientists can use the technique of polymerase chain reaction (PCR) to generate a large number of copies of the DNA for analysis. This technique is so sensitive that the DNA isolated from a single cell is sufficient to obtain the desired nucleotide sequence.

In PCR, the DNA to be copied is placed in a plastic test tube along with nucleotides, DNA polymerase, and two or more short sequences of single-stranded DNA (called primers) that define where DNA polymerase will begin working. DNA polymerase is an enzyme that adds nucleotides to the growing strand. The test tube is then placed in a machine that can vary the temperature of the test tube and its contents.

The test tube is first heated so that the two strands of the original DNA molecule will separate from each other. Each strand will act as a template for production of a new strand. The temperature is then slightly decreased to allow the DNA primers to bind to each template adjacent to the start of nucleotide addition. The temperature is then slightly increased, and DNA polymerase adds nucleotides until a new strand is synthesized on each template, producing two double-stranded DNA molecules after the first cycle that are identical to the original DNA molecule. Each time the steps are repeated, the DNA doubles: Four DNA molecules are produced after two cycles, eight DNA molecules are produced after three cycles, and so on. Repeating these steps thirty times results in about one billion copies of the amplified DNA region.

—*Jason J. Schwartz*

FURTHER READING

McPherson, Michael, and Simon Møller. *PCR (The Basics)*. 2d ed. New York: Taylor & Francis, 2006.

Rudin, Norah, and Keith Inman. *An Introduction to Forensic DNA Analysis*. 2d ed. Boca Raton, Fla.: CRC Press, 2002.

PROKARYOTES

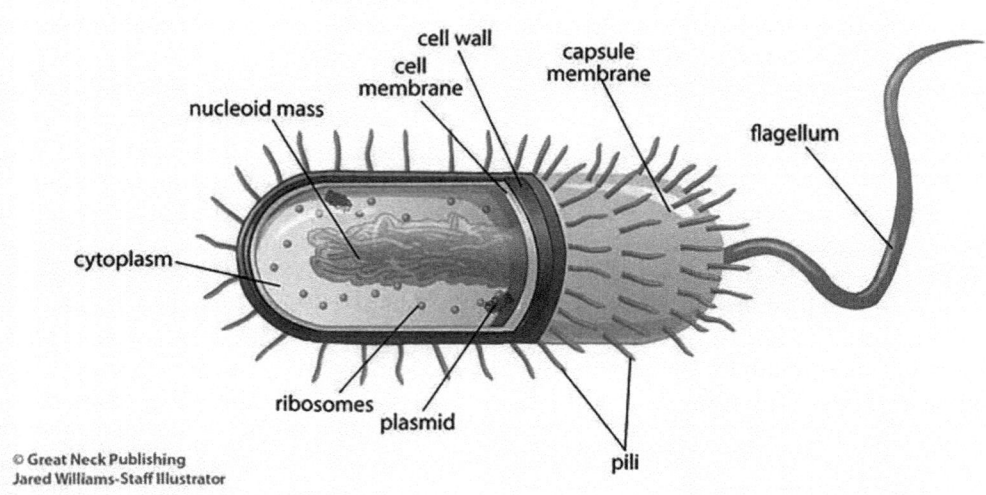

Prokaryotes: Prokaryotes are microscopic organisms which have been on Earth for millions of years. They differ from eukaryotes in that they are very simple and do not have a nucleus to enclose DNA, nor specialized cell organelles. EBSCO

FIELDS OF STUDY

Biotechnology; Environmental biotechnology (or Green biotechnology); Geology; Microbiology

ABSTRACT

Prokaryotes are primitive, one-celled organisms that have left an extensive fossil record in the form of sedimentary structures produced by physiological activity of cell communities. For 80 percent of the earth's history, communities of prokaryotes made up the biosphere of the earth. They are a well-defined group of organisms and occupy a highly diverse variety of habitats.

CHARACTERISTICS OF PROKARYOTES

From about 3.5 to 1 billion years ago, life on Earth, as determined from the fossil record, consisted entirely of one-celled organisms that have a cell morphology and a metabolism different from those of all other life-forms. These organisms, the prokaryotes, are characterized by their lack of a cell nucleus, their lack of sexual reproduction (meiosis), the small size of the prokaryotic cell, and their distinctive biochemistry. Prokaryotes are neither plants nor animals, although the aerobic photosynthetic forms, often called blue-green algae, have in the past been placed with the plants. Eukaryotes, organisms with a cell nucleus and a larger, more complex cell, make up the animals, plants, fungi, and protists. Prokaryotes are thus quite separate from all other life-forms in terms of their cell biology. Prokaryotes and eukaryotes are the two most basic categories of living things, exhibiting basic differences in their biologic processes that are greater than those that exist between animals or plants or between any of the other kingdoms.

Prokaryotes are mainly single-celled organisms, usually found living together in "colonies" consisting of immense numbers of cells. Their deoxyribonucleic acid (DNA) is distributed throughout the cell; it is not, as in the case of the eukaryotes, localized in a cell nucleus surrounded by a nuclear membrane. The prokaryotic cell is smaller by a factor of ten than the average eukaryotic cell. It lacks chloroplasts and mitochondria and consequently is considered primitive when compared with the eukaryotic cell.

Prokaryotes constitute the kingdom Monera, one of five kingdoms in modern taxonomy. (The other kingdoms are the protists, fungi, animals, and plants, all of which have more complex eukaryotic-type cells.) Phyla, or categories, within the kingdom Monera include the bacteria, the cyanobacteria (or blue-green algae), the archaebacteria, and the prochlorophytes. The bacteria, as well as the other moneran phyla, are further subdivided into a number of classes. Bacterial classes of the Monera include the eubacteria, photosynthetic bacteria, myxobacteria (slime bacteria), actinomycetes (moldlike bacteria), and other groups, each characterized by its own distinctive metabolism and biochemistry. The bacteria consist of obligate or strict anaerobes and facultative anaerobes; the former include the photosynthetic bacteria, which differ from the cyanobacteria not only in their ability to function, if required, under anaerobic conditions and low light levels, but also in their different photosynthetic pigment.

The archaebacteria are considered by some to be the most primitive and ancient of the monerans. Archaebacteria have a number of biochemical and metabolic characteristics that allow them to live under very adverse conditions—conditions such as those that appear to have existed during the early history of the earth. The archaebacteria are defined from their ribosomal ribonucleic acid (RNA), which in sequencing is quite different from that of all other monerans. The archaebacteria differ fundamentally from the other bacteria classes in structural and biochemical aspects as well.

GEOLOGICAL SIGNIFICANCE

Fossil prokaryote cells of great antiquity have become widely known from the fossil record. They were first reported in the 1910's by C. D. Walcott from 1.5-billion-year-old strata of western Montana (Belt series). However, the authenticity of these fossils was doubted until the discovery, in 1954, of one of the oldest known paleontological "windows" on life of the past, the 2-billion-year-old Gunflint biota. Since then, many occurrences of prokaryote cell fossils have been reported, most from very fine-textured flinty cherts associated with stromatolites of the Proterozoic (latter part of the Precambrian) eon and dating from as far back as 3.5 billion years.

The geologic significance of the prokaryotes is great: Not only do they (at present as well as in the geologic past) play an important part in the recycling of many chemical elements, but they also have a role in basic geologic processes such as weathering and other alteration of rocks. For example, prokaryotes are involved in the formation of stromatolites. Stromatolites are layered organosedimentary structures, frequently found fossilized in rock strata of many different geologic ages. Stromatolites come in a considerable variety of shapes and sizes; the different types have often been given Linnaean biological names because, when originally discovered, they were thought to be fossil organisms like corals or sponges. Most stromatolites are dome-shaped, finger-shaped, or laminar structures that have a characteristic "signature." They can form significant parts of rock strata, particularly in limestone and dolomites. Stromatolites are found in rock strata as ancient as the Archean (former part of the Precambrian) eon and are particularly diverse and abundant in strata of the Proterozoic. Locally, they can be quite common in early Paleozoic marine strata as well.

The origin of stromatolites was debated for many years; as late as the 1950's, many paleontologists seriously doubted their biogenic origin. This doubt stemmed, at least in part, from the fact that stromatolites occur so much further in the geologic past than do any other fossils. Through thousands and thousands of meters of Precambrian strata, they are the only fossils that can be found. Early workers on stromatolites, such as Walcott, suggested a cyanobacterial origin for them. The discovery of the well-preserved cells of prokaryotic type in digitate (fingerlike) stromatolites of the Gunflint Chert of Ontario in the 1950's led to a gradual acceptance by most geologists and paleontologists of the organic origin of the majority of stromatolites. It became clear that, under the right conditions, small, fragile cells could be preserved in very ancient strata.

During the 1970's and 1980's, studies on Precambrian stromatolites and the prokaryotic organisms responsible for them became widespread. Stromatolite occurrences going as far back as 3.5 billion years have been documented. These ancient stromatolites yield not only morphological information but also carbon isotope ratios indicative of a biogenic origin. They sometimes supply biochemical information in the form of hydrocarbons, amino acids, and porphyrins (the latter is apparently a degradation product of original photosynthetic pigment). In 1999, studies of Proterozoic rocks from western

Australia confirmed the presence of chemicals that could only have been synthesized by cyanobacteria.

The morphology of a prokaryotic organism is simple. Unlike fossils of eukaryotic organisms, fossils of most prokaryotes provide little specific information about the actual living organism. Prokaryotic cells can be single coccoid (spherical) forms, or they can be elongate chains of cells, as with the filaments, or trichomes, of the cyanobacteria.

STUDY OF PROKARYOTES

A standard petrographic thin section mounted on a glass slide is the common mount for observing cells preserved in a stromatolite. Oil immersion is usually required if fossil prokaryote cells are to be observed. Thin slivers of a stromatolite can also be examined under oil immersion; however, the best results are generally with well-made thin sections. Often considerable trial and error is involved in finding stromatolites that preserve cells and then in actually locating those cells; different parts of a particular stromatolite specimen usually have varying degrees of cell preservation. Very fine-grained sediments, such as those that occur with stromatolites preserved by black cherts or finely crystalline limestones, generally give the best results.

In this section under high optical magnification, a stromatolite may exhibit fossil cyanobacterial cells as either filaments or rod-shaped forms. If preservation of these small prokaryote cells is excellent, as in the stromatolites of the Gunflint formation, the biogenic origin of the cells will be clear, and distinct cell types can be observed. When most stromatolites are examined in thin section under high magnification, however, the biogenicity of the small objects seen is usually not so certain. Often, small black globules of carbon, suggestive of macerated cells, are evident, but their origin usually cannot be proved. Contaminants such as spores, pollen grains, bacteria cells, and fungi fragments can be a problem, particularly in examination of suspected fossiliferous rocks when thin sections are not used. Even with most thin sections, the unequivocal verification of a biogenic origin for fossil cells is rare. In the case of the Gunflint prokaryotes, the detail preserved in these fossil cells is highly remarkable; some of them show internal cell structure and cells in the process of division.

The earliest stromatolites that yield these fossil cells are generally either broad domes or laminar forms. Associated cells either are single-cell coccoid forms or consist of probable chains of photosynthetic bacteria. Chains of cells of filamentous cyanobacteria generally first appeared about 2.3 billion years ago, and this appearance of filaments agrees fairly well with the first appearance of branched or digitate stromatolites, for which filamentous cyanobacteria seem to be responsible.

CHEMICAL SIGNATURES

Often more significant than single-cell morphology or the megascopic morphology of a stromatolite is the chemical signature left by a group of prokaryotes as a consequence of their metabolic activity. Prokaryotes are classified according to their type of metabolism; some prokaryotes have a metabolism that enables them to occupy a wider variety of ecological niches than do eukaryotic organisms. Anaerobic and aerobic forms are the two fundamental forms of prokaryotic metabolism. In these two categories are the autotrophs and the heterotrophs; heterotrophic prokaryotes require previously formed organic material on which to live, while autotrophs do not. The autotrophs obtain their energy from their environment either in the form of sunlight (photoautotrophs) or through chemical reactions such as oxidation, as in the sulfur-oxidizing bacteria; such bacteria are called chemoautotrophs. This type of metabolism is unique in the organic world, for all other life-forms obtain their energy from photosynthesis or through utilization of the chemical energy contained in previously formed organic compounds. The cyanobacteria are photoautotrophs and are responsible for the formation of the various types of stromatolites. The process of photosynthesis changes the microenvironment around the photosynthesizing prokaryote; the mineral precipitation that results is responsible for the formation of stromatolite layers.

Some stromatolites contain oxidized manganese, cobalt, or other "transitional" elements, possibly incorporated into these fossil communities by oxidative metabolism of bacteria. Chemoautotrophic prokaryotes, which are various types of bacteria, may leave a chemical signature in the form of these oxides and precipitate their production of a layered stromatolite-like structure containing these oxidized metals. A number of bacteria oxidize manganese to higher oxidation states so that it is precipitated; deep-sea manganese nodules presently being formed are believed to have such an origin. Sectioning of these nodules shows a finely layered, stromatolite-like structure. Some of the heavy-metal-bearing stromatolites of the early Precambrian may reflect a similar chemoautotrophic metabolism. Analysis of organic residues present in many stromatolites in small quantities can sometimes shed light upon the specific organisms responsible for forming them. This technique, however, has met with only limited success, although degradation products of the photosynthetic pigment present in cyanobacteria have been identified, supporting the cyanobacterial origin of many ancient stromatolites.

The earliest stromatolites, those of Archean age (about 3.5 billion years old), exhibit certain distinctive morphological and chemical aspects. Some of these early stromatolites may be products of anaerobic assemblages of photosynthetic bacteria rather than of cyanobacteria communities. Geochemical evidence suggests that the atmosphere in the Archean may have been anoxygenic (oxygen-free) and that the photosynthetic bacteria, not being obligate aerobes, would have been favored by such an environment.

——*Bruce L. Stinchcomb*

Further Reading

Broadhead, T. W., ed. *Fossil Prokaryotes and Protists: Notes for a Short Course.* Knoxville: University of Tennessee, 1987. Information on a broad range of fossil prokaryotes and protist microfossils is presented. This work is the text of one of a series of "short courses" sponsored by the Paleontological Society, but it can be useful to anyone interested in the various fossil groups covered.

Fedonkin, Mikhail A., James G. Gehling, Kathleen Grey, Guy M. Narbonne, and Patricia Vickers-Rich. *The Rise of Animals: Evolution and Diversification of the Kingdom Animalia.* Baltimore: Johns Hopkins University Press, 2007. An up-to-date account of all the known Ediacaran sites that is extensively and beautifully illustrated. Accessible to all interested readers.

Gunde-Cimerman, Nina, Aharon Oren, and Ana Plemenita. *Adaptation to Life at High Salt Concentrations in Archaea, Bacteria, and Eukarya.* New York: Springer, 2011. This text provides information on high-salinity habitats and the halophilic organisms found there. The environment overview is followed by sections devoted specifically to Archaea, bacteria, fungi, protozoa, and viruses.

Margulis, Lynn, and Michael J. Chapman. *Kingdoms and Domains: An Illustrated Guide to the Phyla of Life on Earth.* 4th ed. Boston: Elsevier, 2009. Provides a comprehensive overview (current at the time of publication) of life's classification and phylogeny. An improvement on Margulis's previous work, *Five Kingdoms*, this text includes more recent discoveries and systematics. The classification is a modified version of Linnaean taxonomy.

McMenamin, Mark. *Discovering the First Complex Life: The Garden of the Ediacara.* New York: Columbia University Press, 1998. This entertaining study of the earliest complex life-forms on the planet details the author's work on these organisms. Written for the interested student but understandable by the general reader.

Nisbet, Evan G. *The Young Earth: An Introduction to Archean Geology.* Boston: Unwin Hyman, 1987. A highly comprehensive coverage of geologic phenomena of the earth's earliest geologic time span, the Archean eon. Included in this work is information on both crustal evolution and biosphere. The book's sections vary considerably in technical coverage and terminology, some parts being readily comprehensible to the lay reader while others are quite technical and require considerable background in trace element geochemistry, isotope geochemistry, and petrology.

Schopf, J. William, ed. *Major Events in the History of Life.* Boston: Jones and Bartlett, 1992. An excellent overview of the origin of life, the oldest fossils, and the early development of plants and animals. Written by specialists in each field but at a level that is suitable for high school students and undergraduates. Although technical language is used, most terms are defined in the glossary.

Yaacov, Davidov, and Eduard Jurkevitch. "Predation Between Prokaryotes and the Origin of Eukaryotes." *BioEssays* 31 (2009): 738-757. Discusses how eukaryotes evolved as a result of parasitism or predation occurring between prokaryotes. Covers the topic of mitochondrial symbiosis.

PROTEOMICS AND PROTEIN ENGINEERING

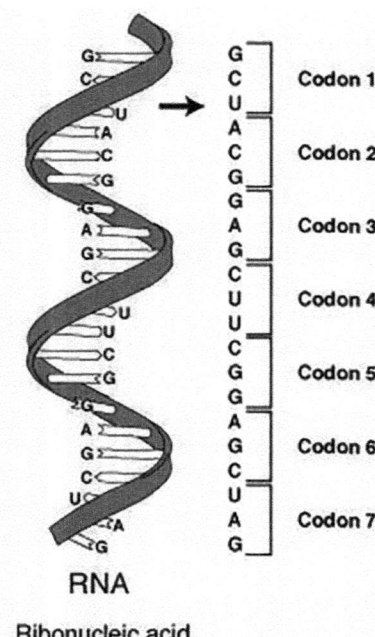

A series of codons in part of a messenger RNA (mRNA) molecule. Each codon consists of three nucleotides, usually corresponding to a single amino acid. The nucleotides are abbreviated with the letters A, U, G and C. This is mRNA, which uses U (uracil). DNA uses T (thymine) instead. This mRNA molecule will instruct a ribosome to synthesize a protein according to this code. [Public domain], via Wikimedia Commons

FIELDS OF STUDY

Biotechnology; Environmental biotechnology (or Green biotechnology); Genetics; Medical biotechnology (or Red biotechnology); Microbiology

ABSTRACT

Proteomics is the study of an organism's complete set of proteins. The term "proteome" is commonly used to describe all the proteins that are made by an organism's cells; it can also be described as all the proteins that are synthesized by a particular cell at a particular time. Protein engineering is the process of developing useful or valuable proteins for practical use. Protein engineering uses two strategies for engineering proteins: rational design and directed evolution. Both techniques have been developed to synthesize and manipulate proteins. To study proteins, one needs to determine the sequence of each protein's amino acids and its corresponding three-dimensional structure.

DEFINITION AND BASIC PRINCIPLES

Proteomics is the study of proteins: their structure and function and their interactions with other proteins. Proteins are made from the primary sequence of deoxyribonucleic acid (DNA), which is then transcribed into a messenger RNA (mRNA) molecule, which in turn is translated into polypeptide chains that will form a three-dimensional functional protein product. The study of proteins is complicated—more so than the study of genomics—because of the complexity of modifications that take place for the DNA sequence to become a protein product and because the proteome changes from cell to cell and at different time periods in a cell or organism's life.

Once the primary sequence of DNA forms an mRNA molecule, it is then translated into a polypeptide chain, in which posttranslational modifications (chemical modifications after translation) take place, which change the functional nature of that protein product. Specific posttranslational modifications include such processes as phosphorylation, ubiquitination, methylation, acetylation, glycosylation, oxidation, and nitrosylation. Although these processes may not be familiar to most people, all are processes that have the capability of modifying the organism's DNA to create various protein products. Once these changes in the protein take place, each cell may or may not express that protein because of variables such as the time in the organism's life, whether a functioning protein is needed in the type of cell in which the protein resides, or whether the specific conditions of the cell are conducive for a functioning protein. Also, proteins may have to communicate or interact with other proteins to become functional.

With all these possibilities to express certain proteins, the study of proteomics and the proteome is quite extensive and complex. However, understanding proteins and their functions will make it possible to design new drugs for the treatment of diseases and will provide researchers with a better knowledge of how cells work, how they interact with other cells, and how these interactions relate to a cell's ability to survive.

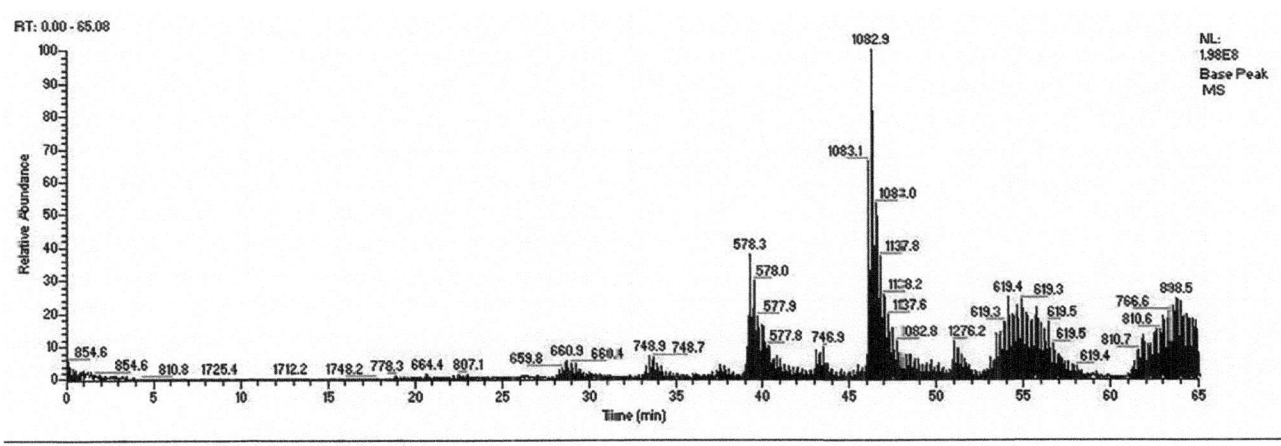

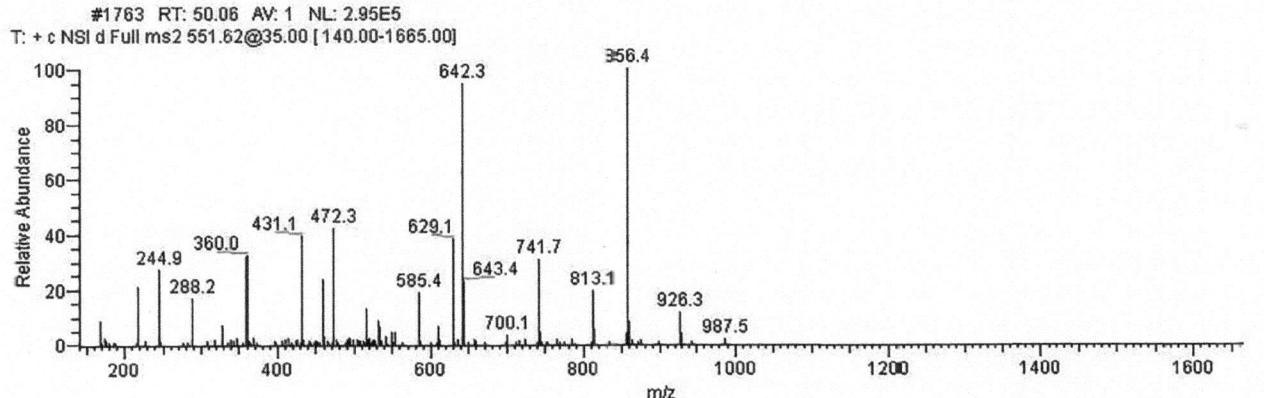

Chromatography trace and MS/MS spectra of a peptide. By Florian H. Tirk (Tirkfl.) [GFDL (http://www.gnu.org/copyleft/fdl.html) or CC-BY-SA-3.0 (http://creativecommons.org/licenses/by-sa/3.0/)], via Wikimedia Commons

BACKGROUND

The term "proteome" was coined by Marc Wilkins in 1994 as part of his doctoral thesis work on two-dimensional gel electrophoresis on proteins. Gel electrophoresis is a procedure that allows researchers to visualize the presence or absence of a protein in a specialized medium. Wilkins used this term to describe the entire complement of proteins expressed by the genome, which can be described as either all the proteins in a cell or an entire organism. The term "proteomics" was coined in 1997 to describe the study of proteins in the genome, thereby combining the two terms "protein" and "genomics" into one. When the Human Genome Project was completed in 2003, the entire human genome had been mapped. With the knowledge of the human DNA sequence, the next most logical step was to turn to the set of proteins in a cell or organism.

As the discipline of proteomics began, proteins were first studied by analyzing messenger RNA. Messenger RNA (mRNA) is a particular form of RNA that transcribes the genetic instructions of DNA to proteins for later expression. In early proteomic research, scientists took DNA, converted it into mRNA through transcription, and then studied the resulting mRNA to look for its protein product. It was found, however, that mRNA does not always correlate directly to protein content. First, mRNA may not always translate into a functional protein; second, the amount of protein produced from any mRNA differs based on the gene from which it is transcribed and how much protein that cell may need at any one time; third, the protein product may differ extensively from the original mRNA message based on posttranslational modifications; and fourth, proteins may need to interact with other proteins to be functional. All these complex features in studying the proteome have made new methods of study necessary, and new approaches to studying proteins have become available.

HOW IT WORKS

With the advent of new molecular and cell biology technologies, which began with the Human Genome Project in the 1990's, the study of proteins in the early 2000's advanced significantly. Different technologies were developed, and the data generated from these studies grew so large that bioinformatic systems were developed to store all the data for interpretation and sharing with the scientific community at large. Protein studies have numerous steps and various technologies for isolating the proteins and analyzing them.

Isolation of Proteins. The first approach to the study of proteins is to disrupt cells and separate out the proteins from the other particles in the cell. Cells are first treated with a detergent or urea so that enzymes will not degrade the proteins. These steps can be accomplished using kits that will perform a whole-cell extraction, producing small pieces from which one can isolate the proteins of interest, for example, isolating the hydrophobic proteins that are involved as membrane proteins.

Separation and Purification. Once the proteins have been isolated, one can then begin to separate and purify them. A common method is to use affinity or liquid chromatography to separate out specific proteins or a family of related proteins through the binding of proteins to specific substrates, which will elute out specific proteins from a mixture. Certain technologies have made it possible to extract very small quantities of proteins from a mixture.

Proteins can also be labeled with an isotope to measure the quantity of protein separated by using mass spectrometry. Tags can also be peptide derived rather than isotopes; peptide-tagged protein can be isolated and purified directly when the proteins to be studied are already known. One can then gather information on the abundance of proteins in cells.

Two-Dimensional Gel Electrophoresis. Two-dimensional gel electrophoresis was developed in the 1970's and has enhanced the ability to separate out specific proteins. This method puts a mixture of proteins through two gels or dimensions. The first dimension separates proteins by isoelectric identification using a pH (acidity-alkalinity) gradient, and the second dimension is SDS-PAGE (sodium dodecylsulfate-polyacrylamide gel electrophoresis), which separates proteins by their molecular weight (larger molecules move slowly through a gel and smaller proteins move quickly and farther along the gel). A number of companies sell kits to perform two-dimensional gel electrophoresis, making these steps easier than in the past.

Protein Identification and Mass Spectrometry. Once the gels have been run, interpreting the data may be challenging. The data appear as spots for each of the protein products that have been separated. To identify these spots, certain dyes—fluorescent dyes or blue or silver stains—are used to stain specific proteins.

Analyzing different proteins is generally accomplished with mass spectrometry. Proteins are usually ionized by different processes. One then can use computer software to analyze the data to compare each of the proteins expressed.

Microarrays. Protein microarrays take a mixture of proteins and add them to antibodies, antigens, enzymes, substrates, membrane receptors, or ligands, which can recognize the protein with which they associate or interact. These specific proteins will light up on an array, or chip, that identifies specific proteins from the thousands of proteins in the mixture. This technique, which is being used but is still in development, is a powerful tool and has the potential of eliminating numerous laborious steps being used to identify proteins.

APPLICATIONS AND PRODUCTS

Proteomics has immense potential for a broad number of practical applications in medicine and the pharmaceuticals industry.

Basic Research. One major area of proteomic research is to understand how the amino acid primary sequence specifies the stability and dynamics of protein conformation. This research would provide information on how to design novel functionalities of proteins and on disorders that work by changing a protein's three-dimensional structure, such as amyloidosis and prion diseases. Studying the folding and unfolding of proteins will help scientists understand the three-dimensional structures of proteins.

Disease Detection. One major application of proteomics is for disease detection. The National Cancer Institute is working on ways that proteomic technologies can be used for the detection of important proteins seen in disease. This technology would use biomarkers and target proteins that can be used to detect known diseases, including cancers, autoimmune disorders, and inflammatory diseases, as well as to screen for allergies.

Biomarkers that can be used for early detection of disease include gene mutations, gene transcription and translation modifications, and alterations in protein products. By looking at free DNA in serum, clinical testing has developed serum screens, and with the addition of biomarkers, testing has been expanded to use oncogene mutations, microsatellite instability regions in the DNA, and hypermethylation of promoter regions in DNA for the detection of cancer.

Clinical testing techniques using proteins as the basis for detection of disease include western blotting, immunohistochemical staining, enzyme-linked immunosorbent assay (ELISA), and mass spectrometry.

One such approach to disease detection is the use of proteomic pattern diagnostics to detect cancer. The first report of using this technique coupled with mass spectrometry was to identify ovarian cancer. More than two-thirds of ovarian cancer is not detected until it is in an advanced disease stage. Early detection is essential if this disease is to be treated and cured successfully.

One detection method is to identify discriminatory patterns of proteins that indicate ovarian cancer. Serum (obtained through a noninvasive procedure) from both normal women and those with ovarian cancer shows distinct patterns of proteins. This approach can be used for other cancers and diseases once the protein patterns are established.

Disease Treatment Though Novel Drugs. The development of novel drugs using proteomics for disease therapy holds tremendous potential. The major step forward is to identify proteins associated with disease, which can then be used as targets for drug development. First, a protein that is implicated in a disease must be identified. Then, drugs can be designed based on the protein's three-dimensional structure to interfere with the action of the protein. This can be achieved by developing molecules that fit into the protein at the active site to stop the enzymatic reaction that would normally occur. Inactivation of enzyme function may also be a means of developing personalized drugs based on different individuals' genetic, and thereby, protein makeup. With the advent of computer databases of protein structures, computer techniques can fit different molecules into a three-dimensional structure called virtual ligand screening.

For cancer diseases alone, drug development is a major area of scientific endeavor, and the industry continues to grow. For example, the development of novel proteins through protein engineering has led to new drugs. Top7, a fusion protein, was used to create an interleukin 1 blocker, Arcalyst (rilonacept), which was approved by the U.S. Food and Drug Administration in 2008 for the treatment of cryopyrin-associated periodic syndromes.

Protein Engineering. The field of protein engineering involves developing and creating useful or important proteins. Protein engineering encompasses two main strategies: rational design and directed evolution.

In rational design, mutations are induced to make changes to a protein to change the structure or function of that protein. Because mutagenesis techniques are well established, this technique has great potential; however, the difficulty in its success is that a protein's structure and function may not be known in detail. For proteins that are not well defined, it may be impossible to determine what mutations to incorporate.

Random mutations may be induced in a protein, and the variant proteins with the desired qualities sorted out from the protein mix. Once the variant proteins have been isolated, further mutations and selection of those variants are performed. This process of mutations and selection is called directed evolution because it simulates natural selection and may produce more fit or successful proteins. Another technique used in directed evolution is DNA shuffling, which mixes and matches pieces of variants to produce better protein products. This process is similar to the recombination process that occurs naturally in individual cells during sexual reproduction. One advantage of directed evolution is that it does not require prior knowledge of a protein or the knowledge of which mutation to induce. Rather, mutations are randomly induced, and those mutations are monitored to see what effect they have on the protein expression in a cell. The difficulty of directed evolution is that it requires high throughput, in which large amounts of DNA must be mutated and the protein products monitored for the desired qualities. This process may not be feasible for all proteins.

SOCIAL CONTEXT AND FUTURE PROSPECTS

With the coupling of advanced instrumentation and new technologies and informatics tools, the field of proteomics has great potential in disease detection, drug development, and basic research on proteins.

Another major aspect to the future of proteomics is whether a patent system that facilitates research and can effectively change with the dynamics of the field of genomics and proteomics can be developed. A definite conflict of interest exists between for-profit industries and nonprofit research institutes. This makes research studies difficult to finish and publish so that other researchers can freely access the results. This dilemma is worldwide and leads to higher costs of doing research and difficulties in converting research results to clinical useful applications. Intellectual property laws and patenting are difficult issues and becoming more complex and burdensome with time.

The issue of how to use systems biology, including proteomic research, to improve the health of individuals is a major priority. It is becoming increasingly apparent that proteomics will have a major role in creating a predictive, preventative, and personalized approach to medicine. This raises the question of how individualized medicine will be handled by insurance companies and whether there will be disparity between individuals in their access to individualized medical procedures and therapies. The idea that genomic and proteomic research may outpace societal change must be recognized and steps must be taken to address all issues.

——*Susan M. Zneimer, PhD*

FURTHER READING

Alberghina, Lilia. *Protein Engineering in Industrial Biotechnology*. Amsterdam: Hardwood Academics, 2000. An overview of the applications of proteins and how protein engineering can be used to solve problems in industrial biotechnology. Describes how protein engineering enhances purification of recombinant proteins and how it is applied to health care, including the development of new vaccines.

Hamacher, Michael, et al., eds. *Proteomics in Drug Research*. Weinham, Germany: Wiley-VCH, 2006. Contains information on technologies and applications, particularly in pharmaceuticals.

Liebler, Daniel. *Introduction to Proteomics*. Totowa, N.J.: Humana Press, 2002. Discusses key concepts in proteomics, including the methodology and instrumentation used. Describes the applications of mass spectrometry and provides an excellent introduction and overview of proteomics.

Mishra, Nawin C., and Günter Blobel. *Introduction to Proteomics: Principles and Applications*. Hoboken, N.J.: John Wiley & Sons, 2010. Introduces the field of proteomics and examines the role proteomics plays in the study of biological systems in general and disease in particular. Provides an understanding of the structure, function, and interactions of proteins and how they are used for identifying diseases and developing new drugs.

Pennington, Stephen, and Michael Dunn, eds. *Proteomics: From Protein Sequence to Function*. London: Garland, 2001. Gives an overview of how proteins can be identified, and used for clinical procedures and research. Discusses the technologies used in the field and protein structure and function.

Wulfkuhle, D. J., et al. "Proteomic Applications for the Early Detection of Cancer." *Nature Reviews* 3 (2003): 267-275. Describes cancer detection with a great review of techniques used in proteomics.

R

Radiocarbon dating

Measuring 14C is now most commonly done with an accelerator mass spectrometer. [Public domain], via Wikimedia Commons

FIELDS OF STUDY

Archaeology; Environmental biotechnology (or Green biotechnology); Genetics

ABSTRACT

Radiocarbon dating is a means of determining the approximate time at which biological processes ceased in a once-living organism or in related organic substances. It allows scientists to estimate the ages of organic materials and the formations in which they occur.

RADIOACTIVE CARBON-14

Carbon compounds are among the most abundant in nature. Carbon, in the form of atmospheric carbon dioxide, is continuously moved through major environmental systems, a process known as the carbon cycle. The photosynthesis of carbon dioxide establishes carbon in various compounds in plants, which in turn may be consumed by animals. Most carbon is absorbed as carbon dioxide by the oceans or appears there as dissolved carbonate or bicarbonate compounds. Carbon occurs naturally in three isotopes: carbon-12, the most common; carbon-13, also a stable isotope; and the radioactive isotope carbon-14, which exists in minute quantities. Radiocarbon dating draws on several assumptions concerning the natural production of carbon-14, its presence in various environmental cycles, and its rate of decay, or half-life.

Radioactive carbon-14 is formed in the upper atmosphere as the result of bombardment by energetic cosmic radiation emanating from deep space. Statistically, it is most likely that free neutrons, which result from collisions between proton cosmic radiation and atmospheric gases, will shortly collide with molecules of stable nitrogen, nitrogen-14, by far the most abundant gaseous element in the atmosphere. The resulting reaction normally expels a proton from the nitrogen nucleus, producing an atom which now behaves chemically like carbon but is heavier than the stable carbon isotopes. Free carbon does not remain long in the upper atmosphere. It quickly combines with oxygen molecules to form carbon dioxide, whereupon it enters into various geological and biological processes.

BETA DECAY

The half-life of carbon-14 usually is expressed as $5,568 \pm 30$ years, though it is more likely to be on the order of $5,730 \pm 30$ years. (Once large numbers of dates had been calculated on the basis of the former figure, leading scientific journals generally preferred to stay with it.) The particular process of radioactive decay of carbon-14 is called beta decay, in which the "extra" neutron in the nucleus emits a beta particle (essentially an electron) and a neutrino (an uncharged particle), thus changing itself, in effect, from a neutron into a proton. The result is once again an atom of stable nitrogen.

As long as an organism is alive and normal biological processes are occurring, the rate of accumulation of carbon-14 is in approximate equilibrium with the rate of radioactive decay of the carbon-14 already in the organism. This level of equilibrium is extremely small, on the order of one atom of carbon-14 to every

1 trillion atoms of carbon-12. The moment that biological processes cease, however, the equilibrium is broken, and the quantity of carbon-14 in the once-living organism begins to decrease at the predictable rate of radioactive decay. By measuring the rate of beta radiation from the residual carbon-14, one may estimate the age of a material, or, more precisely, when the biological processes of the organism from which the material derives ceased to operate. In general, the age of material which gives off only half as much beta radiation as a living organism would be 5,730 years; the age of material emitting only 25 percent as much radiation would be 11,460 (5,730 × 2) years; and so on.

EVOLUTION OF MEASUREMENT TECHNIQUES

After fifteen years of research, Willard F. Libby, an American physicist, introduced the radiocarbon dating method in 1949. Since Libby's pioneering work, radiocarbon dating has evolved to include several counting techniques. Normally the substance under study must be destroyed by combustion or other processes to produce gaseous carbon dioxide or a hydrocarbon gas, whose carbon-14 content is measured by a gas counter. Liquid counting techniques have also been developed, in which the carbon dioxide from the substance under study is synthesized into more complex liquid hydrocarbons, such as benzene. After 1980, direct measurement techniques, using particle accelerators and mass spectrometers, increasingly replaced gas and liquid counting.

The evolution of measurement techniques has greatly enhanced the usefulness of radiocarbon dating. The minimum acceptable mass of a sample substance is much smaller than that required in the years immediately following Libby's introduction of the method. Refined techniques and improved instrumentation also have extended the effective chronological limits of the method. At first, only materials less than about 20,000 years old could be dated with any confidence, but by the fortieth anniversary of Libby's initial dating experiments, there was wide agreement that the effective limit was about 40,000 years and, when supporting data could be gleaned from other dating methods, possibly 70,000 years.

RISK OF CONTAMINATION

The extremely small quantities of carbon-14, even in living organisms, demand that substances undergoing counts be prepared most carefully and that every possible source of contamination be considered. For example, fallout from atmospheric testing of thermonuclear weapons in the 1950s interfered with some of Libby's early experiments. Numerous chemical processes affect archaeological artifacts and other substances which might be candidates for radiocarbon analysis. Carbon compounds from associated soil layers find their way into dating samples. Carbonate encrustations may develop around other samples, especially in locations with abundant groundwater. In the 1980s, some radiocarbon laboratories had to consider their local environments, where, in many cases, the buildup of carbon dioxide and other carbon compounds in the atmosphere from fossil fuel combustion threatened to interfere with sample integrity and analysis. (Fossil fuels, because of their great geological age, contain virtually no carbon-14.)

Some substances are at a greater risk of contamination than others. Woody plant remains, such as charcoal or building materials carbonized by fire, produce the best results. Less dependable are textiles, fibers, and the remains of nonwoody plants, which do not live as long as woody species and, as a result, may reflect short-term or local carbon-14 anomalies. Bone is notorious for its ability to absorb carbon compounds from its surroundings. Inorganic substances which contain carbon compounds, such as eggshell and marine shell, also present serious problems of carbonate contamination. Of these substances, marine shell is preferred, since the extraneous carbonate levels can be determined on the basis of relatively constant values present in ocean water.

ADJUSTING FOR VARIABILITY

In reality, radiocarbon dating is subject to numerous variable factors and is not nearly so accurate. Radiocarbon dates are expressed as years BP or "before present." For example, increases in atmospheric carbon dioxide since the Industrial Revolution have diluted the concentration of carbon-14 in what is known as the Suess effect, named after Austrian chemist Hans Suess. Observations by the THEMIS spacecraft fleet have shown that twenty times more solar particles cross the earth's magnetic shield when it is aligned with the sun's magnetic field, compared to when the two magnetic fields are oppositely directed. "Present" is actually a zero base date of 1950.

Scientists chose this year because it was close to the first experimental application of the method and because the buildup of atmospheric radioactivity from nuclear weapons tests in later years introduced complications into the measurement process. Each date is expressed with a standard error, or standard deviation—in essence, a "confidence level"—indicated by a plus-or-minus sign. A typical radiocarbon date of, say, 2750 BP ± 60 indicates that there is a 66 percent probability that the true date falls within sixty years of 2750 BP and a 95 percent probability that the true date falls within twice the standard error—in this case, 120 years.

Another source of variability in test results was only dimly suspected before about 1970. Libby assumed that the formation rate of carbon-14 in the upper atmosphere had a constant value over time, since the process depended on cosmic radiation. Scientists now know that the formation rate varies over time and space, in part according to the strength and contours of the earth's magnetic field. The field deflects a certain portion of cosmic radiation, so some less energetic particles never reach the atmosphere. At any time, for example, more carbon-14 is formed in the atmosphere near the poles, where the magnetic field is relatively weak, than near the equator, where it is stronger.

The strength of the magnetic field itself is affected by cyclical changes in solar activity. The sun has roughly an eleven-year cycle of sunspot activity; the earth's magnetic field is most energetic when sunspots are most numerous. Longer-term and even more significant changes apparently have occurred in more recent decades of solar history, as in the case of the so-called Maunder minimum (1645-1715), when sunspot activity may have been almost nonexistent. During such a period, the earth's magnetic field would be less energetic, carbon-14 formation levels would be abnormally high, and materials from the era would appear "younger" in radiocarbon terms than they otherwise would.

Dendrochronology (the study of tree rings) provides a means of identifying these anomalies and thus calibrating radiocarbon dates more closely with absolute chronology. In principle, each growth ring in a tree lives only for one year. By comparing the known age of tree rings with their radiocarbon ages, scientists can locate anomalies in the carbon-14 absorption rate and, therefore, the formation rate. Wood samples from the extremely long-lived bristlecone pine provide invaluable data. The oldest living bristlecone is more than 4,500 years old, and the remains of dead specimens have yielded a calibration matrix extending almost nine thousand years into the past.

Several other dating systems have been developed which utilize the half-life of radioactive isotopes, and some provide corroboration of radiocarbon chronologies. Further support comes from closely related techniques, such as thermoluminescence. These methods, as well as the established practices of classical archaeology, strengthen the credibility of the radiocarbon technique.

APPLICATION TO ARCHAEOLOGY

Through radiocarbon dating applications, scientists have amassed a wealth of information on such matters as sea-level fluctuation, climatic change, glaciation, habits of marine life, volcanic activity, and early forms of atmospheric pollution from the combustion of fossil fuels. From time to time, results of carbon-14 analysis also attract attention by establishing the authenticity of artistic works or the ages of religious relics.

By far the most significant application of radiocarbon dating, however, has been in providing archaeologists with the means of constructing chronological relationships independently of traditional archaeological assumptions about cultural processes. The technique is especially valuable in prehistoric archaeology, for which little or no documentation is available and artifact interpretation previously could allow only relative chronological sequences. In the Western Hemisphere, Africa south of the Sahara, and other parts of the world where "prehistory"—in the sense of absence of documentation—extends nearly into the modern era and chronologies had been mere guesswork, carbon-14 dating was a revolutionary technological breakthrough.

The results obtained from applying carbon-14 dating to prehistoric materials, however, made the technique immediately controversial. For example, early dates for the origins of agriculture in the Near East—generally referred to as the Neolithic Revolution—were dramatically earlier than archaeologists had expected. At Jericho, one of the first sites to provide such material, scholars previously had placed the start of the Neolithic Revolution at around 4000 BCE, but carbon-14 dates were on the order of four thousand years earlier. Further research in the Near

East has established a chronological frontier for the Neolithic Revolution at 10,000 BCE. or even earlier.

A second major achievement of radiocarbon dating was the establishment of the first outlines of an absolute chronology for developments similar to the Neolithic Revolution in Mexico and Central America. Supported by pollen analysis, carbon-14 dating of the remains of early domesticated maize suggested an astonishingly long continuum for agriculture in the region extending back to 5000 BCE. This finding was the first real indication of the depth and complexity of Mesoamerican cultural heritage.

In sub-Saharan Africa, the immense potential of charcoal for providing carbon-14 dates led to reconstruction of the prehistory of a region which, before 1950, was unknown. Charcoal is especially plentiful at the large numbers of iron-smelting sites in Africa. Slag from the smelting operation, sometimes even iron remains, may be dated using carbon-14; carbon migrates as an impurity into the iron. Before 1950, conventional wisdom attributed the presence of iron-smelting technology in sub-Saharan Africa to cultural borrowing, either from Phoenician traders or from the Egyptians. Carbon-14 dates clearly established a thriving iron technology in West Africa as early as 1000 BCE, several hundred years before the earliest known incidence of smelting in Egypt itself.

CONTROVERSIAL RESULTS

Perhaps the most disturbing revelation of the first decade or so of carbon-14 dating was its support for an extended Neolithic chronology in Europe. Although archaeologists already were debating the merits of a "long" versus "short" European Neolithic, most scholars and cultural opinion favored the latter, since it accommodated the notion of cultural "diffusion" from the ancient Near East, relegating Europe to a state of barbarism until stimulated by classical civilizations. The famous megalithic structure of Stonehenge in southern England, for example, was thought to have been inspired by Mycenaean influence and was therefore dated to around 1400 BCE Yet, carbon-14 results obtained from reindeer bone fragments at the site—the bones probably were digging implements used in construction—suggested dates for vital parts of the Stonehenge complex that were several centuries earlier, making it, in effect, pre-Mycenaean. These findings threatened to sever the "diffusionist" link with the Near East and forced a reassessment of the state of prehistoric European culture.

The most severe test of the radiocarbon technique occurred when it was used to corroborate dates for Egyptian historical artifacts that already had been fairly accurately dated with conventional archaeological methods. Early results were not reassuring. The method repeatedly generated dates for materials from the second millennium BCE that were several centuries too recent. Since these discrepancies were more or less consistent in scope and did not alter the *sequences* of Egyptian history, they raised serious questions about the integrity of the whole chronological framework of ancient history patiently constructed over decades by archaeological research. Conversely, many traditional scholars, confident of their work and suspicious of a technique derived from nuclear physics, preferred to reject the radiocarbon system out of hand.

By the early 1970s, these difficulties, together with the developing precision of sample handling techniques and instrumentation, had led to the realization that there were variables in the carbon-14 cycle unaccounted for by Libby's earlier formulations—principally the magnetic field fluctuations, which, as noted earlier, result in periodic changes in the carbon-14 formation rate. Once radiocarbon dates could be calibrated on the basis of tree-ring data, they agreed very closely with previously established dates for Egyptian artifacts and therefore confirmed the work of traditional archaeologists. With that, the last serious barrier to acceptance of the technique disappeared. Some of the prehistoric dates for Europe, however, when corrected according to tree-ring data, actually pushed the Neolithic horizon even further into the past. The earliest portions of Stonehenge are now believed to be separated from what once were thought to be their Mycenaean origins by what one archaeologist has called a "yawning millennium."

In the 1980s, investigators from England, Switzerland, and the United States applied radiocarbon dating techniques to fibers from the "Shroud of Turin." Held in reverence by many as the burial shroud of Jesus, this piece of linen retains the image of a bearded man with marks consistent with crucifixion. Measurements by separate laboratories agreed that the flax from which the linen was produced grew sometime in the thirteenth or fourteenth centuries—far too recent to have been the burial shroud of Jesus. Some scholars continue to dispute the findings, however, including criticizing technical aspects of the dating process.

INTERFACE BETWEEN SCIENCE AND TRADITION

Radiocarbon dating, together with similar techniques using isotopes of other elements and a variety of methods drawn from the physical and life sciences since 1950, has elaborated a picture of human history and earth history to a degree that could not have been conceived by earlier scholars. Among the method's practical benefits is a much more sophisticated knowledge of the scope of natural climatic change, without which it would not be possible to make useful scientific or political decisions on matters that may affect future, human-induced climatic and environmental change. Radiocarbon dating also has shattered many long-standing notions about European prehistory, Europe's historical relationship with the ancient Near East, and the antiquity and complexity of non-European civilizations, thereby undermining fundamental assumptions about the centrality of Western civilization in the human saga.

The history of the application and results of carbon-14 dating require that one look carefully at the conditions and assumptions associated with certain dates and at the stage in the technique's development from which those dates derive. A prime example of how these matters can fuel controversy was the confusion generated by some carbon-14 dates which seemed to suggest that established dates for Egyptian artifacts of the second millennium BCE. were several centuries too early. Biblical scholars had long worried about an unexplained gap between the dates, derived from scientific genealogical and textual studies, for the Hebrew Exodus from Egypt and the establishment of the ancient kingdom of Israel. In terms of regarding the Old Testament as historically accurate, it would be convenient if several centuries of Egyptian history could be "erased" and earlier events moved up to fill the gap. Ironically, that is just what early carbon-14 dating of Egyptian artifacts suggested. Some biblical scholars were ecstatic that science seemed to verify the accounts of the Bible. However, later calibration of these Egyptian dates using tree-ring data reinstated traditional Egyptian chronology and nullified this temporary congruence. Clearly, the lay reader who encounters results of radiocarbon dating encounters a complex interface between modern science and the most fundamental issues in the Western religious and historical tradition.

——*Ronald W. Davis*

FURTHER READING

Agrawal, D. P., and M. G. Yadava. *Dating the Human Past*. Pune: Indian Society for Prehistoric and Quaternary Studies. 1995. An interesting look at the use of radiocarbon dating technology in relation to anthropology. Suitable for the nonscientist. Illustrations.

Bard, E., F. Rostek, and Guillemette Menot-Combes. "A Better Radiocarbon Clock." *Science* 303 (2004): 178-179. A strong discussion of the need for a radiocarbon curve that extends beyond that of INTCAL98. This article is written in a manner that is accessible to undergraduates as well as graduate students.

Bard, Edouard, and Wallace S. Broecker, eds. *The Last Deglaciation: Absolute and Radiocarbon Chronologies*. Berlin: Springer-Verlag, 1992. This book deals with the radiocarbon dating processes and techniques that have been used to unravel the mysteries of glaciers and massive changes in climate. This book provides a simple explanation of radiocarbon dating and illustrates its applications.

Currie, L. A. "The Remarkable Metrological History of Radiocarbon Dating." *Journal of Research of the National Institute of Standards and Technology* 109 (2004): 185-217. This article accounts the discovery and progression of radiocarbon dating as a tool in geological and archaeological studies. It is extremely detailed and provides an excellent background for anyone using carbon dating techniques.

Fleming, Stuart. *Dating in Archaeology: A Guide to Scientific Techniques*. New York: St. Martin's Press, 1976. Places the methods and results of radiocarbon dating in the context of other dating techniques, such as dendrochronology, thermoluminescence, fission track dating, pollen analysis, and chemical methods. Discusses how several of these techniques may corroborate one another in establishing chronologies. Excellent bibliography.

Lowe, J. John, ed. *Radiocarbon Dating: Recent Applications and Future Potential*. Chichester, N.Y.: John Wiley and Sons, 1996. This college-level book offers a comprehensive overview of the techniques and protocols of radiocarbon dating. Several of the essays explore possible future usage and applications. Illustrations and maps help to clarify difficult concepts. Bibliographical references.

Reimer, Paula J., et al. "IntCal104 Terrestrial Radiocarbon Age Calibration, 0-26 cal kyr BP." *Radiocarbon*. 46 (2004): 1029-1058. This article provides a technical review of the past state of the radiocarbon calibration curve. More recent data are used to update and refine the curve. Written in a highly technical manner for practicing geologists and graduate students.

Renfrew, Colin. *Before Civilization: The Radiocarbon Revolution and Prehistoric Europe*, 2d ed. London: Penguin Books, 1990. A comprehensive discussion of the development of radiocarbon dating, the early discrepancies, and their correction using tree-ring data. Offers a thorough analysis of some alternative approaches to prehistory based on carbon-14 results from Europe.

Rick, T. C., R. L. Vellanoweth, and J. Erlandson. "Radiocarbon Dating and the 'Old Shell' Problem: Direct Dating of Artifacts and Cultural Chronologies in Coastal and Other Aquatic Regions." *Journal of Archaeological Sciences* 32 (2005): 1641-1648. An interesting and easy-to-follow discussion of a common problem in sample selection for carbon dating. This article addresses the problem of "old wood" and compares it to the sampling of shells at archaeological sites. Provides the basic principles of radiocarbon dating in a manner accessible to the layperson with some science background.

Scott, E. M., M. S. Baxter, and T. C. Aitchison. "A Comparison of the Treatment of Errors in Radiocarbon Dating Calibration Methods." *Journal of Archaeological Science* 11 (1984): 455-466. Discusses several methods of calibration with respect to the varying degrees of accuracy required in specific applications.

Taylor, R. E. "Fifty Years of Radiocarbon Dating." *American Scientist* 88 (January/February 2000): 60-67. This paper reviews the entire five decades of development of this remarkable technique. Taylor reviews the basic technique, deviation of carbon-14 dates from true dates due to variation in the magnetic field, means of correction, and attaining increased sensitivity by the application of mass spectrometry.

Taylor, R. E. *Radiocarbon Dating: An Archaeological Perspective*. New York: Academic Press, 1987. An excellent treatment of the procedures and complexities involved in measuring radiocarbon content. Discusses instrumentation, sources of contamination, case studies using various substances, and the historical development of radiocarbon methodology. The bibliography covers a broad range of sources.

Vita-Finzi, Claudio. *Recent Earth History*. New York: Halsted Press, 1974. An account of the physical changes undergone by the earth during the Holocene (modern) geologic era, presented in the form of a stratigraphical narrative based throughout on radiocarbon dates.

Wagner, Gunther A., and S. Schiegl. *Age Determination of Young Rocks and Artifacts: Physical and Chemical Clocks in Quaternary Geology and Archaeology*. New York: Springer, 2010. The authors cover various materials and dating methods. Well organized, accessible to advanced undergraduates and graduate students.

Walker, Mike. *Quaternary Dating Methods*. New York: Wiley, 2005. This text provides a detailed description of current dating methods, followed by content on the instrumentation, limitations, and applications of geological dating. Written for readers with some science background, but clear enough for those with no prior knowledge of dating methods.

Walther, John Victor. *Essentials of Geochemistry*, 2d ed. Jones & Bartlett Publishers, 2008. Contains chapters on radioisotope and stable isotope dating and radioactive decay. Geared more toward geology and geophysics than toward chemistry, this text provides content on thermodynamics, soil formation, and chemical kinetics.

"Willard Libby and Radiocarbon Dating." *American Chemistry Society*, 2016, www.acs.org/content/acs/en/education/whatischemistry/landmarks/radiocarbon-dating.html. Accessed 31 Aug. 2017.

Wilson, David. *The New Archaeology*. New York: New American Library, 1974. Perhaps the best account for the general reader of the development and subsequent refinement of radiocarbon dating techniques. Discusses their enormous impact on the field of prehistoric archaeology and perceptions of prehistory.

Refuse-derived fuel

FIELDS OF STUDY

Archaeology; Biochemistry; Biology; Biophysics; Biosystems engineering; Civil engineering; Environmental biotechnology (or Green biotechnology); Genetics

ABSTRACT

The burning of refuse-derived fuel has some advantages over the burning of coal in terms of reducing damage to the environment, but this alternative energy source is not completely clean.

Raw municipal solid waste (MSW) is a notoriously poor fuel because of its high moisture and low heat content. In addition, mass-burn incineration of MSW produces a broad range of atmospheric pollutants, and the ash produced may become concentrated in potentially toxic elements, such as cadmium or arsenic. One goal of the production of refuse-derived fuel (RDF) is to improve the combustion of MSW by producing a fuel of lower moisture content, more uniform size, greater density, and lower ash content than raw MSW. Another goal is to reduce the amount of material in landfills.

HOW IT WORKS

A typical process for the production of RDF involves passage of raw wastes through a screen to remove small, inert materials (such as stones, soil, and glass), pulverization of larger particles in a shredding device, separation of ferrous metals by magnetic extraction, and segregation of the lightweight, mostly organic fraction in an upward airstream (this step is known as air classification). The shredded organic waste fraction can be used directly as a fuel (fluff RDF), or it can be compressed into high-density pellets or cubettes (densified RDF). The latter material is popular because it is easy to transport and store and because of its adaptability in handling and combustion.

RDF can be utilized as a cofuel with coal or fired separately. A major advantage of RDF over coal is that its sulfur content is markedly lower (0.1 to 0.2 percent compared with 5 percent or more for some coal samples), as is its nitrogen content. Both sulfur and nitrogen are among the more notorious atmospheric precursors to acid rain. Also, as a result of processing, RDF contains smaller amounts of potentially toxic metals than does MSW.

RDF possesses only 50-60 percent of the calorific value and 65-75 percent of the density of typical bituminous coals. As a consequence, considerably larger weights of RDF must be burned to obtain performance similar to that of coal. The use of RDF in a boiler may therefore have an adverse impact on the performance of the boiler's systems for air-pollution control and ash removal. In addition, the chlorine content of RDF is higher than typical coal. Other problems sometimes associated with the use of RDF concern odors and dust production in storage and particulate, carbon monoxide, and hydrogen chloride discharges during combustion.

—*John Pichtel*

FURTHER READING

Hickman, H. Lanier, Jr. "Refuse-Derived Fuel and Energy Recovery: Fulfilling the Resource Recovery Promise." In *American Alchemy: The History of Solid Waste Management in the United States*. Santa Barbara, Calif.: Forester Communications, 2003.

Niessen, Walter R. "Refuse-Derived Fuel Systems." In *Combustion and Incineration Processes*. 3d ed. New York: Marcel Dekker, 2002.

Renewable and nonrenewable resources

A coal mine in Wyoming, United States. Coal, produced over millions of years, is a finite and non-renewable resource on a human time scale. [Public domain], via Wikimedia Commons

FIELDS OF STUDY

Biochemistry; Biology; Biophysics; Biosystems engineering; Civil engineering; Environmental biotechnology (or Green biotechnology); Genetics; Political Science

ABSTRACT

Nature provides numerous energy resources. Nonrenewable resources were the primary source of energy for the twentieth century. However, with the depletion of nonrenewables, interest in renewable forms of energy has generated increasing research and development of renewables.

BACKGROUND

Nonrenewable resources cannot be readily replaced after consumption. A renewable resource is one that is continuously available, such as solar energy, or one that can be replaced within several decades, such as wood.

NONRENEWABLE ENERGY SOURCES

Nonrenewable resources may be subdivided into four categories: metals (such as copper and aluminum), industrial minerals (such as lime and soda ash), construction materials (sand and gravel), and energy resources (coal, oil, and uranium). Of the nonfuel substances, metals are most prone to depletion by overproduction, but recycling can prolong their useful lifetime almost indefinitely.

Construction materials, although not readily recyclable, are abundant and ubiquitous in the Earth's crust, rendering them a virtually unlimited resource. Although less plentiful, the most widely used industrial minerals are unlikely to be depleted in the near future; on the scale of centuries, however, they are an endangered resource if current levels of production are maintained. It is probable that environmental concerns will reduce future production.

The major forms of nonrenewable energy production are fossil fuel combustion (using oil and coal) and nuclear fission (using uranium). Of the total energy consumed by Americans, only 7 percent is from renewable resources, while 85 percent is from fossil fuels, predominantly oil.

Coal was the first fossil fuel to be used extensively, and it remains the most abundant. Coal can be burned directly or converted into petroleum or petroleum products, through the expenditure of additional energy. When used as fuel, coal creates many problems. Mines are environmentally destructive, and coal is the most difficult fossil fuel to transport. When coal is burned, vast quantities of sulfur compounds (which form sulfuric acid in the atmosphere) are released, while the carbon in the coal becomes carbon dioxide, a greenhouse gas believed to contribute to global warming. The carbon in coal also has many other valuable (nonpolluting) uses in the chemical industry.

Oil is the world's major source of energy because it is abundant and relatively inexpensive. Its high rate of use will result in its depletion during the twenty-first century. When burned as gasoline in cars, it releases carbon dioxide; various dangerous air pollutants, such as carbon monoxide and nitrogen oxides; and uncombusted hydrocarbons (a major cause of photochemical smog). Natural gas, formed when organic materials decompose, is usually found with petroleum reservoirs. Its supply, rate of consumption, and probable future are comparable to those of petroleum. It is widely used because it is relatively inexpensive, clean, and nonpolluting (although it does add carbon to the atmosphere).

Tar sands, principally found in Canada, are a low-grade source of petroleum that is feasible to mine and process only when oil prices are relatively high. Two additional problems limit this source: About as much energy is required to extract usable oil as is created when it is combusted, and the process has raised environmental concerns. Oil shales, abundant in the western United States, appear theoretically to be a major source of future petroleum products. The amount of oil tied up in shale exceeds the remaining total world reserve of oil. To extract oil, however, the shale must be mined and heated by processes requiring large quantities of water in regions where water is scarce. Additionally, the total energy required for extraction exceeds the energy created when the oil is burned.

Nuclear reactors produce energy through controlled fission of uranium 235. No air pollution is produced, the mining operations are relatively small and safe, and the resource being consumed has no other known use. On the other hand, reactor technology is sophisticated and elaborate, and complicated devices are prone to breakdowns. A reactor breakdown can have disastrous consequences if radioactive materials are released into the environment. Of equal or greater concern is how the by-products of nuclear power production—nuclear waste—should be disposed of over the long term.

RENEWABLE ENERGY SOURCES

The most abundant renewable energy resource is solar energy, the source of most other renewables as well as the original source of fossil fuels. The supply is enormous and inexhaustible, but most is wasted because it occurs in a dilute form that requires expensive hardware to concentrate. Also, it reaches Earth in its most dilute form during the winter, when it is most needed for heating. In cloudy regions it is not even available when demand for it is greatest.

Like solar energy, wind represents a large and potentially inexhaustible source of energy. However, when wind energy is used to generate electricity, expensive collectors are required. Wind energy is not feasible everywhere, and even when feasible it is not always available. Power derived from moving water, such as that provided by hydroelectric dams, makes an important contribution to the world's energy supply. Many of the best sites have already been dammed, however, and development of a number of other sites is unwise because of ecological reasons or the sites' scenic beauty.

Tidal energy utilizes the ebb and flow of tides to create electricity by trapping seawater at the extremes of high and low tide and releasing it through turbines. Although a potentially large energy source, it is economically feasible only where there are naturally high tides (4.5 meters or more) and where a narrow inlet encloses a large bay.

Geothermal energy uses the heat from natural hot springs to create steam to power turbines, which are used to create electricity. Because the heat must be close to the surface, there are few known sites from which geothermal electrical energy can be extracted economically. Also, because pipelines must be run over many hectares to collect steam, the power-generating stations tend to be ugly and noisy.

Vegetation (biomass) energy uses plants or animal products derived from plants as a source of fuel. This source includes wood, organic wastes, ethanol, and methane gas from biodigestion. This type of renewable resource is renewable only if harvesting is controlled and if resources exist to cultivate the source. Thus, trees must be given sufficient time to mature, and corn must be cultivated before ethanol can be produced. Although vegetation has a long history as a source of fuel, efficient and sustainable techniques have yet to be introduced.

——*George R. Plitnik*

FURTHER READING

Boyle, Godfrey, ed. *Renewable Energy*. 2d ed. New York: Oxford University Press in association with the Open University, 2004.

Cassedy, Edward S., and Peter Z. Grossman. *Introduction to Energy: Resources, Technology, and Society*. 2d ed. New York: Cambridge University Press, 1998.

Evans, Robert L. *Fueling Our Future: An Introduction to Sustainable Energy*. Cambridge, England: Cambridge University Press, 2007.

González, Pablo Rafael. *Running Out: How Global Shortages Change the Economic Paradigm*. New York: Algora, 2006.

Greiner, Alfred, and Willi Semmler. *The Global Environment, Natural Resources, and Economic Growth*. New York: Oxford University Press, 2008.

Hinrichs, Roger A., and Merlin Kleinbach. *Energy: Its Use and the Environment.* 4th ed. Belmont, Calif.: Thomson, Brooks/Cole, 2006.

Kozlowski, Ryszard, Gennady Zaikov, and Frank Pudel, eds. *Renewable Resources: Obtaining, Processing, and Applying.* Hauppauge, N.Y.: Nova Science, 2009.

Kruger, Paul. *Alternative Energy Resources: The Quest for Sustainable Energy.* Hoboken, N.J.: John Wiley & Sons, 2006.

Pimentel, David, ed. *Biofuels, Solar, and Wind as Renewable Energy Systems: Benefits and Risks.* New York: Springer, 2008.

Twidell, John, and Tony Weir. *Renewable Energy Resources.* 2d ed. New York: Taylor & Francis, 2006.

REPRODUCTIVE SCIENCE AND ENGINEERING

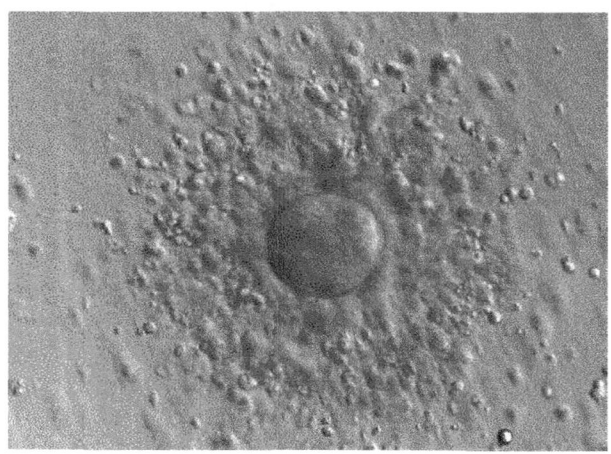

Oocyte with surrounding granulosa cells. By Ekem at en.wikipedia [Public domain], via Wikimedia Commons

FIELDS OF STUDY

Biotechnology; Genetics; Microbiology; Philosophy and Religious Studies; Reproductive technology

ABSTRACT

Reproductive science and engineering is concerned with the examination and regulation of the physiological mechanisms involved in human reproduction, such as conception and birth, and with diagnosing and treating disorders of reproduction, such as male and female infertility. Because the ability to successfully bear offspring is the core objective driving the success of all animal species, reproductive research is of deep importance on a purely scientific level. In addition, by providing methodologies that enable infertile couples to conceive, such as artificial insemination and in vitro fertilization, this discipline has profound effects on both individual human lives and the population trends of societies, nations, and the world as a whole.

DEFINITION AND BASIC PRINCIPLES

Reproductive science and engineering is the study of the physical and chemical processes that underlie human reproduction. It involves the application of scientific technologies to treat disorders of reproduction, such as infertility. It also includes the development and use of methods to interfere with or prevent impregnation, such as contraception and sterilization. Among the many approaches reproductive scientists take to these issues, three of the most significant involve mechanical, chemical, and genetic strategies for engaging with reproduction. Micromanipulation is an umbrella term for any reproductive assistance technique that involves the physical handling of sperm, oocytes, or embryos on a microscopic scale, using specialized tools. Another of the key tools that is widely used by reproductive scientists is a set of pharmaceutical products that mimic or interfere with the chemical signals naturally produced by the body. These artificial hormones can be used to manipulate the human reproductive system in a variety of ways. Reproductive genetics is a rapidly expanding subfield of reproductive science that applies the tools of genetic research and DNA-based technologies to issues of conception, childbirth, and inheritance.

BACKGROUND

Written records indicate that humans have struggled with infertility since ancient times. For example, infertility is mentioned in texts from ancient Greece and Persia. For much of European history, infertility was generally attributed to women, not men, and was considered a sign of impiety because bearing children was considered a blessing from God. The first experiments with artificial insemination were carried out in the middle of the nineteenth century by the American gynecologist J. Marion Sims, who also carried out surgical procedures designed to widen the cervix and thus facilitate the entry of sperm. In the late nineteenth century, scientists began to acknowledge the potential role of male infertility, using microscopes to test the potency of sperm.

The first half of the twentieth century witnessed the discovery of the three most important hormones involved in reproduction, the female hormones estrogen and progesterone, and the male hormone testosterone. Soon after, companies began to manufacture the first synthetic hormones for the treatment of infertility. In 1944, the first laboratory test showing that human oocytes could be fertilized in vitro was carried out. Thirty-four years later, the first so-called test-tube baby was born in England, and sperm banks became more common. The late twentieth century saw two more milestones in infertility treatment: the first successful implantation and pregnancy with an egg that had been cryopreserved and the development of intracytoplasmic sperm injection technology.

HOW IT WORKS

Micromanipulation. The basic setup of a micromanipulator is a microscope connected to robotic arms that are powered by electric motors and moved by hydraulic or pneumatic controls that may require foot pedals or joysticks, or both. The robotic arms are in turn connected to incredibly tiny glass tools. By looking through the microscope, which magnifies the cells hundreds of times, and manipulating the controls, the operator is able to tinker with the gametes and embryos with great precision—in essence performing a kind of microsurgery. Some micromanipulation techniques involve lasers, which can move segments of a cell from place to place and slice open the thick membrane around an oocyte. This membrane, known as the zona pellucida, is often slit open to facilitate the entry of sperm. Other techniques involve the use of electric currents that can cause the membranes of two different cells to join together or transfer genetic material from one cell to another. Intracytoplasmic sperm injection is the most common micromanipulation procedure.

Reproductive Pharmacology. Reproductive pharmacology makes use of synthetic hormones to produce a desired effect—whether contraceptive in nature or intended to increase fertility. For example, most birth control pills use some combination of artificial forms of the female hormones estrogen and progesterone. These substances interfere with the normal cycle of ovulation and menstruation, thus suppressing a woman's ability to conceive. For example, the chemical signals sent by the pill may prevent a woman's pituitary gland from releasing a chemical signal that induces ovulation, so that the ovaries do not release any oocytes. Or it may prevent the lining of the uterus from thickening, thus inhibiting the implantation of a fertilized egg. Other ways in which artificial hormones may prevent pregnancy include thickening the mucus found in the cervix so that it is more difficult for sperm to travel through it and reducing the rate at which oocytes migrate from the ovaries toward the uterus. Artificial hormones can also be used to treat infertility. For example, injections of follicle-stimulating hormone (FSH) and human chorionic gonadotropin (hCG) stimulate the process of ovulation, while gonadotropin-releasing hormone alters the timing of ovulation, making a woman's fertile period more regular and ensuring that an oocyte is not released into the uterus until it has developed properly. These and other pharmaceutical tools can help physicians correct problems associated with common female fertility disorders. The most common disorder leading to infertility in women is polycystic ovary syndrome (PCOS), which results in excessively high levels of androgen (a male hormone) and ovulation that is irregular or entirely absent.

Genetics. Some infertility disorders are associated with a genetic defect of one kind or another. Male infertility has been linked with microdeletions, or tiny missing parts, in the Y chromosome, and with mutations in particular genes. To identify these genetic components of infertility, scientists often conduct what is known as a genome-wide association study, or whole-genome association study. This is a technique by which the entire set of genetic material belonging to each member of a group of subjects is scanned and compared in order to pinpoint specific genetic variations that are more prevalent in people with a certain trait, such as infertility.

Screening Tests. Another common tool of reproductive genetics is the use of screening tests to identify genetic traits in embryos produced by in vitro fertilization before they are implanted in a woman's uterus. After a cell sample has been retrieved from the blastocyst or embryo, various techniques can be used to screen its DNA for possible genetic abnormalities associated with diseases such as Down syndrome or cystic fibrosis. For example, short pieces of DNA can be artificially produced that are specially designed to bind to and mark mutated DNA in the sample, if it exists. Alternatively, the DNA in the sample can be directly examined to look for known mutations. Tests can also be carried out that reveal enzymes and proteins produced by specific genes.

APPLICATIONS AND PRODUCTS

In Vitro Fertilization. In vitro fertilization (IVF) is one of the most common applications of assisted reproduction technology. The Latin term *in vitro* literally means "in glass." In vitro fertilization is a medical procedure in which egg cells are fertilized not inside the body (in vivo) but within an artificial laboratory environment. Because this procedure is both complex and expensive, it is often employed to help couples for whom other infertility treatments have already failed.

The first step in an IVF cycle involves the use of artificial hormones, such as follicle-stimulating hormone and human chorionic gonadotropin, in drug form. During normal ovulation, a woman's ovaries produce a single egg; this treatment, known as superovulation, causes the ovaries to produce multiple oocytes. Next, the oocytes, along with some follicular fluid, are retrieved from the patient's ovaries via a needle. They are allowed to incubate under controlled laboratory conditions, with sperm collected from the patient's partner or with donor sperm. After a day or two, the eggs are examined to determine which, if any, have been successfully fertilized. If necessary, intracytoplasmic sperm injection may be used to inseminate the eggs. This is a micromanipulation procedure that can be used to artificially induce fertilization when sperm have low motility (do not move well). In this process, a micromanipulator with a thin glass pipette on the end is used to pick up and inject a single sperm directly into an oocyte. Next, the fertilized eggs—now known as embryos—are either frozen for later use or transferred into the uterus using a speculum and a catheter. Typically, multiple embryos are inserted into the uterus so as to increase the chances of at least one successful implantation. This also increases the possibility of a multiple birth.

Because the success of any given cycle of IVF treatments is by no means guaranteed, many couples choose to use cryopreservation techniques to freeze embryos produced in one cycle for future use. This streamlines the process a couple must go through if treatment does, in fact, have to be repeated. It also reduces the need for performing invasive procedures on the female patient.

Intrafallopian Transfer. Gamete intrafallopian transfer (GIFT) is a procedure that resembles in vitro fertilization. The main difference between the two applications is that with GIFT, fertilization takes place not within a laboratory setting but rather inside the female patient's body. GIFT is a minimally invasive surgical procedure in which a catheter is placed through a small keyhole incision. Oocytes and semen are inserted through the catheter into the Fallopian tubes. At this point, fertilization and implantation of the embryo may or may not occur.

Zygote intrafallopian transfer (ZIFT) is a procedure that combines elements of both in vitro fertilization and GIFT. First, gametes are extracted and fertilized under controlled laboratory conditions. Next, they are inserted into the Fallopian tubes using the same method as in GIFT. Because GIFT and ZIFT are more invasive and expensive than in vitro fertilization, they are much less commonly performed. They may be recommended for couples whose struggles with infertility are more severe or who have not responded positively to previous cycles of IVF treatment.

Artificial Insemination. Artificial insemination is a widely used, minimally invasive procedure used to help couples with a variety of infertility problems, such as a male partner with low sperm count or motility, the existence of natural antibodies in either the male or female partner that attack sperm, or characteristics of the cervix shape that make fertilization difficult. Artificial insemination can be carried out using either the intended father's sperm or that of a donor obtained from a sperm bank. In either case, once the sperm has been collected, it is physically inserted, via a catheter, into the woman's cervix. Though the technique is simple, timing is extremely important—artificial insemination must take place either just before or on the day of ovulation to be successful. In practice, it is often carried out on two consecutive days to increase the chances of fertilization.

Surrogacy. Surrogacy is an attractive option for women who are infertile because their uterus is abnormal or has been removed or who are believed to be at high risk for miscarriage or other complications of pregnancy. In gestational surrogacy, an embryo fertilized though in vitro fertilization is implanted into the uterus of a woman who is healthy and fertile, and has agreed to act as a surrogate. In many cases, the surrogate donates her own egg to be fertilized via in vitro fertilization with the biological father's sperm or that of a donor, then carries the baby to term. Some surrogates perform this service out of pure altruism or generosity; these women are usually close friends or relatives of the intended mother. Others are unrelated strangers who are compensated financially by the parents.

Contraception and Sterilization. Some applications of reproductive science are intended to prevent, not facilitate, impregnation. Most of the available contraceptive methods are designed for use by the female partner alone, although a few are meant for shared use or use by the male partner alone. Barrier methods, such as the sponge, the diaphragm, the male and female condoms, and the cervical cap, prevent sperm from reaching the egg inside the Fallopian tube. Hormonal methods, such as the pill, the vaginal ring, injected or implanted devices, and certain intrauterine devices, release artificial hormones into a woman's body to interfere with the process of ovulation and prevent eggs from being released into the Fallopian tubes. Emergency contraception, which can be used up to three days after intercourse, uses high doses of estrogen to prevent a fertilized egg from being implanted in the uterus.

Other contraceptive methods are less reliable. These include the rhythm method, in which partners carefully monitor the woman's body temperature and menstrual cycle to determine the date of ovulation and avoid intercourse during this time. Sterilization is the most decisive method of preventing impregnation. A vasectomy is a reversible surgical procedure that results in sterility for the male partner. It involves cutting or otherwise sealing both the right and the left vas deferens, the tubes through which sperm travel into the penis. A tubal ligation is a nonreversible surgical procedure that results in sterility for the female partner. It involves sealing the Fallopian tubes so that eggs are unable to pass from them into the uterus.

SOCIAL CONTEXT AND FUTURE PROSPECTS

Perhaps the most significant social impact of reproductive science and engineering is the way it has transformed the opportunities available to women in the workplace and, more broadly, the shape of families themselves. Because technologies such as in vitro fertilization enable women to bear children successfully later in their lives, many choose to delay parenthood until they have established themselves fully within their careers. Artificial insemination not only has enabled infertile couples to fulfill their desire to bear children but also—because sperm can be acquired through donor banks—has facilitated the rise of the modern phenomenon of single parenthood by choice. Also, assisted reproductive technologies have provided a means for same-sex couples to become biological parents.

Some very useful technologies created by reproductive science and engineering have the potential to be turned into what some observers fear are unethical applications. For example, preimplantation genetic diagnosis has profound benefits because it enables couples to raise their chances of having healthy babies. However, it has also generated a certain amount of controversy because the same techniques could, in theory, be used to allow couples to select embryos with certain very specific traits. For example, embryos could be chosen or engineered to have genes encoding for eye or hair color or perhaps traits such as intelligence or physical beauty—the so-called designer baby concept. Other ethical questions provoked by assistive reproductive technologies include the question of what to do with leftover frozen embryos and whether and how much women should be compensated for egg or embryo donation. Finally, cloning is a highly controversial and often misunderstood area in biomedical research. Reproductive cloning, which involves creating a precise genetic copy of an existing organism through a process known as somatic cell nuclear transfer, is banned from being done with humans in most countries.

—*M. Lee, MA*

FURTHER READING

Elder, Kay, Doris Baker, and Julie Ribes. *Infections, Infertility, and Assisted Reproduction.* 2004. Reprint. New York: Cambridge University Press, 2010. A detailed, illustrated examination of the microbiology of assisted reproductive technologies. Each chapter includes references, further reading suggestions, and frequently appendixes outlining procedures and protocols.

Green, Ronald Michael. *Babies by Design: The Ethics of Genetic Choice.* 2007. Reprint. New Haven, Conn.: Yale University Press, 2009. A bioethicist tackles moral dilemmas provoked by emerging genetic engineering technologies. Contains a glossary of relevant technical terms.

Jones, Richard E., and Kristin H. Lopez. *Human Reproductive Biology.* Rev. ed. London: Academic Press, 2010. A comprehensive introductory textbook, heavily illustrated with diagrams and photographs. Each chapter includes a summary, further reading, and advanced reading list.

Romundstad, Liv Bente, et al. "Effects of Technology or Maternal Factors on Perinatal Outcome After Assisted Fertilisation: A Population-based Cohort Study." *The Lancet* 372, no. 9640 (August, 2008): 737–743. Finds that adverse outcomes associated with births following assisted reproduction are not caused by technological factors. Includes several tables.

Spar, Debora. *The Baby Business: How Money, Science and Politics Drive the Commerce of Conception.* Boston: Harvard Business School Press, 2006. A critical overview of the fertility industry and how reproductive technologies are used in the marketplace. Includes numerous tables and extensive end notes listing sources.

Scanning Probe Microscopy

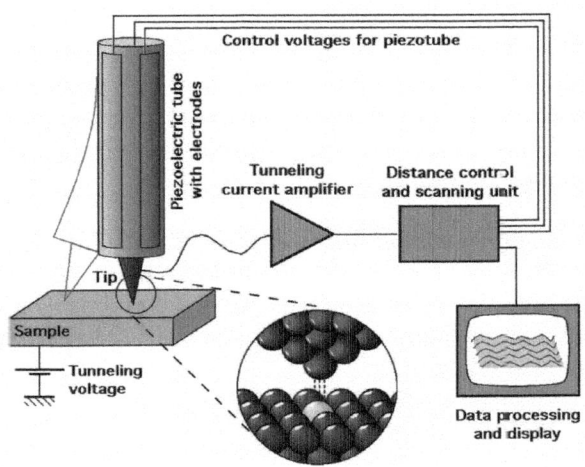

Schematic view of an STM (scanning tunneling microscope) By Michael Schmid. [CC BY-SA 2.0 at (https://creativecommons.org/licenses/by-sa/2.0/at/deed.en)], via Wikimedia Commons

FIELDS OF STUDY

Biotechnology; Biophysics

ABSTRACT

Scanning probe microscopy is a methodology that allows direct observation of structures and properties at the atomic and molecular scales. The techniques are applicable to a wide variety of purposes, providing information that is otherwise inaccessible. Scanning probe microscopy is particularly appropriate to nanotechnology, permitting the direct construction of nanoscale objects in an atom-by-atom manner.

DEFINITION AND BASIC PRINCIPLES

Scanning probe microscopy is a methodology that interfaces the macroscale of human observation (1–10^{-2} meters) to the atomic scale of the physical world, providing direct observation of features in the range of 100 micrometers to 10 picometers (10^{-4}–10^{-11} m).

Though the upper range of resolution for scanning probe microscopy is well within the range of other methods of microscopy, it is the lower range that is of most interest, because it is at this range that direct observation of atomic and molecular scale properties is possible.

The basic principles of scanning probe microscopy are founded in the quantum mechanical properties of atoms. The methods use the measurement of electronic properties (current, voltage, or "atomic force") as the means of observing the nature of surfaces and surface phenomena.

Quantum mechanics describes and defines the behavior of electrons in atoms. One of the rules of quantum behavior is that electrons are constrained to specific locales known as orbitals within the structure of an atom and are not allowed to exist at the boundaries of those locales. An observable property called quantum mechanical tunneling occurs, however, which permits electrons to move from one locale to another across the orbital boundaries. In scanning probe microscopy the miniscule electrical current due to quantum mechanical tunneling between the atoms at a surface and the atoms at the tip of an atomic-scale probe is measured as a function of their relative positions. This provides a corresponding atomic scale map of the surface structure.

BACKGROUND AND HISTORY

The scanning tunneling microscope (STM) was invented in 1982 by German physicist Gerd Binnig and Swiss physicist Heinrich Rohrer. Their device could be conceived of as a sort of quantum mechanical phonograph, in which an exceedingly sharp metallic needle scans a surface in a manner similar to the way a phonograph needle scans the groove of a vinyl phonograph record. The needle would ideally taper down to a single atom at the point, enabling atom-to-atom interaction at the surface. Sensitive digital-electronic measurement devices would measure the electronic tunneling current between the tip and the surface. The devices would then relay that data directly to a computer that would then correlate the values according to the relative dimensions and spatial relationships of the probe tip and the surface. The result would be displayed as an image having resolution of atomic scale features.

In 1986, Binnig, along with American electrical engineer Calvin Quate and Swiss physicist Christoph Gerber, introduced the scanning force microscope (SFM), which is also known as the atomic force microscope, or AFM, a variation on the STM that maintained a constant force between the scanning tip and the surface. This allowed any surface to be scanned, whereas the STM could only be used with electrically conductive surfaces. More recently, scanning near-field optical microscopy (SNOM) was developed, using measurement of short-range components of electromagnetic fields (a very small light source) between tip and surface to produce the equivalent of a photographic representation of the surface features.

HOW IT WORKS

To appreciate the operation of scanning probe microscopy, it is necessary to understand the scale on which it operates. The unaided human eye can discern detail as small as approximately 0.1 millimeter. Optical microscopes can extend this to a resolution of about 0.0001 meter. Scanning electron microscopes can typically produce images with a resolution as fine as 10 micrometers (0.00001 meter). This is the range at which scanning probe microscopes only begin to work, and they typically provide information with a resolution of as little as 10 picometers (0.00000000001 meter).

At this scale, the operation is in the realm of quantum mechanical physics rather than classical physics, and the effects associated with that scale are very different from those that occur on a larger scale. The most important difference lies in what is meant by the word "surface."

Quantum Mechanical Physics. On scales that are significantly larger than atomic and molecular diameters, a "surface" is solid matter, analogous to a smooth tabletop. At the atomic scale of quantum mechanics, however, there is no such thing as a hard surface in that sense.

Quantum mechanics describes the structure of atoms as a very small, dense nucleus of massive protons and neutrons, surrounded by a cloud of electrons that is 100,000 times greater in diameter than the nucleus. The electron cloud is therefore very diffuse. The electrons in a neutral atom are equal in number to the protons contained in the nucleus, and are confined to specific three-dimensional regions, called orbitals, around the nucleus. and are allowed to have only very specific energies according to the orbitals they occupy. At this scale of operation, a scanning probe microscope measures the electromagnetic interaction of the electron clouds in the atoms of the probe tip and atoms of the surface being scanned.

Scanning Tunneling Microscopes (STMs). The STM operates by moving the atoms-wide point of the scanning tip across a metallic, and therefore electrically conducting, surface at a distance of less than one nanometer (10^{-9} meter). The device measures the magnitude of the "tunneling current" that arises between the probe tip and the surface atoms. This current is exceedingly small, and its measurement requires extremely sensitive digital sampling and amplification electronics, and computers to process the measured data according to the relative geometries of the tip and the surface. As the probe scans, the angle between the tip and the atomic surface changes, as does the distance between them. The tunneling current measurement thus has three-dimensional vector-field properties, and because the probe tip maintains a constant orientation, variations in the tunneling current are presumably caused by the three-dimensional shape of the atoms being scanned.

Atomic Force Microscopes (AFMs). The AFM operates in essentially the same manner as the STM, except that its function is to maintain a constant measured electrical force between the probe tip and the atomic surface being scanned. In this function, the probe tip follows the shape of the atomic surfaces directly, rather than measuring a property difference that changes according to the shape of the surface. Several different modes of operation are available within this context, such as constant contact, noncontact, intermittent contact, lateral force, magnetic force, and thermal scanning. Each mode provides a different type of information about the surface atoms.

Scanning Near-Field Optical Microscopes (SNFOMs). The SNFOMs use an extremely small-point light source as the probe, rather than a physical tip. Measurement of the effect that the surface has on the light provides the image data. The technique can employ a broad range of wavelengths to investigate different properties of the atoms being scanned.

APPLICATIONS AND PRODUCTS

The field of scanning probe microscopy is a high technology research practice. Its direct applications are limited to the analytical study of surface phenomena and structures. The nature of the techniques of scanning probe microscopy makes it the method of choice for examination and study of surfaces that are not amenable to any other means of close examination, especially those of certain biological materials. Scanning probe microscopy, with its ability to probe and manipulate single atoms and so to form molecule-sized structures, is invaluable in chemistry and in the development of nanotechnology.

Surface Chemistry. Chemical reactions take place at the level of the outermost electronic orbitals of atoms, or at the electronic surface of the atoms involved. In catalyst-mediated reactions, the interaction among chemical species takes place on the surface of the catalyst material, where the atoms of the reactants have interacted electronically with the atoms of the catalytic material. This lowers the energy barriers that must be overcome for the reaction to occur, with the result that the desired reaction is facilitated.

Scanning probe microscopy has a spectroscopy mode that allows the direct measurement of electron energies in single atoms and of single atomic bonds between atoms in a molecule. The methodology provides a better understanding of the chemistry that takes place at surfaces, in turn contributing to the development of new and improved chemical processes. Perhaps the most important aspect of this is the enhanced knowledge of the mechanisms of surface phenomena such as oxidation and chemical corrosion.

Electronic Materials and Integrated Circuits. How small can a functional transistor be? Technology enables the construction of several million transistors on the surface of the small silicon chip that is the central processing unit (CPU) of a computer. Scanning probe microscopy enables the physical study of such structures in minute detail. It also enables the construction and study of transistor structures that are orders of magnitude smaller. The research in this field examines possible ways to construct integrated circuits and computers that exceed existing capabilities of production.

The available methods of production of integrated circuits have essentially reached the physical limit of their capabilities to construct viable semiconductor transistors. Atomic force microscopy is being used to investigate the construction of transistor-like structures based on quantum dots rather than on semiconductor junctions.

Another area of research is the construction of molecule-sized transistors made from graphene or carbon nanotubes and other materials. These technologies, when fully developed, will completely change the nature of computing by enabling the construction of a quantum computer, a device that could carry out in seconds calculations that existing computers would require possibly billions of years to complete.

Data storage would also be revolutionized by these innovations. Research indicates that data storage will soon reach capacities measured in terabits per square centimeter. The ultimate binary data storage density would have each bit stored in the space of one atom, a density that can be envisioned only with scanning probe microscope technology.

Nanostructures and Nanodesign. Nanotechnology works at the nanometer scale of 10^{-9} meters. To appreciate this scale, imagine the length of one millimeter divided into one million segments, each of which would be one nanometer. The concept of nanotechnology is to produce physical machines constructed to that scale. Because scanning probe microscopy can manipulate single atoms, it can be used to construct nanoscale, and even picoscale, physical mechanisms. The latter are essentially individual molecules whose physical structures imitate those of much larger devices and mechanisms, such as gears.

In September, 2011, researchers at Tufts University reported the successful use of low-temperature scanning tunneling microscopy to construct a working electric motor consisting of a single molecule. As can be imagined, this is a complex field of research, because quantum effects play a significant role in the interoperability of such small devices.

One application of scanning probe microscopy that is of immediate importance is the study of friction and abrasion at the atomic level, which is where those processes take place. Atomic force microscopy can be used to literally scratch the surface of a material, providing detailed information about how friction and abrasion actually work and about what might be done to lessen or prevent those effects.

Biological Studies. Scanning electron microscopy (SEM) has been the workhorse of biological research since its invention, providing detailed images of extremely small biological structures. The technology has some practical limits, however, because of the principles on which it functions. Many biological materials that are of interest cannot be studied in detail using SEM, but are amenable to study using scanning probe microscopy. The methods are useful in measuring the forces that exist among functional groups in biological and organic chemical structures.

SOCIAL CONTEXT AND FUTURE PROSPECTS

Scanning probe microscopy is a field that will have very little direct social context because of the extremely small scale of its subject matter. However, the secondary effects of the knowledge and technology derived from research and development in this field could have a large social impact, primarily because of the economic benefits from control of friction and from new technologies for integrated circuits and magnetic memory media. Any real predictions for the future prospects of the field of scanning probe microscopy are entirely conjectural.

—*Richard M. Renneboog, MSc*

FURTHER READING

Bhushan, Bharat, Harald Fuchs, and Masahiko Tomitori, eds. *Applied Scanning Probe Methods VIII: Scanning Probe Microscopy Techniques (NanoScience and Technology)*. Berlin: Springer, 2008. Provides the most up-to-date information on this rapidly evolving technology and its applications.

Howland, Rebecca, and Lisa Benatar. "A Practical Guide to Scanning Probe Microscopy." ThermoMicroscopes. March 2000. Web. Accessed September, 2011. This introductory guide to scanning probe microscopy describes several techniques and operating modes of the devices, as well as their structure and principles of operation, and discusses the occurrence of image artifacts in their use.

Meyer, Ernst, Hans Josef Hug, and Roland Bennewitz. *Scanning Probe Microscopy: The Lab on a Tip*. Berlin: Springer, 2004. This book provides an excellent overview of scanning probe microscopy before delving into more detailed discussions of the various techniques.

Mongillo, John. *Nanotechnology 110*. Westport, Conn.: Greenwood, 2007. Demonstrates the essential relationship between and value of scanning probe microscopy to nanotechnology.

SCIENCE OF CLONING

FIELDS OF STUDY

Biotechnology; Genetics; Microbiology; Philosophy and Religious Studies; Reproductive technology

ABSTRACT

Cloning is any type of biological reproduction that produces offspring that are genetically identical to their parents. Cloning occurs naturally, since many organisms routinely reproduce through natural cloning processes. Artificial cloning technologies include molecular cloning, which reproduces large quantities of discrete segments of DNA; reproductive cloning, which uses assisted reproductive technologies to produce animals that share the same desirable genetic characteristics as another living or previously existing organism; and therapeutic cloning, which uses the same techniques as reproductive cloning but instead derives useful cell lines from cloned embryos.

DEFINITION AND BASIC PRINCIPLES

Cloning is a means of producing biological organisms, cells, or DNA molecules that are genetically identical to their progenitors. There are natural forms of cloning and three main types of artificial cloning: molecular, reproductive, and therapeutic cloning.

Natural mechanisms of cloning occur in organisms such as bacteria that simply split or fragment into identical copies of themselves. In other organisms, reproductive cells, or gametes, undergo a process called parthenogenesis, in which they initiate development without the benefit of fertilization. Cloning is uncommon in mammals, but rarely, early mammalian embryos undergo a form of cloning called twinning, in which the embryo splits into two embryos, which develop into genetically identical twins.

Molecular cloning, also known as recombinant DNA technology or DNA cloning, involves the transfer of an isolated fragment of DNA from an organism of interest to a host cell that replicates it. Such isolated DNA fragments are known as cloned DNA or genes.

Reproductive cloning uses assisted reproductive technologies to generate animals with the same nuclear genome as another animal. The particular procedure used during reproductive cloning is called somatic cell nuclear transfer (SCNT). Cloned embryos are gestated in the womb of a surrogate mother until they come to term. Cloned organisms are not genetically modified organisms but are simply produced through a type of assisted reproduction.

Therapeutic cloning uses the same procedures as reproductive cloning; however, instead of transferring the cloned embryo into the womb of a surrogate mother, the embryo is further manipulated in the laboratory to make cell cultures of embryonic cells for basic or clinical research.

BACKGROUND

Sea urchins were the first animal cloned in the laboratory. In 1894, Hans Dreisch isolated sea urchin embryo cells and watched them develop into small, separate larvae. In 1902, Hans Spemann used the same procedure of embryo splitting to isolate cells from salamander embryos, which also developed into identical adult salamanders. In 1903, U.S. Department of Agriculture employee Herbert Webber coined the word "clon" for asexually produced cells or organisms, which later evolved into "clone." This term comes from the Greek *klon*, which means 'trunk" or "branch." Horticulturists have used this term for more than a century, since an entire new plant can grow from a cutting, resulting in a plant that is genetically identical to the plant from which the cutting was taken.

In 1928, Spemann cloned salamanders by transferring the nucleus, the subcellular compartment that houses the chromosomes, from one salamander embryo into the egg of another. Since Spemann's seminal experiments, scientists have adapted nuclear transfer technology to clone other organisms. In 1952, frogs were cloned, and in 1963, the Chinese embryologist Tong Dizhou cloned a carp to produce the first cloned fish. During the 1980s and 1990s, sheep, cows, and mice were cloned. However, all these animals were cloned by using nuclei from embryos. In 1996, Ian Wilmut and his team at the Roslin Institute in Edinburgh, Scotland, cloned a sheep from an adult cell, demonstrating that adult cells could serve as the source of genetic material for animal clones. This technological feat was followed by the cloning of goats, mules, gaurs (an endangered species), horses, pigs, mouflons (a wild sheep), mice, rats, dogs, cats, water buffalos, camels, rabbits, deer, wolves, and African wildcats, and even embryos from nonhuman primates and humans.

HOW IT WORKS

Molecular Cloning. To clone a gene, the DNA of the model organism is selectively fragmented by enzymes called restriction endonucleases (REs) and inserted into another piece of DNA called a cloning vector. Cloning vectors are either small circles of DNA called plasmids, bacterial viruses, or bacterial or yeast artificial chromosomes. They ferry the DNA fragments from the genome of the model organism into a host cell (either a bacterium or yeast). This population of host cells collectively carries the entire genome of the model organism in small fragments, and is called a gene library.

To isolate a gene from a gene library requires a probe, which is a fragment of DNA or RNA of any length that has a sequence that is complementary to the sequence of the gene that is to be isolated. Probes can be made synthetically or can come from the genes of closely related organisms. By screening the gene library with the probe, the gene of interest is cloned, which simply means to isolate it from all the other sequences found in the genome of the model organism.

Alternatively, scientists can synthesize small strands of DNA called primers, whose sequences are complementary to different locations in the gene. These primers can be used to specifically amplify the gene from the library by means of a polymerase chain reaction (PCR). A polymerase chain reaction makes large quantities of the gene of interest from a very small amount of starting material, and the amplified DNA can also be cloned into a cloning vector or analyzed directly.

Reproductive Cloning. To clone an animal, mature eggs are isolated from females of the animal species that is to be cloned. The egg is enucleated by piercing it with a microscopically narrow (0.0002-inch-wide) glass tube that is used to vacuum out the egg nucleus. The enucleated egg is fused with a cell from the body of the animal to be cloned and activated with either chemicals or an electric current. This procedure is called somatic cell nuclear transplantation (SCNT).

After activation, the egg divides and grows like a newly formed embryo. However, if the animal is a mammal, the embryo can survive only for a limited period of time before it must implant into the inner layer of the mother's womb. Therefore, a surrogate female from the same species of the animal to be cloned, or a closely related species, is made pseudopregnant by feeding her hormones, and the embryo is released into her receptive womb where it then implants. Barring any technical or biological mishap, the cloned embryo will develop, and the process will result in a live birth.

Therapeutic Cloning. To make embryonic cell cultures, cloned embryos are made by means of somatic cell nuclear transplantation. They are then either disassembled in the laboratory and used to establish embryonic cell cultures or gestated in a surrogate mother to the fetal stage, at which time the fetus is aborted, and cells from the fetus are used to establish fetal cell cultures.

By culturing specific cells from cloned embryos, scientists can make embryonic stem cell (ESC) cultures. During mammalian development, two distinct cell populations form after the first few days of embryonic development. The trophoblast, or the flattened, outer layer of cells, will eventually form the placenta and its associated structures. The inner cell mass (ICM) is the round, inner clump of cells that develop to form the embryo proper and a few structures associated with the placenta. If ICM cells are isolated and cultured on feeder cells, a layer of nondividing skin cells that secrete a cocktail of growth-promoting chemicals, the ICM cells will grow and spread over the surface of the culture dish. Such a culture is an embryonic stem cell culture, and these cells are pluripotent, which means that they can differentiate into any cell type in the adult body.

APPLICATIONS AND PRODUCTS

Molecular Cloning. Organisms that express cloned genes make many useful pharmaceuticals such as human insulin, growth hormone, clotting factors, fertility drugs, and vaccines. Cloned genes are also used to genetically screen individuals for genetic diseases. Pharmacologists even use cloned genes for pharmacogenetics, which screens patients for the presence of gene variants that can profoundly affect the efficacy and toxicity of particular drugs. This allows clinicians to tailor treatment to the exact genetic makeup of the patient to maximize treatment efficacy and minimize side effects. Such a strategy is called personalized medicine. Cloned genes are also used in gene therapy, which delivers cloned genes into the bodies of patients who suffer from genetic diseases in an attempt to cure them. Patients with cancer and inherited deficiencies of the immune system, blindness, and blood-based defects have been treated with gene therapy protocols.

In agriculture, the introduction of cloned genes into plants that are used as food crops has generated transgenic crops. These crops display several advantageous traits: reduced dependence on agrochemical applications (for example, Bt-corn and herbicide-resistant crops), increased nutritional value (for example, Golden Rice), increased resistance to environmental stresses, and reduced spoilage (for example, the Flavr Savr tomato).

Despite safety and ethics concerns regarding genetically modified organisms (GMOs) in agricultural products, the International Service for the Acquisition of Agri-Biotech Applications (ISAAA) reported in 2013 that genetically modified (GM) crops were planted in twenty-seven countries by approximately 18 million farmers worldwide. Over 60 percent of the world's population live in the twenty-seven countries that are planting GM crops. In 2006,

U.S. government statistics showed that 87 percent of the global genetically modified crops were grown in developed countries. By 2013, however, ISAAA reported that Latin American, Asian, and African farmers grew 54 percent of the global GM crops compared to the 46 percent grown in developed countries worldwide. This trend continued into 2016, when the ISAAA reported increased soybean plantings in countries such as Argentina and Brazil. Corn, soybeans, cotton, alfalfa, and canola were the major crops, often modified for insect resistance. Rice has been genetically enhanced for more iron and vitamins to alleviate malnutrition in Asia. Other plants have been modified to survive weather variances.

Reproductive Cloning. When farmers identify food animals with desirable traits, they typically breed those animals as much as possible to improve the genetic quality of their herds and flocks. However, such prize animals inevitably die. Propagating these animals by reproductive cloning and mating them to as many animals as possible preserves the exceptional genetic content of a prize animal and allows it to produce far more offspring. This significantly raises the genetic quality of the flock or herd, and commercial dissemination of such cloned animals to other farmers raises the overall genetic quality of food animals. Reproductive cloning also eliminates the need for artificial insemination, which is often expensive and inconvenient.

Cloning effectively maintains high-quality animal stocks. Reproductive cloning of only the healthiest and most productive animals increases their numbers and improves the gene pool (sum total of genetic diversity) and overall health of food animals. This results in safer and healthier food and reduces the use of growth hormones, antibiotics, and other chemicals in the raising of animals.

In the field of conservation biology, the numbers of endangered species are often increased by captive breeding programs. However, not all endangered species can effectively breed in captivity. Reproductive cloning can aid in the preservation of those organisms that do not reproduce in captivity. Cloning can also resurrect genetic material from dead animals and potentially expand the gene pool of endangered species. In 2001, scientists at the University of Teramo, Italy, cloned the European mouflon, an endangered sheep, from cells sampled from a dead animal. When combined with other reproductive technologies, cloning can help save endangered species.

Cloned animals also serve as excellent research models. Because each cloned animal is genetically identical, experiments on cloned animals are devoid of differences caused by heterogeneous genetic backgrounds. Genetic manipulation of cloned animals allows researchers to modify genes of interest and more completely analyze their contribution to development and disease. Modifying particular genes of cloned animals also generates model systems for particular genetic diseases. Cloned, transgenic mice and cloned knockout mice, which have had a specific gene inactivated, are examples of the vast usefulness of such model systems.

Of enormous interest is modifying the genomes of cloned animals so that they can produce clinically and pharmaceutically significant products. By genetically modifying pigs, it is possible to make cloned pigs that contain organs that are fit for transplantation into humans (xenotransplantation). Also, producing antibodies, clotting factors, or even vaccines in the blood or milk of farm animals provides a means to mass-produce potentially expensive pharmaceutical agents at a fraction of the normal cost. This process is called pharming.

Therapeutic Cloning. Therapeutic cloning has tremendous potential for numerous clinical applications. Embryonic stem cells (ESCs) made from therapeutic cloning procedures are pluripotent. Therefore, injured, diseased, or failing tissues or organs could potentially be replaced by tissues or organs manufactured from embryonic stem cells in the laboratory or fetal cells from cloned fetuses. Furthermore, embryonic stem cells made from cloned embryos, or any tissues or organs fashioned from these cells, would not be regarded by the patient's body as foreign. Experiments in laboratory animals have shown that such scenarios are possible. Therapeutic cloning coupled with embryonic stem cells technology, could christen a new era of regenerative medicine.

Embryonic stem cells from cloned embryos have toxicological applications. Toxicologists typically use laboratory animals or cultured cells to gauge the biological effects of natural or industrially produced molecules on human beings. Unfortunately, laboratory animals show limited utility as a model for human toxicology, and cultured cells do not represent the response of an organ or tissue to foreign molecules. Furthermore, neither of these model systems can assess the individual responses people will have to such molecules, because the genetic variation between individual humans causes differential responses to drugs, toxins, or environmental pollutants. However, cultured embryonic stem cells from cloned embryos can test the biological effects of drugs or environmental pollutants on cells made from a specific person. In addition, because these cells can be differentiated into various tissues and even organs, they can be used to evaluate the individual and tissue-specific responses people might have to particular drugs or pollutants.

SOCIAL CONTEXT AND FUTURE PROSPECTS

Despite the reservations of some people, cloning is a part of everyday life. Many of the foods Americans consume contain some genetically engineered products. Physicians prescribe medicines, give vaccines, and apply other biological products made by genetically engineered microorganisms on a quotidian basis.

Nevertheless, many people have raised concerns over cloning technologies. First, conservation biologists have suggested that cloning endangered species does not address the habitat destruction and environmental degradation that pushed these species to near extinction in the first place. Second, cloning only makes one species and does not re-create an ecosystem. For example, cloning cannot recapitulate a coral reef or an old growth forest. Thus, many argue that it is the wrong solution for the problem.

Genetically modified organisms have become the focal point of concern for several environmental activism groups. Such groups oppose GMOs because they believe that the cloned genes inserted into them can spread to other species and cause severe environmental disruption and that genetically engineered foods have not been sufficiently tested and are potentially dangerous to human health.

The most contentious aspect of cloning technologies is human genetic engineering and reproductive cloning. Until the cloning of the sheep Dolly in 1997, it was thought that adult specialized cells could not be made to revert to nonspecialized cells that can give rise to any type of cell. However, Dolly was created from a specialized adult cell from a ewe's udder. U.S. president Bill Clinton asked the National Bioethics Advisory Commission to form recommendations about the ethical, religious, and legal implications of human cloning. In June 1997, the commission concluded that attempts to clone humans are "morally unacceptable" for safety and ethical reasons. There was then a moratorium placed on using federal funds for human cloning. In January 1998, the U.S. Food and Drug Administration (FDA) declared that it had the authority to regulate human cloning and that any human cloning must have FDA approval.

In 2013, the science of cloning experienced another major breakthrough when researchers at the Oregon Health & Science University successfully cloned human embryonic stem cells. Using nuclear transfer, the scientists implanted the skin of a fetus into the nucleus of an egg before prompting division, creating a ball of cells that includes embryonic stem cells with the same genetic material as the original skin cell. The main purpose of the research was not to further the prospect of human cloning, however, but to establish a process to consistently clone stem cells for use in the treatment of diseases and other human ailments.

China, a country that has always striven to become a global leader in the advancement of cloning technology, announced in late 2015 that one of its major companies, BoyaLife, would be opening a facility designated to mass commercialization of cloned cattle largely in response to the country's increased need for beef, which represented a marked departure from other countries' focus on cloning solely for the purpose of breeding. Earlier that year, the European Parliament had voted to institute a ban on the cloning of any farm animals, their offspring, and derivative products.

Transhumanists are some of the most energetic proponents of human cloning and genetic enhancement. As a movement, Transhumanism regards infirmity, disease, aging, and death as undesirable and unnecessary and views science and technology as the means to defeat human limitations. Transhumanists' main argument for human cloning is that reproductive freedoms extend to everyone, and therefore, every human being has an inherent right to clone himself or herself.

Opponents of human cloning object to the manufacturing of human beings. Cloned children are made to be identical to someone else and therefore will always live in the shadow of the original person and never be completely the person they choose to be. These unreasonable expectations can psychologically damage them and violate their human dignity and individuality. Cloning would also alter the concept of human nature and therefore undermine the very foundation of liberal democracy.

In the future, the argument and debate over cloning will not dissipate, but cloning research will certainly advance and provide more and more examples of the utility of this remarkable technology.

———*Michael A. Buratovich, PhD*

FURTHER READING

Alexander, Brian. *Rapture: A Raucous Tour of Cloning, Transhumanism, and the New Era of Immortality.* Basic, 2004.

Fukuyama, Francis. *Our Posthuman Future: Consequences of the Biotechnology Revolution.* Picador, 2003.

Hanson, Charles, et al. "Transplantation of Human Embryonic Stem cells onto a Partially Wounded Human Cornea In Vitro." *Acta Ophthalmologica*, vol. 91, no. 2, 2013, pp. 127–30.

"ISAAA Brief 46-2013: Executive Summary." *ISAAA*, 25 Mar. 2014, www.isaaa.org/resources/publications/briefs/46/executivesummary/. Accessed 28 July 2014.

Jensen, Eric A. *The Therapeutic Cloning Debate: Global Science and Journalism in the Public Sphere.* Ashgate, 2014.

Lynch, Colum. "UN Backs Human Cloning Ban." *Washington Post*, 9 Mar. 2005, www.washingtonpost.com/wp-co/hotcontent/index.html?section=nation/specials/science/cloning. Accessed 28 July 2014.

Mitchell, C. Ben, et al. *Biotechnology and the Human Good.* Georgetown UP, 2007.

Park, Alice. "Scientists Report First Success in Cloning Human Stem Cells." *CNN*, 16 May 2013, www.cnn.com/2013/05/15/health/time-cloning-stem-cells. Accessed 24 Oct. 2016.

Phillips, Tom. "Largest Animal Cloning Factory Can Save Species, Says Chinese Founder." *Guardian*, 24 Nov. 2015, www.theguardian.com/world/2015/nov/24/worlds-largest-animal-cloning-factory-can-save-species-says-chinese-founder. Accessed 24 Oct. 2016.

"Pocket K No. 16: Biotech Crop Highlights in 2015." *ISAAA*, June 2016, isaaa.org/resources/publications/pocketk/16/default.asp. Accessed 24 Oct. 2016.

Prado, José Rafael, et al. "Genetically Engineered Crops: From Idea to Product." *Annual Review of Plant Biology*, vol. 65, 2014, pp. 769–90.

Semple, Kirk. "UN to Consider Whether to Ban Cloning of Human Embryos." *The New York Times*, 3 Nov. 2003, www.nytimes.com/2003/11/03/world/un-to-consider-whether-to-ban-some-or-all-forms-of-cloning-of-human-embryos.html. Accessed 28 July 2014.

Shanks, Pete. *Human Genetic Engineering: A Guide for Activists, Skeptics, and the Very Perplexed.* New York: Nation Books, 2005. A helpful explication of the science behind cloning, coupled with stern warnings against it, by a noted social activist.

Silver, Lee. *Challenging Nature: The Clash Between Biotechnology and Spirituality.* New York: Harper Perennial, 2006. A Princeton stem cell scientist explains the science behind biotechnology and stem cells. He offers some rather harsh critiques of more conservative thinkers who do not agree with his optimistic views of genetic enhancement and embryonic stem cells.

———. *Remaking Eden: How Genetic Engineering and Cloning Will Transform the American Family.* New York: Harper Perennial, 2007. A very readable introduction to the science of cloning and genetic engineering by a noted mammalian embryologist, who believes that humans should be cloned and that people should welcome the profound changes that it will invoke within human societies.

Wilmut, Ian, Keith Campbell, and Colin Tudge. *The Second Creation: Dolly and the Age of Biological Control.* New York: Farrar, Straus and Giroux, 2000. The two researchers who made Dolly team up with a noted British science writer to give a personal but rigorous explanation and thoughtful examination of cloning. Contains a helpful glossary of terms.

Seed banks

Seedbank at the USDA Western Regional Plant Introduction Station [Public domain], via Wikimedia Commons

FIELDS OF STUDY

Agronomy; Biotechnology; Environmental biotechnology (or Green biotechnology); Forestry; Horticulture; Microbiology; Plant pathology

ABSTRACT

Because modern agricultural practices focus on only a limited number of crops, seed banks have been established to help promote biodiversity and protect the world's future food supply by preserving the seeds, or genetic material, of thousands of plant species that might otherwise become extinct. Seed banks also provide a backup of genetic plant material in the event of wars, accidents, or environmental disasters on a local or global scale.

HOW IT WORKS

Modern-day seed banks, also referred to as gene banks, are based on an agricultural tradition that has persisted for thousands of years: saving seeds from one season to ensure a supply that can be used to plant crops the next season. Seeds have also been passed down from one generation of farmers to the next, but as agricultural mass production has increased around the world, thousands of plant species have been irrevocably lost simply because no farmers or agricultural corporations chose to grow them over a period of time.

By the early twenty-first century, approximately 1,400 organizations identifying themselves as seed or gene banks were in operation worldwide. These banks can range from small, informal facilities that specialize in certain types of seeds to larger organizations such as the Seed Savers Exchange, a nonprofit organization established in 1975 to preserve not only the genetic but also the cultural and historical heritages of plants. Located on 360 hectares (890 acres) in Decorah, Iowa, the Seed Savers Exchange maintains more than 25,000 varieties of fruit, flower, vegetable, and herb seeds. This group goes beyond the activities traditionally conducted by seed banks in that it makes many uncommon seed varieties available for sale, and it plants its seed stock on a rotating basis in order to generate new seeds and ensure that the plant species remain part of the planet's active ecosystem.

One of the world's most impressive seed banks is the vast Svalbard Global Seed Vault, a facility opened by the Norwegian government in 2008 with assistance from the Global Crop Diversity Trust. Built inside a mountain on an island within the Arctic Circle, this high-tech seed bank has the capacity to store 4.5 million seed samples; it maintains a constant interior climate in spite of temperature variations in the surrounding permafrost. Nicknamed the "Doomsday Vault" by the media, it is designed to withstand nuclear warfare and terrorist attacks as well as natural disasters such as flooding, and is therefore considered to be a global "insurance policy" in the event of agricultural disaster. The Svalbard Global Seed Vault is actually a bank for banks; most of its "depositors" are smaller seed banks around the world that are located in areas that are geographically or politically less stable. It thus represents a vital step in efforts to protect the earth's plant biodiversity. The Global Crop Diversity Trust has the further goal of providing funding and other support for multiple crop repositories around the world.

—*Amy Sisson*

FURTHER READING

Fenner, Michael, and Ken Thompson. "Soil Seed Banks." In *The Ecology of Seeds*. New York: Cambridge University Press, 2005.

Fowler, Cary. "The Svalbard Seed Vault and Crop Security." *BioScience* 58, no. 3 (2008): 190-191.

Rosner, Hillary. "The Gatherers." *Popular Science*, January, 2008, 60-64, 66, 91.

Shea, Neil. "Norway's Ark." *National Geographic*, June, 2007, 14-21.

STEM CELL RESEARCH AND TECHNOLOGY

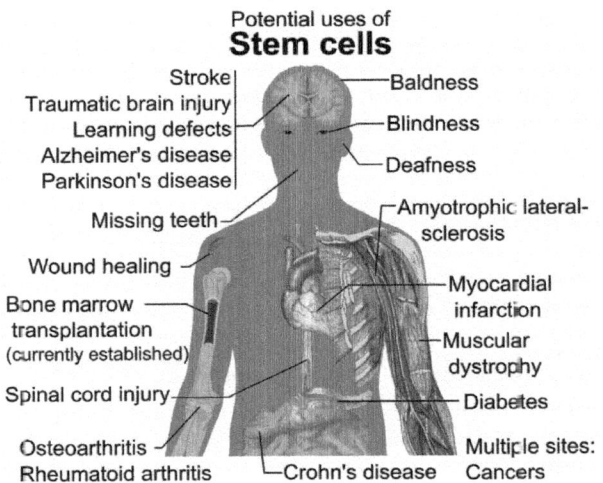

Diseases and conditions where stem cell treatment is promising or emerging.[62] Bone marrow transplantation is, as of 2009, the only established use of stem cells. By Mikael Häggström. [Public domain], via Wikimedia Commons

FIELDS OF STUDY

Biotechnology; Genetics; Medical biotechnology (or Red biotechnology); Philosophy and Religious Studies

ABSTRACT

Stem cell research is the field of science that examines specific cells that have the ability to divide indefinitely in culture and that give rise to specialized cells in order to provide therapy for diseases. There are two main types of stem cells: embryonic and somatic stem cells. Embryonic stem cells are formed in the early stages of embryonic development, and somatic stem cells are adult stem cells found in various tissues in the body. Stem cells have the potential to be used as therapy to replace or repair a person's cells or tissues that are damaged or dysfunctional in the treatment or cure of diseases.

DEFINITION AND BASIC PRINCIPLES

Stem cells have the basic properties of being undifferentiated cells that can divide indefinitely and have the potential to develop into many different types of cells in a body during early embryogenesis and during growth of an individual. Stem cells are different from other cells in the body in that they can renew themselves through cell division, allowing them to act as a repair mechanism and to replenish cells that are damaged or that die. When each stem cell divides, it has the potential of either remaining a stem cell or becoming another cell type with a more specialized function

There are two main types of stem cells: embryonic and somatic (also called adult stem cells). Embryonic stem cells are pluripotent, in that they have the capability of becoming any type of cell in the body. This is because these cells arise from the blastocyst, early in embryogenesis, making up the inner mass of cells. The inner cell mass gives rise to the entire body of an organism, including all the specialized cell types and organs, such as the heart, muscle, brain, skin, and other tissues. Somatic stem cells are considered to be multipotent and are found only in specialized tissues in the body, which are specific populations of cells that are used to generate replacements for cells that are damaged or die through the normal aging process of cells and because of injury or disease.

Stem cell therapy uses stem cells to replace or repair a patient's cells or tissues that are damaged or missing. Stem cell therapy is still experimental, in that it has not yet proven to be effective or safe, but stem cells have the potential to treat many diseases.

BACKGROUND

Embryonic stem cells were first studied in the mouse in 1981, when scientists discovered ways to derive embryonic stem cells from mouse embryos. This led to the discovery, in 1998, of a method to derive stem cells from human embryos and grow them in the laboratory. However, the use of human embryonic stem cells—taken from embryos that were originally created for reproductive purposes—has been limited because a number of stem cell lines have been allowed to be grown in the laboratory for research purposes. This constraint led to further discoveries of how to derive stem cells from somatic tissues. In 2006, scientists discovered conditions that would allow these specialized tissue stem cells to be reprogrammed genetically to become pluripotent. These stem cells, reprogrammed to express certain genes or maintain these cells in a stem cell-like state, are called induced pluripotent stem cells. In March 2009, the ban on generating new stem cell lines was lifted by President Barack Obama, making federal funding for embryonic stem cell research available without the previous limits on the stem cell lines generated.

HOW IT WORKS

To identify stem cells, cells first have to be grown in the laboratory, or cultured. The first step in isolating stem cells is to transfer the inner cell mass of a blastocyst into a cultural medium in a laboratory dish. The culture medium contains nutrients that cells need to grow and divide. Stem cells do not always grow, but when the cells continue to grow and divide, they are then divided into other culture dishes, called subculturing, so that millions of copies of the same stem cell (cloning) can be used for research.

Embryonic Stem Cells. Embryonic stem cells are the easiest of stem cells to divide and reproduce in culture, and they have been shown to live for months without differentiating. When these cells continue in their stem cell state, they are considered pluripotent and have the same genetic makeup as the original stem cells from the inner cell mass. These cells are referred to as an embryonic stem cell line. These cells may be frozen and shipped to other laboratories for further culturing and experimentation.

Somatic Stem Cells. Somatic (adult) stem cells are undifferentiated cells that are found in a tissue or organ that can renew themselves and differentiate to become specialized cells of that tissue or organ. Adult stem cells are used to regenerate or repair the tissue in which they are located. Known somatic stem cells are located in the brain, bone marrow, peripheral blood and blood vessels, muscles, skin, teeth, heart, liver, ovarian epithelium, and testes. To be used as a somatic stem cell, these cells need to demonstrate that they can generate a line of genetically identical cells that can give rise to all the differentiated cell types of that tissue. Once these cells are identified, they can be used to regenerate and repair cells within that tissue. Experiments are ongoing in transdifferentiation, in which certain somatic stem cells are reprogrammed into other cell types or even to become like embryonic stem cells, called induced pluripotent stem cells, with the introduction of embryonic cells.

APPLICATIONS AND PRODUCTS

There are several reasons why stem cells are important in science and the advancement of health care.

Cell Specialization and Development. Pluripotent stem cells help scientists understand the complexity of human development and how genes work to make decisions so that cells differentiate to become specialized cells. As development proceeds from an embryo to an individual human, genes turn on and off to give rise to protein expression and cell differentiation. These decision-making genes control the expression of pluripotent stem cells. Scientists know that certain diseases, such as cancer and birth defects, are caused by abnormal cell division and cell specialization. Understanding normal cell development will allow scientists to determine the errors that cause debilitating and often lethal diseases.

Medical Drug Testing. Stem cell research potentially may change the way new medical drugs are developed and tested for safety. These new drugs can be tested on stem cell lines first. Using pluripotent stem lines will expand the cell types that can be tested in the laboratory, before a drug is tested on animals and humans, streamlining the process for drug development.

Cell Therapies. Stem cells have the potential to be used to generate cells and tissues that could replace or regenerate damaged cells and tissues in humans. Such cell therapies could help treat disorders that disrupt cell function or destroy tissues, such as cancer, heart disease, diabetes, spinal cord injury, arthritis, Parkinson's disease, and Alzheimer's diseases. Modern medicine relies on donated organs and tissues to replace destroyed tissue in heart, bone marrow, and kidney transplants. However, the number of people suffering from these disorders far outnumbers the organs and cells available. Stem cells offer a unique opportunity to create a renewable source of replacement cells and tissues to treat these diseases. Another problem in the transplant process is that the recipient's body tends to reject the foreign cells from the donor. With stem cells, research could focus on developing modifications to these cells to minimize tissue incompatibility or to create tissue banks with common tissue type profiles that would be accepted by a large number of individuals.

Somatic Cell Nuclear Transfer. The technique called somatic cell nuclear transfer is still in the research stage. In somatic cell nuclear transfer, the nucleus of virtually any somatic cell is taken from an individual patient and fused with a donor egg cell from which the nucleus has been removed. That cell is then stimulated to develop into a blastocyst and the inner cell mass is taken to create a culture of pluripotent stem cells. These stem cells can be stimulated to develop into specialized cells that are needed to repair damaged tissues or organs. Because the genetic information is taken from the individual patient, these cells theoretically would not be rejected by the patient as they are genetically identical to those of the individual. This type of transplantation would not require immune-suppressing drugs to be successful, and patients would have a far greater chance of survival.

Somatic Stem Cell Therapies. There are disadvantages and advantages to using somatic stem cells for therapies. One disadvantage is that these stem cells are multipotent but not pluripotent, and the types of cells that can be developed are limited. Previously, it was thought that somatic stem cells could develop into only the specialized cells from which they were derived, making it necessary to use only bone marrow stem cells for bone marrow transplantation, liver stem cells for liver diseases, and so on. However, experiments on mice have shown that, for example, when neural stem cells were placed into bone marrow, a variety of blood cell types were produced. So it is possible that even specialized stem cells may be manipulated to be wider reaching in their potential than previously thought. However, the biggest limitation of somatic stem cells to date is that they have not been isolated from all the tissues of the body.

Transplantation. One advantage of somatic stem cells is in transplantation. If these cells could be isolated from a patient and directed to divide and specialize in a manner that conveys normal cell function, they could then be transplanted back into the patient without immune rejection. This would also reduce or avoid the need for embryonic stem cells from human embryos or human fetal tissue. However, isolating somatic stem cells and growing them in culture has been difficult. Even if it becomes possible, growing and manipulating them quickly enough to correct a disease state may be impossible. Rigorous research will be required to overcome the obstacles of this type of cell therapy.

SOCIAL CONTEXT AND FUTURE PROSPECTS

Acceptance of stem cell research has been greatly expanded with publicity regarding the potential benefits of this type of research. Also, there have been numerous scientific publications leading to advances in the field. However, considerable controversy remains regarding the ethical implications of using embryonic stem cells. In the United States, much debate has centered on the use of human embryos and fetal tissue created for reproductive use. Although these embryos are no longer needed, using them for research means they no longer can be used to produce a viable individual. The morals and ethics of this continue to be debated.

However, stem cell research is not limited by the availability of embryonic stem cells. Alternative stem cells, such as somatic stem cells and induced stem cells, have been developed, and researchers may be able to use these instead. The competitive nature of scientific endeavors has led to the advancement of all fields of science, and continued work in this field has produced further success in the use and potential of stem cells as a source of eliminating the threat of some of the most deadly human diseases. The uses for these cells is continually expanding with research; for example, in 2015 the Harvard Stem Cell Institute (working with several other institutions) was able to grow neurons from stem cells to help with several disorders.

—*Susan M. Zneimer, PhD*

FURTHER READING

Fox, Cynthia. *Cell of Cells: The Global Race to Capture and Control the Stem Cell.* New York: Norton, 2007. Print.

Greene, Alexander L. *Encyclopedia of Stem Cell Research.* Hauppauge: Nova Science, 2012. Digital file.

Haerens, Margaret, ed. *Embryonic and Adult Stem Cells.* Detroit: Greenaven, 2009. Print.

Humber, James M., and Robert F. Almeder, eds. *Biomedical Ethics Reviews: Stem Cell Research.* Totowa: Humana, 2004. Print.

Lanza, R, P., and Anthony Atala. *Essentials of Stem Cell Biology.* 3rd ed. San Diego: Elsevier, 2014. Digital file.

Panno, Joseph. *Stem Cell Research: Medical Applications and Ethical Controversy.* New York: Facts On File, 2004. Print.

Sell, Stewart, ed. *Stem Cells Handbook.* 2nd ed. Totowa: Humana, 2013. Print.

Srivastava, Rakesh K., and Sharmila Shankar. *Stem Cells and Human Diseases.* New York: Springer, 2012. Digital file.

Ungar, Laura. "Stem-Cell Advances May Quell Ethics Debate." *Courier-Journal* [Louisville]. USA Today, 22 June 2014. Web. 20 Aug. 2014.

Wobus, A. M., and K. R. Boheler, eds. *Stem Cells.* New York: Springer, 2006. Print.

SYNTHETIC FUELS

Staff Sergeant Rusty Jones prepares to fuel an A-10C Thunderbolt II March 25, 2010, with a 50/50 blend of Hydrotreated Renewable Jet and JP-8. The A-10 then flew what was the first flight of an aircraft powered solely by a biomass-derived jet fuel blend. By Samuel King Jr. (U.S. Air Force Public Affairs.) [Public domain], via Wikimedia Commons

FIELDS OF STUDY

Biophysics; Biotechnology; Chemical engineering; Chemistry; Environmental biotechnology (or Green biotechnology)

ABSTRACT

Given the finite nature of the world's stores of natural petroleum, the development of economically viable, environmentally safe, and renewable synthetic fuels is important for human survival.

Synthetic fuels are normally produced from abundantly occurring natural resources such as coal, tar sands, oil shale, and biomass. One of the main objectives in the production of a synthetic fuel is to eliminate sulfur and nitrogen from the fuel compound, thereby creating an environmentally clean energy source. Oxides of nitrogen and sulfur dioxide are among the most undesirable of common air pollutants. Sulfur dioxide is one of the major causes of acid rain, which is created when sulfur dioxide combines with water vapor in the atmosphere to form sulfuric acid. Similarly, oxides of nitrogen produce nitric acid. These acids fall back to earth in rain and are detrimental to aquatic life as well as botanical life. Synthetic fuel manufacturers thus strive to eliminate these pollutants, as well as others such as carbon monoxide, hydrocarbons, particulates, and photochemical oxidants, from the fuel supply.

PRINCIPLES OF SYNTHETIC FUEL MANUFACTURE

The manufacture of liquid and gaseous synthetic fuels normally involves transforming naturally occurring carbonaceous raw material through a suitable conversion process. The techniques employed include hydrogenation, devolatilization, decomposition, and fermentation. The principal aim in the manufacture of synthetic fuel is to achieve a low carbon-to-hydrogen atomic mass ratio, or a high hydrogen-to-carbon atomic ratio, whenever possible. This results in a clean-burning fuel that releases by-products that are harmless to the environment. For example, pure methane (CH_4), with a molecular weight of 16, has a high hydrogen-to-carbon ratio of 4:1. Methane gas is a common component that is absorbed into coal. The method used to release the gas involves fracturing the coal and exposing it to low pressures. Coal-bed methane is one of the cleanest-burning fossil fuels; the by-products of burning it are simply carbon dioxide and water. Synthetically generated substitute natural gas is more than 90 percent methane. Natural gas (of which methane is the chief constituent) has a hydrogen-to-carbon ratio of approximately 3.4:1, which is also quite high. The ratios for liquefied petroleum gas and for naphtha lie between 2:1 and 3:1. (In comparison, the ratios for gasoline and fuel oil are less than 2:1. Bituminous coal has one of the lowest values, with ratio of much less than 1:1.)

COAL GASIFICATION AND LIQUEFACTION

Although coal is among the most abundant natural energy sources, it is also among the dirtiest. The composition of this solid fossil fuel is a major disadvantage; it consists of about 70 percent carbon and about 5 percent hydrogen, translating to a highly undesirable carbon-to-hydrogen mass ratio of 14:1. Coal-burning power-generating stations thus spew out large quantities of gases that are harmful to the environment. Despite the use of such emission-reducing devices as electrostatic precipitators, the levels of pollutants emitted by coal-burning plants remain high. Techniques such as coal gasification and coal liquefaction yield synthetic fuels that are safer for the environment.

The process of coal gasification involves making coal react with steam at very high temperatures (in the range of 1,000 degrees Celsius, or 1,832 degrees Fahrenheit). This process produces synthetic gas. Three types of synthetic gas are in common use. Low-calorific-value gas (also called producer gas) is used in turbines. Medium-calorific-value gas (also called power gas) is used as a fuel gas by various industries. High-calorific-value gas (also called pipeline gas) is a very good substitute for natural gas and is well suited to economical pipeline transportation. Pipeline gas contains more than 90 percent methane; as a result, it has a high hydrogen-to-carbon ratio.

The process of coal liquefaction is employed to generate a liquid fuel with a high hydrogen-to-carbon ratio; it is also used to obtain low-sulfur fuel oil. Several methods are employed to accomplish coal liquefaction, including direct catalytic hydrogenation, indirect catalytic hydrogenation, pyrolysis, and solvent extraction. All of these methods produce fuels that are much safer for the environment than the original coal.

TAR SANDS AND OIL SHALE

Naturally occurring tar sands contain grains of sand, water, and bitumen. Bitumen, a member of the petroleum family, is a high-viscosity crude hydrocarbon. A method known as hot water extraction is used to procure bitumen from tar sands. The bitumen is subsequently upgraded to synthetic crude oil in refineries. Synthetic crude oil (also called syncrude) is similar to petroleum and can be obtained through coal liquefaction as well as from tar sands and oil shale.

Large deposits of tar sands are found in Alberta, Canada; the United States has huge reserves of oil shale in Utah, Wyoming, and Colorado. Oil shale is probably the most abundant form of hydrocarbon on earth. Oil shale is a sedimentary rock that contains kerogen, which is not a member of the petroleum family. A popular method known as retorting is used to produce oil from shale. The process involves the method of pyrolysis, which reduces the carbon content in the raw hydrocarbon through distillation. Because the process is costly, however, the production of shale oil has not provided an economically feasible alternative to petroleum.

BIOMASS FUELS AND GASOHOL

Like oil and coal, biomass is derived from plant life. Oil and coal, however, are considered nonrenewable resources, as it takes vast periods of time for geologic processes to produce these materials naturally. Because biomass consists of any material that is derived from plant life, it is produced in far shorter spans—one hundred years or less—and is thus considered renewable. Wood is the most versatile biomass resource; farm and agricultural wastes, municipal wastes, and animal wastes are also considered to be biomass. Biomass can be processed into fuel using a variety of methods. Fermentation, for example, yields ethanol, or ethyl alcohol (sometimes called grain alcohol). Other methods include combustion, gasification, and pyrolysis.

Gasohol is a mixture of gasoline and small quantities of ethanol. The mixture burns cleaner than conventional gasoline; however, it can cause damage to plastic and rubber materials used in automobile engines. In the United States, therefore, the Environmental Protection Agency (EPA) permits the addition of only 10 percent ethanol by volume to gasoline to create gasohol. Methanol, or methyl alcohol (also called wood alcohol), can also be combined with conventional gasoline to produce cleaner fuel; however, the EPA limits the amount of methane in such mixtures to 3 percent.

OTHER FUELS

A nonpolluting rocket fuel based on alcohol and hydrogen peroxide has been developed by U.S. Navy research engineers at China Lake, California. This nontoxic homogeneous miscible fuel (NHMF) can be modified and used to drive turbines, which in turn drive alternators that produce electricity. Further developments based on what has been learned about this fuel may permit its use in automobiles. During World War II, moreover, Germany produced synthetic fuels in large quantities to meet its energy demands, employing coal gasification and also creating diesel oil and aviation kerosene using a reconstitution process; this process is still in use in many places.

Although the present abundance of natural petroleum limits the economic competitiveness of most synthetic fuels, the finite nature of the world's oil supply virtually ensures that synthetic fuels will become increasingly important energy sources. The U.S. Department of Energy and governmental agencies in many other countries thus provide funding to encourage research into the creation of economically viable, environmentally safe, and renewable synthetic fuels.

—*Mysore Narayanan*

FURTHER READING

Deutch, John M., and Richard K. Lester. "Synthetic Fuels." In *Making Technology Work: Applications in Energy and the Environment.* New York: Cambridge University Press, 2004.

Lorenzetti, Maureen Shields. *Alternative Motor Fuels: A Nontechnical Guide.* Tulsa, Okla: PennWell, 1996.

Manahan, Stanley E. "Adequate, Sustainable Energy: Key to Sustainability." In *Environmental Science and Technology: A Sustainable Approach to Green Science and Technology.* 2d ed. Boca Raton, Fla.: CRC Press, 2007.

Miller, G. Tyler, Jr., and Scott Spoolman. "Nonrenewable Energy." In *Living in the Environment: Principles, Connections, and Solutions.* 16th ed. Belmont, Calif.: Brooks/Cole, 2009.

Speight, James G. *Synthetic Fuels Handbook: Properties, Process, and Performance.* New York: McGraw-Hill, 2008.

Tularemia as a Bioweapon

A scientist at the Rocky Mountain Laboratory of the U.S. Public Health Service performs necropsies on tularemia-infected guinea pigs during the early 1940's. The animals were injected with water from mountain streams suspected as sources of the disease. (Library of Congress.) [Public domain], via Wikimedia Commons

FIELDS OF STUDY

Biophysics; Biotechnology; Chemical engineering; Chemistry; Medical biotechnology (or Red biotechnology); Philosophy and Religious Studies; Political Science

ABSTRACT

Commonly known as rabbit fever, tularemia is a disease endemic in North America as well as parts of Europe and Asia. Its relevance to forensic science lies chiefly in its potential for use as a bioweapon.

HOW IT WORKS

Tularemia is a naturally occurring disease. Its primary hosts are rabbits, prairie dogs, muskrats, and other small mammals, but it can also be transmitted by ticks and deerflies. After infection, onset is rapid. Symptoms include headache, fatigue, dizziness, and nausea. If untreated, tularemia may result in death.

The U.S. Centers for Disease Control and Prevention (CDC) regards *Francisella tularensis* as a viable bioweapon agent because tularemia is highly infective and incapacitating yet has relatively low lethality, a consideration in its possible deployment near a civilian population. The bacterium is easy to distribute both as an aerosol and in municipal drinking water supplies. Aerosol release would have the most widespread effect on public health, especially if done in urban settings. *F. tularensis* is classified as a Category A agent, which means it has serious potential for inducing terror in a population (other Category A agents include *Yersinia pestis*, the bacterium that causes plague; *Variola major*, the virus that causes smallpox; *Bacillus anthracis*, the bacterium that causes anthrax; and *Clostridium botulinum*, the bacterium that causes botulism). Japan, the Soviet Union, and the United States have all stockpiled *F. tularensis* in the form of offensive weapons at different times in their histories. It is now known that the Soviet army used the pathogen against the Germans during World War II in the Battle of Stalingrad.

Because the early symptoms of tularemia are similar to those of many ordinary or seasonal infections, an attack using *F. tularensis* on the general population in any given area in the United States could easily take health authorities by surprise. With an incubation range of one to fourteen days and average onset of symptoms taking from three to five days, an attack might not be immediately detected. Security measures that have been taken against this possibility include the installation in thirty U.S. cities of sensors that constantly monitor the air for deadly pathogens. If epidemiologists suspect the deliberate or unexplained release of the tularemia organism, standard practice is for them to contact the appropriate law-enforcement agencies immediately.

One of the things that makes the possibility of the use of the tularemia pathogen as a weapon particularly worrisome is that no vaccine against the disease is available to the general public, in contrast to other possible bioterror agents such as anthrax and smallpox. Some comfort is provided by the availability of potent and effective antibiotics against tularemia.

—*Robert Klose*

FURTHER READING

Dembek, Zygmunt F., Ronald L. Buckman, Stephanie K. Fowler, and James L. Hadler. "Missed Sentinel Case of Naturally Occurring Pneumonic Tularemia Outbreak: Lessons for Detection of Bioterrorism." *Journal of the American Board of Family Practice* 16 (July/August, 2003): 339-342.

Dennis, David T., et al. "Tularemia as a Biological Weapon: Medical and Public Health Management." *Journal of the American Medical Association* 285 (June 6, 2001): 2763-2773.

Siderovski, Susan Hutton. *Tularemia*. New York: Chelsea House, 2006.

Zygomycetes

FIELDS OF STUDY

Biophysics; Biotechnology; Genetics; Medical biotechnology (or Red biotechnology); Reproductive technology

ABSTRACT

Zygomycetes are a group of fungi that constitute the phylum Zygomycota. Also called zygote fungi, zygomycetes include about 750 species. Most are saprobes, living on decaying plant and animal matter in the soil; some are parasites of plants, of insects, or of small soil animals; some cause the familiar soft fruit rot and black bread mold; and a few occasionally cause severe infections in humans and farm animals.

BACKGROUND

Zygomycetes share many common features with members of other phyla in kingdom *Fungi*. They are rapidly growing, nonphotosynthetic organisms that characteristically form filaments called hyphae. Hyphae are highly branched to form an interwoven network mass called mycelium. All zygomycetes are terrestrial and reproduce by means of spores. No motile cells are formed at any stage of their life cycle. The primary component of their cell wall is chitin, and the primary storage polysaccharide in the cytoplasm is glycogen. Most zygomycetes have coenocytic hyphae, within which the cytoplasm can frequently be seen streaming rapidly. Both sexual and asexual reproduction occurs in zygomycetes.

Members of *Zygomycota* play important roles both ecologically and economically. Some species (such as *Rhizopus stolonifer*) cause soft fruit rot, posing a problem for transport and storage of many fruits. The same fungi may also feed on bread and other bakery foods, a potentially serious health hazard. Others, such as *Glomus versiforme*, may form intimate and mutually beneficial symbiotic associations with plant roots called mycorrhizae (literally, "fungus roots").

Yet another group of zygomycetes, the *trichomycetes*, form a fascinating relationship with arthropods. Trichomycetes are found in the larvae of aquatic insects, millipedes, crayfish, and even crustaceans living at the bottom of the ocean near hydrothermal vents. They usually reside in the guts of these animals and are thought to provide vitamins to their hosts. Members of zygomycetes in the order of *Entomophthorales* have great ecological significance based on their parasitic relation with insects and other small pest animals. They are being increasingly used in the biological control of insect pests of crops.

REPRODUCTION AND LIFE CYCLE

Even though by appearance all haploid hyphae of zygomycetes look identical, they are actually of two different mating types. When the two hyphae are in close proximity, hormones are released that cause an outgrowth near their hyphal tips to come together and develop into gametangia. Although some species are homothallic (self-fertilizing), most *Zygomycota* species are heterothallic, requiring a combination of + and − strains for sexual reproduction. The two strains "mate" sexually through the combination of two gametangia. In the process, the walls between the two touching gametangia dissolve, fusing their haploid nuclei to form diploid zygospores (hence the name *Zygomycota* for this phylum). Zygospores have very thick walls and thus are very hardy, able to tolerate extreme environment conditions. Zygospores are dispersed through the air and can remain dormant until conditions are favorable for growth. Zygospores then undergo meiosis and germinate, producing structures on which sporangia (spore cases) are formed. The sporangia produce and disperse numerous haploid spores, marking the beginning of the asexual part of the reproductive cycle.

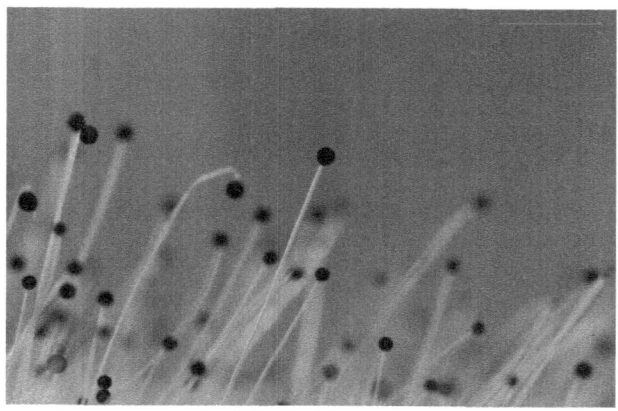

Detail of Sporangia of a Zygomycota species growing on a peach. By Zephyris (Own work.)

During asexual reproduction, haploid spores released by sporangia germinate on food such as fruits, bread, and dung, producing haploid hyphae. These hyphae in turn may produce more hyphae or additional spores within sporangia through mitosis, and the cycle begins again. Asexual reproduction via haploid spores of sporangia is universal among all species of zygomycetes. Two examples illustrate the important role of zygomycetes in human lives: *Rhizopus stolonifer* and *Glomus versiforme*.

RHIZOPUS STOLONIFER

This is one of the best-known and most familiar members of phylum *Zygomycota*. *Rhizopus stolonifer* is a black mold that forms cottony masses on the surfaces of moist, carbohydrate-rich foods and similar substances that are exposed to air. This organism is a serious pest for stored fruits and vegetables, bread, and other types of staple food. Many people are familiar with rotten fruits or aged bread that are covered by *R. stolonifer*.

The life cycle of *R. stolonifer* is similar to those of other species of *Zygomycota*. The mycelium of *R. stolonifer* is composed of several distinct types of haploid hyphae. Most of the mycelium consists of rapidly growing, coenocytic hyphae, which grow through the substrate (such as orange or bread), absorbing nutrients. From coenocytic hyphae, arching hyphae called stolons are formed. The stolons form rhizoids wherever their tips come into contact with the substrate. From each of these points, a sturdy, erect branch arises, which is called a sporangiophore. Each sporangiophore produces a spherical sporangium at its apex. A sporangium begins as a swelling sac, into which a number of nuclei flow. The sporangium is eventually isolated from other hyphae by the formation of a structure called a septum. The protoplasm within is cleaved, and a cell wall forms around each spore. The sporangium wall becomes black as it matures, giving the mold its characteristic color. Each mature spore, upon dispersal, is capable of germinating under adequate conditions to produce a new mycelium. Each year *R. stolonifer* causes an estimated loss of millions of dollars to farmers, fruit growers, and consumers.

GLOMUS VERSIFORME AND *MYCORRHIZAE*

As one of the most important groups of zygomycetes, *Glomus versiforme* and related genera grow in intimate associations with the roots of plants, forming mycorrhizae. Mycorrhizae not only dramatically increase the surface area of roots for absorption but also help convert nutrients in soil into forms usable by plants. For many forest trees, if seedlings are grown in a sterile nutrient solution and then transplanted to grassland soil, they grow poorly and may eventually die from malnutrition. However, if a small amount of forest soil containing the appropriate fungi (including *G. versiforme*) is added to the soil around the roots of the seedlings, normal growth is restored. Studies have found that in forest soil *G. versiforme* and related fungi ensure the formation of mycorrhizae and restore the normal growth of seedlings.

Mycorrhizae occur in most groups of vascular plants. The fungal partner *G. versiforme* helps plant roots to absorb and transfer essential nutrients such as phosphorus, zinc, manganese, and copper. By extending several centimeters out from colonized roots in all directions, the plants are able to obtain nutrients from a much larger volume of soil than would be possible otherwise. In return, *G. versiforme* obtains carbohydrates from the host plants. Some fungi may simply attach to the outer surface of the root to form a sheath of hyphae around the root called ectomycorrhizae. In addition to surface extension, other fungi may penetrate into the root to form endomycorrhizae. Of the two major types of mycorrhizae, endomycorrhizae occur in about 80 percent of all vascular plants. The *G. versiforme* hyphae penetrate the cortical cells of the plant root, where they form either minute, highly branched, treelike structures called arbuscules or swellings called vesicles. Such endomycorrhizae are particularly important in the tropics, where soils tend to be positively charged and thus retain phosphates so tightly that this nutrient is available only in very limited supplies for plant growth. Since the impoverished farmers there are often unable to afford fertilizers, endomycorrhizae play a critical role in making phosphates available to crops in these regions. The commercial applications of endomycorrhizae to crops in other regions to reduce fertilizer use and increase yields appear to be an increasingly attractive possibility as well.

———*Yujia Weng*

FURTHER READING

Ayres, P., and P. Nigel. "Weeding with Fungi." *New Scientist*, September 1, 1990, 36-39. Describes an herbicide made from fungi, which can devastate weeds but not crops. Could be an ideal form of weed control.

Christensen, C. M. *Molds, Mushrooms, and Mycotoxins*. Minneapolis: University of Minnesota Press, 1975. Well-organized and thoughtful discussions on some fungi and their ecological and economic importance.

Dix, N. J., and J. Webster. *Fungal Ecology*. London: Chapman & Hall, 1995. A easy-to-read yet comprehensive account of fungal diversity in various ecological communities. Includes informative discussions on the critical roles fungi play in maintaining the health of ecosystems.

Matossian, M. K. *Poisons of the Past: Molds, Epidemics, and History*. New Haven, Conn.: Yale University Press, 1989. A very good summary on the effects of food poisoning resulting from bread molds on human history. A useful reference book for public health.

Smith, D. C., and A. E. Douglas. *The Biology of Symbiosis*. Baltimore, Md.: Edward Arnold, 1987. An outstanding, concise account of biological symbiosis, including several chapters on symbiotic relationships between fungi and plants.

Important Figures in Biotechnology

David Baltimore

Dr. David Baltimore, 2014. By Bob Paz.

American molecular biologist

- **Born:** March 7, 1938; New York, New York
- **Primary field:** Biology
- **Specialties:** Molecular biology; virology; microbiology

Molecular biologist and Nobel laureate David Baltimore has achieved great influence as a researcher, scientific policymaker, and leader of academic institutions. He is most celebrated for his molecular virology discoveries in recombinant DNA research.

Early Life

David Baltimore was born in New York City on March 7, 1938, to parents Gertrude Lipschitz, an experimental psychologist, and Richard Baltimore, who worked in the garment industry. His decision to become a molecular biologist began in 1955 when he was a junior in high school and his mother arranged for him to spend a summer working on mouse genetics with a group of research biologists at the Jackson Memorial Laboratory in Bar Harbor, Maine. That summer Baltimore also met future colleague and friend Howard Martin Temin, who had just graduated from college and served as a mentor.

Baltimore joined the faculty of MIT in 1968, where he would remain for much of the next thirty years, and conducted research that was ahead of its time in its interdisciplinary nature. He envisioned an environment in which researchers approached complex scientific challenges from a multifaceted perspective. His laboratory thus relied not only on the tools of molecular biology, but also on those of virology and immunology.

Life's Work

In 1970, Baltimore codiscovered reverse transcriptase with Howard Temin, his 1955 summer mentor, who was working independently of Baltimore at the time. Baltimore and Temin discovered separately, and within days of each other, that reverse transcriptase is an uncommon enzyme used by retroviruses to synthesize DNA from RNA. Baltimore showed that retroviruses have encoded in them genetic instructions for the manufacture of an enzyme (reverse transcriptase) capable of copying DNA from an RNA template. The newly-created viral DNA is then able to transform an infected cell into a malignant cell.

Baltimore's codiscovery of reverse transcriptase had its roots in his early research on the ways in which poliovirus manages to penetrate cellular defenses and then replicate. He had conducted this work at the Salk Institute in California for three years beginning in 1965. This research led Baltimore to conduct experiments focusing on two particular RNA tumor viruses: one that causes leukemia in mice and another that causes sarcoma in chickens.

At the time that Baltimore began his experiments, scientists already knew that under certain conditions some viruses caused tumors in chickens and that some viruses could also cause different types of cancer in other animals such as mice. Researchers eventually discovered that both viruses containing DNA as their genetic material and viruses containing the kind of genetic material known as RNA were capable of triggering a normal cell to undergo growth characteristics typical of a tumor cell. A central mystery remained: How did the genetic material of a tumor-causing RNA virus apparently manage to infiltrate the DNA of a healthy cell?

Up until that time, most molecular biologists believed transfers of genetic information to be strictly one-way from DNA to RNA. Baltimore, who had been studying virus-specific enzymes that copy RNA from RNA, began using similar techniques to investigate whether some viruses were capable of copying DNA from RNA. Baltimore's experiments clarified the molecular process and the role of reverse transcriptase by which cancer-causing RNA viruses succeed in infecting and permanently changing healthy cells.

In 1975, Baltimore shared the Nobel Prize in Physiology or Medicine with Howard Temin and Renato Dulbecco for groundbreaking insights into the interaction between tumor viruses and the genetic material of the cell. He continued at MIT as a professor of biology as well as a member of the staff at the MIT Center for Cancer Research until 1990.

In 1982, Baltimore established the Whitehead Institute for Biomedical Research, an independent entity affiliated with MIT. Baltimore served as the institute's first director from 1982 until 1990.

In 1990, Baltimore left the Whitehead Institute to become president of Rockefeller University but he stepped down the following year due to a controversy rooted in a 1986 scientific paper he had coauthored and published in the journal *Cell* with several colleagues, including MIT faculty member Thereza Imanishi-Kari. After a researcher in Imanishi-Kari's laboratory was unable to replicate the results reported in the paper, she accused Imanishi-Kari of falsifying the data. Baltimore was never accused of misconduct, but as a coauthor of the paper, he was nonetheless associated with it.

After a lengthy investigation that lasted several years, the Office of Scientific Integrity at the National Institutes of Health, which had funded the research, determined in 1994 that Imanishi-Kari had falsified her data, and it was recommended that she be barred from receiving federal funding for ten years.

Baltimore stood behind Imanishi-Kari and defended her research and integrity; the paper, however, was eventually retracted. Baltimore remained on the Rockefeller faculty, but he left the university in 1994 and returned to MIT. In 1996, the appeals board at the Department of Health and Human Services, which had been appointed by the federal government to review the case, found that Imanishi-Kari was not guilty of misconduct, and the ruling was overturned. Baltimore was appointed president of the California Institute of Technology (Caltech) in 1997, a position he retained until September 2006.

As of 2017, he remained Robert Andrews Millikan Professor of Biology at Caltech.

Baltimore has continued his research in using retrovirus vectors to modify the immune system. One of his major areas of inquiry is the potential use of gene therapy to ward off certain types of cancer and to ultimately protect against HIV infection. The therapy consists of introducing genetically reprogrammed immune cells into a subject in order to manufacture antibodies that might serve as a vaccine. The model has shown some promise in animal subjects, but it has not yet been perfected for human patients.

Although Baltimore has stated publicly that an effective HIV vaccine may never be found, he has also expressed a commitment to continue the quest. Baltimore has served on and presided over numerous boards including as president of the American Association for the Advancement of Science (AAAS) in 2008 and then as chairman of the board of directors for the organization.

Impact

Baltimore's codiscovery helped lay the groundwork for recombinant DNA technology, which scientists have used to produce more disease-resistant food crops as well as a variety of medical products including insulin, growth hormone, and certain vaccines. In 1975, in response to early concerns about potential hazards or abuses his work helped unleash, Baltimore helped organize the International Congress on Recombinant DNA Molecules, better known as Asilomar (after the California location where the group first met), which created what amounted to the first ethical and safety guidelines for the use of recombinant DNA technology. The voluntary standards that were then adopted by the scientific community represented a landmark attempt at self-regulation in the face of rapidly advancing new technologies.

Baltimore's unexpected discovery that RNA can be transformed into DNA also made possible new approaches to fighting HIV, the virus that causes AIDS. His work has had profound implications for understanding the role viruses play in the development of cancer and for understanding the molecular basis of human immune response.

Baltimore's expertise has shaped U.S.-policy responses to the AIDS pandemic. An early proponent of federal funding for AIDS-related research, Baltimore

> **DAVID BALTIMORE: TUMOR-GROWTH THROUGH VIRAL RNA**
>
> In the 1950s, virologist Renato Dulbecco's research helped to explain how DNA tumor viruses change normal cells into cancerous cells. His work essentially revealed that the viral DNA gets into the nucleus of a cell, at which point it combines with the cellular DNA. Thereafter, when the cell divides, the characteristics of the viral DNA, and not the cellular DNA, are passed down. Accordingly, the disease spreads. Subsequently, virologist Harold Temin hypothesized in the 1960s that an RNA tumor virus—one that would cause cancer—could infect healthy cells by making a DNA copy of itself.
>
> In 1970, David Baltimore was studying leukemia and sarcoma viruses at the Massachusetts Institute of Technology (MIT) and was able to demonstrate the occurrence of a certain enzyme in viral RNA called transcriptase. This enzyme allowed the RNA virus to make a DNA copy of itself. In other words, a molecule of RNA in a cancer-causing virus could change into viral DNA, and then splice itself to the DNA of a host cell. Simultaneously, Temin independently proved the same hypothesis at the University of Wisconsin–Madison, utilizing an avian myoblastosis virus, a type of leukemia found in birds. This process came to be known as reverse transcriptase.
>
> The trio waited five years to receive their Nobel Prize for their discoveries, which was considered a short waiting period in the scientific community; Baltimore was only thirty-seven years old when he received the Nobel Prize. Though they were awarded the Nobel Prize nearly forty years ago, the research and work of Baltimore, Dulbecco, and Temin on the reverse transcriptase has had significant implications for the understanding of and research into viruses and diseases. Specifically, discovery of reverse transcriptase led to a better understanding of retroviruses, most importantly human immunodeficiency virus (HIV), which causes acquired immune deficiency syndrome (AIDS). In fact, the scientists who discovered that HIV causes AIDS were able to make that discovery by using the reverse transcriptase reaction.

cochaired the 1986 National Academy of Sciences committee charged with developing a national strategy for responding to AIDS. In 1996, he was also appointed to direct the National Institutes of Health AIDS Vaccine Research Committee. In 2006, he cofounded Calimmune, a private company dedicated to investigating gene therapies for HIV. He was chair of the company's board until 2015, when he stepped down, remaining a board member. As of 2016, the company was also investigating a vaccine against HIV.

Baltimore has published over 600 peer-reviewed papers. In addition to winning the Nobel Prize, he was presented the Gustave Stern Award in Virology (1970) and the National Medal of Science (1999).

—*Beverly Ballaro*

FURTHER READING

Baltimore, David. "7 Questions for Nobel Laureate David Baltimore." Interview by Mary Engel. *Hutch News*, 6 Aug. 2015, www.fredhutch.org/en/news/center-news/2015/08/questions-for-nobel-laureate-david-baltimore.html. Accessed 21 Mar. 2017.

———. "The Medicine of the Future." Town Hall, Los Angeles. 11 Feb. 2003. Address.

———. "Limiting Science: A Biologist's Perspective." *Daedalus* 104.4 (2005): 7–15.

Boyer, Lauren. "Fighting the Fight: Dr. David Baltimore." *National Science & Technology Medals Foundation*, 19 Dec. 2016, www.nationalmedals.org/stories/baltimore-and-the-cure. Accessed 21 Mar. 2017.

Crotty, Shane. *Ahead of the Curve: David Baltimore's Life in Science*. Berkeley: U of California P, 2001.

Kruglinski, Susan. "The *Discover* Interview: David Baltimore." *Discover* Sept. 2006: 50–53.

Françoise Barré-Sinoussi

French virologist

- **Born:** July 30, 1947; Paris, France
- **Also known as:** Francoise Barré-Sinoussi; Francoise Barre-Sinoussi
- **Primary field:** Biology
- **Specialty:** Virology

French virologist Françoise Barré-Sinoussi received the 2008 Nobel Prize in Physiology or Medicine for her codiscovery, with Luc Montagnier, of the human immunodeficiency virus (HIV), the retrovirus that causes acquired immune deficiency syndrome (AIDS).

Early Life

Françoise Barré-Sinoussi (frahn-SWAHZ ba-RAY-see-noo-see) was born Françoise Sinoussi to Roger Sinoussi and Jeanine Fau Sinoussi on July 30, 1947, in Paris, France. Though a native Parisian, she spent her childhood summers in rural Auvergne, where she became a devoted observer of the natural world.

As a student, Barré-Sinoussi showed a strong predilection for scientific subjects. After graduating high school in 1966, she was torn between pursuing a college degree in science or medicine. Reasoning that medical studies would place a financial burden on her parents, she opted for coursework in the natural sciences at the University of Paris. As college graduation approached, Barré-Sinoussi wondered whether a research career working in a laboratory was right for her. To find out, she decided to volunteer in one. At first, no lab was willing to take her. Only after months of searching and a tip from a friend did she find a lab willing to accept her services.

Located in the Paris suburb of Marnes-la-Coquette, the lab was a branch of the Pasteur Institute. Barré-Sinoussi joined a team led by Jean Claude Chermann. The focus of their study was retroviruses and cancer in mice. Inspired by Chermann, Barré-Sinoussi spent so much time at the lab that she neglected her university studies and stopped attending classes. She only appeared on campus to pass her final exams.

Impressed with Barré-Sinoussi's efforts, Chermann encouraged her to pursue a PhD. Without quitting the lab, she completed her doctoral studies at the University of Paris in 1974. Her thesis focused on how, in a cultured sample, a synthetic molecule could be used to control leukemia caused by the Friend virus.

Following a year of postdoctoral studies at the National Institutes of Health (NIH), in Bethesda, Maryland, Barré-Sinoussi returned to Paris and rejoined Chermann's lab, working on a team lead by Luc Montagnier. On October 7, 1978, she married Jean-Claude Barré, a man she had met while working toward her PhD.

Life's Work

In the early 1980s, Willy Rozenbaum, a Paris doctor, confronted a disturbing trend. Many of his patients were coming in suffering from unusual illnesses. The variety of symptoms almost defied logic, but the end result was uniformly horrific. Despite Rozenbaum's best efforts, his patients were dying, the mysterious illness destroying their immune systems. Other than the shared immunodeficiency, the one commonality that connected most of the early cases was that the affected patients were predominantly male and homosexual. A similar outbreak in the same demographic had first been reported in the United States in 1981. Across the globe, more cases were being observed; the disease, now known as acquired immunodeficiency syndrome (AIDS), was spreading at an alarming rate.

Françoise Brun-Vézinet, a virologist who worked with Rozenbaum, wondered if a retrovirus might be the culprit. He sought out Luc Montagnier's assistance in late 1982. Montagnier, Barré-Sinoussi, and their team at the Pasteur Institute had spent much of their time in the late 1970s and early 1980s studying the connection between retroviruses and certain forms of cancer. A retrovirus is distinct from other types of virus based on its genetic material and its method of replication. A retrovirus's genome is stored in a single strand of ribonucleic acid (RNA). Like other viruses, retroviruses can only reproduce in an infected host. Using an enzyme called reverse transcriptase, the retrovirus accomplishes this by creating deoxyribonucleic acid (DNA) from its own RNA. With the help of another enzyme, this DNA then bonds with the genetic material of the host cell.

Though the project departed somewhat from their usual area of expertise, Montagnier agreed to take it on and asked Barré-Sinoussi to lead the investigation. Provided with samples from a patient diagnosed with the illness, Barré-Sinoussi and the Pasteur Institute team isolated and identified the deadly retrovirus in early 1983.

The team announced its discovery in the May 1983 issue of *Science*, and Barré-Sinoussi traveled to the United States to present the team's data. Molecular biologists at the Pasteur Institute started constructing the virus's genome. As their findings and those of other researchers piled up, the medical community soon reached the conclusion that the virus, later renamed human immunodeficiency virus (HIV), is the cause of AIDS. Though Barré-Sinoussi had spent most of her career investigating retroviruses and cancer, she shifted her focus. HIV/AIDS was killing people by the tens of thousands and in time would go on to kill tens of millions more. She recognized the scope of the crisis and dedicated her professional life solely to HIV research.

Though Chermann, her long-time mentor, left the Pasteur Institute in 1987, Barré-Sinoussi has stayed on throughout the decades. In 1992, she was placed in charge of her own laboratory, the Institute's Biology of Retroviruses Unit. Renamed the Regulation of Retroviral Infections Unit in 2005, Barré-Sinoussi's lab continued to operate at the forefront of HIV research well into the 2010s.

Since 1983, a major aspect of Barré-Sinoussi's work has been a commitment to sharing her skills and insights with other scientists as well as in countries where a lack of wealth and infrastructure make combating the HIV/AIDS epidemic especially difficult. A seminal trip to the Central African Republic (CAR) in 1985 convinced her of the importance of cultivating relationships in this and other countries. Connecting with researchers, leaders, and average citizens would facilitate the exchange of ideas and expertise across continents. Though Barré-Sinoussi's travels have taken her throughout the world, much of her time has been spent working in sub-Saharan Africa and Southeast Asia, in particular.

In 2008, Barré-Sinoussi and Montagnier were awarded the Nobel Prize in Physiology or Medicine for their discovery of HIV. Also receiving a share of the award was Harald zur Hausen for proving that human papilloma viruses cause cervical cancer. The first French woman ever to receive such a Nobel Prize, Barré-Sinoussi downplayed the honor but expressed hope that it would help promote further HIV/AIDS research.

Responding to a 2009 statement by Pope Benedict XVI that condoms were not an effective method of HIV/AIDS prevention, Barré-Sinoussi cosigned an open letter to the pontiff outlining the scientific case for their efficacy.

The author of over 220 articles, Barré-Sinoussi also holds seventeen patents. She received the French Légion d'Honneur in 2006. A co-chair of the United Nations Commission on AIDS Prevention (UNAIDS), she was appointed president of the International AIDS Society (IAS) in 2012.

Impact

The identification of HIV by the Montagnier–Barré-Sinoussi team was the first crucial step in the fight against HIV/AIDS, setting in motion a number of subsequent advances that would lead to a better understanding of the disease, its prevention, and its treatment. Having identified the virus, doctors could now develop ways to test for it. This proved crucial in eliminating blood transfusions as a means of HIV transmission and ensuring the safety of blood banks. Transmission rates have declined thanks to other preventive measures, such as increased condom use and needle exchanges.

While no cure has been found, what was once a death sentence has now become in many cases a manageable illness. Antiretroviral therapies developed in the years since the Pasteur Institute's discoveries are helping those with HIV to live out their natural lives. Nevertheless, as Barré-Sinoussi's commitment demonstrates, the fight against HIV/AIDS is far from over. The global infection rate was nearly one percent as of 2010, an alarming figure, and in certain countries it has been much higher. Often these nations lack the institutions necessary for sustained countermeasures.

As one of the major human health crises of modern times, the HIV/AIDS epidemic has illuminated a host of societal shortcomings. Nevertheless, one sector did distinguish itself. In an abbreviated period of time and in an atmosphere of widespread global panic, the medical community isolated and identified the cause of HIV/AIDS and quickly developed strategies to beat it back. Barré-Sinoussi and the Pasteur Institute team's achievement thus stands among the most important medical breakthroughs of the twentieth century.

——*Paul McCaffrey*

Further Reading

Bell, Sigall T., Courtney L. McMickens, and Kevin Selby. *AIDS (Biographies of Disease)*. Santa Barbara: Greenwood, 2011. Print. Presents a history of HIV/AIDS, from the first reported outbreak to the latest prevention and treatment methods, written by three Harvard-trained physicians, cataloging various efforts to decipher and combat the disease. Bibliography, index, glossary.

Engel, Jonathan. *The Epidemic: A Global History of AIDS*. New York: Smithsonian/Harper Collins, 2006. Print. Chronicles the HIV/AIDS crisis and the world's response to it, approaching the issue from a number of vantage points—scientific, medical, political, and cultural. Bibliography; index.

Montagnier, Luc. *Virus: The Co-Discoverer of HIV Tracks Its Rampage and Charts the Future*. Trans. Stephen Sartarelli. New York: Norton, 2000. Print. Describes the Pasteur Institute's work to isolate and identify HIV, offering an insider's view of the early years of the HIV/AIDS epidemic. Bibliography, index.

Shilts, Randy. *And the Band Played On: Politics, People, and the AIDS Epidemic*. New York: St. Martin's, 2007. Print. Examines the initial stages of the AIDS/HIV crisis and the response to it by medical researchers, politicians, and the larger culture. Describes the dispute between the Pasteur Institute and the American Dr. Robert Gallo over who first identified HIV. Bibliography, index.

BARRÉ-SINOUSSI AND THE PASTEUR INSTITUTE TEAM IDENTIFY THE HIV VIRUS

In January 1982, Françoise Barré-Sinoussi and her research team at the Pasteur Institute were tasked with determining whether the global epidemic then unfolding was caused by a retrovirus. Barré-Sinoussi expected it to be a fairly simple experiment. Using a lymph node from a biopsy performed on a patient thought to have the illness, the team harvested lymphocytes, a type of white blood cell, from the sample. They then cultured the lymphocytes and tested the culture for reverse transcriptase activity. Such activity would confirm the presence of a retrovirus.

The use of lymphocytes for the culture was not unintentional. Among the doctors fighting HIV/AIDS in the first years of the epidemic, the conventional wisdom was that the disease targeted the immune system. White blood cells are one of the body's main instruments in fighting off infection. In blood samples taken from patients suffering from HIV/AIDS, there were hardly any to be found. Whatever the infecting agent, it seemed to be attacking white blood cells.

The results during the first week of testing were inconclusive. The sample showed no sign of reverse transcriptase activity. The next week, however, weak activity was reported. Soon the pace sped up considerably. Unfortunately, just as the deadly virus was about to reveal itself, the T lymphocyte cells within the sample began dying off and with them the enzyme activity. Alarmed, the team started searching for new lymphocyte cells to add to the culture. After acquiring lymphocyte cells from a blood donor, they successfully integrated the cells into the culture. Once they did, the reverse transcriptase activity spiked. In February 1983, the deadly retrovirus, initially named lymphadenopathy associated virus (LAV), was photographed under a microscope.

The team's examination of HIV revealed an especially clever virus. By targeting the T lymphocytes, it sabotages the host's ability to fight it off while weakening the host's defenses against other potential infections. Concealment is also one of HIV's strengths. Unlike other cancer-causing retroviruses, HIV does not immediately set off rapid cell growth. Rather, this process requires an activation trigger. Individuals can unknowingly carry the infection for years without showing any symptoms and can potentially spread the disease. Meanwhile, by bonding with its host's DNA, the virus hides its own genetic information and makes it difficult, if not impossible, to root out all traces of it from the host.

Even with its powers of concealment, Barré-Sinoussi and her team had identified HIV. Their discovery enabled researchers to begin studying the inner workings of HIV, probing it for weakness and developing methods to counteract it.

Paul Berg

Paul Berg - 1980 Albert Lasker Basic Medical Research Award Winner. [Public domain], via Wikimedia Commons

Microbiologist and educator

- **Born:** June 30, 1926; Brooklyn, New York
- **Area of achievement:** Science and technology

Berg won the Nobel Prize in Chemistry in 1980 for his pioneering technique of splicing together deoxyribonucleic acid (DNA) from different types of organisms, which revolutionized the study of viral chromosomes and launched the field of genetic engineering.

Early Life

Paul Berg was born to Sarah and Harry Berg on June 30, 1926, in Brooklyn. Paul Berg graduated from the New York City public schools in 1943 with a great interest in microbiology, fostered by Mrs. Wolf, a high school teacher who ran an after-school science club. Berg entered Pennsylvania State University in 1943, served in the U.S. Navy from 1944 to 1946 and returned to school to graduate with a degree in biochemistry in 1948. He then moved to Case Western Reserve University in Cleveland, Ohio, graduating with a doctorate in biochemistry in 1952 and taking a yearlong American Cancer Society postdoctoral fellowship with Herman Kalckar at the Institute of Cytophysiology in Copenhagen, Denmark, and then a second fellowship with Arthur Kornberg at Washington University in St. Louis, Missouri. In 1956, he became an assistant professor of microbiology. He accepted the position of professor of biochemistry at Stanford University's School of Medicine in 1959.

Life's Work

Throughout the 1950s Berg patiently conducted a series of experiments to determine how amino acids, which combine to form proteins, join together. In 1956, he determined that a special molecule specifically joined the amino acid methionine to the ribonucleic acid (RNA) during replication. This molecule was one of a class of similar molecules, each specific to a unique amino acid that ultimately became known as transfer RNA (tRNA). His important discovery increased his interest in the study of genes, so he spent a sabbatical year in 1967 learning about deoxyribonucleic acid (DNA) tumor viruses at the Salk Institute with Renato Dulbecco. He returned to Stanford and began working with the monkey tumor virus SV40 to figure out how mammalian genes operate. Patient accumulation of data from many experiments enabled him to map out where on the DNA molecule the various viral genes occurred and the relationships among the various specific sequences of nucleotides in the genes and how they affect the DNA of the host organism that they infect. This information was important in ascertaining how cells became cancerous, that is, exhibit abnormal reproduction and growth.

Berg realized that he might be able to combine the DNA of the SV40 with the DNA of a bacteriophage (a type of virus) that would infect the intestinal bacterium *Escherichia coli*. In this way it would be possible to study a gene from one species in isolation from the usual set of genes with which it naturally interacted in its original host, since it was in another host that lacked such genes. This was the first time that anyone had artificially engineered at the genetic level what was in effect an organism possessing genes from two different organisms. Berg's methodology also would enable scientists to quickly replicate particular proteins or other desirable materials using the second organism as a convenient host, a technique widely applied in agricultural, pharmaceutical, and chemical industries.

The problem with these newly created organisms was that they contained foreign genes, and the harmful effects of such artificial manipulation could not be reasonably predicted. This suggested to several scientists who were following these developments that caution should be exercised. A group of scientists in Boston first raised the issue. Berg published a letter in the July 26, 1974, issue of *Science* that listed a series of recommendations from a quickly convened Committee on Recombinant DNA Molecules Assembly of Life Sciences that Berg chaired for the National Academy of Sciences. It raised the possibility of the biological hazards posed by the new genetic engineering and called for a temporary moratorium on further work until such time as a meeting of experts could be assembled to "discuss appropriate ways to deal with the potential biohazards of recombinant DNA molecules."

A group of 140 scientists with a few lawyers and science journalists gathered for a three-day meeting at the Asilomar Conference Center in Pacific Grove, California, on February 27, 1975, to take up these issues. A set of safeguards was created that was further refined by the Recombinant DNA Advisory Committee of the National Institutes of Health. Among many provisions was a requirement that special facilities be created for work with recombinant DNA to ensure that organisms undergoing genetic manipulation did not escape from laboratory facilities. Initially there was considerable controversy over these rules and their implementation on university and research campuses across the nation. Ultimately the standards for work with many organisms were relaxed, but certain pathogenic organisms require high standards of biosecurity. Berg became a professor emeritus at Stanford; he ceased conducting research in 2000 but still serves on many important advisory boards.

J. Michael Bishop

American molecular biologist

- **Born:** February 22, 1936; York, Pennsylvania
- **Primary field:** Biology
- **Specialties:** Molecular biology; biochemistry; virology

Significance

Berg's work over many years to elucidate the biochemistry of nucleic acids, especially in regards to recombinant DNA, resulted in his winning the Nobel Prize in Chemistry in 1980 (with biochemists Walter Gilbert and Frederick Sanger), the Albert Lasker Basic Medical Research Award in 1980, the National Medal of Science in 1983, the Biotechnology Heritage Award in 2005, and numerous other awards. His careful attention to the ethical dimensions of his scientific work, motivated by a precautionary principle, has been heralded as a model for others to emulate. He chaired the advisory board of the Human Genome Project of the National Institutes of Health as well as numerous other important boards and panels over the years.

—*Dennis W. Cheek*

Further Reading

Berg, Paul. "Paul Berg—Biographical." *NobelPrize.org*. Nobel Prize AB, Mar. 2004. Web. 25 Apr. 2016.

Berg, Paul, and Maxine Singer. *Dealing with Genes: The Language of Heredity*. Herndon: University Science Books, 2008. Print.

Frederickson, Donald S. *The Recombinant DNA Controversy: A Memoir. Science, Politics, and the Public Interest, 1974–1981*. Washington, D.C.: Amer. Society for Microbiology, 2001. Print.

Morange, Michel. *A History of Molecular Biology*. Cambridge: Harvard UP, 1998. Print.

Wade, Nick. *The Ultimate Experiment*. New York: W. H. Freeman, 1977. Print.

Wright, Susan. *Molecular Politics: Developing American and British Regulatory Policy for Genetic Engineering, 1972–1982*. Chicago: U of Chicago P, 1994. Print.

Oncogenes, genes which when mutated transform a normal cell into a malignant one, had been discovered in tumor viruses during the early 1970s. J. Michael Bishop isolated the first known human oncogene and demonstrated its origin in the human genome.

Early Life

John Michael Bishop was born in York, Pennsylvania, then a small town near the Susquehanna River, on February 22, 1936, as one of John and Carrie Bishop's three children. His father was the Lutheran minister to two local parishes. Bishop's education during his grade school years occurred in a two-room schoolhouse; four of his school years were spent with a teacher who provided the discipline Bishop would need later in life. Although he received a rigorous education in subjects such as history and penmanship, he noted that the science curriculum was somewhat lacking.

Bishop was valedictorian of his small high school class and earned a varsity letter in track. It was during his high school years that Bishop began to study under Dr. Robert Kough, a local physician who instilled an interest in human biology in his protégé. After graduation Bishop entered Gettysburg College, a small private college in Gettysburg, Pennsylvania, where he majored in chemistry in preparation for medical school. Bishop graduated with a degree in chemistry in 1957, having been elected to the Phi Beta Kappa honor society. He also met his future wife, Kathryn Putman; their marriage in 1959 produced two sons.

Bishop was admitted to medical schools at both the University of Pennsylvania and Harvard University and entered the latter in 1957. It was there that Bishop had his first opportunity to participate in scientific research. Two pathologists, Benjamin Castleman and Edgar Taft, allowed him access to their laboratories during his second year, resulting in some expertise in that scientific area. It was during Bishop's third year in school that he had his first exposure with viral research in animal virology under the auspices of researcher and Harvard professor Elmer Pfefferkorn. Bishop spent his fourth year working with Pfefferkorn outside of the classroom and learning early techniques in the nascent field of molecular biology. Bishop graduated with his medical degree in 1962.

Life's Work

Following two years of clinical training as house physician at Massachusetts General Hospital, Bishop joined the laboratory of Leon Levintow at the National Institutes of Health in Bethesda, Maryland, where he was trained in research in the medical sciences. Levintow's area of research involved studying the physical and chemical nature of poliovirus genomic ribonucleic acid (RNA) and proteins; Bishop's research addressed the replication strategy of poliovirus RNA and produced his first professional publications.

Following Levintow's departure for the University of California, San Francisco (UCSF) in 1967, Bishop joined the laboratory of Gebhard Koch in Hamburg, Germany. Bishop and Koch had collaborated on their polio work at Bethesda; the year overseas significantly expanded Bishop's world view. Upon returning to the United States in 1968, Bishop received two offers for faculty positions; one with Levintow at the then little-known medical school at UCSF as Assistant Professor of Microbiology and Immunology. He accepted this position and stayed there for the remainder of his academic career.

Bishop had planned on continuing his research into the replication of poliovirus when an element of serendipity entered his career. In the laboratory next to his at UCSF was researcher Warren Levinson, who was studying the growth of Rous sarcoma virus (RSV), which causes some cancers. The virus had been discovered as a filterable agent early in the twentieth century, and had been the subject of significant research for two primary reasons: RSV was capable of transforming normal cells into malignant ones, and it was the first of what were subsequently called retroviruses to be characterized. In 1970 virologists David Baltimore and Howard Temin discovered the presence of an enzyme (reverse transcriptase) in the capsid of these viruses, which could copy the RNA genome into DNA. The DNA would then integrate into the genome of infected cells. The puzzle was to determine the mechanism of cell transformation by the virus.

In 1970 Bishop was joined by postdoctoral fellow Harold Varmus and the two developed a professional relationship with the goal of studying how RSV transforms cells. By this time scientists had focused on a specific viral gene, named the *src* gene for its ability to induce sarcomas. The initial hypothesis among scientists was that expression of the *src* gene by RSV following infection induced the malignant transformation. Bishop and Varmus discovered it was the opposite situation; the *src* gene originated within the cell, and was picked up by the virus following integration. The normal cell gene, now referred to as a proto-oncogene, functions to

> ## BISHOP DISCOVERS ONCOGENE PRESENCE IN NONMALIGNANT CELLS
>
> During the 1960s, in studying the role of RNA tumor viruses as prospective etiological agents of cancer in animals, University of Wisconsin researcher Howard Temin discovered an enzyme called reverse transcriptase. The function of the enzyme was to copy the RNA genome of the virus agent, known as the Rous sarcoma virus (RSV), into DNA, allowing the virus to integrate within the chromosome of the host cell. The relatively small genome of the viral agent, which encoded only four proteins, allowed researchers to pinpoint the particular gene that was instrumental in transforming cells from a normal to a malignant state: the gene, referred to as the src gene in the virus, was subsequently termed an oncogene, reflecting its role in cell transformation.
>
> Bishop, joined in 1970 by postdoctoral fellow Harold Varmus, began a study into the evolutionary history of the src gene: Was the src gene specific to the virus? Its presence made little evolutionary sense. While the other genes proved necessary for replication and survival of the virus, the src gene was completely superfluous, providing no necessary function. Was there a homologue in the cell, a gene that exhibited a function similar to the viral oncogene?
>
> Using a technique called DNA hybridization, based on the premise that fragments of DNA that are complementary to each other will bind—or hybridize—when incubated together, Bishop and Varmus began a search for a cellular counterpart to the src gene. Using a probe prepared from the viral genetic material, Bishop, Varmus, and their assistant Dominique Stehelin incubated the viral probe with chicken DNA, knowing if the two hybridized this would mean the cell contained a counterpart to the viral src gene. The unanticipated result of the experiments was that not only did the cell chromatin contain a homologous gene, the cellular gene was nearly identical to the viral oncogene. Using DNA obtained from a variety of birds, Bishop and Varmus found the same results; the src gene was present in all cells. When they repeated their work using cells from mammals, including humans, they found the same results; identical counterparts to the viral src gene were present in all cells. Subsequent to the discovery of oncogenes in tumor viruses, scientists had been working with the hypothesis that oncogenes such as src had originated in viruses and were transferred to normal cells following infection. The reality was exactly the opposite. Oncogenes, later termed proto-oncogenes, actually originated within the cell and were picked up by the virus as a result of integration of the viral DNA into the cellular chromosome.
>
> It was subsequently determined that the viral oncogene and cellular proto-oncogene did exhibit differences in regulation. The viral src gene was much like the short circuit in a light switch, constantly in an "on" position. The cell proto-oncogene was regulated and expressed only at the appropriate time in cell division. Nevertheless Bishop and Varmus's discovery provided a step in understanding the role of genes in development of cancer.

regulate cell division. Only when the gene undergoes a mutation does it result in cell transformation. The *src* gene was merely the first of the so-called oncogenes to be discovered in the cell. The discovery by Bishop and Varmus subsequently led to the discovery of over one hundred such proto-oncogenes, some by Bishop but most by colleagues entering the field opened by Bishop and Varmus. Bishop's later research addressed a more in-depth understanding of the roles played by proto-oncogenes in cells from a wide variety of species, ranging from fruit flies to humans. In 1989 Bishop and Varmus were awarded Nobel Prizes in Physiology or Medicine for their discovery.

In 1972 Bishop was promoted to full professor in both the microbiology and immunology department. He also became a professor in the department of biochemistry and biophysics. He was director of both the G.W. Hooper Research Foundation and the Program in Biological Sciences at UCSF. In 2003, Bishop was awarded the National Medal of Science. In 1998 he was appointed chancellor of UCSF, retiring in 2009. During his period in office, longer than any of the previous seven people in that position, Bishop guided what was at the time the largest academic expansion in the biomedical sciences in the United States.

IMPACT

Bishop's discovery and characterization of cellular proto-oncogenes set the stage for an understanding of molecular processes that regulated cell division and also provided an understanding of the role played by mutations in transforming normal cells into malignant cells. The conservation of proto-oncogenes through evolution carried an inherent explanation of their normal function: a critical role in the regulation of cell replication. Once the first proto-oncogene known as *src* (short for "sarcoma") was discovered and characterized, it was only a brief period before additional proto-oncogenes were identified, some by Bishop himself. Within twenty years after Bishop's work with Varmus, over one hundred such genes were identified and characterized. All play some role in initiating or regulating cell division.

Identification of such genes was only the initial step in understanding their role in cell regulation. By studying the effects of mutations in these genes, Bishop was instrumental in the developing story of the role played by gene products in the regulation of cell replication. That role also provided answers to the question of how mutagens such as chemicals or radiation transform a cell into one that is malignant. They had discovered that by inducing a mutation in the appropriate proto-oncogene, cell regulation is disrupted.

The significance of Bishop's discoveries lies in the realization that malignancies are not generally associated with infectious agents such as viruses, which had been the working hypothesis during the 1960s. Rather, the nature of cancer lies within the cell genome itself.

———*Richard Adler*

Further Reading

Bishop, J. Michael. *How to Win the Nobel Prize: An Unexpected Life in Science.* Cambridge: Harvard UP, 2003. Print. Describes Bishop's early training in the medical field and his transition to molecular biology.

Bunz, Fred. *Principles of Cancer Genetics.* New York: Springer, 2008. Print. Presents an examination of the roles played by Bishop and Varmus in their discoveries of cellular oncogenes.

Mukherjee, Siddhartha. *The Emperor of All Maladies: A Biography of Cancer.* New York: Scribner, 2010. Print. Provides a historical account of the earliest descriptions of cancer, assisting in developing readers' understanding of the molecular basis of the disease.

Pecorino, Lauren. *Molecular Biology of Cancer: Mechanisms, Targets, and Therapeutics.* New York: Oxford UP, 2008. Print. An introductory text presenting the role played by oncogenes and describing events that can trigger a malignancy.

Weinberg, Robert A. *The Biology of Cancer.* New York: Garland, 2006. Print. An in-depth study of the disease. The discoveries by Bishop and Varmus of the roles played by cellular oncogenes represent a significant portion of the story.

Herbert Wayne Boyer

Photo of a molecular biologist Herbert Boyer. By Jane Gitschier. [CC-BY-2.5 (http://creativecommons.org/licenses/by/2.5)], via Wikimedia Commons

American biochemist

- **Born:** July 10, 1936; Darry, Pennsylvania
- **Primary fields:** Biology; genetics
- **Primary inventions:** First recombinant DNA organism; human insulin

Boyer, along with Stanley Norman Cohen, developed the basic techniques used in genetic engineering. Boyer and Cohen were the first to use restriction endonucleases to cut DNAs from two different sources, splice them together to make a recombinant DNA molecule, and express the recombinant DNA molecule after insertion into E. coli *cells.*

Early Life

Herbert Wayne Boyer was born in 1936 in the western Pennsylvania town of Derry. After graduating from Derry High School, where he played football, Boyer commuted to St. Vincent College in Latrobe,

Pennsylvania, where he majored in premed biology, receiving a B.S. in biology and chemistry in 1958. He then attended graduate school at the University of Pittsburgh. In 1959, he married Marigrace Hensler. Boyer completed his PhD work in 1963, after which he did three years of postgraduate work focusing on biochemistry at Yale University in the laboratories of Edward Adelberg and Bruce Carlton. While conducting postdoctoral research, Boyer was active in the Civil Rights movement.

LIFE'S WORK

In 1966, Boyer accepted an assistant professorship of biochemistry and biophysics at the University of California, San Francisco (UCSF). His research focused on the isolation and characterization of EcoR1, a restriction endonuclease enzyme from *Escherichia coli* (*E. coli*) that cuts molecules of deoxyribonucleic acid (DNA) at very specific nucleotide sequences. Boyer discovered that EcoR1 creates DNA molecules with short (four-nucleotide), single-stranded, complementary, "sticky" overhang ends.

In 1972, Boyer teamed up with Stanley Norman Cohen, a professor, research scientist, and physician at Stanford University who had been studying the insertion of small, circular, nonchromosomal DNA molecules called plasmids, which reside and replicate in a variety of bacteria, into *E. coli*. Cohen had realized that two DNAs cut with Boyer's EcoR1 restriction endonuclease could easily be spliced together because of the complementary overhangs. In 1973, Boyer, Cohen, Annie Chang, and Robert Helling cut, combined, and ligated two different plasmids and inserted them into *E. coli*, marking the construction of the first recombinant DNA using restriction endonucleases as well as the beginning of the era of genetic engineering, also known as gene cloning. Now scientists could recombine DNAs at will and insert them into other organisms. In 1974, Boyer and Cohen, along with Chang, Helling, John Morrow, and Howard Goodman, succeeded in constructing a recombinant DNA molecule from the DNA of a plasmid and the ribosomal RNA-coding DNA of *Xenopus laevis*, the African clawed frog. After insertion of the recombinant plasmid/*Xenopus* DNA into *E. coli*, the recombinant DNA was transcribed into *Xenopus laevis* ribosomal ribonucleic acid (rRNA). This was the first demonstration of a eukaryotic DNA molecule (the rRNA gene from *Xenopus laevis*) being expressed in a foreign organism, *E. coli*. Boyer and Cohen next patented their gene-splicing technique.

In 1976, Boyer was promoted to professor of biochemistry and biophysics at UCSF, a position he would hold until his retirement in 1991. In April, 1976, Boyer and Robert Swanson, a financier and venture capitalist who studied chemistry and management at the Massachusetts Institute of Technology (MIT), incorporated Genentech (Genetic Engineering Technology), a company whose first focus was to synthesize the human insulin gene, recombine it with a plasmid, and insert it into *E. coli* with the hope that the *E. coli* would synthesize human insulin. Swanson conceived of the project because the incidence of diabetes in the United States was increasing while the availability of bovine and porcine insulin was decreasing. Swanson and Boyer joined forces with Arthur Riggs, a geneticist at the City of Hope, a clinical research hospital in Duarte, California. By 1978, the Genentech/City of Hope team had successfully cloned and expressed the human insulin gene in *E. coli*. Genentech licensed the production and marketing of human insulin to Eli Lilly. Genentech, one of the leading biotechnology companies in the world, became a publicly traded company in 1980. In 1985, it became the first biotechnology company to produce and market its own medicinal product, human growth hormone (hGH), marketed under the name Protropin.

Boyer was vice president of Genentech until 1990, when he became a member of the board of directors. In 1994, he became a member of the board of directors of Allergan, Inc., a company that focuses on the discovery and development of innovative pharmaceuticals. He served as chairman from 1998 to 2001 and was elected vice chairman in 2001.

Boyer is the recipient of numerous awards, including the Albert Lasker Award for Basic Medical Research (1980), the National Medal of Science (1990), the Lemelson-MIT Prize for Invention and Innovation with Stanley Cohen (1996), the Albany Medical Prize (2004), and the Shaw Prize in Life Sciences and Medicine (2004). He was elected to the American Academy of Sciences in 1979 and to the National Academy of Sciences in 1985. In 1991, after a substantial gift by Boyer and his family, Yale University dedicated the Boyer Center for Molecular Medicine. In 2007, St. Vincent College renamed the School of Natural Science, Mathematics, and Computing the Herbert W. Boyer School. After

> **PRODUCTION OF HUMAN INSULIN**
>
> Although Herbert Wayne Boyer, Stanley Norman Cohen, and their colleagues developed the basic techniques of gene splicing and genetic engineering, it was Boyer who orchestrated the development of these techniques into ones that could be exploited in the production of medicinally important proteins and enzymes, leading to the birth of the biotechnology and biopharmaceutical industries.
>
> Financier Robert Swanson thought that the techniques could be used to produce human insulin, a protein that would be needed in the near future since the number of diabetics in the United States was increasing while the supply of bovine and porcine pancreases from which diabetics received their insulin was decreasing. Boyer and Swanson founded Genentech to produce human insulin. The Swanson-Boyer Genentech team elicited the assistance of Arthur Riggs and his colleagues at the City of Hope clinical research hospital in Duarte, California, in the project. Since the use of human DNA in recombinant DNA experimentation was not permitted by the National Institutes of Health guidelines involving recombinant DNA research, the Genentech-City of Hope team thought that the best approach was to synthesize the human insulin "gene" rather than to isolate it. Under a loophole in the guidelines, the use of a synthetic human gene in constructing recombinant DNA molecules would not be banned. The strategy was to synthesize, clone, and express in *Escherichia coli* (*E. coli*) bacteria cells a human gene that would code for human insulin. Riggs suggested that he and his colleagues Keiichi Itakura, Herb Heyneker, and John Shine first synthesize, clone, and express the human somatostatin gene since somatostatin was considerably smaller—fourteen amino acids compared to insulin's fifty-one amino acids. This project was completed in August, 1977, and published in December, 1977.
>
> Soon after the somatostatin project was completed, the Genentech-City of Hope team, with the assistance of Itakura, David Goeddel, Dennis Kleid, and Roberto Crea, began the human insulin project. Since insulin is composed of two polypeptide chains, the strategy was to synthesize, clone, and express separately in *E. coli* a gene coding for each chain. The chains would then be isolated, mixed, joined, and folded into a functioning insulin molecule. On September 6, 1978, Genentech and City of Hope announced that they had successfully cloned and expressed the human insulin gene in *E. coli*. Twelve days earlier on August 25, Genentech licensed the production of human insulin to Eli Lilly. In 1980, a small-scale clinical trial involving fourteen patients was begun in England, followed by a much larger clinical trial in the United States in 1982. Human insulin became the first human protein of medicinal value to be produced by genetic-engineering techniques and approved for use by the U.S. Food and Drug Administration (FDA). The FDA approved the use of human insulin, marketed under the name Humulin, on October 29, 1982.

retiring, Boyer became professor emeritus of biochemistry and biophysics at UCSF.

IMPACT

Boyer and Cohen's collaboration that led to the development of gene-splicing and genetic-engineering techniques had a profound impact on biology by revolutionizing the study of genetics and molecular biology. Their development of these techniques allowed for the construction, insertion, cloning, and expression of recombinant DNA molecules in foreign hosts and spawned the development of the biotechnology and biopharmaceutical industries.

Although the polymerase chain reaction (PCR) eventually replaced cloning as a method to amplify specific DNA molecules, the cloning of recombinant DNA molecules gave scientists a method to amplify and isolate specific DNA molecules to provide enough copies for DNA sequencing and analysis of gene structure and function. These analyses provided scientists with basic knowledge of the structure of the gene, the nature of mutations, and the control of gene expression.

The expression of recombinant DNA molecules in foreign hosts provided a mechanism for the production by foreign hosts of a variety of human proteins and enzymes to treat human disease. The first medicinally valuable protein that became commercially available was human insulin, in 1982. Insulin was quickly followed by the production of other recombinant human proteins, including hGH, interferon, tissue plasminogen activator (t-PA), and factor VIII clotting factor.

The Boyer-Cohen techniques also led to the development of a variety of transgenic strains of animals and plants. Today, animals such as goats, sheep, pigs, and cows are engineered to produce a variety of human proteins and enzymes of medicinal value, including t-PA, lactoferrin, factor VIII, factor IX, and alpha

1-antitrypsin. Plants are engineered to be insect-resistant, virus-resistant, and pesticide-resistant, allowing for increased yield and a reduction in the use of pesticides. Plants are also engineered to increase nutritional value. Rice, for example, has been engineered to produce beta-carotene, the precursor of vitamin A.

——*Charles L. Vigue*

FURTHER READING

Drlica, Karl. *Understanding DNA and Gene Cloning: A Guide for the Curious.* 4th ed. New York: John Wiley & Sons, 2004. Addresses all aspects of gene cloning, beginning with the structure and expression of DNA and ending with the construction of recombinant DNA and its subsequent cloning and expression.

Hall, Stephen S. *Invisible Frontiers: The Race to Synthesize a Human Gene.* New York: Atlantic Monthly Press, 1987. An well-referenced account of Genentech's success in synthesizing the human insulin gene and its subsequent expression in *E. coli*.

Martineau, Belinda. *First Fruit: The Creation of the Flavr Savr Tomato and the Birth of Biotech Food.* New York: McGraw-Hill, 2001. An authoritative account of the development and subsequent marketing and demise of Calgene's Flavr Savr tomato, the first genetically engineered food to come to market. The author was an employee of Calgene and intimately involved with the development of the tomato. Excellent references.

ERWIN CHARGAFF

TABLE II
Purine and Pyrimidine Contents of Salmon Sperm DNA
The results are expressed in moles per mole of P in the hydrolysate.

Experiment No.*	Preparation No.	Hydrolysis procedure	Nitrogenous constituent				Recovery of nitrogenous constituents		Total
			Adenine	Guanine	Cytosine	Thymine	Purines	Pyrimidines	
1	1	1	0.27	0.18			0.45		
2		1	0.26	0.19			0.45		
3		1			0.17	0.28		0.45	
4		1			0.18	0.28		0.46	
5		2	0.28	0.20	0.21	0.27	0.48	0.48	0.96
6		2	0.30	0.22	0.20	0.29	0.52	0.49	1.01
7		2	0.27	0.18	0.19	0.25	0.45	0.44	0.89
8		2	0.28	0.21	0.20	0.27	0.49	0.47	0.96
9	2	1	0.25	0.18			0.43		
10		1	0.29	0.20			0.49		
11		2	0.29	0.18	0.20	0.27	0.47	0.47	0.94
12		2	0.28	0.21	0.19	0.26	0.49	0.45	0.94
13		2	0.30	0.21	0.20	0.30	0.51	0.50	1.01

* In each experiment between twelve and twenty-four determinations of individual purines and pyrimidines were performed.

Chargaff results on salmon DNA By Erwin Chargaff, Rakoma Lipshitz, Charlotte Green, and M. E. Hodes (http://www.jbc.org/content/192/1/223.full.pdf) [Public domain], via Wikimedia Commons

UKRAINIAN AMERICAN BIOCHEMIST

- **BORN:** August 11, 1905; Czernowitz, Austria-Hungary (now Chernovtsy, Ukraine)
- **DIED:** June 20, 2002; New York, New York
- **PRIMARY FIELD:** Biology
- **SPECIALTIES:** Biochemistry; molecular biology

Twentieth-century Ukrainian American biochemist Erwin Chargaff was a pioneer in the field of genetics. The discoveries he made about the base ratios of DNA, which became known as Chargaff's rules, led to the determination of the structure of DNA.

EARLY LIFE

Erwin Chargaff was born on August 11, 1905, in Czernowitz, a provincial capital in what was then the Austro-Hungarian Empire. He was the son of Rosa Silberstein Chargaff and Hermann Chargaff, the owner of a small bank in Czernowitz. During the early part of World War I, Czernowitz and its surrounding region were occupied by the Russian army. The Chargaff family was forced to flee to Vienna, where they remained following the armistice in 1918.

Though the Chargaff family was relatively well-to-do prior to World War I, like much of former Austria-Hungary, they were financially ruined after the empire collapsed. Despite the family's financial troubles, Chargaff was able to attend secondary school at the prestigious Maximilian Gymnasium in Vienna. The school gave Chargaff a solid grounding in the classics: Latin, Greek, mathematics, and history.

In 1924, Chargaff enrolled at the University of Vienna. Unsure of which program to study, he decided upon chemistry almost by default; he knew little of the subject, but nothing else seemed of interest. Chemistry also offered a better possibility of employment after graduation. Chargaff's dissertation presented his work on organic silver complexes and the action of iodine on azides—compounds that form the conjugate base of the highly volatile hydrazoic acid—for which he was awarded a doctorate in 1928. He had produced two research publications prior to his dissertation, and he decided to remain in research following graduation. The same year Chargaff earned his doctorate, he married Vera Broido, a woman he had met at the university. They had one son, Thomas, born in 1938.

LIFE'S WORK
There were few research positions available in Austria after Chargaff earned his PhD, so he decided to go to the United States. A colleague who had recently completed a lecture tour in the United States informed Chargaff that Rudolph Anderson, a researcher studying the tubercle bacillus (tuberculosis) at Yale University in New Haven, Connecticut, was interested in hiring a student to study lipids of the tubercle bacillus. Chargaff applied, and along with his acceptance he was awarded the Milton Campbell Research Fellowship in Organic Chemistry, which included a stipend for expenses. Chargaff spent two years in Anderson's laboratory, publishing seven papers while there. Among his discoveries was the unusual branching of fatty acid chains on the surface of the tubercle bacillus, which accounted for scientists' difficulty in staining the bacterium for identification.

Chargaff returned to Europe in 1930 and joined the bacteriology department at the University of Berlin. He considered the three years he spent there as a private lecturer among the happiest of his life. During his time in Berlin, Chargaff continued his research on tubercle bacilli. He also became involved in the Lübeck case, in which several physicians were accused of killing infants by giving them live tuberculosis bacilli instead of the Bacillus Calmette-Guérin (BCG) tuberculosis vaccine.

Chargaff helped to prove that the vaccine preparations were not at fault.

In January 1933, Adolf Hitler assumed the government of Germany. Being Jewish and told he was no longer welcome at the university, Chargaff left Berlin for the Pasteur Institute in Paris, France. Despite his acceptance at the institute, Chargaff recognized that its lack of resources, money, and equipment would hamper his ability to carry out meaningful research. Remaining only until the end of 1934, Chargaff returned to the United States in 1935, where he was offered a position as research associate in the Department of Biochemistry at Columbia University in New York City. He remained at Columbia for the remainder of his professional career.

Chargaff's initial research at Columbia dealt with plant proteins as well as the components of blood clotting. Two factors led Chargaff to move his research in a different direction. In 1944, Austrian physicist Erwin Schrödinger published a series of his lectures in a book, *What Is Life?*, in which he speculated that the source of genetic information in a cell was some form of a chemical crystal. While Schrödinger's ideas were premature, their publication led a number of researchers, including Chargaff, to investigate the molecular basis for genetics. That same year, Canadian American physician and bacteriologist Oswald Avery and his colleagues demonstrated that deoxyribonucleic acid (DNA) was the genetic material in cells.

Intrigued by the role of DNA, Chargaff began an investigation into the chemical composition of the molecule. The components of DNA had been established earlier by Lithuanian American biochemist Phoebus Levene in 1929. Levene determined that DNA is composed of four different nucleotides. Each nucleotide contains a five-carbon deoxyribose sugar and a phosphate group, but they differ in the nitrogen base attached—adenine, guanine, thymine, and cytosine. Levene had proposed that the bases were present in equal amounts, which was the foundation for his 1929 hypothesis regarding tetranucleotide.

To determine the quantities of nucleotide bases in DNA, Chargaff used biochemical techniques that became available in the late 1940s, including the separation and quantification of molecules through paper chromatography and spectrophotometry. He discovered that contrary to the tetranucleotide hypothesis, the bases were not present in equal

ERWIN CHARGAFF: BASE RATIOS IN DNA

Once the role of DNA as the hereditary material in cells was established by Oswald Avery in 1944, Erwin Chargaff began focusing his research on the molecular structure of DNA, a subject that was still relatively new in the 1940s. Beyond the knowledge that DNA was a polymer and was composed of molecular classes of purines (adenine and guanine) and pyrimidines (thymine, cytosine, and uracil)—ringed structures known to be constituents of nucleic acids—scientists had largely avoided research into what had previously been thought of as simply storage molecules for phosphate.

Chargaff began his work by addressing the question of how much molecules must differ before a scientist recognizes the significance of that difference. Were all nucleic acids identical in their structure, and if not, how subtle were those differences? Biochemical techniques necessary to analyze this had yet to be developed, and variations in the size of DNA for different organisms or within the tissues of the same organism were unknown.

In order to determine if there were variations in the structure of nucleic acids, Chargaff had to develop procedures for the purification of DNA. Some methods were already routine; others he developed through trial and error. Using these procedures, Chargaff isolated and purified DNA from a wide variety of organisms, including yeast, bacteria, human sperm, and from different tissues within the organisms themselves.

The next steps for analysis required separation and visualization procedures for the purines and pyrimidines. One of Chargaff's choices for separation was the use of paper chromatography. The technique is analogous to that of placing a drop of ink on a piece of paper and allowing water to migrate through the paper; the constituents of the ink will separate on the basis of their water solubility. To separate purines and pyrimidines, a variety of solvents were allowed to migrate through a paper upon which the nucleic acid components were spotted, or marked. Next, Chargaff applied mercury compounds to the purines and pyrimidines to distinguish them. The concentrations of each component were then determined through a process called spectrophotometry, which measures a material's relative absorption of wavelengths of light. The results of Chargaff's experiments were codified in what became known as Chargaff's rules.

Chargaff observed that regardless of the origin of the DNA—whether from bacteria, yeast, or human sperm—the concentration of adenine always equaled the concentration of thymine (AT), while the concentration of guanine always equaled the concentration of cytosine (GC). While these base ratios of DNA never changed, the relative amount of each could vary within different organisms. For example, DNA from a human always contained 27 percent adenine and 27 percent thymine, while DNA from the tuberculosis bacterium contained 12 percent adenine and 12 percent thymine.

The equivalence of DNA base ratios became the first of Chargaff's rules. The variation of their relative proportions between organisms became his second.

---

amounts, but varied depending on any given species. Furthermore, the quantity of adenine always equaled that of thymine, and the quantity of guanine always equaled that of cytosine. This equality of DNA base ratios became part of what is known as Chargaff's rules.

In 1950, Chargaff published his findings in a paper entitled, "Chemical Specificity of Nucleic Acids and Mechanism of Their Enzymatic Degradation." The significance of the ratios was missed by Chargaff at the time, since the actual structure of DNA was still unknown. However, his discovery played a key role two years later in enabling American molecular biologist James D. Watson and British molecular biologist Francis Crick to produce their model for the structure of DNA.

Chargaff spent many of his later years mired in controversy. Despite being one of the pioneers in the field of genetics, he objected to genetic engineering, considering it a greater threat to the world than nuclear technology. Still, he was the recipient of numerous honors, including election to the National Academy of Sciences in 1965 and being awarded the National Medal of Science in 1975. Chargaff served as chair of Biochemistry at Columbia University from 1970 to 1974, after which he retired as professor emeritus. He died on June 20, 2002, at the age of ninety-six.

IMPACT

Chargaff is recognized for his contributions to several areas of biochemistry, and he published numerous papers; it was his discovery of the base ratios of DNA, Chargaff's rules, for which he is most remembered. The significance of his observation was largely overlooked until Watson and Crick began building models of prospective structures of DNA in 1952. The ratios provided a clue toward that structure when Watson and Crick realized that not only did adenine bond with thymine (AT), and guanine with cytosine (GC), but that the space of the AT base pair was exactly the same as with the GC pair. Once they determined the relative positions of the components of DNA, with the phosphate-sugar backbones held together by the AT or GC pairings, the structure of DNA became apparent.

In their 1953 paper describing the structure of DNA, Watson and Crick acknowledged the role Chargaff's discovery played in their work. While Chargaff was recognized for his discovery by Watson, Crick, and others, he was never awarded a Nobel Prize. When the 1962 Nobel Prize in Physiology or Medicine was awarded to Watson, Crick, and British biochemist Maurice Wilkins for their work on DNA, Chargaff responded with critical letters sent to scientists worldwide. He considered his contribution to the study of DNA equal to theirs, arguing that it was he who provided the critical information necessary for determining the molecular structure of DNA. Although the Nobel Prize committee overlooked Chargaff's work, his contributions to the fields of biochemistry and genetics are still recognized.

—*Richard Adler*

Further Reading

Chargaff, Erwin. *Heraclitean Fire: Sketches from a Life Before Nature.* New York: Rockefeller UP, 1978. Print. Offers an autobiography of Chargaff's life and philosophical approach to science, featuring anecdotes and stories that provide insight into the field of science and the behaviors of scientists.

Hausmann, Rudolf. *To Grasp the Essence of Life: A History of Molecular Biology.* Norwell: Kluwer Academic, 2010. Print. Includes the history of DNA, covering its discovery, function, and determination of structure. The contributions of Chargaff are described.

Olby, Robert C. *Francis Crick: Hunter of Life's Secrets.* Cold Spring Harbor: Cold Spring Harbor Laboratory P, 2009. Print. Presents a biography of Francis Crick, covering the work he did with James D. Watson leading to the determination of the structure of DNA. Includes the confrontation between Chargaff and Crick.

Witkowski, Jan A., ed. *Inside Story: DNA to RNA to Protein.* Cold Spring Harbor: Cold Spring Harbor Laboratory P, 2005. Print. Collected perspectives on major figures in molecular biology and their work, including Chargaff.

Stanley Cohen

American biochemist

- **Born:** November 17, 1922; Brooklyn, New York
- **Areas of achievement:** Medicine; science and technology

Cohen, with Rita Levi-Montalcini, discovered, isolated, and characterized nerve growth factor. His research led to a better understanding of how wounds heal and of how to treat many serious ailments, such as cancer, neurodegenerative diseases, and cardiovascular disease.

Early Life

Stanley Cohen was born on November 17, 1922, in Brooklyn, New York. He was the son of Russian Jewish immigrants—his father, Louis, was a tailor, and his mother, Fannie, a homemaker. As a child, Cohen suffered from polio, which left him with a permanent limp. Cohen describes himself in these years as being extremely interested in how things worked—from the interior workings of a telephone to the gears on his bicycle. While his parents had a high regard for education, money was tight, and all four Cohen children attended the local public school, James Madison High School. His grades were good enough to earn his entrance to Brooklyn College, a city school that charged no tuition fees. Following his graduation in 1943 with a double major in chemistry and biology, Cohen worked in a milk-processing plant doing bacteriology work. At this point, his ambition was to become a laboratory technician, but one of his former professors at Brooklyn College offered him a scholarship in biology to Oberlin College in Ohio. Because of his keen interest in the subject, Cohen enthusiastically accepted the opportunity.

Cohen graduated from Oberlin with a master's degree in zoology in 1945. He was then awarded a PhD fellowship in biochemistry at the University of

Michigan. His thesis was on earthworm metabolism. While his days were spent working in the laboratory, his nights were often spent collecting thousands of earthworms from the university grounds. Upon completion of his PhD in 1948 he accepted a postdoctoral fellowship at the University of Colorado. In 1952, Cohen decided to study the emerging technique of using radioisotopes in metabolic studies, and he obtained an American Cancer Society Fellowship to work at Washington University in the radiology department. A year later he moved to the zoology department, where he embarked on a lifelong study of growth factors, for which he and Rita Levi-Montalcini were later to share the Nobel Prize in Physiology or Medicine in 1986.

Life's Work

Cohen joined the laboratory of Viktor Hamburger and Levi-Montalcini at Washington University in 1957. Hamburger and Levi-Montalcini were both Jewish refugees from Nazi Europe who had built a vibrant research laboratory investigating why nerves in chick embryo eggs grew when implanted with a mouse tumor. They recruited Cohen, a biochemist, to identify the active substance. He isolated the substance, nerve growth factor (NGF), and then created antibodies that blocked the growth factor's activity.

By this time, Cohen had a young family to support, and in 1959 he accepted his first teaching position as assistant professor of biochemistry at Vanderbilt University. There, he continued to study growth factors, focusing on epidermal growth factor (EGF), which he discovered, isolated, purified, and sequenced. He also identified the receptor for EGF and discovered its mechanism of action. This led to a new understanding of the mechanisms whereby a cell responds to signals from its surroundings.

In 1976, Cohen was appointed American Cancer Society Research Professor at Vanderbilt Together with Levi-Montalcini, he was awarded the Nobel Prize in Physiology or Medicine in 1986, for their discovery of growth factors. He was also named Distinguished Professor at Vanderbilt in that same year. Until his retirement in 2000 he was a familiar, unassuming figure in the research halls of Vanderbilt, usually seen smoking his pipe. Upon his retirement he became Distinguished Professor Emeritus at Vanderbilt. Over the course of his career he received many other honors and awards, including the Vanderbilt University Earl Sutherland Prize for Achievement in Research in 1977, the National Medal of Science in 1986, and the Albert Lasker Basic Medical Research Award in 1986. Cohen was elected to the National Academy of Science in 1980 and to the American Academy of Arts and Sciences in 1984. He was recognized with induction into the National Institute of Child Health and Human Development Hall of Honor in 2007.

Significance

The discovery of growth factors had a major impact on understanding how cell growth in the body is stimulated in normal and abnormal circumstances. This had enormous implications for cancer therapy and led directly to the development of targeted anticancer drugs such as Herceptin (trastuzumab) and Gleevec (imatinib mesylate). EGF is also currently used to treat corneal ulcers and in burn healing. Growth factors may also hold important clues to a variety of other diseases, including Alzheimer's disease. Applications of growth factors in the areas of aging and regenerative medicine, cardiovascular biology, cancer, and metabolic disorders have also been actively pursued.

——*Rosemary Whelan*

Further Reading

Cohen, Stanley N. "The Manipulation of Genes." *Scientific American* 233 (July, 1975): 24-33. Eminently readable and profusely illustrated, this article provides the general reader with an excellent historical perspective on the development of recombinant DNA technology and a sound description of the scientific processes involved. Discussion of the use of plasmids as vectors is particularly useful.

Cohen, Stanley. "Origins of Growth Factors: NGF and EGF." *Journal of Biological Chemistry* 283, no. 49 (2008): 33793-33797. Print.

Cohen, Stanley. "Stanley Cohen—Autobiography" *The Nobel Prizes 1986.* Stockholm: Nobel Foundation, 1987. Print.

Drlica, Karl. *Understanding DNA and Gene Cloning: A Guide for the Curious.* 4th ed. New York: John Wiley & Sons, 2004. A good introduction to the impact of DNA technology on human lives.

COHEN AND BOYER DEVELOP RECOMBINANT DNA TECHNOLOGY

- DATE 1973
- LOCALE Stanford, California
- KEY FIGURES: *Stanley Norman Cohen* (b. 1935), physician and molecular geneticist; *Herbert Wayne Boyer* (b. 1936), bacteriologist and molecular geneticist; *Paul Berg* (b. 1926), biochemist; *Hamilton Oliver Smith* (b. 1931), molecular geneticist

Stanley Norman Cohen and Herbert Wayne Boyer pioneered techniques that now allow scientists to insert DNA from any source into bacteria and to detect the expression of the foreign genes in these simple cells.

Recombinant DNA (deoxyribonucleic acid) technology—known also in various guises as "genetic engineering," "genetic modification," and "gene cloning"—is an area of scientific investigation and applied biology that has, since its inception in 1973, revolutionized molecular biology, allowing scientists to address questions in cell biology that could not be addressed by earlier methods. Recombinant DNA methods allow molecular biologists to add one or a small number of genes from essentially any organism to simple bacterial cells. These foreign genes can be made to become an integral part of the bacterium, replicating along with the bacterial genetic material and thus stably transmitted from one bacterial generation to the next. The foreign genes can also be made to be functional in their bacterial host—that is, they can be induced to make their normal gene products.

Bacteria are very simple single-celled organisms that are ubiquitous in nature. Although some are capable of causing disease, most bacteria are harmless to humans. Some, like the common intestinal bacterium *Escherichia coli* (*E. coli*), are normal inhabitants of the human body that are essential to human life. Each *E. coli* cell has a single circular DNA molecule, or chromosome, containing between two thousand and three thousand genes. In addition, some cells have one or more additional small circular DNA molecules called plasmids. A typical plasmid contains on the order of five to ten genes and is therefore much smaller than the *E. coli* chromosome. These plasmids are semiautonomous, meaning that while they are incapable of leading a cell-free existence, they generally remain separate from the larger chromosome and control and direct their own replication and transmission to each daughter cell at cell division. Plasmids that contain genes for resistance to certain antibiotics, viruses, and so forth can provide the host cell with useful properties.

The "basic experiment" of recombinant DNA technology involves four essential elements: a method of generating pieces of DNA from different sources and splicing them back together; a "vector" molecule (often a plasmid) that can replicate both itself and any foreign DNA linked to it; a way to get this composite, or recombinant, DNA molecule back into a suitable bacterial host; and a means to separate those bacterial cells that have picked up the desired recombinant plasmid from those cells that have not.

As part of the process, the recombinant plasmids are then reintroduced back into *E. coli* host cells in a process called "transformation." An essential feature of transformation is treatment of the host cells with calcium chloride, which weakens the cell walls and membranes, allowing the reconstituted plasmid DNA to be taken up inside the cells. If all has gone well, these genetically engineered clones of bacterial cells will then stably replicate the foreign DNA, along with the rest of the chromosomal and plasmid DNA of each cell generation; the products of the foreign genes—ribonucleic acid (RNA) or protein—will be made as well.

By the early 1970's, the stage was set for the advent of recombinant DNA technology. DNA "ligases" (enzymes that play a significant role in the process) had been discovered and purified independently in five separate laboratories in 1967. Hamilton Oliver Smith described the first restriction endonuclease enzyme in 1970, and shortly thereafter Herbert Wayne Boyer described the isolation of EcoRI, a restriction endonuclease that became extremely important in the development of cloning methods. Paul Berg and his group described the construction of the first recombinant DNA molecules in a test tube, and at about the same time researchers in Stanley Norman Cohen's laboratory reported on the first successful transformation experiments in *E. coli*.

In the fall of 1973, Cohen and Boyer were the first researchers to describe successful completion of a recombinant DNA experiment. Their report detailed the mixing and subsequent reconstitution of DNAs from two separate plasmids in *E. coli*. Shortly thereafter, they described experiments in which DNA from a plasmid found in an unrelated bacterium was successfully cloned in *E. coli*, and one year later they reported on the first successful cloning of animal genes in *E. coli*.

Recombinant DNA technology is widely considered to be the most significant advance in molecular biology since the elucidation of the molecular structure of DNA in 1953 by biophysicists James D. Watson and Francis Crick. It soon became apparent, however, that the technology had opened a Pandora's box of social, ethical, and political issues unprecedented in scientific history. The research held the potential of addressing biological problems of fundamental theoretical and practical importance, yet it generated real concerns also, because some experiments might present new and unacceptable dangers. Even in the course of scholarly research with the best intentions, there was concern that a laboratory accident or an unanticipated experimental result might introduce dangerous genes into the environment, with *E. coli* carrying them.

Soon after the scientific concerns were first voiced, a conference was planned to allow many of the leading researchers in molecular biology to try to assess the potential dangers of recombinant DNA technology. The conference was held at the Asilomar Conference Center in February of 1975. Six months earlier, however, eleven respected authorities in molecular biology, including Cohen, Boyer, Berg, and others who helped develop recombinant DNA techniques, signed a letter that was simultaneously published in three English and American scientific journals. This letter called for a voluntary moratorium on recombinant DNA experiments until questions about potential hazards could be resolved. The development of a set of guidelines for recombinant DNA research, a modification of which was later adopted by the National Institutes of Health, was discussed at the Asilomar Conference. Levels of both biological and physical "containment" were defined, and each type of recombinant DNA experiment was assigned to an appropriate level. Some types of experiments were banned. In the years that followed the initial furor, guidelines have been modified accordingly, as many of the initial fears about possible dangers have proved to be groundless.

As predicted, recombinant DNA technology has proved to have extensive practical applications, particularly in the fields of medicine and agriculture. Virtually all insulin-dependent diabetics now take human insulin made by genetically engineered bacteria. Human growth hormone, prolactin, interferon, and other human gene products with specific therapeutic uses in medicine are available only because they can be made in quantity by using cloning. In agriculture, improved species of genetically modified crop plants have been designed to help address problems in global food supplies. Of particular note is the effort to clone the bacterial genes for nitrogen fixation into crop plants, thus obviating the need for most fertilizers.

Fredrickson, Donald S. *The Recombinant DNA Controversy: A Memoir—Science, Politics, and the Public Interest, 1974-1981*. 2d ed. Washington, D.C.: ASM Press, 2001. The author, National Institutes of Health director from 1975 to 1981, offers a glimpse into the heated controversy surrounding the new technology.

Grobstein, Clifford. "The Recombinant-DNA Debate." *Scientific American* 237 (July, 1977): 22-33. Written at the height of the public controversy over the risks and benefits of recombinant DNA research, this thoughtful article provides a unique perspective to the history of the phenomenon. Colorful, helpful illustrations complement Grobstein's balanced treatment of the sensitive issues.

Jackson, David A., and Stephen R. Stich, eds. *The Recombinant DNA Debate*. Englewood Cliffs, N.J.: Prentice-Hall, 1979. A collection of seventeen essays, written by recognized leaders in the fields of molecular biology, ethics, and philosophy, covering all aspects of the controversy. Of particular note are essays by Robert L. Sinsheimer and George Wald, two of the most vocal scientists who pushed for a halt to recombinant DNA research.

Knowles, Richard V. *Genetics, Society, and Decisions*. Columbus, Ohio: Charles E. Merrill, 1985. This broad-based college text was written for nonbiology majors with an interest in science and the social issues raised by the new advances in biology. Good treatment of basic principles of genetics, particularly as applied to humans. Chapter 19 provides a straightforward presentation of recombinant DNA, including the relevant science, history, applications, and controversial issues. Many useful illustrations.

Kresge, N., R. D. Simoni, and R. Hill. "Precocious Newborn Mice and Epidermal Growth Factor: The Work of Stanley Cohen." *Journal of Biological Chemistry* 281, no. 10 (2006): e10-e11. Print.

Snyder, Bill. "Stanley Cohen's Nobel Prize: 25 Years of Progress." *ResearchNews@Vanderbilt*. Vanderbilt University, 9 Dec. 2011. Web. 25 Apr. 2016.

Vigue, Charles L., and William G. Stanziale. "Recombinant DNA: History of the Controversy." *American Biology Teacher* 41 (November, 1979): 480-491. Written primarily for teachers of secondary school biology, this short article summarizes the history and controversy surrounding the recombinant DNA debate. Should be readily accessible to the average reader and a good first choice for further reading.

Francis S. Collins

NIH Director Francis Collins, 2009. By Bill Branson, NIH. [Public domain], via Wikimedia Commons

American geneticist

- **Born:** April 14, 1950, Staunton, Virginia, VA
- **Primary field:** Biology
- **Specialty:** Genetics

A pioneer in the field of human genetics, physician Francis Collins served as the director of the National Human Genome Research Institute at the National Institutes of Health (NIH) in Washington, D.C., for fifteen years. His groundbreaking work on mapping the complete human genome led to his being named director of the NIH by U.S. president Barack Obama in 2009.

Early Life

Francis Sellars Collins was born on April 14, 1950, in Staunton, Virginia. His parents, Fletcher and Margaret Collins, were Yale-educated activists who were commissioned by First Lady Eleanor Roosevelt to rejuvenate a mining town in West Virginia during the Great Depression. Fletcher went on to work as a folksong collector, a supervisor in an aircraft factory during World War II, and a drama teacher at a women's college. In the 1940s, the Collinses bought a farm in Virginia's Shenandoah Valley and attempted to work the land without the benefit of machinery.

Margaret homeschooled her four sons, of whom Collins was the youngest. When he was ten, he entered the public school system. His exposure to institutionalized religion was limited to a stint in the choir of a local Episcopal church. When he was fourteen, his interest in science was sparked by an enthusiastic chemistry teacher. He graduated from Robert E. Lee High School at sixteen and entered the University of Virginia. During his freshman year, Collins became an agnostic and later an atheist.

After earning a bachelor's degree in chemistry in 1970, Collins entered Yale University to pursue a PhD in physical chemistry. While preparing his dissertation on theoretical quantum mechanics, he began to have doubts about his chosen field. The isolation of spending long hours in the laboratory conflicted with his growing desire to serve people. As he explored other areas of science, he became interested in the genetic code during a course in biochemistry.

In 1974, he graduated from Yale and entered medical school at the University of North Carolina at Chapel Hill, earning an MD in 1977. During his third year of medical school, he worked with terminally ill patients who exhibited a strong faith in God, which caused him to reconsider his atheism. He applied the same intellectual rigor to his spiritual search as he did to his scientific endeavors. English writer and Christian apologist C. S. Lewis's *Mere Christianity* (1952) was a major influence on Collins's thinking. During a hike in the mountains, Collins had a conversion experience and became a committed Christian.

Life's Work

Collins served as a resident in internal medicine at North Carolina Memorial Hospital in Chapel Hill from 1978 to 1981 and then returned to Yale as a Fellow in Human Genetics. While at Yale, he and Sherman Weissman published a paper on chromosome jumping, a genome-mapping procedure that bypasses areas of DNA that are difficult to clone. In 1984, Collins accepted a post at the University of Michigan as professor of internal medicine and human genetics, where he developed a gene-hunting approach that he called positional cloning, a gene-identification strategy that became an integral part of molecular genetics.

Collins's main interest was to identify and isolate genes that caused serious hereditary diseases. During the late 1980s, he joined Lap-Chee Tsui and a team of researchers at Toronto's Hospital for Sick Children who were working to find the genes responsible for cystic fibrosis. Using the chromosome jumping method he pioneered, Collins and his colleagues identified the gene in 1989. Their work helped spur discoveries of genes that cause Huntington's disease and neurofibromatosis, among other conditions.

In 1993, Collins's growing reputation as a premier gene hunter led to his appointment as director of the National Center for Human Genome Research. He succeeded Nobel laureate James D. Watson, who had discovered the spiral structure of DNA with Francis Crick. In 1997, the center was renamed the National Human Genome Research Institute. The goal of the organization was to map the human genome in order to understand the hereditary nature of common diseases such as breast and prostate cancer. In June of 2000, U.S. president Bill Clinton announced that a working draft of the first genetic blueprint of the human species was complete. Joining the president on the podium were Collins and biochemist Craig Venter, the founder of Celera Genomics, a private-sector company that was also working on sequencing human DNA.

In 2007, Collins founded the BioLogos Foundation, which is dedicated to exploring the compatibility between science and religion. The foundation promotes theistic evolution, a controversial belief that God is the author of the physical universe and has used evolution as a vehicle to create all life on Earth. Collins's notable achievements, as well as his interest in the harmony between science and faith, led Pope Benedict XVI to appoint him to the Pontifical Academy of Sciences in 2009.

On July 8, 2009, Collins was nominated by President Barack Obama to be the director of the National Institutes of Health (NIH). Kathleen Sebelius, the secretary of Health and Human Services, announced that his nomination had been unanimously confirmed by the Senate. His appointment was praised by many of his colleagues, but some were not as enthusiastic. In 2006, Collins published *The Language of God,* a memoir that details his path to faith, his advocacy of theistic evolution, and his belief that religion and science are compatible. Critics were afraid that his evangelical Christianity would negatively influence his decisions on controversial topics. He sought to reassure his detractors, however, by stating that his personal beliefs would have no impact on his leadership of the NIH.

Occupying one of the most powerful positions in Washington, the director of the NIH oversees 27 institutes and research centers, more than 20,000 onsite staff members, and 325,000 offsite researchers. Collins's experience as director of the National Human Genome Research Institute prepared him well for the formidable challenge of running a sprawling government agency.

In addition to his management responsibilities, Collins dealt with ethical questions and political controversies, including embryonic stem cell research. His critics were afraid that his religious beliefs would cause him to roll back the advances in the field. Collins, however, believed that embryonic stem cell research would be crucial in developing new treatments for a variety of medical conditions, including restoring movement in people who are paralyzed. Although he was opposed to creating embryonic stem cells for research purposes, he fought for congressional funding to use existing lines, as well as adult cells. He also advocated for keeping genetic information private and spoke against discrimination by employers and health insurance agencies based on one's genetic profile. In 2013 he outlined plans to reduce use of chimpanzees in NIH-funded studies, and in 2015 he announced that the NIH would no longer be funding any research involving chimpanzees. In 2016, with the help of an additional $680 million contributed to the project by the federal government, he launched the Cancer Moonshot Initiative to accelerate cancer research.

For his contributions to genetic research, Collins received the Kilby International Award (1993) the Presidential Medal of Freedom (2007), and the National Medal of Science (2009). He was also elected to the Institute of Medicine and the National Academy of Sciences.

Impact

When Collins became director of the National Genome Research Institute in 1993, the Human

> **FRANCIS COLLINS INVENTS THE TECHNIQUE OF CHROMOSOME JUMPING**
>
> Francis S. Collins first became interested in DNA sequencing when he was a research fellow at Yale during the 1980s. At the time, the process was laborious. During one study on hereditary fetal hemoglobin, which is associated with sickle-cell anemia, it took Collins eighteen months to discover the single-letter alteration in the DNA code that causes fetal hemoglobin production. During the same period, geneticists were talking about sequencing the entire human genome, which was thought to consist of three billion base pairs of chromosomes. Given the slow pace at which geneticists were able to discover even one sequence, it seemed unlikely that the entire human genome would be mapped within Collins's lifetime.
>
> Several years later, Collins became interested in the genetic basis of cystic fibrosis (CF). The challenge of finding one genetic alteration among three billion chromosome pairs was daunting. In 1985, Collins and his team whittled down the possible location of the CF gene to within two million base pairs on chromosome 7. To further isolate the gene, Collins and his colleagues invented a procedure called chromosome jumping. Chromosome jumping was used to skip over regions of DNA that are difficult to clone and move along the chromosome more quickly than the traditional method of chromosome walking. The technique enables geneticists to conduct gene searches in multiple areas of the chromosome, thereby speeding up the process. In 1984 Collins and his mentor Sherman Weissman jointly published a paper on the topic titled "Directional cloning of DNA fragments at a large distance from an initial probe: a circularization method."
>
> Though intrigued, the scientific community was skeptical that the technique would have any practical impact on alleviating human disease. Financial constraints and research roadblocks frustrated Collins and his team. In 1989, they joined Lap-Chee Tsui at the Hospital for Sick Children in Toronto to pursue CF research using chromosome jumping. In May of that year, they discovered that cystic fibrosis was caused by three deleted letters of the DNA code on a previously unknown gene. Collins's discovery, pinpointing the location of the CF on a specific chromosome, had huge implications for treating and possibly finding a cure for cystic fibrosis, as well as other conditions.

Genome Project had been in existence for three years. It was regarded as one of the most important scientific projects ever undertaken. Because he was a physician specializing in genetics, Collins was seen as the perfect choice to head the institute. Initially he refused the appointment, but the challenge of sequencing the human genome proved irresistible, and he accepted the directorship.

Not only was Collins attracted to cutting-edge genetic research, he was also committed to the principle that scientists should have free access to genetic information. This conviction was not shared by Craig Venter at Celera Genomics, a private enterprise also dedicated to sequencing human DNA. Venter was determined to patent genes, which would give his company a monopoly on the information. Celera also planned to charge subscription fees for access to the data. Eventually, Venter bowed to Collins's advocacy for open access to the sequencing information.

The public availability of the data has transformed scientists' understanding of how disease is caused. Discoveries made during Collins's tenure as director have the potential to revolutionize medicine. These discoveries include the identifications of genes associated with Parkinson's disease and prostate cancer, alterations of BRCA 1 and BRCA 2 (which are associated with breast, ovarian, and prostate cancers), the identification of genetic factors influencing late-onset Alzheimer's disease, and others. In combination with traditional diagnostic tools such as personal medical history, family history, magnetic imaging, and laboratory tests, information on specific hereditary risk factors could be used to tailor personal treatments and drug therapies.

In 2014, Johns Hopkins University named an award for Collins. The Francis S. Collins Scholar Award is given to researchers studying neurofibromatosis.

—*Pegge Bochynski*

Further Reading
Boyer, Peter, J. "The Covenant." *New Yorker* 6 Sept. 2010: 60–67. Print.
Collins, Francis S. "Why I'm a Man of Science—and Faith." Interview. *National Geographic*, 19 Mar. 2015, news.nationalgeographic.com/2015/03/150319-three-questions-francis-collins-nih-science. Accessed 24 Mar. 2017.
Collins, Francis S. "Q and A with NIH Director Dr Francis S Collins: The Value of Global Health

Research and Training." Interview by Roger Glass. *Global Health Matters*, vol. 15, no. 6, 2016, p. 5. Fogarty International Center, www.fic.nih.gov/News/GlobalHealthMatters/november-december-2016/Documents/fogarty-nih-global-health-matters-newsletter-november-december-2016.pdf.

Collins, Francis S. *The Language of God: A Scientist Presents Evidence for Belief.* New York: Free, 2006.

Collins, Francis S. *The Language of Life: DNA and the Revolution in Personalized Medicine.* New York: Harper, 2010. P

"Francis S. Collins, MD, PhD." *National Human Genome Research Institute*, 25 Sept. 2015, www.genome.gov/10001018/former-nhgri-director-francis-collins-biography. Accessed 24 Mar. 2017.

Giberson, Karl W., and Francis S. Collins. *The Language of Science and Faith: Straight Answers to Genuine Questions.* Downers Grove: InterVarsity, 2011. Print.

CARL F. CORI

Carl Ferdinand Cori, By Nobel Foundation. [Public domain], via Wikimedia Commons

CZECH AMERICAN BIOCHEMIST

- **BORN:** December 5, 1896; Prague, Austria-Hungary (now Czech Republic)
- **DIED:** October 20, 1984; Cambridge, Massachusetts
- **PRIMARY FIELD:** Biology
- **SPECIALTIES:** Biochemistry; physiology

Carl Cori and his wife, Gerty Cori, were Czech American biochemists who discovered the eponymous Cori cycle, the mechanism by which lactic acid produced during metabolism is converted to glucose in the liver.

EARLY LIFE

Carl Ferdinand Cori was born in Prague, then part of the Austro-Hungarian Empire, on December 5, 1896. He was the second of three children born to Dr. Carl Isidor Cori, who had earned both a medical degree and a doctorate in zoology, and Maria Lippich Cori, daughter of Ferdinand Lippich, a professor of mathematical physics and the developer of the polarimeter. When Cori was two years old, the family moved to Trieste, Italy, where his father had been appointed director of the Marine Biological Station. In addition to the scientific opportunities in marine biology and oceanography provided by his father, the young Cori was exposed to the geology and history of the region, resulting in a lifelong interest in archeology. Living in Trieste, Cori was also exposed to a wide variety of ethnic and religious communities. He became fluent in Italian, one of eight languages he would learn to speak over the course of a lifetime.

From 1906 to 1914, Cori was enrolled in the Trieste Gymnasium (secondary school), where he was educated primarily in the classics. Graduating in 1914, Cori entered the German University of Prague with the intent to study medicine. Among his fellow students was Gerty Radnitz, with whom he shared both a love of science and of the outdoors. They soon began collaborating on their research.

In 1916, with Austria-Hungary engaged in World War I, Cori was drafted into the army, where he first served with the ski troops before being assigned to a bacteriological laboratory with the sanitary corps.

After surviving bouts of typhoid fever and several incidents in dealing with military bureaucracy, Cori was assigned to a hospital on the Piave River near the front lines in Italy. Fluent in Italian, Cori also dealt with illness and health concerns among the local civilian population.

LIFE'S WORK

At the end of the war in 1918, Cori returned to Prague. In 1920, he finished his medical degree and moved to Vienna, Austria, for his postdoctoral studies. In August of that year, he married Gerty Radnitz. They would have one son, Carl Thomas, born in 1936.

While in Vienna, Cori carried out research at the Pharmacological Institute at the University of Vienna while continuing his clinical work. Meanwhile, Gerty Cori carried out her own postdoctoral studies at Vienna's Carolinen Children's Hospital. The near-starvation conditions in Vienna—the result of the outcome of the war and the refusal of physicians to accept the dietary supplements provided by the American Relief Administration—resulted in her developing xerophthalmia, a vitamin A deficiency. She was able to overcome the condition only by returning to Prague.

Cori remained in Vienna to continue his research, a rarity given the economic crisis at the time. He was able to continue thanks to his father, who sent him a shipment of frogs to study. Moving both the frogs and his research to the University of Graz in Austria and working with German pharmacologist Otto Loewi, Cori was able to study the mechanism of seasonal variation of vagus nerve action on the heart. His data came to the attention of another German pharmacologist, Hans Horst Meyer, a retired member of the institute. Meyer recommended Cori to Dr. Harvey Gaylord, director of the State Institute for the Study of Malignant Diseases (now the Roswell Park Memorial Institute) in Buffalo, New York. Gaylord had been trying to hire a biochemist for the institute and Cori accepted his offer, arriving early in 1922.

Cori's wife Gerty joined him shortly thereafter, where she was appointed to the pathology department at the institute. They would remain there for nine years, becoming naturalized citizens of the United States in 1928 and continuing the collaborative research that marked their professional careers—despite the resistance they encountered at having a woman on an equal level with men in the laboratory. While the Coris performed some research on the origins and causes of cancer, it was here that they began their lifelong work on carbohydrate metabolism. They published some eighty papers during their time in Buffalo.

The Coris' interest in carbohydrate chemistry stemmed from several sources. Gerty's father suffered from diabetes, a fatal affliction at the time. It was said he had asked Gerty to investigate the cause of abnormal glucose levels, which characterized the disease. In 1923, German physiologist Otto Warburg had observed abnormal levels of lactic acid built up in tumor cells, which coincided with Coris' study of cancer. The Coris wanted to investigate how glucose metabolism was regulated and, in particular, how the hormones insulin and epinephrine functioned in regulation. They observed that insulin increased the conversion of glucose to glycogen in muscle, but decreased conversion in the liver. Epinephrine produced the opposite effect. Their results also suggested an intermediate to the pathway would be in the blood; lactic acid was found to be that intermediate. The interconversion of glucose and glycogen became known as the lactic acid cycle, or Cori cycle. The Coris' subsequent 1929 publication of an extensive review of carbohydrate metabolism placed them in the forefront of the field.

In 1931, the Coris moved to St. Louis, Missouri, where Carl subsequently became chair of the Department of Pharmacology at the Washington University School of Medicine. In addition to his research, Cori was also expected to deliver thirty to forty lectures to the medical students. The Coris' carbohydrate research in Buffalo had produced an unusual finding: the conversion of glycogen to lactic acid is also accompanied by a loss of inorganic phosphate in the tissue. Continuing this area of research in St. Louis, the Coris determined that a missing intermediate was produced: a hexose (six-carbon) sugar phosphate (glucose-1-phosphate). The sugar phosphate became known as the Cori ester, and the enzyme that catalyzed its formation was termed a phosphorylase. In 1947, the Coris were awarded the Nobel Prize in Physiology or Medicine for their discovery of the Cori cycle.

Following World War II, Cori resigned as chair of the Department of Pharmacology to accept a similar position in the Department of Biochemistry, where he remained for twenty-one years. His research during the remainder of his academic career

CARL F. CORI: THE METABOLIC RELATIONSHIP BETWEEN GLUCOSE AND GLYCOGEN

During their research into carbohydrate chemistry in Buffalo, New York, Carl and Gerty Cori developed a procedure for measuring the level of glucose phosphate in muscle tissues. Exposure of the tissue to the hormone epinephrine produced a significant increase in the concentration of the carbohydrate. Concurrently there was a decrease in the level of free phosphate in that tissue. The question that was to be addressed was whether there was a relationship between the two events. Adding to the observation was the discovery by a colleague of Carl Cori's, Polish biochemist Jakub Parnas, that the reaction observed by the Coris was accompanied by a decrease in glycogen concentration in tissue. The entire process was termed phosphorolysis, reflecting the reduction in phosphate levels. The Coris and Parnas decided to establish a collaborative relationship to investigate the discovery.

The nature of the results observed by the Coris suggested several metabolic steps were taking place during the reactions. First, they found that the rate of the reactions significantly increased if adenosine monophosphate (AMP) was added to the reaction mixture. The significance was initially unclear. The calculations addressing the changes in concentrations of the reactants and products of the reactions also produced discrepancies, which could only be explained if they assumed an unknown intermediate was being produced.

Using a variety of purification steps, the Coris were able to crystallize the intermediate and determined it was an unusual form of modified glucose: glucose-1-phosphate. The molecule was eventually synthesized in the laboratory and was named the Cori ester.

The isolation of the intermediate was critical in explaining the results observed in previous experiments, and it provided an interpretation for the conversion of glycogen into glucose. When glycogen is broken down into its constituent glucose, free phosphate is added to glucose by an enzyme the Coris named phosphorylase. This explained the reduction in levels of phosphate that the Coris had observed in the muscle tissue. The phosphorylase activity in turn is boosted in the presence of the AMP, which explained the earlier results as well. The reactions that the Coris had discovered were critical in understanding the interconversion of glucose and glycogen in the liver and muscles of the body.

---

focused on the enzymes of carbohydrate metabolism. Most of this work dealt with enzymes that function in the oxidation of sugars, such as glucose, in what is known as the glycolytic pathway, as well as enzymes involved in the metabolism of glycogen. Gerty Cori was particularly interested in what were known as glycogen-storage diseases, which she continued to work on in collaboration with her husband until her death in 1957. In 1960, Cori married Anne Fitzgerald-Jones.

In 1966, Cori retired from Washington University and moved to Harvard University in Cambridge, Massachusetts, where he was appointed Visiting Professor of Biological Chemistry at the School of Medicine. He continued his research at Massachusetts General Hospital during his last years, remaining active until his death on October 20, 1984.

Impact

The Coris' work opened a new area of medical research, that of carbohydrate chemistry, focusing on the interconversion between glucose and glycogen. The Coris also discovered the cellular functions of the enzyme and its analogous counterparts. Prior to their research, the prevailing hypothesis was that large biological polymers could only be synthesized within a living cell. Using three purified enzymes, the Coris demonstrated that it was possible to synthesize large molecules, such as glycogen, in a test tube.

The Coris also found that a second molecule, adenosine monophosphate (AMP), is required to activate one form of the phosphorylase. This was the first example of an activator that functions in what became known as an allosteric mechanism. After identifying the first allosteric control, the Coris were able to identify other enzymes that metabolized carbohydrates.

The publications that grew out of the Coris' collaborative research not only characterized the primary enzymes of carbohydrate metabolism but also provided the background for understanding the structure and synthesis of carbohydrates in the body. The Coris' legacy was beyond that of their own research. The biochemical techniques they developed for purification and analysis of enzymes were adapted by their colleagues carrying out analogous research in metabolic processes.

—*Richard Adler*

Further Reading

Cori, Carl F. "The Call of Science." *Annual Review of Biochemistry* 38 (1969): 1–21. Print. Autobiography of the author. Cori describes his life in science and the research that led to receiving the 1947 Nobel Prize.

McGrayne, Sharon Bertsch. *Nobel Prize Women in Science: Their Lives, Struggles, and Momentous Discoveries*. Secaucus, NJ: Carol, 1993. Print. Includes a biography of Gerty Cori and her professional and personal collaboration with her husband, Carl Cori. Emphasizes the Coris' contributions in biochemistry.

Salway, J. G. *Medical Biochemistry at a Glance*. 3rd ed. Malden: Wiley-Blackwell, 2012. Print. Covers the relevance of biochemistry to clinical medicine. Summarizes metabolic pathways, including the Cori cycle and its regulation. Illustrations, index.

Worek, Michael, ed. *Nobel: A Century of Prize Winners*. Ontario: Firefly, 2008. Print. Profiles two hundred Nobel Prize winners, including Carl and Gerty Cori. Includes illustrations to explain scientific concepts and photographs of the laureates.

Erasistratus

Greek physician

- **Born:** c. 304 BCE; Chios, Greece
- **Died:** c. 250 BCE; Mycale, Ionia, Asia Minor
- **Primary field:** Biology
- **Specialties:** Anatomy; physiology

The third-century BCE physician Erasistratus helped establish a scientific basis for studies of human anatomy and physiology. Together with the anatomist Herophilus, Erasistratus carried out the first recorded systematic dissections of the human body, which yielded information that undermined many misconceptions of the body.

Early Life

Erasistratus was born at Iulis on the Greek island of Chios around 304 BCE. His father, uncle, and brother were all physicians. He began his study of medicine first in Athens and then at the university on the island of Cos, where Hippocrates, widely regarded as the father of Western medicine, had founded his prestigious medical school.

For a part of his career, Erasistratus served as a physician at the royal court of Seleucus I Nicator, a former Macedonian general under Alexander the Great who ruled Syria after Alexander's death. In his later life, Erasistratus stopped practicing medicine to concentrate exclusively on research. He carried out most of this work in the Egyptian city of Alexandria, then part of the Greek empire.

Alexandria thrived during Erasistratus's lifetime as a hub of intellectual, cultural, and scientific progress. Indeed, it was the scientific center of the ancient world, where scholars were fusing the best knowledge and methods generated by Greek, Egyptian, Macedonian, Babylonian, and other ancient civilizations. The kings of the Ptolemaic dynasty, who succeeded Alexander the Great as rulers of Egypt, were ambitious and generous patrons of the arts and sciences. When Ptolemy Soter ascended the throne, he commissioned the establishment of a library and museum system intended to outshine its counterpart at Athens. Erasistratus undertook his scientific investigations aided by access to this library, which housed more than half a million of the most important manuscripts in Western culture.

Anatomy, in particular, benefited from the fusion of learning in Alexandria. The death rituals of embalming and mummification had created an Egyptian expertise on the internal structures of both animal and human bodies. Erasistratus was able to combine this traditional knowledge, available from the library at Alexandria, with the more recent conceptions of Greek physicians such as Hippocrates. Egyptian embalming practices also created a favorable climate for experiential, rather than speculative, research methods; the Ptolemies permitted the public dissection of human cadavers. A number of ancient writers claimed that Erasistratus also carried out vivisections of inmates taken from the royal prisons. Scholars have not been able to confirm this, however. It is clear, though, that to understand better the physical changes death triggers, Erasistratus frequently conducted autopsies on the bodies of people immediately after they had died.

Life's Work

Erasistratus used the knowledge he gained from the vivisection and dissection of animals to make inferences about human anatomy. By comparing rabbit, stag, and human brains, for example, Erasistratus deduced a correspondence between the quantity and complexity of cerebral convolutions—the "coils" that make up the surface of the brain—and the level of intelligence; the greater the surface area and volume of an organism's brain, he discovered, the greater that organism's intellectual capacity will be. His extensive dissections also enabled him to map the brain and nervous system more accurately than any of his predecessors. He correctly distinguished between the cerebrum and cerebellum and described the ventricles within the brain and the membranes that cover it. Most significantly, he broke ranks with those ancient philosophers, including Aristotle, who believed the heart to be the seat of human intelligence; he agreed with Plato that this function belonged to the brain alone. Moreover, Erasistratus understood the brain as the starting point for all nerves and recognized the differences between sensory and motor nerves.

Despite the accuracy of his descriptions of human anatomy, Erasistratus's investigations into human physiology yielded mixed results. In the case of some functions, such as digestion, Erasistratus was remarkably ahead of his time in proposing a mechanical, or cause-and-effect, explanation. In other cases, such as circulation and cardiovascular function, his theories were less accurate.

Erasistratus based his understanding of all the workings of the human body on a theory of vapors he called *pneuma*, which according to Erasistratus, existed in the air outside of the body, were inhaled into the lungs, and were then carried through the pulmonary vein to the heart. Once inside the body, these pneuma made possible three basic physiological functions.

The heart, Erasistratus believed, transformed these pneuma into "vital spirits" dispersed via blood vessels to the brain and the rest of the body. The brain converted the pneuma it received into "animal spirits" that, when conducted by the nerves, made possible muscle movements and sensory perceptions. The liver manufactured blood, which was then dispersed to various organs that consumed it as a form of fuel.

While the reasoning behind pneumatic theory led Erasistratus to make fundamental errors, his theories nonetheless reveal a crude grasp of the concepts of oxygen exchange and blood circulation. This pattern of understanding physiological processes as a whole while getting many of the mechanical details wrong repeated itself in Erasistratus's studies of other bodily systems.

Erasistratus was far off the mark, for example, in many of his beliefs about cardiovascular functions. He was convinced that blood was made in the liver and distributed throughout the body only by the veins. Because the arteries in the bodies on which he performed postmortem examinations were typically empty of blood, Erasistratus concluded that they must routinely be full of air and mistakenly thought that the arteries collected pneuma from the lungs and distributed it throughout the body. This common mistaken belief is immortalized in the word "artery" itself, coming from the Greek *aer* (air) and *aeirein* (to lift or raise).

At the same time, some of his views on the circulatory system were remarkably modern. Many ancient thinkers conceived of the heart as the seat of human emotions, but Erasistratus was among the earliest to realize that the heart functions as a pump. He also described in detail the structure and role of the bicuspid and tricuspid valves of the heart, which open and close to permit the flow of blood in one direction only. He also theorized the existence of capillaries many centuries before the late Renaissance physician William Harvey famously did.

Although Erasistratus is famous for his contributions to anatomy and physiology, he is equally well known for what is perhaps the first psychiatric diagnosis and cure in recorded Western medicine Plutarch, writing after Rome had reduced Greece to a conquered territory, recorded an incident that occurred while Erasistratus served as royal physician under Seleucus I Nicator. The king begged Erasistratus to save the life of his son Antiochus, who was wasting away from a mysterious ailment. Erasistratus, noting the young man's symptoms, quickly deduced that he was suffering from unrequited love. Carefully observing the prince's reaction whenever a beautiful woman of the court came to visit him, Erasistratus soon realized that the presence of only one woman, Antiochus's stepmother Stratonice, caused the lovesick prince's face to flush, voice to crack, and pulse to race. Erasistratus, Plutarch recounts, informed the king that there was no hope

> **ERASISTRATUS: ERASISTRATUS AND THE MECHANICS OF METABOLISM**
>
> Erasistratus was born around 304 BCE at Iulis on the Greek island of Chios. Although none of his original work has survived, he was clearly very influential in his time and is quoted and referenced extensively by later authors, especially the second-century CE physician and philosopher known as Galen. Erasistratus's main contributions were in the study of medicine, specifically in describing the workings of the body. Dissection was an important factor in his observations. Some later authors accused him of performing vivisection (dissections on living people), but it is not known whether this is true.
>
> Among Erasistratus's crucial observations was that digestion was a mechanical process. He understood correctly that the closing of the larynx during swallowing prevented liquids from invading the lungs and accurately described the contractions of the gastric muscles that help churn food into bits in the stomach. Although his details were not perfect, he contradicted the prevailing idea that food was processed by the heat of the body. The stomach was widely thought to be like an oven, cooking the food and thereby getting energy from it. Erasistratus claimed that food was ground up by contractions in the stomach. In his famous experiment on birds, he found that a bird that was confined and starved would lose weight after a few days. The weight loss could be accounted for by the amount of waste the bird produced in the same amount of time, directly linking food, digestion, and waste.
>
> Erasistratus was one of the first to roughly intuit the process we now know as metabolism, although he could not understand its chemical nature. He believed that the organs of the body received nourishment from the bloodstream through a process of absorption. However, he enjoyed a lesser degree of accuracy when he attempted a teleological approach. In contrast to explanations that rely on mechanical principles of causation, a teleological perspective tries to explain events in terms of ends or purposes; thus Erasistratus jumped to the mistaken conclusion that some of the broken-down food in the stomach passed via the blood vessels to the liver, which somehow transformed it into blood.
>
> Despite some inaccuracies in his assumptions, Erasistratus had a solid grasp on the overall systemic workings of the body and laid the groundwork for later anatomists and physiologists.

of saving Antiochus's life, that a cure was impossible because Antiochus was in love with Erasistratus's own wife. When the king begged Erasistratus to give up his wife to save Antiochus's life, proclaiming that he would do the same if he could, Erasistratus finally revealed the true object of Antiochus's passion. Seleucus divorced Stratonice and permitted his son to marry her, naming the couple king and queen.

Erasistratus died at Mycale in Asia Minor around 250 BCE. The circumstances of his death remain unknown. One legend claims that, out of despair over an incurable foot ulcer, Erasistratus committed suicide by drinking hemlock.

IMPACT

Through his work on the human body, Erasistratus helped lay the foundation for later studies in anatomy and physiology. Although his own writings are no longer extant in modern times, Erasistratus's observations were noted and commented upon by other ancient scholars whose work has survived. Among these was the third-century CE physician-philosopher Galen, who, by extensively critiquing Erasistratus's conclusions, preserved a record of his predecessor's work for ages to come. Flawed though Erasistratus's theories were, they represented an early systematic approach to studying the human body, contributed to Greek knowledge about medicine, and influenced physicians and philosophers in subsequent generations.

—*Beverly Ballaro*

FURTHER READING

Galen. *Galen on Bloodletting*. Trans. Peter Brain. New York: Cambridge University Press, 2009. Print. Includes translations of Galen's works against Erasistratus and the Erasistrateans, with extensive annotations.

———. *Galen on Diseases and Symptoms*. Ed. Ian Johnson. Cambridge University Press, 2006. Print. Discusses Erasistratus's concept of disease transmission along with Galen's opposing perspective.

Jackson, Michael, and Amy Norrington. "An A to Z of Medical History: Part 1." *Student British Medical Journal* 10 (Sept. 2002): 317. Print. An exploration of the history of medicine covering Erasistratus and his dissection techniques.

Lutz, Peter L. "The Alexandrian Period." *The Rise of Experimental Biology: An Illustrated History.* Totowa: Humana, 2002. Print. Discusses Erasistratus's thought and achievements within the context of Alexandrian science of the time.

Alexander Fleming

Scottish biologist

- **Born:** August 6, 1881; Lochfield, Scotland
- **Died:** March 11, 1955; London, England
- **Primary field:** Biology
- **Specialty:** Bacteriology

Scottish biologist Sir Alexander Fleming is best known for his role in the discovery of penicillin, the world's first cheap, effective antibiotic. Although he insisted that his discovery was merely the result of luck and a keen eye, he shared the 1945 Nobel Prize in Physiology or Medicine for his achievement.

Early Life

Born in rural Lochfield, Scotland, on August 6, 1881, Alexander Fleming was the third child of sheep farmer Hugh Fleming and his second wife, Grace Sterling Morton. The large family also included four children from Hugh Fleming's first marriage. The young Fleming (known to his family and friends as Alec) divided his time between working on his family's farm and roaming the countryside with his siblings. Fleming was only seven years old when his father died, leaving Fleming and his siblings to help his mother tend the farm.

While in Scotland, Fleming went to local country schools and the Kilmarnock Academy. He moved to London when he was fourteen to live with his stepbrother Tom, a doctor. For two years, Fleming attended the Polytechnic Institute in Regent Street before finding work as a clerk at a shipping company. When the Boer War broke out between the United Kingdom and the Boers of South Africa, Fleming left this job and, in 1900, joined a Scottish regiment with two of his brothers. Shortly after the war, Fleming's uncle died and bequeathed Fleming two hundred pounds. His stepbrother encouraged Alec to use the money to go into medicine.

Fleming enrolled at London's St. Mary's Hospital Medical School in 1901, beginning a lifetime relationship with the institution. His early medical career was influenced by Sir Almroth Wright, a professor at St. Mary's. Wright had been researching vaccines, which were still controversial in the early twentieth century, and his ideas intrigued Fleming.

When Fleming graduated in 1906, he planned to become a surgeon, which would have meant leaving St. Mary's. However, the captain of the school's rifle club didn't want lose one of his best shooters, so he convinced Fleming to change his focus to bacteriology, allowing him to stay at St. Mary's and work with Wright.

Life's Work

Fleming was an unusual doctor. A skilled glassblower, he was known for making his own test tubes. He enjoyed the research that many other doctors saw as a necessary evil. He was also one of the only doctors in London to use the controversial syphilis treatment known as Salvarsan, which had been developed by German chemist-physicist Paul Ehrlich and his team. Salvarsan was also known as 606 because it was the six hundred and sixth arsenical compound that Ehrlich's team devised. Fleming eventually became known as "Private 606" because he administered the treatment so often.

When World War I broke out and devolved into a stalemate, the trenches became notorious for their unsanitary conditions. In fact, soldiers were more likely to die from infection than from enemy fire. Fleming and Wright, who both served as medical officers during the war, sought to solve this problem. By studying blood and tissue samples from wounded or dead soldiers, Fleming found that most of the soldiers had been exposed to the *Clostridium welchii* bacterium, which causes gas gangrene. The bacterium thrived in manure deposited by soldiers' horses on the battlefields. In 1915, Fleming married Sarah Marion McElroy. McElroy was an Irish nurse Fleming had met in London years earlier. The couple's son, Robert, was born in 1924.

After the war ended, in 1918, Fleming returned to St. Mary's to continue his research. He was determined to find a way to kill bacteria, and thus prevent death from minor infections like those he encountered in the war. Fleming was convinced that there were bacteria-killing properties present in human tears, saliva, and mucus, and he spent most of the 1920s looking for them. Fleming saw this theory proved partially true in 1921, when some of his own mucus dripped onto a culture of bacteria, dissolving it. Through his research in this area, he discovered the antibacterial enzyme lysozyme, which he named in 1922 in his paper "On a Remarkable Bacteriolytic Element Found in Tissues and Secretions." Fleming's attempts to synthesize lysozyme were unsuccessful, but his discovery led to a better understanding of the inner workings of the human body and of the purpose of mucus.

Though he had abandoned the idea of making a powerful antibiotic from bodily fluids, Fleming did not give up trying to find ways to prevent the spread of disease and infection. He began studying a type of bacteria called *Staphylococcus*, which causes the respiratory disease pneumonia, as well as septicemia, or blood poisoning, and numerous other types of infection. Despite being promoted to the position of chair of bacteriology at St. Mary's in 1926, Fleming made little headway in his research until 1928. That year, he accidentally discovered that mold growing in a petri dish had killed part of a *Staphylococcus* culture. After studying the properties of the mold, which he identified as a strain of *Penicillium*, Fleming was able to distill a small amount of it into a liquid, which he called "penicillin." Fleming suspected that this powerful

---

FLEMING DISCOVERS PENICILLIN

In 1928, Alexander Fleming was conducting experiments on strains of Staphylococcus bacteria. In September of that year, Fleming returned to his London laboratory after a vacation and noticed that mold spores had contaminated a petri dish containing a Staphylococcus culture, and that the mold had killed the bacteria colonies growing around it. Fleming identified the mold as belonging to the genus Penicillium (later confirmed as Penicillium notatum), and subsequently found that other strains of Penicillium produced a similar bacteria-killing (or "antibiotic") substance. The substance, which Fleming dubbed "penicillin," after the name of the mold, was effective in destroying not just Staphylococci but other kinds of pathogenic (disease-causing) bacteria as well. It was also nontoxic and did not interfere with the action of white blood cells.

In 1929, Fleming presented his findings in a paper, "On the Antibacterial Action of Cultures of a Penicillium," published in the British Journal of Experimental Pathology. He continued experimenting with penicillin, hoping to prove that it could be used to treat bacterial disease without harming patients. However, while he was able to make a penicillin solution that that could control infections in superficial wounds, he was not able to isolate enough of it to successfully conduct the key experiment of injecting penicillin into mice infected with bacteria and halting the infection.

A few years later, pharmacologist Howard Florey, biochemist Ernst Chain, and their team at Oxford University began their own experiments with penicillin. By 1940 they were able to isolate and concentrate a penicillin solution that was one thousand times as strong as Fleming's. They were then able to conduct successful experiments in mice, which led to clinical trials that proved that penicillin could be used effectively and safely in human patients to treat systemic bacterial infections. The drug was able to be mass-produced in time to be of enormous service during World War II, earning Fleming, Florey, and Chain the 1945 Nobel Prize in Physiology or Medicine.

Although Fleming had concluded most of his own work on penicillin by 1932, he continued to think about its impact as a drug. Penicillin works by inhibiting the growth of bacterial cell walls, and Fleming realized that if it were used in insufficiently large doses, penicillin-resistant bacteria would develop. He knew this because he had created mutant bacteria in his own lab, simulating the type of incomplete treatment he was concerned about. The surviving bacteria strains in his lab grew stronger and built thicker cell walls that eventually were able to withstand the effects of penicillin. This ill effect would be more pronounced, and more immediate, he said in a 1945 interview with the New York Times, if people began taking the drug orally instead of intravenously, because they would be able to medicate themselves without a doctor's supervision. His concerns were subsequently borne out, and antibiotic resistance is a growing problem in the twenty-first century.

Nonetheless, Fleming's isolation of penicillin was a major landmark in the history of modern medicine, ushering in the era of antibiotics and bringing under control a huge range of potentially fatal infections.

antibiotic was capable of killing many of the deadliest bacteria and could, if properly administered, save thousands of lives.

Fleming was unable to produce penicillin in large enough doses to make it practical for medical use, but Oxford scientists Howard Florey, Ernst Chain, and Norman Heatley, who had been trying to unlock the secrets of lysozyme, took notice of Fleming's discovery and obtained samples of his *Penicillium* specimen. The three scientists began working together in the late 1930s to find the practical applications for penicillin, eventually setting up a factory at Oxford and helping to spur mass production of the drug in the United States. In the early 1940s, they were able to mass produce the new and powerful antibiotic for use during World War II, and the world suddenly became very interested in penicillin. By the end of the war, it was credited with saving the lives of millions of people who otherwise would have died from bacterial infections. For their work, Florey, Chain, and Fleming shared a Nobel Prize in Physiology or Medicine in 1945. Fleming went on to receive worldwide fame and honors for his discovery of penicillin.

In 1932, Fleming switched his focus from penicillin to other projects at St. Mary's, including the study of sulfonamides as antibacterial agents and oversight of the hospital's vaccine production program. In 1946 he became the principal of St. Mary's inoculation department, which eventually became known as the Wright-Fleming Institute of Microbiology. After retiring from the institute in 1948, he was an emeritus professor until 1954. Fleming's research was interrupted in 1949 when his wife died. He worked alone for several years before remarrying in 1953, this time to a young doctor named Amalia Koutsouri-Voureka. Fleming died of a heart attack about two years later, on March 11, 1955.

IMPACT

Fleming was celebrated around the world for his role in the discovery of penicillin, which became known as a wonder drug for its ability to fight infectious disease. The discovery of penicillin's antibacterial properties helped inaugurate the development of modern antibiotics—along with the pharmaceutical industry that mass produced these and other drugs. In the mid-twentieth century, synthetic penicillin would be used to help treat diseases such as gangrene, pneumonia, syphilis, and tuberculosis.

Later assessments of Fleming's work would challenge what Fleming himself called the "Fleming myth," that penicillin was the result of one scientist's lucky discovery rather than teams of scientists working systematically, and would place more emphasis on the work done by Florey, Chain, and others at Oxford. However, Fleming's work on penicillin is recognized as enabling the Oxford team's investigation into penicillin to be as efficient and focused as it was.

——*Alex K. Rich*

FURTHER READING

Brown, Kevin. *Penicillin Man: Alexander Fleming and the Antibiotic Revolution*. 2004. Stroud: Sutton, 2005. Print. Tells the story of Fleming and his discovery, as well as the history of the antibiotic revolution.

Bud, Robert. *Penicillin: Triumph and Tragedy*. New York: Oxford UP, 2007. Print. Describes the discovery and development of penicillin, considered a miracle drug until it proved nonresistant to "superbugs."

Lax, Eric. *The Mold in Dr. Florey's Coat: The Story of the Penicillin Miracle*. New York: Holt, 2004. Print. Argues that scientists who continued Fleming's research should be credited with the discovery of penicillin.

CAMILLO GOLGI

ITALIAN PHYSICIAN

- **BORN:** July 7, 1843; Corteno, Italy
- **DIED:** January 21, 1926; Pavia, Italy
- **PRIMARY FIELD:** Physiology
- **SPECIALTIES:** Microbiology; histology

Nineteenth-century physician and histologist Camillo Golgi was the first Italian scientist to be awarded the Nobel Prize. He is best known for his work on the human nervous system, including the discovery of a tendon sensory organ called the Golgi receptor. He is also known for formulating a method of staining nerve cells and tissues in order to observe their behavior.

Camillo Golgi, Nobel Prize in Physiology or Medicine 1906. [Public domain], via Wikimedia Commons

Early Life

Camillo Golgi was born in Corteno, Lombardy, Italy on July 7, 1843. He studied medicine at the University of Pavia, intending to become a physician like his father. Golgi received his medical degree in 1865 but never practiced medicine, choosing instead to focus on medical research. His early interest was in psychiatry, and he became a student intern at the Institute of Psychiatry at Pavia under criminologist Cesare Lombroso. As an assistant to Lombroso, Golgi studied mental illness, working to incorporate the family history of patients with available clinical data. This work inspired Golgi to search for connections between mental illness and abnormalities in the nervous system.

Golgi's introduction to histology, the study of cells and tissue with a microscope, began under the tutelage of pathologist Giulio Bizzozero. (Golgi's wife, Lina Aletti, was Bizzozero's niece.) In the late nineteenth century, doctors did not understand that the brain is comprised of of individual neurons. Although scientists made use of the cell theory developed by German botanist Matthias Schleidon and German physiologist Theodor Schwann in 1839, the theory had not yet been applied to the nervous system. At the time, differentiating individual brain cells under a microscope was impossible.

Many types of human cells are roughly spherical in shape, but nerve cells, or neurons, are long and narrow. When scientists of Golgi's era attempted to section these tissues in preparation for viewing under a microscope, they destroyed these fragile structures. Scientists theorized that nervous tissue might actually be continuous cytoplasm (the gel-like substance inside cells), and not be composed of cells at all.

Life's Work

Golgi developed a staining technique that proved nervous system tissue is made up primarily of cells. This breakthrough occurred after he exposed nervous system tissue hardened by potassium bichromate to silver nitrate. Hardening the tissue protected the long, delicate neurons. Silver particles adhered to the neurilemma (the membrane around the neuron), staining it black. Under the microscope, the finest details stood out against a contrasting yellow background. Scientists were able to observe the structure of the neuron for the first time.

Golgi's black reaction was a crucial development in the field of neuroscience. His work helped scientists discover that each neuron has a soma, or cell body, with long, narrow extensions called axons and dendrites. These extensions, or processes, reach out to interact with other neurons. Electrical signals are transmitted cell-to-cell along networks of neurons, sending the messages necessary for thought and movement.

Densely populated neurons are difficult to see clearly, but when only a few of them are stained, pathways between interconnected neurons become clear. Golgi's selective staining technique was the key to understanding the organization of this complex tissue. For the first time, scientists realized that signals were being communicated between neurons in the brain and spinal cord. Golgi's black reaction stains certain neurons in brain tissue, but not others. Histologists still do not understand this phenomenon.

Golgi applied the black reaction to the study of brain structures including the olfactory lobes (the control center in the brain for the sense of smell), gray matter (nervous tissue of the brain and spinal cord), and the cerebellum (a portion of the brain

> **CAMILLO GOLGI: CAMILLO GOLGI DEVELOPS METHOD FOR STAINING NERVE CELLS**
>
> After graduating from the University of Pavia in Italy in 1865, Golgi began his clinical work in medicine. However, he became increasingly interested in studying brain functions, particularly the pathological basis of mental illnesses. He accepted a position in the psychiatric clinic at the hospital of Cesare Lombroso, an Italian criminologist and physician who believed criminality was an inherited trait. Golgi hoped to determine whether the physical nature of the brain differed between criminals, the mentally ill, and the average person.
>
> During this time, Golgi also participated in lectures and demonstrations provided by Italian physician Giulio Bizzozero with the Institute of General Pathology at the University of Turin. Bizzozero was a professor of pathology known for his study of blood cells. Bizzozero instructed Golgi in the use of the microscope for observation and study of cells and tissues, a technique then still in its infancy. Among the problems was that of fixing and staining tissues for improved observations. Hematoxylin, a dye prepared by the boiling of logwood, had been used in staining tissues since the eighteenth century, and when combined with eosin, a technique developed in the 1870s, had become the standard method of staining cells and tissues. However, H & E stain, as the combined technique was called, was not useful in staining nerve cells.
>
> In the early 1870s Golgi began experimenting with metal impregnation using silver nitrate as a means to observe nerve cells in a procedure he termed the "black reaction." The tissue was first fixed in potassium dichromate and then stained with the silver compound. The product, silver chromate, would only stain nerve cells and not cells in surrounding tissues.
>
> Among Golgi's observations was the revelation that each nerve cell consists of a long structure called an axon, with shorter branched components known as dendrites, which are associated with the cell body. The ability to stain nerve cells opened up an entirely new area of pathology: the study of the brain at the cellular level. Golgi continued his work studying the various portions of the brain, including the pathological lesions present in the brain of a patient with what is now termed Huntington's disease. He was unable, however, to observe any pathological differences between the brains of the criminally insane, obtained after death, and average brains.
>
> Golgi was appointed Professor of Histology at the University of Pavia in 1876 where he continued his work on neurological functions. He is given credit for the discovery of the Golgi organ, a nerve organ proprioreceptor, which in part monitors the movement of skeletal muscles, and the Golgi-Mazzoni corpuscles, which envelop nerve endings. In his later years, Golgi also centered his observations on the study of kidney histology.
>
> In the 1890s, he reported the presence of an internal membrane system within cells, which later became known as the Golgi apparatus, a series of internal membrane organelles that function in the modification and secretion of proteins.

that coordinates movement and balance). His published results of these studies, along with his anatomical drawings, earned him the Nobel Prize in Physiology or Medicine in 1906. The black reaction is still in use today.

By 1872, financial problems had forced Golgi from the University of Pavia. He became chief medical officer at a hospital in Abbiategrosso, Italy In 1876, Golgi returned to the University of Pavia, this time as a professor of histology. There he directed his own laboratory, taking on students and assistants to continue his research.

Around 1886, Golgi became interested in malaria. Malaria is an infection caused by protozoan (single-celled eukaryotic) parasites that results in repeated cycles of fever and chills. The disease is transmitted by the bite of infected *Anopheles* mosquitoes and is common in tropical climates where mosquitoes are plentiful. Mosquitoes become infected with malaria by biting a person or bird with malaria. They transmit the disease by going on to bite an uninfected person. When a person is infected, the parasites reproduce in the liver and then penetrate the red blood cells. They continue to reproduce inside the red blood cells, breaking out at predictable intervals to cause bouts of fever.

Golgi investigated the periodic nature of fevers in malaria patients, concluding that there were actually two forms of malaria: one that causes fever every other day, and one that causes fever every third day. Golgi discovered differences in the reproductive rates between the two forms of the disease. Whenever new parasites broke out of red blood cells and into the bloodstream, the patient suffered another bout of fever. Golgi was the first scientist to notice that the timing of fevers corresponded to the reproductive cycle of the parasite.

Golgi discovered the intracellular organelle known as the Golgi apparatus by accident while observing partially stained Perkinje cells (neurons from the cerebellar cortex). The Golgi apparatus is described as resembling a stack of pita bread, because it is comprised of stacked, flattened sacs of membrane.

Golgi named the organelle the "*apparato reticolare interno*" ("internal reticular apparatus"). This later became known as the Golgi apparatus, or, more simply, the Golgi. He published the discovery in 1898, but some scientists dismissed it as an artifact of the staining process. Artifacts of staining include thickened spots of stain or other microscopic irregularities that could be confused with real cellular structures. It was not until the invention of the electron microscope in the 1930s that the discovery was confirmed. Electron microscopes produce images with much better magnification and resolution than the light microscopes available to Golgi.

We now know that the Golgi apparatus functions in the cell like a post office does in a city, sorting and shipping proteins to their destinations. Newly synthesized proteins enter the apparatus, where polysaccharide (carbohydrate) chains bound to their surfaces are added to, or trimmed. These polysaccharide chains act as the address labels, showing the destination of each kind of protein. Just as a business letter has the name and address both on the envelope and the letter itself, matching polysaccharide chains are attached both to the protein and to the membranous vesicle that contains it.

When the proteins are ready, vesicles, or bubbles of membrane, pinch off the Golgi apparatus in much the same way as a bubble is blown from bubble gum. These vesicles travel to their destinations inside the cell, or to the plasma membrane of the cell to be used elsewhere in the body.

Golgi and his wife were never able to have children of their own. They adopted Golgi's niece, Carolina, and raised her as their own daughter. In recognition of their contributions to the study of the nervous system, Golgi and the Spanish histologist Santiago Ramón y Cajal shared the 1906 Nobel Prize in Physiology or Medicine. Golgi was sixty-three at the time. In his later years, he became an influential dean and faculty member at the University of Pavia. In 1900, he became a politician, where his focus remained on public health. Italy joined World War I in 1915. Despite his advanced age, Golgi headed up a military hospital in Pavia that treated wounded soldiers. Golgi retired in 1918 and died in Pavia, Italy, on January 21, 1926.

IMPACT

Golgi helped to improve our understanding of the human brain and nervous system. Researchers worldwide still use his method of nervous tissue staining. The neuron doctrine, which holds that individual cells make up the nervous system, was established as a direct result of his work. The Golgi Hall at the Museum of History at the University of Pavia houses an exhibit on Golgi's life and accomplishments. His Nobel diploma is on display there.

—*Courtney Farrell*

FURTHER READING

Alberts, Bruce, et al. *Essential Cell Biology*. New York: Garland Science, 2009. Print. Offers an introduction to the basic concepts of cell biology. Illustrated.

Mazzarello, Paolo. *Golgi: A Biography of the Founder of Modern Neuroscience*. Oxford: Oxford UP, 2009. Print. Presents a detailed review of Golgi's life and work, including his achievements in neuroscience, cell biology, and microbiology.

Mironov, Alexander A. *The Golgi Apparatus: State of the Art 110 Years after Camillo Golgi's Discovery*. New York: Springer, 2010. Print. Details the function of the Golgi apparatus and reviews scientific developments related to the fields of molecular biology and microscopy.

Carol W. Greider

American molecular biologist

- **Born:** April 15, 1961; San Diego, California
- **Also known as:** Carolyn Widney Greider
- **Primary field:** Biology
- **Specialties:** Molecular biology; biochemistry; cellular biology

Carol Greider was awarded the Nobel Prize in Physiology or Medicine in 2009 for her work in discovering the enzyme telomerase and identifying the role of telomeres in the maintenance of linear chromosomes.

Early Life

Carolyn Widney Greider was born April 15, 1961, in San Diego, California, the second child of Kenneth and Jean Foley Greider. Both parents were scientists, her father having earned a PhD in nuclear physics and her mother a PhD in botany. Her parents were completing postdoctoral fellowships in San Diego at the time of her birth; they moved to New Haven, Connecticut, in 1962. In 1965, Greider's father accepted a position in the Department of Physics at the University of California, Davis, and the family returned to California, where Greider spent her formative years.

In 1967, Greider's mother committed suicide, exacerbating the difficulties Greider was facing in school. Among her challenges was an inability to sound out words, resulting in her placement in a remedial spelling class. It was only later that the problem would be recognized as a form of dyslexia. In 1971, Greider's father took a sabbatical leave from Davis and moved to Heidelberg, Germany, where Greider and her brother, Mark, were enrolled in the Englisches Institut and where, by necessity, she learned German. Her dyslexia created problems at the school, particularly in her English classes, as she was unable to properly spell English words.

When she returned to California, Greider became increasingly interested in biology as she moved through both junior and senior high school, graduating from Davis Senior High School in 1979. Her decision to study science was particularly motivated by her twelfth-grade biology teacher. Greider's father, who supported her love of reading, reminded her that a strong academic background would make available many opportunities in the future. Following high-school graduation, Greider decided to enroll at the University of California, Santa Barbara (UCSB) in the College of Creative Studies.

Life's Work

A friend of Greider's mother's, Beatrice "Beazy" Sweeney, was on the UCSB faculty and encouraged Greider to enroll in laboratory courses. In addition to aiding Sweeney with her studies of the movement of chloroplasts in a dinoflagellate, Greider also had the opportunity to work with other researchers in their studies of microtubule proteins.

Greider spent her junior year at the University of Göttingen in Germany, working with Klaus Weber at the Max Planck Institute. The biology class in which she enrolled provided the opportunity to study polytene chromosomes in *Chironomus*, an organism related to the fruit fly. In addition to giving her the chance to publish, Greider believed the work introduced her to an "appreciation for the beauty of chromosomes." She graduated with a BS in biology in 1983.

Despite an outstanding undergraduate career, when it came time to apply for graduate admission, only two universities offered Greider interviews: the University of California, Berkeley and the California Institute of Technology. Once again, her dyslexia had proved a challenge when taking standardized tests, and her low scores precluded additional schools from accepting her applications. Greider met with biologist Elizabeth Blackburn at Berkeley during her interview and accepted an opportunity to work in Blackburn's laboratory.

Blackburn had been attempting to characterize the sequence of nucleotide bases found on the tips of linear chromosomes, a region known as the telomeres, in the protozoan *Tetrahymena*. Blackburn and geneticist Jack Szostak had found evidence that the telomere sequences consisted of a series of bases repeated in tandem. Whether there was a specific enzyme that synthesized these units was unclear. Blackburn was convinced that a heretofore unidentified enzyme was involved, and Greider decided to pursue the question as part of her doctoral work once she became part of Blackburn's group in May 1984.

CAROL W. GREIDER DISCOVERS TELOMERASE, THE ENZYME THAT SYNTHESIZES TELOMERES

By the spring of 1984, Elizabeth Blackburn had spent several years attempting to explain a mechanism by which telomeres could be incorporated and maintained on the ends of linear chromosomes. Several models existed, including one in which recombination between different strands of DNA might account for their elongation. None of the models were satisfactory, which left the hypothesis that telomeres were being newly synthesized and then added. Blackburn and Jack Szostak had proposed the existence of an enzyme that could accomplish this synthesis, but they had no firm data to support their idea. This was the situation when Carol Greider joined the laboratory, and it was her task to determine whether such an enzyme did exist and, if so, to isolate it.

The main challenge during the first months was to develop a test for the suspected enzyme. Initial results were inconclusive; fragments of DNA both with and without telomeres could be equally elongated with material extracted from Tetrahymena cells. Unknown at the time was that internal nucleases were creating single-stranded fragments, and what Greider was observing was just the repair of those fragments. In December, Greider decided upon a new approach. Since the sequence of Tetrahymena's telomeres was known by then, and consisted in part of a string of guanine residues (TTGGGG), Greider added an artificial telomere to the cell extract and observed the results. The extract did indeed elongate the artificial telomere, indicating the presence within it of the enzyme responsible.

The next six months were spent further characterizing the enzyme, dubbed telomerase, as well as confirming that the activity being observed was not the result of other cell enzymes. Other experiments pointed to a unique feature of the enzyme. The sequence of repeat bases from Tetrahymena telomeres consisted of the nucleotide sequence TTGGGG, while the equivalent se-quence of bases obtained from the yeast Saccharomyces consisted of the sequence TTGGG—three guanine residues instead of four. When Blackburn and Greider combined the telomeric sequence obtained from the yeast with the telomerase extract from Tetrahymena, they observed that instead of copying the three-guanine sequence (GGG), the enzyme from Tetrahymena initially added a fourth G residue before copying the sequence; in other words, the telomerase somehow recognized the absence of the fourth G. One possible explanation was that the telomerase enzyme includes an RNA template, which it copies to produce the repeated sequence of the telomere. Greider and Blackburn later confirmed that telomerase does in fact consist of both a protein portion, which carries out the actual catalysis, and an RNA template that it copies, thus making the enzyme a reverse transcriptase.

While Greider and Blackburn quickly found indirect evidence of enzymatic activity, part of the challenge lay in differentiating activity associated with a DNA repair protein from that of the suspected telomere-producing enzyme. On December 25, 1984, they finally obtained definitive results: the enzyme that synthesized telomeres was a newly recognized protein not previously described. The enzyme was initially referred to as telomere terminal transferase, shortly renamed simply as telomerase. The work was published the following December.

Greider received her PhD in molecular biology in 1987, after which she joined the Cold Spring Harbor Laboratory in Cold Spring Harbor, New York, on a postdoctoral fellowship. There, she continued with her research on telomeres and successfully cloned the gene that encodes telomerase. Following the successful application for a National Institutes of Health (NIH) grant, Greider was promoted to assistant investigator at the laboratory and was able to hire her first graduate student.

She also began collaborating with Calvin Harley from McMaster University in Ontario, Canada, who had been studying cellular senescence (cell aging) and its application to what was called a mitotic clock. Some twenty years earlier, cell biologist Leonard Hayflick had observed that cells could only replicate some fifty times before aging and death; cells from embryos could replicate dozens of times, while those obtained from an adult would barely replicate at all. The molecular basis for the clock was unknown. In contrast, transformed or cancer cells did not appear to be bound by such a clock, seeming to achieve immortality. Greider and Harley were able to demonstrate a relationship between telomere length and senescence: as chromosomes are replicated, the telomeres become progressively shorter, leading to senescence and death. They discovered that the gene-encoding telomerase is activated in cancer cells, suggesting a mechanism by which cancer cells become immortal.

Harley subsequently left academia for the biotechnology company Geron, headquartered in Menlo

Park, California. For a time, Greider maintained her relationship with Harley, even joining the advisory board of Geron. However, she saw a conflict between the business side of research and the importance of public relation, as exemplified by Geron, and the academic pursuit of research for its own sake. Eventually she broke with Harley and resigned from the advisory board.

In 1997, Greider accepted a position as an associate professor at the Johns Hopkins University School of Medicine in Baltimore, Maryland, where she continued her work on the characterization of telomerase and the implications of telomere formation in malformation of stem cells. She has since become the Daniel Nathans Professor and Director of Molecular Biology and Genetics at the institute. In 2009, she, Blackburn, and Szostak were awarded the Nobel Prize in Physiology or Medicine for their work in characterizing telomeres and telomerase and identifying how they protect and maintain chromosomes. Greider and her husband, historian Dr. Nathaniel Comfort, have two children.

IMPACT

The initial impact following the December 1985 publication of Greider's work with Blackburn and Szostak was minimal. Many scientists, including both Greider and Blackburn, were not yet convinced their interpretation of telomere formation was correct. Ann Pluta and Virginia Zakian at the Fred Hutchinson Cancer Research Center in Seattle, Washington, for example, observed recombination events in telomere formation in yeast, an alternative explanation of synthesis. Further, telomeres were not yet sequenced in human cells at the time, so what had been observed in *Tetrahymena* might not have applied to human cells. It was only when others made similar observations, and when the presence of telomeres on the tips of chromosomes from other eukaryotic species, including humans, was observed, that the real significance of the discovery was recognized.

As Greider noted in the lecture she presented upon receipt of the Nobel Prize, the presence of telomeres in their repeating form across all ranges of species suggests that they arose early in evolution and have an important function within the cell. The question that remained was how the telomeres were produced, and it was here that Greider's discovery was particularly important. The ability of *Tetrahymena* cell extracts to elongate telomeres in the yeast *Saccharomyces* suggested an unknown enzyme was involved; Greider's discovery of that enzyme finally explained the mechanism of telomere formation.

—*Richard Adler*

FURTHER READING

Brady, Catherine. *Elizabeth Blackburn and the Story of Telomeres.* Cambridge: MIT P, 2009. Print. A biography of Blackburn that includes the work carried out by her graduate student, Carol Greider, and their discovery of telomerase.

Lewis, Ricki. *Discovery: Science as a Window to the World.* Malden: Blackwell, 2001. Print. A summary of scientific research, beginning with the debunking of spontaneous generation. One chapter addresses the discovery and characterization of telomeres.

Wayne, Tiffany. *American Women of Science since 1900.* Santa Barbara: ABC-CLIO, 2011. Print. A two-volume summary of prominent American women scientists. Includes life statistics (birth, education) of each woman, along with a brief summary of her scientific contributions.

ALFRED D. HERSHEY

AMERICAN BIOLOGIST

- **BORN:** December 4, 1908; Owosso, Michigan
- **DEATH:** May 22, 1997
- **PRIMARY FIELD:** Biology
- **SPECIALTIES:** Molecular biology; virology; microbiology

A founder of modern molecular biology; experimenting with bacteriophages, demonstrated that DNA (deoxyribonucleic acid) carries genetic information and is the molecular basis of heredity; for that achievement, shared with Max Delbrück and Salvador Luria (both of whom had worked independently of him) the 1969 Nobel Prize for medicine or physiology for their "discoveries concerning the replication mechanism and the genetic structure of viruses," which contributed indirectly to *"an increased understanding of the mechanism of inheritance and of those mechanisms that control the development, growth, and function of tissues and organs"*; *did his research at the Carnegie Institution of Washington Genetics Research Unit at Cold Spring Harbor, New York, with the assistance of Martha Chase; joined the Carnegie research unit in 1950 and became its director in 1962; retired in 1972; died at his home in Syosset, New York.*

EARLY LIFE

The American biologist Alfred Day Hershey was born in Owosso, Michigan, to Alma (Wilbur) and Robert D. Hershey. He attended public schools in Owosso and Lansing before entering Michigan State College (now Michigan State University), where he obtained a B.S. in chemistry in 1930. Remaining at Michigan State for graduate studies, he received a PhD in bacteriology in 1934 and was then appointed to the staff of the Department of Bacteriology at Washington University in St. Louis, Missouri. After serving as an assistant bacteriologist for two years, he became an instructor; he was appointed assistant professor in 1938 and associate professor in 1950.

During his early years at Washington University, Hershey worked under J. J. Bronfenbrenner, who had been investigating bacteriophages since their discovery in 1915. The bacteriophage, a type of virus that infects bacterial cells, is the simplest form of life and, like other viruses, consists of protein and nucleic acid.

LIFE'S WORK

Early in the twentieth century, scientists demonstrated that the inheritance of physical characteristics is governed by genes, which reside on chromosomes in the nucleus of each cell. Chromosomes contain proteins associated with nucleic acids, large molecules composed of units of sugar, phosphate, and nitrogenous bases known as purines or pyrimidines. Biochemical studies revealed two kinds of nucleic acids: ribonucleic acid (RNA) and deoxyribonucleic acid (DNA). It had been believed that only proteins, consisting of amino acids linked together in chains, were sufficiently complex to carry the genetic information; DNA molecules were considered too uniform and repetitive. In the 1940s, however, it was discovered that genes are made of DNA and that DNA directs the biosynthesis of cellular proteins, enzymes, and coenzymes (the heat-stable, water-soluble portion of an enzyme), thus establishing the role DNA plays in controlling the biochemical processes of the cells.

Between 1940 and 1947 Max Delbruck at Vanderbilt University in Nashville, Tennessee, was analyzing the life cycle of the bacteriophage. In collaboration with Salvador Luria at Columbia University, he demonstrated that bacterial cells undergo spontaneous mutation to resist destruction by bacteriophages. Their findings, published in 1943, became a standard for the analysis and presentation of experimental results in bacteriological research. Delbruck, Luria, and Hershey formed the cadre of the Phage Group, an informal group dedicated to bacteriophage research–specifically, the mechanisms of phage replication. Encouraging free exchange of ideas among independent investigators, the Phage Group urged other scientists to concentrate on seven strains of bacteriophage that infect the colon bacillus *Escherichia coli* strain B, so that experimental results could be easily compared.

In 1946 Hershey and Delbruck, working independently, discovered that different strains of bacteriophage may exchange genetic material if more than one strain infects the same bacterial cell. Delbruck was a brilliant theoretician, whereas Hershey was a better experimentalist, and it was Hershey who obtained unequivocal evidence of genetic exchange, which he called genetic recombination. It was the first laboratory demonstration of recombination of genetic material in viruses.

Hershey left Washington University in 1950 to join the staff of the Genetics Research Unit of the Carnegie Institute at Cold Spring Harbor on Long Island, New York. Electron microscopic analysis had revealed the bacteriophage structure to be a protein head, encapsulating a DNA core, and a slender protein tail. In 1952, together with Martha Chase, a geneticist, Hershey discovered how bacteriophages infect bacterial cells. The method employed by Hershey and Chase was based on the fact that phage protein contains no phosphorus and DNA has no sulfur. After cultivating two batches of bacteriophage–one with radioactive phosphorus, the other with radioactive sulfur–the two researchers traced the isotopes during the process of infection and determined that the bacteriophage first attached itself to the bacterial-cell membrane by its protein tail; the nucleic acid core was then injected into the bacterial cell. To separate the empty bacteriophage coats (containing the sulfur isotope) from the bacterial cells (labeled with phosphorus), the suspension was spun in a blender to break the attachment of the viral tails to the bacterial membranes. The suspension was finally centrifuged to separate cellular and fluid fractions. These blender experiments confirmed that DNA is the genetic material of the bacteriophage and, by inference, of all other organisms as well.

During the 1950s and 1960s Hershey continued to investigate the biochemical structure and function of bacteriophage DNA. His work established that bacteriophage DNA is single-stranded–unlike the DNA of higher organisms–and that some bacteriophagic DNA is circular. Moreover, he demonstrated that DNA differs from one species to another.

Impact

The 1969 Nobel Prize for Physiology or Medicine was awarded to Hershey, Luria, and Delbruck "for their discoveries concerning the replication mechanism and the genetic structure of viruses." On presenting the award, Sven Gard of the Karolinska Institute noted the significance of these discoveries for biochemistry, genetics, and other fields of research, adding that the three laureates "must ... be regarded as the original founders of the modern science of molecular biology."

From 1962 until his retirement in 1974, Hershey served as director of the Genetics Research Unit at Cold Spring Harbor. "Although it would be difficult to imagine three personalities more unlike than those of Delbruck, Luria, and Hershey," wrote Gunther Stent of the Harvard Medical School, "they have one trait in common–total incorruptibility–and it is just this trait ... that these three men managed to impose on an entire scientific discipline."

After their first meeting in 1943, Max Delbruck described Hershey in a note to Luria: "Drinks whiskey but not tea, simple, to the point, likes living in a sailboat for three months, likes independence." In 1946 Hershey married Harriet Davidson, with whom he had a son.

Hershey has received an honorary degree from the University of Chicago as well as the Lasker Award of the American Public Health Association (1958) and the Kimber Genetics Award of the National Academy of Sciences (1965). He is a fellow of the American Academy of Arts and Sciences and a member of the National Academy of Sciences.

—*Richard Adler*

Further Reading

"Alfred D. Hershey - Biographical". *Nobelprize.org*. Nobel Media AB 2014. Web. 15 Dec 2017. http://www.nobelprize.org/nobel_prizes/medicine/laureates/1969/hershey-bio.html

Stahl, Franklin W. *Alfred Day Hershey: December 4, 1908-May 22, 1997*. Washington, D.C: National Academy Press, 2001. Print.

Watson, James D. A *Passion for DNA: Genes, Genomes, and Society*. Oxford: Oxford Univ. Press, 2000. Print.

Edward B. Lewis

American geneticist

- **Born:** May 20, 1918; Wilkes-Barre, Pennsylvania
- **Died:** July 21, 2004; Pasadena, California
- **Primary field:** Biology
- **Specialties:** Genetics; evolutionary biology

Twentieth-century American geneticist Edward B. Lewis shared the Nobel Prize in Medicine for discovering the genetic aspects of embryo development. His work laid the foundation for developmental genetics and demonstrated the relationship between radiation and cancer at the genetic level.

Early Life

Edward B. Lewis was born on May 20, 1918, in Wilkes-Barre, Pennsylvania, the younger of two sons. His father was forced to close his jewelry store during the Great Depression. While the family struggled financially, Lewis's parents were committed to making sure their sons stayed in school. By the time Lewis was ready to go to college, his older brother was able to provide financial help to the family so that he could attend school.

As a young man, Lewis had two passions: music and biology. He was so proficient at the flute that he played with the Wilkes-Barre Symphony while he was still in high school, earning a music scholarship to Bucknell University, which he attended for one year before transferring to the University of Minnesota. Lewis's love of music continued throughout his life, and he continued to take music lessons while doing graduate work at the California Institute of Technology (Caltech).

During his youth, Lewis also developed a love of animals, and he kept snakes, toads, and fish as pets. While in high school, he became interested in working with *Drosophila* (fruit flies), and he and his friend Edward Novitski ordered cultures so that they could perform experiments on these insects. Novitski later became an award-winning geneticist. Following high school, Novitski attended Purdue University, while Lewis enrolled at Bucknell. In 1939, Lewis transferred to the University of Minnesota, partly to work in the laboratory of geneticist Clarence Paul Oliver on *Drosophila melanogaster*. He completed his work at Minnesota quickly and proceeded to Caltech to work with geneticist Alfred Sturtevant. Lewis graduated with a PhD from Caltech in 1942. Following his doctoral work, Lewis completed a master's degree in meteorology and did additional work in oceanography.

Life's Work

After spending four years in the U.S. Army Air Corps, Lewis returned to Caltech as an instructor in 1946. Not long after his return, he met genetics student Pamela Harrah while working with *Drosophila*. Harrah discovered a mutant that would become important to the study of gene regulation. Lewis and Harrah married in 1946 and continued to collaborate in the lab. The couple had three sons. Lewis was named full professor at Caltech in 1956. He was appointed Thomas Hunt Morgan Professor of Biology in 1966, a post he held until his retirement in 1988.

Lewis's work at Minnesota followed Oliver's work closely. In his *Drosophila* experiments, Oliver discovered that a single gene might have two components or might actually be two genes. Lewis worked with Oliver on these experiments, setting the stage for decades of work on recombinant genes and demonstrating that several recognized genes could well be duplicates.

At Caltech, Lewis made his first major discoveries in his work with fruit flies. At the time, it was not believed that recombination could occur within one gene; such processes only occurred between genes.

Through his experiments, Lewis found that the star-recessive component of the gene was actually a second locus within the gene. Thus, rather than this component being yet another allele in a series of genes, it turned out to be another group of genes within the same gene. Lewis used the word "pseudoalleles" to describe what was originally thought to be a single gene. Building on earlier experiments, he was also able to show that if pairs of genes come from one ancestral gene, then their functions will be similar and the genes will interact. Lewis's theory builds on the notion that after duplication, one pair of genes can mutate and perform new functions, while the other can retain older functions. In this way, the number of genes increases and the acquisition of new functions occurs in heterozygotes.

Lewis also discovered a process he called transvection. Because of the rearrangement of chromosomes, chromosomes were sometimes prevented from working closely together. He experimented with these inhibited chromosomes to conclude that any chromosome break within a certain number of experiments could inhibit pairing enough to indicate this transvection phenotype. Lewis determined that such patterns can be recognized in the generation immediately succeeding the break, saving the amount of time needed to investigate the effects of chromosome breakage. Building upon this discovery, Lewis developed several technical methods that became standard for experimenting with fruit flies. He discovered that neutrons were much more effective than X-rays or gamma rays in producing chromosome breaks. He also developed a procedure for treating fruit flies with the potent mutagen EMS that soon became standard procedure. Lewis also developed a machine that aided scientists in counting large numbers of flies. The fruit flies were suspended in a liquid and passed through a narrow tube, enabling experimenters to count large numbers of flies more easily.

Lewis's experiments demonstrated that a deeply ordered set of regulatory genes underlay the various segments of a fruit fly's body. A skilled experimenter and an expert at introducing mutations, Lewis persistently introduced duplications into the flies as he attempted to map the structure of each body segment and to confirm that tandem genes were responsible for the structure of these segments. In later experiments, he made connections between these mutant flies and their evolutionary ancestors. In the early

EDWARD B. LEWIS ON RADIATION AND CANCER

Ever since the advent of X-ray technology and the development of radiation for use in various medical and scientific purposes, many scientists have raised questions about the effects of radiation on its subjects. Does radiation cause mutations? Does it have genetic effects on its subjects, causing mutations that genes carry over generations of development? Are low doses of radiation safe to use with minimal genetic effects? Do high doses of radiation have adverse genetic effects? What effects do high and low doses of radiation have on the body? Are the effects on the body the same or different than the effects on the genes? Is there a way to protect against adverse effects of radiation? What is the relation between radiation and cancer?

Lewis had been using radiation to induce mutations in his experiments for years, but he found eventually that neutrons provided a more reliable way to induce mutations than either X-rays or gamma rays. Still, it was not until the mid-1950s that Lewis was drawn into the conversation about the effects of radiation on the body and its function as a cancer-causing agent.

At the time, several scientists thought that cancer developed out of somatic (bodily) mutations rather than genetic ones. If this was true, Lewis thought, then scientists should be able to draw the line between radiation and cancer in the same way that they could draw the line if they were focusing on the effect of radiation on the genes. Lewis publicly challenged the wisdom of the time and argued that, in the absence of any hard data, a straight-line relationship holds in the case of somatic effects as it does in the case of genetic effects. Such was the case, he argued, for low-dose as well as high-dose exposure. Lewis was able to obtain data from studies done in Nagasaki and Hiroshima and published an article in Science in which he used this evidence to demonstrate that the effect of low doses of radiation on the body was the same as such doses on the genes. While numerous scientists were arguing that there was no threshold for that could be set for the effect of doses of radiation on the body, Lewis argued that the same threshold that exists for genetic effects must be used for somatic effects.

Lewis wrote several papers devoted to this subject. In one of them, he proposed that increases in leukemia could be traced to radiation treatment for hyperthyroidism. His conclusions corresponded to later discoveries related to radiation from nuclear tests and its relation to the thyroid and cancer.

Although Lewis never opposed the Atomic Energy Commission, he was responsible for contemporary thinking about developing standards for radiation protection. Lewis posited that there is a linear relationship between radiation and cancer and that there is no threshold (low or high) for doses of radiation and their effects on genes or bodies; these views have influenced generations of scientists' procedures for dealing with radiation.

1960s, Lewis produced a four-winged fly that looked like a dragonfly. His experiments demonstrated that mutations cause very simple changes in an insect's body structure. Lewis was able to construct a triple-mutant chromosome (anterobithorax, bithorax, postbithorax) that provided a map of the complex structure of the fly's body. Lewis was awarded the Nobel Prize in Physiology or Medicine in 1995 for his work on the bithorax function of *Drosophila*, and he is often credited with founding the field of developmental genetics.

After his retirement in 1988, he continued to work on issues related to developmental genetics and medicine. Lewis died in 2004 at the age of eighty-six.

IMPACT

Lewis's work on mutations, developmental genetics, evolution, radiation, and cancer won him numerous prizes and awards, including the Thomas Hunt Morgan Medal (1983), the Gairdner Foundation International Award (1987), the Wolf Foundation Prize in Medicine (1989), the Rosenstiel Award

(1990), the National Medal of Science (1990), the Albert Lasker Award for Basic Medical Research (1991), and the Louisa Gross Horwitz Prize (1992). For his groundbreaking work on *Drosophila*, he was a corecipient of the 1995 Nobel Prize in Physiology or Medicine, along with Christiane Nusslein-Volhard and Eric Wieschaus. Because of Lewis's work with *Drosophila* and his expert experiments on mutations, decades of subsequent researchers have been able to build on his findings regarding specific genes or pairs of genes that influence various evolutionary changes in both wild and domestic *Drosophila* populations.

Lewis developed new laboratory methods for working with *Drosophila* populations, such as a device for counting large numbers of fruit flies in experiments, and he pioneered the use of EMS, a highly potent mutagen that became a standard tool for introducing mutations in populations. Moreover, his insights into the properties and functions of regulatory genes were soon embraced by geneticists and had a tremendous impact on the ways that scientists understood gene regulation and development and evolution.

Although Lewis's writings and research about the relationship between levels of radiation and cancer are not well known, his writing in this area caused a shift in public policy regarding radiation protection. His studies indicated that the effect of either low or high doses of radiation could result not only in changes to genetic makeup but in changes to the body. Lewis was called several times to testify before the Atomic Energy Commission, and his writings about these matters can be found in *Genes, Development, and Cancer* (2004), a collection of his writings drawn from his lifelong work.

Lewis's important work in developmental genetics changed the direction of a number of scientific fields and influenced a large number of scientists and their work.

——*Henry L. Carrigan, Jr.*

FURTHER READING

Crow, James F., and Welcome Bender. "Edward B. Lewis, 1918–2004." *Genetics* 168: 4 (Dec. 2004): 1773–83. Print. Presents an overview of Lewis's life and work. Contains several remembrances of Lewis from colleagues and former students.

Lipshitz, H. D. *Genes, Development, and Cancer: The Life and Work of Edward B. Lewis.* Boston: Kluwer Academic, 2004. Print. Complete collection of Lewis's papers divided into sections that reflect the many areas of research in which Lewis was engaged. Includes summaries.

Ruse, Michael, and Joseph Travis, ed. *Evolution: The First Four Billion Years.* Cambridge, MA: Harvard UP, 2009. Print. Presents an introduction to the theory of evolution that contains a small overview of Lewis and his work.

Konrad Lorenz

Lorenz as a Soviet POW in 1944. [Public domain], via Wikimedia Commons

Austrian zoologist

- **Born:** November 7, 1903; Vienna, Austria-Hungary (now Austria)
- **Died:** February 27, 1989; Altenburg, Austria
- **Primary field:** Biology
- **Specialties:** Ethology; zoology

A key figure in the study of animal behavior, Konrad Lorenz was one of the first biologists to combine behavioral science with evolutionary theory, and his controversial views on human behavior are among the earliest examples of evolutionary psychology.

Early Life

Konrad Zacharias Lorenz was born on November 7, 1903, in Vienna, Austria. His parents, both physicians, owned a large estate near a wooded area. Young Lorenz was fascinated by animals, and he cared for an extensive menagerie that included reptiles, amphibians, birds, dogs, cats, and monkeys. He treated his animal collection as a research project, keeping written records of feeding regimens and behaviors.

At the request of his father, Lorenz studied medicine, first at Columbia University, then at the University of Vienna, where he received an MD degree in 1928. Though his parents hoped he would become a physician, Lorenz studied anatomy and embryology under Ferdinand Hochstetter, with the goal of understanding the biology of animal behavior. Lorenz remained at the University of Vienna, where he worked as an assistant professor and, in 1933, completed a PhD in zoology.

Life's Work

In the early twentieth century, the study of animal behavior was not considered a serious scientific discipline, something that Lorenz wanted to change. Lorenz collaborated with German biologist Oskar Heinroth, who was using morphology (the physical appearance, or developing appearance, of an animal) as the basis for comparative studies of behavior. Lorenz used Heinroth's research as the foundation for several papers on the behavior of birds and fish. Hochstetter supported Lorenz's research as a type of comparative anatomy, thereby allowing him to publish in high-profile biological journals.

In Altenburg during the mid-1930s, Lorenz continued to breed and raise captive birds, his primary research subjects. He was interested in imprinting, the learning process by which animals follow and imitate the first object or creature they see after birth. Imprinting was first observed and recorded in 1873 by naturalist Douglas Spalding but not formally investigated until Heinroth's 1910 paper on the subject. Lorenz conducted extensive research on imprinting, including numerous experiments with hybrid graylag geese. In these experiments, Lorenz found that, by using himself as the surrogate parent and modeling behavior for the goslings' imprinting, he could induce them to imitate and court humans rather than geese. Lorenz also noted that, after imprinting, the altered behavior appeared to be permanently fixed.

Lorenz also worked with ornithologist Wallace Craig, who published an important series of papers on what he called the "behavioral sequences" of doves in 1918. Lorenz named these sequences "fixed-action patterns," which he described as motor patterns initiated by an environmental stimulus but carried to completion with the removal of the stimulus.

During the 1930s, Lorenz gave lectures detailing his theories on instinctive behavior and his method of using anatomy and physiology as the basis for behavioral study. Lorenz asserted that an animal's behavior depended on evolutionary adaptations, and that the comparative study of closely related species could illuminate the evolutionary process. Lorenz's approach to behavior was classified as ethology, now understood as the morphological and physiological study of animal behavior. The ethologists of Lorenz's time were opposed by scientists from the school of comparative psychology, which held that animal behavior and human psychology should be investigated using the same methods. Comparative psychologists conducted many of their experiments in controlled, laboratory environments, which they believed were necessary to evaluate behavior quantitatively. Ethologists, on the other hand, preferred to study animal behavior in natural settings, and preferred observation to laboratory analysis.

In 1936, Lorenz became coeditor of *Zeitschrift für Tierpsychologie*, a leading journal of ethology, while working as a lecturer in comparative anatomy at the University of Vienna. That year, Lorenz began collaborating with Nikolaas Tinbergen, a Dutch ornithologist who shared many of Lorenz's views on instinctive behavior. Lorenz and Tinbergen conducted a series of experiments involving the egg-rolling behavior of graylag geese, using Lorenz's pets as their study subjects.

In 1939, Lorenz was appointed chair of the psychology department at Albertus University in Königsberg, Germany, where his experiences breeding animals led him to develop theories about the effects of domestication on behavior. Lorenz noted that domestic geese have an increased drive for feeding and mating behavior. Lorenz used the evidence in his controversial theories about the effects of domestication on the human species.

After Nazi Germany annexed Austria in 1939, Lorenz joined the Nazi Party, a decision that would lead later generations to doubt the quality of his research. Nazi scientists supported Lorenz's theories on domestication and sponsored his research until 1941. In 1940, Lorenz wrote a now-famous paper in which he used Nazi terminology of racial purity to argue his theories of humanity's self-domestication. In 1941, Lorenz joined the German military; the following year, he was transferred to the front lines and, later, taken prisoner by the Russians. Lorenz was forced to work at a hospital in Armenia treating patients suffering from traumatic disorders. He remained a captive for six years, during which time he wrote his first book. Later published under the title *The Natural Science of the Human Species* (also called *The Russian Manuscript*), the book dealt with instinct, evolution, and comparative animal research.

When Lorenz was repatriated in 1948, he obtained a position at the Max Planck Institute in Seewiesen, Germany. With ethologist Erich von Holst, he co-founded the institute's department of comparative ethology in 1950. In 1952, Lorenz published an English translation of the widely popular *King Solomon's Ring*, a memoir of his experiences as a naturalist and a popular account of his theories on animal behavior.

In 1953, American psychologist Daniel Lehrmann criticized Lorenz, Tinbergen, and other ethologists for their definition of innate knowledge. This criticism started a series of debates between Lorenz and Lehrmann about the causes of behavior and the relative importance of learned versus instinctual, or innate, behavior patterns (this is often called the "nature versus nurture" debate). Eventually, however, Lorenz and Lehrmann became friends, and Lorenz credited Lehrmann as instrumental in the creation of his 1965 book *Evolution and Modification of Behavior*.

Lorenz's experience in medicine helped him to clarify his opinions on the modification of human behavior through domestication and civilization. In his 1963 book *On Aggression*, Lorenz proposed that aggression may be best understood by examining the evolutionary and instinctual drives that lead to aggressive behavior in animals. Lorenz first became interested in aggression following his observations of fish, and after reading research on the behavior of bees published by his colleague, Austrian ethologist Karl von Frisch. Much of *On Aggression* is presented in a philosophical rather than biological form, but the book inspired tremendous debate within the scientific community.

Lorenz retired from the Max Planck Institute in 1973, and became the director of the newly-formed department of animal sociology in the Comparative Ethology Institute of the Austrian Academy of Sciences. That year, Lorenz, Tinbergen, and Karl von Frisch were jointly awarded the Nobel Prize in Physiology or Medicine for their work in founding the field of ethology. In his acceptance speech, Lorenz openly addressed and apologized for his involvement with the Nazis, and expressed his intention to become involved in efforts to sustain natural resources and to further conservation.

Later that year, Lorenz published *Behind the Mirror: A Search for a Natural History of Human Knowledge*, an attempt to examine the origins of intelligence and consciousness in humans. In 1974, Lorenz captured the public imagination again when he published *Civilized Man's Eight Deadly Sins*, which addressed environmental destruction and the dangers of overpopulation. Both books became national best sellers and paved the way for modern research on conservation and consciousness.

Konrad Lorenz died of kidney failure on February 27, 1989, at his family home in Altenberg.

Impact

Lorenz was a pioneer in the study of animal behavior. His ethological investigations presaged the fields of sociobiology and evolutionary psychology, which use evolutionary adaptations as a foundation for investigating many facets of human behavior, including child abuse and rape. His comparative methods for studying animals played a crucial role in the development of early ethology. The awarding of the Nobel Prize to Lorenz and his colleagues in 1973 marked a significant moment in ethology's legitimation as a scientific discipline. His theories helped to define the evolutionary origins of behaviors that span different corners of the animal kingdom. He argued that animals are genetically predisposed to learn knowledge that is advantageous to their survival. Later in his career, his work concerned the behaviors and the ecological costs of human civilization. The aggressive nature of humanity, he argued, could be mitigated through environmental changes and provisions for basic instincts.

—*Micah L. Issitt*

LORENZ ADVANCES THE STUDY OF HUMAN AND ANIMAL BEHAVIOR

Konrad Lorenz was a prominent Austrian zoologist and ethologist. He was awarded the Nobel Prize for Physiology or Medicine in 1973 for pioneering work in animal behavior; he received the prize alongside fellow ethologists Nikolaas Tinbergen and Karl von Frisch. Much of Lorenz's work was based on studies of bird behavior, but his theories helped to define the relationship between evolution and behavior across the animal kingdom. He is now recognized as one of the founders of ethology, the comparative study of animal behavior, used to investigate the evolutionary origin of behavioral patterns.

Born in Vienna, Austria, Lorenz developed an interest in animals at a young age. Biographical accounts indicate that he kept a wide variety of pets and often spent time nursing injured wildlife, including numerous species of birds and small mammals. Lorenz earned his MD in 1928 and his PhD in zoology in 1933, both from the University of Vienna. During this time, Lorenz published his first scientific paper, a detailed analysis of the behavior of a pet jackdaw (*Coloeus monedula*), which was published in a prominent German ornithological journal. Over the course of his career, Lorenz taught comparative behavior at the University of Vienna and the University of Königsberg.

Lorenz bred and maintained captive colonies of ducks, geese, and various songbirds, which he used to study behavior between parents and offspring. Studies of captive bird colonies led to Lorenz's discovery of imprinting, his term for a critical bonding period that occurs in young ducklings, goslings, and a variety of other animals. Lorenz found that during imprinting a duckling or gosling is evolutionarily primed to respond to behaviors observed in their environments, such as the movements and vocalizations of its parents. Under normal circumstances, the bird will imprint on its parents and thereafter follow and learn by imitating the parent animal's behaviors. Lorenz, acting as a parental figure himself, found that a young bird could be induced to imprint onto another species and would follow this animal, modeling its behavior.

Lorenz dedicated most of his life to studying instinctive behavior. He helped to demonstrate how conflicting and opposing instinctual drives could be reconciled to give rise to various behavioral patterns. He showed that the observed behavioral patterns of an organism are largely governed by evolutionary optimization. Animals are genetically predisposed to learn behaviors that will advance and protect the survival of the species.

Toward the end of his life, Lorenz attempted to use lessons from studying instinctual behavior to better understand human behavior and the formation of human society. In his book *On Aggression* (1963), Lorenz attempts to trace the evolution of aggression in human behavior to the origins of similar behaviors in ancestor species. Until his death in 1989, Lorenz continued to analyze the human condition, relating the failings of human society to an underlying difficulty in reconciling instinctual drives with utilitarian virtues.

Further Reading

Burkhardt, Richard W. *Patterns of Behavior: Konrad Lorenz, Niko Tinbergen, and the Founding of Ethology.* Chicago: U of Chicago P, 2005. Print. Examines the institutions, scientists, and contexts by which ethology became a respected scientific discipline. Discusses Lorenz at length in context of contemporary ethological scientists.

Lorenz, Konrad. *King Solomon's Ring: New Light on Animal Ways.* 1952. Trans. Marjorie Kerr Wilson. New York: Routledge, 2002. Print. Collects observations, anecdotes, and facts on animal psychology. Humorous, eloquent writing is appropriate for general readers. Includes illustrations by the author.

———. *On Aggression.* 1963. Trans. Marjorie Kerr Wilson. New York: Routledge, 2007. Print. An inquiry into aggression as a means of survival among a variety of animals. Examines humanity and proposes solutions to the problem of human aggression.

Barbara McClintock

American geneticist

- **Born:** June 16, 1902; Hartford, Connecticut
- **Died:** September 2, 1992; Huntington, New York
- **Primary field:** Biology
- **Specialty:** Genetics

A pioneer in classical and molecular genetics, McClintock won the Nobel Prize in Physiology or Medicine in 1983. Her theories that patterns of genetic traits caused by mutations do not follow the accepted rules of genetics, and that sections of chromosomes detach and move to a new location during development, were far ahead of contemporary genetic research.

Early Life

Barbara McClintock was the third daughter of Sara Handy and Thomas Henry McClintock. Shortly after their marriage, Thomas McClintock finished medical school at Boston University, and after a few relocations, the couple moved to Hartford, Connecticut, where Barbara was born. Less than two years after McClintock's birth, her younger brother Malcolm was born. McClintock's mother, unable to cope with four small children and financial issues, decided to send McClintock to live with her aunt and uncle in Massachusetts. McClintock came back to her parents' house when it was time for her to start school. In the interim, the McClintocks had moved to the Flatbush section of Brooklyn, New York, which in those days was a semirural area.

In high school, McClintock discovered science. She loved information and problem solving. Her mother began to worry that McClintock would pursue knowledge to the detriment of her "feminine development" and tried to dissuade her, but McClintock was set on attending Cornell University, a school that was known for accepting and supporting the education of women.

Beginning her studies at Cornell in 1919, McClintock soon blossomed both intellectually and socially. Elected president of the women's freshman class, she began to date, played banjo in local cafes, and managed a heavy class schedule. After a couple of years, however, she began to be disillusioned with the social whirl. Devoting more and more time to academics, she was allowed to take a graduate class in genetics during her junior year, and her lifelong fascination with the subject began.

Life's Work

At the time of McClintock's matriculation at Cornell, genetics was a new science, scarcely more than a few decades old. Cytology (the branch of biology that concerns cells) was relatively new, and many cellular structures were still a mystery. Evolutionary biologist Thomas Hunt Morgan, working with *Drosophila* (fruit flies) at Columbia, had proposed that genes were located on the chromosomes like beads on a string, but many scientists did not accept that idea.

McClintock started her career with the successful identification of maize (Indian corn) chromosomes, distinguishing the ten individual chromosomes that constitute each individual kernel. After completing her bachelor's degree, McClintock began graduate work in the Botany Department; the Department of Plant Breeding, which included genetics, did not accept women as graduate students. She received her doctorate in 1927, when she was twenty-four. Fascinated by the work being done by Morgan and others, she stayed on at Cornell to try to parallel Morgan's *Drosophila* work with maize. Two other researchers who came to Cornell at this time were Marcus Rhodes and George Wells Beadle, both of whom later became prominent geneticists. In this environment, McClintock's research thrived. By 1931, she had published nine papers on maize chromosomal morphology and attracted renown within her field.

During these years, Harriet Creighton, a new graduate student, arrived at Cornell. McClintock convinced her to enroll as a cytology and genetics major, after which Creighton began working as McClintock's assistant. Toward the end of the first year, McClintock suggested that Creighton attempt to prove the commonly assumed correlation between chromosomal crossover (an actual physical exchange of chromatids during meiosis) and genetic crossover (an organism's display of a combination of parental traits that are normally linked). McClintock believed that this could be proved through a series of experiments involving a particular maize chromosome, and she had identified and isolated kernels that involved the traits needed. Creighton and McClintock worked on the problem until the spring of 1931, when, at the urging of Morgan, they published their successful results. It became a landmark essay in classical genetics.

As faculty positions were simply not open to women at Cornell, McClintock decided to leave. For a number of years, McClintock moved between Cornell, the California Institute of Technology, and the University of Missouri, where with Lewis Stadler she was researching the existence of ring

McCLINTOCK DISCOVERS TRANSPOSABLE GENETIC ELEMENTS IN MAIZE

Barbara McClintock's work in genetics began during her time at Cornell University, where she majored in botany. In 1921, McClintock enrolled in a genetics course taught by Claude B. Hutchinson, a professor in the Department of Plant Breeding, and in a cytology course taught by Lester Sharp in the Department of Botany. These courses stimulated McClintock's growing interest in genetics and, particularly, in the role of chromosomes.

McClintock remained at Cornell as an instructor of botany until 1931. During this period, her research addressed the functions of the ten chromosomes found in maize, the ancestor of modern corn. Among her associates were future Nobel laureate George Beadle, who was studying for his PhD alongside geneticist Rollins Emerson and Marcus Rhoades, who would become a major figure in the field as well. McClintock's research during this period included her development of a staining technique for the visualization of maize chromosomes. This technique allowed her to demonstrate linkage among phenotypic traits (that is, traits that are related to the properties produced by an organism's interaction with its environment). She was also able to demonstrate "crossing-over," or the chromosomal exchange of DNA fragments, during gamete production. In 1931, her demonstration allowed her to publish the first genetic map of maize.

After a brief time in Germany, McClintock became an assistant professor at the University of Missouri, where she would remain until 1941. During this time, McClintock observed that exposing maize to X-rays resulted in the breakage and reunion of chromosomal elements, an event that she later found could occur spontaneously as well.

Leaving Missouri, McClintock joined the genetics laboratory at Cold Spring Harbor, New York. She would remain with the laboratory until 1967, and it was here that she would carry out her most notable research. McClintock's discovery of transposable elements began with breeding experiments of maize in the summer of 1944. Each of 450 plants was fertilized with a parent that contained a broken end on one of its chromosomes. McClintock expected mutations to be associated with the regions of the broken chromosomes. Instead of the expected pattern, however, she observed a large number of bizarre mutations, particularly mutations associated with chlorophyll production. Drawing upon previous genetic studies by Hermann Muller on the effects of X-rays on chromosomal breakage, McClintock ultimately found that she was observing the breakage and repair/reunion of chromosomes. The repair mechanism, she suggested, is dependent upon the stage of cell division. The results of these breeding experiments accounted for the mottled pattern of colors in kernels of corn.

By 1948, McClintock had developed a working hypothesis to explain her results. Two genetic elements, termed Ac (Activator) and Ds (Dissociator), were responsible. Each genetic element represented a transposable element, with Ac regulating the transposition of Ds. In the 1970s, the Ac locus was found to encode a transposase, a catalytic enzyme that regulates the transposition of Ds. The position of Ds determines the extent of pigment production in the kernel; the time of transposition determines the size of the pigment spot. In 1983, McClintock was awarded the Nobel Prize in Physiology or Medicine for her work.

chromosomes. In 1933, McClintock received a fellowship to go to Germany and work with the famous geneticist Richard B. Goldschmidt. She found the reality of Nazi Germany unbearable and returned within the year. In 1935, Stadler convinced the University of Missouri to offer McClintock a position as assistant professor, which she accepted. Her unconventional ways and outspokenness did not fit in well at Missouri, and while her research went well, she was passed over for promotion in favor of men whose credentials were far less impressive. In 1941, she relocated to the Carnegie Institute of Washington at Cold Spring Harbor, New York.

In 1944, the National Academy of Sciences elected McClintock to its membership. She was only the third woman to receive this honor. Between 1944 and 1951, McClintock studied patterns of genetic traits caused by mutations, traits that did not seem to follow the accepted rules of genetics. These traits did seem to occur with some regularity, however, implying that some form of controlling factors were involved. McClintock proposed that, at some point during development, sections of chromosomes actually detached and moved to a new location, contrary to Morgan's "beads on a string" concept. Furthermore, these transpositions seemed to be controlled by a factor on the chromosome itself. In 1951, she presented her findings at the annual Cold Spring Harbor Symposium. While her work received a cool reception in the scientific community, Jacques Monod and François Jacob published a paper that confirmed some of her findings in 1961. McClintock immediately submitted an article for publication and presented another at Cold Spring Harbor, but still her work was not accepted.

In the mid-1960s, McClintock's situation began to improve. She received the Kimber Genetics Award in 1967 and the National Medal of Science in 1970, as more of her findings were confirmed by other researchers and studies again started citing her work. This trend continued in the 1970s, and McClintock started publishing again. She gained new prominence, winning awards and prizes that brought her prestige and money. She won the Nobel Prize in Physiology or Medicine in 1983, becoming the first woman to win an unshared Nobel Prize in this category and the third woman to do so in any science category.

Still at Cold Spring Harbor, McClintock continued her research, maintaining her independence and isolation despite the belated acceptance of her work. In later years, she studied Tibetan Buddhism and biofeedback techniques, which fascinated her in their incorporation of the same holistic approach she had brought to the study of maize for so many years. McClintock died on September 2, 1992, at the age of ninety.

IMPACT

McClintock's work was influential in the field of genetics. Her discoveries on the behavior of chromosomes during genetic transposition earned her the Nobel Prize for Physiology or Medicine in 1983. Her research on the incidence and distribution of chromosomal features and abnormalities helped account for the evolution of the varieties of corn crops.

Throughout her life, McClintock wanted gender to be disregarded in assessments of a scientist's worth. She wanted to be accepted and respected on her own merit, as a researcher and expert in her field. She confronted and overcome many barriers throughout her career. Others followed her example. Geneticist Evelyn Witkin, for example, established the Barbara McClintock Chair at Rutgers University to acknowledge the enormous debt she owed McClintock for her inspiration and encouragement in their ten years together at Cold Spring Harbor. Creighton, who owed the inception of her career to McClintock, commented, "It was the best steering anyone could have given me." McClintock paved a path that continues to be followed by many scientists, both men and women. Her complete dedication and unique vision serve as an example to women in every field.

——*Margaret Hawthorne*

FURTHER READING

Bennetzen, Jeff L., and Sarah C. Hake. *Handbook of Maize: Genetics and Genomics.* New York: Springer, 2009. Print. Examines maize as a significant food and fuel crop and a subject for scientific research. Includes a chapter devoted to McClintock's research.

Comfort, Nathaniel C. *The Tangled Field: Barbara McClintock's Search for the Patterns of Genetic Control.* Cambridge, Mass.: Harvard UP, 2001. Print. Examines biographer Evelyn Fox Keller's depiction of McClintock's milieu and McClintock's own description of her role in science. Maintains that

McClintock was always a respected scientist and a distinguished figure.

Cullen, J. Heather. *Barbara McClintock*. New York: Chelsea House, 2003. Print. Chronicles McClintock's life and career. Appropriate for high-school-level readers.

Fedoroff, Nina, and David Botstein, eds. *The Dynamic Genome*. Cold Spring Harbor: Cold Spring Harbor Laboratory P, 1992. Print. A collection of essays written as tributes and remembrances for McClintock's ninetieth birthday. Includes technical articles on genetics and McClintock's impact, as well as reprints of four of McClintock's most influential articles.

Lewin, Roger. "A Naturalist of the Genome." *Science* 222 (28 Oct. 1983): 402–405. Print. Published shortly after McClintock received the Nobel Prize. Includes a concise and clear overview of the work for which the prize was awarded as well as some discussion of where McClintock's discoveries have led since.

Christiane Nüsslein-Volhard

German biologist

- **Born:** October 20, 1942; Magdeburg, Germany
- **Primary field:** Biology
- **Specialty:** Genetics

German biologist Christiane Nüsslein-Volhard is best known for her study of the genes of Drosophila melanogaster, also known as the fruit fly. In recognition of her genetics research, Nüsslein-Volhard shared the Nobel Prize in Physiology or Medicine in 1995.

Early Life

Christiane Nüsslein-Volhard was born on October 20, 1942, in Magdeburg, Germany, the second of five children born to Rolf Volhard and Brigitte Haas. Both of Nüsslein-Volhard's parents were skilled musicians and painters, as were her siblings. Nüsslein-Volhard never had the same knack for visual arts that her brother and sisters had, though she was a proficient flutist. Her father also encouraged Nüsslein-Volhard's academic development, discussing mathematics and science with his daughter in his spare time.

Nüsslein-Volhard learned the value of hard work early on, as her family did not have much money. She and her siblings made their own clothing and other possessions, and Christiane committed herself to scientific projects of her own design, based on books she had read. Much of Nüsslein-Volhard's childhood was spent on her grandparents' farm, where she was allowed to interact with and learn about animals. By the age of twelve, she knew she wanted to be a biologist. In high school, Nüsslein-Volhard performed well in subjects that she enjoyed and ignored those she did not, such as English. Her teachers recognized, however, that young Nüsslein-Volhard possessed a unique scientific mind and a propensity for critical thought. She had several strong female role models in her academic life, as many of her teachers, including her biology professor, were women.

Animal behavior and biology had always interested the young Nüsslein-Volhard, but for a long time she had considered studying medicine and becoming a doctor. Shortly before she graduated from high school, her father died suddenly, renewing her interest in medicine. After working as a nurse for a month, however, Nüsslein-Volhard lost interest in the medical profession and decided to devote herself fully to biological research.

After graduating from high school, Nüsslein-Volhard entered the Johann-Wolfgang-Goethe University in Frankfurt. Having studied biology for years on her own, Nüsslein-Volhard was bored with many of the discoveries discussed in her biology lectures, though she was occasionally interested in botany. She felt that she had already learned about all of the exciting aspects of biology on her own and was subsequently drawn to physics. Then, a chemistry class she had taken reminded her of her love of biology. In 1964, Nüsslein-Volhard left Frankfurt to study chemistry at the University of Tübingen. Five years later, she graduated with a degree in biochemistry and a growing interest in genetics, moving on to doctoral work at Tübingen's Max-Planck-Institut für Virusforschung (Max Planck Institute for Virus Research).

Life's Work

By 1973, as Nüsslein-Volhard was finishing her doctoral thesis on DNA and RNA sequences, she became increasingly interested in embryology. Alfred Gierer, who was in charge of the Tübingen School, encouraged Nüsslein-Volhard to study developmental biology, then a fairly new discipline. Later that year, she met Walter Gehring, who had begun studying the development of the common fruit fly, or *Drosophila melanogaster*. Gehring invited Nüsslein-Volhard to conduct research at his laboratory in Basel, Switzerland, which she did for several years. In 1978, Nüsslein-Volhard became interested in the work of Eric Wieschaus, an American biologist who was studying the development of fly embryos.

Nüsslein-Volhard eventually moved to Heidelberg to work with Wieschaus at the European Molecular Biology Laboratory. The pair began researching the embryonic genetic development of fruit flies and of animals in general. Unlike many of their contemporaries, Nüsslein-Volhard and Wieschaus were interested in how genes were related, rather than how they were different.

Nüsslein-Volhard and Wieschaus wanted to figure out how animals grow from single cells into complex organisms, a process called embryogenesis. Expanding on the earlier work of Edward B. Lewis (with whom they would later share the Nobel Prize), Nüsslein-Volhard and Wieschaus isolated the genes that cause flies to grow from eggs into adults.

Nüsslein-Volhard and Wieschaus exposed the flies to chemicals that were known to cause genetic mutations. Though she did not know precisely what kinds of mutations would occur, Nüsslein-Volhard assumed that one of the genes affected would likely be the gene for which she was searching. When mutations occurred, the affected flies were mated with normal flies. Nüsslein-Volhard analyzed the offspring of these couplings so that she could see which genes were causing the mutations.

While the specific variables of the experiments were somewhat haphazard, Nüsslein-Volhard pioneered several important techniques for studying the fruit flies and eventually cemented *D. melanogaster* as the prototypical species for genetic experimentation.

Nüsslein-Volhard found that oil could be used to make the outer layer of an embryo, the chorion, transparent, allowing her to examine the embryo without removing the membrane. She also discovered that she could divide up eggs from a single fly and place them in test tubes, meaning that she could study the effects of many chemicals at once. Apart from the discoveries she would eventually make about the specific genes that affect development, Nüsslein-Volhard's techniques completely redefined the methods used in genetics and biochemistry research.

After performing thousands of trials with *D. melanogaster*, Nüsslein-Volhard and Wieschaus eventually located about 140 genes (out of the fruit fly's 20,000 total genes) that seemed to be essential to the process of embryogenesis. They gave the isolated genes names such as oskar, gurken, and hedgehog. Hedgehog ended up being a very important gene, responsible for limb development. Its analog in humans, called sonic hedgehog, is particularly important in the development of the spinal cord and the brain. Nüsslein-Volhard and Wieschaus published their initial findings in 1980, to relatively little fanfare outside of the scientific community. Nüsslein-Volhard extrapolated her results with flies to determine that genetic mutation is often the cause of abnormalities and miscarriages in humans and other animals.

Genetics had been a taboo subject in Germany since World War II. Many Germans considered Nüsslein-Volhard an enemy of science because of her interest in genetic engineering. Eventually, however, Nüsslein-Volhard's discoveries in genetics were hailed as a breakthrough. Her findings led to a better understanding of the behavior of genes and their role in human disorders, including cancer and, eventually, Parkinson's, Huntington's, and Alzheimer's diseases, all of which result in damage to and degradation of the brain. Following Nüsslein-Volhard's studies, it was eventually discovered that humans share 44 percent of their genetic makeup with fruit flies, and that 61 percent of known human disorders are related to those shared genes. Far from merely explaining how animals grow, Nüsslein-Volhard helped to explain the basic behavior of genes in all animals.

Nüsslein-Volhard continued her work with fruit flies for many years, returning to Tübingen in 1985 to become the director of the prestigious Max Planck Institute for Developmental Biology (originally the Max Planck Institute for Virus Research, where she had done her PhD work). In 1995, Nüsslein-Volhard shared the Nobel Prize in Physiology or Medicine with Edward Lewis and Eric Wieschaus. Despite her longstanding fascination with flies and her prominence

NÜSSLEIN-VOLHARD DISCOVERS MECHANISM OF EMBRYONIC GENE REGULATION

Nüsslein-Volhard's doctoral work at the University of Tübingen centered on understanding gene regulation. However, she became increasingly interested in applying her work on developmental regulation to animal systems. At the time, numerous scientists at the university were studying developmental biology and utilizing the latest techniques of molecular biology. Nüsslein-Volhard decided the fruit fly Drosophila would be the ideal organism for the study of developmental biology as a postdoctoral project.

Drosophila presented a number of advantages. It was easily manipulated and easy to mutate. Some bicaudal mutants (in which the embryos had mirror-image duplicated abdomens) had already been discovered. While Drosophila encoded some 20,000 genes, they were distributed on only four easily visible sets of chromosomes. Nüsslein-Volhard happened to meet another postdoctoral student, Eric Wieschaus, who was also interested in the subject of developmental biology, and the two developed a collaboration that ultimately carried out the successful study.

Nüsslein-Volhard and Wieschaus were each offered the opportunity to work at the European Molecular Biology Laboratory in Heidelberg in 1978, and it was there they began research aimed at understanding embryonic development, or embryogenesis. The project began with the screening for lethal mutations in Drosophila, particularly those that affected the segmentation patterns formed during the immediate period after embryogenesis. They quickly determined possible roles for a number of newly recognized genes.

Regulation of embryonic development in Drosophila begins with several "morphogens," gene products known as "maternal effect genes" from the female that form a gradient in the embryo. Sites of higher concentration produce the anterior (head) portion through activation of specific early genes; the thorax and abdomen form at lower concentrations. Eventually the embryo will develop fifteen segments; each segment will subsequently produce other features of the fly such as wings or legs.

Nüsslein-Volhard's work focused on regulation of segment production. She and Wieschaus determined that early development is regulated by three classes of genes: (1) gap genes, mutations of which result in missing segments; (2) pair-rule genes, mutations of which result in the disappearance of alternate segments; and (3) segmentation-polarity genes, mutations of which result in missing segments or segments that become mirror images of each other. Nüsslein-Volhard and Wieschaus discovered that regulatory genes actually encode transcription factors, each set of which activates the next class of genes.

The observations and discoveries found by Nüsslein-Volhard and Wieschaus were subsequently found to have a far more important impact than initially thought. The regulatory genes in fruit flies have precise counterparts in more evolved organisms, including humans. Understanding genetic development in Drosophila translates to a better understanding of human embryonic development as well.

in that specialized field, Nüsslein-Volhard has since focused her work on zebra fish.

Impact

Nüsslein-Volhard has also worked to encourage equality for women scientists. Despite her Nobel win, she recognizes the disparity between women and men in the sciences. Indeed, she was only the eleventh woman to be awarded a Nobel Prize in the sciences. To help close this gap, Nüsslein-Volhard set up a foundation in Germany to provide support to scientists who are also mothers, claiming that the traditional roles assigned to mothers can often interfere with a scientist's research projects.

Nüsslein-Volhard continues her studies of genetic development in her role as director of the Max Planck Institute. In 2005, she was awarded an honorary doctor of science degree from Oxford University.

—Alex K. Rich

Further Reading

Dreifus, Claudia. "A Conversation with Christiane Nüsslein-Volhard: Solving a Mystery of Life, Then Tackling a Real-Life Problem." *New York Times*, 4 July 2006: 2. Print. Presents an interview with Nüsslein-Volhard in which she answers questions about gender discrimination among professional scientists, her Nobel Prize–winning research on fruit flies, and her parents' experiences during World War II.

Nüsslein-Volhard, Christiane. *Coming to Life: How Genes Drive Development.* San Diego: Kales, 2006. Print. Presents an introduction to the genetic development of bilaterian embryos (embryos of animals characterized by left-right symmetry).

Wade, David. "Nobel Women." *Science* 18 Jan. 2002: 439. Print. Discusses the awarding of the Nobel Prize to women scientists between 1977 and 1995, including Nüsslein-Volhard.

Frederick Sanger

British biochemist

- **Born:** August 13, 1918; Rendcomb, England
- **Died:** November 19, 2013; Cambridge United Kingdom
- **Primary field:** Chemistry
- **Specialties:** Biochemistry; molecular biology; genetics

Twentieth-century biochemist Frederick Sanger was awarded the Nobel Prize in Chemistry in 1958 for his discovery of the structure of the insulin molecule and again in 1980 for his invention of a technique for determining the sequence of DNA.

Early Life

Frederick Sanger was born on August 13, 1918, in Rendcomb, England, to a Quaker family. He was the son of a physician, also named Frederick Sanger, and his wife Cicely, heiress to a cotton-manufacturing business. Sanger attended St. John's College at the University of Cambridge, where he obtained a bachelor's degree in natural sciences in 1939. After his undergraduate degree he received first class examination scores in advanced biochemistry. He stayed on at Cambridge to begin his PhD in biochemistry.

In keeping with his Quaker tradition, Sanger was a conscientious objector during World War II. Instead of joining the military he worked in the laboratory of Albert Neuberger, where he studied the metabolism of the amino acid lysine. Sanger was awarded his PhD in 1943, after which he won a Beit Memorial Fellowship for Medical Research. This income enabled him to work in the laboratory of A. C. Chibnall at Cambridge. Chibnall, whose research group had been working with insulin, suggested that Sanger explore the structure of the molecule.

Life's Work

Insulin is a hormone manufactured in the pancreas. It brings glucose (sugar) from the blood into cells, where it is used for energy. Because some diabetics require injections of insulin, a purified form of bovine insulin was available to Sanger for his research.

All proteins are made of amino acids linked together in a specific order. There are twenty common amino acids. Just as the letters of the alphabet can be rearranged to make infinite words, these twenty amino acids form many different proteins. When Sanger began his work, scientists knew proteins were made of amino acids, but they had not yet discovered that the amino acids in each protein were arranged in a specific sequence.

Sanger used the chemical dinitrophenol to bind to one amino acid at a time in the insulin molecule, removing each amino acid from the rest of the chain. He identified the individual amino acids using paper chromatography, a technique for identifying components in liquids. In paper chromatography a solvent is drawn up by filter paper and amino acids dissolved in the solvent will move different distances up the paper. Some amino acids are more strongly attracted to the paper and others to the solvent. When the paper dries, the amino acid leaves a smear behind. Sanger examined these smears to identify each amino acid.

The process was slow, and it took until 1955 to determine the whole sequence. Sanger discovered that the insulin molecule is actually made of two separate chains of amino acids, or polypeptides, linked together. Insulin was the first protein ever to be sequenced, but the work was important in other ways too. Since scientists knew that DNA (deoxyribonucleic acid) coded for proteins, the fact that proteins were ordered molecules meant that DNA must have a sequence that is important too. It was later discovered that the sequence of nitrogenous bases in a gene determines the sequence of amino acids in a protein. In this way, DNA carries heritable traits from one generation to the next.

In 1958, Sanger earned the Nobel Prize in Chemistry for his sequencing work on insulin. Diabetics have benefited from his research because it enabled pure human insulin to be produced using techniques of genetic engineering. Before this advancement, people with diabetes had to rely on insulin purified from animals. Animal insulin was not as effective, and sometimes caused allergic reactions.

Sanger was appointed to the Medical Research Council at Cambridge in 1951. By 1962 the council moved to a new location in the Laboratory of Molecular Biology, also at Cambridge, and Sanger was promoted to head of the Division of Protein Chemistry there. Sanger was surrounded by colleagues who were interested in DNA, and he began working on a method to sequence the DNA molecule.

Nucleotides are the building block molecules that make up DNA. Each nucleotide includes one of four nitrogenous bases: adenine (A), thymine (T), guanine (G), or cytosine (C). The sequence of these bases codes for the sequence of amino acids in a protein. Sanger developed a technique to read the sequence of bases in DNA—a technique based on DNA's ability to replicate itself. The double helical DNA molecule looks like a spiral staircase in which the "stairs" are made of pairs of bases bound together, A with T and C with G. Because of this precise pairing, each strand can serve as a template to synthesize a new strand.

During synthesis of a growing strand, the new nucleotide to be added binds to an oxygen atom on the last nucleotide in line. Sanger came up with the idea of using dideoxynucleotides, which are nucleotides with no oxygen, to bind on to. When these get incorporated into a growing strand, no new nucleotides can bind, and the chain is terminated.

The reaction mixture includes normal nucleotides as well, so it produces a mixture of different-length pieces of DNA. They get separated on a slab of gel, leaving a pattern that looks like the bar code on a price tag. That bar code reveals the bases in the DNA sequence.

Sanger and his research group eventually sequenced the genome (all the DNA) of the bacteriophage phi-X 174. A bacteriophage is a virus that attacks bacteria. Bacteriophages are convenient to study because they can grow in petri dishes with bacteria, and because they cannot infect people. This particular virus had a genome of 5,375 nucleotides, which is small compared to genomes other organisms. Sanger won the 1980 Nobel Prize in Chemistry for developing the technique and sequencing the genome of bacteriophage phi-X 174. This was the first intact genome ever to be sequenced from any organism. Sanger shared the prize with Paul Berg and Walter Gilbert.

More ambitious sequencing efforts followed, including the first sequencing of the genome of human mitochondria. Mitochondria are subcellular organelles that provide humans with energy. Now comparisons of mitochondrial DNA from different populations are revealing paths of migration taken by ancient peoples as they spread out of Africa.

IMPACT

The sequencing of bacteriophage (lambda) was Sanger's largest effort, at over 48,000 nucleotides. DNA sequencing was a tremendous achievement because it allows us to compare a gene with its protein product. Some inherited diseases are due to mutations in a gene for a crucial protein. Sanger's work laid the foundation for curing inherited diseases using genetic engineering.

For example, some children are born with SCIDS (severe combined immunodeficiency syndrome), an inherited defect in the immune system that prevents them from fighting off disease. These children once had to be kept in sterile plastic chambers just to stay alive. Using Sanger's sequencing method, researchers identified the "misspelling" in the DNA sequence. Experimental genetic engineering has corrected the problem for some children with SCIDS, who are now freed from confinement and living normal lives.

FREDERICK SANGER DETERMINES THE STRUCTURE OF INSULIN

Proteins are large molecules that contain carbon, hydrogen, oxygen, nitrogen, and sulfur. They transport molecules, provide support and protection, and control chemical reactions. By the early 1940s, it was well established that proteins were made of some twenty different amino acids, but little was known about their arrangement and order. Determining the building plan for these complicated molecules was one of the greatest research problems of the twentieth century.

In 1943, Frederick Sanger began a twelve-year study that established the unique amino acid sequence of insulin—the hormone needed for the treatment of diabetes. Working with fellow British biochemist A. C. Chibnall, Sanger showed that the building blocks of proteins have a definite, specified order. Sanger developed strategies and techniques for uncovering this sequence that have been applied to many other proteins.

First, Sanger determined how many chains there were on the insulin molecule. The first amino acid in any chain has a free amino end. Sanger attached a colored dye, fluorodinitrobenzene (FDNB), to the insulin molecule. The dyed insulin was then broken into individual amino acids. Sanger found that the dye had reacted with two different amino acids. Therefore, he concluded that insulin had two chains. Sanger separated the two chains by oxidizing (combining with oxygen) the sulfurs that joined them.

When a protein is treated with strong acid, it comes apart into its individual amino acids. Sanger, however, gently treated each insulin chain with a weaker acid solution. The chain came apart but broke into bigger pieces of three, four, or five amino acids. Sanger separated these fragments and used his dye to find the identity of the first amino acid in each piece. He also determined the amino acid composition of each small piece and began to put the pieces together.

Besides using the acid solution that broke the chain in various places, Sanger also used digestive enzymes—chymotrypsin and trypsin—to cut the amino acid chains in specific places. The order of the chain was determined by overlapping the various fragments.

Once Sanger had successfully found the order of the two chains of the insulin molecule, he had to figure out how they were joined. The smaller chain, the amino acid with a sulfur group, had four cysteines (crystalline amino acids derived from cystine). The larger chain had two cysteines. Therefore, there were several different ways in which the chains could be linked.

By 1955, Sanger had determined the amino acid sequence of the 21-unit A chain and the 30-unit B chain of bovine insulin. He also found that insulin from other animals differed very slightly in its composition. He showed that these two chains were linked by sulfur bridges joining the cysteines at position 7 in both chains. A second disulfide bridge linked position 20 of the first chain with 19 of the second. This landmark work provided conclusive evidence that a protein had a very specific sequence of amino acids.

Frederick Sanger's work in determining the sequence of the amino acid units in insulin established conclusively that proteins have a definite order in the arrangement of their amino acids. This finding suggested to other scientists that there must be a genetic code that provides the information for the amino acid sequence. For his work Sanger was awarded the 1958 Nobel Prize in Chemistry.

An automated version of the Sanger method was also used in the international Human Genome Project. The sequencing of the entire genome from a volunteer was recently completed and the sequence is still being analyzed. Questions about the ethical implications of the technology have also been raised, and continue to be discussed.

Frederick Sanger retired in 1983. The Sanger Centre was established in his honor in 1992, funded by the Wellcome Trust. Located in Cambridge, England, it is primarily focused on investigating genomes, with an eye toward health and medical applications of the work. The Sanger Centre was one of the sequencing centers involved in the Human Genome Project.

—*Courtney Farrell*

FURTHER READING

García-Sancho, Miguel. "A New Insight into Sanger's Development of Sequencing: From Proteins to DNA, 1943–1977." *Journal of the History of Biology* 43.2 (2010): 265–323. Print. Discusses Sanger's scientific career related to his research on RNA and DNA sequencing, genomics, and bioinformatics and biomedicine.

Manchester, Keith. "Protein Sequencing Fifty Years Ago: Fred Sanger and the Amino Acid Sequence of Insulin." *South African Journal of Science* 101 7–8 (2005): 327–30. Print. Presents information on Sanger's work with the amino acid and protein sequencing and his study of the molecular structure of insulin.

Sanger, Frederick, and Margaret Dowding. *Selected Papers of Frederick Sanger: With Commentaries.* Singapore: World Scientific, 1996. Print. Includes publications from Sanger's research on RNA and DNA sequencing, amino acids, and insulin, as well as Sanger's reflections on his papers and why they were included in the collection.

HANS SPEMANN

Hans Spemann (June 27, 1869 – September 9, 1941). [Public domain], via Wikimedia Commons

GERMAN BIOLOGIST

- **BORN:** June 27, 1869; Stuttgart, Germany
- **DIED:** September 9, 1941; Freiburg, Germany
- **PRIMARY FIELD:** Biology
- **SPECIALTIES:** Zoology; anatomy

In a career that spanned the late nineteenth and early twentieth century, German biologist Hans Spemann taught zoology and comparative anatomy for forty years and conducted influential research at several German universities. In the course of transplantation experiments, he was the first scientist to create a clone. Spemann received the Nobel Prize in Physiology or Medicine for his discovery of the principle of embryonic induction.

EARLY LIFE

Hans Spemann was born on June 27, 1869 in Stuttgart, Germany. The son of prominent publisher Johann Wilhelm Spemann and his wife Lisinka Hoffman Spemann, he was raised in a wealthy family. After graduating from the local gymnasium, Hans worked for his father for one year and served a mandatory year of military service. Spemann then enrolled in a medical studies program at the University of Heidelberg, where he worked with anatomist Karl Gegenbaur, who inspired his interest in zoology. After earning his bachelor's degree in medicine in 1892, Spemann married Klara Binder. The couple had three sons and one daughter. Not long after his marriage, Spemann enrolled at the University of Munich for additional clinical training and laboratory work.

In 1894, Spemann relocated to the University of Würzburg to begin teaching zoology, where he took graduate courses at the university's Zoological Institute under such professors as cytologist Theodor

Boveri, physiologist and cell biologist Otto Bütschli, plant physiologist Julius Sachs, and physicist Wilhelm Röntgen, a Nobel laureate. In 1895, Spemann earned his PhD in anatomical studies with concentrations in zoology, botany, and physics. His dissertation focused on the cell lineage of nematodes. The following year, Spemann was stricken with tuberculosis.

Life's Work

Following his recovery, Spemann returned with new purpose to Würzburg to continue his duties as a lecturer. Shortly after the turn of the century, he began a period of intense laboratory research and experimentation on the embryos of salamanders, newts, and frogs, concentrating on the division and transplantation of the cells of fertilized eggs. He published his first paper related to his studies in 1901.

In 1908, Spemann became professor of zoology and comparative anatomy at the University of Rostock, one of the oldest and largest institutions of higher education in northern Europe. He continued his work experimenting with amphibian embryos, investigating embryological induction in the development of particular tissues.

Between 1914 and 1919, Spemann served as chair of the department of experimental embryology and developmental mechanics at the Kaiser Wilhelm Institute (KWI) for Biology. From 1915 to 1918, Spemann served as codirector of the institute with geneticist Carl Correns.

In 1919, Spemann moved to Freiburg, a city on the French border. As a zoology professor at the University of Freiburg, Spemann set up a Department of Embryology (later renamed the Spemann Graduate School of Biology and Medicine). At Freiburg, Spemann and his colleagues and students carried out numerous experiments in transplantation. The department attracted many students who would become well-known scientists in Germany or abroad, including embryologist Viktor Hamburger and biologist Johannes Holtfreter.

Spemann's doctoral student, Hilde Mangold, conducted work of particular importance. She carried out a series of experiments that Spemann designed, adroitly using microsurgery tools and techniques that Spemann pioneered. During 1921 and 1922, in support of her advanced degree, Mangold performed nearly five hundred transplanted grafts that demonstrated the "organizer effect" (later called embryonic induction), a key principle in the cell division of fertilized eggs. In 1924, Mangold and Spemann coauthored a paper entitled "Induction of Embryonic Primordia by Implantation of Organizers from Different Species."

The paper served as the primary basis for the awarding of the 1935 Nobel Prize in Physiology or Medicine to Spemann. Mangold died of injuries she sustained after a gasoline heater in her home exploded. The Nobel Prize committee does not give awards posthumously, and Mangold was never formally recognized for her contributions, though Spemann always fully credited her important work in public and in his papers.

Spemann retired from his teaching work soon after winning the Nobel Prize, the first embryologist so honored. In 1938, he published a book—*Embryonic Development and Induction*—detailing his earlier experiments. The book also discusses an experiment involving the replacement of one egg nucleus with another nucleus, a concept that laid the foundation for nuclear-transfer cloning in the early 1950s. Spemann died in 1941 at age seventy-two.

Impact

Throughout his career, Spemann was preoccupied with the practical and theoretical aspects of experimental embryology, which was a brand-new field of research when he began his career in the late nineteenth century.

Much of Spemann's work involved the manual division of the microscopic cells of fertilized eggs taken from such animals as newts, salamanders, and frogs in order to observe what happens when cells are split at various places and at different stages of development. To do so, Spemann had to invent the tools necessary to perform such experiments. He created needle-like knives from extruded glass to cut the embryos, and made miniscule pipettes to suck up fragments of cells. He also formed almost invisible glass rods with blunt ends to make small impressions in wax to hold embryos in place and built infinitesimal bridges made of glass to support delicate transplanted tissues. Always innovative, Spemann even used fine hairs from his infant daughter's head to make tiny loops and nooses with which to divide cells.

Spemann spent his entire career exploring the subject of cell development and evolution. In the process, he established many of the principles and techniques of modern embryology and set the stage for the science of cloning.

———*Courtney Farrell*

FURTHER READING
Rheinberger, Hans-Jörg. *An Epistemology of the Concrete: Twentieth-Century Histories of Life.* Durham: Duke UP Books, 2010. Print. Examines the lives, work, and ultimate impact of twentieth-century experimental biologists and life scientists, including Spemann.
Shubin, Neil. *Your Inner Fish: A Journey into the 3.5 Billion-Year History of the Human Body.* New York: Vintage, 2009. Print. Illustrated work incorporating Spemann's experiments in embryological development into an interesting discussion of animal evolution in general and human evolution in particular.
Slack, Jonathan M W. *Essential Developmental Biology.* Hoboken: Wiley-Blackwell, 2005. Print. Explains the processes involved in the development of an embryo on molecular and cellular levels, from fertilization to maturity; contains many full-color drawings, a glossary, and a bibliography.

JACK W. SZOSTAK

CANADIAN AMERICAN BIOLOGIST

- **BORN:** November 9, 1952; London, England
- **PRIMARY FIELD:** Biology
- **SPECIALTIES:** Genetics; biochemistry; molecular biology

Biologist Jack W. Szostak has led many groundbreaking investigations into the origins of life and genetic solutions to disease. His scientific contributions have earned him many honors, including the Nobel Prize for Physiology or Medicine in 2009.

EARLY LIFE
Jack William Szostak is the son of Bill and Vi Szostak. His father, an aeronautical engineer with the Royal Canadian Air Force, was studying at Imperial College when Jack was born in London, England. He was still an infant when his parents returned to Canada. Jack and his two sisters grew up in Germany and Canada as their father was transferred to different posts during a twenty-year career. Eventually, the family settled in Canada near Montreal, Quebec.

An excellent student, Szostak became interested in science at an early age, particularly chemistry and biology. As a boy, he had a basement lab, stocked with substances his mother brought home from the chemical company where she worked, and produced several spectacular explosions. In high school, he built a hydroponic garden. One summer, he worked at his mother's chemical-testing laboratory.

At the age of fifteen, Szostak graduated from Riverdale High School in the Pierrefonds-Roxboro section of Montreal. In 1968, he entered McGill University to begin his undergraduate education in cellular biology. Two years later, he joined the summer program at the Jackson Laboratory, a biomedical research institution in Bar Harbor, Maine. As part of the laboratory's mission to investigate genetics in order to prevent, treat, and cure human diseases, Szostak worked with mice, dissecting thyroid glands to analyze mutant hormones.

Upon returning to McGill, Szostak concentrated on plant biology with a focus on algae and published his first paper on peptide hormones. He graduated with a Bachelor of Science degree in 1972, then used a fellowship to enter Cornell University to study the DNA sequencing of genomes of yeasts, microscopic single-celled fungi. He earned his PhD in biochemistry in 1977 and stayed at Cornell for two additional years to conduct further genetic research.

LIFE'S WORK
In 1979, Szostak was appointed as a teacher of biological chemistry at Harvard Medical School and simultaneously became an independent researcher at the Dana-Farber Cancer Institute in Boston, Massachusetts. He continued to explore genetics, particularly the process of recombination, in which double-stranded DNA is broken and then repaired by combining it with other genetic material.

The following year, while attending a conference on nucleic acids, Szostak met Elizabeth Blackburn. A biological researcher at the University of California, San Francisco, Blackburn was conducting genetic experiments with freshwater protozoa that were similar in nature to Szostak's studies of yeasts. In subsequent discussions, Blackburn and Szostak focused on the function of telomeres, regions at the ends of DNA molecules that protect chromosomes from degrading. The two scientists decided to collaborate to test whether Blackburn's protozoa's telomeres would work with Szostak's yeasts.

Blackburn soon added former postdoctorate student Carol Greider, later a professor of molecular biology and genetics at Johns Hopkins University School of Medicine, to the collaborative team. In 1983, Szostak successfully became the first scientist to clone yeast telomeres. Meanwhile, Blackburn and Greider, applying the results of Szostak's research, were able to isolate and identify telomerase, the enzyme that creates telomeres in DNA.

The recombination and telomere breakthroughs attracted many postgraduate students to Szostak's lab and resulted in new opportunities. In 1984, when Szostak began teaching genetics at Harvard Medical School, he also accepted an appointment to the Department of Molecular Biology at Massachusetts General Hospital. By the late 1980s, when Szostak became a full professor of genetics at Harvard Medical School, his research team and Blackburn's had expanded knowledge of telomere genetics. Szostak's group produced mutated yeast cells that shortened telomeres and prematurely aged the cells; Blackburn's group experienced similar results in mutating protozoan telomerase RNA.

Having exhausted the possibilities of yeasts, Szostak began to gravitate toward the study of ribozymes, which are ribonucleic acid (RNA) molecules that were discovered during the 1980s. Ribozymes catalyze chemical reactions and are key components in the study of how life originated billions of years ago, as well as potential agents in the creation of genetic therapies to combat diseases.

In the early 1990s, Szostak's lab concentrated mainly on RNA. He developed a method of in vitro selection of biological molecules that would allow individual molecules to be screened for specific functions. Essentially an adaptation of the forces of natural selection, the test-tube technique allowed Szostak and his colleagues to evolve RNA aptamers (a term he coined), engineered acid-based molecules that bind to particular molecular targets like cells or tissues. Aptamers show great promise in their ability to zero in on specific diseases; in the early twenty-first century, the U.S. Food and Drug Administration approved an aptamer-based drug for the treatment of age-related macular degeneration, a major cause of visual impairment.

Late in the 1990s, in addition to his other duties, Szostak joined the Howard Hughes Medical Institute, a private nonprofit organization headquartered in Chevy Chase, Maryland, that funds biological and medical research. Between 2000 and 2007, Szostak was also associated with the National Aeronautic and Space Administration (NASA) and was an active participant in the Astrobiology Institute in California. A principal investigator of the NASA's exobiology and evolutionary biology program, he contributed his expertise to studies of the origin, evolution, and distribution of life in the universe.

In the 2000s, Szostak's laboratory research began to center on the creation of protocells, which are replications of ancient inorganic matter that as a result of chemical processes organized into biological life, established a metabolism, became self-reproducing, and evolved into higher forms. In working to unravel the question of how life began, Szostak hopes ultimately to produce genetically generated and chemically based treatments of diseases to help preserve and prolong human life.

Impact

Since the early 1970s, Szostak has worked in tandem with his colleagues in attempting to understand the intricacies of genetics. His research in a number of related areas of molecular biology has far-ranging implications across several disciplines.

Szostak's collaboration with Elizabeth Blackburn and Carol Greider, which resulted in the discovery of the telomere-creating enzyme telomerase and earned the trio the 2009 Nobel Prize in Physiology or Medicine, offers tremendous promise for combating aging and degenerative diseases. Szostak's technique to develop aptamers has considerably shortened the evolutionary period of RNA molecules and simplified the screening process, opening the door for the genetic targeting of particular diseases. His contributions to NASA programs may pay dividends in the future, if humans ever explore other planets and strange new worlds beyond the solar system, and his investigations into the origins of life ultimately may reveal one of nature's longest-held secrets and could pave the way toward human immortality.

Szostak's recognitions for his innovative research include memberships in the National Science Foundation, the National Institutes of Health, the National Research Council, and the American Academy of Arts and Sciences. He was elected to the National Academy of Sciences in 1998 and to the New York Academy of Sciences in 1999. In addition to the Nobel Prize, he shared the 2006 Lasker Award with Elizabeth Blackburn and Carol Greider for their joint discovery of telomerase. Szostak has also been honored with the National Academy of Sciences Award in Molecular Biology (1984), the Genetics Society of America Medal (2000), and the Heineken Prize for Medicine (2008).

—*Jack Ewing*

FURTHER READING

Atkins, John F., Raymond F. Gesteland, and Thomas R. Cech, eds. *RNA Worlds: From Life's Origins to Diversity in Gene Regulation.* Cold Spring Harbor: Cold Spring Harbor Lab, 2010. Print. A close look at the microscopic world of ribonucleic acid (RNA).

Kauffman, Stuart. *Investigations.* New York: Oxford UP, 2000. Print. An examination of the origins and basis of life. Includes discussion in chapter 2, 'The Origins of Life,' of Szostak's work with aptamers and his attempts to produce self-replicating RNA.

Szostak, Jack W. *The Origins of Life.* Cold Spring Harbor: Cold Spring Harbor Lab, 2010. Print. An illustrated discussion encapsulating Szostak's primary area of study: how prehistoric chemical reactions caused organic molecules to coalesce and replicate into what is known as life, and the possibility that similar reactions have or will occur in extraterrestrial settings.

J. CRAIG VENTER

Picture of J. Craig Venter from an article by Liza Gross (no photo credit): "A New Human Genome Sequence Paves the Way for Individualized Genomics." Gross L PLoS Biology Vol. 5, No. 10, e266 doi:10.1371/journal.pbio.0050266. [CC BY 2.5 (http://creativecommons.org/licenses/by/2.5)], via Wikimedia Commons

AMERICAN BIOLOGIST

- **BORN:** October 14, 1946; Salt Lake City, Utah
- **PRIMARY FIELD:** Biology
- **SPECIALTIES:** Genetics; biochemistry

Prominent genomics researcher and entrepreneur, J. Craig Venter is recognized for a wide array of accomplishments that includes the first complete sequencing of the human genome, as well as his 2010 creation of the first self-replicating bacterial cell with synthetic DNA.

EARLY LIFE

John Craig Venter was born on October 14, 1946, in Salt Lake City, Utah, but he spent his childhood in the city of Millbrae, California. He was the second of four children born to John and Elizabeth Venter, who were both members of the United States Marine Corps during World War II.

As a child, Venter was a self-described mischief-maker who loved playing around the railroad tracks near his house. He displayed a strong mechanical and engineering ability, excelling at such projects. Venter was an exceptionally competitive swimmer but did not shine academically. At Mills High School, where he enrolled in 1960, he was rebellious, bored, and frequently in trouble. (In 2007, when his genome became the first individual human genome to be completely sequenced, Venter would discover a possible explanation for his behavior: a genetic anomaly associated with Attention Deficit Hyperactivity Disorder.)

At seventeen, Venter dropped out of high school and moved alone to Costa Mesa, California, but was soon drafted into the military. He enlisted in the United States Navy and began serving as a senior corpsman in a field hospital in Da Nang, South Vietnam. The exposure to death and critical injury had a profound impact on him, and when Venter returned to the United States in 1968, he began taking steps toward a medical career. Venter first took classes at the College of San Mateo, a community college in California, and later he was accepted as a student at the University of California, San Diego (UCSD).

In 1972, he received his BS degree in biochemistry and remained at UCSD to pursue graduate studies. He earned a PhD in physiology and pharmacology in 1975. By this time, his ambitions had turned from practicing medicine to laboratory research. His doctoral dissertation focused on how the hormone adrenaline interacts with cells.

Life's Work

After completing his doctorate, Venter accepted a position as a junior faculty member at the medical school of the State University of New York. In 1982, he became the deputy director of the molecular immunology department at the Roswell Park Cancer Institute. Venter joined the National Institutes of Health (NIH), a government research body based in Bethesda, Maryland, in 1984. It was during his tenure there that the focus of his work turned toward genetic sequencing.

Genetic sequencing is the attempt to map out the structure and function of the chemical building blocks, or base pairs, that form strands of DNA (deoxyribonucleic acid). DNA is the carrier of the genetic instructions that drive the development of most living organisms. A DNA base pair consists of either a molecule of cytosine paired with one of guanine, or a molecule of adenine paired with one of thymine. These bases are abbreviated with the letters A, T, C, and G; a sequenced genome looks like a series of letters, such as "CCAAGTAC."

One of Venter's early projects was the development of so-called expressed sequence tags (ESTs). An EST consists of a small, unique sequence of DNA, a few hundred base pairs in length. This sequence is a portion of a gene that expresses, or contains instructions for producing, a particular protein. By matching ESTs to a longer stretch of genomic information, researchers can quickly identify full-length genes that serve known functions.

In 1992, Venter left the NIH to start his own nonprofit research foundation, the Institute for Genomic Research (TIGR). TIGR made the news in 1995 when it published the first complete genome sequence belonging to a free-living organism—that of *Haemophilus influenzae*, a bacterium that causes various human infections. Although others had managed to sequence virus genomes completely, *Haemophilus influenzae* was a much more complex organism. Venter's success in mapping all of its base pairs in one year was considered remarkable.

Venter had sequenced the bacterium genome using a technique known as "shotgun sequencing." At the time, the traditional method of DNA-sequencing relied on sequencing large chunks of a genome, cloning them in bacterial artificial chromosomes (BACs), a type of engineered DNA molecule, and breaking those down again into smaller pieces. Only after every piece has been sequenced could they be fitted back together. This technique is highly accurate but expensive and time-consuming. Venter's much faster and cheaper shotgun technique breaks a genome into much smaller pieces, which can be sequenced and aligned simultaneously.

In 1998, Venter founded a private company called Celera Genomics. Celera's main goal was to speed up the completion of the sequencing of the entire human genome. This information would provide a database for medical researchers in developing diagnostic tests for diseases based on genetic markers, as well as personalized therapies targeting specific genetic anomalies. The publicly funded Human Genome Project, which consisted of a consortium of international research groups, had been working on this endeavor since 1990 but continued to use the BAC-to-BAC technique. Venter believed the shotgun method would provide a quicker and less expensive solution.

In 2000, what had been a very visible and sometimes contentious race between the public and private sectors to map the complete human genome ended on a collaborative note, with Venter and Francis Collins of the Human Genome Project jointly announcing that both groups had finished draft sequences. It was generally accepted that Celera "won" the race, since its draft was completed several months before that of the Human Genome Project. In the years that followed, Celera scientists would go on to publish the genomes of other organisms, including the mouse and the fruit fly.

Leaving Celera in 2002, Venter threw himself into the field of synthetic biology: the attempt to build

VENTER CREATES THE FIRST SELF-REPLICATING CELL WITH SYNTHETIC DNA

In May 2010, the J. Craig Venter Institute announced its successful creation of the first self-replicating bacterial cell with an entirely synthetic genome. While Venter's team had replicated the bacterium Myocoplasma genitalium in 2008, this new genome represented the longest set of DNA yet to be produced. Venter's team had been working on this line of research since 1995. The release of the paper in which Venter described the achievement received a great deal of media attention. Venter had designed and built a complete artificial bacterial chromosome, inserted it into a living bacterial cell that already possessed its own DNA, and watched the synthetic DNA replace the cell's original genetic code. The synthetic DNA provided the instructions that the bacterium used as it replicated. This replication formed new copies of the cell, copies that also featured the synthetic DNA.

Venter assembled the design for the artificial chromosome on a computer, using the actual DNA of the bacterium Mycoplasma mycoides as a reference. He made several changes to the natural genome, deleting several genes and adding new segments of genetic code that served no practical function but stood for the names of the researchers, their contact information, and a series of famous quotations from such figures as physicist Richard Feynman and author James Joyce. These changes served as watermarks—identifying markers that would distinguish the synthetic genome from that of the original cell.

The first practical problem Venter and his colleagues had to overcome was the question of how they could build a complete bacterial chromosome, which would consist of over one million base pairs of DNA. Existing techniques only enabled the synthesis of partial genomes, consisting of a few thousand base pairs. Venter solved this problem by cutting the code for the synthetic DNA into over a thousand smaller pieces. He added additional overlapping base pairs at the ends of each piece that would later be used to stitch them back together as well as providing a sequence that would enable them to survive in yeast cells. Once these pieces were synthesized, Venter began inserting them into yeast cells. The yeast took in and recombined the DNA pieces into larger wholes until the entire synthetic genome formed a single continuous chromosome.

Finally, the completed synthetic genome was inserted into a living bacterium—not Mycoplasma mycoides, but the related Mycoplasma capricolum. After several unsuccessful trials, Venter's team finally managed to grow a colony of bacteria that possessed the synthetic genome. The cells were producing Mycoplasma mycoides proteins instead of those typical of Mycoplasma capricolum. Their native genetic code had been completely replaced by Venter's synthetic code.

In spite of the breakthrough, Venter admitted that the synthetic genome was a mere "plagiarism" of nature, a copy of a naturally occurring organism. As of 2012, the J. Craig Venter Institute aspired to develop more advanced synthetic organisms for ecological and industrial purposes—organisms, for example, that could repair the environmental damages of pollution and supply sources of fuel.

living biological systems using standardized parts, such as DNA sequences, proteins, and other organic compounds. Among other projects, Venter has been sailing around the world collecting undersea microbes to help build a library of organic building blocks. In 2010, he made the headlines again for creating the first living cell with an entirely synthetic genome. He continues to work on cutting-edge genomics research at the J. Craig Venter Institute, the merged incarnation of several organizations he had founded previously.

Venter has been married three times. He has one child, Christopher Emrys Rae Venter, with his former wife Barbara Rae. He lives in San Diego, California

IMPACT

Venter has received numerous scientific prizes and awards, including the National Medal of Science in 2008. He has twice been named as one of *Time* magazine's most influential people in the world. Many of the tools he has worked on developing or promoting are now widely used throughout biological research. The EST technique for rapid gene identification, for instance, has been used to pinpoint the genes involved in several human diseases, including Alzheimer's and a particular form of colon cancer. The initially controversial technique of shotgun sequencing is, more recently, a conventional tool of genomics. Most current approaches to whole-genome sequencing use some hybrid form of shotgun and BAC-to-BAC sequencing.

Although Venter's bacterium with a synthetic genome was not, as it was sometimes described, the first "synthetic life," it represented a step toward a host of potential applications. The ability to design and manipulate pieces of genetic code like building blocks could enable the creation of clean, renewable sources of alternative energy, self-repairing materials, or highly effective vehicles for drug-delivery. Because of the potential for the tools of synthetic biology to be used in harmful ways, Venter's announcement prompted President Barack Obama to create a bioethics commission to study the regulatory implications of this emerging field.

—*M. Lee*

FURTHER READING

Biello, David, and Katherine Harmon. "Tools for Life: What's Next for Cells Powered by Synthetic Genomes?" *Scientific American* 303.2 (2010): 17–18. Print. Brief commentary on the impact that synthetic biology, such as Venter's artificial bacterial genome, may have on the study of life.

Lesk, Arthur M. *Introduction to Genomics*. New York: Oxford UP, 2007. Print. Student-focused volume covering human genomics, comparative genomics, evolution and genomic change, sequencing techniques, proteomics, and systems biology. Includes extensive full-color illustrations, sidebars, and end-of-chapter problems and exercises.

Pennisi, Elizabeth. "Synthetic Genome Brings New Life to Bacterium." *Science* 21 (2010): 958–59. Print. Explanation of the steps involved in Venter's 2010 creation of the first entirely artificial (bacterial) genome and successful insertion into a recipient cell.

Shreeve, James. *The Genome War: How Craig Venter Tried to Capture the Code of Life and Save the World*. New York: Alfred A. Knopf, 2004. Describes the competition between Venter and Collins to decode the human genome.

Venter, J. Craig. *A Life Decoded: My Genome, My Life*. New York: Penguin, 2007. Print. Provides a look at Venter's personal story, scientific discoveries, and the politics that surround genomic research.

JAMES D. WATSON

James Watson in 1992. By National Cancer Institute (NCI). [Public domain], via Wikimedia Commons

AMERICAN BIOLOGIST

- **BORN:** April 6, 1928; Chicago, Illinois
- **ALSO KNOWN AS:** James Dewey Watson
- **PRIMARY FIELD:** Biology
- **SPECIALTIES:** Genetics; virology

James Watson, working with British biologist Francis Crick, helped identify the double-helix structure of the deoxyribonucleic acid (DNA) molecule, and co-discovered the process of replication responsible for heredity. He also conducted significant research on protein synthesis and the role of viruses in cancer.

EARLY LIFE

James Dewey Watson was born on April 6, 1928, in Chicago, Illinois, and was named after his father, a businessman. Although the Great Depression impacted the family financially, the Watson home was filled with books. As a child, Watson read classic Russian and English literature and became absorbed in almanacs and other reference books. He spent his early years in Chicago and attended the University of Chicago Nursery School, Horace Mann Elementary School, and South Shore High School.

A gifted student, Watson received a scholarship to an experimental program at the University of Chicago for students who had completed two years of high school. He chose to major in zoology and graduated with a Bachelor of Science degree in 1947, at the age of nineteen.

Watson became interested in genetics during his senior year of college. Like Francis Crick, he was inspired by the book *What Is Life?* (1946) by Nobel Prize–winning physicist Erwin Schrödinger. In his book, Schrödinger proposes that genetic material may be found in a molecule, meaning that molecular biology holds the answers to the question of the evolution of life on Earth. Instead of becoming an ornithologist or a naturalist, Watson began entertaining thoughts about studying DNA.

Watson obtained a fellowship to attend graduate school at the University of Indiana at Bloomington, where he studied genetics with Hermann J. Muller. His adviser for his thesis on bacteriophages was Salvatore Luria, a prominent microbiologist. Watson spent one summer at Cold Spring Harbor Laboratory on Long Island, New York, with Luria, learning about bacteriophages.

In 1950, Watson received his PhD in zoology and obtained a Merck Fellowship from the National Research Council to study bacteriophage-infected cells at the University of Copenhagen. He analyzed DNA in bacterial viruses with biochemist Herman Kalckar, who took Watson to a symposium on micromolecules held at the Zoological Station in Naples, Italy, in the spring of 1951. It was there that he first saw an X-ray diffraction photograph of DNA that depicted a helical structure, taken by Maurice Wilkins, a physicist and X-ray crystallographer at King's College in London. It was well known that Wilkins and Rosalind Franklin, his colleague at King's College, had been researching DNA, and suddenly Watson became interested in the research.

LIFE'S WORK

In October 1951, Watson joined a group of scientists researching proteins at Cambridge University's Cavendish Laboratory, not far from King's College. There, he met Francis Crick, a British physicist who had returned to graduate school after World War II and was still working on his PhD.

Crick and Watson struck up a friendship based on their mutual interest in DNA. Crick shared everything he had heard about the King's College team, which had been making considerable progress in cracking the genetic code. Watson mentioned his experience at the symposium in Naples earlier that year and raved about Wilkins's X-ray diffraction photograph. Despite the fact that neither scientist was assigned to research heredity, they decided to enter the race to determine the structure of the DNA molecule. Watson's background in biology complemented Crick's background in physics. However, it would take time for the pair to reach the necessary level of expertise. In addition, Crick had to teach Watson about X-ray crystallography, which was the chief means of recording and investigating the structure of molecules at the time.

Watson and Crick had a number of problems to solve before the basis of human heredity could be revealed. Even with the aid of X-ray diffraction techniques, molecular analysis is very difficult. The technique Watson and Crick used was to make physical models of molecular structures. In the spring of 1952, Watson and Crick saw another X-ray diffraction photograph taken by Rosalind Franklin, the chemist and expert X-ray crystallographer working with Wilkins at King's College. Her photograph clearly depicted the double-helix structure of DNA. Based on her photograph, Watson and Crick were able to put together a model showing the correct structure of DNA.

Crick and Watson submitted their findings, titled "A Structure for Deoxyribose Nucleic Acid," to the scientific journal *Nature*. They were published on April 25, 1953, followed by another article on May 30. Watson and Crick had beaten the King's College team, as well as the distinguished American chemist Linus Pauling, who had also been close to solving the DNA puzzle.

In 1962, Watson, Crick, and Wilkins were awarded the Nobel Prize, although many scientists believe that Rosalind Franklin, had she not died prior to 1962 would have been named instead, for it was her X-ray diffraction photograph that depicted the double helix.

After his work at Cambridge, Watson returned to the California Institute of Technology (Caltech) as a senior research fellow. He worked with students on bacteriophages and conducted additional research in X-ray diffraction studies of ribonucleic acid (RNA).

In 1955, he returned to Cambridge, where he studied tobacco mosaic virus using X-ray diffraction and worked with Crick on a theory about the structure of DNA in viruses. In 1956, Harvard University offered Watson an assistant professorship in the biology department; he became a full professor in 1961. He wrote the first undergraduate textbook on genetics, *Molecular Biology of the Gene* (1965). After several revisions, it is still in use more than forty years later.

Watson remained a bachelor until 1968, when he married Elizabeth Lewis, a sophomore at Radcliffe College. That year, Watson was named director of the Cold Spring Harbor Laboratory. At the time, the institution was suffering from financial problems. Watson turned it into a premier academic institution specializing in cancer research. In 1994, Watson was named president of the institution and worked to obtain accreditation for its doctoral programs. The Watson School of Biological Sciences is now named for him, and he watched the first graduate students obtain their doctorates in June 2004.

In 1968, Watson published his memoir of the DNA discovery, *The Double Helix: A Personal Account of the Discovery of DNA*. The book was controversial for its depiction of Watson's colleagues, and women's groups disapproved of his portrayal of Rosalind Franklin. In his memoir *Genes, Girls, and Gamow: After the Double Helix* (2002), Watson gives Franklin credit for her important role in solving the DNA puzzle.

In 1989, Watson was asked to direct the National Center for Human Genome Research at the National Institutes of Health. For three years, he oversaw the initial development of the Human Genome Project, which led to the draft sequencing of the human genome in 2003 and the finished sequencing in 2006. Concerned about the project's ethical implications, Watson created the Ethical, Legal, and Social Issues Research Program (ELSI).

In addition to the Nobel Prize, Watson has been awarded many other honors, including the Presidential Medal of Freedom in 1977, the National Medal of Science in 1997, and a number of honorary degrees. In 2007, the 454 Life Science Corporation presented Watson with his personalized, full sequence genome—or specific genetic makeup. Watson published the genome online.

Impact

The discovery of DNA's double-helix structure brought about a revolution in biochemistry because it allowed scientists to analyze the genetic makeup of all living organisms. Without the discovery of the double helix, sequencing the human genome would have been impossible. Having a complete sequence of the human genome is a major scientific achievement. However, researchers have much to learn about the number, exact location, functions, and expressions of human genes. As scientists increase their knowledge of how genetic variations affect individuals, new genetic therapies and technologies will become possible. Watson's own work in genetics and molecular biology has had direct implications for poliomyelitis research and the fight against cancer.

In anticipation of these and other breakthroughs made possible by genetic research, Watson and others have begun to address the host of often complex ethical, legal, and social issues that have already arisen, including fair use, privacy and confidentiality, reproductive rights, medical and public health education, environmental concerns, commercialization, and human responsibility.

—*Sally Driscoll*

Further Reading

Ridley, Matt. *Francis Crick: Discoverer of the Genetic Code*. New York: Harper, 2006. Print. Eminent Lives series. Provides a succinct account of Crick's work on DNA with Watson.

Watson, James D. *Avoid Boring People: Lessons from a Life in Science*. New York: Knopf, 2007. Print. Discusses Watson's own recollections of his early life and scientific achievements up to the mid-1970s.

———, and Andrew Berry. *DNA: The Secret of Life*. New York: Knopf, 2003. Print. Traces the evolution of the study of DNA.

WATSON AND CRICK DISCOVER DNA'S DOUBLE HELIX STRUCTURE

In the fall of 1951, when James D. Watson began work at the Cavendish Laboratory at Cambridge University in England, there was no agreed upon understanding of how human genetic information was stored and transmitted. Some biologists believed that genes were protein molecules; others thought that DNA (deoxyribonucleic acid), which was already known to exist in the chromosomes of all cells, was the carrier of genetic information. Watson and his colleague Francis Crick became convinced that research into DNA was most likely to provide the answer to this central question. However, the structure of the DNA molecule was unknown.

The DNA molecule consists of long chains of sugar and phosphate bases. The four different types—adenine (A), cytosine (C), guanine (G), and thymine (T)—are attached to the sugar and sugar phosphates in chemically complex ways. The length of these chains is astonishing; the largest human chromosome contains about three billion base pairs.

Watson and Crick made physical models of the DNA molecule using parts representing these different chemical components. These parts were constructed so that they could not be physically connected to other parts if the chemical bond could not exist in nature. The structure of the model was verified by comparing the angles of the connections to the angles revealed by X-ray diffraction photographs of the actual DNA molecule.

Watson and Crick built their first model out of cardboard and wire, relying on Watson's memory of rival Maurice Wilkins's X-ray diffraction photograph. This first model depicted three strands, or spirals, rather than the two-strand, double helix model that they would later determine to be correct. In addition, the bases were positioned incorrectly and were lacking the right chemical structures.

After they saw rival Rosalind Franklin's X-ray diffraction photograph that clearly showed the double helix structure in 1952, Watson and Crick set about determining the correct placement and pairing of the A, C, G, and T bases. While visiting Cambridge, Austrian chemist Erwin Chargaff took a look at the model and explained his recent determination regarding ratios between the bases, which stated that base A was equivalent to base T and that base C was equivalent to base G. These are now known as Chargaff's rules. Another chemist, Jerry Donohue, assisted Watson and Crick by showing them the correct chemical structure of the bases. The bases form what are now called the "rungs of the ladder," whereas in the first model Watson and Crick had placed the bases along the backbone, or the "rails of the ladder."

From his work on proteins, Crick guessed that the two strands of DNA move in opposite directions and then determined correctly that DNA replicates when the two strands "unzip" and copy themselves. The partners had put together the final, correct model of DNA by March 7, 1953.

EDMUND BEECHER WILSON

AMERICAN ZOOLOGIST

- **BORN:** October 19, 1856; Geneva, Illinois
- **DIED:** March 3, 1939; New York, New York
- **PRIMARY FIELD:** Biology
- **SPECIALTIES:** Cellular biology; zoology; genetics

American zoologist Edmund Beecher Wilson is considered one of most influential cell biologists in history. During a career that spanned the late nineteenth and early twentieth centuries, he conducted multiple studies on the cellular division of eggs in insects and marine creatures, and helped define the vital role of chromosomes in determining gender across the animal kingdom.

EARLY LIFE

Edmund Beecher "Eddy" Wilson was descended from a New England family whose ancestors came to

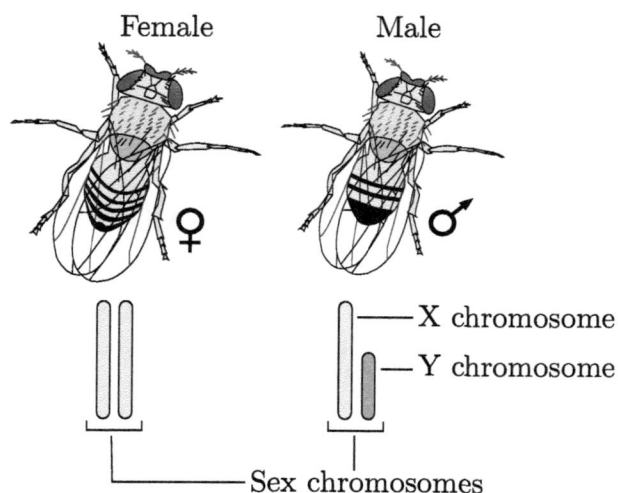

Drosophila XY sex-chromosomes. By GYassineMrabetTalk. (Own work.) [GFDL (http://www.gnu.org/copyleft/fdl.html) or CC BY-SA 4.0-3.0-2.5-2.0-1.0 (https://creativecommons.org/licenses/by-sa/4.0-3.0-2.5-2.0-1.0)], via Wikimedia Commons

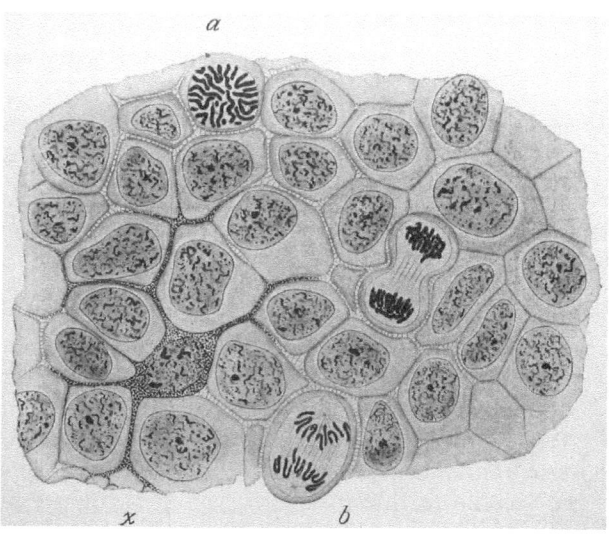

Figure 1 from Wilson, Edmund B. (1900) *The Cell in Development and Inheritance* (2nd ed.), New York: The Macmillan Company. [Public domain], via Wikimedia Commons

America from England on the *Mayflower*. His father, Isaac Grant Wilson, was a lawyer and judge. His mother's name was Caroline Louisa Clarke Wilson. Wilson's parents relocated from Massachusetts to Geneva, Illinois, in the early 1840s. One of five children, Wilson developed an interest in music and natural history at an early age, studying birds, reptiles, and insects.

At the age of sixteen, Wilson became a salaried teacher at a one-room schoolhouse in the village of Oswego, Illinois, just south of Geneva. He presided over twenty-five students, ranging in age from six to eighteen, instructing them in reading, writing, arithmetic, and history. In 1872, he took an entrance exam for West Point. Despite earning a top score, he was too young to be admitted. The following year, Wilson entered Antioch College in southern Ohio, where he took a challenging course of study that included Latin, Greek, zoology, botany, chemistry, geometry, and trigonometry. To help support himself, he worked as an assistant in a geographical survey of Lake Ontario and Lake Erie.

In 1875, Eddy transferred to Yale University. He worked the summer of 1877 with the U.S. Commission of Fish and Fisheries (also known as the Fish Commission, established in 1871). Sailing out of Gloucester, Massachusetts, aboard the *Speedwell*, a naval steamship, he participated in dredging operations and collected marine animals. Wilson graduated from Yale in 1878, majoring in biology. That same year, he published the first of more than one hundred zoological papers. He stayed on for another year at Yale, taking graduate courses in embryology and heredity. Wilson was then granted a fellowship to study at Johns Hopkins University, where he remained for three years while conducting research at the university's marine station on Chesapeake Bay. He earned his PhD in 1881.

Life's Work

In 1882, Wilson traveled abroad to further his education. He first studied at Cambridge University, taking courses from some of the leading British scientists of the day, including zoologist and comparative anatomist Adam Sedgwick, physiologist and embryologist Walter Heape, zoologists William Hay Caldwell and William Bateson, and physiologist Michael Foster. He then moved to Germany, where he conducted laboratory research under zoologist Rudolf Leuckart and took classes from physiologist and comparative anatomist Carl F. W. Ludwig at the University of Leipzig. Wilson then worked for a year with German scientist Anton Dohrn, founder and director of the Zoological Station in Naples, Italy, conducting experiments and making observations of marine animals. At the station, Wilson also worked alongside marine

WILSON'S KEY GENETIC DISCOVERY: XX AND XY CHROMOSOMES

Edmund Beecher Wilson began his scientific investigations as a traditional biologist, following the conventional precept of "ontogeny recapitulates phylogeny." The theory, since discredited, was concocted by influential German biologist-naturalist Ernst Haeckel, who hypothesized that animals in the process of embryonic development (ontogeny) pass through or recapture stages that reflect the evolution of ancient ancestors (phylogeny).

Wilson's early research centered on observation and examination of the structure and activity of the hydra, a fresh-water animal with tentacles that has the ability to regenerate missing limbs. He later studied annelids (earthworms) and platyhelminthes (flatworms), which can also recreate lost segments and are self-propagating hermaphrodites. Wilson's investigations led him to concentrate on embryology, with a particular focus on cytology— defined as the study of the structure, function and chemistry of cells—especially as related to the cellular division that occurs in fertilized eggs.

In the late 1880s and early 1890s, in the process of attempting to draw connections between heredity and cellular development, he conducted research at the Marine Biological Laboratory and elsewhere. Wilson experimented with the cell division in the eggs of marine worms (Nereis), studied germination in mollusks (Dentalium), compared cleavage stages and cellular organization in the eggs of sea snails (Patella), and induced artificial parthenogenesis—reproduction without fertilization—in the eggs of sea urchins (Echinoidea). By the end of the nineteenth century, Wilson had concluded that chromosomes, packages of genes found in the nuclei of cells, were the primary carriers of hereditary information in any organism.

To further develop his idea, Wilson focused his research on the embryology of insects, paying particular attention to chromosomes. Using the fruit fly (Drosophilidae) as his initial subject of study, and later extending his research with investigation of arachnids (Scorpiones), Wilson painstakingly separated and counted the chromosomes within fertilized eggs. Among the pairs of chromosomes contained in cell nuclei—half contributed by the male, half by the female of the species—that provide an organism's genetic blueprint (now termed the genome), he sometimes found an additional set of chromosomes. Through careful observation, Wilson noticed that the extra pair of chromosomes was present only in germ cells produced by the male. Furthermore, when the fertilized egg began to divide, the extra chromosome was found in only one of the two new cells. When the cell divided again, two of the four cells contained the extra chromosome; this process continued throughout cell division.

On the basis of his research, Wilson concluded that—unlike the other chromosomes, called autosomes, which govern other characteristics of the offspring—the extra chromosome was solely concerned with determining the sex of the new individual. Females always contribute an XX chromosome to the sexual development of an egg. Males, however, can either contribute the usual XX or an extra XY chromosome. When both male and female contribute XX chromosomes, the result is female. When the male contributes the extra XY chromosome, the progeny is male.

Wilson's discovery, which could be applied to other species, including humans—had far-reaching implications, particularly in animal breeding, and was highly influential in the further development of genetics as a specialized science. Later research conducted by other scientists would reveal anomalies within the X-Y chromosome system, and slightly different systems in certain organisms—butterflies, frogs, and some fish—for determining sex. Remarkably, one of Wilson's former students, Nettie Stevens, simultaneously conducted similar research and independently reached the same conclusions at the same time as Wilson, and today both Wilson and Stevens are jointly credited with the discovery of the XX and XY sex-determinant chromosome system.

biologist Hugo Eisig and comparative invertebrate anatomist Arnold Lang.

Wilson returned to the United States in 1883 to teach biology at Williams College. After spending a year as lecturer at the Massachusetts Institute of Technology (MIT), he took a position at Bryn Mawr College, where he served as head of the biology department for six years.

Following another year abroad in Munich, Naples, and Sicily, Wilson was appointed to a post at Columbia University, where he helped establish the school's new department of zoology. He began spending summers researching at the Marine Biological Laboratory in Woods Hole, Massachusetts.

Wilson remained at Columbia for the rest of his professional career, teaching biology and zoology, including specialized courses on heredity, cytology, chromosomes, variation, and evolution. Under Wilson's influence, Columbia's department of zoology became internationally renowned as a leading institution in the study of modern genetics. Many of Wilson's students became leaders in the fields of biology and zoology. His students Thomas Hunt Morgan and Hermann Muller later won Nobel Prizes.

Throughout his time at Columbia, Wilson conducted extensive research on cell organization, structure, and function, with a particular emphasis on studies of eggs through every stage of development. In the course of his research, he became an expert in the use of microscopes and was known for his proficiency in tissue sectioning and slide staining.

While at Columbia, Wilson published numerous papers based on his research related to chromosomes, spermatogenesis, and experimental embryology that helped to refine existing concepts of evolution. During his tenure, he published an authoritative book, *The Cell in Development and Inheritance* (1896), which became a standard college text and went through second (1915) and third (1925) editions before advancements in the field rendered it obsolete. Part of the work's long popularity was due to the clarity of the illustrations that Wilson, an excellent draftsman, had included.

In 1904, Wilson married Anne Maynard Kidder of Washington, D.C., whose family was associated with the U.S. Fish Commission and with the Marine Biological Station. The couple had a daughter, Nancy, who became a professional cellist. In 1906, the couple took a long working vacation, traveling through Arizona, Wyoming, and California to collect insects for Wilson's investigations into chromosomes.

The first decade of the twentieth century was a productive time for Wilson. Between 1905 and 1912—during a revival in interest in the genetic studies of late Austrian scientist Gregor Mendel—he conducted extensive research into heredity, chromosomes, and cell division. Inducted into the American Academy of Arts and Sciences in 1902 and into the American Association for the Advancement of Science in 1913, Wilson continued his work until the end of his life. He published his final paper in 1937, nine years after retiring as a professor at Columbia. Wilson died in 1939 at age eighty-two.

IMPACT

During his four decades of teaching, Wilson influenced numerous students who worked to advance scientific understanding of biology. Morgan (Nobel Prize, 1933) and Muller (Nobel Prize, 1946) are perhaps the best known of Wilson's former students, but his classes also inspired other such leading evolutionary biologists and geneticists as Nettie Stevens, Gary N. Calkins, James McGregor, Alfred Henry Sturtevant, and Marcella O'Grady Boveri.

Considered America's first cytologist (cell biologist), Wilson also pioneered the disciplines of embryology, heredity, and genetics through his research on marine animals and insects. Among his many contributions to the study of evolution was his coining of the term "stem cell" in 1896, and his discovery of the XX and XY chromosomes.

Wilson received many honors and awards for his work, both during his lifetime and posthumously. In 1925, the National Academy of Sciences presented him with the Daniel Giraud Elliot Medal for his contributions to zoology. In 1928, he received the Gold Medal of the Linnean Society of London. In 1936, he garnered the U.S. National Academy of Sciences John J. Carty Medal and Award. Over the years, he collected honorary degrees from numerous universities. In his memory, the American Society for Cell Biology established the E. B. Wilson Medal and Lecture as its highest honor for accomplishments in the field.

—*Jack Ewing*

Further Reading

Carlson, Elof Axel. *Mutation: The History of an Idea from Darwin to Genomics.* Long Island: Cold Spring Harbor Lab P, 2011. Print. Discusses the birth and development of mutation, a key concept in the theory of evolution that has occupied geneticists—including Edmund Beecher Wilson—for more than 150 years.

Laubichler, Manfred D., and Jane Maienschein, eds. *From Embryology to Evo-Devo: A History of Developmental Evolution.* Cambridge: MIT P, 2007. Print. Presents an overview of the modern field of evolutionary developmental biology, which has grown out of earlier research into embryology, heredity, and genetics.

Wilson, Edmund Beecher. *The Supernumerary Chromosomes and their Relation to the Odd or Accessory Chromosome.* Whitefish: Kessinger, 2010. Print. Reprint of Wilson's groundbreaking 1909 work on chromosomes as determinants of sex.

Norton David Zinder

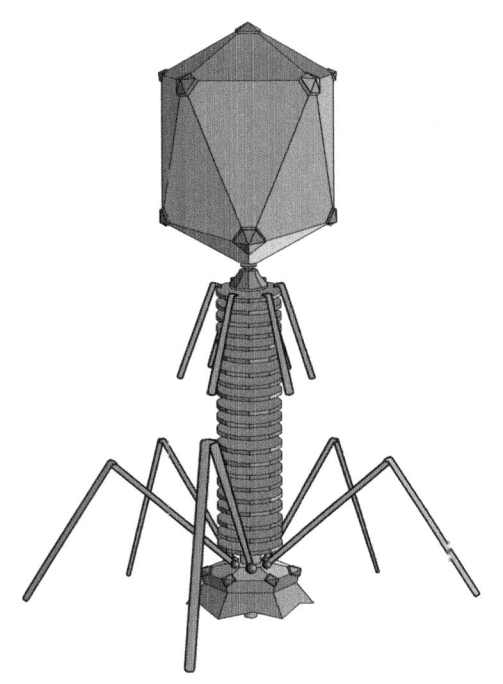

Artistic rendering of a T4 bacteriophage. By Adenosine (Own work,) [CC BY-SA 3.0 (https://creativecommons.org/licenses/by-sa/3.0) or GFDL (http://www.gnu.org/copyleft/fdl.html)], via Wikimedia Commons

American virologist

- **Born:** November 7, 1928; New York, New York
- **Died:** February 3, 2012; New York, New York
- **Primary field:** Biology
- **Specialties:** Genetics; virology; microbiology

As a graduate student, twentieth-century American virologist Norton Zinder discovered transduction, the process by which genetic material is transferred between bacteria by bacteriophage (bacterial viruses). Later in his career, he discovered the first bacteriophage in which RNA was the genetic material.

Early Life

Norton David Zinder was born November 7, 1928. He was the first of two sons of Harry Zinder, a local manufacturer, and Jean Gottesman Zinder. Following his early education in the New York public school system, he attended the Bronx High School of Science, founded in 1938 to provide education for gifted students in science and mathematics. He graduated in 1944 at the age of fifteen.

Zinder enrolled at Columbia University in New York City. He received his bachelor's degree in biology at the age of eighteen in 1947. While an undergraduate at Columbia, Zinder spent time in the zoology laboratory of biologist Francis J. Ryan. Among Ryan's former students was molecular biologist Joshua Lederberg, who by then was on the faculty at the University of Wisconsin, and it was Ryan who helped Lederberg apply techniques while working with the bread mold *Neurospora* to a system utilizing *Escherichia coli*. Ryan encouraged Zinder to pursue his postgraduate work at the University of Wisconsin in order to work with Lederberg.

Life's Work

Zinder joined Lederberg's laboratory in 1948. Lederberg had recently completed his graduate work with geneticist Edward Tatum and had co-discovered bacterial conjugation, the transfer of genetic material between cells. Lederberg had carried out this work using the bacterium *Escherichia coli* (*E. coli*), and was interested in determining whether the

ZINDER DISCOVERIES: TRANSDUCTION AND THE TRANSFER OF GENETIC INFORMATION BY BACTERIOPHAGE

In 1948, Norton Zinder joined Joshua Lederberg's laboratory as a first-year graduate student. Two years earlier Lederberg and Edward Tatum had discovered recombination as a means of transferring genetic material among strains of the bacterium *Escherichia coli*. Zinder planned to study the same phenomenon among a different species of bacteria, *Salmonella typhimurium*. The choice of Salmonella was not fortuitous. It was a known pathogen—certain strains were the etiological agents of typhoid fever—and a large number of mutant strains were already available in the Lederberg laboratory. The existence of so many strains suggested they might be useful in applying the recombination studies previously used for *E. coli*.

Zinder began by refining the procedure routinely used in creating auxotrophs (mutants), the addition of mutagens to the bacterial growth media, followed by screening for bacteria that had become the desired mutants. The challenge was the difficulty in the selection of mutants against the background of wild-type (normal) bacteria. Zinder correctly assumed that if penicillin was added to media that lacked the nutrient the mutant required, any (wild-type) bacteria that grew would be killed, leaving only mutants to be removed, washed, and isolated.

The initial experiments carried out by Zinder in which various combinations (crosses) of nutritional mutant auxotrophs were incubated together proved negative; he found no evidence for conjugation. Finally, after trying over one hundred combinations of crosses, he observed the presence of some recombinants. One explanation was these could potentially represent spontaneous reversion to the wild-type rather than the result of conjugation. The problem was addressed by using individual auxotrophs with combinations of mutations, making spontaneous reversion of several genes simultaneously unlikely.

Conjugation between bacteria required direct contact. However, when Zinder placed the two different strains on opposite sides of a filter he again observed recombination, making conjugation unlikely to have been the explanation. Since the agent that caused the recombination must have passed through the filter, Zinder and Lederberg referred to it as a "filterable agent." Identification of the agent as a virus began with elimination of what it was not. Since it was not affected by the enzyme DNase, it was not likely free DNA, as would be the case with transformation. Its sensitivity to heat also ruled out DNA and pointed to a protein component. When the agent was observed with an electron microscope, it was identified as a bacteriophage. The biologic characteristics of the agent, such as the limitations in the quantities of the agent that could infect bacteria, also helped identify the agent as a bacteriophage. The timing of the discovery also played a role: Several years earlier Andre Lwoff had discovered lysogeny, the ability of bacteriophage to integrate into the bacterial genome.

The term transduction ("to lead across") in defining the role of bacteriophage as vector was coined by Lederberg at a scientific meeting in 1951. Lederberg applied the term in a more general sense, in part because he was not convinced of the unique role played by bacteriophage. Zinder, however, used the term in the sense in which it was subsequently defined, the view being widely accepted by the mid-1950s.

process could also take place among other species of bacteria.

Zinder began this work with the bacterium *Salmonella*, chosen in part because of the large number of available mutants. The experimental method was similar to that previously used by Lederberg for *E. coli*, crossing two different strains that had different amino acid nutritional requirements and observing whether recombinants were produced that were capable of synthesizing those amino acids. In the process of conjugation, it was thought that the bacteria must come into contact with each other for the genetic information to pass. However, Zinder found in his experiments that contact was not necessary; the two different strains could even be kept separate using a filter between them. Zinder's interpretation was that something physical was passing through the filter from one cell to another. The agent was shortly identified as a bacteriophage, a bacterial virus that was incorporating fragments of DNA from one strain and passing them to the second strain upon infection. Lederberg termed the process *transduction*. The same phenomenon was subsequently demonstrated in other species of bacteria.

In December 1949, Zinder married Marilyn Estreicher. Their fifty-five year marriage lasted until her death in 2004. The couple had two sons.

After receiving his PhD at Wisconsin in Medical Microbiology in 1952, Zinder accepted an offer as Assistant Professor at the Rockefeller Institute in New York, where he remained for the rest of his professional career. In 1960, Timothy Loeb joined Zinder's laboratory as a graduate student. Loeb was interested in whether bacteriophage existed that were "male" specific and were capable of infecting the pilus on *E. coli* bacteria, the means by which conjugation took place. Since *E. coli* was found in sewage, a logical place to search for such bacteriophage was at the sewage plant in New York. Loeb quickly found seven types of such phage, which he and Zinder termed f1 (for fertility) through f7. F1 was found to contain single-stranded DNA as its genetic material, but f2 through f7 all contained RNA as genetic material, the first such bacteriophage demonstrated to contain RNA. Zinder found that the viral RNA was not only the genome, but that it served as its own messenger RNA, associating with ribosomes for translation into proteins.

In 1964, Zinder was promoted to professor of genetics at the institute, and a decade later he was promoted to John D. Rockefeller, Jr. Professor of Molecular Genetics. That same year saw his appointment as chair of the committee for evaluation of the Virus Cancer Program. The Zinder Report's recommendation for altering the funding process resulted in a significant reorganization of the cancer program. In 1974, Zinder was also among those expressing concern about hazards associated with the new technology of recombinant DNA. The 1975 Asilomar Conference provided guidelines for such work.

During the late 1980s, scientists began work on what became known as the Human Genome Project, the goal of which was to sequence the entire human genome. Two figures stood at the forefront of the proposed project: James Watson, Nobel laureate for his role in determining the structure of DNA, and J. Craig Venter, biochemist and physiologist who entered the project as a private entrepreneur. The concern among many scientists was that if the project was publicly funded through research grants, the funding would occur at the expense of other equally worthwhile projects. At the time it was believed that relatively little of the human genome was relevant and that most consisted of unread sequences considered "junk." Zinder repeatedly pushed for serious consideration by funding committees and was ultimately successful. The presence of Venter as a private citizen duplicating the work that Watson hoped to lead was also a concern. In 2000, Zinder negotiated an understanding between the different parties that allowed the project to continue.

Zinder died February 3, 2012 in New York at the age of eighty-three.

Impact

Zinder's discovery of transduction represented the third mechanism of transfer of genetic information between bacteria. As noted by other scientists in the field of genetics, several aspects common to each of these mechanisms involving movement of genetic elements in bacteria set them apart from genetic recombination in eukaryotic organisms. First, recombination did not involve the entire genome but only portions of the DNA. The discovery of transduction altered the thinking applied to the role of viruses. In addition to possibly killing the bacterial cells they infect, viruses could also serve as vectors for the transfer of genetic material.

Within several years, two forms of transduction were shown to take place. Generalized transduction, as first described by Zinder, involved the transfer of random fragments of DNA, largely anything that was packaged during bacteriophage assembly. The discovery of lysogeny in bacteria, the integration of bacteriophage into the host genome usually at nonrandom sites, allowed for what became known as specialized transduction, the transfer of only those genes adjacent to the site of integration in the host.

The discovery of transduction provided one more technique for fine-structure genetic mapping of bacteria. By determining which genes co-transduced in the same fragment, it became possible to observe which particular genetic markers were adjacent to each other in the bacterial genome. Among the first of these discoveries was that genes that regulate specific metabolic pathways or regulate the expression of genes within a pathway are often closely linked to each other on the bacterial genome in an area that became known as the operon. For example, in the early 1960s, French molecular biologists François Jacob and Jacques Monod produced their model of regulation of the *lactose operon* by means of transducing specific genes within the operons of *Escherichia coli* and *Shigella*. The model they produced became the basis for Jacob and Monod being awarded Nobel Prizes in Physiology or Medicine in 1965.

——*Richard Adler*

FURTHER READING

Brock, Thomas. *The Emergence of Bacterial Genetics*. Woodbury, New York: Cold Spring Harbor Lab. P, 1990. Print. Reviews the history behind understanding bacterial genetics. Zinder's role in the discovery of transduction is discussed.

Crotty, Shane. *Ahead of the Curve: David Baltimore's Life in Science*. Berkeley: U of California P, 2001. Print. Biography of Nobel laureate David Baltimore. Addresses the significance of Zinder's role in the discovery of RNA phage.

Witkowski, Jan. *The Inside Story: DNA to RNA to Protein*. Woodbury, New York: Cold Spring Harbor Lab. P, 2005. Print. Includes a discussion of Zinder's discovery of RNA phage, which was among the first indications RNA could also serve as genetic material.

TIME LINE OF INVENTIONS AND SCIENTIFIC ADVANCEMENTS IN BIOTECHNOLOGY

Before Common Era (BCE)

7000 BCE	Chinese discover fermentation through beer making.
6000 BCE	Yogurt and cheese made with lactic acid-producing bacteria by various people.
4000 BCE	Egyptians bake leavened bread using yeast.
500 BCE	Moldy soybean curds used as an antibiotic.
250 BCE	The Greeks practice crop rotation for maximum soil fertility.
100 CE	Chinese use chrysanthemum as a natural insecticide.

Pre-20th Century

1663	First recorded description of dying cells by Robert Hooke.
1675	Antonie van Leeuwenhoek discovers and describes vagina and protozoa.
1798	Edward Jenner uses first viral vaccine to inoculate a child from smallpox.
1802	The first recorded use of the word biology.
1824	Henri Dutrochet discovers that tissues are composed of living cells.
1838	Protein discovered, named and recorded by Gerardus Johannes Mulder and Jöns Jacob Berzelius.
1862	Louis Pasteur discovers the bacterial origin of fermentation.
1863	Gregor Mendel discovers the laws of inheritance.
1864	Antonin Prandtl invents first centrifuge to separate cream from milk.
1869	Friedrich Miescher identifies DNA in the sperm of a trout.
1871	Ernst Hoppe-Seyler discovers invertase, which is still used for making artificial sweeteners.
1877	Robert Koch develops a technique for staining bacteria for identification.
1878	Walther Flemming discovers chromatin leading to the discovery of chromosomes.
1881	Louis Pasteur develops vaccines against bacteria that cause cholera and anthrax in chickens.
1885	Louis Pasteur and Emile Roux develop the first rabies vaccine and use it on Joseph Meister.

20th Century

1919	Károly Ereky, a Hungarian agricultural engineer, first uses the word biotechnology.
1928	Alexander Fleming notices that a certain mould could stop the duplication of bacteria, leading to the first antibiotic: penicillin.
1933	Hybrid corn is commercialized.
1942	Penicillin is mass-produced in microbes for the first time.

Time Line of Inventions and Scientific Advancements in Biotechnology

1950	The first synthetic antibiotic is created.
1951	Artificial insemination of livestock is accomplished using frozen semen.
1952	L.V. Radushkevich and V.M. Lukyanovich publish clear images of 50 nanometer diameter tubes made of carbon, in the Soviet Journal of Physical Chemistry.
1953	James D. Watson and Francis Crick describe the structure of DNA.
1958	The term bionics is coined by Jack E. Steele.
1964	The first commercial myoelectric arm is developed by the Central Prosthetic Research Institute of the USSR, and distributed by the Hangar Limb Factory of the UK.
1972	The DNA composition of chimpanzees and gorillas is discovered to be 99% similar to that of humans.
1973	Stanley Norman Cohen and Herbert Boyer perform the first successful recombinant DNA experiment, using bacterial genes.
1974	Scientist invents the first biocement for industrial applications.
1975	Method for producing monoclonal antibodies developed by Köhler and César Milstein.
1978	North Carolina scientists Clyde Hutchison and Marshall Edgell show it is possible to introduce specific mutations at specific sites in a DNA molecule.
1980	The U.S. patent for gene cloning is awarded to Cohen and Boyer.
1982	Humulin, Genentech's human insulin drug produced by genetically engineered bacteria for the treatment of diabetes, is the first biotech drug to be approved by the Food and Drug Administration.
1983	The Polymerase Chain Reaction (PCR) technique is conceived.
1990	First federally approved gene therapy treatment is performed successfully on a young girl who suffered from an immune disorder.
1994	The United States Food and Drug Administration approves the first GM food: the "Flavr Savr" tomato.
1997	British scientists, led by Ian Wilmut from the Roslin Institute, report cloning Dolly the sheep using DNA from two adult sheep cells.
1999	Discovery of the gene responsible for developing cystic fibrosis.
2000	Completion of a "rough draft" of the human genome in the Human Genome Project.

21st Century

2001	Celera Genomics and the Human Genome Project create a draft of the human genome sequence. It is published by Science and Nature Magazine.
2002	Rice becomes the first crop to have its genome decoded.
2003	The Human Genome Project is completed, providing information on the locations and sequence of human genes on all 46 chromosomes.

2008	Japanese astronomers launch the first Medical Experiment Module called "Kibo", to be used on the International Space Station.
2009	Cedars-Sinai Heart Institute uses modified SAN heart genes to create the first viral pacemaker in guinea pigs, now known as iSANs.
2012	Thirty-one-year-old Zac Vawter successfully uses a nervous system-controlled bionic leg to climb the Chicago Willis Tower.
2012	Researchers at the University of Washington in Seattle announce the successful sequencing of a complete fetal genome using nothing more than snippets of DNA floating in its mother's blood.
2013	Two research teams announced a fast and precise new method for editing snippets of the genetic code. The so-called CRISPR system takes advantage of a defense strategy used by bacteria.
2013	Researchers in Japan developed functional human liver tissue from reprogrammed skin cells.
2013	Researchers published the results of the first successful human-to-human brain interface.
2013	Doctors announced that a baby born with HIV had been cured of the disease.
2014	Researchers showed that blood from a young mouse can rejuvenate an old mouse's muscles and brain.
2014	Researchers figured out how to turn human stem cells into functional pancreatic ß cells—the same cells that are destroyed by the body's own immune system in type 1 diabetes patients.
2014	A woman gave birth to a baby after receiving a womb transplant.
2014	An international team of scientists reconstructed a synthetic and fully functional yeast chromosome, potentially leading to custom-built organisms (human organisms included).
2015	Scientists from Singapore's Institute of Bioengineering and Nanotechnology designed short strings of peptides that self-assemble into a fibrous gel when water is added for use as a healing nanogel.
2015	Researchers in Sweden developed a blood test that can detect cancer at an early stage from a single drop of blood.
2015	Scientists discovered a new antibiotic, the first in nearly 30 years, that may pave the way for a new generation of antibiotics and fight growing drug-resistance. The antibiotic, teixobactin, can treat many common bacterial infections, such as tuberculosis, septicaemia, and C. diff.
2015	A team of geneticists finished building the most comprehensive map of the human epigenome, mapping more than 100 types of human cells to help researchers better understand the complex links between DNA and diseases.
2015	Stanford University scientists revealed a method that may be able to force malicious leukemia cells to change into harmless immune cells, called macrophages.
2015	Using cells from human donors, doctors, for the first time, built a set of vocal cords from scratch. The cells were urged to form a tissue that mimics vocal fold mucosa —vibrating flaps in the larynx that create the sounds of the human voice.

2016	CRISPR, the revolutionary gene-editing tool that promises to cure illnesses and solve environmental calamities, took a major step forward this year when a team of Chinese scientists used it to treat a human patient for the very first time.
2016	Researchers found that an ancient molecule, GK-PID, is the reason single-celled organisms started to evolve into multicellular organisms approximately 800 million years ago.
2016	For the first time, bioengineers created a completely 3D-printed 'heart on a chip.'
2017	Researchers at the National Institute of Health discovered a new molecular mechanism that might be the cause of severe premenstrual syndrome known as PMDD.
2017	Scientists at the Salk Institute in La Jolla, CA, said they're one step closer to being able to grow human organs inside pigs.
2017	First step taken toward epigenetically modified cotton.
2017	Research reveals different aspects of DNA demethylation involved in tomato ripening process.
2017	Sequencing of green alga genome provides blueprint to advance clean energy, bioproducts.
2017	Scientists engineer disease-resistant rice without sacrificing yield.
2017	Blood stem cells grown in lab for the first time.
2017	Researchers at Sahlgrenska Academy – part of the University of Gothenburg, Sweden – generated cartilage tissue by printing stem cells using a 3D-bioprinter.
2017	Two-way communication in brain-machine interface achieved for the first time.

GLOSSARY

animal breeding: The practice of selecting and mating domesticated animals to enhance their contributions to humans.

animal testing: Use of nonhuman animals for research purposes, particularly medical research

anthrax: Deadly disease caused by the soil bacterium Bacillus anthracis.

anthropogeomorphology: the alteration of the Earth's surface by human activity, has profound effects on the environment, including land, air, and water, and has the potential to alter local climates as well.

antibiotic-resistant bacteria: Ability of some bacteria to resist or entirely withstand the effects of antimicrobial agents.

antibiotics as defense against biological warfare: Therapeutic agents that kill infectious microorganisms.

artificial organs: complex systems that assist or replace failing organs.

audio engineering: also known as sound engineering and audio technology, is the recording, manipulation, and reproduction of sound, especially music.

bioassays: Tests that use biological organisms to detect the presence of given chemical substances or determine the biological activity of known substances in particular environments

biochemical engineering: involves designing and building those industrial processes that use catalysts, feedstocks, or absorbents of biological origin.

biodetectors: Devices comprising highly specific sensing components—such as biolayers of DNA, proteins, or enzymes— immobilized on surfaces that serve as transducers that measure electrical signals produced by interactions between the biomolecules of interest and the biolayers.

bioengineering: interdisciplinary field of applied science that deals with the application of engineering methods, techniques, design approaches, and fundamental knowledge to solve practical problems in the life sciences, including biology, geology, environmental studies, and agriculture.

biofertilizers: Fertilizers consisting of either naturally occurring or genetically modified microorganisms

biofuels: renewable fuels generated from or by organisms. They can be manufactured from this organic matter and, unlike fossil fuels, do not require millennia to be produced.

biostratigraphy: the method of identifying and differentiating layers of sedimentary rock (strata) by their fossil content.

biosynthetic: to any type of material produced via a biosynthetic process. A biosynthetic process uses enzymes and energetic molecules to transform small molecules into larger molecules within the cells of organisms.

biotechnology: the use of living organisms, or substances obtained from such organisms, to produce products or processes of value to humankind

biotoxins: Toxic substances that originate from biological sources, including viruses, **bacteria**, fungi, algae, and plants.

botany: an old branch of science that began with early humankind's interest in the plants around them. In modern society, plant science extends beyond that interest to cutting-edge biotechnology

botulinum toxin: Highly toxic substance produced by the Clostridium botulinum bacterium that targets nerve tissue and blocks neuromuscular transmission of impulses in the body, causing the paralytic disease botulism.

bubonic plague: Highly contagious human bacterial disease with a very high rate of mortality.

correlation: the determination of the equivalence of age or stratigraphic position of two strata in separate areas, or, more broadly, determination of the geological contemporaneity of events in the geologic histories of two areas

eubacteria: minute, prokaryotic organisms that inhabit a range of habitats considerably greater in diversity than those occupied by other organisms; exclusive of archaebacteria, they constitute the majority of prokaryotes

eukaryotic cell: the cell type present in all animals, plants, fungi, and protists; each cell has a distinct nucleus and mitochondria, chloroplasts, and other subcellular structures absent in prokaryotic cells

fossils: remains or traces of animals and plants preserved by natural causes in the earth's crust

genetic engineering: the manipulation of deoxyribonucleic acid (DNA) and the transfer of genes or gene components from one species to another

index fossil: a fossil that can be used to identify and determine the age of the stratum in which it is found

industrial processes: used in food, waste-management, pharmaceutical, and agricultural plants are often called unit operations. Those unit operations used in combination with biological organisms or molecules include heat and mass transfer, bioreactor design and operation, filtration, cell isolation, and sterilization.

microbialite: a biogenic sedimentary structure that is found fossilized in sedimentary rock strata of various ages and that is attributed to the life activities of prokaryotes

polders: Dutch land areas reclaimed from sea or marsh

prokaryotes: generally, single-celled organisms that often grow in colonies and that are made of prokaryotic cells

prokaryotic cell: the cell type found in the Kingdom Eubacteria and Archaebactera, characterized by a number of criteria, including the absence of a cell nucleus, mitochondria, and chloroplasts

ribosome: a large multienzyme complex made up of protein and ribonucleic acid (RNA) molecules that carry and process information stored in deoxyribonucleic acid (DNA); this information is also carried by RNA to synthesize proteins

sedimentary rocks: rocks formed by the accumulation of particles of other rocks or of organic skeletons or of chemical precipitates or some combination of these

stratigraphy: the study of layered rocks, especially of their sequence and correlation

stratum (plural, strata): a single bed or layer of sedimentary rock

stromatolite: a biogenic sedimentary layered structure produced by sediment trapping, binding, or precipitation as a result of the photosynthesis of microorganisms, principally cyanobacteria (blue-green algae)

GENERAL BIBLIOGRAPHY

"A Citizen's Guide to Bioremediation." *EPA*, Sept. 2012. Accessed from https://clu-in.org/download/Citizens/a_citizens_guide_to_bioremediation.pdf.

"Alfred D. Hershey - Biographical". *Nobelprize.org* Nobel Media AB 2014. Web. 15 Dec 2017. http://www.nobelprize.org/nobel_prizes/medicine/laureates/1969/hershey-bio.html

"Animal Research Numbers Continue Downward Trend According to Newly-Released Report." *National Anti-Vivisection Society*. Natl. Anti-Vivisection Soc., 6 July 2015. Web. 9 July 2015.

"Animal Testing & Cosmetics." *US Food & Drug Administration*, 29 July 2014, www.fda.gov/Cosmetics/ScienceResearch/ProductTesting/ucm072268.htm. Accessed 3 Nov. 2016.

"Animals in Laboratories." *The Humane Society of the United States*, www.humanesociety.org/about/departments/animals_research.html. Accessed 3 Nov. 2016.

"Antimicrobial Resistance." *WHO*. World Health Organization, Apr. 2014. Web. 10 Mar. 2013.

"Biomedical Engineers." *Occupational Outlook Handbook*. Bureau of Labor Statistics, 17 Dec. 2015, www.bls.gov/ooh/architecture-and-engineering/biomedical-engineers.htm. Accessed 27 Oct. 2016.

"Ending Cosmetics Testing on Animals." *Cruelty Free International*. Cruelty Free International, n.d. Web. 14 June 2016.

"Fingerprint Examiners Found to Have Very Low Error Rates." *PR Newswire US*, February 2 2015.

"Francis S. Collins, MD, PhD." *National Human Genome Research Institute*, 25 Sept. 2015, www.genome.gov/10001018/former-nhgri-director-francis-collins-biography. Accessed 24 Mar. 2017.

"ISAAA Brief 46-2013: Executive Summary." *ISAAA*, 25 Mar. 2014, www.isaaa.org/resources/publications/briefs/46/executivesummary/. Accessed 25 July 2014.

"Opening the Doors on Animal Testing." *EBD*. BBC, 15 May 2014. Web. 15 June 2016.

"Plant Biotechnology Timeline." *National Institute of Food and Agriculture*. USDA, 18 Mar. 2009. Web. 24 Sept. 2013.

"Pocket K No. 16: Biotech Crop Highlights in 2015." *ISAAA*, June 2016, isaaa.org/resources/publications/pocketk/16/default.asp. Accessed 24 Oct. 2016.

"Summary Report for: Biochemical Engineers." *O*Net OnLine*, www.onetonline.org/link/summary/17-2199.01. Accessed 27 Oct. 2016.

"Willard Libby and Radiocarbon Dating." *American Chemistry Society*, 2016, www.acs.org/content/acs/en/education/whatischemistry/landmarks/radiocarbon-dating.html. Accessed 31 Aug. 2017.

"Data." *UNOS*, 2016, www.unos.org/data/. Accessed 26 Oct. 2016.

"Largest WTG Yet Makes Its Debut." *Modern Power Systems* 34.3 (2014): 39–40. *Energy & Power Source*. Web. 15 Jan. 2015.

"Tidal Power." *Chemical Business* 28.10 (2014): 53. *MasterFILE Premier*. Web. 15 Jan. 2015.

Acquaah, George. *Principles of Plant Genetics and Breeding*. 2nd ed. Hoboken: Wiley, 2012. Print.

Addis, Bill. *Building: Three Thousand Years of Design Engineering and Construction*. New York: Phaidon, 2007. Print.

AE Biofuels. http://www.alternative-energy-news.info/technology/biofuels/

Ager, Derek V. *The Nature of the Stratigraphical Record*. 3rd ed. Hoboken, N.J. John Wiley & Sons Inc., 1993. A witty and unabashedly curious look at stratigraphy; some of the discussion centers on biostratigraphy. An extensive bibliography, index, and a few well-chosen illustrations illuminate the text.

Agrawal, D. P., and M. G. Yadava. *Dating the Human Past*. Pune: Indian Society for Prehistoric and Quaternary Studies. 1995. An interesting look at the use of radiocarbon dating technology in relation to anthropology. Suitable for the nonscientist. Illustrations.

Akinyemi, Okoro M. *Agricultural Production: Organic and Conventional Systems*. Enfield, N.H.: Science Publishers, 2007.

Alberghina, Lilia. *Protein Engineering in Industrial Biotechnology*. Amsterdam: Hardwood Academics, 2000. An overview of the applications of proteins and how protein engineering can be used to solve problems in industrial biotechnology. Describes how protein engineering enhances purification of recombinant proteins and how it is applied to health care, including the development of new vaccines.

Alberts, Bruce, et al. *Essential Cell Biology*. New York: Garland Science, 2009. Print. Offers an introduction to the basic concepts of cell biology. Illustrated.

Aldridge, Susan. *The Thread of Life: The Story of Genes and Genetic Engineering*. 1996. Reprint. New York: Cambridge University Press, 2000.

Alexander, Brian. *Rapture: A Raucous Tour of Cloning, Transhumanism, and the New Era of Immortality*. New York: Basic Books, 2004. A reporter examines the fringe groups that support human cloning and genetic enhancement and finds people who want to defeat the effect of entropy and live forever.

Alexander, Martin. *Biodegradation and Bioremediation*. 2d ed. San Diego, Calif.: Academic Press, 1999.

Alibek, Ken, with Stephen Handelman. *Biohazard: The Chilling True Story of the Largest Covert Biological Weapons Program in the World—Told from Inside by the Man Who Ran It*. London: Hutchinson, 1999. Print.

Allhoff, Fritz, et al., eds. *Nanoethics: The Ethical and Social Implications of Nanotechnology*. Hoboken, N.J.: Wiley-Interscience, 2007.

American Academy of Ophthalmology. *Clinical Optics*. San Francisco: American Academy of Ophthalmology, 2006. This volume covers the fundamental concepts of optics as it relates to lenses, refraction, and reflection. It also covers the basic optics of the human eye and the fundamental principles of lasers.

American Society for Cell Biology. *Exploring the Cell*. Available on the World Wide Web at http://www.ascb.org/pubs/exploring.pdf. Features excellent images of cells and cellular processes.

Anderson, Burt, Herman Freedman, and Mauro Bendinelli, eds. *Microorganisms and Bioterrorism*. New York: Springer, 2006. Print.

Andrews, Lori B. *The Clone Age: Adventures in the New World of Reproductive Technology*. Henry Holt, 1999.

Aretha, David. *Steroids and Other Performance-Enhancing Drugs*. Berkeley Heights: Enslow, 2005. Print.

Arnold, Frances H. *Directed Evolution Library Creation: Methods and Protocols*. Totowa, N.J.: Humana Press, 2010. Encyclopedic collection of protocols for generating libraries of randomly mutagenic enzyme genes in bacteria, with tables, graphs, and some figures.

Arnold, Frances H., and George Georgiou, eds. *Directed Enzyme Evolution: Screening and Selection Methods*. Totowa, N.J.: Humana Press, 2010. Laboratory protocol book that describes, in great detail with figures and graphs, some rather ingenious techniques for screening mutant clones of enzyme genes.

Artmann, Gerhard M., and Shu Chien, editors. *Bioengineering in Cell and Tissue Research*. Springer, 2008.

Arya, Dev. *Aminoglycoside Antibiotics: From Chemical Biology to Drug Discovery*. New York: Wiley-Interscience, 2007. Describes the design and synthesis of antibiotics and the process of antibiotic resistance.

Atkins, John F., Raymond F. Gesteland, and Thomas R. Cech, eds. *RNA Worlds: From Life's Origins to Diversity in Gene Regulation*. Cold Spring Harbor: Cold Spring Harbor Lab, 2010. Print. A close look at the microscopic world of ribonucleic acid (RNA).

Atlas of Israel: Cartography, Physical and Human Geography. New York: Macmillan, 1985. Documents and maps all geophysical data available in Israel for over forty years of quantitative measurements from meteorology, oceanography, demography, and other disciplines.

Atlas, Ronald M., and Jim Philp, eds. *Bioremediation: Applied Microbial Solutions for Real-World Environmental Cleanup*. Washington, D.C.: ASM Press, 2005.

Aubrey, Allison. "Outrage over Government's Animal Experiments Leads to USDA Review." *NPR*. NPR, 6 Feb. 2015. Web. 9 July 2015.

Avery, Alex, and Dennis Avery. "High-Yield Conservation: More Food and Environmental Quality through Intensive Agriculture." In *Agricultural Policy and the Environment*, edited by Roger E. Meiners and Bruce Yandle. Lanham, Md.: Rowman & Littlefield, 2003.

Avise, John C. *The Hope, Hype, and Reality of Genetic Engineering: Remarkable Stories from Agriculture, Industry, Medicine, and the Environment*. New York: Oxford UP, 2004. Print.

———. *Molecular Markers, Natural History, and Evolution*. New York: Chapman and Hall, 1994. Examines relationships between genetic changes, taxonomy, and evolution.

Ayres, P., and P. Nigel. "Weeding with Fungi." *New Scientist*, September 1, 1990, 36-39. Describes an herbicide made from fungi, which can devastate weeds but not crops. Could be an ideal form of weed control.

Bahrke, Michael S., and Charles E. Yesalis, eds. *Performance-Enhancing Substances in Sport and Exercise.* Champaign: Human Kinetics, 2002. Print.

Bailey, James E., and David F. Ollis. *Biochemical Engineering Fundamentals.* 2d ed. New York: McGraw-Hill, 2006. Covers all aspects of biochemical engineering in an understandable manner.

Bainbridge, William Sims, editors. *Berkshire Encyclopedia of Human-Computer Interaction: When Science Fiction Becomes Fact.* Berkshire, 2004. 2 vols.

Balkin, Karen F., ed. *Food-Borne Illnesses.* San Diego, Calif.: Greenhaven Press, 2004. Collection of essays presents a variety of perspectives on food safety issues.

Baltimore, David. "7 Questions for Nobel Laureate David Baltimore." Interview by Mary Engel. *Hutch News*, 6 Aug. 2015, www.fredhutch.org/en/news/center-news/2015/08/questions-for-nobel-laureate-david-baltimore.html. Accessed 21 Mar 2017.

———. "Limiting Science: A Biologist's Perspective." *Daedalus* 104.4 (2005): 7–15.

———. "The Medicine of the Future." Town Hall. Los Angeles. 11 Feb. 2003. Address.

Barbaro, Anna, Patrizia Cormaci, and Aldo Barbaro "DNA Analysis from Mixed Biological Materials." *Forensic Science International* 146, supp. 1 (Fall 2004): S123-S125.

Barbosa-Cánovas, Gustavo, et al. *Global Issues in Food Science and Technology.* Burlington: Academic, 2009. Print.

Bard, E., F. Rostek, and Guillemette Menot-Combes. "A Better Radiocarbon Clock." *Science* 303 (2004): 178-179. A strong discussion of the need for a radiocarbon curve that extends beyond that of INTCAL98. This article is written in a manner that is accessible to undergraduates as well as graduate students.

Bard, Edouard, and Wallace S. Broecker, eds. *The Last Deglaciation: Absolute and Radiocarbon Chronologies.* Berlin: Springer-Verlag, 1992. This book deals with the radiocarbon dating processes and techniques that have been used to unravel the mysteries of glaciers and massive changes in climate. This book provides a simple explanation of radiocarbon dating and illustrates its applications.

Barnum, Susan R. *Biotechnology: An Introduction.* 2nd ed. Belmont: Wadsworth, 2005. Print.

Barrows, Geoffrey, Steven Sexton, and David Zilberman. "Agricultural Biotechnology: The Promise and Prospects of Genetically Modified Crops." *Journal of Economic Perspectives* 28, no. 1 (2014): 99–120.

Barry, W. B. N. *Growth of a Prehistoric Time Scale.* Rev. ed. Palo Alto, Calif.: Blackwell Scientific Publications, 1987. Largely devoted to the history of how the global geologic time scale was formulated, much of this book is a history of biostratigraphy. Well illustrated, with a good bibliography and an index.

Bart, Jan C. J., and Natale Palmeri. *Biodiesel Science and Technology: From Soil to Oil.* Cambridge, England: Woodhead 2010.

Bartelt, Margaret A. *Diagnostic Bacteriology. A Study Guide.* Philadelphia: Davis, 2000. Print.

Basl, John. "The Ethics of Creating Artificial Consciousness." *American Philosophical Association Newsletters: Philosophy and Computers* 13.1 (2013): 25–30. *Philosophers Index with Full Text.* Web. 25 Feb. 2015.

Basra, A. S., ed. *Heterosis and Hybrid Seed Production in Agronomic Crops.* Binghamton, N.Y.: Food Products Press, 1999. This book discusses current research in some of the most important crops of the world.

Baudrillard, Jean. *The Vital Illusion.* Edited by Julia Witwer. Columbia UP, 2000.

Baura, Gail D. *Engineering Ethics: An Industrial Perspective.* Boston: Elsevier, 2006. Print.

Baxevanis, Andreas D., and B. F. Francis Ouellette, eds. *Bioinformatics: A Practical Guide to the Analysis of Genes and Proteins.* 3d ed. Hoboken, N.J.: John Wiley & Sons, 2005. Covers bioinformatics from the database and searching perspective. Contains chapters on various biological databases and their search interfaces.

Bayne, Kathryn A. L., et al. *Laboratory Animal Welfare.* London: Academic Press, 2014. Print.

Beasley, Val Richard, et al. "Diagnostic and Clinically Important Aspects of Cyanobacterial (Blue-Green Algae) Toxicoses." *Journal of Veterinary Diagnostic Investigation* 1 (October, 1989): 359-365. Scholarly article focuses on the diagnosis of biotoxins in animals.

Behnisch, Peter A. "Biodetectors in Environmental Chemistry. Are We at a Turning Point?" *Environment International* 27 (December, 2001): 441-442.

Belgrader, P., et al. "Automated DNA Purification and Amplification from Blood-Stained Cards

Using a Robotic Workstation." *BioTechniques* 19 (September, 1995): 426-432.

Bell, Sigall T., Courtney L. McMickens, and Kevin Selby. *AIDS (Biographies of Disease)*. Santa Barbara: Greenwood, 2011. Print. Presents a history of HIV/AIDS, from the first reported outbreak to the latest prevention and treatment methods, written by three Harvard-trained physicians, cataloging various efforts to decipher and combat the disease. Bibliography, index, glossary.

Bennetzen, Jeff L., and Sarah C. Hake. *Handbook of Maize: Genetics and Genomics*. New York: Springer, 2009. Print. Examines maize as a significant food and fuel crop and a subject for scientific research. Includes a chapter devoted to McClintock's research.

Berg, Paul, and Maxine Singer. *Dealing with Genes: The Language of Heredity*. Herndon: University Science Books, 2008. Print.

———. "Paul Berg—Biographical." *NobelPrize.org*. Nobel Prize AB, Mar. 2004. Web. 25 Apr. 2016.

Berk, Zeki. *Food Process Engineering and Technology*. Academic Press, 2009.

Berlatsky, Noah. *Artificial Intelligence*. Detroit: Greenhaven, 2011. Print.

———. *Nanotechnology*. Greenhaven, 2014.

Bertheau, Yves. *Genetically Modified and Non-Genetically Modified Food Supply Chains: Coexistence and Traceability*. Chichester: Wiley-Blackwell, 2013. Print.

Bhushan, Bharat, Harald Fuchs, and Masahiko Tomitori, eds. *Applied Scanning Probe Methods VIII: Scanning Probe Microscopy Techniques (NanoScience and Technology)*. Berlin: Springer, 2008. Provides the most up-to-date information on this rapidly evolving technology and its applications.

Biello, David, and Katherine Harmon. "Tools for Life: What's Next for Cells Powered by Synthetic Genomes?" *Scientific American* 303.2 (2010): 17–18. Print. Brief commentary on the impact that synthetic biology, such as Venter's artificial bacterial genome, may have on the study of life.

———. "Whatever Happened to Advanced Biofuels?" *Scientific American*, 26 May 2016, www.scientificamerican.com/article/whatever-happened-to-advanced-biofuels/. Accessed 25 Oct. 2016.

Binnewies, Tim T., et al. "Ten Years of Bacterial Genome Sequencing: Comparative-Genomics-Based Discoveries." *Functional and Integrative Genomics* 6 (July, 2006): 165-185.

Binns, Chris. *Introduction to Nanoscience and Nanotechnology*. Hoboken: Wiley, 2010. Print.

Birke, Lynda, et al. *The Sacrifice: How Scientific Experiments Transform Animals and People*. Purdue UP, 2007.

Bishop, J. Michael. *How to Win the Nobel Prize: An Unexpected Life in Science*. Cambridge: Harvard UP, 2003. Print. Describes Bishop's early training in the medical field and his transition to molecular biology.

Biswas, Abhijit, et al. "Advances in Top-Down and Bottom-Up Surface Nanofabrication: Techniques, Applications & Future Prospects." *Advances in Colloid and Interface Science* 170.1–2 (2012): 2–27. Print.

Black, C. A. *Soil-Plant Relationships*. New York: John Wiley & Sons, 1988.

Blume, Stuart. *The Artificial Ear: Cochlear Implants and the Culture of Deafness*. New Brunswick, N.J.: Rutgers University Press, 2010. Historical study of implant development and implementation.

Boopathy, Raj, Sara Shields, and Siva Nunna. "Biodegradation of Crude Oil from the BP Oil Spill in the Marsh Sediments of Southeast Louisiana, USA." *Applied Biochemistry And Biotechnology* 167, no. 6 (July 2012): 1560–1568.

Bostom, Nick, and Eliezer Yudkowsky. "The Ethics of Artificial Intelligence." *Machine Intelligence Research Institute*. MIRI, n.d. Web. 23 Sept. 2016.

———. "Ethical Issues in Advanced Artificial Intelligence." *NickBostrom.com*. Nick Bostrom, 2003. Web. 23 Sept. 2016.

Bougaze, David, Thomas R. Jewell, and Rodolfo G. Buiser. *Biotechnology. Demystifying the Concepts*. San Francisco: Benjamin/Cummings, 2000. Classical book on biotechnology and bioprocessing.

Bourdon, Richard M. *Understanding Animal Breeding*. 2d ed. Upper Saddle River, N.J.: Prentice Hall, 2000.

Bourgaize, David, Thomas R. Jewell, and Rodolfo G. Buiser. *Biotechnology: Demystifying the Concepts*. San Francisco: Benjamin Cummings, 2000. Excellent introduction to biotechnology.

Bourne, Joel K. "Green Dreams." *National Geographic* 212, no. 4 (October, 2007): 38–59.

Bowman, John L., ed. *Arabidopsis: An Atlas of Morphology and Development*. New York: Springer, 1994. Print.

Boyer, Lauren. "Fighting the Fight: Dr. David Baltimore." *National Science & Technology Medals*

Foundation, 19 Dec. 2016, www.nationalmedals.org/stories/baltimore-and-the-cure. Accessed 21 Mar. 2017.

Boyer, Peter, J. "The Covenant." *New Yorker* 6 Sept. 2010: 60–67. Print.

Boyle, Godfrey, ed. *Renewable Energy*. 2d ed. New York: Oxford University Press in association with the Open University, 2004.

Brady, Catherine. *Elizabeth Blackburn and the Story of Telomeres*. Cambridge: MIT P, 2009. Print. A biography of Blackburn that includes the work carried out by her graduate student, Carol Greider, and their discovery of telomerase.

Braga, Newton C. *Bionics for the Evil Genius: Twenty-five Build-It-Yourself Projects*. McGraw-Hill, 2006.

Breeze, Roger G., Bruce Budowle, and Steven E. Schutzer, eds. *Microbial Forensics*. Burlington, Mass.: Elsevier Academic Press, 2005. Reviews the relationships between microbe physiology and forensics.

Brenner, R. L., and T. R. McHargue. *Integrative Stratigraphy Concepts and Applications*. Englewood Cliffs, N.J.: Prentice-Hall, 1988. Chapter 11 of this college-level textbook provides a detailed look at biostratigraphic concepts, methods, and applications. Well illustrated, with extensive reference lists and an index.

Broadhead, T. W., ed. *Fossil Prokaryotes and Protists: Notes for a Short Course*. Knoxville: University of Tennessee, 1987. Information on a broad range of fossil prokaryotes and protist microfossils is presented. This work is the text of one of a series of "short courses" sponsored by the Paleontological Society, but it can be useful to anyone interested in the various fossil groups covered.

Brock, Thomas. *The Emergence of Bacterial Genetics*. Woodbury, New York: Cold Spring Harbor Lab. P, 1990. Print. Reviews the history behind understanding bacterial genetics. Zinder's role in the discovery of transduction is discussed.

Bronzino, Joseph D., and Donald R. Peterson. *Biomechanics: Principles and Practices*. Boca Raton: CRC, 2014. *eBook Collection (EBSCOhost)*. Web. 25 Feb. 2015.

Brookfield, Michael E. *Principles of Stratigraphy*. Hoboken, N.J.: Wiley-Blackwell, 2004. Written for undergraduate students, this text provides an overview of the principles and applications of stratigraphy. Includes stratigraphic techniques and case studies. Organized into three sections, beginning with foundational material, followed by data collection and research topics, and completed with interpretations and analysis.

Brown, Kevin. *Penicillin Man: Alexander Fleming and the Antibiotic Revolution*. 2004. Stroud: Sutton, 2005. Print. Tells the story of Fleming and his discovery, as well as the history of the antibiotic revolution.

Brubaker, Bob. "*Yersinia pestis* and the Bubonic Plague." In *The Prokaryotes*, edited by Martin Dworkin et al. 3d ed. Vol. 6. New York: Springer, 2006.

Buckleton, John, Christopher M. Triggs, and Simon J. Walsh, eds. *Forensic DNA Evidence Interpretation*. Boca Raton, Fla.: CRC Press, 2005.

Bud, Robert. *Penicillin: Triumph and Tragedy*. New York: Oxford UP, 2007. Print. Describes the discovery and development of penicillin, considered a miracle drug until it proved nonresistant to "superbugs."

Budiansky, Stephen. *If a Lion Could Talk*. New York: Free Press, 1998. A good collection of contemporary and historical cases of animal intelligence. The stories cover a wide range of examples seen in various animal species.

Budowle, Bruce, et al. "Genetic Analysis and Attribution of Forensic Evidence." *Critical Review of Microbiology* 31 (October, 2005): 233-254.

———. "Quality Sample Collection, Handling, and Preservation for an Effective Microbial Forensics Program." *Applied and Environmental Microbiology* 72 (October, 2006): 6431-6438.

Bunz, Fred. *Principles of Cancer Genetics*. New York: Springer, 2008. Print. Presents an examination of the roles played by Bishop and Varmus in their discoveries of cellular oncogenes.

Burkhardt, Richard W. *Patterns of Behavior: Konrad Lorenz, Niko Tinbergen, and the Founding of Ethology*. Chicago: U of Chicago P, 2005. Print. Examines the institutions, scientists, and contexts by which ethology became a respected scientific discipline. Discusses Lorenz at length in context of contemporary ethological scientists.

Butler, John M. *Forensic DNA Typing: Biology, Technology, and Genetics of STR Markers*. 2d ed. Burlington, Mass.: Elsevier Academic Press, 2005. Provides a detailed examination of DNA fingerprinting analysis using STR markers. Intended for readers with background in the sciences.

———, et al. "Allele Frequencies for Fifteen Autosomal STR Loci on U.S. Caucasian, African American, and Hispanic Populations." *Journal of Forensic Sciences* 48 (Summer, 2003): 908-911.

Capriccioso, Richard P. "Genetic Testing." *Cancer*, edited by Jeffrey A. Knight, Salem Press, 2009. Salem Health.

Carbone, Larry. *What Animals Want: Expertise and Advocacy in Laboratory Animal Welfare Policy.* New York: Oxford UP, 2004. Print.

Cardell, Carolina, Isabel Guerra, and Antonio Sánchez-Navas. "SEM-EDX at the Service of Archaeology to Unravel Historical Technology." *Microscopy Today* 17, no. 14 (August, 2009): 28-33. An overview of the use of scanning electron microscopy to analyze archaeological materials. Includes diagrams and full-color photomicrographs.

Cardwell, Diane. "JetBlue Makes Biofuels Deal to Curtail Greenhouse Gases." *New York Times*, 19 Sept. 2016, www.nytimes.com/2016/09/20/business/energy-environment/jetblue-makes-biofuels-deal-to-curtail-greenhouse-gases.html. Accessed 25 Oct. 2016.

Carlson, Elof Axel. *Mutation: The History of an Idea from Darwin to Genomics.* Long Island: Cold Spring Harbor Lab P, 2011. Print. Discusses the birth and development of mutation, a key concept in the theory of evolution that has occupied geneticists—including Edmund Beecher Wilson—for more than 150 years.

Carmichael, D. G. *Infrastructure Investment: An Engineering Perspective.* Boca Raton: Taylor, 2015. Print.

Carracedo, Angel. *Forensic DNA Typing Protocols.* Totowa, N.J.: Humana Press, 2005.

Carroll, Pamela M., et al., eds. *Model Organisms in Drug Discovery.* Chichester: Wiley, 2003. Print.

Cassedy, Edward S., and Peter Z. Grossman. *Introduction to Energy: Resources, Technology, and Society.* 2d ed. New York: Cambridge University Press, 1998.

Castilho, Leda R., et al., eds. *Animal Cell Technology: From Biopharmaceuticals to Gene Therapy.* New York: Taylor & Francis, 2008.

Cathomen, Toni, Matthew Hirsch, and Matthew H. Porteus. *Genome Editing: The Next Step in Gene Therapy.* Springer, 2016.

Chargaff, Erwin. *Heraclitean Fire: Sketches from a Life Before Nature.* New York: Rockefeller UP, 1978. Print. Offers an autobiography of Chargaff's life and philosophical approach to science, featuring anecdotes and stories that provide insight into the field of science and the behaviors of scientists.

Charlier, R. H., and C. W. Finkl. *Ocean Energy: Tide and Tidal Power.* London: Springer, 2009.

Chien, Shu, Peter C. Y. Chen, and Y. C. Fung, eds. *An Introductory Text to Bioengineering.* Hackensack, N.J.: World Scientific Publishing, 2008.

Chisti, Yusuf. "Biodiesel from Microalgae." *Biotechnology Advances* 25, no. 3 (2007): 294-306.

Chrispeels, Maarten J., and David E. Sadava. *Plants, Genes, and Agriculture.* Sudbury, Mass.: Jones and Bartlett, 1994.

———. *Plants, Genes, and Crop Biotechnology.* 2nd ed. Sudbury: Jones & Bartlett, 2003. Print.

Christensen, C. M. *Molds, Mushrooms, and Mycotoxins.* Minneapolis: University of Minnesota Press, 1975. Well-organized and thoughtful discussions on some fungi and their ecological and economic importance.

Churchill, B. W. *Biological Control of Weeds with Plant Pathogens.* Edited by R. Charudattan and H. Walker. New York: Wiley, 1982. Print.

Coats, William Sloan, et al. *The Practitioner's Guide to Biometrics.* Chicago: American Bar Association Publishing, 2007.

Cohen, Stanley N. "The Manipulation of Genes." *Scientific American* 233 (July, 1975): 24-33. Eminently readable and profusely illustrated, this article provides the general reader with an excellent historical perspective on the development of recombinant DNA technology and a sound description of the scientific processes involved. Discussion of the use of plasmids as vectors is particularly useful.

———. "Origins of Growth Factors: NGF and EGF." *Journal of Biological Chemistry* 283, no. 49 (2008): 33793-33797. Print.

———. "Stanley Cohen—Autobiography." *The Nobel Prizes 1986.* Stockholm: Nobel Foundation, 1987. Print.

Cole, Leonard A. *The Anthrax Letters: A Medical Detective Story.* Washington, D.C.: Joseph Henry Press, 2003.

Collins, Francis S. "Q and A with NIH Director Dr Francis S Collins: The Value of Global Health Research and Training." Interview by Roger Glass. *Global Health Matters*, vol. 15, no. 6, 2016, p. 5. *Fogarty International Center*, www.fic.nih.gov/News/GlobalHealthMatters/november-december-2016/

Documents/fogarty-nih-global-health-matters-newsletter-november-december-2016.pdf.

———. "Why I'm a Man of Science—and Faith." Interview. *National Geographic*, 19 Mar. 2015, news.nationalgeographic.com/2015/03/150319-three-questions-francis-collins-nih-science. Accessed 24 Mar. 2017.

———. *The Language of God: A Scientist Presents Evidence for Belief*. New York: Free, 2006.

Comfort, Nathaniel C. *The Tangled Field: Barbara McClintock's Search for the Patterns of Genetic Control*. Cambridge, Mass.: Harvard UP, 2001. Print. Examines biographer Evelyn Fox Keller's depiction of McClintock's milieu and McClintock's own description of her role in science. Maintains that McClintock was always a respected scientist and a distinguished figure.

Conger, B. V., ed. *Cloning Agricultural Plants via In Vitro Techniques*. Boca Raton: CRC, 1981. Print.

Conkin, Paul K. *A Revolution Down on the Farm: The Transformation of American Agriculture Since 1929*. Lexington, Ky.: University Press of Kentucky, 2008.

Cooper, Jon, and Tony Cass, eds. *Biosensors: A Practical Approach*. 2d ed. New York: Oxford University Press, 2004.

Cooper, Marion R., Anthony W. Johnson, and Elizabeth A. Dauncey. *Poisonous Plants and Fungi: An Illustrated Guide*. 2d ed. London: TSO, 2003. Comprehensive volume describes the many varieties of poisonous plants and fungi.

Coors, J. G., and S. Pandey. *Genetics and Exploitation of Heterosis in Crops*. Madison, Wis.: American Society of Agronomy and Crop Science Society of America, and Soil Science Society of America, 1999. Provides an account of the various issues related to hybrid vigor, or heterosis.

Cordesman, Anthony H. *Terrorism, Asymmetric Warfare, and Weapons of Mass Destruction: Defending the US Homeland*. Westport: Praeger, 2002. Print.

Cori, Carl F. "The Call of Science." *Annual Review of Biochemistry* 38 (1969): 1–21. Print. Autobiography of the author. Cori describes his life in science and the research that led to receiving the 1947 Nobel Prize.

Cowan, Marjorie Kelly, and Kathleen Park Talaro. *Microbiology: A Systems Approach*. 2d ed. Boston: McGraw-Hill, 2008. General microbiology text focuses on the health sciences. Includes a chapter devoted to description of microbial identification techniques.

Cox, Margaret, et al. *The Scientific Investigation of Mass Graves*. New York: Cambridge University Press, 2008.

Creager, Angela N. H. *The Life of a Virus: Tobacco Mosaic Virus as an Experimental Model, 1930–1965*. Chicago: Chicago UP, 2002. Print.

Crommelin, Daan J. A., Robert D. Sindelar, and Bernd Meibohm, eds. *Pharmaceutical Biotechnology: Fundamentals and Applications*. 3d ed. New York: Informa Healthcare, 2008.

Crompton, Malcolm. "Biometrics and Privacy: The End of the World as We Know It or the White Knight of Privacy?" In *Biometrics: Security and Authentication*. Sydney: Biometrics Institute, 2003.

Crotty, Shane. *Ahead of the Curve: David Baltimore's Life in Science*. Berkeley: U of California P, 2001. Print. Biography of Nobel laureate David Baltimore. Addresses the significance of Zinder's role in the discovery of RNA phage.

Crow, James F., and Welcome Bender. "Edward B. Lewis, 1918–2004." *Genetics* 168: 4 (Dec. 2004): 1773–83. Print. Presents an overview of Lewis's life and work. Contains several remembrances of Lewis from colleagues and former students.

Cullen, J. Heather. *Barbara McClintock*. New York: Chelsea House, 2003. Print. Chronicles McClintock's life and career. Appropriate for high-school-level readers.

Cummings, Craig A., and David A. Relman. "Microbial Forensics: When Pathogens Are 'Cross-Examined.'" *Science* 296 (2002): 1976–79. Print.

Cummings, Stephen, ed. *Bioremediation: Methods and Protocols*. New York: Humana Press, 2010. Experts in the field of environmental biotechnology present innovative and imaginative bioremediation techniques in pollution removal.

Currie, L. A. "The Remarkable Metrological History of Radiocarbon Dating." *Journal of Research of the National Institute of Standards and Technology* 109 (2004): 185–217. This article accounts the discovery and progression of radiocarbon dating as a tool in geological and archaeological studies. It is extremely detailed and provides an excellent background for anyone using carbon dating techniques.

Cutler, Stephen J., and Horace G. Cutler, eds. *Biologically Active Natural Products: Pharmaceuticals*. Boca

Raton, Fla.: CRC Press, 2000. Demonstrates the connections between agrochemicals and pharmaceuticals and explores the uses of plants and plant products in the formulation and development of pharmaceuticals.

Cyranoski, David. "CRISPR Gene-Editing Tested in a Person for the First Time." *Nature News*, Nature Publishing Group, 15 Nov. 2016, www.nature.com/news/crispr-gene-editing-tested-in-a-person-for-the-first-time-1.20988.

Dahms, A. Stephen. "Biotechnology: What It Is, What It Is Not, and the Challenges in Reaching a National or Global Consensus." *Biochemistry and Molecular Biology Education*, vol. 32, no. 4, 2004, pp. 271–278., doi:10.1002/bmb.2004.494032040375.

Das, Ravindra. *Biometric Technology: Authentication, Biocryptography, and Cloud-Based Architecture.* CRC Press, 2015.

Davies, Kevin. *The $1,000 Genome: The Revolution in DNA Sequencing and the New Era of Personalized Medicine.* New York: Free Press, 2010. Looks at how less expensive, faster means of obtaining a person's genome will change medicine and make it more tailored to the individual.

Davis, Rowland H. *Neurospora: Contributions of a Model Organism.* New York: Oxford UP, 2000. Print.

De Gray, Aubrey, and Michael Rae. *Ending Aging: The Rejuvenation Breakthroughs That Could Reverse Aging in Our Lifetime.* New York: St. Martin's Griffin, 2008.

Deacon, J. W. *Microbial Control of Plant Pests and Diseases.* Research Triangle Park: Instrumentation Systems & Automation, 1983. Print.

Decker, Janet. *Anthrax.* New York: Chelsea, 2003. Print.

Deedrick, Douglas W. "Hairs, Fibers, Crime, and Evidence: Part 1—Hair Evidence." *Forensic Science Communications* 2 (July, 2000).

Dembek, Zygmunt F., Ronald L. Buckman, Stephanie K. Fowler, and James L. Hadler. "Missed Sentinel Case of Naturally Occurring Pneumonic Tularemia Outbreak: Lessons for Detection of Bioterrorism." *Journal of the American Board of Family Practice* 16 (July/August, 2003): 339-342.

Demetzos, Costas. *Pharmaceutical Nanotechnology: Fundamentals and Practical Applications.* Adis, 2016.

Dennis, David T., et al. "Tularemia as a Biological Weapon: Medical and Public Health Management." *Journal of the American Medical Association* 285 (June 6, 2001): 2763-2773.

DeSalle, Michael, and Michael Yudell. *Welcome to the Genome: A User's Guide to the Genetic Past, Present, and Future.* Hoboken, N.J.: Wiley-Liss, 2005. Starts with a brief history of genetics and description of the science before examining how the genome was sequenced. Analyzes the likely medical and agricultural applications.

Deutch, John M., and Richard K. Lester. "Synthetic Fuels." In *Making Technology Work: Applications in Energy and the Environment.* New York: Cambridge University Press, 2004.

Dewick, Paul. *Medicinal Natural Products: A Biosynthetic Approach.* New York: John Wiley & Sons, 2009. Comprehensive textbook describing biosynthetic methods and processes, including new techniques in genetic engineering and isolation of genes.

Dieiter, George E., and Linda C. Schmidt. *Engineering Design.* 5th ed. Boston: McGraw, 2013. Print.

DiLorenzo, Daniel J., and Joseph D. Bronzino, eds. *Neuroengineering.* Boca Raton, Fla.: CRC Press, 2008. Essential review of neuroengineering developments written by leaders in the field.

Dittmar, Tim. *Audio Engineering 101: A Beginner's Guide to Music Production.* Waltham: Focal, 2012. Print.

Dix, N. J., and J. Webster. *Fungal Ecology.* London: Chapman & Hall, 1995. An easy-to-read yet comprehensive account of fungal diversity in various ecological communities. Includes informative discussions on the critical roles fungi play in maintaining the health of ecosystems.

Dixon, Richard A., and Robert A. Gonzales, eds. *Plant Cell Culture: A Practical Approach.* 2nd ed. New York: Oxford UP, 1994. Print.

Dodds, John H., and Lorin W. Roberts. *Experiments in Plant Tissue Culture.* 3rd ed. New York: Cambridge University Press, 1995. Print.

DOE Joint Genome Institute. http://www.jgi.doe.gov

Dolby, Victoria. *All about Soy Isoflavones and Women's Health.* New York: Avery, 1999. Frequently asked questions, and answers to them, written for the nonscientist.

Doran, Pauline M. *Bioprocess Engineering Principles.* London: Academic Press, 2009. A solid, basic textbook for students entering the field.

Dott, Robert H., Jr., and Donald R. Prothero. *Evolution of the Earth.* 8th ed. New York: McGraw-Hill,

2009. This basic textbook on historical geology is aimed at students of geology. However, it is very readable by anyone with a background in science. Presents an up-to-date account of the earth's history from the viewpoint of plate tectonics. Includes a glossary.

Doudna, Jennifer A. and Samuel H. Sternberg. *A Crack in Creation: Gene Editing and the Unthinkable Power to Control Evolution*. Houghton Mifflin Harcourt, 2017. Written by one of the scientists who discovered this earthshaking technology; focuses on whether or not to actually use this method to change our DNA and the promises and perils of this gene-editing tool.

Dreifus, Claudia. "A Conversation with Christiane Nüsslein-Volhard: Solving a Mystery of Life, Then Tackling a Real-Life Problem." *New York Times*, 4 July 2006: 2. Print. Presents an interview with Nüsslein-Volhard in which she answers questions about gender discrimination among professional scientists, her Nobel Prize–winning research on fruit flies, and her parents' experiences during World War II.

Drexler, K. Eric. *Engines of Creation: The Coming Era of Nanotechnology*. New York: Anchor, 1986. Print.

———. *Radical Abundance: How a Revolution in Nanotechnology Will Change Civilization*. New York: Public-Affairs, 2013. Print.

Drlica, Karl. *Understanding DNA and Gene Cloning: A Guide for the Curious*. 4th ed. New York: John Wiley & Sons, 2004. A good introduction to the impact of DNA technology on human lives.

Drug and Alcohol Testing Industry Association http://www.datia.org

Durand, Dominique M. "What Is Neural Engineering?" *Journal of Neural Engineering* 4, no. 4 (September, 2006). Written by the editor in chief of the journal, who defines NE and its scope.

Dykstra, Michael J., and Laura E. Reuss. *Biological Electron Microscopy: Theory, Techniques, and Troubleshooting*. 2d ed. New York: Kluwer Academic, 2003. A guide to using microscopic instrumentation in cytological research. Covers conventional light microscopy, transmission electron microscopy, scanning electron microscopy, and photomicroscopy.

Edwards, Brian K. *The Economics of Hydroelectric Power*. Northampton, Mass.: Edward Elgar, 2003.

Edwards, Steven A. *The Nanotech Pioneers: Where Are They Taking Us?* Weinheim, Germany: Wiley-VCH, 2006.

Eggins, Brian R. *Biosensors: An Introduction*. New York: John Wiley & Sons, 1996.

Einsiedel, Edna. "Brave New Sheep: The Clone Named Dolly." In *Biotechnology: The Making of a Global Controversy*, edited by Martin W. Bauer and George Gaskell. New York: Cambridge University Press, 2002.

Elder, Kay, Doris Baker, and Julie Ribes. *Infections, Infertility, and Assisted Reproduction*. 2004. Reprint. New York: Cambridge University Press, 2010. A detailed, illustrated examination of the microbiology of assisted reproductive technologies. Each chapter includes references, further reading suggestions, and frequently appendixes outlining procedures and protocols.

Enderle, John D., et al., editors. *Introduction to Biomedical Engineering*. 3rd ed., Amsterdam: Elsevier/Academic Press, 2012. Print.

Engebretson, Monica. "Will South Korea Be the Next Country to Beat the United States in Ending Cosmetics Testing on Animals?" *Huffington Post*. HuffingtonPost.com, 6 Jan. 2015. Web. 9 July 2015.

Engel, Jonathan. *The Epidemic: A Global History of AIDS*. New York: Smithsonian/Harper Collins, 2006. Print. Chronicles the HIV/AIDS crisis and the world's response to it, approaching the issue from a number of vantage points—scientific, medical, political, and cultural. Bibliography; index.

Enriquez, Juan and Steve Gullans. *Evolving Ourselves: Redesigning the Future of Humanity—One Gene at a Time*. Current, 2016. Discusses the rapidly changing field of altering human evolution; shows the inner workings of innovative molecular biology and how this will affect who humans become in the future.

Espejo, Roman. *Performance-Enhancing Drugs*. Farmington Hills: Greenhaven, 2015. Print.

Esquinas-Alcazar, José. "Protecting Crop Genetic Diversity for Food Security: Political, Ethical, and Technical Challenges." *Nature Reviews: Genetics* 6, no. 12 (December, 2005): 946-953.

Ethier, C. Ross, and Craig A. Simmons. *Introductory Biomechanics: From Cells to Organisms*. Cambridge, England: Cambridge University Press, 2007. Provides an introduction to biomechanics and also discusses clinical specialties, such as cardiovascular, musculoskeletal, and ophthalmology.

Evans, Gareth, and Judith Furlong. *Environmental Biotechnology: Theory and Application*. New York: John

Wiley & Sons, 2003. A detailed examination of environmental biotechnology, focusing on present-day practices, the potential for biotechnological interventions, and microbial techniques and methods.

Evans, Robert L. *Fueling Our Future: An Introduction to Sustainable Energy.* Cambridge, England: Cambridge University Press, 2007.

Evans, William Charles. *Trease and Evans Pharmacognosy.* 15th ed. New York: W. B. Saunders, 2002.

Faber, Kurt. *Biotransformations in Organic Chemistry: A Textbook.* 5th ed. New York: Springer-Verlag, 2004. A very clear, useful textbook on the uses of enzymes in chemistry that includes a chapter on engineered enzymes.

Fairbanks, Daniel J. *Relics of Eden: The Powerful Evidence of Evolution in Human DNA.* Amherst, N.Y.: Prometheus Books, 2007. An examination of the field of evolutionary genomics that asserts that there is no dichotomy between religion and science.

Falconer, D. S., and Trudy F. C. Mackay. *Introduction to Quantitative Genetics.* 4th ed. New York: Longman, 1996.

Federal Bureau of Investigation. "Next Generation Identification (NGI)." *Fingerprints & Other Biometrics,* 2016.

Fedonkin, Mikhail A., James G. Gehling, Kathleen Grey, Guy M. Narbonne, and Patricia Vickers-Rich. *The Rise of Animals: Evolution and Diversification of the Kingdom Animalia.* Baltimore: Johns Hopkins University Press, 2007. An up-to-date account of all the known Ediacaran sites that is extensively and beautifully illustrated. Accessible to all interested readers.

Fedoroff, Nina, and David Botstein, eds. *The Dynamic Genome.* Cold Spring Harbor: Cold Spring Harbor Laboratory P, 1992. Print. A collection of essays written as tributes and remembrances for McClintock's ninetieth birthday. Includes technical articles on genetics and McClintock's impact, as well as reprints of four of McClintock's most influential articles.

———, and Nancy Marie Brown. *Mendel in the Kitchen: A Scientist's View of Genetically Modified Food.* Joseph Henry Press, 2004.

Fehr, W. R., and H. H. Hadley. *Hybridization of Crop Plants.* Madison, Wis.: American Society of Agronomy and Crop Science Society of America, 1980. Brings together the experience of plant breeders and scientists in a form that can be used by the layperson.

Fellows, P. J. *Food Processing Technology: Principles and Practice.* 3rd ed., CRC Press, 2009.

Fenichell, Stephen. *Plastic: The Making of a Synthetic Century.* New York: HarperCollins, 1996. A well-researched account of the plastics industry, focusing on the social and historical contexts of plastics, their technical development, and the many uses for synthetic fibers.

Fenner, Michael, and Ken Thompson. "Soil Seed Banks." In *The Ecology of Seeds.* New York: Cambridge University Press, 2005.

Fernandez-Cornejo, Jorge, et al. "Genetically Engineered Crops in the United States." US Department of Agriculture / Economic Research Service, Feb. 2014.

Ferrer, Manuel, et al. "Metagenomics for Mining New Genetic Resources of Microbial Communities." *Journal of Molecular Microbiology and Biotechnology* 16, nos. 1/2 (2009): 109-123.

Fetrow, Charles W. *The Complete Guide to Herbal Medicines.* Springhouse, Pa.: Springhouse, 2000. Accessible information available on more than three hundred herbal medicines.

Field, Thomas G., and Robert E. Taylor. *Scientific Farm Animal Production: An Introduction to Animal Science.* 10th ed. Upper Saddle River, N.J.: Prentice Hall, 2011.

Filson, Glen C., ed. *Intensive Agriculture and Sustainability: A Farming Systems Analysis.* Vancouver: University of British Columbia Press, 2004.

Fingas, Merv. *The Basics of Oil Spill Cleanup.* 2d ed. Boca Raton, Fla.: CRC Press, 2001.

Fischman, Josh. "Merging Man and Machine: The Bionic Age." *National Geographic,* vol. 217, no. 1, 2010, pp. 34–53.

Fisher, Elizabeth. "Why We Should Accept Animal Testing." *Huffington Post.* AOL (UK), 17 July 2013. Web. 14 Aug. 2014.

Fleming, Stuart. *Dating in Archaeology: A Guide to Scientific Techniques.* New York: St. Martin's Press, 1976. Places the methods and results of radiocarbon dating in the context of other dating techniques, such as dendrochronology, thermoluminescence, fission track dating, pollen analysis, and chemical methods. Discusses how several of these techniques may corroborate one another in establishing chronologies. Excellent bibliography.

Food and Agriculture Organization and World Health Organization. *Safety Assessment of Foods Derived from Genetically Modified Microorganisms.* Geneva: World Health Organization, 2001.

Foster, George T., ed. *Focus on Bioterrorism.* New York: Nova Science, 2006. Print.

Foster, Steven, and Rebecca L. Johnson. *Desk Reference to Nature's Medicine.* Washington, D.C.: National Geographic Society, 2006.

Fowler, Cary. "The Svalbard Seed Vault and Crop Security." *BioScience* 58, no. 3 (2008): 190-191.

Fox, Cynthia. *Cell of Cells: The Global Race to Capture and Control the Stem Cell.* New York: Norton, 2007. Print.

Fox, Renée C., and Judith P. Swazey. *Spare Parts: Organ Replacement in American Society.* Oxford UP, 1992.

Frankenberger, William, and Sally Benson, eds. *Selenium in the Environment.* Boca Raton, Fla.: CRC Press, 1994.

Franklin, Sarah. *Dolly Mixtures: The Remaking of Genealogy.* Durham, N.C.: Duke University Press, 2007.

Fredrickson, Donald S. *The Recombinant DNA Controversy: A Memoir—Science, Politics, and the Public Interest, 1974-1981.* 2d ed. Washington, D.C.: ASM Press, 2001. The author, National Institutes of Health director from 1975 to 1981, offers a glimpse into the heated controversy surrounding the new technology.

Freeman, Scott, and Jon Herron. *Evolutionary Analysis.* New York: Prentice Hall, 2000. Covers the connections between evolution and taxonomy.

Friedman, Dan. *Sound Advice: Voiceover from an Audio Engineer's Perspective.* Bloomington: AuthorHouse, 2010. Print.

Fritz, Sandy, comp. *Understanding Germ Warfare.* New York: Warner Books, 2002. Collection of materials describes twenty-first century bioterrorism and germ weapons, including anthrax, smallpox, plague, viral fevers, and toxins. Also discusses methods of delivery of biological agents and their identification, symptoms, and treatment.

Frozen Ark. Frozen Ark Project, 2011. Web. 12 Mar. 2015.

Fukuyama, Francis. *Our Posthuman Future: Consequences of the Biotechnology Revolution.* New York: Picador, 2003. A historian's admonition of the consequences of the biotechnology revolution and its potential to abolish human rights and erode the foundations of liberal democracy.

Fung, Wing K. "User-Friendly Programs for Easy Calculations in Paternity Testing and Kinship Determinations." *Forensic Science International* 136 (Fall, 2003): 22-34.

Gad, Shayne Cox, ed. *Handbook of Pharmaceutical Biotechnology.* Hoboken, N.J.: John Wiley & Sons, 2007.

Galen. *Galen on Bloodletting.* Trans. Peter Brain. New York: Cambridge University Press, 2009. Print. Includes translations of Galen's works against Erasistratus and the Erasistrateans, with extensive annotations.

———. *Galen on Diseases and Symptoms.* Ed. Ian Johnson. Cambridge University Press, 2006. Print. Discusses Erasistratus's concept of disease transmission along with Galen's opposing perspective.

Gamborg Oluf L., and Gregory C. Phillips, eds. *Plant Cell, Tissue and Organ Culture: Fundamental Methods.* New York: Springer, 1995. Print.

García-Sancho, Miguel. "A New Insight into Sanger's Development of Sequencing: From Proteins to DNA, 1943–1977." *Journal of the History of Biology* 43.2 (2010): 265–323. Print. Discusses Sanger's scientific career related to his research on RNA and DNA sequencing, genomics, and bioinformatics and biomedicine.

Garrett, Laurie. *The Coming Plague: Newly Emerging Diseases in a World Out of Balance.* New York: Farrar, Straus and Giroux, 1994. Discusses the increase in outbreaks of infectious diseases in the late twentieth century as well as ways to prevent such outbreaks.

Garrett, Roger A., and Hans-Peter Klenk, eds. *Archaea: Evolution, Physiology, and Molecular Biology.* Malden, Mass. Wiley-Blackwell, 2007. A compilation of scientific reviews and specialist articles discussing the Archaea domain. Discusses the phylogenetic research conducted on these organisms as well as the physiology and specifically metabolic pathways of Archaea. Written for a scientific audience with a biology background.

Gee, Henry. *Jacob's Ladder: The History of the Human Genome.* New York: W. W. Norton, 2004. Examines what human genome sequencing reveals and how this information may be used in the future.

Geller, Howard. *Energy Revolution: Policies for a Sustainable Future.* Washington, D.C.: Island Press, 2003.

Genetic Testing Registry. National Center for Biotechnology Information, National Institutes of Health,

2016, www.ncbi.nlm.nih.gov/gtr/. Accessed 28 Oct. 2016.

Genomic Science Program, U.S. Department of Energy http://genomicscience.energy.gov/index.shtml

Giarratano, Joseph, and Peter Riley. *Expert Systems: Principles and Programming.* 4th ed. Boston: Thomson, 2005. Print.

Gibbs, Samuel, and Alex Hern. "Apple Watch 2 Brings GPS, Waterproofing and Faster Processing." *The Guardian*, 8 Sept. 2016, www.theguardian.com/technology/2016/sep/07/iphone-7-launch-apple-watch-2-gains-gps-longer-battery-life. Accessed 28 Oct. 2016.

Giberson, Karl W., and Francis S. Collins. *The Language of Science and Faith: Straight Answers to Genuine Questions.* Downers Grove: InterVarsity, 2011. Print.

Gibson, D. G., et al. "Creation of a Bacterial Cell Controlled by a Chemically Synthesized Genome." *Science Express* (May 20, 2010). Announces the creation of a self-replicating bacterial cell governed by a synthetic genome.

———. *Sanford Guide to Antimicrobial Therapy 2015.* 45th ed. Sperryville: Antimicrobial Therapy, 2015. Print.

Glazer, Alexander N., and Hiroshi Nikaido. *Microbial Biotechnology: Fundamentals of Applied Microbiology.* New York: Cambridge University Press, 2007. In-depth analysis of the application of microorganisms in bioprocessing.

Gliessman, Stephen R., and Martha Rosemeyer, eds. *The Conversion to Sustainable Agriculture: Principles, Processes, and Practices.* Boca Raton, Fla.: CRC Press, 2010.

González, Pablo Rafael. *Running Out: How Global Shortages Change the Economic Paradigm.* New York: Algora, 2006.

Goodman, Justin R., et al. "Mounting Opposition to Vivisection." *Contexts* 11.2 (2012): 68–69. Print.

Grace, Eric S. *Biotechnology Unzipped: Promises and Realities.* Rev. 2d ed. Washington, D.C.: National Academies Press, 2006.

Graffy, Elizabeth A., and David R. Foran. "A Simplified Method for Mitochondrial DNA Extraction from Head Hair Shafts." *Journal of Forensic Sciences* 50 (September, 2005): 1119–22.

Graham, Linda E., James M. Graham, and Lee W. Wilcox. *Algae.* 2nd ed. San Francisco: Pearson, 2009. Print.

Greeley, Henry T. "The Uneasy Ethical and Legal Underpinnings of Large-Scale Genomic Biobanks." *Annual Review of Genomics and Human Genetics* 7 (2007): 343-364. Discusses the various issues related to the existence of private databases that contain genetic information on individuals.

Green, Ronald Michael. *Babies by Design: The Ethics of Genetic Choice.* 2007. Reprint. New Haven, Conn.: Yale University Press, 2009. A bioethicist tackles moral dilemmas provoked by emerging genetic engineering technologies. Contains a glossary of relevant technical terms.

Greene, Alexander L. *Encyclopedia of Stem Cell Research.* Hauppauge: Nova Science, 2012. Digital file.

Greenspoon, S. A., et al. "Application of the BioMek 2000 Laboratory Automation Workstation and the DNA IQ System to the Extraction of Forensic Casework Samples." *Journal of Forensic Sciences* 49 (2004): 29-39.

Greiner, Alfred, and Willi Semmler. *The Global Environment, Natural Resources, and Economic Growth.* New York: Oxford University Press, 2008.

Grobstein, Clifford. "The Recombinant-DNA Debate." *Scientific American* 237 (July, 1977): 22-33. Written at the height of the public controversy over the risks and benefits of recombinant DNA research, this thoughtful article provides a unique perspective to the history of the phenomenon. Colorful, helpful illustrations complement Grobstein's balanced treatment of the sensitive issues.

Grotzinger, John, and Tom Jordan. *Understanding Earth*, 6th ed. New York: W. H. Freeman, 2009. This comprehensive physical geology text covers the formation and development of the earth. Readable by high school students as well as by general readers. Includes an index and a glossary of terms.

Groves, M. J., ed. *Pharmaceutical Biotechnology.* 2d ed. Boca Raton, Fla.: Taylor & Francis, 2006.

Gu, Jenny, and Phillip E. Bourne, eds. *Structural Bioinformatics.* 2d ed. Hoboken, N.J.: John Wiley & Sons, 2009. Combination textbook and manual covering all aspects of protein bioinformatics, including the major protein databases, visualization, mass spectrometry, and protein modeling.

Guerrini, Anita. *Experimenting with Humans and Animals: From Galen to Animal Rights.* Baltimore: Johns Hopkins UP, 2003. Print.

Guillemin, Jeanne. *Anthrax: The Investigation of a Deadly Outbreak.* Berkeley: U of California P, 2001. Print.

Gunde-Cimerman, Nina, Aharon Oren, and Ana Plemenita. *Adaptation to Life at High Salt Concentrations in Archaea, Bacteria, and Eukarya.* New York: Springer, 2011. This text provides information on high-salinity habitats and the halophilic organisms found there. The environment overview is followed by sections devoted specifically to Archaea, bacteria, fungi, protozoa, and viruses.

Guzman, Carlos Alberto, and Giora Z. Feuerstein, eds. *Pharmaceutical Biotechnology.* New York: Springer, 2009.

Haerens, Margaret, ed. *Embryonic and Adult Stem Cells.* Detroit: Greenaven, 2009. Print.

Haley, James, and Tamara Roleff. *Performance-Enhancing Drugs.* San Diego: Greenhaven, 2003. Print.

Hall, Elizabeth A.H. *Biosensors.* Englewood Cliffs, N.J.: Prentice Hall, 1991.

Hall, Stephen S. *Invisible Frontiers: The Race to Synthesize a Human Gene.* New York: Atlantic Monthly Press, 1987. An well-referenced account of Genentech's success in synthesizing the human insulin gene and its subsequent expression in *E. coli*.

Hall, Susan J. *Basic Biomechanics.* 5th ed. New York: McGraw-Hill, 2006. A good introduction to biomechanics, regardless of one's math skills.

Hamacher, Michael, et al., eds. *Proteomics in Drug Research.* Weinham, Germany: Wiley-VCH, 2006. Contains information on technologies and applications, particularly in pharmaceuticals.

Hamill, Joseph, and Kathleen Knutzen. *Biomechanical Basis of Human Movement.* 4th ed. Philadelphia: Lippincott, 2015. Print. Integrates anatomy, physiology, calculus, and physics and provides the fundamental concepts of biomechanics.

Hampton, Dave. *So, You're an Audio Engineer: Well, Here's the Other Stuff You Need to Know.* Parker: Outskirts, 2005. Print.

———. *The Business of Audio Engineering.* 2nd ed. New York: Hal Leonard, 2013. Print.

Hanson, Bryan. *Understanding Medicinal Plants: Their Chemistry and Therapeutic Action.* New York: Haworth Herbal Press, 2005.

Han, Lei. "Genetically Modified Microorganisms: Development and Applications." In *The GMO Handbook: Genetically Modified Animals, Microbes, and Plants in Biotechnology*, edited by Sarad R. Parekh. Totowa, N.J.: Humana Press, 2004.

Hanson, Charles, et al. "Transplantation of Human Embryonic Stem cells onto a Partially Wounded Human Cornea In Vitro." *Acta Ophthalmologica*, vol. 91, no. 2, 2013, pp. 127–30.

Hartwell, Leland H., et al. *Genetics: From Genes to Genomes.* 5th ed., McGraw-Hill Education, 2015.

Hatze, H. "The Meaning of the Term 'Biomechanics.'" *Journal of Biomechanics* 7.2 (1974): 89–90. Print.

Haugen, David M., editor. *Animal Experimentation.* Greenhaven Press, 2007. Opposing Viewpoints.

Hausmann, Rudolf. *To Grasp the Essence of Life: A History of Molecular Biology.* Norwell: Kluwer Academic, 2010. Print. Includes the history of DNA, covering its discovery, function, and determination of structure. The contributions of Chargaff are described.

Hay, James G., and J. Gavin Reid. *Anatomy, Mechanics, and Human Motion.* 2d ed. Englewood Cliffs, N.J.: Prentice Hall, 1988. A good resource for upper high school students, this text covers basic kinesiology.

———. *The Biomechanics of Sports Techniques.* 4th ed. Englewood Cliffs: Prentice, 1993. Print.

Hayes, Allyson E., ed. *Cryogenics: Theory, Processes and Applications.* Hauppauge: Nova, 2010. Print.

He, Bin, ed. *Neural Engineering.* New York: Kluwer Academic/Plenum Publishers, 2005. Introductory overview of research in neural engineering.

Hedberg, H. D., ed. *International Stratigraphic Guide.* New York: John Wiley & Sons, 1976. The international "rule book" for stratigraphy. It sets procedures and standards to be met when naming stratigraphic units. It also defines many terms used in stratigraphy and has an extensive bibliography. Chapter 6 is devoted to biostratigraphy.

Heinzle, Elmar, Arno P. Biwer, and Charles L. Cooney. *Development of Sustainable Bioprocesses: Modeling and Assessment.* Hoboken, N.J.: John Wiley & Sons, 2007. Looks at making bioprocesses sustainable by improving them. Includes case studies on citric acid, biopolymers, antibiotics, and biopharmaceuticals.

Hekimi, Siegfried, editor. *The Molecular Genetics of Aging.* Springer, 2000. Results and Problems in Cell Differentiation.

Helander, Martin. *A Guide to Human Factors and Ergonomics.* 2nd ed., CRC Press, 2006.

Hench, Larry L., and Julian R. Jones, eds. *Biomaterials, Artificial Organs, and Tissue Engineering.* CRC Press, 2005.

Herrick, James W., and Dean R. Snow, eds. *Iroquois Medical Botany.* Syracuse, N.Y.: Syracuse University Press, 1997. A fascinating look at one Native American body of knowledge of herbal medicines. Important not only for those interested in herbal medicine but also for those studying American Indian cosmology as it relates to material culture. Illustrated, with references, index.

Hickman, H. Lanier, Jr. "Refuse-Derived Fuel and Energy Recovery: Fulfilling the Resource Recovery Promise." In *American Alchemy: The History of Solid Waste Management in the United States.* Santa Barbara, Calif.: Forester Communications, 2003.

Hill, Walter E. *Genetic Engineering: A Primer.* The Netherlands: Harwood Academic Publishers, 2000. Provides a concise overview. Illustrations, glossary, appendix, and index.

Hinrichs, Roger A., and Merlin Kleinbach. *Energy: Its Use and the Environment.* 4th ed. Belmont, Calif.: Thomson, Brooks/Cole, 2006.

Hobson, Art. *Physics: Concepts and Connections.* 5th ed. Boston: Pearson Addison-Wesley, 2010. Includes chapters on light, geometric optics, wave nature of light, and a section on night vision imaging.

Hochberg, Robert, and Kathleen Gabric. "A Provably Necessary Symbiosis." *The American Biology Teacher* 72, No. 5 (2010): 296-300. This article describes some mathematics that can be taught in biology classrooms.

Hodge, Russ. *Genetic Engineering: Manipulating the Mechanisms of Life.* New York: Facts On File, 2009. Print.

Holmes, Chris. *Spores, Plague, and History: The Story of Anthrax.* Dallas: Durban, 2003. Print.

Howe, Christopher. *Gene Cloning and Manipulation.* 2nd ed., Cambridge: Cambridge University Press, 2007. Print.

Howland, John L. *The Surprising Archaea: Discovering Another Domain of Life.* New York: Oxford University Press, 2000. Provides an all-encompassing knowledge of Archaea, from their evolution to their ecology. The author captures the excitement of their discovery within the scientific community.

Howland, Rebecca, and Lisa Benatar. "A Practical Guide to Scanning Probe Microscopy." ThermoMicroscopes. March 2000. Web. Accessed September, 2011. This introductory guide to scanning probe microscopy describes several techniques and operating modes of the devices, as well as their structure and principles of operation, and discusses the occurrence of image artifacts in their use.

Hrazdina, Geza, ed. *The Use of Agriculturally Important Genes in Biotechnology.* Washington, D.C.: NATO Scientific Affairs Division, 2000.

Huffman, Wallace E., and Robert E. Evenson. *Science for Agriculture: A Long-Term Perspective.* 2nd ed., Blackwell, 2006.

Hughes, Brian. "Technology Becomes Us: The Age of Human-Computer Interaction." *The Huffington Post,* 20 Apr. 2016, www.huffingtonpost.com/brian-hughes/technology-becomes-us-the_b_9732166.html. Accessed 28 Oct. 2016.

Humber, James M., and Robert F. Almeder, eds. *Biomedical Ethics Reviews: Stem Cell Research.* Totowa: Humana, 2004. Print.

Hummel, Susanne. *Ancient DNA Typing: Methods, Strategies, and Applications.* New York: Springer, 2002.

Hung, George K. *Biomedical Engineering: Principles of the Bionic Man.* World Scientific, 2010.

Hvistendahl, Mara. "China's Three Gorges Dam: An Environmental Catastrophe?" *Scientific American,* March 25, 2008. Assessment of China's initial failure to consider the dam's ecological impacts and recent acknowledgment of its possible consequences for the environment.

Illman, Walter, and Pedro Alvarez. "Performance Assessment of Bioremediation and Natural Attenuation." *Critical Reviews in Environmental Science and Technology,* 39, no. 4 (April, 2009): 209-270. A critical review of the state-of-the-art in performance assessment methods. Discusses future research directions in bioremediation, natural attenuation, chemical fingerprinting, and molecular biological tools.

International Food Information Council Foundation. "Food Biotechnology Timeline." *Food Biotechnology: A Communicator's Guide to Improving Understanding.* Food Insight, n.d. PDF. Web. 24 Sept. 2013.

Jabr, Ferris. "Building Tastier Fruits & Veggies." *Scientific American* 311.1 (2014): 56–61. Print.

Jackson, David A., and Stephen R. Stich, eds. *The Recombinant DNA Debate.* Englewood Cliffs, N.J.: Prentice-Hall, 1979. A collection of seventeen essays, written by recognized leaders in the fields of molecular biology, ethics, and philosophy, covering all aspects of the controversy. Of particular note are essays by Robert L. Sinsheimer and George Wald, two of the most vocal scientists who pushed for a halt to recombinant DNA research.

Jackson, Michael, and Amy Norrington. "An A to Z of Medical History: Part 1." *Student British Medical Journal* 10 (Sept. 2002): 317. Print. An exploration of the history of medicine covering Erasistratus and his dissection techniques.

Jain, Anil K., Arun Russ, and Sharath Pankanti. "Biometrics: A Tool for Information Security." *IEEE Transactions on Information Forensics and Security* 1, no. 2 (2006): 125–143.

James, Stuart H., and Jon J. Nordby, eds. *Forensic Science: An Introduction to Scientific and Investigative Techniques*. 2d ed. Boca Raton, Fla.: CRC Press, 2005.

Jeffreys, A. J., V. Wilson, and S. L. Thein. "Individual-Specific 'Fingerprints' of Human DNA." *Nature* 316 (1985): 76-79. Landmark paper that introduced the concept of DNA fingerprinting as a method of identification.

Jenkins, Amanda J., and Bruce A. Goldberger, eds. *On-Site Drug Testing*. Totowa, N.J.: Humana Press, 2002. Discusses on-site methods of testing for drugs in hospital, criminal, workplace, and school settings. Looks at many specific tests, discussing their efficacy and their underlying principles.

Jensen, Eric A. *The Therapeutic Cloning Debate: Global Science and Journalism in the Public Sphere*. Ashgate, 2014.

Jha, A. R. *Cryogenic Technology and Applications*. Burlington: Elsevier, 2006. Print.

Jokinen, Jussi P. P. "Emotional User Experience: Traits, Events, and States." *International Journal of Human-Computer Studies*, vol. 76, 2015, pp. 67–77.

Jones, Richard E., and Kristin H. Lopez. *Human Reproductive Biology*. Rev. ed. London: Academic Press, 2010. A comprehensive introductory textbook, heavily illustrated with diagrams and photographs. Each chapter includes a summary, further reading, and advanced reading list.

Jorde, Lynn B., et al. *Medical Genetics*. 5th ed. Elsevier, 2016.

Jördening, Hans-Joachim, and Josef Winter. *Environmental Biotechnology: Concepts and Applications*. Weinheim, Germany: Wiley-VCH, 2005. A solid foundation for students wishing to study environmental biotechnology. Examines in detail the microbiological treatment of waste and pollution in water, soil, and air.

Kakalios, James. *The Physics of Superheroes*. 2d ed. New York: Gotham Books, 2009. Uses comic-book references to cover basic physics theory. Includes chapters on mechanics, energy (heat and light), and modern physics.

Kalloo, G., and J. B. Chowdhury, eds. *Distant Hybridization of Crop Plants*. Berlin: Springer-Verlag, 1992.

Kandler, Otto, and Wolfram Zillig, eds. *Archaebacteria Eighty-five: Proceedings of the EMBO Workshop on Molecular Genetics of Archaebacteria*. Forestburgh, N.Y.: Lubrecht and Cramer, 1987. A proceedings volume on molecular genetics, biology, and biochemistry of archaebacteria.

Kar, Ashutosh. *Pharmacognosy and Pharmacobiotechnology*. 2d ed. Tunbridge Wells, England: Anshan, 2008.

Karch, Stephen B., ed. *Workplace Drug Testing*. Boca Raton, Fla.: CRC Press, 2008. Examines regulations and mandatory guidelines for federal workplace drug testing and describes techniques. Provides sample protocols from the nuclear power and transportation industries.

Karunakaran, Chandran, et al. *Biosensors and Bioelectronics*. Elsevier, 2015.

Katoh, Shigeo, and Fumitake Yoshida. *Biochemical Engineering: A Textbook for Engineers, Chemists, and Biologists*. John Wiley & Sons, 2009.

Katz, Bruce F. *Neuroengineering the Future: Virtual Minds and the Creation of Immortality*. Hingham, Mass.: Infinity Science Press, 2008. Fascinating introduction to this field, describing the state of the art and speculating on long-term developments.

Katz, Linda B., ed. *Agroterrorism: Another Domino?* New York: Novinka, 2005. Print.

Kauffman, Stuart. *Investigations*. New York: Oxford UP, 2000. Print. An examination of the origins and basis of life. Includes discussion in chapter 2, "The Origins of Life," of Szostak's work with aptamers and his attempts to produce self-replicating RNA.

Kerr, Andrew. *Introductory Biomechanics*. London: Elsevier, 2010. Print.

Khan, Arshad Hassan. *Desalination Processes and Multistage Flash Distillation Practice*. New York: Elsevier, 1986.

Khan, Muhammad Sarwar, Iqrar A. Khan, and Debmalya Barh. *Applied Molecular Biotechnology: The Next Generation of Genetic Engineering.* CRC Press, 2016.

Khudyakov, Yury E., and Paul Pumpens. *Viral Nanotechnology.* CRC Press, 2016.

Kieleczawa, Jan. *DNA Sequencing: Optimizing the Process and Analysis.* Sudbury, Mass.: Jones & Bartlett, 2005.

King, Robert C., et al. *A Dictionary of Genetics.* 8th ed., Oxford UP, 2013.

Klass, Donald L. *Biomass for Renewable Energy, Fuels, and Chemicals.* San Diego, Calif.: Academic Press, 1998.

Knowles, Richard V. *Genetics, Society, and Decisions.* Columbus, Ohio: Charles E. Merrill, 1985. This broad-based college text was written for nonbiology majors with an interest in science and the social issues raised by the new advances in biology. Good treatment of basic principles of genetics, particularly as applied to humans. Chapter 19 provides a straightforward presentation of recombinant DNA, including the relevant science, history, applications, and controversial issues. Many useful illustrations.

Kobilinsky, Lawrence F., Louis Levine, and Henrietta Margolis-Nunno. *Forensic DNA Analysis.* New York: Chelsea House, 2007. Presents a comprehensive introduction to the use of STRs in DNA fingerprinting. Includes discussion of future directions, including mitochondrial and Y chromosome analyses.

Kohler, Robert E. *Lords of the Fly: Drosophila Genetics and the Experimental Life.* Chicago: Chicago UP, 1994. Print.

Kolata, Gina Bari. *Clone: The Road to Dolly, and the Path Ahead.* New York: HarperCollins, 1998.

Komolprasert, Vanee, and Kim M. Morehouse, editors. *Irradiation of Food and Packaging: Recent Developments.* American Chemical Society, 2004.

Kozlowski, Ryszard, Gennady Zaikov, and Frank Pudel, eds. *Renewable Resources: Obtaining, Processing, and Applying.* Hauppauge, N.Y.: Nova Science, 2009.

Kozubek, James. *Modern Prometheus: Editing the Human Genome with Crispr-Cas9.* Cambridge University Press, 2016. Discusses the potential for gene editing, including ethical and legal implications; tells the story across a 50-year timeline, including stories of the scientists involved in the process.

Kresge, N., R. D. Simoni, and R. Hill. "Precocious Newborn Mice and Epidermal Growth Factor: The Work of Stanley Cohen." *Journal of Biological Chemistry* 281, no. 10 (2006): e10-e11. Print.

Kress-Rogers, Erika, ed. *Handbook of Biosensors and Electronic Noses: Medicine, Food, and the Environment.* Boca Raton, Fla.: CRC Press, 1997.

Kreuzer, Helen, and Adrianne Massey. *Recombinant DNA and Biotechnology: A Guide for Students.* Oxford: Blackwell Science, 2000. Introductory overview. Illustrations and index.

Krishnan, K. N., with D. R. Berwick. *Developing a Police Perspective and Exploring the Use of Biometrics and Other Emerging Technologies as an Investigative Tool in Identity Crimes.* Payneham, S. Aust.: Australasian Centre for Policing Research, 2004.

Kruger, Paul. *Alternative Energy Resources: The Quest for Sustainable Energy.* Hoboken, N.J.: John Wiley & Sons, 2006.

Kruglinski, Susan. "The *Discover* Interview: David Baltimore." *Discover* Sept. 2006: 50–53.

Kumar, Srinibas. "Biotechnology: Scope and Branches of Biotechnology." *Biology Discussion*, 26 Oct. 2015, www.biologydiscussion.com/biotechnology/branches-biotechnology/biotechnology-scope-and-branches-of-biotechnology/15653.

Kunzig, Robert. "Scraping Bottom: The Canadian Oil Boom." *National Geographic*, March, 2009. Fair but controversial report on the Canadian oil industry's bringing about both an economic boom and environmental degradation.

Kyte, Lydiane, and John Kleyn. *Plants from Test Tubes: An Introduction to Micropropagation.* 3rd ed. Portland: Timber, 1996. Print.

LaFreniere, Gilbert. *The Decline of Nature: Environmental History and the Western Worldview.* Bethesda, Md.: Academica Press, 2008. Examines human policies that have affected nature through history, detailing both philosophical attitudes underlying those policies and their possible catastrophic results.

Laidlaw, Stuart. *Secret Ingredients: The Brave New World of Industrial Farming.* Toronto: McClelland & Stewart, 2004.

Lamoreux, M. Lynn, et al. *The Colors of Mice: A Model Genetic Network.* Hoboken: Wiley, 2010. Print.

Lanza, R, P., and Anthony Atala. *Essentials of Stem Cell Biology.* 3rd ed. San Diego: Elsevier, 2014. Digital file.

Larkin, Marilynn. "Microbial Forensics Aims to Link Pathogen, Crime, and Perpetrator." *The Lancet Infectious Diseases* 3.4 (2003): 180–81. Print.

Laubichler, Manfred D., and Jane Maienschein, eds. *From Embryology to Evo-Devo: A History of Developmental Evolution.* Cambridge: MIT P, 2007. Print. Presents an overview of the modern field of evolutionary developmental biology, which has grown out of earlier research into embryology, heredity, and genetics.

Lauer, William C., ed. *Desalination of Seawater and Brackish Water.* Denver, Colo.: American Water Works Association, 2006.

Lax, Eric. *The Mold in Dr. Florey's Coat: The Story of the Penicillin Miracle.* New York: Holt, 2004. Print. Argues that scientists who continued Fleming's research should be credited with the discovery of penicillin.

Lazer, David, ed. *DNA and the Criminal Justice System: The Technology of Justice.* Cambridge, Mass.: MIT Press, 2004. Collection of essays explores the ethical and procedural issues related to DNA evidence. Includes a chapter by Associate Justice Stephen G. Breyer of the U.S. Supreme Court.

Lazo, John, and Peter Wipf. "Combinatorial Chemistry and Contemporary Pharmacology." *The Journal of Pharmacology and Experimental Therapeutics* 293, no. 3 (February, 2000): 705-709. Describes the process of combinatorial chemistry. Includes experimental strategies and flow charts describing the screening of compounds.

Lesk, Arthur M. *Introduction to Bioinformatics.* 3d ed. New York: Oxford University Press, 2008. Comprehensive overview of genomes, proteomics, protein structure, databases, phylogenetics, programming languages, and more.

———. *Introduction to Genomics.* New York: Oxford UP, 2007. Print. Student-focused volume covering human genomics, comparative genomics, evolution and genomic change, sequencing techniques, proteomics, and systems biology. Includes extensive full-color illustrations, sidebars, and end-of-chapter problems and exercises.

Levetin, Estelle, and Karen McMahon. *Plants and Society.* Boston: WCB/McGraw-Hill, 1999. Chapter 13, "Legumes," and the unit on plants and human health are relevant.

Lewin, Benjamin. *Genes VIII.* San Francisco: Benjamin Cummings, 2003. In-depth look at genes and molecular biology.

Lewin, Roger. "A Naturalist of the Genome." *Science* 222 (28 Oct. 1983): 402–405. Print. Published shortly after McClintock received the Nobel Prize. Includes a concise and clear overview of the work for which the prize was awarded as well as some discussion of where McClintock's discoveries have led since.

Lewis, Ricki. *Discovery: Science as a Window to the World.* Malden: Blackwell, 2001. Print. A summary of scientific research, beginning with the debunking of spontaneous generation. One chapter addresses the discovery and characterization of telomeres.

———. *Human Genetics: Concepts and Applications.* 11th ed., McGraw-Hill Education, 2015.

Lewis, Walter H., and Memory P. F. Elvin-Lewis. *Medical Botany: Plants Affecting Human Health.* 2d ed. Hoboken, N.J.: J. Wiley, 2003. An excellent in-depth (544-page) study of plants and the medicines they produce, examining plants' effects on human health as injurious, remedial, or psychoactive. Includes bacteria, fungi, and seaweeds as well as flowering plants.

Liang, George H., and Daniel Z. Skinner. *Genetically Modified Crops: Their Development, Uses, and Risks.* New York: Food Products Press, 2004.

Liebler, Daniel. *Introduction to Proteomics.* Totowa, N.J.: Humana Press, 2002. Discusses key concepts in proteomics, including the methodology and instrumentation used. Describes the applications of mass spectrometry and provides an excellent introduction and overview of proteomics.

Liebsch, Manfred, et al. "Alternatives to Animal Testing: Current Status and Future Perspectives." *Archives of Toxicology* 85 (2011): 841–58. Print.

Lindler, Luther E., Frank J. Lebeda, and George W. Korch, eds. *Biological Weapons Defense: Infectious Diseases and Counterbioterrorism.* Totowa, N.J.: Humana Press, 2005. Prominent experts in biodefense research—many from the U.S. Army Medical Research Institute of Infectious Diseases—describe how to identify the presence of biological weapons through proteomic and genomic analysis.

Lipkin, Steven Monroe and John Luoma. *The Age of Genomes: Tales from the Front Lines of Genetic Medicine.* Beacon Press, 2016. Focuses on the real-life stories of patients who may be helped by this type of gene editing in an easy-to-read and accessible way.

Lipshitz, H. D. *Genes, Development, and Cancer: The Life and Work of Edward B. Lewis.* Boston: Kluwer

Academic, 2004. Print. Complete collection of Lewis's papers divided into sections that reflect the many areas of research in which Lewis was engaged. Includes summaries.

Liska, Ken. *Drugs and the Human Body with Implications for Society.* Upper Saddle River, N.J.: Pearson/Prentice Hall, 2004. Simply describes the various classes of drugs and drug testing methods.

Lodish, Harvey, et al. *Molecular Cell Biology.* 4th ed. New York: W. H. Freeman, 2000. One of the most authoritative cell biology reference texts available. Provides excellent graphics, CD-ROM animations of cellular processes, and primary literature references.

Lorenz, Konrad. *King Solomon's Ring: New Light on Animal Ways.* 1952. Trans. Marjorie Kerr Wilson. New York: Routledge, 2002. Print. Collects observations, anecdotes, and facts on animal psychology. Humorous, eloquent writing is appropriate for general readers. Includes illustrations by the author.

———. *On Aggression.* 1963. Trans. Marjorie Kerr Wilson. New York: Routledge, 2007. Print. An inquiry into aggression as a means of survival among a variety of animals. Examines humanity and proposes solutions to the problem of human aggression.

Lorenzetti, Maureen Shields. *Alternative Motor Fuels: A Nontechnical Guide.* Tulsa, Okla: PennWell, 1996.

Lowe, J. John, ed. *Radiocarbon Dating: Recent Applications and Future Potential.* Chichester, N.Y.: John Wiley and Sons, 1996. This college-level book offers a comprehensive overview of the techniques and protocols of radiocarbon dating. Several of the essays explore possible future usage and applications. Illustrations and maps help to clarify difficult concepts. Bibliographical references.

Lucy, David. *Introduction to Statistics for Forensic Scientists.* Hoboken, N.J.: John Wiley & Sons, 2005.

Lurquin, Paul F. *High Tech Harvest: Understanding Genetically Modified Food Plants.* Westview, 2002.

Lutz, Peter L. "The Alexandrian Period." *The Rise of Experimental Biology: An Illustrated History.* Totowa: Humana, 2002. Print. Discusses Erasistratus's thought and achievements within the context of Alexandrian science of the time.

Lydersen, Bjorn K., Nancy A. D'Elia, and Kim L. Nelson, eds. *Bioprocess Engineering: Systems, Equipment and Facilities.* New York: John Wiley & Sons, 1994. Describes equipment and facilities for industrial fermentation.

Lynch, Colum. "UN Backs Human Cloning Ban." *Washington Post,* 9 Mar. 2005, www.washingtonpost.com/wp-co/hotcontent/index.html?section=nation/specials/science/cloning. Accessed 28 July 2014.

Lynch, J. M. *Soil Biotechnology: Microbiological Factors in Crop Production.* Malden, Mass.: Blackwell, 1983.

Madhavan, Guruprasad, et al., editors. *Career Development in Bioengineering and Biotechnology.* Springer, 2008.

Madigan, Michael T., et al. *Brock Biology of Microorganisms.* 12th ed. San Francisco: Benjamin Cummings, 2008. Several chapters of this popular textbook describe microbial metabolism and the application of microorganisms in industry.

Madigan, Michael T., John M. Martinko, Paul V. Dunlap, and David P. Clark. *Brock Biology of Microorganisms.* 12th ed. Upper Saddle River, N.J.: Pearson Prentice Hall, 2008. Widely respected basic microbiology textbook includes information about biological weapons and methods of microbial identification.

Malhotra, Bansi D., et al. "Recent Trends in Biosensors." *Current Applied Physics* 5 (February, 2005): 92-97.

Manahan, Stanley E. "Adequate, Sustainable Energy: Key to Sustainability." In *Environmental Science and Technology: A Sustainable Approach to Green Science and Technology.* 2d ed. Boca Raton, Fla.: CRC Press, 2007.

Manchester, Keith. "Protein Sequencing Fifty Years Ago: Fred Sanger and the Amino Acid Sequence of Insulin." *South African Journal of Science* 101.7–8 (2005): 327–30. Print. Presents information on Sanger's work with the amino acid and protein sequencing and his study of the molecular structure of insulin.

Mann, J. *Murder, Magic, and Medicine.* Rev. ed. New York: Oxford University Press, 2000. An interesting and readable book on the use of natural plant products for medicinal purposes.

Mansfield, A. J., and J. L. Wayman. *Best Practices in Testing and Reporting Performance of Biometric Devices: Version 2.01.* Teddington, Middlesex, England: National Physical Laboratory, 2002.

Marguet, Philippe, et al. "Biology by Design: Reduction and Synthesis of Cellular Components and

Behavior." *Journal of the Royal Society Interface* 4, no. 15 (2007): 607–623. Review on metabolic engineering and synthetic biology written for the general public.

Margulis, Lynn, and Michael J. Chapman. *Kingdoms and Domains: An Illustrated Guide to the Phyla of Life on Earth*. 4th ed. Boston: Elsevier, 2009. Provides a comprehensive overview (current at the time of publication) of life's classification and phylogeny. An improvement on Margulis's previous work, *Five Kingdoms*, this text includes more recent discoveries and systematics. The classification is a modified version of Linnaean taxonomy.

Martineau, Belinda. *First Fruit: The Creation of the Flavr Savr Tomato and the Birth of Biotech Food*. New York: McGraw-Hill, 2001. An authoritative account of the development and subsequent marketing and demise of Calgene's Flavr Savr tomato, the first genetically engineered food to come to market. The author was an employee of Calgene and intimately involved with the development of the tomato. Excellent references.

Mataigne, Fen. *Medicine by Design: The Practice and Promise of Biomedical Engineering*. Baltimore: The Johns Hopkins University Press, 2006.

Matossian, M. K. *Poisons of the Past: Molds, Epidemics, and History*. New Haven, Conn.: Yale University Press, 1989. A very good summary on the effects of food poisoning resulting from bread molds on human history. A useful reference book for public health.

Maytal, Ben-Zion, and John M. Pfotenhauer. *Miniature Joule-Thomson Cryocooling: Principles and Practice*. New York: Springer, 2013. Print.

Mazzarello, Paolo. *Golgi: A Biography of the Founder of Modern Neuroscience*. Oxford: Oxford UP, 2009. Print. Presents a detailed review of Golgi's life and work, including his achievements in neuroscience, cell biology, and microbiology.

McClellan, Marilyn. *Organ and Tissue Transplants: Medical Miracles and Challenges*. Enslow, 2003.

McClintock, J. Thomas. *Forensic DNA Analysis: A Laboratory Manual*. CRC Press, 2008.

McGavin, George. *Endangered: Wildlife on the Brink of Extinction*. Richmond Hill: Firefly, 2006. Print.

McGovern Institute for Brain Research at MIT, Genome Editing with CRISPR-Cas9, https://www.youtube.com/watch?v=2pp17E4E-O8.

McGowran, Brian. *Biostratigraphy: Microfossils and Geological Time*. New York: Cambridge University Press, 2008. This text expands the application of biostratigraphy from the original use in paleontology to the fields of petroleum exploration and deep-ocean drilling. Addresses the relatively new methodology of studying microfossils and absolute aging.

McGrayne, Sharon Bertsch. *Nobel Prize Women in Science: Their Lives, Struggles, and Momentous Discoveries*. Secaucus, NJ: Carol, 1993. Print. Includes a biography of Gerty Cori and her professional and personal collaboration with her husband, Carl Cori. Emphasizes the Coris' contributions in biochemistry.

McMenamin, Mark. *Discovering the First Complex Life: The Garden of the Ediacara*. New York: Columbia University Press, 1998. This entertaining study of the earliest complex life-forms on the planet details the author's work on these organisms. Written for the interested student but understandable by the general reader.

McMenamin, Mark. *Discovering the First Complex Life: The Garden of the Ediacara*. New York: Columbia University Press, 1998. This entertaining study of the earliest complex life-forms on the planet details the author's work on these organisms. Written for the interested student but understandable by the general reader.

McNamee, Gregory. *Careers in Renewable Energy: Get a Green Energy Job*. PixyJack, 2008.

McPherson, Michael, and Simon Møller. *PCR (The Basics)*. 2d ed. New York: Taylor & Francis, 2006.

Medicinal Plant Working Group http://www.nps.gov/plants/MEDICINAL/index.htm

Metz, Matthew, ed. *Bacillus Thuringiensis: A Cornerstone of Modern Agriculture*. Binghamton: Haworth, 2003. Print.

Meyer, Ernst, Hans Josef Hug, and Roland Bennewitz. *Scanning Probe Microscopy: The Lab on a Tip*. Berlin: Springer 2004. This book provides an excellent overview of scanning probe microscopy before delving into more detailed discussions of the various techniques.

Meyerowitz, Elliot M., ed. *Arabidopsis*. Plainview: Cold Spring Harbor, 1994. Print.

Mikityuk, Andrey. "Mr. Ethanol Fights Back." *Forbes*, November 24, 2008, 52–57.

Miller, G. Tyler, Jr., and Scott Spoolman. "Nonrenewable Energy." In *Living in the Environment: Principles, Connections, and Solutions*. 16th ed. Belmont, Calif.: Brooks/Cole, 2009.

Miller, Judith, Stephen Engelberg, and William Broad. *Germs: Biological Weapons and America's Secret War.* New York: Simon, 2001. Print.

Mills, Kelly. "University Opts for Biometric Security." *Computerworld,* January 25, 2002, 3–4.

Minsky, Marvin, and Seymour Papert. *Perceptrons: An Introduction to Computational Geometry.* Rev. ed. Boston: MIT P, 1990. Print.

Mironov, Alexander A. *The Golgi Apparatus: State of the Art 110 Years after Camillo Golgi's Discovery.* New York: Springer, 2010. Print. Details the function of the Golgi apparatus and reviews scientific developments related to the fields of molecular biology and microscopy.

Mishra, Nawin C., and Günter Blobel. *Introduction to Proteomics: Principles and Applications.* Hoboken, N.J.: John Wiley & Sons, 2010. Introduces the field of proteomics and examines the role proteomics plays in the study of biological systems in general and disease in particular. Provides an understanding of the structure, function, and interactions of proteins and how they are used for identifying diseases and developing new drugs.

Misra, J. C., ed. *Biomathematics: Modelling and Simulation.* Hackensack, N.J.: World Scientific, 2006. This book provides an in-depth guide to several modern applications of biomathematics and includes many helpful illustrations.

Mitchell, C. Ben, et al. *Biotechnology and the Human Good.* Washington D.C.: Georgetown University Press, 2007. A distinctly Christian assessment of the application of biotechnology to humans that remains optimistic but cautious and concerned.

Monamy, Vaughan. *Animal Experimentation: A Guide to the Issues.* 2nd ed. New York: Cambridge University Press, 2009. Print.

Mongillo, John. *Nanotechnology 110.* Westport, Conn.: Greenwood, 2007. Demonstrates the essential relationship between and value of scanning probe microscopy to nanotechnology.

Monroe, Judy. *Steroids, Sports, and Body Image: The Risks of Performance-Enhancing Drugs.* Berkeley Heights: Enslow, 2004. Print.

Montagnier, Luc. *Virus: The Co-Discoverer of HIV Tracks Its Rampage and Charts the Future.* Trans. Stephen Sartarelli. New York: Norton, 2000. Print. Describes the Pasteur Institute's work to isolate and identify HIV, offering an insider's view of the early years of the HIV/AIDS epidemic. Bibliography, index.

Montaigne, Fen. *Medicine by Design: The Practice and Promise of Biomedical Engineering.* Baltimore: The Johns Hopkins University Press, 2006. Bioengineering (including neuroengineering) applications made accessible to the nonspecialist through vignettes and portraits of researchers.

Morange, Michel. *A History of Molecular Biology.* Cambridge: Harvard UP, 1998. Print.

Moreno, Lilliana I., and Bruce McCord. "Separation of DNA for forensic Applications Using Capillary Electrophoresis." In *Handbook of Capillary and Microchip Electrophoresis and Associated Microtechniques,* edited by James P. Landers. 3d ed. Boca Raton, Fla.: CRC Press, 2008.

Morgan, Rose M. *The Genetics Revolution: History, Fears, and Future of a Life-Altering Science.* Westport, Conn.: Greenwood Press, 2006.

Morrison, Adrian R. *An Odyssey with Animals: A Veterinarian's Reflections on the Animal Rights and Welfare Debate.* Oxford UP, 2009.

Morrison, Robert Thornton, and Robert Nielson Boyd. *Organic Chemistry.* 5th ed. Newton, Mass.: Allyn & Bacon, 1987. Provides one of the best and most readable introductions to organic chemistry and polymerization.

Mosier, Nathan S., and Michael R. Ladisch. *Modern Biotechnology: Connecting Innovations in Microbiology and Biochemistry to Engineering Fundamentals.* John Wiley & Sons, 2009.

Moss, Cynthia. *Elephant Memories.* New York: William Morrow, 1988. An interesting account of thirteen years of field observations concerning the behavior of elephants in the Amboseli National Park in Kenya.

Moulton, Benjamin W. "DNA Fingerprinting and Civil Liberties." *Journal of Law, Medicine and Ethics* 34 (Summer, 2006): 147-148. Presents an overview of the issues addressed at a symposium on the topic of DNA fingerprinting and civil liberties. Contributions to the symposium appear as articles in the same issue of the journal.

Mukherjee, Siddhartha. *The Emperor of All Maladies: A Biography of Cancer.* New York: Scribner, 2010. Print. Provides a historical account of the earliest descriptions of cancer, assisting in developing readers' understanding of the molecular basis of the disease.

Mur, Cindy, ed. *Drug Testing.* Farmington Hills, Mich.:Greenhaven Press/Thomson Gale, 2006. A collection of essays on drug testing in schools and the workplace, discussing efficacy and ethical issues such as privacy.

Murphy, Kenneth M., Paul Travers, and Mark Walport. *Janeway's Immunobiology*. 7th ed., Taylor & Francis, 2007.

Nagy, M., et al. "Optimization and Validation of a Fully Automated Silica-Coated Magnetic Beads Purification Technology in Forensics." *Forensic Science International* 152, no. 1 (2005): 13–22.

Nakamura, Yusuke. "DNA Variations in Human and Medical Genetics: Twenty-Five Years of My Experience." *Journal of Human Genetics*, vol. 54, 2009, pp. 1–8.

Nanavati, Samir, Michael Thieme, and Raj Nanavati. *Biometrics: Identity Verification in a Networked World*. New York: John Wiley & Sons, 2002.

National Academies Press *Desalination: A National Perspective*. http://books.nap.edu/openbook.php?record_id=12184&page=R1

National Center for Food Protection and Defense. *Food Defense Education: Post 9/11*. Minneapolis: Author, 2007. Report on a three-year study explores food safety education programs in the United States, with emphasis on the work of criminal justice professionals.

National Human Genome Research Institute. http://www.genome.gov

National Research Council of the National Academies. *Desalination: A National Perspective*. Washington, D.C.: National Academies Press, 2008.

———. *Science, Medicine, and Animals: A Circle of Discovery*. National Academies Press, 2004. The National Academies Press, www.nap.edu/catalog/10733/science-medicine-and-animals. Accessed 3 Nov. 2016.

Nebel, Bernard J., and Richard T. Wright. *Environmental Science: Towards a Sustainable Future* 10th ed. Englewood Cliffs: Prentice Hall, 2008. Describes several bioprocesses used in waste treatment and pollution control.

Nei, Masatoshi, and Suhir Kumar. *Molecular Evolution and Phylogenetics*. New York: Oxford University Press, 2000. Covers the mathematics of phylogenetic analysis.

Nelson, Gerald C., ed. *Genetically Modified Organisms in Agriculture: Economics and Politics*. San Diego, Calif.: Academic Press, 2008.

Nelson, Lewis S., Richard D. Shih, and Michael J. Balick. *Handbook of Poisonous and Injurious Plants*. 2d ed. New York: Springer, 2007. Provides useful information on many different plant biotoxins.

Nemerow, Nelson L., et al., eds. *Environmental Engineering: Environmental Health and Safety for Municipal Infrastructure, Land Use and Planning, and Industry*. Hoboken: John Wiley & Sons, 2009. 3 vols.

New England Bio Labs, CRISPR/Cas9 and Targeted Genome Editing: A New Era in Molecular Biology, https://www.neb.com/tools-and-resources/feature-articles/crispr-cas9-and-targeted-genome-editing-a-new-era-in-molecular-biology.

Newell, Frank W. *Ophthalmology: Principles and Concepts*. 5th ed. St. Louis: Mosby, 1982. Covers basic eye anatomy, optics, and retinal physiology and biochemistry.

Newton, David E. *GMO Food: A Reference Handbook*. Santa Barbara, California: ABC-CLIO, 2014.

———. *The Animal Experimentation Debate*. Santa Barbara, California: ABC-CLIO, 2013. Print.

Nicholl, Desmond S. T. *An Introduction to Genetic Engineering*. New York: Cambridge University Press, 2008. Print.

Niessen, Walter R. "Refuse-Derived Fuel Systems." In *Combustion and Incineration Processes*. 3d ed. New York: Marcel Dekker, 2002.

Nikiforuk, Andrew. *Tar Sands: Dirty Oil and the Future of a Continent*. Berkeley, Calif.: Greystone Books, 2008. Examines potential long-term environmental effects of Canada's oil sands industry in light of possible climatic changes and the ethics of fossil fuel consumption.

Nisbet, Evan G. *The Young Earth: An Introduction to Archean Geology*. Boston: Unwin Hyman, 1987. A highly comprehensive coverage of geologic phenomena of the earth's earliest geologic time span, the Archean eon. Included in this work is information on both crustal evolution and biosphere. The book's sections vary considerably in technical coverage and terminology, some parts being readily comprehensible to the lay reader while others are quite technical and require considerable background in trace element geochemistry, isotope geochemistry, and petrology.

Nunnally, Brian K. *Analytical Techniques in DNA Sequencing*. New York: Taylor & Francis, 2005.

Nüsslein-Volhard, Christiane. *Coming to Life: How Genes Drive Development*. San Diego: Kales, 2006. Print. Presents an introduction to the genetic development of bilaterian embryos (embryos of animals characterized by left-right symmetry).

O'Mathúna, Dónal P. *Nanoethics: Big Ethical Issues with Small Technology*. London: Continuum, 2009.

Ogg, James G., Gabi Ogg, and Felix M. Gradstein. *The Concise Geologic Time Scale*. New York: Cambridge University Press. 2008. This book is a complete overview of the geological time scale, including stratigraphy topics such as chronostratigraphy and magnetic stratigraphy. It is organized by geological time periods. Also contains a reference appendix, the geological time scale table, and indexing.

Ohkawa, H., H. Miyagawa, and P. W. Lee, eds. *Pesticide Chemistry: Crop Protection, Public Health, Environmental Safety*. New York: Wiley-VCH, 2007.

Olby, Robert C. *Francis Crick: Hunter of Life's Secrets*. Cold Spring Harbor: Cold Spring Harbor Laboratory P, 2009. Print. Presents a biography of Francis Crick, covering the work he did with James D. Watson leading to the determination of the structure of DNA. Includes the confrontation between Chargaff and Crick.

Old, R. W., and S. B. Primrose. *Principles of Gene Manipulation: An Introduction to Genetic Engineering*. Boston: Blackwell Scientific, 1994. Focuses on methods of cloning. Illustrations, appendices, references, and index.

Orent, Wendy. *Plague: The Mysterious Past and Terrifying Future of the World's Most Dangerous Disease*. New York: Free Press, 2004.

Ortega-Rivas, Enrique, editor. *Processing Effects on Safety and Quality of Foods*. CRC Press, 2010.

Ostergaard, Simon, Lisbeth Olsson, and Jens Nielsen. "Metabolic Engineering of *Saccharomyces cerevisiae*." *Microbiology and Molecular Biology Reviews* 64, no. 1 (2000): 34–50. Describes metabolic engineering techniques using *S. cerevisiae* as an example.

Owen, Meran R. L., and Jan Pen, eds. *Transgenic Plants: A Production System for Industrial and Pharmaceutical Proteins*. New York: Wiley, 1996. Print.

Page, George. *Inside the Animal Mind*. New York: Doubleday, 1999. The author begins with a historical account of the popular and scientific views about animal intelligence. He provides good details about the various attempts to communicate with primates by teaching them sign language or through the use of keyboards.

Page, Roderic, and Edward Holmes. *Molecular Evolution: A Phylogenetic Approach*. Malden, Mass.: Blackwell Science, 1998. Concentrates on molecular evolution in populations and how species arise.

Pahl, Greg. *Biodiesel: Growing a New Energy Economy*. 2nd ed., Green, 2008.

Panno, Joseph. *Stem Cell Research: Medical Applications and Ethical Controversy*. New York: Facts On File, 2004. Print.

Parekh, Sarad R. *The GMO Handbook: Genetically Modified Animals, Microbes, and Plants in Biotechnology*. Totowa, N.J.: Humana Press, 2004.

Park, Alice. "Scientists Report First Success in Cloning Human Stem Cells." *CNN*, 16 May 2013, www.cnn.com/2013/05/15/health/time-cloning-stem-cells. Accessed 24 Oct. 2016.

Park, Sheldon J., and Jennifer R. Cochran, eds. *Protein Engineering and Design*. Boca Raton, Fla.: CRC Press, 2010. Covers the broader field of protein engineering—methods of developing altered proteins for novel applications—in two sections: one on experimental protein engineering and the other on computational design. Includes discussion of enzyme engineering using both rational and combinatorial approaches.

Parker, Philip M., and James N. Parker. *Bubonic Plague: A Medical Dictionary, Bibliography, and Annotated Research Guide to Internet References*. San Diego, Calif.: ICON Health Publications, 2003.

Pascal, Kintz. *Analytical and Practical Aspects of Drug Testing in Hair*. Boca Raton, Fla.: CRC Press, 2006. Looks at advances in the use of strands of hair for drug testing in the workplace and in forensic crime laboratories and techniques for detecting specific drugs.

Pasternak, Jack J. *An Introduction to Human Molecular Genetics: Mechanisms of Inherited Diseases*. 2nd ed., Wiley-Liss, 2005.

Paul, Ellen Frankel, and Jeffrey Paul, editors. *Why Animal Experimentation Matters: The Use of Animals in Medical Research*. Transaction Publishers, 2001.

Pavese, Franco, and Gianfranco Molinar Min Beciet. *Modern Gas-Based Temperature and Pressure Measurements*. 2nd ed. New York: Springer, 2013. Print.

Pecorino, Lauren. *Molecular Biology of Cancer: Mechanisms, Targets, and Therapeutics*. New York: Oxford UP, 2008. Print. An introductory text presenting the role played by oncogenes and describing events that can trigger a malignancy.

Pennington, Stephen, and Michael Dunn, eds. *Proteomics: From Protein Sequence to Function*. London: Garland, 2001. Gives an overview of how proteins can be identified, and used for clinical procedures and research. Discusses the technologies used in the field and protein structure and function.

Pennisi, Elizabeth. "Synthetic Genome Brings New Life to Bacterium." *Science* 21 (2010): 958–59. Print. Explanation of the steps involved in Venter's 2010 creation of the first entirely artificial (bacterial) genome and successful insertion into a recipient cell.

Pereira, Filipe, et al. "Identification of Species with DNA-Based Technology: Current Progress and Challenges." *Recent Patents on DNA and Gene Sequence*, vol. 2, 2008, pp. 187–200.

Peruski, Anne Harwood, and Leonard F. Peruski, Jr. "Immunological Methods for Detection and Identification of Infectious Disease and Biological Warfare Agents." *Clinical and Diagnostic Laboratory Immunology* 10 (July, 2003): 506-513. Technical article describes immunological methods of biological weapon identification.

Peterson, Donald R., and Joseph D. Bronzino, eds. *Biomechanics: Principles and Applications*. 2d ed. Boca Raton, Fla.: CRC Press, 2008. A collection of twenty articles on various aspects of research in biomechanics.

Petroski, Henry. *Success through Failure: The Paradox of Design*. 2006. Rpt. Princeton: Princeton UP, 2008. Print.

Pettit, George. *Biosynthetic Products for Cancer Chemotherapy*. Vol. 5 London: Elsevier Science, 1985. A discussion of the fundamental processes involved with screening for antitumor agents.

Phillips, Tom. "Largest Animal Cloning Factory Can Save Species, Says Chinese Founder." *Guardian*, 24 Nov. 2015, www.theguardian.com/world/2015/nov/24/worlds-largest-animal-cloning-factory-can-save-species-says-chinese-founder. Accessed 24 Oct. 2016.

Pierce, Benjamin A. *Genetics: A Conceptual Approach*. 4th ed. New York: W. H. Freeman, 2012. Print.

Pierik, R. L. M. *In Vitro Culture of Higher Plants*. Dordrecht: Martinus Nijhoff, 1987. Print.

Pilch, Richard F., and Raymond A. Zilinskas, eds. *Encyclopedia of Bioterrorism Defense*. Hoboken: Wiley, 2005. Print.

Pimentel, David, ed. *Biofuels, Solar, and Wind as Renewable Energy Systems: Benefits and Risks*. New York: Springer, 2008.

Plotkin, Mark J. *Medicine Quest: In Search of Nature's Healing Secrets*. New York: Viking, 2000.

Pollack, Andrew. "FDA Approves Drug from Gene-Altered Goats." *The New York Times*, 6 Feb. 2009, www.nytimes.com/2009/02/07/business/07goatdrug.html. Accessed 7 Nov. 2016.

Pollan, Michael. *The Omnivore's Dilemma: A Natural History of Four Meals*. New York: Penguin Press, 2007.

Powell, John. *How Music Works: The Science and Psychology of Beautiful Sounds, from Beethoven to the Beatles and Beyond*. New York: Little, 2010. Print.

Powell, Russell. "The Evolutionary Biological Implications of Human Genetic Engineering." *Journal of Medicine & Philosophy* 37.3 (2012): 204–25. Print.

Prado, José Rafael, et al. "Genetically Engineered Crops: From Idea to Product." *Annual Review of Plant Biology*, vol. 65, 2014, pp. 769–90.

Prendergast, Patrick, ed. *Biomechanical Engineering: From Biosystems to Implant Technology*. London: Elsevier, 2007. One of the first comprehensive books for biomechanical engineers, written with the student in mind.

Primrose, Sandy B., et al. *Principles of Gene Manipulation*. 6th ed., Blackwell, 2003.

Probstein, Ronald F., and Edwin R. Hicks. *Synthetic Fuels*. Mineola, N.Y.: Dover Publications, 2006.

Prothero, Donald R. *Bringing Fossils to Life*. 2d ed. Boston: McGraw-Hill, 2004. This well-illustrated and entertaining text covers a broad range of paleontological topics, including biostratigraphy. Glossary, bibliography, and index.

Purchase, Helen C. *Experimental Human-Computer Interaction: A Practical Guide with Visual Examples*. Cambridge University Press, 2012.

Ramsden, Jeremy. *Nanotechnology*. Elsevier, 2016.

Rand, Gary M., ed. *Fundamentals of Aquatic Toxicology: Effects, Environmental Fate, and Risk Assessment*. 3d ed. Boca Raton, Fla.: CRC Press, 2008.

Ratner, Daniel, and Mark A. Ratner. *Nanotechnology and Homeland Security: New Weapons for New Wars*. Upper Saddle River: Prentice, 2004. Print.

———. *Nanotechnology: A Gentle Introduction to the Next Big Idea*. Upper Saddle River: Prentice, 2003. Print.

Raven, Peter H., Ray F. Evert, and Susan E. Eichhorn. *Biology of Plants*. 6th ed. New York: W. H. Freeman/

Worth, 1999. Introductory botany text provides a chapter introducing eukaryotic cells in general and plant cells specifically.

Ray, Paresh Chandra, Hongtao Yu, and Peter P. Fu. "Toxicity and Environmental Risks of Nanomaterials: Challenges and Future Needs." *Journal of Environmental Science and Health*, Part C, Environmental Carcinogenesis and Ecotoxicology Reviews 27, no. 1 (2009): 1-35.

Reece, Jane B., and Neil A. Campbell. *Biology*. 6th ed. Menlo Park, Calif.: Benjamin Cummings, 2002. Introductory textbook. Illustrations, problems sets, glossary, index.

Regan, Tom. *The Case for Animal Rights*. Updated ed., U of California P, 2004.

Rehbinder, E., et al. *Pharming: Promises and Risks of Biopharmaceuticals Derived from Genetically Modified Plants and Animals*. Berlin: Springer, 2009.

Reid, Walter V. "Gene Co-ops and the Biotrade: Translating Genetic Resource Rights into Sustainable Development." *Journal of Ethnopharmacology* 51, nos. 1-3 (April, 1996): 75-92.

Reimer, Paula J., et al. "IntCal104 Terrestrial Radiocarbon Age Calibration, 0-26 cal kyr BP." *Radiocarbon*. 46 (2004): 1029-1058. This article provides a technical review of the past state of the radiocarbon calibration curve. More recent data are used to update and refine the curve. Written in a highly technical manner for practicing geologists and graduate students.

Reitdorf, Jens, et al., eds. *Microscopy Techniques*. New York: Springer, 2005. A technical reference book designed for those with a biomedical background, including numerous tables and diagrams, plus appendixes detailing mathematical formulas.

Renfrew, Colin. *Before Civilization: The Radiocarbon Revolution and Prehistoric Europe*, 2d ed. London: Penguin Books, 1990. A comprehensive discussion of the development of radiocarbon dating, the early discrepancies, and their correction using tree-ring data. Offers a thorough analysis of some alternative approaches to prehistory based on carbon-14 results from Europe.

Resnik, David B. *Owning the Genome: A Moral Analysis of DNA Patenting*. Albany: State University of New York Press, 2004.

Rheinberger, Hans-Jörg. *An Epistemology of the Concrete: Twentieth-Century Histories of Life*. Durham: Duke UP Books, 2010. Print. Examines the lives, work, and ultimate impact of twentieth-century experimental biologists and life scientists, including Spemann.

Richards-Kortum, Rebecca. *Biomedical Engineering for Global Health*. Cambridge University Press, 2010.

Rick, T. C., R. L. Vellanoweth, and J. Erlandson. "Radiocarbon Dating and the 'Old Shell' Problem: Direct Dating of Artifacts and Cultural Chronologies in Coastal and Other Aquatic Regions." *Journal of Archaeological Sciences* 32 (2005): 1641-1648. An interesting and easy-to-follow discussion of a common problem in sample selection for carbon dating. This article addresses the problem of "old wood" and compares it to the sampling of shells at archaeological sites. Provides the basic principles of radiocarbon dating in a manner accessible to the layperson with some science background.

Ridley, Matt. *Francis Crick: Discoverer of the Genetic Code*. New York: Harper, 2006. Print. Eminent Lives series. Provides a succinct account of Crick's work on DNA with Watson.

Rimmer, Matthew. *Intellectual Property and Biotechnology: Biological Inventions*. Northampton, Mass.: Edward Elgar, 2008.

Rogers, Ben, Jesse Adams, and Sumita Pennathur. *Nanotechnology: The Whole Story*. Boca Raton: CRC, 2013. Print.

———. *Nanotechnology: Understanding Small Systems*. 2nd ed. Boca Raton: CRC, 2011. Print.

Romundstad, Liv Bente, et al. "Effects of Technology or Maternal Factors on Perinatal Outcome After Assisted Fertilisation: A Population-based Cohort Study." *The Lancet* 372, no. 9640 (August, 2008): 737–743. Finds that adverse outcomes associated with births following assisted reproduction are not caused by technological factors. Includes several tables.

Roper, Stephan M., and Owatha L. Tatum. "Forensic Aspects of DNA-Based Human Identity Testing." *Journal of Forensic Nursing*, vol. 4, 2008, pp. 150–56.

Rose, Nickolas. *The Politics of Life Itself: Biomedicine, Power, and Subjectivity in the Twenty-first Century*. Princeton, N.J.: Princeton University Press, 2006.

Rosillo-Calle, Frank, et al., eds. *The Biomass Assessment Handbook: Bioenergy for a Sustainable Environment*. Sterling, Va.: Earthscan, 2008.

Rosner, Hillary. "The Gatherers." *Popular Science*, January, 2008, 60-64, 66, 91.

Rudacille, Deborah. *The Scalpel and the Butterfly: The War between Animal Research and Animal Protection.* Farrar, Straus and Giroux, 2000.

Rudin, Norah, and Keith Inman. *An Introduction to Forensic DNA Analysis.* 2d ed. Boca Raton, Fla.: CRC Press, 2002. Provides a good introduction to the use of biological evidence in forensics as well as the history and application of DNA fingerprinting in forensic investigations.

Rumelhart, David E., James L. McClelland, and the PDP Research Group. *Parallel Distributed Processing: Explorations in the Microstructure of Cognition.* 1986. Rpt. 2 vols. Boston: MIT P, 1989. Print.

Ruse, Michael, and Joseph Travis, ed. *Evolution: The First Four Billion Years.* Cambridge, MA: Harvard UP, 2009. Print. Presents an introduction to the theory of evolution that contains a small overview of Lewis and his work.

Russell, Stuart, and Peter Norvig. *Artificial Intelligence: A Modern Approach.* 3rd ed. Upper Saddle River: Prentice, 2010. Print.

Ryder, Oliver A., et al. "DNA Banks for Endangered Animal Species." *Science* 288.5464 (2000): 127–77. Print.

Saboowala, Hakim. *CRISPR Cas 9: An Enzymatic Scissor for Specific Site Modification of Genome.* Amazon Digital Services, 2016. Discusses how the components of CRISPR-Cas9 can be combined in multiple ways to edit the genome.

Sachs, Jessica Snyder. *Good Germs, Bad Germs: Health and Survival in a Bacterial World.* New York: Hill, 2007. Print.

Salisbury, F. B., and C. W. Ross. *Plant Physiology.* Pacific Grove, Calif.: Brooks/Cole, 1985.

Salway, J. G. *Medical Biochemistry at a Glance.* 3rd ed. Malden: Wiley-Blackwell, 2012. Print. Covers the relevance of biochemistry to clinical medicine. Summarizes metabolic pathways, including the Cori cycle and its regulation. Illustrations, index.

Sandøe, Peter, and Stine B. Christiansen. *Ethics of Animal Use.* Oxford, England: Blackwell, 2008.

Sanger, Frederick, and Margaret Dowding. *Selected Papers of Frederick Sanger: With Commentaries.* Singapore: World Scientific, 1996. Print. Includes publications from Sanger's research on RNA and DNA sequencing, amino acids, and insulin, as well as Sanger's reflections on his papers and why they were included in the collection.

Savageau, Michael. *Biochemical Systems Analysis: A Study of Function and Design in Molecular Biology.* New York: CreateSpace, 2010. Detailed textbook describing the immune system and gene regulation.

Savage-Rumbaugh, Sue, Stuart G. Shanker, and Talbot J. Taylor. *Apes, Language, and the Human Mind.* New York: Oxford University Press, 1998. A book written from an academic and scientific perspective about the ability of chimpanzees to communicate by means of symbols and a keyboard display.

———. *Kanzi: The Ape at the Brink of the Human Mind.* New York: John Wiley & Sons, 1994. This book presents an apparent breakthrough in the communication with chimpanzees using symbols and a keyboard. It includes a number of incidences of the spontaneous construction of sentences by the primates.

Schatten, Heide, and Gheorghe M. Constantinescu, eds. *Comparative Reproductive Biology.* Ames, Iowa: Blackwell, 2007.

Scheindlin, Stanley. "Clinical Enzymology: Enzymes As Medicine." *Molecular Interventions* 7, no. 1 (February, 2007): 4-8 An absorbing and readable summary of the use of engineered enzymes in clinical diagnoses and treatments.

Schnell, Santiago, Ramon Grima, and Philip Maini. "Multiscale Modeling in Biology." *American Scientist* 95 (March-April, 2007): 134-142. This article gives an overview of how biological models are created and provides several modern examples of biomathematical applications.

Scholar, Eric M., and William B. Pratt, eds. *The Antimicrobial Drugs.* New York: Oxford UP, 2000. Print.

Schopf, J. William, ed. *Major Events in the History of Life.* Boston: Jones and Bartlett, 1992. An excellent overview of the origin of life, the oldest fossils, and the early development of plants and animals. Written by specialists in each field but at a level that is suitable for high school students and undergraduates. Although technical language is used, most of the terms are defined in the glossary.

Schopf, J. William, ed. *Major Events in the History of Life.* Boston: Jones and Bartlett, 1992. An excellent overview of the origin of life, the oldest fossils, and the early development of plants and animals. Written by specialists in each field but at a level that is suitable for high school students and undergraduates. Although technical language is used, most terms are defined in the glossary.

Schuster, Brian G. "A New Integrated Program for Natural Product Development and the Value of an Ethnomedical Approach." *Journal of Alternative and Complementary Medicine* 7, no. 1 (2001): S61-S72.

Scott, Celicia. *Doping: Human Growth Hormone, Steroids, and Other Performance-Enhancing Drugs.* Broomall: Mason Crest, 2015. Print.

Scott, E. M., M. S. Baxter, and T. C. Aitchison. "A Comparison of the Treatment of Errors in Radiocarbon Dating Calibration Methods." *Journal of Archaeological Science* 11 (1984): 455-466. Discusses several methods of calibration with respect to the varying degrees of accuracy required in specific applications.

Scott, Elizabeth, and Paul Sockett. *How to Prevent Food Poisoning: A Practical Guide to Safe Cooking, Eating, and Food Handling.* Hoboken, N.J.: John Wiley & Sons, 1998.

Scott, Elizabeth, and Paul Sockett. *How to Prevent Food Poisoning: A Practical Guide to Safe Cooking, Eating, and Food Handling.* Hoboken, N.J.: John Wiley & Sons, 1998. Provides thorough information on food poisoning's causes and symptoms. Includes a chapter devoted to the science of food poisoning.

Scragg, Alan. *Biofuels: Production, Application, and Development.* Cambridge, Mass.: CAB International, 2009.

———. *Environmental Biotechnology.* 2d ed. New York: Oxford University Press, 2005. Examines the multitude of ways in which environmental biotechnology is applied in pollution control, environmental management, and removal of oil and minerals.

Sears, Andrew, and Julie A. Jacko, editors. *The Human-Computer Interaction Handbook: Fundamentals, Evolving Technologies, and Emerging Applications.* 2nd ed., Erlbaum, 2008.

Selinger, Ben. *Chemistry in the Marketplace.* 5th ed. Sydney: Allen & Unwin, 2002. The seventh chapter of this book provides a concise overview of many fiber materials and their common uses and properties.

Sell, Stewart, ed. *Stem Cells Handbook.* 2nd ed. Totowa: Humana, 2013. Print.

Semple, Kirk. "UN to Consider Whether to Ban Cloning of Human Embryos." *The New York Times*, 3 Nov. 2003, www.nytimes.com/2003/11/03/world/un-to-consider-whether-to-ban-some-or-all-forms-of-cloning-of-human-embryos.html. Accessed 28 July 2014.

Service, Robert F. "The Hydrogen Backlash." *Science* 305, no. 5686 (August 13, 2004): 958–961.

Shanks, Pete. *Human Genetic Engineering: A Guide for Activists, Skeptics, and the Very Perplexed.* New York: Nation Books, 2005. A helpful explication of the science behind cloning, coupled with stern warnings against it, by a noted social activist.

Shapiro, Stewart, ed. *Encyclopedia of Artificial Intelligence.* 2nd ed. New York: Wiley, 1992. Print.

Sharp, Heken, et al. *Interaction Design: Beyond Human-Computer Interaction.* 2nd ed., John Wiley & Sons, 2007.

Sharp, Lesley A. *Bodies, Commodities, and Biotechnologies: Death, Mourning, and Scientific Desire in the Realm of Human Organ Transfer.* Columbia UP, 2008.

Shea, Neil. "Norway's Ark." *National Geographic*, June, 2007, 14-21.

Shilts, Randy. *And the Band Played On: Politics, People, and the AIDS Epidemic.* New York: St. Martin's, 2007. Print. Examines the initial stages of the AIDS/HIV crisis and the response to it by medical researchers, politicians, and the larger culture. Describes the dispute between the Pasteur Institute and the American Dr. Robert Gallo over who first identified HIV. Bibliography, index.

Shostak, Stanley. *Becoming Immortal: Combining Cloning and Stem-Cell Therapy.* State U of New York P, 2002.

Shreeve, James. *The Genome War: How Craig Venter Tried to Capture the Code of Life and Save the World.* New York: Alfred A. Knopf, 2004. Describes the competition between Venter and Collins to decode the human genome.

Shubin, Neil. *Your Inner Fish: A Journey into the 3.5 Billion-Year History of the Human Body.* New York: Vintage, 2009. Print. Illustrated work incorporating Spemann's experiments in embryological development into an interesting discussion of animal evolution in general and human evolution in particular.

Siderovski, Susan Hutton. *Tularemia.* New York: Chelsea House, 2006.

Silveira, Semida, ed. *Bioenergy: Realizing the Potential.* San Diego, Calif.: Elsevier, 2005.

Silver, Lee. *Challenging Nature: The Clash Between Biotechnology and Spirituality.* New York: Harper Perennial, 2006. A Princeton stem cell scientist explains the science behind biotechnology and stem cells.

He offers some rather harsh critiques of more conservative thinkers who do not agree with his optimistic views of genetic enhancement and embryonic stem cells.

———. *Remaking Eden: How Genetic Engineering and Cloning Will Transform the American Family.* New York: Harper Perennial, 2007. A very readable introduction to the science of cloning and genetic engineering by a noted mammalian embryologist, who believes that humans should be cloned and that people should welcome the profound changes that it will invoke within human societies.

Simon, Paul. *Tapped Out: The Coming World Crisis in Water and What We Can Do About It.* New York: Welcome Rain, 1998.

Sims, Ralph. *The Brilliance of Bioenergy: In Business and Practice.* London: James and James, 2002.

Singer, Peter. *Animal Liberation: The Definitive Classic of the Animal Movement.* Updated ed., Harper Perennial, 2009.

Singh, Om V., and Steven P. Harvey, eds. *Sustainable Biotechnology: Sources of Renewable Energy.* London: Springer, 2009.

Singh, Sheo B., and Fernando Pelaez. "Biodiversity, Chemical Diversity, and Drug Discovery." *Progress in Drug Research* 65, no. 141 (2008): 142-174.

Singh, V. P., and R. D. Stapleton, Jr., eds. *Biotransformations: Bioremediation Technology for Health and Environmental Protection.* New York: Elsevier Science, 2002.

Skipper, H. D., and R. F. Turco, eds. *Bioremediation: Science and Applications.* Madison, Wis.: American Society of Agronomy, 1995.

Slack, Jonathan M. W. *Essential Developmental Biology.* Hoboken: Wiley-Blackwell, 2005. Print. Explains the processes involved in the development of an embryo on molecular and cellular levels, from fertilization to maturity; contains many full-color drawings, a glossary, and a bibliography.

Slater, Adrian, Nigel W. Scott, and Mark Fowler. *Plant Biotechnology: The Genetic Manipulation of Plants.* New York: Oxford University Press, 2008.

Sluder, Greenfield, and D. E. Wolf, eds. *Digital Microscopy.* 3d ed. Boston: Elsevier Academic Press, 2007. A guide to coordinating microscopes with digital cameras to capture and analyze microscopic images. Includes detailed laboratory exercises to demonstrate principles in action.

Smith, D. C., and A. E. Douglas. *The Biology of Symbiosis.* Baltimore, Md.: Edward Arnold, 1987. An outstanding, concise account of biological symbiosis, including several chapters on symbiotic relationships between fungi and plants.

Smith, Frederick P., ed. *Handbook of Forensic Drug Analysis.* Burlington: Elsevier, 2005. Print.

Smith, Louis D. S., and Hiroshi Sugiyama. *Botulism: The Organism, Its Toxins, the Disease.* 2d ed. Springfield, Ill.: Charles C Thomas, 1988.

Smith, Marquard, and Joanne Morra, editors. *The Prosthetic Impulse From a Posthuman Present to a Biocultural Future.* MIT, 2007.

Sneader, W. *The Evolution of Modern Medicines.* New York: Wiley, 1985. Provides excellent coverage of how plants contributed to the development of many pharmaceuticals.

———. *Drug Discovery: A History.* Hoboken, N.J.: Wiley, 2005.

Snyder, Bill. "Stanley Cohen's Nobel Prize: 25 Years of Progress." *ResearchNews@Vanderbilt.* Vanderbilt University, 9 Dec. 2011. Web. 25 Apr. 2016.

Soegaard, Mads, and Rikke Friis Dam, editors. *The Encyclopedia of Human-Computer Interaction.* 2nd ed., Interaction Design Foundation, 2014.

Spar, Debora. *The Baby Business: How Money, Science and Politics Drive the Commerce of Conception.* Boston: Harvard Business School Press, 2006. A critical overview of the fertility industry and how reproductive technologies are used in the marketplace. Includes numerous tables and extensive end notes listing sources.

Speight, James G. *Synthetic Fuels Handbook. Properties, Process, and Performance.* New York: McGraw-Hill, 2008.

Spentzos, Dimitri. "Gene Expression Signature with Independent Prognostic Significance in Epithelial Ovarian Cancer.' *Journal of Clinical Oncology* 22, no. 23 (December, 2004): 4648-4658. The research article describes the diagnosis of ovarian cancer and the use of biomarkers for detection.

Spiegler, K. S., and A. D. K. Laird, eds. *Principles of Desalination.* 2d ed. New York: Academic Press, 1980.

Srivastava, Rakesh K., and Sharmila Shankar. *Stem Cells and Human Diseases.* New York: Springer, 2012. Digital file.

Stahl, Franklin W. *Alfred Day Hershey: December 4, 1908-May 22, 1997.* Washington, D.C: National Academy Press, 2001. Print.

Stallard, Brian. "Oil Eaters: How Nature Cleans Up the Deepwater Horizon Spill." *Nature World News*, August 11, 2014.

Stanforth, Stephen. *Natural Product Chemistry at a Glance*. New York: Wiley-Blackwell, 2006. An introductory textbook that describes much of the organic chemistry involved in biosynthesis.

Stanley, S. M. *Exploring Earth and Life Through Time*. 2d ed. New York: W. H. Freeman, 1989. An excellent introductory-level college textbook on historical geology. It reviews the history of life and the many fossil forms found in strata in the earth's crust. Chapter 5 includes a discussion of biostratigraphy. Lavishly illustrated, with extensive references, glossaries, appendices on fossil groups, and an index.

Stannard, Jerry, Katherine E. Stannard, and Richard Kay, eds. *Pristina Medicamenta: Ancient and Medieval Medical Botany*. Brookfield, Vt.: Ashgate, 1999. Articles on premodern texts on plants.

Stephanopoulos, Gregory N., Aristos A. Aristidou, and Jens Nielsen. *Metabolic Engineering: Principles and Methodologies*. San Diego: Academic Press, 1998. Classic text on metabolic engineering.

Stemke, Douglas J. "Genetically Modified Organisms: Biosafety and Ethical Issues." In *The GMO Handbook: Genetically Modified Animals, Microbes, and Plants in Biotechnology*, edited by Sarad R. Parekh. Totowa, N.J.: Humana Press, 2004.

Stewart, Eileen. "Food Irradiation: More Pros than Cons? Part 1." *The Biologist*, vol. 51, no. 2, 2004, pp. 91–96.

———. "Food Irradiation: More Pros than Cons? Part 2." *The Biologist*, vol. 51, no. 3, 2004, pp. 141–44.

Stine, Keith J. *Carbohydrate Nanotechnology*. Wiley, 2016.

Stockwell, Christine. *Nature's Pharmacy: A History of Plants and Healing*. London: Century, 1988. An excellent discussion of medicinal products from plants.

Stone, Richard. *Mammoth: The Resurrection of an Ice Age Giant*. New York: Basic, 2002. Print.

Strom, Stephanie. "F.D.A. Takes Issue with the Term 'Non-G.M.O.'" *The New York Times*, 20 Nov. 2015, www.nytimes.com/2015/11/21/business/fda-takes-issue-with-the-term-non-gmo.html. Accessed 28 Oct. 2016.

Sumner, Judith, and Mark J. Plotkin. *The Natural History of Medicinal Plants*. Portland, Oreg.: Timber Press, 2000. An accessible introduction to the world of medicinal plants by a Harvard University botanist, from Europe in the Middle Ages to the modern pharmacopeia.

Substance Abuse and Mental Health Services Administration http://www.samhsa.gov

Substance Abuse Program Administrators Association http://www.sapaa.com

Swede, Helen, Carol L. Stone, and Alyssa R. Norwood. "National Population-Based Biobanks for Genetic Research." *Genetics in Medicine* 9, no. 3 (2007): 141-149. Focuses on the ethics issues related to the establishment of national genetic databases.

Syngellakis, S., ed. *Biomass to Biofuels*. WIT Press, 2015.

Szostak, Jack W. *The Origins of Life*. Cold Spring Harbor: Cold Spring Harbor Lab, 2010. Print. An illustrated discussion encapsulating Szostak's primary area of study: how prehistoric chemical reactions caused organic molecules to coalesce and replicate into what is known as life, and the possibility that similar reactions have or will occur in extraterrestrial settings.

Talbot-Smith, Michael, ed. *Audio Engineer's Reference Book*. 2nd ed. Woburn: Focal, 1999. Print.

———. *Sound Engineering Explained*. 2nd ed. Woburn: Focal, 2001. Print.

Taylor, R. E. "Fifty Years of Radiocarbon Dating." *American Scientist* 88 (January/February 2000): 60-67. This paper reviews the entire five decades of development of this remarkable technique. Taylor reviews the basic technique, deviation of carbon-14 dates from true dates due to variation in the magnetic field, means of correction, and attaining increased sensitivity by the application of mass spectrometry.

Taylor, R. E. *Radiocarbon Dating: An Archaeological Perspective*. New York: Academic Press, 1987. An excellent treatment of the procedures and complexities involved in measuring radiocarbon content. Discusses instrumentation, sources of contamination, case studies using various substances, and the historical development of radiocarbon methodology. The bibliography covers a broad range of sources.

Thakur, Indu Shekhar. *Environmental Biotechnology: Basic Concepts and Applications*. New Delhi: I. K. International, 2006. A comprehensive examination of environmental processes and the many possible applications of environmental biotechnology,

General Bibliography

Thatcher, Jim, et al. *Web Accessibility: Web Standards and Regulatory Compliance*. Springer, 2006.

The Human Genome Project. http://www.ornl.gov/sci/techresources/Human_Genome/home.shtml

Thieman, William J., and Michael A. Palladino. *Introduction to Biotechnology*. 3rd ed. Boston: Pearson, 2013. Print.

Triere, Detlef, and Peter Hemmersbach. *Doping in Sports*. Berlin: Springer, 2010. Examines sports doping from its beginning, covering the use of anabolic steroids, erythropoietin, human growth hormone, and gene doping in humans and the doping of race horses. Effects of the drugs, detection methods, and regulations are also discussed.

Tipler, Paul A., and Gene Mosca. *Physics for Scientists and Engineers*. 6th ed. New York: W. H. Freeman, 2008. Paul Tipler's physics text has been a staple for introductory university physics courses for many years. Chapters cover basic physics concepts including the basic physics of optics and the dual wave and particle nature of light.

Traynor, Ann J., and Reed J. Jensen. "Direct Solar Reduction of CO_2 to Fuel: First Prototype Results." *Industrial and Engineering Chemistry Research* 41, no. 8 (2002): 1935-1939.

Trease, G. E., and W. C. Evans. *Trease and Evans' Pharmacognosy*. 15th ed. W. B. Saunders, 2002. One of the most complete treatises on the production of drugs from plants. At nearly 600 pages, covers all scientific aspects of the topic, from taxonomy, cellular biology, and phytochemistry through genetics. Drugs are examined in chapters that group them by chemical class. The scope is broad, including vitamins and hormones and even alternative therapies such as homeopathic medicine and aromatherapy. Professionals will appreciate the chapters on investigative methodologies. Appendices, index.

Trestrail, John Harris, III. *Criminal Poisoning: Investigational Guide for Law Enforcement, Toxicologists, Forensic Scientists, and Attorneys*. 2d ed. Totowa, N.J.: Humana Press, 2007. Focuses on intentional poisonings and the techniques used to investigate poisoning crimes.

Treuting, Piper M., and Suzanne M. Dintzis, eds. *Comparative Anatomy and Histology: A Mouse and Human Atlas*. Boston: Elsevier, 2012. Print.

---

such as bioremediation, bioprocessing, and bioteaching.

Tucker, Jonathan B., ed. *Toxic Terror: Assessing Terrorist Use of Chemical and Biological Weapons*. Cambridge, Mass.: MIT Press, 2000.

Tufte, Edward R. *The Visual Display of Quantitative Information*. 2nd ed. Graphics, 2007.

Tuelyan, Victor. *Genetically Modified Food Sources: Safety Assessment and Control*. Elsevier 2013.

Twidell, John, and Tony Weir. *Renewable Energy Resources*. 2d ed. New York: Taylor & Francis, 2006.

U.S. Department of Agriculture. Animal Breeding, Genetics, and Genomics. www.csrees.usda.gov/animalbreedinggeneticsgenomics.cfm

Ungar, Laura. "Stem-Cell Advances May Quell Ethics Debate." *Courier-journal* [Louisville]. *USA Today*, 22 June 2014. Web. 20 Aug. 2014.

Ursano, Robert J., Anne E. Norwood, and Carol S. Fullerton, eds. *Bioterrorism: Psychological and Public Health Intervention*. New York: Cambridge University Press, 2004. Print.

"Need Continues to Grow." *Organ Procurement and Transplantation Network*, US Dept. of Health and Human Services, optn.transplant.hrsa.gov/need-continues-to-grow/. Accessed 27 Feb. 2015.

U.S. Department of Labor Drug-Free Workplace Adviser. http://www.dol.gov/elaws/drugfree.htm

Vacca, John R. *Biometric Technologies and Verification Systems*. Amsterdam: Elsevier - Butterworth Heinemann, 2007. Print.

Vaitheeswaran, Vijay. *Power to the People: How the Coming Energy Revolution Will Transform an Industry, Change Our Lives, and Maybe Even Save the Planet*. New York: Farrar, Straus and Giroux, 2003.

Valentinuzzi, Max. *Understanding the Human Machine: A Primer for Bioengineering*. Hackensack, N.J.: World Scientific Publishing, 2004.

Van der Werf, Julius, Hans-Ulrich Graser, Richard Frankham, and Cedric Gondro, eds. *Adaptation and Fitness in Animal Populations: Evolutionary and Breeding Perspectives on Genetic Resource Management*. London: Springer, 2009.

Van Sciver, Steven W. *Helium Cryogenics*. 2nd ed. New York: Springer, 2012. Print.

Vasic-Racki, Durda. "History of Biotransformations: Dreams and Realities." *Industrial Biotransformations*. Edited by Andreas Liese, Karsten Seelbach, and Christian Wandrey, Wiley, 2000.

Venter, J. Craig. *A Life Decoded: My Genome, My Life*. New York: Penguin, 2007. Print. Privides a look at

455

General Bibliography

Venter's personal story, scientific discoveries, and the politics that surround genomic research.

Ventura, Guglielmo, and Lara Risegari. *The Art of Cryogenics: Low-Temperature Experimental Techniques*. Burlington: Elsevier, 2008. Print.

Vigneshvar, S., et al. "Recent Advances in Biosensor Technology for Potential Applications—An Overview." *Frontiers in Bioengineering and Biotechnology*. 16 Feb. 2016. journal.frontiersin.org/article/10.3389/fbioe.2016.00011/full.

Vigue, Charles L., and William G. Stanziale. "Recombinant DNA: History of the Contorversy." *American Biology Teacher* 41 (November, 1979): 480-491. Written primarily for teachers of secondary school biology, this short article summarizes the history and controversy surrounding the recombinant DNA debate. Should be readily accessible to the average reader and a good first choice for further reading.

Vita-Finzi, Claudio. *Recent Earth History*. New York: Halsted Press, 1974. An account of the physical changes undergone by the earth during the Holocene (modern) geologic era, presented in the form of a stratigraphical narratuve based throughout on radiocarbon dates.

Wade, David. "Nobel Women." *Science* 18 Jan. 2002: 439. Print. Discusses the awarding of the Nobel Prize to women scientists between 1977 and 1995, including Nüsslein-Volhard.

Wade, Nick. *The Ultimate Experiment*. New York: W. H. Freeman, 1977. Print.

Wagner, Gunther A., and S. Schiegl. *Age Determination of Young Rocks and Artifacts: Physical and Chemical Clocks in Quaternary Geology and Archaeology*. New York: Springer, 2010. The authors cover various materials and dating methods. Well organized, accessible to advanced undergraduates and graduate students.

Wagner, Viqi, ed. *Do Infectious Diseases Pose a Serious Threat?* New Haven: Greenhaven, 2005. Print.

Wald, Matthew L. "Is Ethanol for the Long Haul?" *Scientific American* 296, no. 1 (January, 2007): 42-49.

Walker, John F., and Nicholas Jenkins. *Wind Energy Technology*. New York: John Wiley & Sons, 1997.

Walker, Mike. *Quaternary Dating Methods*. New York: Wiley, 2005. This text provides a detailed description of current dating methods, followed by content on the instrumentation, limitations, and applications of geological dating. Written for readers

Principles of Biotechnology

with some science background, but clear enough for those with no prior knowledge of dating methods.

Walker, Sharon. *Biotechnology Demystified*. McGraw-Hill, 2006.

Wall, Judy, ed. *Bioenergy*. Washington, D.C.: ASM Press, 2008.

Walsh, Christopher. *Antibiotics: Actions, Origins, Resistance*. Washington: ASM, 2003. Print.

Walsh, Gary. *Biopharmaceuticals: Biochemistry and Biotechnology*. 2d ed. New York: John Wiley & Sons, 2003.

Walsh, P. S., D. A. Metzger, and R. Higuchi. "Chelex 100 as a Medium for Simple Extraction of DNA for PCR-Based Typing from Forensic Material." *BioTechniques* 10 (April, 1991): 506-13.

Walter, Lynne Paige, and Ellen Hodgson Brown. *Nature's Pharmacy: Break the Drug Cycle with Safe, Natural Treatments for Two Hundred Everyday Ailments*. Upper Saddle River, N.J.: Prentice Hall, 1999. Typical of a wave of similar publications that began to appear after the 1996 law that removed "food supplements" from FDA regulatory responsibility, this 400-plus-page reference catalogs common ailments, from acne to whooping cough, offering signs, symptoms, and suggestions for alternative treatments in addition to traditional Western medicine.

Walther, John Victor. *Essentials of Geochemistry*, 2d ed. Jones & Bartlett Publishers, 2008. Contains chapters on radioisotope and stable isotope dating and radioactive decay. Geared more toward geology and geophysics than toward chemistry, this text provides content on thermodynamics, soil formation, and chemical kinetics.

Watkins, James. *Introduction to Biomechanics of Sport and Exercise*. London: Elsevier, 2007. Print.

Watson, James D. *A Passion for DNA: Genes, Genomes, and Society*. Oxford: Oxford Univ. Press, 2000. Print.

Watson, James D. *Avoid Boring People: Lessons from a Life in Science*. New York: Knopf, 2007. Print. Discusses Watson's own recollections of his early life and scientific achievements up to the mid-1970s.

Watson, James D., and Andrew Berry. *DNA: The Secret of Life*. New York: Knopf, 2003. Print. Traces the evolution of the study of DNA.

———, Michael Gilman, Jan Witkowski, and Mark Zoller. *Recombinant DNA*. 2d ed. New York: W. H.

456

Freeman, 1992. Emphasizes applications. Illustrations, reading lists, and index.

Wayne, Tiffany. *American Women of Science since 1900.* Santa Barbara: ABC-CLIO, 2011. Print. A two-volume summary of prominent American women scientists. Includes life statistics (birth, education) of each woman, along with a brief summary of her scientific contributions.

Weasel, Lisa H. *Food Fray: Inside the Controversy over Genetically Modified Food.* New York: Amacom-American, 2009. Print.

Weaver, Robert F., and Philip W. Hedrick. *Genetics.* 3d ed. Dubuque, Iowa: W. C. Brown, 1997.

Weinberg, Robert A. *The Biology of Cancer.* New York: Garland, 2006. Print. An in-depth study of the disease. The discoveries by Bishop and Varmus of the roles played by cellular oncogenes represent a significant portion of the story.

Weinberger, Charles B. "*Instructional Module on Synthetic Fiber Manufacturing.*" Gateway Engineering Education Coalition: 30 Aug. 1996. This article presents an introduction to the chemical engineering of synthetic fiber production, giving an idea of the sort of training and specialization required for careers in this field.

Weiss, Marcia J. "Beware! Uncle Sam Has Your DNA: Legal Fallout from Its Use and Misuse in the U.S." *Ethics and Information Technology* 6, no. 1 (2004): 55-63. Discusses the constitutionality of DNA profiling.

Wenisch, A., R. Kromp, and D. Reinberger. *Science or Fiction: Is There a Future for Nuclear?* Vienna: Austrian Ecology Institute, 2007.

Wheelis, Mark, Lajos Rózsa, and Malcolm Dando, eds. *Deadly Cultures: Biological Weapons Since 1945.* Cambridge: Harvard UP, 2006. Print

Wilmut, Ian, and Roger Highfield. *After Dolly: The Uses and Misuses of Human Cloning.* New York: W. W. Norton & Company, 2006.

———, Keith Campbell, and Colin Trudge. *The Second Creation: Dolly and the Age of Biological Control.* New York: Farrar, Straus and Giroux, 2000. The two researchers who made Dolly team up with a noted British science writer to give a personal but rigorous explanation and thoughtful examination of cloning. Contains a helpful glossary of terms.

Wilson, David. *The New Archaeology.* New York: New American Library, 1974. Perhaps the best account for the general reader of the development and

subsequent refinement of radiocarbon dating techniques. Discusses their enormous impact on the field of prehistoric archaeology and perceptions of prehistory.

Wilson, Edmund Beecher. *The Supernumerary Chromosomes and their Relation to the Odd or Accessory Chromosome. Whitefish:* Kessinger, 2010. Print. Reprint of Wilson's groundbreaking 1909 work on chromosomes as determinants of sex.

Wilson, Edward O. *On Human Nature.* 1978. Harvard UP, 2004.

Winchester, Simon. *The Map That Changed the World.* London: Penguin, 2001. An entertaining study of William Smith, who developed the idea of biostratigraphy and used it to produce the first geologic map of Great Britain. Written for the interested reader but valuable to students and professionals also.

Witkowski, Jan. *The Inside Story: DNA to RNA to Protein.* Woodbury, New York: Cold Spring Harbor Lab. P, 2005. Print. Includes a discussion of Zinder's discovery of RNA phage, which was among the first indications RNA could also serve as genetic material.

Wobus, A. M., and K. Boheler, eds. *Stem Cells.* New York: Springer, 2006. Print.

Woese, Carl R. "Archaebacteria." *Scientific American* 244 (June 1981): 98-122. One of the most comprehensive articles available on the archaebacteria. Distinctive attributes characteristic of the archaebacteria as determined through molecular biology are enumerated. The author was one of the workers originally involved in the discovery of the biochemical uniqueness of archaebacteria.

Wolfensohn, Sarah, and Maggie Lloyd. *Handbook of Laboratory Animal Management and Welfare.* 4th ed. Oxford: Wiley, 2013. Print.

Woodward, John D., Jr., Nicholas M. Orlans, and Peter T. Higgins. *Biometrics.* New York: McGraw-Hill, 2003.

Worek, Michael, ed. *Nobel: A Century of Prize Winners.* Ontario: Firefly, 2008. Print. Profiles two hundred Nobel Prize winners, including Carl and Gerty Cori. Includes illustrations to explain scientific concepts and photographs of the laureates.

World Anti-Doping Agency. http://www.wada-ama.org

World Health Organization *WHO Guidelines on Good Agricultural and Collection Practices (GACP) for Medicinal Plants.* http://whqlibdoc.who.int/publications/2003/9241546271.pdf

Wright, Richard T., and Dorothy F. Boorse. *Environmental Science: Toward a Sustainable Future*. 11th ed. Boston: Benjamin/Cummings, 2011. This textbook describes several bioprocesses used in waste treatment and pollution control.

Wright, Susan. *Molecular Politics: Developing American and British Regulatory Policy for Genetic Engineering 1972–1982*. Chicago: U of Chicago P, 1994. Print.

Wulfkuhle, D. J., et al. "Proteomic Applications for the Early Detection of Cancer." *Nature Reviews* 3 (2003): 267–275. Describes cancer detection with a great review of techniques used in proteomics.

Yacov, Davidov, and Eduard Jurkevitch. "Predation Between Prokaryotes and the Origin of Eukaryotes." *BioEssays* 31 (2009): 738–757. Discusses how eukaryotes evolved as a result of parasitism or predation occurring between prokaryotes. Covers the topic of mitochondrial symbiosis.

Yadav, A. K., S. Ray Chaudhuri, and M. R. Motsara, eds. *Recent Advances in Biofertilizer Technology*. New Delhi: Society for Promotion and Utilisation of Resources and Technology, 2001.

Yang, Shang-Tian. *Bioprocessing for Value-Added Products from Renewable Resources: New Technologies and Applications*. Amsterdam: Elsevier, 2007. Reviews the techniques for producing products through bioprocesses and lists suitable organisms,

proteomics, and protein structure and modeling.

including bacteria and algae, and describes their characteristics.

Yao, Nan, and Zhong Lin Wang, eds. *Handbook of Microscopy for Nanotechnology*. New York: Kluwer Academic, 2005. An overview of microscopy applications in nanotechnology. Each of the twenty-two chapters contains a discussion of a specific microscopic instrument or technique by nanotechnology specialists working in different fields.

Yesalis, Charles E. *Anabolic Steroids in Sport and Exercise*. 2d ed. Champaign: Human Kinetics, 2000. Print.

Young, Tomme R. *Genetically Modified Organisms and Biosafety: A Background Paper for Decision-Makers and Others to Assist in Consideration of GMO Issues*. International Union for Conservation of Nature, 2004.

Yount, Lisa. *Biotechnology and Genetic Engineering*. 3d ed. New York: Facts On File, 2008. Print.

Your Genome. Facts, What is CRISPR-Cas9? http://www.yourgenome.org/facts/what-is-crispr-cas9.

Zenios, Stefanos, Josh Makower, and Paul Yock, eds. *Biodesign: The Process of Innovating Medical Technologies*. New York: Cambridge University Press, 2010.

Zvelebil, Markéta, and Jeremy Baum. *Understanding Bioinformatics*. New York: Garland Science, 2008. Intermediate text with detailed descriptions on sequence alignments, phylogenetics, genomics,

SUBJECT INDEX

A

abzymes 185
acquired immunodeficiency syndrome (AIDS) 203, 351, 352
adalimumab 43
adsorption 42
adult stem cells 337, 338
adulteration 171–172
Advanced Cell Technologies (ACT) 131
aerobic fermentation 60, 235
aerosolized plague 122
aerospace engineering 176
agarose gel 157, 165, 195
agglutination test 76
agricultural biotechnology 131, 288
agricultural engineering 55, 134
Agrobacterium tumefaciens 96, 213, 259, 295
agroterrorism 72
A-efacept 43
algae 3, 50, 56, 60, 79, 102, 148, 236
alpha carbon 300
alternative energy sources 1–3
Alzheimer's disease 54
amino acids 110, 183, 221, 222, 236, 300, 400–402
aminoglycosides 20
amylases 44
anabolic steroids 285
anaerobic fermenters 235
animal breeding 4–5, 425
animal cell cultures 99
animal medical research 9–10
animal testing 6–8, 425
Animal Welfare Act of 1966 (AWA) 7, 10
anthrax 11–13, 73, 425
anthropogeomorphology 14–16, 425
antibacterials 18, 243
antibiotic-resistant bacteria 17–19, 425
antibiotics 19–21, 98–99, 235–236, 425
   mode of action 20
antibody production 94, 109, 110
antigen preparation 110
anti-inflammatory agents 243
antimycobacterials 20
AquaBounty Technologies 207
Arabidopsis thaliana 259–260
aramid fibers 191
archaebacteria 21–24, 305

architectural engineering 176
artificial bacterial chromosome 409
artificial hybridization 231
artificial insemination 113, 324–325
artificial intelligence 25–28
   intelligent tutor systems 29
artificial organs 30–34, 425
artificial retinas (AR) 273
aspirin 243, 301
atomic force microscope (AFM) 252, 254, 265, 328
atomic-scale probe 327
Atryn 9
audio engineering 34–38, 425
autogamous 231
automated detection systems (ADSs) 12
avastin 43

B

Bacillus anthracis 11–13, 73, 75, 284
Bacillus thuringiensis (*B.t.*) toxins 96, 114, 206, 236
bacterial artificial chromosomes (BACs) 408
bacterial resistance 18–19
bacteriophages 354, 386, 387, 401, 411, 417–419
Basic Local Alignment Sequence Tools (BLAST) 67, 68, 78
batch cultures 41
beta-lactam antibiotics 184
bevacizumab 43
bioartificial liver (BAL) 126
bioartificial organs 125–126
bioassays 39–40, 425
bioaugmentation 180–181
biobutanol 50–51
biochemical engineering 40–45, 53, 425
biochronology 106, 107
biodetectors 46–47, 181, 425
biodiesel 3, 50, 59, 61, 63, 64
biodiversity 68–69, 296
bioenergy technologies 47–51
bioengineering 51–56, 176, 425
bioethanol 44, 60
biofertilizers 57–58
biofilms 181
biofuels 51, 58–62, 98, 236–237, 249, 425
   production 44
   and synthetic fuels 62–65
biogas 3, 50, 59, 60, 64, 237

Subject Index

bovine somatotropin 114
bovine serum albumin (BSA) conjugation 110
botulism 117, 120, 281
botulinum toxin 73, 120–121, 425
bottom-up nanofabrication 266
botany 118–119, 425
bone ingrowth 82
B lymphocytes 43
blue-green algae 22, 304
blood doping 286
bitumen 341
birth defects 203, 338
biotoxins 116–118, 425
BioThrax 12
biotechnology 112–116
biosynthetics 108–111, 425
biosynthetic temporary skin substitute 111
biostaugraphy 105–107, 425
biostimulation 180–181
biosequence databases 68
biosensors 47, 103–104, 110, 181
bioremediation 100–102, 180
airlift 98
bioreactors 40, 41–42, 45, 64, 98, 101, 235
bioprocess engineering 97–100, 181
biopolymers 189, 190
biopesticides 95–96
bionics, 91–94
bionanotechnology 54–55
biometric identification systems 88–90
biometric eye scanners 87–88
biomedical engineering 41, 51, 54, 91–94
biomechanics 84–86
biomechanical engineering 80–83
biomathematics 77–80
biomaterials 32, 33, 94
biomass power 3
production 99, 236
biomass 49, 50, 59, 61, 65, 341
biomarkers 111, 311
identification 75–76
bubonic plague 121–122
botulinum toxin 120–121
biological weapons 71–74
biological warfare 11–13
biological containment 115
bioinstrumentation 54, 70
bioinformatics 66–70, 223
biohydrogen 59, 60, 62

C

Caenorhabditis elegans 258, 260
callus tissue 113, 133
carbapenems 20
carbohydrates 298, 372, 373
carbon fiber 190, 191
carbon nanotubes 265, 268, 269
cardiac glycosides 244, 302
cardiac resynchronization therapy (CRT) devices 82
carotenoids 235, 248, 288
catalytic antibodies 184, 285
cell engineering 123–126
cell matrices 125
cellulose fibers 190
Centers for Disease Control and Prevention (CDC) 117, 143, 343
cephalopods 7
cetuximab 43
charge-coupled device (CCD) sensors 254
Chelex extraction 153, 157
chemical engineering 41, 176
chemosynthetic 109
chemotaxonomic strategies 202
chinese hamster ovary (CHO) cells 42, 43
chromatography 42, 309
cephalosporins 20
chemical treatment 290
Chlamydomonas reinhardtii 260
Chlorella pyrenoidosa 260
cloning 93, 94, 127–134, 165, 168, 221
Clostridium acetobutylicum 60, 234
Clostridium botulinum 117, 120, 143, 343
ciprofloxacin 12
civil engineering 176–177
clustered regularly interspaced short palindromic repeats (CRISPR) 195–196
coal-bed methane 341
coal gasification 341
coal liquefaction 341
cochlear implants (CI) 82, 272, 273, 275
cognitive ethology 238
Combinatorial Active-Site Saturation Test (CASTing) 183
combinatorial chemistry 110, 111

butanol 59, 60, 62
bubonic plague 121–122, 425
BSM-2000 (Universal Detection Technology) 12
brain-computer interfaces (BCIs) 54, 274

Down syndrome 145, 324
double helix structure, DNA 299, 412, 413
Dolly, sheep 166–168
DNA typing 163–164
DNA structure 161, 164–165
DNA sequencing 162–163
DNA recognition instruments 160–161
DNA profiling 158–160
DNA isolation methods 156–158
DNA fingerprinting 154–156
DNA extraction 152–154
DNA database controversies 150–151
DNA banks 149–150
DNA analysis 145–149
disposable micropumps 110
disaccharide 298
direct solar conversion 1
dipicolinic acid (DPA) 12
dideoxyribonucleic acids (ddNTPs) 162
differential extraction 157
Diamond v. Chakrabarty 144–145, 194
dendrochronology 315
dehydration synthesis 297
deep brain stimulation (DBS) 272, 274
daidzein 187
dew point 426
desalination plants 140–141
DENDRAL 26

D
daidzein 187
cystic fibrosis (CF) 370
cyanobacteria 22, 57, 236, 305–307
cutaneous anthrax 11, 12
cryonics 137
cryogenics 136–139
cross-pollinated plants 230
CRISPR-Cas9 135, 196
Creutzfeldt-Jakob disease (CJD) 203
coumestrol 187
cosmetics testing 7
correlation 106, 117, 426
conventional plant breeding 206, 295
contraceptive methods 325
continuous culture systems 41
computer engineering 177
computational fluid dynamics (CFD) 31
complementary DNA (cDNA) 165
Combined DNA Index System (CODIS) 151, 159

doxycycline 12, 19
Drosophila melanogaster 260, 388, 390, 398
drug testing 169–173

E
electrical engineering 53–54, 177
electron microscopy 251–253
electroporation 290
embryonic gene regulation 399
embryonic induction 404
embryonic stem cell (ESC) 129, 130, 332, 333, 337, 338
embryo transfer 113
Energy Policy Act 60
environmental biotechnology 179–182
environmental engineering 177, 182
environmental genomics 216
Environmental Protection Agency (EPA) 10, 12, 213, 270, 342
enzyme-linked immunosorbent assay (ELISA) 76, 110
epidermal growth factor (EGF) 365
equilibrium 81, 85, 160, 313
Erasistratus 374–376
erbitux 43
Escherichia coli 18, 247, 417
estradiol 187
estriol 187
estrogens from plants 186–188
estrone 187
ethanol 50, 59, 60, 63, 237
fermentation 44
eubacteria 22, 426
eukaryotic cells 23, 23, 299, 426
evolutionary genomics 216
expressed sequence tags (ESTs) 408
extreme halophiles 23–24

F
familial issues 151
fed-batch operation 41
fermenters 41, 53, 64, 97, 98, 99, 234–236
fiber technologies 189–192
Fischer-Tropsch process 63, 64
flavonoids 248
fluorescence microscopy 251, 253
fluoroquinolones 20
Foodborne Diseases Active Surveillance Network (FoodNet) 143

462

glycosaminoglycan (GAG) 43
glycopeptide antibiotics 20
glycerol 248, 298
Glomus versiforme 346–347
Global Positioning System (GPS) 177
Gleevec 365
germ-line therapy 224
geothermal energy 3, 321
geographic information system (GIS) technology 241
geoengineering 56
Genzyme Transgenics 194
genistein 187
genomics 214–219
genome decoding 217
genetic resources 199–202
Genetic Information Nondiscrimination Act 218
genetic erosion 199–200
genetic engineering 53, 112–116, 118–121, 181, 193–198, 288, 295, 426
genetically modified organisms (GMOs) 134, 185, 209–212, 288, 332, 334
genetically modified food 196, 205–208
genetically modified crops 52, 134, 194, 196, 198, 206, 333
genetically modified animals 196, 207, 209, 211
genetically engineered mice 175
genetically engineered pharmaceuticals 203–205
genetically altered bacteria 212–213
gene therapy 115, 224
gene testing 223
Gene Logic's BioXpress System 111
gene library 128, 165, 331
GenBank, 67, 223
gel electrophoresis 262
gastrointestinal anthrax 11, 12
gasohol 341–342
gas chromatography/mass spectrometry (GC/MS) 170–171
gamete intrafallopian transfer (GIFT) 324

G

functional electrical stimulation (FES) 272–274
Frankenfood 45
Francisella tularensis 343
fragile X syndrome 148
Fossils 23, 24, 48, 49, 51, 105–107, 305, 306, 320, 426
food poisoning, detection and prevention of 142–143
food engineering 44

structure 402
359, 360, 400, 401
insulin 92, 93, 97, 114, 197, 210, 213, 222, 223, 236,
insect resistance 206, 207, 333
inorganic molecules 298
inner cell mass (ICM) 129, 332, 337
inhalation anthrax 11, 13
industrial fermentation 233–237
industrial farming 241–242
industrial engineering 177
index fossil 105, 106, 426
inbreeding 5, 231
imatinib mesylate 365

I

hydrolysis 298
hydrogen 59, 61, 63, 64, 248, 298, 299
hydroelectric power 2
hydrocarbon gases 3
hybrid vigor 5, 231
hybridization 230–232
humira 43
Human Microbiome Project 69
human immunodeficiency virus (HIV) 147, 197, 352
human growth hormone (HGH) 43, 114, 285
Human Genome Project 67, 69, 111, 195, 225, 309, 408
human genetic engineering 220–225
human-computer interaction (HCI) 226–230
human cloning 94, 131, 220, 225, 334
homozygous 160, 163, 231
Hodgkin-Huxley equations 78
high-throughput screening method 110
heterosis 231, 232
hercepun 203, 365
herbicide tolerance 207, 211
hemodynamics 31–32
Health Research Extension Act 7
Health Insurance Portability and Accountability Act (HIPAA) 152
Hardy-Weinberg principle 159–160
Haemophilus influenzae 21, 283, 408

H

Green Revolution 134, 241, 295, 296
greenhouse gases 1, 3, 44, 56, 60, 61, 320
gray goo scenario 267, 269
Gore-Tex 33
Gödel's incompleteness theorem 28

Subject Index

I
intelligence 237–240
intensive farming 241–242
interferon 43, 114, 213
intergeneric crosses 232
interspecific crosses 232
in vitro engineering 124–125
in vitro fertilization (IVF) 324
in vivo engineering 125
ion exchange methods 141
irradiation 280–282
isoflavonoids 187
isoprenoids 248

J
Jatropha plants 5
Joule-Thomson effect 138

K
karyotype 145
keyhole limpet hemocyanin (KLH) 110

L
lactic acid fermentation 44
landfill gas 49, 50, 61
light-emitting diode (LED) technology 266–267
light microscopes 251
linezolid 20
lipids 298–299
liquid counting techniques 314
living modified organisms (LMOs) 116
low-throughput bioinformatics 67

M
macrolide antibiotics 20
macromolecules 298
magnetic resonance imaging (MRI) 54, 93, 138
manufacturing engineering 177–178
marker-assisted selection 5
mass transfer efficiency 32–33
materials science 53
mechanical engineering 52, 91, 175, 178
media, fermentation 234
medicinal plants 243–244
membrane methods 141
messenger ribonucleic acid (mRNA) 221, 300, 308, 309
metabolic engineering 245–249
metagenomics 69
metal fibers 190
methane 3, 59–61, 63, 64
methanogens 24, 101
microalgae 44, 60, 99
microarray 217, 310
microbial forensics 13, 75, 118
microbialite 22–24, 426
micromanipulation 323, 324
microparticle bombardment 289, 290
micropropagation 289
microsatellites 155, 263
microscopy 250–255
microsurgery 253, 254, 323, 404
microsystems 273
minisatellites 155
mitochondrial DNA (mtDNA) 257–258
model organisms 258–261
molecular cloning 127, 128, 129, 331–333
molecular farms 296
molecular hydrogen 59, 63, 248
molecular systematics 261–263
monobactams 20
monoclonal antibodies 43
monoculture 134
monosaccharides 298
multilayer insulation (MLI) technique 137
MYCIN 26
mycorrhizae 346–347
mycotoxins 117

N
nanobots 269
nanoimprint lithography 266
nanoparticles 111, 264, 265, 268–270
nanotechnology 104, 254, 264–267, 329
environment 268–270
National Association for Biomedical Research (NABR) 9
National Center for Biotechnology Information (NCBI) 67, 68
National Institute on Drug Abuse (NIDA) 169
natural resources 1
near-field scanning optical microscopes (NSOM) 252
needle-free drug delivery systems 111
nematodes 96
Neolithic Revolution 315, 316
nerve growth factor (NGF) 365
neural engineering 270–275
neural prostheses (NP) 272–273
neuroaugmentation 272

N

neuromodulation 271, 272
neuromuscular simulation 272
Newton's laws of motion 85
night vision technology 275–278
nonrenewable resources 320–321
nontoxic homogeneous miscible fuel (NHMF) 342
Norwalk viruses 143
nuclear energy sources 1
nuclear engineering 178
nucleic acid(s) 299–300
nucleic acid amplification testing (NAAT) 147
nucleic acid hybridization 165
nucleotides 299, 401

O

Online Mendelian Inheritance of Animal (OMIA) 68
Online Mendelian Inheritance of Man (OMIM) 68
opiate alkaloids 244, 302
optical microscopy 251, 328
organic extractions 153, 154, 157

P

pacemakers 54, 82, 93
paper chromatography, 362, 400
paradigm shift 200–201
paralytic shellfish poisoning (PSP) toxins 118
parthenogenesis 127, 331
pasteurization 279–282
pathogen genomic sequencing 283–284
penicillin 9, 12, 20, 235, 243
People for the Ethical Treatment of Animals (PETA) 9
perfluorocarbon fluids 33
performance-enhancing drugs 284–287
pharmaceuticals 42–43, 131, 248
pharmacogenomics 224–225
PharmGKB 68
phenylketonuria (PKU) 145, 215
phospholipids 299
physical containment 115
phytoestrogens 187–188
phytoremediation process 100, 102
pico hydel 2
plant biotechnology 287–291
time line of 292–293
plant breeding 294–297
plant cell 119, 297–300
plant cloning 133–134
plant-derived medicines 301–302
plant propagation 294–297
plant tissue cultures 113, 133, 289, 290
plant transformation 289–290
pluripotent stem cells 338, 339
pneuma 375
pneumonic plague 73, 122
polders 15, 426
polycystic ovary syndrome (PCOS) 323
polyethylene glycol (PEG) 185
polygenic traits 4
polymerase chain reaction (PCR) 12, 76, 146, 155, 157, 165–166, 283, 303
limitations of 148
polysaccharide 298, 345, 382
primers 76, 128, 146, 147, 162, 165, 166, 303, 332
probes 128, 146, 155, 165, 265, 331, 357
prokaryotes 304–307, 426
protein biomarker assays 110–111
Protein DataBank 68
protein engineering 311–312
Protein Information Resource (PIR) 67
Protein Sequence-Activity Relationship (ProSAR) 183
protein structure 184, 185, 300
proteomics 223, 308–311
protoplasts 290
protozoa 96
Public Health Service Policy on Humane Care and Use of Laboratory Animals (PHS Policy) 7
PulseNet 143, 144
pyrolysis 49, 341

Q

quinine 243, 244, 301, 302

R

rabbit fever 343–344
radiation 79, 92, 276, 279–282, 389
radiocarbon 313–317
RasMol 68
Rebif 43
recombinant DNA technology 92, 93, 113–114, 164–166, 204, 366–367
Recombivax HB 115
red biotechnology 213
reforestation 14, 296
refuse-derived fuel (RDF) 319
renewable resources 321
reporter genes 290, 291
reproductive cloning 127–130, 331, 332, 333

Subject Index

reproductive pharmacology 323
reproductive science and engineering 322-326
restriction endonucleases (REs) 128
restriction enzymes 165
restriction fragment length polymorphism (RFLP) 146, 147, 155, 157
restriction mapping 263
retina 88
retinal bioengineering 273
Rhizobium 57, 58
rhizopus stolonifer 346
ribosomal RNA (rRNA) 300
ricin 74, 117
Roundup Ready soybeans 134
Rous sarcoma virus (RSV) 356

S

Saccharomyces cerevisiae 247-249
Saizen 43
salicylic acid 243
saturated fatty acids 299
scaffolds 122
scanning electron microscope (SEM) 252, 330
scanning force microscope (SFM) 328
scanning near-field optical microscopy (SNOM) 328, 329
scanning probe acoustic microscopes (SPAM) 252
scanning probe microscopy 252, 327-330
scanning tunneling microscope (STM) 252, 265, 327, 328
secondary metabolites 187, 201
sedimentary rocks 426
seed banks 336
selectable markers 290
self-pollinated plants 231
semen 113, 153, 161, 324
semi-batch operation 41
semisynthetic enzymes 183
sensors 42, 254, 266
separation techniques 42
septicemic plague 122
Serostim 43
severe acute respiratory syndrome (SARS) 72
severe combined immunodeficiency syndrome (SCIDS) 401
shed skin cells 153
short tandem repeats (STRs) 147, 152, 153, 155, 156, 159
shotgun sequencing 408, 410
simple sequence repeats (SSRs) 155
single nucleotide polymorphisms (SNPs) 13, 146-147
smallpox 73, 117
smart materials 266
Smith-Waterman algorithm 78
solar photovoltaic power (PV) technology 1-2
solar power 1-2
somatic cell nuclear transfer (SCNT) 127, 128, 331
somatic stem cells 337-339
stem cells 94, 337-339
sterilization 42
steroids 99, 173, 186, 285
sterols 249
Streptococcus infections 19
stromatolite 22, 23, 305-307, 426
Substance Abuse and Mental Health Services Administration (SAMHSA) 170
sulfanilamides 20
superovulation 113, 324
surrogacy 325
syngas 50, 64
synthetic biology 216-217
synthetic fuels, 62-65, 340-342
synthetic polymers 53, 189, 190

T

tar sands 321, 341
Tay-Sachs disease 145
Taxol 201, 244, 248
telomere terminal transferase 384
telomeres 167, 383-385, 406
Tetrahymena 383-385
tetracyclines 20, 235
therapeutic cloning 127-130, 331-334
thermoacidophiles 23
tidal power 2-3
tissue engineering 40, 43-44, 123-126
tissue plasminogen activating factor (TPA) 184
transduction 418, 419, 420
transfer RNA (tRNA) 221, 222, 300
transformation 195-196
transgenic plant 291, 296, 302
transplantation 33, 34, 43, 94, 123, 124, 129, 333, 339
transposable genetic elements in maize 395
trastuzumab 365
trichloroethylene (TCE) 101
triglycerides 299
trimethoprim 20

465

Subject Index

triticale 232
top-down nanofabrication 264, 266
tositumomab 43
totipotency 133
toxicogenomics 223
transcranial magnetic stimulation 274
transcutaneous electrical nerve stimulation (TENS) 274
transformed plant cells 290–291
transmission electron microscope (TEM) 252, 264–265
trichomycetes 345
tularemia 343–344
turbine efficiency 2
two-dimensional gel electrophoresis 310

U

United Nations Convention on Biological Diversity (CBD) 200, 201
unsaturated fatty acids 299
U.S. Department of Agriculture (USDA) 6, 10
U.S. Humane Cosmetics Act 7

V

variable number tandem repeats (VNTRs) 90, 146, 147
ventricular assist device (VAD) 31, 32
virus-indexed plants 296
virus resistance 207

W

waste management 44–45
waxes 299
wind power 2
World Anti-Doping Agency (WADA) 173

X

Xenotransplantation, 130, 333
XX and XY chromosomes 415

Z

zoaparanol 244, 301
zygomycetes 345–347
zygote intrafallopian transfer (ZIFT) 324